SUSTAINABLE FISHERIES MANAGEMENT: PACIFIC SALMON

SUSTAINABLE FISHERIES MANAGEMENT: PACIFIC SALMON

E. Eric Knudsen,
Cleveland R. Steward,
Donald D. MacDonald,
Jack E. Williams, and
Dudley W. Reiser

Editors

LEWIS PUBLISHERS

Boca Raton   New York

COVER PHOTOGRAPHS: (Graphic Design by Dale Harkness, Harkness Design)
FRONT COVER: (TOP): Mike Galesloot, Nesconlith Indian Band/Shuswap Nation Fisheries Commission: Coho enumeration efforts in the Thompson River watershed. **(BOTTOM):** Main underwater photo of salmon: Neil McDaniel, Prespawning sockeye in the Adams River, B.C.
BACK COVER: (TOP LEFT): Blake Covernton, Pro Plan Services: Commercial salmon troller; **(TOP RIGHT):** Blake Covernton, Pro Plan Services: Martin Andrew of the Boothroyd Band, dipnetting and transport of sockeye salmon past a landslide on the Nahatlatch River; **(CENTER):** George Gatenby, Sports Fisherman: Catch and release of 30 lb King salmon; **(BOTTOM):** Mike Galesloot, Nesconlith Indian Band/Shuswap Nation Fisheries Commission: Coho enumeration efforts in the Thompson River watershed.

Library of Congress Cataloging-in-Publication Data

Sustainable fisheries management : Pacific salmon / E. Eric Knudsen
   ... [et al.].
       p. cm.
   Includes bibliographical references and index.
   ISBN 1-56670-480-4 (alk. paper)
   1. Pacific salmon fisheries--Management. 2. Fishery conservation.
I. Knudsen, E. Eric. II. Title: Pacific salmon.
SH346.S87   1999
333.95′6656153′09795--dc21

99-37599
CIP

Preface

William D. Ruckelshaus

Recently, my wife and I moved back to the Pacific Northwest—something the salmon try to do every year as they live out their inspiring life cycles. Unlike us, the salmon do not always find a hospitable environment either here or elsewhere on their life's journey. There are many reasons. Simply put, there are more People in the salmon's way and they struggle more with the problems that come with expanding human populations. There have been a number of reports issued over the past few years that have chronicled the broad declines and local extinction of many salmon, steelhead and sea-run cutthroat stocks in the region.

The people who fish for a living and the communities in which they live have been hit hard. Our resource agencies are in danger of being overwhelmed by the complexity and magnitude of the problem. These developments and my own personal sense of the loss have heightened my sensitivity to the plight of the salmon and the people whose lives they affect.

Why are salmon faring so poorly? Who is responsible? What can be done to reverse the recent declines in salmon populations? Although I am aware of what is happening to the salmon and consider it a tragedy, I have come to no conclusion about who is at fault, and I don't intend to. The one thing that I am certain of is that the only truly innocent parties in all of this are the salmon and the generations of people yet to come. It seems to me that the responsibility falls upon all of us – fishermen, resource managers and concerned citizens alike – to take the steps necessary to ensure that salmon populations recover to the point that our children will be able to enjoy the quality of life we once took for granted.

We all share the salmon resource here in the Pacific Northwest. This means we will only succeed in maintaining or expanding the resource if we share the responsibility for its recovery. Because of my past history as administrator of the U.S. Environmental Protection Agency, and because of my fisherman's interest in salmon, I accepted an invitation to join the Advisory Board of the Sustainable Fisheries Foundation. The material provided to me by the Foundation indicated a goal of protecting, enhancing, and promoting the wise use of fisheries resources in the Pacific Northwest. The Foundation was little more than three months old when the founders approached me. What they lacked in experience, they made up for in enthusiasm. I asked them, "What is your goal, what are you trying to achieve?" The answer was deceptively simple, "We are trying to put more salmon back in the rivers and lakes of the Pacific Northwest." Determining exactly how to accomplish this goal has defied the efforts of a great many dedicated and talented people. Among the people I've talked with, the prevailing belief is that, as complex as they are, the challenges facing us can be overcome. For one, we have a reasonably good understanding of what salmon need to persist and evolve and what factors influence their well-being at different places and times. We know, for instance, that hatcheries pose special problems but with improvements in technology they will present fewer risks and help stabilize salmon-related economies in the future. The effects of habitat degradation and variable ocean conditions on salmon survival and harvest levels are becoming better understood. The importance of maintaining a diversity of wild stocks and ensuring that weaker stocks are not over-fished when mixed with more abundant runs has become a basic tenet of fisheries management. In general, good science and conservation translate into good management. If our understanding of the problem is adequate and people are committed to improving management, why has the situation not improved? The problem isn't so much with managing fish as it is with managing people having different needs, values, laws, institutions, and accessibility to the fish and the resources upon which they depend. Fisheries problems are fundamentally social problems. What we are dealing with is a classic "tragedy of the commons," as articulated nearly 30 years ago by Garrett Hardin, a professor at the University of California at Santa Barbara. Hardin talked about a meadow or commons where livestock owners would graze cattle. In pursuing their narrow self-interests they would use as much of the pasture as they could, graze as many cows as they could until the whole thing collapsed. Even though it was obvious that the meadow would be over-grazed the individual owners pursued their narrow, short-term interests at the expense of the whole. The problem was stated in another way by William James: the trouble with man is that he cannot have enough without having too much.

In addition to the salmon themselves, we have several commons: the rivers and the lakes in which they spawn and rear, the ocean in which they grow to maturity, and the landscapes affecting the quality of these aquatic ecosystems. Ownership and management of these resources vary across jurisdictions and ranges in scale from individual and corporate owners, to tribal organizations and smaller governmental units, to state, provincial and federal governments. The impulse to accrue wealth and prosperity across all levels of organization tends to foster overuse of the commons on which man's well-being is dependent. We want to use the rivers, oceans and land to maximize our quality of life. For some, salmon are an important determinant of our quality of life; for others, they are not that important.

So what has happened so far? Well, so far we have had rhetoric, a lot of it. More words, in fact, than fish and that is likely to continue. We have proponents and opponents on every issue. We have studies and lawsuits, charges and countercharges. Any day of the week you can pick up the newspaper and find articles regarding the studies and the lawsuits and the disagreements. As a former lawyer and government official, I can tell you who is the principal beneficiary of all this and it isn't the fish.

We know better than to believe that if any one of these things – the imposition of court orders, tighter regulations, or political interference in decision-making – were to succeed, then the problem will be over. They won't solve the larger problem and in fact some actions may even make it worse. Solutions will require broarder public education, involvement, and participation in the decision-making process. Public awareness and acceptance will largely determine management success. And, the key to gaining acceptance is to get all the parties and stakeholders to put their interests on the table and participate in a process that seeks to accommodate them without compromising the resource or the commons. It is important to recognize that the path we are now on means failure for everyone and particularly the salmon.

Ultimately, in a free society it is impossible to gain a lasting solution to a problem without the consent of the people. Unless people are allowed to participate in the decision-making process today, they can effectively block any solution that is proposed tomorrow. As trust in our governmental institutions eroded over the past decades, people have demanded a bigger say in the decisions that affect them. Systems that don't provide that opportunity for public involvement don't work. The public fears hidden or hostile government agendas, mistrusts basic "facts" offered in support of actions or decisions, and feels left out when policy changes are being considered. They must be involved or stagnation will result.

When trust in government institutions seriously erodes, two solutions are possible: one is dictatorship and the other is more democracy. What we are suggesting is the second alternative, more democracy. This solution requires that society and the governmental agencies charged with the management of our natural resources establish a shared vision, common policies, and a process for collaborative management.

Let people in, give them a real role in the decision-making process, give them a piece of the action. Don't suggest they have no knowledge of what is going on, that they are not experts, or that they have no right to be included. Do encourage them to understand the importance of putting more salmon in the lakes and rivers. It is a simple enough message. What is it that we need to do? What is it that their communities need to do? What is it that their industries or institutions, that they feel so much a part of, need to do in order to accommodate the interest of the whole, to preserve the commons? When we have seen people come together around solutions, it works. Solutions can be had; the resource can be enhanced.

The solutions developed at this conference (and in this book) will form the basis of a strategy for sustainable fisheries of West Coast salmon and steelhead and help figure out how to accomplish it. Obviously, our work is cut out for us. We cannot give up; after all, we human beings are supposed to be the most intelligent species on Earth. Collectively, we have put in jeopardy other species completely dependent upon us for survival. Shame on us. Maybe we are not as intelligent as we thought. Now is the time for us to get on with proving that we are. I wish us all good luck and, on behalf of the salmon, my best wishes in the future.

This preface is an excerpt from a keynote address given by William D. Ruckelshaus (former Administrator of the U.S. Environmental Protection Agency) at the "Toward Sustainable Fisheries: Balancing the Conservation and Use of Salmon and Steelhead in the Pacific Northwest" conference, which was held in Victoria, British Columbia, Canada, April 1996.

Contributors

Richard Aho
U.S.D.A. Forest Service, Stikine Area
P.O. Box 309
Petersburg, AK 99833

James J. Anderson
School of Fisheries
Box 358218
University of Washington
Seattle, WA 98195

James L. Anderson
Department of Resource Economics
University of Rhode Island
Kingston, RI 02881

Neil B. Armantrout
U.S.D.I., Bureau of Land Management
P.O. Box 10226
Eugene, OR 97440

Bruce A. Bachen
Northern Southeast Regional Aquaculture
  Association
1308 Sawmill Cr. Rd.
Sitka, AK 99835

Nat Bingham (deceased)
Formerly of Pacific Coast Federation of
  Fishermen's Association
P.O. Box 29910
San Francisco, CA 94129-0910

Matthew Booker
Department of History
Stanford University
Stanford, CA 94305-2024

Gregory J. Bryant
National Marine Fisheries Service
Southwest Region
Protected Species Division
1330 Bayshore Way
Eureka, CA 95501

Mason Bryant
U.S.D.A. Forest Service
Forestry Sciences Laboratory
2770 Sherwood Lane 2A
Juneau, AK 99801

Carl V. Burger
U.S. Geological Survey
Biological Resources Division
Alaska Biological Science Center
1011 East Tudor Rd.
Anchorage, AK 99503
Current Address:
U.S. Fish and Wildlife Service
Abernathy Fish Technology Center
1440 Abernathy Creek Rd.
Longview, WA 98632

Peggy J. Busby
National Marine Fisheries Service
Northwest Fisheries Science Center
Coastal Zone and Estuarine Studies
  Division
2725 Montlake Blvd. East
Seattle, WA 98112

Calvin H. Casipit
U.S.D.A. Forest Service, Alaska Region
P.O. Box 21628
Juneau, AK 99802

C. Jeff Cederholm
Washington Department of Natural
  Resources
Resource Planning and Asset Management
  Division
P.O. Box 47014
Olympia, WA 98504

Richard M. Comstock
U.S. Fish and Wildlife Service
510 Desmond Dr., S.E.
Suite 102
Lacey, WA 98503-1273

Max Copenhagen
U.S.D.A. Forest Service, Alaska Region
P.O. Box 21628
Juneau, AK 99802

Parzival Copes
Department of Economics and Institute of
  Fisheries Analysis
Simon Fraser University
Burnaby, British Columbia V5A 1S6
Canada

Steven P. Cramer
S.P. Cramer & Associates
300 S.E. Arrow Creek Lane
Gresham, OR 97080

Shannon W. Davis
The Research Group
533 SW 4th St.
Corvallis, OR 97333

Paul DeVries
R2 Resource Consultants, Inc.
15250 N.E. 95th St.
Redmond, WA 98052-2518

Larry G. Dominguez
Washington Department of Natural Resources
Resource Planning and Asset Management
  Division
P.O. Box 47014
Olympia, WA 98504

James E. Doyle
Mt. Baker Snoqualmie National Forest
U.S. Forest Service
21905 64th Ave. West
Mountlake Terrace, WA 98043

Karl K. English
LGL Limited, Environmental Research
  Associates
9768 Second St.
Sidney, British Columbia V8L 3Y8
Canada

Tamra Faris
National Marine Fisheries Service
P.O. Box 21668
Juneau, AK 99802

Melody Farrell
Fisheries and Oceans Canada
Fraser River Action Plan
Habitat and Enhancement Branch
360-555 W. Hastings St.
Vancouver, British Columbia V6B 5G3
Canada

Michelle Fisher
Mt. Baker Snoqualmie National Forest
U.S. Forest Service
21905 64th Ave. West
Mountlake Terrace, WA 98043

Tira Foran
Environmental Defense Fund
5655 College Ave.
Oakland, CA 94618

Gary Freitag
Southern Southeast Regional Aquaculture
  Association
1621 Tongass
Ketchikan, AK 99901

Kurt L. Fresh
Washington Department of Fish and Wildlife
Fish Management Program
Natural Resources Bldg.
1111 Washington St. SE
Olympia, WA 98504

Rodney M. Fujita
Environmental Defense Fund
5655 College Ave, Suite 304
Oakland, CA 94618

Harold J. Geiger
Alaska Department of Fish & Game
P.O. Box 25526
Juneau, AK 99802-5526

W. Stewart Grant
National Marine Fisheries Service
Northwest Fisheries Science Center
Conservation Biology Division
2725 Montlake Blvd. East
Seattle, WA 98112

Cornelis Groot
R.R. #3, Barney Rd.
Ladysmith, British Columbia V0R2E0
Canada

Emily Hanson
Marine Resource Management Program
Oregon State University
Corvallis, OR 97331

Jeffrey J. Hard
National Marine Fisheries Service
Northwest Fisheries Science Center
Conservation Biology Division
2725 Montlake Blvd. East
Seattle, WA 98112

Allen Harthorn
Butte Creek Watershed Conservancy
5342 La Playa Ct.
Chico, CA 95928

Gordon F. Hartman
1217 Rose Ann Dr.
Nanaimo, British Columbia V9T 3Z4
Canada

Jeffrey L. Hartman
Alaska Department of Fish and Game
Division of Commercial Fisheries
P.O. Box 25526
Juneau, AK 99802-5526

Kenneth A. Henry
National Marine Fisheries Service
18564 Springdale Court N.W.
Shoreline, WA 98177

Fern Hietkamp
Fisheries and Oceans Canada
Fraser River Action Plan
Habitat and Enhancement Branch
360-555 W. Hastings St.
Vancouver, British Columbia V6B 5G3
Canada

Jeremy M.B. Hume
Department of Fisheries and Oceans
Cultus Lake Laboratory
4222 Columbia Valley Hwy.
Cultus Lake, British Columbia V2R 5B6
Canada

Kim D. Hyatt
Fisheries and Oceans Canada
Science Branch, Stock Assessment Division
Pacific Biological Station
Hammond Bay Rd.
Nanaimo, British Columbia V9R 5K6
Canada

Kathleen Jensen
Alaska Department of Fish and Game
Division of Commercial Fisheries Management
  and Development
P.O. Box 240020
Douglas, AK 99824-0020

Kim K. Jones
Oregon Department of Fish and Wildlife
Research and Development Section
28655 Highway 34
Corvallis, OR 97333

Jeff Kershner
U.S.D.A. Forest Service
Fisheries Habitat Relationships
Utah State University
Logan, UT 84322-5210

Steve Kessler
U.S.D.A. Forest Service, Eastern Region
310 West Wisconsin Ave., Suite 500
Milwaukee, WI 53203

E. Eric Knudsen
U.S. Geological Survey
Biological Resources Division
Alaska Biological Science Center
1011 East Tudor Rd.
Anchorage, AK 99503

Robert G. Kope
National Marine Fisheries Service
Northwest Fisheries Science Center
Fishery Resource Analysis and Monitoring
  Division
2725 Montlake Blvd. East
Seattle, WA 98112

Kirill V. Kuzishchin
Vorobjevy Gory
MGU
Biology Faculty
Ichthyology Department
Moscow, Russia 119899

Otto E. Langer
Fisheries and Oceans Canada
Fraser River Action Plan
Habitat and Enhancement Branch
360-555 W. Hastings St.
Vancouver, British Columbia V6B 5G3
Canada

Peter W. Lawson
National Marine Fisheries Service
2030 S. Marine Science Dr.
Newport, OR 97365

Danny C. Lee
U.S. Forest Service
Rocky Mountain Research Station
316 East Myrtle
Boise, ID 83702

Steven A. Leider
Washington Department of Fish and Wildlife
Fish Management Program
600 Capitol Way North
Olympia, WA 98501-1091

Colin D. Levings
Fisheries and Oceans
Science Branch
West Vancouver Laboratory
4160 Marine Dr.
West Vancouver, British Columbia V7H 1V2
Canada

James A. Lichatowich
Alder Fork Consulting
182 Dory Rd.
Sequim, WA 98382

Michael R. Link
LGL Limited
9768 Second St.
Sidney, British Columbia V8L 3Y8
Canada
Current Address:
Alaska Department of Fish and Game
333 Raspberry Rd.
Anchorage, AK 99518

Timothy J. Linley
Prince William Sound Aquaculture Corporation
P.O. Box 1110
Cordova, AK 99574

Gino Lucchetti
King County, Department of Natural Resources
Water and Land Resources Division
700 5th Ave., Suite 2000
Seattle, WA 98104-5022

Donald D. MacDonald
Sustainable Fisheries Foundation
2376 Yellow Point Road, R.R.#3
Ladysmith, British Columbia V0R 2E0
Canada

Sergei V. Maximov
Vorobjevy Gory
MGU
Biology Faculty
Ichthyology Department
Moscow, Russia 119899

Geoffrey A. McMichael
Washington Department of Fish and Wildlife
Fish Management Program
201 N. Pearl St.
Ellensburg, WA 98926

Kelly M.S. Moore
Oregon Department of Fish and Wildlife
Research and Development Section
28655 Highway 34
Corvallis, OR 97333

Gary S. Morishima
Quinault Management Center
3010 77th Southeast, Suite 104
Mercer Island, WA 98040

Greta Movassaghi
Mt. Baker Snoqualmie National Forest
U.S. Forest Service
Mountlake Terrace, WA 98043

David W. Narver
British Columbia Wildlife Federation
873 Oliver St.
Victoria, British Columbia V8S 4W5
Canada

Roger Nichols
Mt. Baker Snoqualmie National Forest
U.S. Forest Service
Mountlake Terrace, WA 98043

James G. Norris
Marine Resources Consultants
P.O. Box 816
Port Townsend, WA 98368

Thomas G. Northcote
10193 Giant's Head Rd.
R.R. 2, Site 77B, Comp. 10
Summerland, British Columbia V0H 1Z0
Canada

Steve Paustian
U.S.D.A. Forest Service, Chatham Area
204 Siginaka Way
Sitka, AK 99835

Todd N. Pearsons
Washington Department of Fish and Wildlife
Fish Management Program
600 Capitol Way North
Olympia, WA 98501-1091

Hans D. Radtke
P.O. Box 244
Yachats, OR 97498

Guido R. Rahr, III
Wild Salmon Center
813 SW Alder
Portland, OR 97205

Michael P. Ramey
R2 Resource Consultants, Inc.
15250 N.E. 95th St.
Redmond, WA 98052-2518

Dudley W. Reiser
R2 Resource Consultants, Inc.
15250 N.E. 95th St.
Redmond, WA 98052-2518

Brian E. Riddell
Fisheries and Oceans Canada
Science Branch, Stock Assessment Division
Pacific Biological Station
Hammond Bay Rd.
Nanaimo, British Columbia V9R 5K6
Canada

Bruce E. Rieman
U.S. Forest Service
Rocky Mountain Research Station
316 East Myrtle
Boise, ID 83702

Norma Jean Sands
Alaska Department of Fish and Game
Division of Commercial Fisheries
P.O. Box 25526
Juneau, AK 99802-5526

Ksenia A. Savvaitova
Vorobjevy Gory
MGU
Biology Faculty
Ichthyology Department
Moscow, Russia 119899

Lana Shea
Alaska Department of Fish and Game
P.O. Box 240020
Douglas, AK 99824

Ken S. Shortreed
Department of Fisheries and Oceans
Cultus Lake Laboratory
4222 Columbia Valley Hwy.
Cultus Lake, British Columbia V2R 5B6
Canada

William W. Smoker
University of Alaska, Fairbanks
Juneau Center
School of Fisheries & Ocean Sciences
11120 Glacier Hwy.
Juneau, AK 99801

Cleveland R. Steward
Sustainable Fisheries Foundation
P.O. Box 206
Bothell, WA 98041-0206

John G. Stockner
Department of Fisheries and Oceans
Cultus Lake Laboratory
4222 Columbia Valley Hwy.
Cultus Lake, British Columbia V2R 5B6
Canada

Gilbert Sylvia
Coastal Oregon Marine Experiment Station
Oregon State University
Hatfield Marine Science Center
2030 Marine Science Dr.
Newport, OR 97365

David J. Teel
National Marine Fisheries Service
Northwest Fisheries Science Center
Coastal Zone and Estuarine Studies Division
2725 Montlake Blvd. East
Seattle, WA 98112

Angela Thiessen
Asotin County Conservation District
725 6th St., Suite 102
Clarkson, WA 99403

Russell F. Thurow
U.S. Forest Service
Rocky Mountain Research Station
316 East Myrtle
Boise, ID 83702

Peter J. Tschaplinski
Research Branch
British Columbia Ministry of Forests
P.O. Box 9519 Stn. Prov. Govt.
Victoria, British Columbia V8W 9C2
Canada

(The Honorable) Fran Ulmer
Lieutenant Governor, State of Alaska
P.O. Box 110015
Juneau, AK 99801-0015

Benjamin W. Van Alen
Alaska Department of Fish and Game
Division of Commercial Fisheries
P.O. Box 240020
Douglas, AK 99824-0020

Linda Vane
Sustainable Fisheries Foundation
P.O. Box 206
Bothell, WA 98041

Thomas C. Wainwright
NMFS/FAM Division
2030 S. Marine Science Dr.
Newport, OR 97365

Laurie Weitkamp
National Marine Fisheries Service
Northwest Fisheries Science Center
Coastal Zone and Estuarine Studies Division
2725 Montlake Blvd. East
Seattle, WA 98112

Shauna M. Whidden
Oregon Trout
117 SW Front Ave.
Portland, OR 97204

Jack E. Williams
Bureau of Land Management
1387 S. Vinnell Way
Boise, ID 83709

Current Address:
Boise National Forest
1249 S. Vinnell Way, Suite 200
Boise, ID 83709

James C. Woodey
Pacific Salmon Commission
600-1155 Robson St.
Vancouver, British Columbia V6E 1B5
Canada

Acknowledgments

Supporters—This book could not have been completed without the generous contributions of many individuals and their organizations. In particular, the following organizations contributed to the successful completion and publication of the book.

Sustainable Fisheries Foundation
American Fisheries Society, Western Division
Pacific Salmon Foundation
U.S. Geological Survey, Biological Resources Division
U.S. Bureau of Land Management
National Marine Fisheries Service
Fisheries and Oceans Canada

Peer Review Coordinators—The following individuals graciously volunteered to coordinate the three peer reviews for each chapter in this book.

W. Bradbury	D. MacDonald	D. Schmidt	B. Ward
S. Cramer	W. Mavros	C. Steward	I. Williams
R. Fisher	H. Radtke	G. Sylvia	J. Williams
E. Knudsen	D. Reiser	S. Trefy	

Peer Reviewers—The individuals listed below provided peer reviews for one or more of the manuscript chapters of the book.

A. Argue	D. Gaudett	R. LeBresseur	D. Reiser
M. Asit	N. Gayeski	D. Levy	L. Rutter
D. Bayles	A. Goddard	J. Lichatowich	N.J. Sands
R. Bilby	M. Grayum	A. Lill	D. Schmidt
P. Bisson	D. Hanson	R. Lincoln	J. Scott
T. Brown	S. Hare	C. Luecke	C. Scrivener
T.G. Brown	W. Harrower	D. MacDonald	E. Shott
R. Buchannon	G. Hartman	N. Mantua	T. Sibley
C. Burger	J. Hartman	W. Mavros	P. Slaney
P. Busby	S. Hinch	C. McAllister	G. Spain
S. Carlson	D. Hogan	B. McCay	T. Stearns
C. Carter	R. House	M. Miles	P. Stevenson
S. Cramer	P. Howell	P. Moyle	C. Steward
R. Danehy	P. Hussemer	D. Narver	R. Stone
C. Denton	K. Hyatt	W. Nehlsen	K. Sullivan
P. Devries	W. Jaeger	B. Norberg	G. Taylor
T. Dewberry	K. Jensen	J. Norris	M. Taylor
P. Dygert	S. Johnston	T. Northcote	D. Todd
S. Elliot	S. Kantor	L. Oman	A. Viola
J. Emlen	H. Kinnucan	D. Peacock	C. Walters
G. Fandrei	G. Knapp	E. Pinkerton	A. Wertheimer
T. Faris	E. Knudsen	B. Pyper	C. Wiese
R. Fisher	K. Koski	H. Radtke	I. Williams
R. Francis	S. Kuehl	W. Rees	J. Williams
M. Gaboury	G. Kyle	R. Reisenbichler	

The views expressed in this book are those of the authors and do not necessarily represent the views of the agencies listed above.

In Memory of Nat Bingham

This book is dedicated to the memory of Nathaniel S. Bingham, lifelong commercial salmon fisherman and seaman, past President of the Pacific Coast Federation of Fishermen's Association, and PCFFA's Habitat Conservation Director at his untimely death in May 1998. Known affectionately to his friends as "Captain Habitat," Nat worked all his life for the protection and restoration of the habitat that is the biological basis for productive fisheries. A patient, yet persistent, man with a deep sense of history, Nat was also a bridge-builder and a visionary. Nat leaves behind a legacy of stewardship and activism that fishers, farmers, ranchers, and citizens everywhere can build upon in their efforts toward the creation of a more just and sustainable society.

Table of Contents

Section I. Needs and Values for Sustainable Fisheries

Chapter 1
Setting the Stage for a Sustainable Pacific Salmon Fisheries Strategy 3
E. Eric Knudsen, Donald D. MacDonald, and Cleveland R. Steward

Chapter 2
The Needs of Salmon and Steelhead in Balancing Their Conservation and Use 15
Carl V. Burger

Chapter 3
Science and Management in Sustainable Salmonid Fisheries: The Ball is Not in Our Court 31
Gordon F. Hartman, Cornelis Groot, and Thomas G. Northcote

Chapter 4
The Importance of "Stock" Conservation Definitions to the Concept
of Sustainable Fisheries 51
Kim D. Hyatt and Brian E. Riddell

Chapter 5
The Elements of Alaska's Sustainable Fisheries 63
(The Honorable) Fran Ulmer

Chapter 6
Review of Salmon Management in British Columbia: What Has the Past Taught Us? 67
David W. Narver

Chapter 7
Aboriginal Fishing Rights and Salmon Management in British Columbia:
Matching Historical Justice with the Public Interest 75
Parzival Copes

Section II. Stock Status

Chapter 8
The Status of Anadromous Salmonids: Lessons in Our Search for Sustainability 95
Jack E. Williams

Chapter 9
Endangered Species Act Review of the Status of Pink Salmon from Washington,
Oregon, and California 103
Jeffrey J. Hard, Robert G. Kope, and W. Stewart Grant

Chapter 10

Review of the Status of Coho Salmon from Washington, Oregon, and California 111
*Laurie Weitkamp, Thomas C. Wainright, Gregory J. Bryant, David J. Teel,
and Robert G. Kope*

Chapter 11

Status Review of Steelhead from Washington, Idaho, Oregon, and California 119
Peggy J. Busby, Thomas C. Wainwright, and Gregory J. Bryant

Chapter 12

Status and Distribution of Chinook Salmon and Steelhead in the Interior Columbia
River Basin and Portions of the Klamath River Basin 133
Russell F. Thurow, Danny C. Lee, and Bruce E. Rieman

Chapter 13

Status and Stewardship of Salmon Stocks in Southeast Alaska...161
Benjamin W. Van Alen

Chapter 14

Kamchatka Steelhead: Population Trends and Life History Variation 195
Ksenia A. Savvaitova, Kirill V. Kuzishchin, and Sergei V. Maximov

Section III. Existing Management

Chapter 15

International Management of Fraser River Sockeye Salmon 207
James C. Woodey

Chapter 16

The History and Status of Pacific Northwest Chinook and Coho Salmon Ocean Fisheries
and Prospects for Sustainability 219
Gary S. Morishima and Kenneth A. Henry

Chapter 17

Managing Pacific Salmon Escapements: The Gaps Between Theory and Reality 237
E. Eric Knudsen

Chapter 18

Research Programs and Stock Status for Salmon in Three Transboundary Rivers:
The Stikine, Taku, and Alsek...273
Kathleen Jensen

Section IV. Habitat Assessment

Chapter 19

The Effects of Forest Harvesting, Fishing, Climate Variation, and Ocean Conditions
on Salmonid Populations of Carnation Creek, Vancouver Island, British Columbia 297
Peter J. Tschaplinski

Chapter 20
Habitat Assessment in Coastal Basins in Oregon: Implications for Coho Salmon
Production and Habitat Restoration...329
Kim K. Jones and Kelly M. S. Moore

Chapter 21
An Overview Assessment of Compensation and Mitigation Techniques Used to Assist
Fish Habitat Management in British Columbia Estuaries 341
Colin D. Levings

Chapter 22
Human Population Growth and the Sustainability of Urban Salmonid Streams in the
Lower Fraser Valley 349
Otto E. Langer, Fern Hietkamp, and Melody Farrell

Section V. Artificial Production

Chapter 23
Minimizing Ecological Impacts of Hatchery-Reared Juvenile Steelhead Trout
on Wild Salmonids in a Yakima Basin Watershed..365
Geoffrey A. McMichael, Todd N. Pearsons, and Steven A. Leider

Chapter 24
Economic Feasibility of Salmon Enhancement Propagation Programs 381
Hans D. Radtke and Shannon W. Davis

Chapter 25
The New Order in Global Salmon Markets and Aquaculture Development:
Implications for Watershed-Based Management in the Pacific Northwest 393
Gilbert Sylvia, James L. Anderson, and Emily Hanson

Chapter 26
Alaska Ocean Ranching Contributions to Sustainable Salmon Fisheries 407
William W. Smoker, Bruce A. Bachen, Gary Freitag, Harold J. Geiger,
and Timothy J. Linley

Section VI. Modeling Approaches

Chapter 27
The Proportional Migration Selective Fishery Model..423
Peter W. Lawson and Richard M. Comstock

Chapter 28
A Simulation Model to Assess Management and Allocation Alternatives in Multi-Stock
Pacific Salmon Fisheries..435
Norma Jean Sands and Jeffrey L. Hartman

Chapter 29
Defining Equivalent Exploitation Rate Reduction Policies for Endangered Salmon Stocks.......451
James G. Norris

Chapter 30

Decadal Climate Cycles and Declining Columbia River Salmon 467
James J. Anderson

Chapter 31

The Effect of Environmentally Driven Recruitment Variation on Sustainable Yield
from Salmon Populations..485
Steven P. Cramer

Chapter 32

Using Photosynthetic Rates to Estimate the Juvenile Sockeye Salmon Rearing Capacity
of British Columbia Lakes..505
Ken S. Shortreed, Jeremy M. B. Hume, and John G. Stockner

Section VII. Habitat Protection and Restoration

Chapter 33

Protecting and Restoring the Habitats of Anadromous Salmonids in the Lake Washington
Watershed, an Urbanizing Ecosystem 525
Kurt L. Fresh and Gino Lucchetti

Chapter 34

Rehabilitating Stream Channels Using Large Woody Debris with Considerations
for Salmonid Life History and Fluvial Geomorphic Processes...545
Larry G. Dominguez and C. Jeff Cederholm

Chapter 35

Development of Options for the Reintroduction and Restoration of Chinook Salmon
into Panther Creek, Idaho 565
Dudley W. Reiser, Michael P. Ramey, and Paul DeVries

Chapter 36

Effectiveness of Current Anadromous Fish Habitat Protection Procedures
for the Tongass National Forest, Alaska..583
*Calvin H. Casipit, Jeff Kershner, Tamra Faris, Steve Kessler, Steve Paustian, Lana Shea,
Max Copenhagen, Mason Bryant, and Richard Aho*

Chapter 37

Using Watershed Analysis to Plan and Evaluate Habitat Restoration...601
Neil B. Armantrout

Chapter 38

Watershed Restoration in Deer Creek, Washington—A Ten-Year Review 617
James E. Doyle, Greta Movassaghi, Michelle Fischer, and Roger Nichols

Chapter 39

Integrating History into the Restoration of Coho Salmon in the Siuslaw River, Oregon............625
Matthew Booker

Section VIII. Toward Sustainability

Chapter 40
Community Education and Cooperation Determine Success in Watershed Restoration:
The Asotin Creek Model Watershed Plan..639
Angela Thiessen and Linda Vane

Chapter 41
Spring-Run Chinook Salmon Work Group: A Cooperative Approach to Watershed
Management in California...647
Nat Bingham and Allen Harthorne

Chapter 42
Creating Incentives for Salmon Conservation...655
Rodney M. Fujita and Tira Foran

Chapter 43
Long-Term, Sustainable Monitoring of Pacific Salmon Populations Using Fishwheels
to Integrate Harvesting, Management, and Research...667
Michael R. Link and Karl K. English

Chapter 44
Sanctuaries for Pacific Salmon...675
*James A. Lichatowich, Guido R. Rahr, III, Shauna M. Whidden,
and Cleveland R. Steward*

Chapter 45
One Northwest Community—People, Salmon, Rivers, and the Sea:
Toward Sustainable Salmon Fisheries..687
Donald D. MacDonald, Cleveland R. Steward, and E. Eric Knudsen

Index...703

Section I

Needs and Values
for Sustainable Fisheries

1 Setting the Stage for a Sustainable Pacific Salmon Fisheries Strategy

E. Eric Knudsen, Donald D. MacDonald, and Cleveland R. Steward

Abstract.—Salmon and steelhead *Oncorhynchus* spp. have been keystone species for ecosystems and human cultures of the North American Pacific coast for eons. Yet, in the past century, many populations have been greatly diminished and some are now extinct—the result of a combination of factors, including habitat loss and degradation, overfishing, natural variability in salmon production, negative effects of artificial propagation, and weaknesses in institutional and regulatory structures. We argue that a major shift is required, from the egocentric environmental approach (wherein each part of the ecosystem is managed as a unit) to the ecocentric ecosystem approach (wherein all parts are integrated for management). A management framework is proposed that contains—for each management unit such as a watershed—four elements: management goals; management objectives; ecosystem indicators; and a coordinated action plan. We also describe the Sustainable Fisheries Strategy, a consultative process for developing an ecosystem-based approach toward achieving sustainable Pacific salmon and steelhead populations and fisheries. This book is one of three important underpinnings of the Strategy; the other two are the Strategy itself and a manual being developed to guide community-based programs embracing the principles of sustainable fisheries. This book contains important historical perspectives as well as numerous innovative ideas for moving toward ecosystem-oriented, sustainable management of Pacific salmon and steelhead.

INTRODUCTION

Salmon and steelhead *Oncorhynchus* spp. serve as a powerful symbol for the quality of life enjoyed in the Pacific Northwest and generate a wide range of economic, social, and cultural benefits in the region. Commercial fisheries contribute significantly to local, provincial/state, and national economies, both directly through the sale of fish and indirectly through related service and manufacturing industries. Likewise, sport fisheries provide diverse economic benefits through the tourist, manufacturing, and service sectors. First Nations (Canada) and Tribal (U.S.) fisheries revitalize local economies, encourage social stabilization and renewal, and help maintain the rich cultural heritage that characterizes the Pacific Northwest. Considering the nature and extent of benefits that they bestow, the conservation of our shared salmon and steelhead resources merit national and international priority.

Despite the vast quantities of time, money, and effort spent on fisheries management, there has been a widespread and marked decline in the size and number of salmonid populations in the Pacific Northwest over the last century (e.g., Nehlsen et al. 1991; Slaney et al. 1996). Numerous populations have been extirpated, and many that remain are at precariously low levels and trending downward. While natural factors have undoubtedly played a role, the overriding causes of decline have been associated with human activities. The current status and trends of these populations

3

reflect our inability to effectively manage the fishery resource and take the necessary steps to avert further declines in population abundance. The blame, as well as the responsibility for action, falls upon all of us—scientists, managers, resource users, and the public. Together, we must cooperate in the development and implementation of a comprehensive fisheries management strategy to ensure that west coast salmon populations are sustained and enhanced for future generations.

People throughout the Pacific Northwest are recognizing the need for such a strategy and many have agreed to work cooperatively on the development of a ***Sustainable Fisheries Strategy (SFS) for West Coast Salmon and Steelhead Populations***. According to the more than 500 participants who attended the "Toward Sustainable Fisheries" conference held in Victoria, British Columbia April 26–30, 1996, sustainable fisheries can be defined as:

> the conditions that support healthy, diverse, and productive ecosystems, viable aboriginal, sport, and commercial fisheries, and vital and stable communities throughout the historical range of anadromous Pacific salmonids.

This book, *Sustainable Fisheries Management: Pacific Salmon,* represents an important step toward sustainable fisheries because it articulates a common vision for the future, identifies the factors that are currently impeding our ability to sustainably manage salmon and steelhead, and outlines some new and innovative ideas and strategies for overcoming these constraints and moving toward sustainability. This first chapter provides a context for sustainable fisheries management by briefly describing the factors that influence fisheries sustainability, providing a framework for ecosystem-based fisheries management, and describing the overall strategy to support a transition toward sustainable fisheries management.

CHALLENGES TO THE SUSTAINABILITY OF PACIFIC SALMON

Before non-native settlement, salmon and steelhead populations flourished throughout the Pacific Northwest. As a result, humans could easily harvest all the fish they needed without adversely affecting populations (Booker 2000; Copes 2000). In many places, the abundance of salmon eclipsed the populations that exist today. For example, it has been estimated that the total run to the Columbia River was historically about 10–16 million salmon (Johnson et al. 1997). Today, the total run of salmon to this important river system is on the order of 1–3 million fish, 75% of which are of hatchery origin (Johnson et al. 1997). Similar or more dramatic reductions in the abundance of Pacific salmon have been observed throughout much of their range. The severity and extent of these declines are emphasized by the number of stocks that are either extinct or at risk of extinction (Nehlsen et al. 1991; Slaney et al. 1996). These widespread declines in anadromous salmonid abundance have led to a number of listings by the National Marine Fisheries Service (NMFS) under the U.S. Endangered Species Act (ESA), which at the time of this writing includes the species listed in Table 1.1.

Considering the importance of salmon and steelhead to the people of the Pacific Northwest, it is surprising that we have permitted their populations to fall so precipitously without effective intervention. To understand the underlying reasons for these declines, it is important to recognize that anadromous salmonids are subjected to a wide variety of stressors throughout their life history (see Groot and Margolis 1991; Burger 2000; and Hartman et al. 2000 for information on the salmon life cycle). As generally described below, the main obstacles to fisheries sustainability fall into four main categories: habitat, harvest, production, and institutional structures.

Habitat Loss and Degradation

Scientists and natural resource managers have long realized the need to manage terrestrial and aquatic habitats in ways that maintain normal biophysical processes, linkages within and between

TABLE 1.1
U.S. Endangered Species Act Listing Status of Salmon and Steelhead Evolutionarily Significant Units as of July 1999.

Species	Evolutionarily Significant Unit (ESU)	Proposed or actual listing	Status of listing
Chinook	Sacramento Winter	Endangered	Listed
	Snake River Spring/Summer	Threatened	Listed
	Snake River Fall	Threatened	Listed (ESU modification proposed 3/9/98)
	Puget Sound	Threatened	Listed
	Lower Columbia River	Threatened	Listed
	Upper Willamette	Threatened	Listed
	Upper Columbia River Spring	Endangered	Listed
	S. Oregon/California. Coast	Threatened	Deferred until 9/99
	Cal. Central Valley Spring	Endangered	Deferred until 9/99
	Cal. Central Valley Fall	Threatened	Deferred until 9/99
Chum	Hood Canal Summer	Threatened	Listed
	Columbia River	Threatened	Listed
Coho	Central California	Threatened	Listed
	S. Oregon/N. California Coasts	Threatened	Listed
	Oregon Coast	Threatened	Listed
	Puget Sound/Strait of Georgia	Candidate	Assessments due mid-1999
	Southwest WA/Lower Columbia River	Candidate	Assessments due mid-1999
Sockeye	Snake River	Endangered	Listed
	Ozette Lake	Threatened	Listed
Steelhead	Southern California	Endangered	Listed
	South-Central California Coast	Threatened	Listed
	Central California Coast	Threatened	Listed
	Upper Columbia River	Endangered	Listed
	Snake River Basin	Threatened	Listed
	Lower Columbia River	Threatened	Listed
	California Central Valley	Threatened	Listed
	Upper Willamette River	Threatened	Listed
	Middle Columbia River	Threatened	Listed
	Northern California	Candidate	
	Klamath Mountains Province	Candidate	
	Oregon Coast	Candidate	

habitats, and a diversity of living and nonliving entities (see Hartman et al. 2000). Healthy habitat is a fundamental requirement for salmon (Burger 2000). Unfortunately, evidence from many sources strongly suggests that the natural freshwater, estuarine, and marine habitats used by salmon are under extreme stress, and many have already been irreversibly altered (e.g., Fresh and Lucchetti 2000; Jones and Moore 2000; Levings 2000; Langer et al. 2000).

Expanding human populations and associated demands on natural resources have resulted in the degradation of aquatic and riparian habitats throughout much of the Pacific Northwest (NRC 1996). Specifically, forest management activities have been linked to many adverse effects on salmon habitats, including changes in water quality, streambed substrate composition, stream hydrology, and stream morphology (e.g., Gregory and Bisson 1997). In addition, construction of impoundments for hydropower generation and irrigation has blocked access to important spawning and rearing areas, flooded key habitats, altered streamflow conditions, and degraded water quality. Urbanization and industrial developments have also adversely affected salmon habitats by altering streamflow patterns, degrading water quality, and changing stream morphology (e.g., Fresh and Lucchetti 2000; Langer et al. 2000). Anthropogenic activities have also influenced salmon habitats in marine and estuarine systems, both through direct habitat loss and alteration (e.g., due to log storage, diking, etc.; Levings 2000) and indirectly through releases of substances that contribute to global climate change (e.g., carbon dioxide, nitrous oxide, chloroflourocarbons, etc.). Of particular concern is the progressive northward advancement of warm ocean temperatures, a likely factor contributing to reduced marine survival of Pacific salmonids (Welch et al. 1998).

Salmon are remarkably adaptable, frequently colonizing new habitats and surviving major ecological perturbations (e.g., Milner 1989). However, their resiliency is being overwhelmed by a combination of natural and human-caused disturbances of a kind, magnitude, and frequency unlike any that have occurred in the recent past (i.e., post-glaciation). If deliberate steps are not taken to reduce the impact of human activities, the quality and productivity of freshwater and marine habitats will continue to deteriorate.

HARVEST MANAGEMENT

Fisheries around the world are under immense pressure from fishing activity. According to the Food and Agriculture Organization (FAO) of the United Nations (as cited by Christie 1993), all of the world's 17 major marine fishing grounds have been fished to their limits or beyond. Nine of these fishing grounds are now in serious decline due to overfishing. Recent experience with the Atlantic cod *Gadus morhua* fishery, off the east coast of Canada, has demonstrated that a fisheries collapse can occur even in countries with apparently well-developed management systems and access to the best scientific information. The collapse of the cod fishery has had devastating social and economic consequences for the people who live in fishing communities and many others less directly linked to the resource (Spurgeon 1997). The tragedy of this situation is that despite repeated warnings of the impending collapse by inshore fishermen, scientists, and others (Hutchings et al. 1997) the fishing continued unabated until Atlantic cod populations were reduced to less than 1% of their former abundance, requiring curtailment of virtually all harvest of that species (Spurgeon 1997).

Experience has now shown us the disastrous effects of poorly regulated, unsustainable fishing practices on fishery resources. In turn, reduction in the abundance of these resources can have serious impacts on the social, cultural, and economic fabric of the coastal communities that rely on these precious, otherwise renewable resources. Nonetheless, many of the ecological signals of unsustainability, including decreased biodiversity, reduced productivity, diminished habitat carrying capacity, reduced harvests, and declines in indicator species, have become evident in many areas throughout the Pacific Northwest (NRC 1996).

Many challenges face those responsible for managing the harvest of Pacific salmonids. Most importantly there is a lack of clear objectives for harvest management programs: i.e., are we trying to optimize biomass or biodiversity? (Hyatt and Riddell 2000). In addition, even though stock sizes have been reduced and the variability in yields has increased, the demand for salmon harvesting continues to grow. This creates greater political pressure on managers to open fisheries at a time when the stocks can least support the fishing effort. Moreover, the information and science-based tools upon which we rely to make harvest management decisions (such as escapement and run size, age structure, and stock-recruitment models) are often flawed, unavailable, or incomplete (Knudsen 2000). Thus, it is apparent that many of the strategies and procedures currently used to manage Pacific salmon fisheries are unsustainable; however, some rays of hope do shine through this gloom. For example, management of most Alaskan salmon fisheries is still relatively successful because fish are protected in terminal areas, the harvest rate on mixed stocks has been moderated, and most freshwater habitat is still intact (Van Alen 2000).

SALMON PRODUCTION

Natural Production.—No one knows exactly how many naturally produced salmon and steelhead return to most of the rivers, streams, and lakes of the Pacific Coast; nor is it well understood how many salmon should be escaping to each system. There is, however, no doubt that production of wild salmon and steelhead has declined drastically throughout a significant portion of their range over the last century. These declines have coincided with both subtle and profound reductions in the productivity of freshwater, estuarine, and marine habitats.

Salmon production is strongly influenced by variation in climate which occasionally leads to catastrophic perturbations of freshwater, estuarine, and/or marine survival of salmon (see, for example, Welch et al. 1998). Because salmon migrate through a number of different ecosystems during their life cycle, they are particularly vulnerable to the vagaries of natural phenomena in the various habitats. Each run has its own response to the environmental influences it experiences as it migrates. The run's natural survival "bottleneck" may be egg-devastating floods one year and overly warm ocean waters the next. Alternatively, when there are no substantial bottlenecks, salmon can be impressively productive. No salmon populations are immune to these natural variations; witness the 1997 and 1998 "crashes" of Bristol Bay sockeye and other western Alaskan populations following a number of years of record high run sizes (Kruse 1998).

When habitat degradation and harvest are superimposed on this naturally fragile interdependence between environment and production, it is easy to see how salmon are susceptible to decline (e.g., Lawson 1993). While scientists are still limited in their ability to understand the relationship between salmon populations and their environment, the complexities of population dynamics in runs suffering from both overharvest and altered habitats present even greater intellectual and technological challenges. Much work is still needed in our technical capability to account for the dynamic interdependence of natural and anthropogenic factors. Additionally, new ways of thinking about salmon populations indicate a significant need for social and economic acceptance of the natural variations in salmon abundance (NRC 1996; Knudsen 2000).

Artificial Production.—In response to decreasing natural productivity resulting from overexploitation and habitat loss, fisheries managers turned to artificial production techniques as a means of satisfying the growing demand for fish (e.g., Smoker et al. 2000). While releases of juvenile salmonids from production hatcheries, spawning channels, and other facilities dramatically increased productivity of some populations, the fisheries on these enhanced runs increased pressure on weaker stocks (NRC 1996). Recently, concerns have also been raised about the potential impacts of hatchery-produced fish on wild salmonids, in terms of both survival and genetic diversity (Hilborn 1992). In certain locations (e.g., Columbia River Basin), changes in ocean conditions have resulted in greatly reduced survival of hatchery salmonids, which in turn resulted in decreased benefit-to-cost ratios for these facilities (Radtke and Davis 2000). As such, the use of large hatcheries as a production tool is now commonly viewed as a threat to the sustainability of wild salmon populations. In contrast, conservation aquaculture has been identified as a key element of coordinated salmon recovery efforts (NRC 1996). For example, some populations only exist because of continued hatchery propagation.

INSTITUTIONAL AND REGULATORY STRUCTURES

A wide variety of institutions and organizations are involved in the fisheries and environmental management process. In fact, the existing management framework is so complicated that even those involved in the process find it difficult to fully understand. This complexity in the management of Pacific salmonids and their associated habitats has developed for several reasons. Our existing management structures have evolved over a period of more than 100 years. Over that time, our understanding of the resource and its interactions with people and the environment has increased dramatically. However, rather than implementing a holistic ecosystem-based approach to management, we have chosen to compartmentalize the environment and establish agencies that assume the primary responsibility for managing one or more compartments (i.e., fish, water, forests, urban developments, etc.).

While most of the existing *institutions* have adequately fulfilled their stated mandates, the current status of salmon and steelhead populations and their associated habitats suggests that the existing *institutional management structures* do not adequately respond to the challenges we are currently facing. One need only look at the precipitous and widespread declines of salmon populations in the Pacific region to conclude that, together, the institutions responsible for their management have failed to protect them. While many factors have contributed to the decline (e.g., overfishing, habitat degradation, water quality changes, climate change, etc.), the overriding cause is our failure to effectively and holistically manage human activities so as to avoid mass extinction of non-humans (Hartman et al. 2000). Resource agencies have not yet made the shift to an ecosystem perspective nor have they adopted true adaptive management approaches (Lichatowich 1997).

While resource agencies have made valiant attempts to rectify the shortcomings of the existing management structure, they are often constrained by the lack of a clear mandate, insufficient funding, and/or chronic understaffing. When changes in policy and management have occurred, they have often been incremental, rarely deviating sufficiently from the *status quo* to correct the problem at hand (Cone 1995). There has also been a lack of consistency, over time and between agencies, in setting and implementing management priorities. Most agencies are funded separately and operate under separate mandates, leading to uncoordinated and piecemeal management. For example, Hyatt and Riddell (2000) point out that the objectives of harvest managers and habitat managers often conflict with respect to maintaining "no net loss" in habitat productivity, particularly when the latter depends on forgoing catch to ensure sufficient returns of salmon as key agents of nutrient delivery to streams (see also Bilby et al. 1996). Another problem is that fisheries science has often been stifled in favor of politically or economically favorable outcomes (Hutchings et al. 1997). Lacking clear policy direction and incentives for meaningful change, agencies have often found it difficult to establish effective enough regulations and processes to adequately protect salmon and their habitats.

The lack of consistency and effectiveness among our institutions stems in part from the large number of jurisdictions involved. In the Pacific Northwest alone, there are two federal governments—each represented in the natural resource management arena by a host of agencies—at least six states and provinces, a plethora of local jurisdictions, and numerous indigenous peoples' organizations, all with some role to play in the management of the Pacific salmon resource. Many of the agencies' jurisdictions either overlap or are separated by some political boundary naturally crossed by migrating salmon. Furthermore, many of the decisions influencing salmon habitat, such as land use and zoning, are made at the local government level, institutionally far removed from the primary salmon management agencies. All of these organizations have different agendas and different approaches. Given the number and diversity of jurisdictions and institutions involved, it is little wonder that efforts to protect salmon have been thwarted. Despite efforts to cooperate and achieve consensus, our institutions have not yet developed the collective ability to effectively manage land, water, and fisheries to avoid accelerating the extinction rate of salmon populations in North America.

A FRAMEWORK FOR ECOSYSTEM-BASED MANAGEMENT

Despite the best of efforts, the crisis in Pacific salmon management seems to worsen each year. In California, Oregon, Idaho, and Washington, 1998 was marked by a plethora of listings of salmon populations under the ESA. Each listing generated a requirement to develop and implement an appropriate recovery plan. In British Columbia, sweeping changes in the management of commercial, recreational, and traditional fisheries have been implemented to conserve endangered coho salmon populations and rationalize the commercial fleet (Anderson 1998). Even Alaska has been affected by changes in the abundance of Pacific salmon, prompting Governor Knowles to declare the Bristol Bay fishery a disaster in 1998. Such recent challenges underscore the need for a new, science-based approach to the management of Pacific salmonids and the ecosystems upon which both fish and humans depend. The ecosystem approach provides a framework for meeting these challenges.

The ecosystem approach to planning, research, and management is the most recent phase in a historical succession of environmental management approaches. Previously, we considered ourselves to be separate from the environment in which we lived. This *egocentric approach* viewed the external environment only in terms of our uses of natural resources. However, recent experience has shown us that human activities can have significant and far-reaching impacts on the environment (as evidenced by the declines in salmon abundance) and on the humans who reside in these systems (i.e., the coastal communities that have been devastated by losses in fishing opportunities). Therefore, there is a need for a more holistic approach to environmental management, in which humans are considered as integral components of the ecosystem. The ecosystem approach provides this progressive perspective by integrating the *egocentric* with an *ecocentric view* that considers the broader implications of human activities.

Implementation of the ecosystem approach necessitates the development of an integrated set of policies and managerial practices that relate people to ecosystems of which they are a part, instead of to the external resources or environments with which they interact (Vallentyne and Beeton 1988). The essence of the ecosystem approach is that it relates *wholes* at different levels of integration (i.e., humans and the ecosystems containing humans) rather than the interdependent parts of those systems (i.e., humans and their environment; Christie et al. 1986). The identifying characteristics of the ecosystem approach are:

- a synthesis of integrated knowledge on the ecosystem;
- a holistic perspective of interrelating systems at different levels of integration; and
- actions that are ecological, anticipatory, and ethical (Christie et al. 1986; Vallentyne and Hamilton 1987).

The primary distinction between the environmental (i.e., egocentric) and ecosystem (i.e., ecocentric) approaches is whether the system under consideration is external to (in the environmental approach) or contains (in the ecosystem approach) the population under study (Vallentyne and Beeton 1988). The conventional concept of the environment is like that of *house*—external and detached; in contrast, ecosystem implies *home*—something that we feel part of and see ourselves in, even when we are not there (Christie et al. 1986). The change from the environmental approach to the ecosystem approach necessitates a change in the view of the environment from a political or people-oriented context to an ecosystem-oriented context (Vallentyne and Beeton 1988). This expanded view then shapes the planning, research, and management decisions that are made within and pertaining to the ecosystem.

Implementation of the ecosystem approach on a watershed basis requires a framework in which to express the environmental management policies that have been established for the ecosystem. In general, this framework is comprised of four functional elements. The first element is a statement of broad *management goals* for the ecosystem. These goals must reflect the importance of the ecosystem to local area residents and other stakeholders. The second element of the framework is a set of *objectives* for the various components of the ecosystem which clarify the scope and intent of the ecosystem goals. The third element of the framework is a set of *ecosystem indicators* (including specific *metrics and targets*), which provide an effective means of measuring the level of attainment of each of the ecosystem goals and objectives. The final element of an ecosystem-based strategic planning process is the development of a fully coordinated action plan which outlines the steps that are needed and schedule for achieving the desired goals and objectives (Lichatowich et al. 1995; Mobraud et al. 1997).

DEVELOPMENT OF A SUSTAINABLE FISHERIES STRATEGY

Realization of the agreed-to, high priority of sustaining west coast salmonid resources will require implementation of an ecosystem-based, comprehensive, and coordinated plan that is developed cooperatively by affected interests located throughout the Pacific Northwest. One of the major deterrents to the development and implementation of this type of strategy has been the lack of cooperation between the parties involved in the fisheries management process. At the international level, the Canadian and U.S. governments have had difficulty reaching a lasting accord on Pacific salmon management. Political interference from user groups has also prevented government agencies and fisheries commissions from effectively managing the salmonid stocks within their jurisdictions. Even among First Nations and Native American tribes, more cooperation is needed on issues related to fisheries management actions and sharing of the resource. Sustainability will require that all management entities improve cooperation from now on.

While virtually everyone involved recognizes the need to assure the sustainability of our west coast salmonid resources, there has been little agreement on the underlying causes of population declines or on the actions required to protect and restore these populations. For this reason, the Sustainable Fisheries Foundation (SFF) and its partners have initiated a consultative process for developing an ecosystem-based Sustainable Fisheries Strategy (SFS) for west coast salmon and steelhead populations. This SFS is intended to provide a common vision for the future and a framework for action to protect and restore west coast salmonid populations, from Alaska to California.

The SFS is being developed cooperatively by a wide variety of salmon-based interest groups, including federal, provincial, state, and local government agencies, First Nations, Tribal organizations, resource user groups, conservation groups, and concerned citizens. The SFS is evolving from information and technical discussions presented at the "Toward Sustainable Fisheries" conference, which was designed to provide delegates an opportunity to exchange information on a wide range of topics related to Pacific fisheries management, including:

- Status of salmon and steelhead populations;
- Status of freshwater, estuarine, and marine habitats;
- Status of salmon fisheries and related economies;
- Status of salmon and steelhead management;
- First Nations/Tribal fisheries management;
- Opportunities and strategies for attaining sustainable resource use;
- Integration of natural and artificial production;
- Habitat assessment and restoration techniques and initiatives;
- Establishing more effective administrative and regulatory structures;
- Monitoring, assessment, and adaptive fisheries management;
- Community-based and grassroots fisheries management initiatives;
- Addressing uncertainty in fisheries management;
- Shifting social, economic, and cultural priorities;
- Models for sustainable fisheries; and
- Establishing more effective legal and policy frameworks.

In addition to the comprehensive suite of technical sessions, a series of five work group sessions were also convened during the conference. This gave delegates the opportunity for input to the SFS.

The products resulting from the SFS process and its underpinnings have been published in three forms to facilitate access by participants in the process. First, a draft SFS report, which provides an integrated set of principles, guidelines, and actions for restoring salmonid populations to sustainable levels, has been published, distributed for review, and posted on the Internet (SFF 1996). A series of follow-up workshops have also been convened to refine the SFS that was developed at the Victoria conference, to assess its applicability at the watershed level, and to identify key indicators that can be measured to evaluate progress toward sustainable fisheries. The results of the follow-up workshops have been summarized in various documents that have been published by the SFF and its partners (key recommendations are fully articulated in the final chapter of this book).

Second, a guidance manual is being prepared to integrate the input received throughout the project and provide detailed advice on developing community-based programs that embrace the principles of sustainable fisheries management. This manual is being revised in light of experience gained from several community- and watershed-based forums held during 1996–1998 in Washington and British Columbia.

This book represents a third and critical element of the overall SFS development and implementation process because it provides the principal mechanism for disseminating the science-based information that supports the SFS (i.e., the key technical information presented at the Victoria conference). Importantly, this book provides state-of-the-art information on the following topics:

- Needs and values for sustainable fisheries;
- Status of salmon and steelhead stocks;
- Current approaches to fisheries management;
- Habitat assessment;
- Artificial production;
- Modeling approaches;
- Habitat protection and restoration; and
- Striving toward sustainability.

In addition, the book provides a summary of the guiding principles that form the basis of the SFS. We believe this is an excellent compilation of both historical perspective and numerous innovative ideas for moving toward sustainability. As such, we are hopeful that this book will make a lasting contribution to salmon and steelhead management and provide a template for action that will lead to the implementation of ecosystem-based management and fisheries sustainability in the Pacific Northwest.

ACKNOWLEDGMENTS

We appreciate all the participants in the Sustainable Fisheries Conference who, by their enthusiasm and active involvement, helped to set the stage for this chapter and book and, most importantly, the Sustainable Fisheries Strategy. Thanks to P. Busby for assistance with Table 1.1. We are also grateful for the careful reviews of this chapter by C. Burger, K. Hyatt, and D. Reiser.

REFERENCES

Anderson, D. 1998. Announcement of Canada's coho recovery plan and federal response measures. Communications Directorate, Department of Fisheries and Oceans, Vancouver, British Columbia.

Bilby, R. E, B. R. Fransen, and P. A. Bisson. 1996. Incorporation of nitrogen and carbon from spawning coho salmon into the trophic system of small streams. Canadian Journal of Fisheries and Aquatic Sciences 53:164–173.

Booker, M. 2000. Integrating history into the restoration of coho salmon in Siuslaw River, Oregon. Pages 625–636 in E.E. Knudsen, C.R. Steward, D.D. MacDonald, J.E. Williams, and D.W. Reiser, editors. Sustainable fisheries management: Pacific salmon. Lewis Publishers, Boca Raton, Florida.

Burger, C.V. 2000. The needs of salmon and steelhead in balancing their conservation and use. Pages 15–29 in E.E. Knudsen, C.R. Steward, D.D. MacDonald, J.E. Williams, and D.W. Reiser, editors. Sustainable fisheries management: Pacific salmon. Lewis Publishers, Boca Raton, Florida.

Christie, W.J. 1993. Developing the concept of sustainable fisheries. Journal of Aquatic Ecosystem Health 2:99–109.

Christie, W.J., M. Becker, J.W. Cowden, and J.R. Vallentyne. 1986. Managing the Great Lakes Basin as a Home. Journal of Great Lakes Research 12(1):2–17.

Cone, J. 1995. A common fate: endangered salmon and the people of the Pacific Northwest. Henry Holt and Company, New York.

Copes, P. 2000. Aboriginal fishing rights and salmon management in British Columbia: Matching historical justice with the public interest. Pages 75–91 in E.E. Knudsen, C.R. Steward, D.D. MacDonald, J.E. Williams, and D.W. Reiser, editors. Sustainable fisheries management: Pacific salmon. Lewis Publishers, Boca Raton, Florida.

Fresh, K.L., and G. Lucchetti. 2000. Protecting and restoring the habitats of anadromous salmonids in the Lake Washington Watershed, an Urbanizing Ecosystem. Pages 525–543 *in* E.E. Knudsen, C.R. Steward, D.D. MacDonald, J.E. Williams, and D.W. Reiser, editors. Sustainable fisheries management: Pacific salmon. Lewis Publishers, Boca Raton, Florida.

Gregory, S.V., and P.A. Bisson. 1997. Degradation and loss of anadromous salmonid habitat in the Pacific Northwest. Pages 277–314 *in* D.J. Souder, P.A. Bisson, and R.J. Naiman, editors. Pacific salmon and their ecosystems: status and future options. Chapman & Hall, New York.

Groot, C., and L. Margolis, editors. 1991. Pacific Salmon Life Histories. UBC Press, Vancouver, British Columbia.

Hartman, G.F., C. Groot, and T.G. Northcote. 2000. Science and management in sustainable salmonid fisheries: the ball is not in our court. Pages 31–50 *in* E.E. Knudsen, C.R. Steward, D.D. MacDonald, J.E. Williams, and D.W. Reiser, editors. Sustainable fisheries management: Pacific salmon. Lewis Publishers, Boca Raton, Florida.

Hilborn, R. 1992. Hatcheries and the future of salmon in the Northwest. Fisheries 17(1):5–8.

Hutchings, J.A., C.J. Walters, and R.L. Haedrich. 1997. Is scientific inquiry incompatible with government information control? Canadian Journal of Fisheries and Aquatic Science. 54:1198–1210.

Hyatt, K., and B. Riddell. 2000. The importance of "stock" conservation definitions to the concept of sustainable fisheries. Pages 51–62 *in* E.E. Knudsen, C.R. Steward, D.D. MacDonald, J.E. Williams, and D.W. Reiser, editors. Sustainable fisheries management: Pacific salmon. Lewis Publishers, Boca Raton, Florida.

Johnson, T.H., R. Lincoln, G.R. Graves, and R.G. Gibbons. 1997. Status of wild salmon and steelhead stocks in Washington state. Pages 127–144 *in* D.J. Stouder, P.A. Bisson, and R.J. Naiman, editors. Pacific salmon and their ecosystems: status and future options. Chapman & Hall, New York.

Jones, K.K., and K.M.S. Moore. 2000. Habitat assessment in coastal basins in Oregon: Implications for coho salmon production and habitat restoration. Pages 329–340 *in* E.E. Knudsen, C.R. Steward, D.D. MacDonald, J.E. Williams, and D.W. Reiser, editors. Sustainable fisheries management: Pacific salmon. Lewis Publishers, Boca Raton, Florida.

Knudsen, E.E. 2000. Managing Pacific salmon escapements: The gaps between theory and reality. Pages 237–272 *in* E.E. Knudsen, C.R. Steward, D.D. MacDonald, J.E. Williams, and D.W. Reiser, editors. Sustainable fisheries management: Pacific salmon. Lewis Publishers, Boca Raton, Florida.

Kruse, G.H. 1998. Salmon run failures in 1997–1998: a link to anomalous ocean conditions? Alaska Fishery Research Bulletin 5:55–63.

Langer, O.E., F. Hietkamp, and M. Farrell 2000. Human population growth and the sustainability of urban salmonid streams in the lower Fraser Valley. Pages 349–361 *in* E.E. Knudsen, C.R. Steward, D.D. MacDonald, J.E. Williams, and D.W. Reiser, editors. Sustainable fisheries management: Pacific salmon. Lewis Publishers, Boca Raton, Florida.

Levings, C.D. 2000. An overview assessment of compensation and mitigation techniques used to assist fish habitat management in British Columbia Estuaries. Pages 341–347 *in* E.E. Knudsen, C.R. Steward, D.D. MacDonald, J.E. Williams, and D.W. Reiser, editors. Sustainable fisheries management: Pacific salmon. Lewis Publishers, Boca Raton, Florida.

Lichatowich, J. 1997. Evaluating salmon management institutions: the importance of performance measures, temporal scales, and production cycles. Pages 69–87 *in* D.J. Stouder, P.A. Bisson, and R.J. Naiman, editors. Pacific salmon and their ecosystems: status and future options. Chapman & Hall, New York.

Lichatowich, J., L. Mobraud, L. Lestelle, and T. Vogel. 1995. An approach to the diagnosis and treatment of depleted salmon populations in Pacific Northwest watersheds. Fisheries 20(1):10–18.

Milner, A. M., and R. G. Bailey. 1989. Salmonid colonization of new streams in Glacier Bay, Alaska. Aquaculture and Fisheries Management 20:179–192.

Mobraud, L.E., J.A. Lichatowich, L.C. Lestelle, and T.S. Vogel. 1997. An approach to describing ecosystem performance "through the eyes of salmon." Canadian Journal of Fisheries and Aquatic Sciences 54:2964–2973.

NRC (National Research Council). 1996. Upstream: salmon and society in the Pacific Northwest. National Academy Press, Washington, D.C.

Nehlsen, W., J.E. Williams, and J.A. Lichatowich. 1991. Pacific salmon at the crossroads: Stocks at risk from California, Oregon, Idaho, and Washington. Fisheries 16(2):4–21.

Radtke, H.D., and S.W. Davis. 2000. Economic feasibility of salmon enhancement propagation programs. Pages 381–392 *in* E.E. Knudsen, C.R. Steward, D.D. MacDonald, J.E. Williams, and D.W. Reiser, editors. Sustainable fisheries management: Pacific salmon. Lewis Publishers, Boca Raton, Florida.

Slaney, T.L., K.D. Hyatt, T.G. Northcote, and R.J. Fielden. 1996. Status of anadromous salmon and trout in British Columbia and Yukon. Fisheries 21(10):20–35.

Smoker, W.W., B.A. Bachen, G. Freitag, H.J. Geiger, and T.J. Linley. 2000. Alaska ocean ranching contributions to sustainable fisheries. Pages 407–420 *in* E.E. Knudsen, C.R. Steward, D.D. MacDonald, J.E. Williams, and D.W. Reiser, editors. Sustainable fisheries management: Pacific salmon. Lewis Publishers, Boca Raton, Florida.

Spurgeon, D. 1997. Canada's cod leaves science in hot water. Nature 386:108.

SFF (Sustainable Fisheries Foundation). 1996. Toward sustainable fisheries: Building a cooperative strategy for balancing the conservation and use of west coast salmon and steelhead populations. Sustainable Fisheries Foundation, Ladysmith, British Columbia.

Vallentyne, J.R., and A.L. Hamilton. 1987. Managing the human uses and abuses of aquatic resources in the Canadian ecosystem. Pages 513–533 *in* M.C. Healey and R.R. Wallace, editors. Canadian Aquatic Resources. Canadian Fisheries and Aquatic Sciences Bulletin 215.

Vallentyne, J.R., and A.M. Beeton. 1988. The "ecosystem" approach to managing human uses and abuses of natural resources in the Great Lakes Basin. Environmental Conservation 15(1):58–62.

Van Alen, B.W. 2000. Status and stewardship of salmon stocks in Southeast Alaska. Pages 161–193 *in* E.E. Knudsen, C.R. Steward, D.D. MacDonald, J.E. Williams, and D.W. Reiser, editors. Sustainable fisheries management: Pacific salmon. Lewis Publishers, Boca Raton, Florida.

Welch, D.W., Y. Ishida, and K. Nagasawa. 1998. Thermal limits and ocean migrations of sockeye salmon (*Oncorhynchus nerka*): long-term consequences of global warming. Canadian Journal of Fisheries and Aquatic Sciences 55:937–948.

2 The Needs of Salmon and Steelhead in Balancing Their Conservation and Use

Carl V. Burger

Abstract.—Over the past 100 years, Pacific salmon *Oncorhynchus* spp. and steelhead trout *O. mykiss* populations in the Pacific Northwest have experienced dramatic declines as a result of human population growth and associated development of the region's natural resources. Any strategy to reverse those declines will depend on achieving consensus among a diverse group of stakeholders willing to (1) restore damaged habitats and watersheds; (2) consider the biological needs of salmonid fishes; and (3) understand, conserve, and manage existing levels of biodiversity within an ecosystem context. The objective of this chapter is to review and summarize the needs of Pacific salmon and steelhead trout from perspectives that promote their sustainability and perpetuity, and, within that context, provide a framework for the development of a sustainable fisheries strategy. Although volumes of information are available on the life histories and habitat requirements of Pacific salmonids, new concepts have begun to address the importance of habitat complexity, genetic diversity among locally adapted populations, and the need for appropriate units of conservation. Habitat variability and complexity are the templates that produce diverse, locally adapted populations of Pacific salmon and steelhead trout. Molded by conditions in the environments they colonized, salmon and steelhead have unique adaptations that may have taken hundreds—perhaps thousands—of years to evolve. Any strategy to reverse the declines of salmon and steelhead must focus on conserving both current levels of genetic diversity as well as restoration and maintenance of habitats compatible with the specific needs of individual populations. An argument can be made that we know enough already about the needs of our salmon and steelhead to allow us to make appropriate decisions. There is a strong need to integrate what we know about local salmonid adaptations and habitat needs with conservation and management strategies that address the problems we have created. It may be possible to have healthy, sustainable salmonid populations coastwide if we do not wait too long to make up our minds.

INTRODUCTION

Henry David Thoreau in his *A Week on the Concord and Merrimack Rivers* wrote:

> Perchance after a thousand years, if the fishes will be patient and pass their time elsewhere, meanwhile nature will have leveled the Billerica Dam and the Lowell factories, and the Concord River will run clear again, to be explored by migrating shoals.

The environmental impacts on the fish resources addressed by Thoreau in the northeastern U.S. soon worsened, and a similar dilemma began to materialize on the west coast of North America, in direct proportion to growing numbers of settlers who placed ever-increasing demands on anadromous fishes. The demise of fishery resources in the Pacific Northwest has become a biopolitical crisis that directly confronts the region's primary economic interests: its hydropower, agriculture, forest, and commercial fishing industries (Allendorf and Waples 1996).

As we proceed to take on the enormous task of developing a strategy to reverse the declines of certain Pacific Northwest salmonids, and as we continue to build consensus and "buy-ins" among a diverse and growing group of stakeholders who share in common concerns, we must reflect on and consider the biological needs of our fish. What is it that they need for survival and self-perpetuation? What do they have to do to survive? And what is it that we humans have to avoid, in terms of environmental impacts, if we are to assure the health of our fishery resources into the future?

The objective of this chapter is to summarize the needs of Pacific salmon *Oncorhynchus* spp. and steelhead trout *O. mykiss* from perspectives that promote their sustainability and perpetuity and, within that context, provide a foundation for balancing the conservation and use of salmon and steelhead in the Pacific Northwest.

HISTORICAL PERSPECTIVE

Pacific salmon and steelhead constitute spectacular resources that range from northern California, through British Columbia, to Alaska's Aleutian Islands and beyond. These resources sustained aboriginal peoples and were quite abundant through about the mid-1800s, providing substantial cultural, recreational, economic, and symbolic importance to the Pacific Northwest. For example, up to 16 million anadromous salmonids were thought to return to the Columbia River Basin during that pre-development time period (NPPC 1986)—an estimate that gradually dropped to about one eighth (2 to 3 million fish) by the 1990s (NPPC 1994; NRC 1996).

In the late 1800s, hatcheries began to appear in the Pacific Northwest in an attempt to compensate for overfishing and declining numbers of salmon and steelhead. At the turn of the century, hatcheries were deemed to be necessary in the Columbia River drainage to maintain salmon runs. By the 1920s various fishing restrictions (for example, a ban on fishwheels) were imposed on the Columbia River as salmon numbers continued their precipitous decline. Because as much as 80% of the contemporary Pacific Northwest production of anadromous salmon and trout is estimated to be of hatchery origin, natural production is probably less than 5% of what it was historically (up to 16 million) (Allendorf and Waples 1996).

Through the early and mid-1990s, various factors contributed to the ensuing decline of anadromous fishes in the Pacific Northwest, including human population growth and the associated development of the region's natural resources. Turn-of-the-century overfishing and subsequent irrigation and hydroelectric projects were the major impacts leading to declines in both the U.S. and Canada. Despite tremendous investments in hatcheries, runs began to disappear as habitat was developed and as fishing and other pressures were applied to the resource.

Historic fishery declines can be tied directly to impacts from expanded fishing, forestry, agriculture, hydroelectric, and industrial activities by increasing numbers of settlers in the Pacific Northwest. The explosive population growth in several northwestern U.S. states is apparent from census data reviewed by the NRC (1996): about 100,000 inhabitants were present during the mid- to late 1800s, over 1 million people by 1900, and 8.7 million by 1990 (annual growth rate, 2%). Not surprisingly, and largely associated with construction of dams that blocked fish passage, Pacific salmon disappeared from about 40% of their historic range in Washington, Oregon, California, and Idaho, and the numbers of fish in most remaining populations were substantially reduced (NRC 1996). Although major dams were not constructed on the Fraser River in British Columbia, a similar pattern of decline in salmon runs occurred through the 1970s (NRC 1996) from mining, logging, agriculture, and other forms of industrial development. Although Alaska did not escape historic declines, its generally stable or increasing trend of contemporary Pacific salmon production (Burger and Wertheimer 1995) likely results from conservative management since statehood and a comparatively small number of statewide inhabitants (about 600,000). Because of larger salmon population sizes, anadromous salmonids in Alaska also may have a better ability to recover from both natural habitat disruptions as well as those imposed by human development, than populations subjected to decades of environmental impacts in southern portions of the range. Weakened by many years of human-induced pressures, some populations in non-Alaskan portions of the range contain low numbers of spawners that can no longer withstand severe winters, floods, and other natural events. With 214 populations at risk of extinction and more than 100 populations recently extinct in Washington, Oregon, California, and Idaho, we are truly "at the crossroads" (Nehlsen et al. 1991).

LIFE HISTORY AND HABITAT NEEDS OF PACIFIC SALMONIDS

Since the earliest parts of this century, volumes of information have been written and published on the habitat and survival needs of Pacific Northwest salmon and steelhead trout. One only has to look at recent compendiums such as *Pacific Salmon Life Histories* (Groot and Margolis 1991) to find a wealth of data on the life histories and population dynamics of sockeye *O. nerka*, pink *O. gorbuscha*, chum *O. keta*, chinook *O. tshawytscha*, and coho *O. kisutch* salmon throughout their established ranges. *Influences of Forest and Rangeland Management on Salmonid Fishes and Their Habitats* (Meehan 1991) exemplifies a second collection of chapters and articles on how certain human influences affect salmonids, their life histories, and their diverse habitats. Also, countless journal publications and symposia over the past several decades have addressed the life histories, habitats, and needs of Pacific salmon and steelhead trout. Clearly, there is no shortage of material on these subjects that would preclude development and implementation of a sustainable fisheries strategy.

Fishery scientists have spent careers studying salmon and steelhead — they know the needs of these fishes quite well. And whether scientist, policymaker, politician, or layperson, all are familiar *to some extent* with what the various species of salmonid fishes need, in terms of habitat and life history requirements, for survival, reproduction, and perpetuation of their numbers. Nevertheless, and as the strategy for sustainability is developed, it is appropriate to compare and contrast some of the general needs of Pacific salmon and steelhead trout, with examples of some of the things that may adversely affect them. Also, I will try to integrate what we know about their life history and habitat needs with some key concepts — habitat complexity, genetic diversity, and local adaptation — that have implications for sustaining Pacific Northwest Salmonids.

Life History of Pacific Salmon and Steelhead Trout

Pacific salmon and steelhead trout are anadromous: they migrate to the ocean for maturation and return to freshwater to spawn. Although some investigators have suggested a marine origin for modern-day salmonid fishes, others have proposed a primitive origin entirely within freshwater, and that evolution of anadromy was in response to more abundant food resources in the marine, rather than freshwater environment (Gross 1987).

The migration of juvenile salmonids to the ocean can occur within several days of their emergence from spawning gravels (pink and chum salmon) or following one or more years in freshwater (chinook, coho, and sockeye salmon; steelhead trout). The length of time spent in freshwater as juveniles varies considerably among and within species; it is apparently regulated by both environmental and genetic factors that influence behavior as well as certain physiological changes prior to the seaward migration (Randall et al. 1987). Juvenile migration to the sea (smolt stage) frequently occurs at night during spring or early summer (Figure 2.1). Once at sea, initial movements appear to be oriented within estuaries and along coastlines, but subsequent marine movements occur throughout the North Pacific Ocean and Bering Sea, where distances traveled may be extensive (Groot and Margolis 1991). However, as described for sockeye salmon (Burgner 1991), offshore movements are complex and are affected by physical (season, temperature, and salinity) and biological factors (age, size, availability of food, and population-specific genetics).

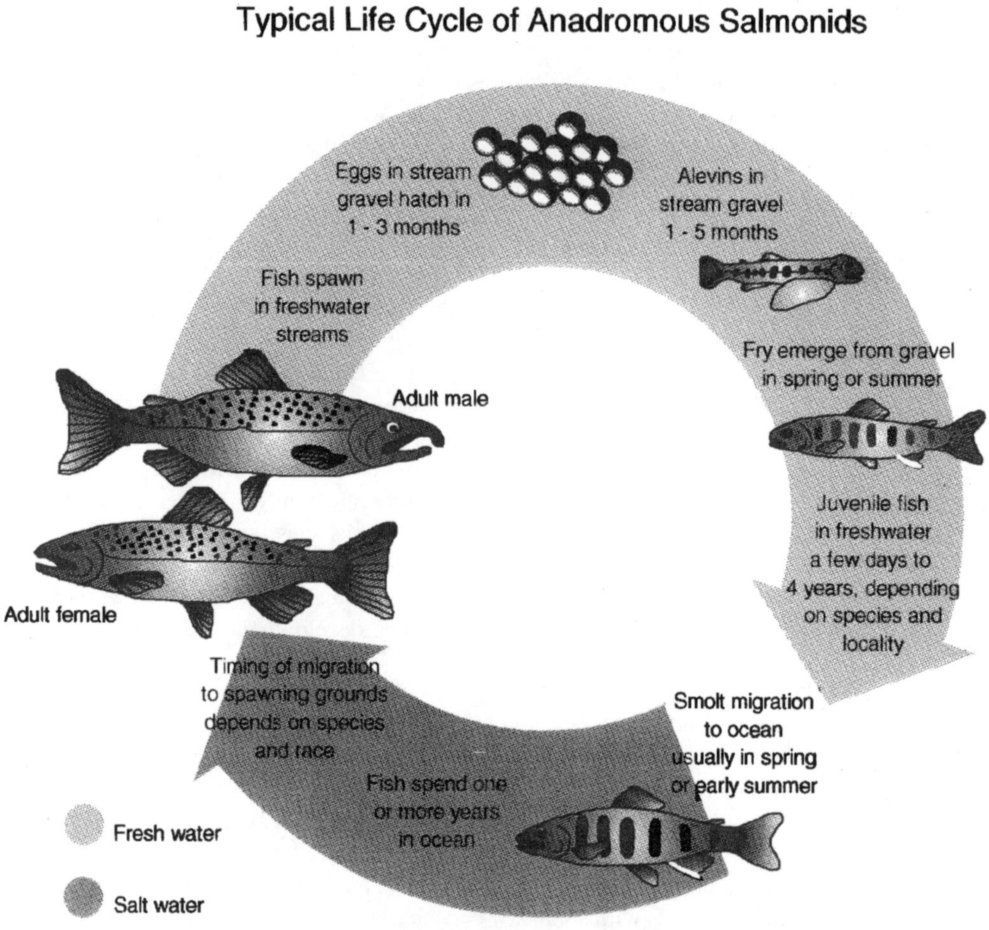

Typical Life Cycle of Anadromous Salmonids

FIGURE 2.1   Typical life cycle of anadromous salmonids (modified from Meehan and Bjornn 1991).

The length of time at sea (usually 1 to 4 years) and the age at which adult salmon and steelhead return to freshwater to spawn varies considerably among and within species (Figure 2.2).

Adult migrations to freshwater spawning areas are also variable and depend on the species, seasonal form, geographical location, and physical and biological factors. The variability in timing of return migrations to freshwater and in spawning activities encompasses virtually all seasons of the year for the five species of Pacific salmon and steelhead trout. It is beyond the scope of this chapter to provide such detail although numerous examples and summaries of species- and population-specific run and spawning times have been published (Groot and Margolis 1991; Meehan and Bjornn 1991).

Adult salmon and steelhead typically spawn during spring, summer, and fall in rivers having substrates suitable for water permeation and oxygenation of eggs that develop in the spawning nest (redd). However, there is substantial variation in the physical features (depths at which eggs are deposited, sizes of substrates used, etc.) of the spawning habitat selected (Bjornn and Reiser 1991), and in the types of environments where spawning occurs. One species (sockeye salmon) is capable of reproducing in a wide array of spawning environments including shorelines and upwelling areas entirely within lakes, lateral tributaries flowing into lakes, and outlet rivers (Burgner 1991). Adult salmon usually guard their redds against predation for one or more weeks

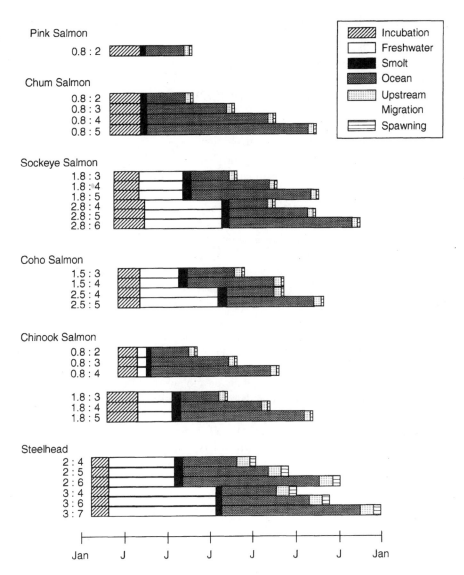

FIGURE 2.2 Lengths of time that representative anadromous salmonids spend in various developmental stages and in fresh water or the ocean. Numbers under species names are age keys indicating the years spent rearing in fresh water from egg to smolt (first column) in relation to the total age of the fish (second column). For example, 1.5 : 3 indicates a fish in its third year of life that had spent 1.5 years incubating and rearing in fresh water before migrating to sea (modified from Meehan and Bjornn 1991).

after spawning. Whereas Pacific salmon all die following spawning-related activities, steelhead trout are capable of surviving, and a certain percentage of steelhead adults returns to the ocean following reproduction.

Eggs of Pacific salmon and steelhead trout hatch within the redd from 1 to 3 months post-fertilization. The resulting alevins (yolk-sac fry) continue to reside within the protective confines of gravel substrates for 1 to 5 additional months (Figure 2.1). The juveniles typically emerge during spring and summer to actively seek food, avoid predators, emigrate to sea (pink and chum salmon) or establish new home ranges within river drainages (chinook and coho salmon; steelhead trout) or lakes (sockeye salmon), to commence the cycle anew.

Needs of Pacific Salmon and Steelhead Trout

Pacific salmon and steelhead trout have varied yet quite specific needs that must be met at different times of the life cycle. When these needs are met, distribution and abundance is maintained and production is often maximized. For example, anadromous salmonids need a sufficient volume and depth of water (at least 18 to 24 cm; Thompson 1972) with unobstructed passage, for adults to reach their spawning grounds, within the constraints of a limited amount of bodily energy. They need flow velocities that do not usually exceed about 2.1 m/s (Thompson 1972) if they are to complete their migrations. Also, they need clean, well-oxygenated, and unpolluted streams and lakes with loose, appropriately sized spawning gravel substrates (range 1.3 to 10.2 cm in diameter; Bell 1986) that are free of silt and fine sediment (Cordone and Kelley 1961). Pacific salmon and steelhead trout need cool or cold temperature regimes to optimize the success of migration, spawning, and egg incubation. Juveniles must have adequate shade and cover to escape predators and they need high aquatic productivity to ensure rapid growth to the smolt stage. A riparian interface between aquatic and terrestrial ecosystems is therefore crucial.

Bjornn and Reiser (1991) provide a very comprehensive review of the various physical and biological habitat requirements of anadromous salmonids by life history stage. There are numerous other publications on the habitat needs of salmonid fishes (for example, Marcus et al. 1990) in the available scientific literature. The point is that, within each species, the life history events and needs of both juveniles and adults are remarkably precise. The events must occur and the needs must be met within very specific windows of time for survival to be maximized.

What do Pacific salmon and steelhead trout have to do, in order to survive? Allendorf and Waples (1996) provide an excellent summary of the steps necessary for an anadromous salmonid to fulfill its life cycle. It must complete embryonic development and emerge at a time appropriate for its natal drainage; find food and avoid predators as a juvenile; locate cool water during rearing or be able to tolerate increased temperatures; undergo the physiological changes associated with the smoltification process at the most opportune time for seaward migration; enter the ocean during a window of time that promotes survival and growth as a subadult; locate food and survive at sea for one or more years; reverse the physiological process of smoltification when returning to freshwater as an adult; home to the natal spawning area and find a mate; and ensure that eggs are fertilized and deposited at an appropriate location, time, and depth. A break in any of the above processes will reduce or prevent survival in the following generation.

The various seasonal runs, races, and ecological forms within several species of Pacific salmon and steelhead trout also must escape a host of commercial, sport, and aboriginal fisheries to complete their life cycles. They need to be able to find their home streams in numbers adequate to ensure high future survival. The large numbers of adults that returned to Northwest watersheds at the turn of the century helped to confer a certain level of population resiliency toward disasters that were natural in origin. That resiliency is no longer present. Hence, salmon and steelhead resources can no longer withstand the combined effects of catastrophes (severe winters, floods, droughts, etc.), harvest rates of 60 to 70% or more in certain populations, and impacts from humans exploiting non-fish resources. Because former levels of population resiliency have been lost, modern adult escapement goals must be based on conservative standards that account for the synergistic effects of natural and human-induced activities, if perpetuation and sustainability are to be achieved.

There is another benefit from large annual escapements of Pacific salmon to home-stream spawning areas. For some time, fishery biologists have realized that the carcasses that remain after salmon spawning and death provide an important source of nutrients in otherwise nutrient-poor environments (Donaldson 1967). Although we have known that these carcasses contribute to the trophic productivity of streams, we have not always known how much. Recent research by Bilby et al. (1996) demonstrates that more than 30% of the nitrogen and carbon in the tissues of juvenile coho salmon and steelhead trout derives from the carcasses of spawned-out salmon. Their study was conducted in tributaries to the Snoqualmie River, Washington, where they also found marine-originating N and C in invertebrate organisms and streamside vegetation, meaning that the carcasses of salmon returning from the sea are contributing to overall health of the aquatic and adjacent terrestrial ecosystems. Such results demonstrate that, as salmon escapements decline, a self-perpetuating cycle of decreasing productivity ensues, leading to fewer and fewer returning adults that can "fertilize" the home stream. An example provided by the World Fisheries Trust (1996) highlights the type of spiraling decline and the unsustainable fishery that can result from too few spawners returning to the home stream. In a population having only about 50 female salmon, perhaps 100,000 eggs will be viable. Of those, about 20,000 will hatch into fry, 1,000 will make it to the sea as smolts, 50 will survive to adulthood, and 15 salmon will escape various commercial, sport, and aboriginal fisheries to return to spawn. Thus, future generations of Pacific salmon are progressively diminished when fewer and fewer fish return to the spawning grounds. The above example underscores the necessity to consider interactions at the ecosystem level, if we want to implement conservation and management strategies that achieve the goal of fisheries sustainability.

Several additional concepts have been used to describe the biological needs and relationships among Pacific Northwest salmonids. These include habitat complexity, the concept of locally adapted populations, and the evolutionary unit concept. Along these lines, it seems appropriate to interject some results from past and present research on Alaska's wild salmonids. Because such research encompasses populations and habitats that have remained relatively pristine, results can provide new insights and implications for restoring depleted populations in more southerly portions of their range.

Habitat Complexity

Pacific salmon and steelhead trout need aquatic habitat with instream complexity that includes deep, unsedimented pools (Bjornn and Reiser 1991), undercut banks with overhanging vegetation (Platts 1991), and an adequate amount of natural woody debris to create resting, feeding, and overwintering areas (Everest et al. 1987). They do not benefit from overly simple, channelized streams, altered by human activities, which offer little or no complexity in which to hide, feed, and rest (Hicks et al. 1991). Nor do they need obstacles to upstream migration or streambed alterations (Atkinson et al. 1967) that can either prevent spawning or suffocate eggs deposited in once-clean gravels (Cordone and Kelley 1961). Good quality soils and forest understory along non-erosive stream banks slowly release stored water and nutrients, from riparian areas (Platts 1991). The root systems of vegetated areas also provide stream bank stability and buffer the erosive effects of water flow (Sedell and Beschta 1991). Stream banks where the forest canopy has been overcut by logging activities or overgrazed by cattle will erode when heavy rain and high flows occur, causing sedimentation of once-productive anadromous fish habitat (Chamberlin et al. 1991; Platts 1991). The result is the loss of a complex reach of stream that formerly produced high numbers of salmonid fishes.

Several studies have confirmed the benefits to juvenile salmonids from adding and simulating various types of cover that increase the complexity of streams. Juvenile salmon and steelhead abundance has been positively correlated with the addition of woody debris to increase habitat complexity in Southeast Alaska streams (Murphy et al. 1984; Heifetz et al. 1986; Johnson et al. 1986). Using unpublished data, Bjornn and Reiser (1991) showed that juvenile steelhead trout were much more abundant in channels having a combination of deep water, undercut banks, large rocks, and brush, than in pools with less cover. Combinations of cover translate into increased complexity and large numbers of fish.

LOCAL ADAPTATION

Habitat variability and complexity are the templates (drivers) that produce diverse, locally adapted populations of Pacific salmon and steelhead trout. Because each population is adapted to its home-stream habitat, conservation and maintenance of habitat complexity promotes conservation of genetic diversity and local adaptation.

In concert with the ability to home to streams of origin after several years at sea, the variation in life history patterns that has evolved within the diverse environments used by anadromous salmonids produces many distinct local populations within a species (Altukhov and Salmenkova 1991; Utter et al. 1993). Using sockeye salmon as an example, both anadromous and non-anadromous (kokanee) forms occur throughout the range (Burgner 1991). Kokanee spend their lives exclusively in freshwater, and this form may have evolved repeatedly, in many locations, from the ocean-going variety (Foote et al. 1989). Another form (residual sockeye salmon; Burgner 1991) also matures without going to sea and is a transition between kokanee and anadromous sockeye salmon. Considerable life history diversity is apparent within the anadromous form. Although a few populations of sockeye salmon spawn in rivers having no nursery lake (Eiler et al. 1992), most anadromous populations return to drainages having lakes adjacent to the spawning grounds (Burgner 1991), where there is additional variability in choice of spawning habitat for early- and late-seasonal forms. Typical of the spawning distribution between early- and late-seasonal forms throughout Alaska, British Columbia, and Russia, early-run sockeye salmon spawn in cold, lateral tributaries of Tustumena Lake, Southcentral Alaska, whereas late-run fish spawn in considerably warmer, lake shoreline areas and in the lake's outlet (Burger et al. 1995). Despite temporal separation in spawning times between the two runs, offspring from both groups can emerge at the most favorable time for lake rearing because their life cycles are thermally adapted to, and in fine synchrony with, the temperature regimes of their specific spawning habitats (e.g., Burger et al. 1985; Brannon 1987).

Within a specific population, the timing of upstream migration and spawning is often remarkably precise from year to year and seems directly related to the temperature regime of the home stream. Spawning times among anadromous salmonids may have evolved in response to the variable temperatures of the home streams and many populations appear to have unique times and temperatures for spawning that maximize survival of fry (Bjornn and Reiser 1991). The actual temperature regime during egg incubation may be the primary evolutionary factor determining time of spawning (Brannon 1987).

A relationship between spawning time and home-stream thermal regime has been observed between the earliest (inland) and latest (coastal) runs of pink salmon in Southeast Alaska (Sheridan 1962), early and late runs of chinook salmon in Southcentral Alaska (Burger et al. 1985), and among populations of sockeye salmon spawning in the Fraser River drainage, British Columbia (Brannon 1987) and Kenai Peninsula, Alaska (Burger et al. 1995). The relationship between time of spawning and temperature promotes synchronous emergence among offspring of early and late forms during favorable periods of aquatic productivity which, in turn, promotes high survival and continued genetic selection for specific run and spawning times. Instances have been noted where life history patterns (migration timing) have changed because of environmental alterations (Atkinson et al. 1967). To the extent that Pacific salmon and steelhead trout are locally adapted to specific home-stream thermal conditions, there are implications for any human-related activities that alter temperature regimes in streams where life history events depend on a predictable annual thermal cycle (Burger et al. 1985).

The migrational behavior of sockeye salmon fry provides additional evidence for local adaptation (Taylor 1991). Fry from riverine spawning areas in lake outlets migrate upstream into the nursery lake after emergence, whereas fry emerging from spawning areas in lateral lake tributaries migrate downstream (Brannon 1972), behavior that promotes favorable feeding, growth, and survival. Similarly, laboratory studies have shown that the compass orientation of other populations of sockeye salmon fry correspond to a direction that would lead them to forage areas (Quinn 1985).

In these cases involving highly variable habitats (templates), genetic selection likely favors those behavioral adaptations that promote high juvenile survival.

Recent studies on Alaskan sockeye salmon (Tustumena Lake) suggest that the ecological differences between early- and late-seasonal forms have a genetic basis. Early-run sockeye salmon spawn in inlet tributaries to Tustumena Lake during early August, whereas late-run fish spawn from mid- to late September in the lake's outlet river. Despite temporal and spatial closeness in reproduction between the two Tustumena runs (spawning times differ by only 4 to 6 weeks; spawning areas by 20 to 30 km or less), 36% of late-run salmon had a mitochondrial DNA haplotype not observed in over 450 samples from early-run tributary spawners (Burger et al. 1997). A similar phenomenon (the most common haplotype in late-run salmon was rare or absent in respective early-run fish) was recently observed across two additional species' lines: chinook salmon from the Kasilof River, Alaska (Adams et al. 1994) and coho salmon that spawn within the Kenai River (C. Burger, unpublished data). These results suggest a high degree of local adaptation between various early- and late-seasonal forms of Pacific salmon in Alaska. The template for selection may have been the different spawning habitats used by the seasonal forms of salmon. In all three cases, late-run fish spawn in rivers downstream of (and warmed by) large lakes whereas early-run salmon spawn in comparatively cooler lake or river tributaries—environments where temperatures are substantially different. In non-Alaskan parts of the Northwest, populations of chinook salmon with different run timing from the same stream are more similar genetically than populations having similar run timing from different areas (Utter et al. 1989). Thus, greater levels of genetic diversity appear to be present among Alaskan populations of Pacific salmon than among those in some southern portions of the range. However, there also appears to be greater variability in within-drainage (home-stream) temperature regimes experienced by early- and late-seasonal forms in Alaska than occurs among seasonal forms that spawn within drainages in more southerly latitudes. Genetically diverse Alaskan populations may have resulted from low numbers of colonizers (genetic "bottlenecks") that evolved under harsh climatic conditions in variable environments.

Clearly, research on Alaska's wild populations confirms that the various seasonal runs of Pacific salmon, adapted for spawning in specific areas, have very specific ecological requirements. Because each stream or river has its own set of complex variables, the salmon inhabiting them are adapted in an equally complex way (Thompson 1965). We may never know the historic richness of the salmon and steelhead resources of the West Coast, because many populations were lost prior to our knowledge of stock structure (Nehlsen et al. 1991). However, a key point can be made: the ecological traits and adaptations (phenotypes) of surviving populations must be considered as we develop strategies to recover and restore depleted populations.

A recently completed genetic evaluation of Frazer Lake (Alaska) sockeye salmon highlights the exacting ecological needs of this species when transplanted to a new environment. Anadromous fish had not colonized Frazer Lake (impassable waterfall) prior to an introduction of sockeye salmon in the 1950s (Blackett 1979). The introduction program used three donor populations from different geographical areas. The donors also represented different ecological forms (early-run lake tributary spawners, late-run lake shoreline spawners, and fish that had reproduced in a lake's outlet river). Other introductions using the anadromous form of *O. nerka* had been largely unsuccessful (Wood 1995). However, contemporary production at Frazer Lake grew to nearly 1 million adults in some years (C. Swanton, Alaska Department of Fish and Game, personal communication). Microsatellite allele frequencies were obtained from tissue samples from the three donor populations and from seven subpopulations presently spawning in Frazer Lake or its tributaries. Genetic analysis showed that the lake outlet donor stock did not contribute to colonization of Frazer Lake—no outlet spawners were observed during spawning ground surveys at Frazer Lake (1995). However, additional genetic results (C. Burger, unpublished data) suggest an affinity between the reproductive adaptations of the donors and habitat they colonized in Frazer Lake: allele frequencies of the lake shoreline donor were most similar to two subpopulations of sockeye salmon that spawn along the lake's shoreline;

the lake tributary donor was most genetically similar to four subpopulations spawning in Frazer Lake tributaries. Apparently, the reproductive adaptations of the donors were maintained in the "new" environment.

Results of the Frazer Lake genetic analyses strongly suggest that populations *and habitats* need to be protected if we are to conserve salmonid genetic diversity and locally adapted populations. Localized populations need protection because they are uniquely adapted to the streams and lakes they inhabit and, consequently, are best able to survive there. Genetic differences among individuals in a local population are the basis for natural selection and adaptive evolution, whereas differences between populations reflect local adaptation to past environmental conditions: the latter represents a pool of genetic variation that is the most valuable component of genetic diversity.

Frazer Lake results may help to explain why some of the historic salmon introduction and restoration efforts have failed. Introduction programs must ensure that donor populations are adapted to the types of conditions to be encountered in the area of release. The only surprise is that we have known this for over a hundred years! Hume (1893; cited by the NRC 1996) stated:

> I firmly believe that like conditions must be had...to bring about like results, and that to transplant salmon successfully they must be placed in rivers where the natural conditions are similar to that from which they have been taken.

The main point is that the parts are not interchangeable. The examples above tell us that populations of anadromous fish are locally adapted to precise environmental conditions. Whenever we lose a specific population or habitat, we often lose a very unique life history pattern and a certain level of ecological and genetic diversity that may be slow to arise again.

Diverse life history strategies are important for population stability and persistence in salmonids (Rieman and McIntyre 1993). Diversity may serve to stabilize populations that inhabit variable environments, and it may help to refound population segments within a species (Mullan et al. 1992; Titus and Mosegaard 1992). The adaptations to localized environments increase the overall diversity of the much larger metapopulation (Hanski and Gilpin 1991; Rieman and McIntyre 1993). An array of small, locally adapted populations (metapopulation) within a species means that these smaller units might not respond to changing environmental conditions as a synchronous group but rather the risk of extinction is spread because the loss of all populations at any one time is lowered (Rieman and McIntyre 1993). These authors state that, in addition, the surviving populations may be able to refound, via straying, weaker subpopulations that decline or become extinct, but that such recolonization depends on a host of factors including suitability of habitat, and it may take considerable time. However, some genetically or ecologically unusual populations may be difficult to replace in human time frames (Allendorf and Waples 1996). Salmonid adaptations have likely taken hundreds if not thousands of years, yet many of the recent human-induced changes and depletions in watersheds have occurred over decades. Despite some apparent evolutionary plasticity among anadromous salmonids, there may be insufficient time for evolutionary compensation, especially when portions of the metapopulation are now missing (NRC 1996).

EVOLUTIONARY UNITS

A publication of the proceedings of an American Fisheries Society symposium held in Monterey, California (Nielsen 1995), presents a wealth of new information about defining evolutionary units of conservation, a subject that cannot be ignored in any discussion of the needs of salmonid fishes.

Although the Evolutionarily Significant Unit concept (Waples 1991) was introduced to describe distinct population segments in a framework for interpreting the meaning of distinct populations under the Endangered Species Act, much controversy remains over how units of conservation are defined. The Monterey Symposium provides a context for examining both the biological species concept (which defines species as being reproductively isolated) and the evolutionary species concept that replaces reproductive isolation with spatial or temporal isolation. Evolutionary units refer to populations that maintain their identity at temporal and spatial scales, in a unique evolutionary lineage.

Defining populations of Pacific salmon and steelhead trout from a unit concept, be it based on genetic parameters or ecological considerations and requirements, is not easy. Just as one set of characteristics does not define the needs of salmonids, one size of unit will not fit all (Behnke 1995). Ecological diversity is the template for biodiversity, and temporal and spatial variability must be considered at scales that range from small sections of streams to entire watersheds (Nielsen 1995). Studies referred to previously provide good examples: even though an early and late run of salmon may spawn within 20 to 30 km of one another in a single drainage, their very different genetic structure and their concordant ecological differences suggest that the individual runs are distinct units. Such a result indicates highly restrictive gene flow between populations that spawn in closely adjacent habitat (Burger et al. 1997); this concept was validated in separate drainages used by three species of salmon. Such findings would not support any definition of a unique unit that was based on overly broad geographic boundaries. Rather, the results suggest that the individual spawning area (tributary, lake shoreline, lake outlet, etc.) may indeed be the appropriate unit of conservation and management for Pacific salmonids (Varnavskaya et al. 1994; Burger et al. 1997).

HABITAT AND POPULATION RESTORATION

Insofar as ecological diversity is the template for biodiversity among Pacific salmon and steelhead trout, the key to sustaining and restoring populations depends on our ability to conserve and provide suitable habitat complementary to life history variation. The distinct genetic and ecological differences among salmonid populations are, after all, a consequence of the unique local environments they inhabit (Gharrett and Smoker 1993). Therefore, the key to sustaining populations of Pacific salmon and steelhead trout is to sustain their habitat using an ecosystem approach that allows for a full expression of biodiversity.

I firmly believe that a watershed approach is necessary to restore the health of the drainages used by Pacific salmonids, and that success will be possible if we focus our energies and activities at this level. We must eliminate activities that degrade habitat at their upland sources, such as erosion from clearcut areas in headwater drainage areas, that sediment downstream spawning areas. Some of the roads used to access timber harvest areas need to be removed or stabilized (stormproofed) for the very same reason. Once the major sources of pollutants and erosion in a watershed are eliminated or stabilized, and when barriers to migration are removed, deep pools and instream complexity can be restored in downstream areas along with stream bank stability. Reeves et al. (1991) provide a wealth of information on techniques to restore and stabilize watersheds and streams. Already practiced or underway with community involvement in both the U.S. and Canada, habitat restoration activities involving broadly based societal decisions are key to any strategy to sustain Pacific salmon and steelhead trout.

What is the role of fish hatcheries in efforts to restore anadromous fish populations? Although much has been published on certain adverse effects of selection and genetic change from artificial propagation (see, for example, Allendorf and Waples 1996), re-designs and newly focused efforts may provide opportunities to reduce the risks of outbreeding depression, domestication selection, and loss of genetic variation between and within populations (Campton 1995). Hatcheries have recently begun to refocus their activities toward conservation and restoration priorities. Hatchery facilities conceptually possess the ability to conserve and maintain specific genetic resources, to assist with wild fish recovery, and to conduct research under controlled conditions to improve our understanding of life history and physiological processes. Fish culture programs and facilities are not only necessary to assist in efforts to achieve a sustainable fisheries strategy, but they may also be necessary to recover and restore depleted populations, particularly in areas impacted by habitat modifications.

Although it is beyond the scope of this chapter to discuss management techniques and options to increase escapements of salmonids to spawning grounds, it should be clear from the material presented that increased escapements are prerequisite to development of a sustainable fisheries strategy (see, for example, Knudsen 2000).

LOSS OF FISH AND HABITATS: ARE THEY REALLY BIOLOGICAL PROBLEMS?

I hope I have accomplished my goal to provide the message that the database on the biological needs of Pacific salmon and steelhead trout is very extensive. Concurrently, I have attempted to demonstrate, using numerous examples, the extreme importance of conserving and restoring salmonid habitat complexity and the importance of variation and diversity within locally adapted populations. The Sustainable Fisheries Conference demonstrated that the declines of our Pacific salmon and steelhead are not really resource problems, but people problems, emanating from growing human needs and demands for natural resources. It was also suggested that we know enough, and that we do not really need more studies to confirm what we already know. From a perspective based on the needs of fish, I echo those sentiments.

We already know enough about the needs of our salmon and steelhead to allow us to make appropriate decisions. We can successfully raise salmonids in hatchery facilities; we can accurately predict how they will respond when we alter their habitats; we know that some of the strongest, healthiest populations exist only in pristine areas; we know their ecosystem functions and interactions; and so on. If we know all of these things, one could surmise that our problems may not relate to biology at all. Our salmonid populations could be declining primarily because we have not fully embraced a willingness to restore and maintain suitable conditions for them.

In this light there are additional needs for Pacific salmon and steelhead trout. We need to learn how to integrate what we know about the local adaptations of salmon and steelhead with management strategies that address the problems we humans have created. To protect and assure the perpetuation of salmonid resources, as well as healthy economies of those who depend on them, we need coordinated action on many fronts. I believe this can be accomplished if society is willing to

- change institutional structures for managing salmon and their habitats;
- refocus the salmon management process toward a community-based, watershed-oriented approach;
- implement true salmon ecosystem management; and
- adequately enforce existing laws and programs.

In summary, we need to act, initiate recovery, protect what we have, and restore what has been damaged. If the public and its elected and appointed decision-makers want healthy, sustainable salmonid populations coast-wide, they can have those resources—if they do not wait too long to make up their minds.

One is reminded of the parable of the wise man approached by a trickster who, motioning to his tightly clasped hands said, "Sir, I hold in my hands a bird. Is it alive or dead?" If the wise man said the bird was dead, the trickster would open his hands to let the bird fly away. If he said it was alive, the trickster would crush the bird and kill it. But the wise man would not thus be fooled. He said to the trickster, "The life of the bird is in your hands."

ACKNOWLEDGMENTS

I thank Jack McIntyre (U.S. Forest Service, retired) for his many thoughtful insights, discussions, and suggestions that substantially improved this manuscript. I also appreciate the reviews and discussions provided by Eric Knudsen and Kim Scribner, Biological Resources Division, U.S.

Geological Survey, Anchorage, Alaska. I thank Ernie Brannon (University of Idaho) and Don MacDonald (Sustainable Fisheries Foundation) for their helpful manuscript reviews. Bill Meehan (U.S. Forest Service, retired) and Ted Bjornn (University of Idaho) conceptualized the material used in the figures, and the American Fisheries Society kindly permitted its use.

REFERENCES

Adams, N. S., and five coauthors. 1994. Variation in mitochondrial DNA and allozymes discriminates early and late forms of chinook salmon (*Oncorhynchus tshawytscha*) in the Kenai and Kasilof rivers, Alaska. Canadian Journal of Fisheries and Aquatic Sciences 51 (Supplement):172–181.

Allendorf, F. W., and R. S. Waples. 1996. Conservation and genetics of salmonid fishes. Pages 238–280 *in* J. C. Avise and J. L. Hamrick, editors. Conservation genetics: case histories from nature. Chapman & Hall, New York.

Altukhov, Y. P., and E. A. Salmenkova. 1991. The genetic structure of salmon populations. Aquaculture 98:11–40.

Atkinson, C. E., J. H. Rose, and T. O. Duncan. 1967. Pacific salmon in the United States. International North Pacific Fisheries Commission Bulletin 23:43–223.

Behnke, R. J. 1995. Overview: morphology and systematics. Pages 41–43 *in* J. L. Nielson, editor. Evolution and the aquatic ecosystem: defining unique units in population conservation. American Fisheries Society Symposium 17, Bethesda, Maryland.

Bell, M. C. 1986. Fisheries handbook of engineering requirements and biological criteria. U.S. Army Corps of Engineers, Office of the Chief of Engineers, Fish Passage Development and Evaluation Program, Portland, Oregon.

Bilby, R. E., B. R. Fransen, and P. A. Bisson. 1996. Incorporation of nitrogen and carbon from spawning coho salmon into the trophic system of small streams: evidence from stable isotopes. Canadian Journal of Fisheries and Aquatic Sciences 53:164–173.

Bjornn, T. C., and D. W. Reiser. 1991. Habitat requirements of salmonids in streams. Pages 83–138 *in* W. R. Meehan, editor. Influences of forest and rangeland management on salmonid fishes and their habitats. American Fisheries Society Special Publication 19, Bethesda, Maryland.

Blackett, R. F. 1979. Establishment of sockeye (*Oncorhynchus nerka*) and chinook (*O. tshawytscha*) salmon runs at Frazer Lake, Kodiak Island, Alaska. Journal of the Fisheries Research Board of Canada 36:1265–1277.

Brannon, E. L. 1972. Mechanisms controlling migration of sockeye salmon fry. International Pacific Salmon Fisheries Commission, Bulletin XXI. New Westminster, British Columbia.

Brannon, E. L. 1987. Mechanisms stabilizing salmonid fry emergence timing. Canadian Special Publication of Fisheries and Aquatic Sciences 96:120–124.

Burger, C. V., R. L. Wilmot, and D. B. Wangaard. 1985. Comparison of spawning areas and times for two runs of chinook salmon (*Oncorhynchus tshawytscha*) in the Kenai River, Alaska. Canadian Journal of Fisheries and Aquatic Sciences 42:693–700.

Burger, C. V., and A. C. Wertheimer. 1995. Status and trends of Pacific salmon in Alaska. Pages 343–347 *in* E. T. LaRoe et al., editors. Our living resources: a report to the Nation on the distribution, abundance, and health of U.S. plants, animals, and ecosystems. U.S. Department of the Interior, National Biological Service, Washington, D.C.

Burger, C. V., J. E. Finn, and L. Holland-Bartels. 1995. Pattern of shoreline spawning by sockeye salmon in a glacially turbid lake: evidence for subpopulation differentiation. Transactions of the American Fisheries Society 124:1–15.

Burger, C. V., W. J. Spearman, and M. A. Cronin. 1997. Genetic differentiation of sockeye salmon subpopulations from a geologically young Alaskan Lake system. Transactions of the American Fisheries Society 126:926–938.

Burgner, R. L. 1991. Life history of sockeye salmon (*Oncorhynchus nerka*). Pages 1–117 *in* C. Groot and L. Margolis, editors. Pacific salmon life histories. UBC Press, Vancouver, British Columbia.

Campton, D. E. 1995. Genetic effects of hatchery fish on wild populations of Pacific salmon and steelhead: what do we really know? Pages 337–353 *in* H. L. Schramm, Jr. and R. G. Piper, editors. Uses and effects of cultured fishes in aquatic ecosystems. American Fisheries Society Symposium 15, Bethesda, Maryland.

Chamberlin, T. W., R. D. Harr, and F. H. Everest. 1991. Timber harvesting, silviculture, and watershed processes. Pages 181–205 *in* W. R. Meehan, editor. Influences of forest and rangeland management on salmonid fishes and their habitats. American Fisheries Society Special Publication 19, Bethesda, Maryland.

Cordone, A. J., and D. W. Kelley. 1961. The influences of inorganic sediment on the aquatic life of streams. California Fish and Game 47:189–228.

Donaldson, J. R. 1967. The phosphorous budget of Iliamna Lake, Alaska, as related to cyclical abundance of sockeye salmon. Doctoral dissertation. University of Washington, Seattle.

Eiler, J. H., B. D. Nelson, and R. F. Bradshaw. 1992. Riverine spawning by sockeye salmon in the Taku River, Alaska and British Columbia. Transactions of the American Fisheries Society 121:701–708.

Everest, F. H., G. H. Reeves, J. R. Sedell, D. B. Hohler, and T. Cain. 1987. The effects of habitat enhancement on steelhead trout and coho salmon smolt production, habitat utilization, and habitat availability in Fish Creek, Oregon, 1983–86. Annual Report for 1986, Bonneville Power Administration, Division of Fish and Wildlife, Project 84–11, Portland, Oregon.

Foote, C. J., C. C. Wood, and R. E. Withler. 1989. Biochemical genetic comparison of sockeye salmon and kokanee, the anadromous and non-anadromous forms of *Oncorhynchus nerka*. Canadian Journal of Fisheries and Aquatic Sciences 46:149–158.

Gharrett, A. J., and W. W. Smoker. 1993. Genetic components in life history traits contribute to population structure. Pages 197–202 *in* J. G. Cloud and G. H. Thorgaard, editors. Genetic conservation of salmonid fishes. Plenum Press, New York.

Groot, C., and L. Margolis, editors. 1991. Pacific salmon life histories. UCB Press, Vancouver, British Columbia.

Gross, M. R. 1987. Evolution of diadromy in fishes. Pages 14–25 *in* M. J. Dadswell and five coeditors. Common strategies of anadromous and catadromous fishes. American Fisheries Society Symposium 1, Bethesda, Maryland.

Hanski, I., and M. Gilpin. 1991. Metapopulation dynamics: brief history and conceptual domain. Biological Journal of the Linnean Society 42:3–16.

Heifetz, J., M. L. Murphy, and K. V. Koski. 1986. Effects of logging on winter habitat of juvenile salmonids in Alaskan streams. North American Journal of Fisheries Management 6:52–58.

Hicks, B. J., J. D. Hall, P. A. Bisson, and J. R. Sedell. 1991. Responses of salmonids to habitat changes. Pages 483–518 *in* W. R. Meehan, editor. Influences of forest and rangeland management on salmonid fishes and their habitats. American Fisheries Society Special Publication 19, Bethesda, Maryland.

Johnson, S. W., J. Heifetz, and K. V. Koski. 1986. Effects of logging on the abundance and seasonal distribution of juvenile steelhead in some southeastern Alaska streams. North American Journal of Fisheries Management 6:532–537.

Knudsen, E. E. 2000. Managing Pacific salmon escapements: the gaps between theory and reality. Pages 237–272 *in* E. E. Knudsen, C. R. Steward, D. D. MacDonald, J. E. Williams, and D. W. Reiser, editors. Sustainable fisheries management: Pacific salmon. Lewis Publishers, Boca Raton, Florida.

Marcus, M. D., M. K. Young, L. E. Noel, and B. A. Mullan. 1990. Salmonid habitat relationships in the Western United States: a review and indexed bibliography. General Technical Report RM-188. U.S. Department of Agriculture, Forest Service, Fort Collins, Colorado.

Meehan, W. R., editor. 1991. Influences of forest and rangeland management on salmonid fishes and their habitats. American Fisheries Society Special Publication 19, Bethesda, Maryland.

Meehan, W. R., and T. C. Bjornn. 1991. Salmonid distributions and life histories. Pages 47–82 *in* W. R. Meehan, editor. Influences of forest and rangeland management on salmonid fishes and their habitats. American Fisheries Society Special Publication 19, Bethesda, Maryland.

Mullan, J. W., K. Williams, G. Rhodus, T. Hillman, and J. McIntyre. 1992. Production and habitat of salmonids in mid-Columbia River tributary streams. Monograph 1, U.S. Department of the Interior, Fish and Wildlife Service, Washington, D.C.

Murphy, M. L., and five coauthors. 1984. Role of large organic debris as winter habitat for juvenile salmonids in Alaska streams. Proceedings of the Western Association of Fish and Wildlife Agencies 64:251–262.

NRC (National Research Council). 1996. Upstream: salmon and society in the Pacific Northwest. National Academy Press, Washington, D.C.

Nehlsen, W., J. E. Williams, and J. A. Lichatowich. 1991. Pacific salmon at the crossroads: stocks at risk from California, Oregon, Idaho, and Washington. Fisheries 16(2):4–21.

Nielsen, J. L., editor. 1995. Evolution and the aquatic ecosystem: defining unique units in population conservation. American Fisheries Society Symposium 17, Bethesda, Maryland.

NPPC (Northwest Power Planning Council). 1986. Compilation of information on salmon and steelhead losses in the Columbia River Basin. Appendix D, 1987 Columbia River Basin Fish and Wildlife Program. Northwest Power Planning Council, Portland, Oregon.

NPPC (Northwest Power Planning Council). 1994. Elements of Bonneville's Fish and Wildlife Revenue Impacts, 1991–1994. Northwest Power Planning Council, Portland, Oregon.

Platts, W. S. 1991. Livestock grazing. Pages 389–423 *in* W. R. Meehan, editor. Influences of forest and rangeland management on salmonid fishes and their habitats. American Fisheries Society Special Publication 19, Bethesda, Maryland.

Quinn, T. P. 1985. Homing and the evolution of sockeye salmon (*Oncorhynchus nerka*). Contributions in Marine Science 27(Supplement):353–366.

Randall, R. G., M. C. Healy, and J. B. Dempson. 1987. Variability in length of freshwater residence of salmon, trout, and char. Pages 27–41 *in* M. J. Dadswell and five coeditors. Common strategies of anadromous and catadromous fishes. American Fisheries Society Symposium 1, Bethesda, Maryland.

Reeves, G. H., J. D. Hall, T. D. Roelofs, T. L. Hickman, and C. O. Baker. 1991. Rehabilitating and modifying stream habitats. Pages 519–557 *in* W. R. Meehan, editor. Influences of forest and rangeland management on salmonid fishes and their habitats. American Fisheries Society Special Publication 19, Bethesda, Maryland.

Rieman, B. E., and J. D. McIntyre. 1993. Demographic and habitat requirements for conservation of bull trout. General Technical Report INT-302. U.S. Department of Agriculture, Forest Service, Fort Collins, Colorado.

Sedell, J. R., and R. L. Beschta. 1991. Bringing back the "Bio" in bioengineering. Pages 160–175 *in* J. Colt and R. J. White, editors. Fisheries Bioengineering Symposium. American Fisheries Society Symposium 10, Bethesda, Maryland.

Sheridan, W. L. 1962. Relation of stream temperatures to timing of pink salmon escapements in Southeast Alaska. Pages 87–102 *in* N. J. Wilimovsky, editor. Symposium on pink salmon. H. R. MacMillan Lectures in Fisheries, University of British Columbia, Vancouver.

Taylor, E. B. 1991. A review of local adaptation in Salmonidae, with particular reference to Pacific and Atlantic salmon. Aquaculture 98:185–207.

Thompson, K. 1972. Determining stream flows for fish life. Pages 31–50 in Proceedings, instream flow requirements workshop. Pacific Northwest River Basins Commission, Vancouver, Washington.

Thompson, W. F. 1965. Fishing treaties and salmon of the North Pacific. Science 150:1786–1789.

Titus, R. G., and H. Mosegaard. 1992. Fluctuating recruitment and variable life history of migratory brown trout *Salmo trutta* in a small, unstable stream. Journal of Fish Biology 41:239–255.

Utter, F. M., G. Milner, G. Stahl, and D. Teel. 1989. Genetic population structure of chinook salmon *Oncorhynchus tshawytscha* in the Pacific Northwest. Fishery Bulletin 85:13–23.

Utter, F. M., J. E. Seeb, and L. W. Seeb. 1993. Complementary uses of ecological and biochemical genetic data in identifying and conserving salmon populations. Fisheries Research 18:59–76.

Varnavskaya, N. V., and six coauthors. 1994. Genetic differentiation of subpopulations of sockeye salmon (*Oncorhynchus nerka*) within lakes of Alaska, British Columbia, and Kamchatka, Russia. Canadian Journal of Fisheries and Aquatic Sciences 51 (Supplement):147–157.

Waples, R. S. 1991. Pacific salmon, *Oncorhynchus* spp., and the definition of "species" under the Endangered Species Act. Marine Fisheries Review 53:11–22.

Wood, C. C. 1995. Life history variation and population structure in sockeye salmon. American Fisheries Society Symposium 17:195–216.

World Fisheries Trust. 1996. The salmon survival guide. World Fisheries Trust, Victoria, British Columbia.

3 Science and Management in Sustainable Salmonid Fisheries: The Ball is Not in Our Court

Gordon F. Hartman, Cornelis Groot, and Thomas G. Northcote

Abstract.—As fisheries scientists and managers, we are but minor players in determining sustainability of Pacific Northwest salmonids and the ecosystem functions that support them. To the extent that we can have a role in achieving "sustainability," the challenges may be greater than we think. The population processes of Pacific salmonids are set within a background of four interacting systems: the forest/land/freshwater, ocean, climate, and human systems. The degree of complexity within these interrelated systems demands that new dimensions be added to our understanding of that complexity, and of its implications in regard to the behavior of ecosystems and biological populations. Beyond these difficult considerations, other challenges to the achievement of sustainability are even more daunting. Whether salmonids and, more importantly, the ecosystems that sustain them, can be maintained, depends on future scenarios of human population and industrial growth—locally, nationally and globally. If the current exponential growth patterns continue, salmonid resources will be lost in a maze of cumulative impacts and a prolonged series of conflicts and compromises. The time horizons at which such losses will occur are not predictable, but their inevitability is indicated by our history and our continuing actions. The public must be educated that fisheries scientists and managers are willing and anxious to fulfill their roles to the greatest extent possible, but the future of salmonids and their habitats will be determined by the direction of society as a whole. The public, political leaders and, last of all, biologists should not allow terms like sustainability to become the "idea-fixes" of the present that replace failed "techno-fixes" of the past. Fisheries scientists and managers have a responsibility to raise these issues as high as possible into public view so that neither society nor politicians can avoid the broader challenges that surround the problem of sustaining salmonids in the growth-dominated Pacific Northwest.

INTRODUCTION

There were two primary purposes of the April 1996 conference, *Toward Sustainable Fisheries: Balancing the Conservation and Use of Salmon and Steelhead in the Pacific Northwest.* The first purpose was to develop a science-based sustainable fisheries strategy that would promote a balanced approach to fisheries resource management and use—one based on sound ecological and economic principles—to ensure that fish populations remain viable, productive, and accessible to future generations. An objective of the conference was to "identify the major biological, economical, social, and political impediments" to the recovery of Pacific salmonid populations. The second purpose was to increase awareness among the public and governments of Canada and the U.S. about the problems that threaten salmonids in the near and distant future and about the efforts that are now underway to save them.

1-56670-480-4/00/$0.00+$.50

Salmonids will not be sustained unless functional environments are maintained for them. In reality, the environment (i.e., habitat) is the resource; the fish are a product from it. Difficulties in maintaining the environments, and the associated natural production of viable and sustainable populations of salmonids within them, occur at two levels. The first level arises from the complexity of the biology, population dynamics, and ecological requirements of salmonids. The outcome of the complex interactions of the many factors involved in salmonid production and survival can change rapidly and is often difficult to predict. The second level is caused by the human system and its effects on the natural environment. Human population size and growth, exploitation of natural resources, production of toxic chemicals and gases, and salmon harvest practices are the major concerns. These factors continuously alter salmonid ecosystems directly or indirectly and challenge the natural sustainability of the fish over time. To sustain salmonids for "future generations," society must respond to the elements that constitute these two levels of difficulty.

The objective of this chapter is to examine whether Pacific Northwest salmonid ecosystems can be maintained in the shadow of human systems that are socially, economically, and politically locked into a growth-dependent system that is coupled with rapid resource exploitation. This chapter will emphasize that, because of globalization of many human system processes, the future of salmonids is connected to the future of the world, for which people are all jointly responsible. Because successful sustainable salmon fisheries is really a matter influenced by global-scale social and political events, a large responsibility of biologists and fisheries managers is to connect salmon sustainability to larger-scale issues. The *Sustainable Fisheries Strategy* should clearly reflect this reality and seek ways to bring these critical issues to public and political attention, in far more effective ways than in the past.

PACIFIC COAST SALMONIDS—COMPLEXITY AND VULNERABILITY

This chapter focuses on Pacific salmonid production and problems along the western part of North America and in the eastern Pacific Ocean. This area is vast and the fish populations within it are complex. Fishery managers must sustain not just the various species of Pacific salmon, but rather the thousands of genetically distinct stocks into which these species are partitioned. This says nothing about the other kinds of fish or the whole web of forage species with which they all interact. The "Pacific Northwest" arena in which these interactions continue to be played out includes an enormous land mass extending from 35° to 65°N latitude, reaching a thousand km inland from the coast. Through this land mass drain five very large rivers (Sacramento-San Joaquin, Columbia, Fraser, Skeena, Yukon), many more moderate-sized rivers, and thousands of smaller streams—most of which once supported salmonid populations. Many still support salmonid populations, but in reduced numbers resulting from cumulative effects of a variety of human activities. Especially hard hit are the Sacramento-San Joaquin and Columbia River systems, but few of the others have been spared.

The arena also extends beyond the continental environment. It extends thousands of kilometers offshore into "international" waters plied, fished, and exploited by fleets whose operators often have little knowledge of, let alone concern for, "sustainability." The offshore marine component of the arena is not just contained within the migratory boundaries of the salmonids of interest, but is subject to uncontrollable and largely unpredictable changes operating on a global scale, such as El Niño events.

A number of aspects of the life histories of Pacific Coast salmonids makes them vulnerable to human impacts and major natural changes. The life histories of the various species in the genus *Oncorhynchus* and their physiological ecology have been reviewed in two recent volumes (Groot and Margolis 1991; Groot et al. 1995). However, there are several characteristics of the Pacific salmon and steelhead trout, fundamental to their biology, that ultimately make them vulnerable.

First, salmonids are anadromous; they spawn and incubate in fresh water, migrate to salt water as juveniles, and return to fresh water as adults. Basic to their anadromous behavior are temporal physiological changes in salinity tolerance (Hoar 1965). The basic pattern of using both fresh and salt water has been modified according to species in various ways during evolution by lengthening or shortening the different life stage components. This has produced a wide spectrum of life history strategies for Pacific salmon and steelhead (Hoar 1965; Groot 1982).

Second, each species has a linked series of discrete life history stages that are tied to a particular habitat or part of it, and there can be a range of ages over which the various stages occur. However, within any population, each stage has to occur within a fixed seasonal time range. Therefore, for each species and populations within it, the life history is carried out through a series of space- and time-dependent stages. For example, the life history of 4-year-old sockeye from the Fraser River carries it through 12 distinct habitat changes (Figure 3.1; Groot and Margolis 1991). The fixed behavior patterns underlying such habitat changes vary between and within species and are related to the specific environments these salmonids depend on during their life cycle (Groot 1982).

During their various life history stages, salmon and steelhead utilize different habitats that are spread over large geographic areas, for various periods of time that range from days to years. With increasing scope and numbers of human activities, the potential for encountering conditions that can adversely affect one or more of these many life history stages increases. For example, the potential for overharvest exists when multiple stocks and species are concentrated in long marine migration routes (e.g., Johnstone Strait), particularly when errors in allocation occur. Similarly, salmon can also be exposed to adverse environmental conditions, such as low dissolved oxygen level and severe parasitism by sea lice in estuaries, when they are concentrated by temperature barriers (e.g., upper end of Alberni Inlet on the west coast of Vancouver Island; Steer and Hyatt 1987; Johnson et al. 1996). Furthermore, a wide range of land and water use activities has the potential to cause adverse effects on salmonids in freshwater systems. Therefore, the combination of anadromy, utilization of multiple habitats, run dominance, homing, death after spawning, and highly variable numbers from year to year, makes it very difficult to effectively manage the mix of species, stocks, and interannual populations that constitute the Pacific salmon resource.

While managing Pacific salmon is difficult, restoration of lost populations is even more difficult. Because Pacific salmon "home," they tend to form subpopulations that are geographically and temporally separated. Such stocks of salmon differ genetically from each other and have distinct life history patterns that are adapted to their home stream. Thus, it may not be possible to restore individual stocks in the event of natural or human-induced population failure. For example, total numbers of sockeye salmon above the Hell's Gate landslide, which occurred in 1913 in the Fraser River, required several decades to even partially restore themselves following fishway construction (Bisson et al. 1992). Some populations, such as the Upper Adams River stock, have never recovered (Roos 1986; Williams 1987). In this case, only a small replacement run has been re-established since restoration work began in the late 1950s (Williams 1987).

SALMONIDS IN A WIDER COMPLEX OF INTER-CONNECTED SYSTEMS

For thousands of years, healthy habitats have supported diverse and productive populations of salmon and steelhead. However, rapidly expanding human populations, both here and elsewhere in the world, have resulted in the alteration of some basic features of salmonid habitats and threaten to cause even greater changes in the future (Figure 3.2).

The importance of human population growth in fisheries matters has been recognized by many within the American Fisheries Society (e.g., Becker 1992; Brouha 1994; Alverson 1995; Eipper 1995; Levy et al. 1996; Hartman 1996; Northcote 1996). Population growth in the whole of British Columbia, the total Fraser River basin, and the Okanagan basin, is exponential and driven largely by immigration from other areas (Hartman 1996; Fraser Basin Council 1997). In the upper Fraser River basin, numbers did not increase in the 1980s, but they are projected to rise by about 50% in the next three decades (Figure 3.3; Hartman 1996; Fraser Basin Council 1997). Currently, about 60,000 people per year are added to the British Columbia population (Anonymous 1990). The

Age	Habitat change	Life cycle stage		Timing
1st year	1. egg to gravel	alevin		winter
	2. gravel to lake	fry	0.0	spring
	3. down lake	fingerling	0.0	summer/ autumn
2nd year	4. lake to estuary	smolt	1.0	spring
	5. estuary north along shore	juvenile	1.0	summer
	6. offshore to wintering grounds	immature	1.0	autumn
3rd year	7. wintering grounds to summer feeding grounds	immature	1.1	spring
	8. summer feeding grounds to wintering grounds	immature	1.1	autumn
4th year	9. wintering grounds southwards to summer feeding grounds	immature	1.2	spring
	10. feeding grounds to home river estuary	maturing		summer
	11. home river estuary to nursery lake	mature adult	1.2	autumn
	12. lake to river spawning grounds	mature adult	1.2	autumn

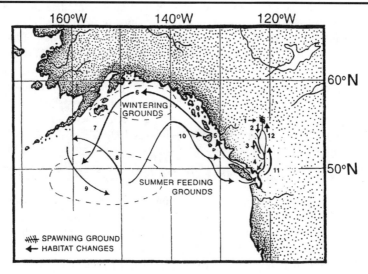

FIGURE 3.1 The habitat changes, life cycle stages, and timing of changes for 4-year-old sockeye salmon. An example of complex life histories, use of many parts of the environment, and time constraints on such use among Pacific salmon (redrawn and modified from Groot 1982).

population of British Columbia's aboriginal people is also increasing sharply (Northcote 1996). In the northwestern U.S., urban population growth is continuous but not exponential, and rural populations are relatively stable (Figure 3.3; Naiman 1992).

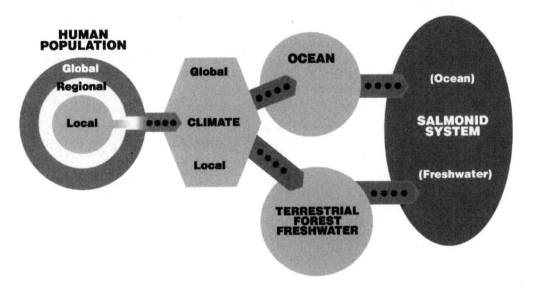

FIGURE 3.2   The salmonid system, a complex of ocean and terrestrial (land/forest/freshwater) environments, is influenced at its core by local and global climate conditions. Climate conditions are being altered by human population impacts.

Significantly, most of the people who reside in the Pacific Northwest live in urban areas. While many urbanites are cognizant of key environmental issues, they may not fully appreciate the impacts they impose on the environment. Nevertheless, it has been shown that the "ecological footprint" of an urban population, such as that in Vancouver, is immense, affecting forests, land, and water over an area about 174 times larger than its political boundary area (Rees 1996).

Considering the nature, extent, and severity of the impacts that are associated with human activities, it is likely that human population growth represents the greatest threat to Pacific salmon. The activities of the rising numbers of people will cumulatively generate increasing impacts on the ecosystems that support salmonid production. Such impacts will vary in nature, range from intensive (e.g., point source discharges) to extensive (e.g., nonpoint source discharges), and range in severity from minor to major. Importantly, such impacts are likely to have cumulative effects on salmon populations. Hence, human population will remain the driving force of environmental degradation, and creatures such as salmonids will act as barometers of such degradation and indicators of risk to ourselves.

Human activities are linked to Pacific salmon at three levels: local (watershed), regional (west coast of North America), and global. At the local level, human activities have direct effects on the lands, forests, and associated freshwater habitats that sustain salmon and steelhead populations (i.e., through land use, water management, and fisheries management). At the regional level, the incremental effects of human activities are likely to be primarily associated with the cumulative effects of land and water use, with fisheries management policies and programs playing a secondary role. At the global level, the cumulative activities of the world's populations are likely to cause or substantially contribute to global climate changes, which in turn will likely affect both marine and terrestrial/freshwater aquatic ecosystems.

EFFECTS OF LOCAL ACTIVITIES ON PACIFIC SALMON

At the local level, human activities can have dramatic effects on the aquatic habitat upon which Pacific salmonids depend. A variety of land use activities, such as logging, mining, urbanization, agriculture, utility corridor, and linear developments (i.e., roads, railways, etc.), have directly affected the quality and quantity of freshwater habitats. Such direct effects result from point and

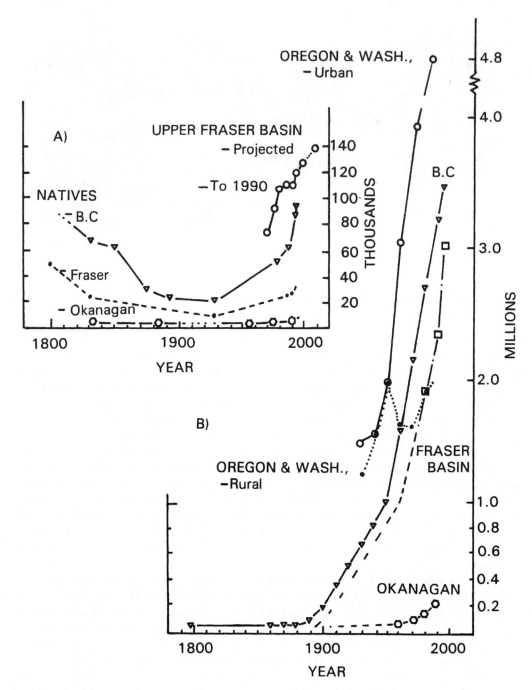

FIGURE 3.3 A. Patterns of population growth for aboriginal people (Natives) in B.C., the Fraser River basin and the Okanagan (redrawn from Northcote 1996), and total population in the upper Fraser basin (redrawn from Hartman 1996). B. Population growth in: Washington and Oregon (redrawn from Naiman 1992), B.C., the Fraser basin, and Okanagan (redrawn from Northcote 1996).

nonpoint source discharges of contaminants (which alter water quality conditions) and the physical destruction of small streams. These developmental activities can also indirectly affect salmonid habitats by removing riparian vegetation (with consequent effects on channel geomorphology and stream temperature regimes), altering upland vegetation (with consequent effects on stream hydrology and water quality), and diverting water for other uses. Reservoir construction for flood control and hydroelectric power generation has also adversely affected the availability and accessibility of fresh-water habitats. Because salmonids utilize freshwater habitats for both spawning and rearing activities, watershed disturbances have significant effects on the size, structure, and stability of salmonid populations (Hall et al. 1987; Holtby 1988; Holtby and Scrivener 1989; Hartman and Scrivener 1990).

Systems for assessing, monitoring, and mitigating the effects of land and water use developments have improved over the past 20 years. In Canada, legislation now exists to require a full environmental assessment of new, major development projects (i.e., Canadian Environmental Assessment Act). While such legislation has definitely improved our ability to protect the environment, politicians still attempt to exempt certain major developments from the environmental assessment process. In addition, historical and existing developments are not subjected to the same assessment process; therefore, we may never know just how much habitat has been altered or permanently lost. Furthermore, many types of human activities are not regulated, or the existing regulations are poorly enforced, which can lead to further degradation of aquatic habitats. Of equal or greater concern, very little information exists to facilitate the assessment of the cumulative effects of the multitude of land use impacts, plus those of climate change, in large drainage basins. Unfortunately, under the present Canadian government approach of severely downgrading research programs, it is unlikely that the capacity to obtain such information will be established.

EFFECTS OF REGIONAL ACTIVITIES ON PACIFIC SALMON

At a regional scale, the cumulative effects of land and water use activities have, and will continue to have, serious impacts on Pacific salmon populations. To illustrate this point, a brief case study has been assembled, which demonstrates cumulative effects in the Fraser River basin. The Fraser is the largest river system in British Columba, draining an area of 209,000 km^2. It is also the most important salmon-producing system in the province.

The Fraser River basin supports a wide range of human activities, which are largely focused on economic development. Environmental change began in the Fraser River basin over 100 years ago with settlement and logging in the lower areas. This was followed by the expansion of placer mining, which began near Hope, British Columbia, and spread to the middle and upper reaches of the Fraser River. The rate of development in the Fraser River basin has been rapid during the last 70 years. The following activities are of greatest concern with respect to the sustainability of salmon and steelhead populations (Figure 3.4):

- Extensive urbanization is evident in the lower reaches of the river. This has resulted in the loss of about 100 small streams or sections of them (Anonymous 1995). It has also resulted in the discharge of 1,181,546 m^3/day of wastewater to the system (Boeckh et al. 1991). Of this total, 313,914 m^3/day were industrial wastes (Schreier et al. 1991).
- Contentious gravel mining operations in major salmon-producing rivers have gone on for years in the lower Fraser River basin. Seventy-six percent of all sand and gravel mined in British Columbia comes from about 70 active pits in the lower Fraser River sub-basin (Schreier et al. 1991). The impacts of gravel mining come from alteration of aquifers, disturbance of river channels, and sediment production from washing operations (Schreier et al. 1991).
- Forestry activities have occurred throughout the Fraser River system. There are currently serious public concerns about forestry effects on water quality in the Quesnel and Horsefly lake areas. Recent expansion of the industry in the Stuart/Takla River drainage has been described by Langer et al. (1992). Logging has occurred over about 2,600 km^2 and was proposed for another 1,800 km^2 in the Stuart/Takla River basin by the end of 1996 (Hartman 1996). To date, 4,400 km^2 will have been logged within an important sockeye salmon production area.

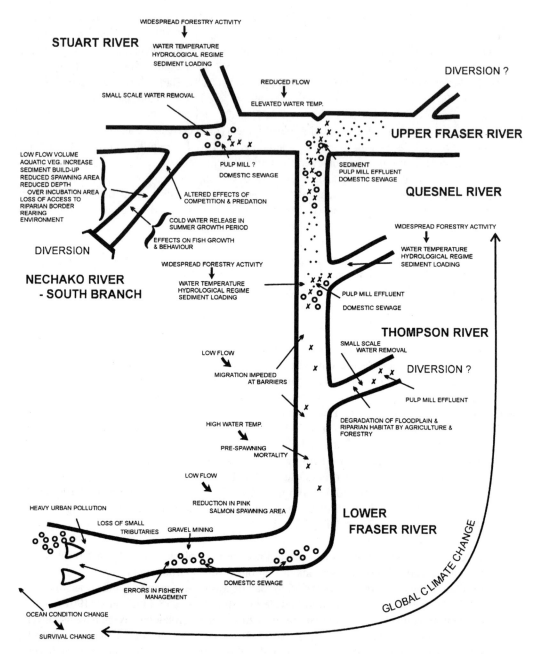

FIGURE 3.4 Schematic representation of the Fraser River drainage with existing or potential environmental impacts. There is high potential for cumulative impacts in the system; however, the combinations of impacts that may be cumulative are not indicated.

- One large water diversion, the Kemano project, has altered the total character of the Nechako River, a major drainage in the upper Fraser River. This diversion project (Figure 3.4) has resulted in elevated water temperatures, loss of spawning space, alteration of rearing area, changes in channel geometry, icing conditions, barriers to migration in the middle Fraser River, and temperature effects on migration in the Nechako River above Prince George (Schouwenburg 1990). Problems surrounding this project are still unresolved. About 65% of the river discharge continues to be diverted (Hartman 1996).

- Extensive water diversion occurs in the Fraser River system. Except for the Kemano project, most diversion is for agriculture. In 1989 there were 17,998 water licenses in the Fraser basin (Boeckh et al. 1991), with 7,463 of these held in the Thompson River drainage alone. Many of these licenses permit only minor abstraction of water. Screening of irrigation intakes now stops the loss of juvenile salmonids in most cases. However, there are still some situations where young salmonids are carried onto fields and lost. Beyond this direct mortality, water diversion results in lower flows and consequently higher temperatures in the affected tributaries and downstream parts of the system.
- Degradation of riparian areas has not been evaluated for the whole Fraser River basin. However, in the Thompson River drainage, concern about damage to riparian habitat led to studies along 56.2, 63.6, 112.9, and 153.8 km of channel in four tributaries. Riparian vegetation was absent along 7 to 44% of the channel, in the four tributaries (Miles 1996). Extensive and severe degradation of riparian habitat has also occurred in the Nicola, Coldwater, and Bonaparte rivers (M. Miles, M. Miles and Associates Ltd., personal communication).
- Water quality in the Fraser River system has been degraded by five pulp mills which, for example, discharged 581,540 m^3/d of wastewater into it in 1985 (Schreier et al. 1991). During the same period, municipalities dumped 972,265 m^3/d of domestic sewage into the river.
- Significant habitat alterations have occurred in the Fraser River estuary. Historical wetland areas have been reduced more than three-fold by dyking and filling of shallow wetlands (Birtwell et al. 1988). Undyked areas have been dredged or used for log storage.
- Placer mining discharges sediment in river systems, but no systematic assessment of operations is available (Schreier et al. 1991). There is also uncertainty about how much acid drainage occurs from old or recent mining in the basin.

This list of impacts in the Fraser River basin, although incomplete, gives cause for concern about salmonid survival for several reasons. First, many of the activities identified are detrimental to salmonid habitat condition even if considered alone. In addition, the impacts associated with many of these activities are likely to be cumulative in their effects on salmonid production (see Hartman 1996). The cumulative effects of land and water use activities in the Fraser River Basin have never been comprehensively evaluated, and the requisite information for conducting such an assessment is unavailable. Furthermore, the existing management regime is incapable of managing cumulative effects, as the institutional processes for responding to habitat issues tend to be structured to deal with single impacts or single project impacts. Finally, the government of Canada is severely reducing its scientific capability to deal with such complex and growing problems. Substantial reductions in support for research have occurred on the west coast of Canada and more reductions are anticipated in the near term. This systematic reduction in scientific capability, especially in the area of habitat research, reflects an abrogation of the government's mandate under a guise of economic restraint.

Superimposed on this backdrop of expanding regional populations and associated environmental degradation is a fisheries management process that has failed to protect the Pacific salmon resource. One of the fundamental shortcomings of this system is the lack of clear management goals and objectives (i.e., management for biomass vs. biodiversity; see Hyatt and Riddell 2000). The lack of such guidance has resulted in the proliferation of production hatcheries and mixed stock fisheries, which are adversely affecting both the diversity and strength of salmon populations. In addition, management decisions are consistently based on incomplete information and erroneous assumptions (see Knudsen 2000). Furthermore, the absence of a workable Pacific Salmon Treaty up to mid-1999 created a climate that enabled fishing interests to exploit salmon populations with little consideration for the future of the resource or those who depend upon it. As such, human activities conducted at the regional level are having serious effects on salmon and populations throughout their range. Displacement of both commercial and sport fishers as a result of reduced salmon abundance erode public support for expensive fisheries management and restoration programs.

EFFECTS OF GLOBAL ACTIVITIES ON PACIFIC SALMON

At a global scale, human activities are increasing the worldwide atmospheric concentrations of carbon dioxide (CO_2), methane (CH_4), nitrous oxide (N_2O), chloro-fluorocarbons (CFCs), and several other trace gases. These gases are essentially transparent to incoming solar radiation, but they absorb and re-radiate the outgoing terrestrial infra-red radiation from the earth's surface, back to the earth. Global climate models predict that a continuing increase in greenhouse gases will raise global temperatures by $3 \pm 1.5°C$ during the next 50 years, during which CO_2 levels will have doubled compared with pre-industrial levels (IPCC 1990). However, there is great uncertainty with respect to the level and timing of warming because the effects of climate change on cloud formation and the role of oceans in cycling and re-absorbing radiative gases are not well understood.

Notwithstanding the uncertainties associated with predicting the magnitude of global climate change, there is ample evidence available to demonstrate that mean air temperatures have recently increased in Canada. From the 1959–1973 period to the 1974–1988 period, air temperatures have increased in the central and western parts of Canada by as much as 1.5 and 2.5°C during spring and winter, respectively (Figure 3.5; Hengeveld 1991). Doubling of CO_2 is predicted to produce further temperature increases up to 4°C during winter along the western part of the continent, where salmon spawning and rearing occur (Figure 3.6; Hengeveld 1991). Increased water run-off at high latitudes and increased summer dryness are expected in association with such temperature increases. Wind, storm, precipitation, and ocean circulation patterns will also be altered, and the ocean level is expected to rise.

Some of the greenhouse gases, such as the chloro-fluorocarbons, affect the ozone layer and ultra-violet light transmission to earth. Up to 50% thinning of the ozone layer over the Antarctic has resulted in increased radiation over the southern ocean (Smith et al. 1992). Not surprisingly, increases in UV-B irradiation have been observed in temperate and polar latitudes in recent decades (Williamson 1995).

Changes in climatic conditions are likely to have significant and long-lasting effects on both the ocean system and on the terrestrial system (including land, forests, and freshwater). Accurately predicting such effects is challenging due to the complex interactions between climate, the physical environment, and Pacific salmon. For instance, the ocean surface conditions are characterized by horizontal and vertical water currents and by temperature and salinity gradients. Water flows are driven by wind and rotation of the earth. Patterns of surface domains, boundary lines, currents, and salinity conditions vary and change seasonally and interannually (Burgner 1991). In the northeastern Pacific Ocean, wind patterns are primarily caused by the Aleutian low pressure field during winter and the California high pressure field during summer. In coastal zones, water flow, temperature, and salinity patterns are strongly influenced by fresh water from large rivers. Oceans, therefore, are non-uniform, changing environments.

Patterns of salmon production are related to changes in ocean conditions in the northern Pacific. A high index number for the Aleutian low pressure field indicates strong winds, increased precipitation, and higher temperatures in the coastal zones of North America (Beamish and Bouillon 1993). This results in a stronger Alaska Gyre, a weaker California Current, and reduced upwelling in the Washington and Oregon coastal zones. The low ocean survival of Washington and Oregon coho and chinook over the past two decades is considered to be the result of this phenomenon.

The work of Beamish and Bouillon (1993) provides a good indication of the connections among climate, oceanographic conditions, and fish production in the Pacific Ocean. Changes in the production of copepods and of pink, chum, and sockeye salmon in the eastern Pacific Ocean were related to changes in meteorological conditions in the north-central Pacific, as indicated by the Aleutian low index (Beamish and Bouillon 1993). The smoothed index rose to a high in the late 1930s and early 1940s when salmon production was relatively high. It was low during the 1950s and 1960s, when salmon production was low and it rose during the 1970s, when salmon production rose again (Beamish and Bouillon 1993).

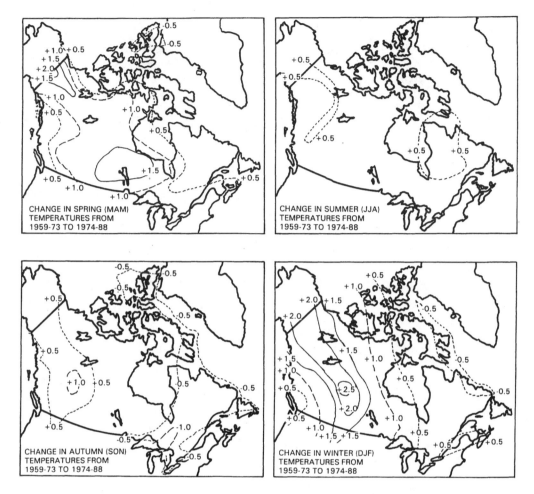

FIGURE 3.5 Changes in air temperature in Canada during the period from 1959–1973 to 1974–1988 (redrawn from Hengeveld 1991).

Global climate change is likely to influence the ocean system in two ways, including increasing surface water temperatures and increasing the delivery of ultraviolet radiation to the ocean surface. The predicted increases in ocean temperature may have extremely important effects on the distribution of Pacific salmon and steelhead. Salmon distributions exhibit sharp step-functions in response to temperature in the oceanic eastern north Pacific in winter and summer (Welch et al. 1995). The critical limits defining boundaries for different species and during specific seasons may limit salmon to a relatively small area of the subarctic Pacific Ocean (Welch et al. 1995). Sockeye salmon are limited to waters with temperatures lower than 7°C in winter in the eastern north Pacific Ocean. Thermal tolerance limits increase from April to August, reach a maximum of 15°C by August, and decline to 7°C by November (Welch et al. 1998a). Thus, sockeye salmon migrate north during autumn and early winter to overwinter, and they move south and remain in water <15°C for summer feeding (Welch et al. 1998a). The distribution of steelhead is also limited by water temperature, with upper and lower temperature thresholds determining both the southern and northern limits of their distribution (Welch et al. 1998a). A climate change associated with doubling of CO_2 levels, and associated ocean surface temperature change, is predicted to change the distribution of steelhead

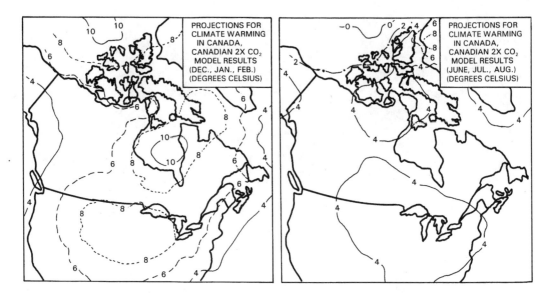

FIGURE 3.6 Projected climate warming, during summer and winter, with a doubling of CO_2 (redrawn from Hengeveld 1991).

and severely reduce the area available for salmon in the northern Pacific Ocean (Welch et al. 1995; Welch et al. 1998a,b).

Salmon and steelhead can also be adversely affected by increased delivery of ultraviolet radiation of the sea surface. Increased UV-B radiation is a concern because it causes direct damage to larval and juvenile fish by decreasing survival and altering other functions. In addition, UV-B can cause indirect damage by changing composition and stability of the phytoplankton communities. While the penetration of UV-B radiation varies with water clarity, up to 10% of the upper marine euphotic zone may be affected by increased levels of ultraviolet radiation. Increased radiation associated with a 25% reduction in ozone concentration would cause a decrease in primary productivity of about 35% near the ocean surface (Smith et al. 1992).

Changes in the climate of the Pacific Northwest are also likely to affect terrestrial, freshwater, and estuarine habitats throughout the range of Pacific salmon. The temperature changes predicted with a doubling of CO_2 will likely alter the distribution of forest types (Figure 3.7; Hengeveld 1991). The area where British Columbia salmon-producing rivers are located is predicted to become semi-arid (Hengeveld 1991). However, within such an area of complex physiography and climate, it is difficult to predict the specific patterns of forest type that will evolve (Northcote 1992). High topographic relief and physiography also cause the same dilemma in predicting the effects of climate warming on river temperatures and flow regimes (Northcote 1992). As forest systems, stream ecology, and salmon production are intimately connected (Hartman and Scrivener 1990; Maser and Sedell 1994), alteration of forest type will necessarily affect salmonid production.

Changes in climatic conditions are likely to significantly alter the freshwater habitats that salmon depend on. Specifically, climate changes will undoubtedly alter stream hydrology, as indicated by changes in the frequency, magnitude, and timing of peak streamflows and extreme low flows. Small streams will be particularly vulnerable because the shift toward warmer, drier conditions in the summer will further reduce streamflow. The presence of water diversions and/or wastewater discharges on such streams will further reduce the availability of suitable salmonid habitats. Changes in the stream hydrology will also affect the delivery of freshwater to estuarine systems, and in so doing adversely affect these important nursery areas and migration corridors.

The increases in air temperatures that have already occurred (Hengeveld 1991), and are predicted in the future (Welch et al. 1998a), will affect surface water temperatures of the river systems

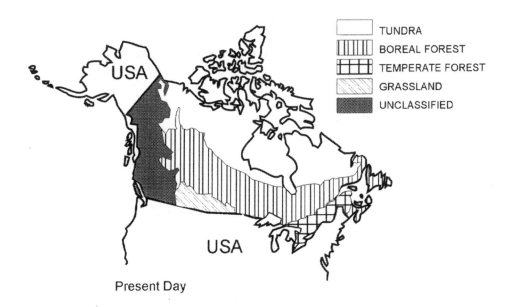

Present Day

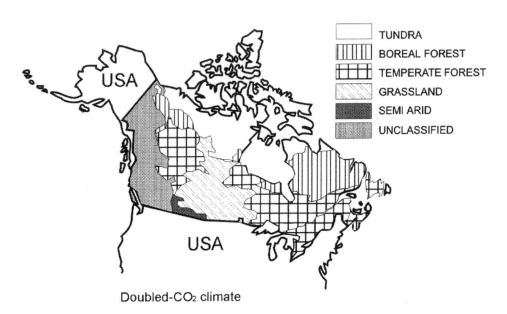

Doubled-CO₂ climate

FIGURE 3.7 Changes in forest and grassland boundaries resulting from typical doubled CO_2 climate (redrawn from Hengeveld 1991).

that produce salmon and steelhead. Importantly, surface water temperatures will increase in many stream systems. Small changes in water temperature during winter exert large impacts on salmon life histories by altering incubation rates, emergence timing, size at critical periods, time of seaward migration, and ultimately survival (Holtby and Scrivener 1989; Hartman and Scrivener 1990).

As various geographic zones are affected by changes in climate, major losses in habitat are likely to occur and certain species of fish may experience conditions beyond their physiological tolerance (Groot and Margolis 1991). For example, climate change will likely shift Shuswap Lake, a major British Columbia sockeye salmon producer, to a more oligotrophic status (i.e., due to the effects of UV-B on phytoplankton production). It is anticipated that climate warming will reduce freshwater production of juvenile sockeye, but may not increase prespawning mortality (Henderson et al. 1992). In the long term, new replacement species may move into areas that undergo change. However, the short-term prospect is for rapid, widespread disruptions of species and possibly loss of whole ecosystems as we know them. It is expected that climate change will have far-reaching effects on the salmonid system.

The conditions that will determine climate regimes of the future (e.g., forest cover, reservoir creation, water diversion, and atmospheric composition) are partly set already. They have been set in motion by local, regional, and global human populations; they have arisen over many decades; and they will be difficult to alter. They will play a large part in determining sustainability of salmonids in the future. However, most aspects of their generation and control are beyond the scope of fisheries scientists and managers. In this sense, the sustainability of anadromous salmonid populations is a broader issue that cannot be managed solely by fisheries managers and scientists.

SALMONID LOSSES IN THE PACIFIC NORTHWEST

Salmonid evolution has permitted species within the group to survive within a background of change and variation over the past ten thousand years in ocean, freshwater, and estuarine environments. However, these changes have occurred on geological time scales. In the future, the fish will have to cope with more rapid environmental changes, wider inter-annual variation in habitat conditions and in their own populations, and other cumulative and increasing impacts arising from higher densities of humans. These changes will occur over decades, rather than centuries or millennia. Salmonid evolution does not occur rapidly enough to allow adaptation to such quick changes.

Habitat destruction, heavy fishing pressure, and changes in ocean conditions have already reduced biodiversity and the number of stocks. Recent reviews of stock status have been completed in Washington, Oregon, and California (Nehlsen et al. 1991; Huntington et al. 1996). Comments by Nehlsen et al. (1991) on the significance of human impacts on salmonids are most relevant here:

> With the loss of so many populations prior to our knowledge of stock structure, the historic richness of the salmon and steelhead resource of the west coast will never be known. However, it is clear that whatever has survived is a small proportion of what once existed, and what remains is substantially at risk.

Based on their review, Nehlsen et al. (1991) listed 54 stocks of special concern, 58 stocks at moderate risk, and 101 stocks at high risk. They stated that 106 or 200 stocks (depending on information source) were locally extirpated. Of the 101 high risk stocks, habitat was associated with the risk level in 96 stocks, fishing in 49 stocks, and other human factors were associated with 39 stocks. Dams and other impacts in the Columbia River basin have reduced natural production to 4 to 7% of predevelopment levels. Ninety-eight healthy stocks were identified in Washington and Oregon, one in California, and none in Montana (Huntington et al. 1996). This health categorization was generous, inasmuch as it considered a stock healthy if its abundance was as little as half what would have occurred without human impacts. Habitat conditions were rated as fair or poor for the majority of watersheds supporting healthy native stocks. More than 85% of the populations surveyed by biologists were perceived to be threatened by some degree of habitat degradation (Huntington et al. 1996).

Two reviews of the status of salmonid stocks have been conducted in British Columbia (Slaney et al. 1996; Hyatt and Riddell 2000). The first review of the status of stocks in British Columbia and Yukon was determined by examining time series of escapement data for increase, decrease, or disappearance (Slaney et al. 1996). The survey was based on DFO escapement estimates, which were the best data available. However, a large fraction of these data are only rough estimates. The results of this review indicate that, at the species aggregate level, salmon stocks other than coho are stable or increasing in British Columbia and Yukon (Slaney et al. 1996). This does not, however, reflect the weak state of chum and wild chinook salmon stocks on the west coast of Vancouver Island. Of the 9,663 stocks identified in British Columbia and Yukon, 5,491 were examined. Of these, 624 were at high risk, 78 were at moderate risk, and 230 were of special concern. One hundred and forty-two stocks were known to have been extirpated this century. The status of 4,172 stocks could not be determined (Slaney et al. 1996).

Hyatt and Riddell (2000) examined the status of sockeye salmon stocks in more detail than was possible in the review of status of all salmon stocks in British Columbia (Slaney et al. 1996). The total production of sockeye salmon has not been reduced by the combination of habitat change, fishing, and ocean conditions. However, there has been a significant loss of small stocks of sockeye salmon. Hyatt and Riddell (2000) stated that, "status classifications of large actively managed stocks diverge from those of the small, passively managed stocks in that virtually all of the high risk and extinct stocks are from the latter group." This is believed to have occurred because the management of sockeye stocks for biomass has been "highly successful." However, major changes are necessary if the maintenance of sockeye salmon biodiversity is to be achieved.

Of 867 stocks of steelhead reviewed by Slaney et al. (1996), eight stocks in the Georgia Strait were classed at high risk of extinction and 143 were deemed of special concern. The only population of summer-run steelhead in the Fraser River, i.e., the one in the Coquihalla River, was virtually gone (Northcote and Atagi 1996). However, changes in fishing pressure during 1995 and 1996 have resulted in an improved escapement of this stock.

SUSTAINABILITY AND MULTI-LEVEL PROCESSES OF CHANGE

Pacific Northwest salmonids will experience broad, relatively rapid effects of climate change during the next few decades. These will exert their influence through effects on both the freshwater and ocean systems (Figure 3.8). They will be driven by local, regional, and global human population increases. The impacts have developed slowly, on a human time scale, and will be reversed slowly (Figure 3.8). The environments of salmonids will also be affected by a range of other factors which are even more directly caused by human activities. These include urbanization, agricultural, forestry, hydroelectric, and transportation development activities (Figure 3.8). The impacts of most of these processes were generated over a short time. Some may be managed and corrected over a short time, but others, such as the restoration of damaged watersheds, may require decades. Impacts on salmonids from hatchery activities may be created quickly, but repaired slowly, if ever (Figure 3.8). Fishing impacts occur over short times so if salmonid populations are not lost, they may be repaired quickly. The cost, technical difficulties, and time frames for response to different impacts may be quite different depending upon the type of activity and impact.

The intractability of this complex of impacts on salmon is discouraging when viewed in its full context. Concern about the non-sustainability of present social processes, on a global scale, has been expressed often and powerfully (e.g., Ehrlich and Ehrlich 1970; Goldsmith et al. 1972; Pirages and Ehrlich 1974; Hardin 1993). In a May 11, 1995 address to the Canadian Society of Zoologists, Dr. P.A. Larkin, touched on several relevant issues:

> On all fronts, progress since 1987 has been discouraging. The root causes of the problems are population growth and poverty, coupled with enhanced capabilities for resource exploitation. More people have been born and more land has been cleared in this century than in all of the previous centuries of human history and the trend continues. World population is projected to be 8 billion by the year 2025, roughly 35% larger than it is now. Education in the imperatives of sustainability still falls far behind the need. Global atmospheric problems are still far from resolution. Perhaps of greatest concern, an increasing number of people either do not care or have lost faith in international efforts to address the issues.

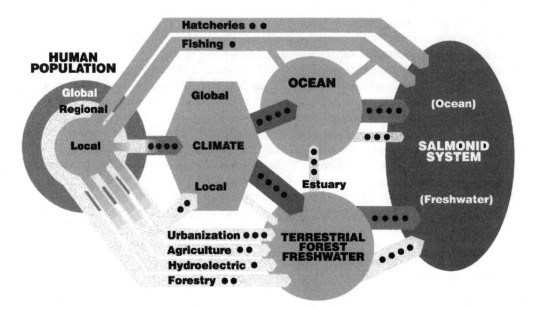

FIGURE 3.8 Schematic representation of population sources, categories and routes of impacts that may affect climate, biological ocean conditions, land-forest-freshwater systems, and "salmonid systems." The core of the figure (see Figure 3.2) indicates connections between local, regional, and global population, and climate, and between climate, ocean and terrestrial environments and salmonid production systems. The overlay of fisheries management elements, hatcheries, and fishing allocation, indicate that these may be cumulative in their impacts on the salmonid system. The additional overlay of hydroelectric, forest management, agricultural and urban development indicates further cumulative impacts. Finally, all of these impacts, in varying combinations according to species, time, and location, may be cumulative in their effect on the salmonid system.

The ideas expressed in this chapter about the direction of society and concern about the future are probably shared by many fisheries biologists, managers, harvesters, and citizens. Population growth and its implications for maintenance of fish resources has been previously recognized (Becker 1992; Brouha 1994; Alverson 1995; Eipper 1995). Like these authors, we are compelled to develop a wide perspective because the globalization of ecological and economic processes is increasing rapidly and the impacts of human population crises flow over national boundaries.

The future sustainability of salmonids will be affected by a broad range of human activities, including climate change, destruction of freshwater ecosystems, hatcheries, and fishing (Figure 3.8). The relative importance of each of these may be different depending upon stock, species, area, or time. However, all of these must be recognized and dealt with simultaneously if the resource is to be sustained. The media, many activists, and some scientists and teachers still tend to dwell, with singular attention, on one part of the complex of issues and ignore the others. Approaches that focus on only one sphere of impacts will not only fail to provide sustainability but, beyond that, they may be counter-productive by either confusing the public or raising its expectations unduly.

A wide spectrum of actions is needed to deal with the total complex of impact issues (Figure 3.8) that challenge the sustainability of salmonids and supporting ecosystems. Some of the problems must be dealt with through more cautious allocation, some through habitat protection and restoration, some through efforts to change global climate trends, and some through efforts to reduce human population growth. There is a requirement for all of these to be supported by continuing research. The groups of people and the ways that they must be educated and involved vary according to the types of processes that affect salmonid ecosystems. The time spans over which effort must be applied are different depending upon the processes involved.

Fisheries scientists and managers traditionally have played only minor roles in influencing major processes in the world (i.e., they have little effect upon the rate, type, and location of human population growth). The processes of industrial development and urbanization will depend upon the things people do, primarily on a regional scale. We may have input into such processes but we alone will not change their overall nature. Fisheries scientists can and should play a role in the development of effective solutions to these daunting challenges. As a first step, the American Fisheries Society should convene a major, international conference to discuss the salient global issues that challenge sustainability of the systems that produce salmon. Such a conference should attempt to establish how tractable these issues are, and to the extent that they may be tractable, what measures should be initiated. Beyond this it should develop a common message for decision-makers and the public.

The processes that drive climate change will depend upon the actions of people all over the world. In the broadest sense, widely diverse groups will influence the achievement of sustainability of Pacific salmon. Within that wide spectrum there are hundreds of millions who do not know where their next meal may come from, let alone anything about salmon. They will go on doing what they have been doing and, through global processes, they will continue to affect salmonid ecosystems. There are many others who may know about salmon, but either do not care about— or lack the time or energy to do anything about—their sustainability. There are even some who would be pleased to see salmon gone from systems like the Fraser River, so the rivers can be more readily exploited for hydroelectric development, industrial use, or water export. Those who care about salmonids represent a small and decreasing part of the total human community. These diverse and often conflicting circumstances will, collectively, determine the sustainability of salmonids.

If there is to be a reprieve for Pacific Northwest salmonids, it must come in the form of initiatives that reach into areas of society beyond fisheries science and management. These must have effects over time scales that differ, depending upon the social or environmental issues involved. Over the past four decades, fisheries management has expanded from allocation control (licensing and law enforcement) to include multiple-use resource management and, most recently, "integrated" management. The latter involves the technical integration of fisheries into a spectrum of other resource management activities. Therefore, the goal of sustainability will require not only local fisheries management remedies, but that fisheries science and management become a part of rational, overall ecological management. To do so, it must expand much further in complexity to influence existing social, economic, and political management at several societal levels. Salmonid ecosystems and the fish populations within them will not be preserved if there is a "sustainability" vacuum among the social, demographic, economic, and political processes that dominate their use. The public and its political representatives should be given this message as strongly as possible. Fisheries scientists and managers should not detract from the potential impact of such a message by offering "sustainability" of salmon when they have, at best, a minor role in delivering it.

ACKNOWLEDGMENTS

Dr. L. Beaudet, L. Duke, L. Hartman, A. Thompson, and G. Utermohle assisted with preparation of figures. W. Hartman spent much time "smoothing out" this chapter. J.C. Scrivener reviewed the manuscript and provided helpful comments. We have discussed various ideas in this chapter with several people. Dr. D. Welch and I. Williams were particularly helpful. We appreciate the efforts of the three peer reviewers who helped us to improve this chapter. We thank these individuals for their assistance.

REFERENCES

Alverson, D.L. 1995. Fisheries management: a perspective over time. Fisheries 20(8):6–7.
Anonymous. 1990. British Columbia population forecast 1990–2016. Province of British Columbia, Ministry of Finance and Corporate Affairs, Planning and Statistics Division, Victoria.
Anonymous. 1995. Lost streams of the lower Fraser River. Canada Department of Fisheries and Oceans, Fraser River Action Plan. Vancouver, British Columbia.

Beamish, R.J., and D.R. Bouillon. 1993. Pacific salmon production trends in relation to climate. Canadian Journal of Fisheries and Aquatic Sciences 50:1002–1016.

Becker, C.D. 1992. Population growth versus fisheries resources. Fisheries 17(5):4–5.

Birtwell, I.K., C.D. Levings, J.S. MacDonald, and I.K. Rogers. 1988. A review of fish habitat issues in the Fraser River system. Water Pollution Research Journal of Canada 23:1–29.

Bisson, P.A., T.P. Quinn, G.H. Reeves, and S.V. Gregory. 1992. Best management practices, cumulative effects, and long-term trends in fish abundance in Pacific Northwest river systems. Pages 189–232 *in* R.B. Naiman, editor. Watershed management: balancing sustainability and environmental change. Springer-Verlag, New York.

Boeckh, I., V.S. Christie, A.H. Dorcey, and H.I. Rueggeberg. 1991. Water use in the Fraser River basin. Pages 181–200 *in* A.H. Dorcey and J.R. Griggs, editors. Water in sustainable development: exploring our common future in the Fraser River basin. Westwater Research Centre, University of British Columbia, Vancouver.

Brouha, P. 1994. Population growth: the real problem. Fisheries 19(9):4.

Burgner, R.L. 1991. Life history of sockeye salmon (*Oncorhynchus nerka*). Pages 1–117 *in* C. Groot and L. Margolis, editors. Pacific salmon life histories. UBC Press, Vancouver, British Columbia.

Ehrlich, P.R., and A.H. Ehrlich. 1970. Population, resources, environment. W.H. Freeman and Company, San Francisco.

Eipper, A.W. 1995. Our quiet crisis. Fisheries 20(9):23–49.

Fraser Basin Council. 1997. Charter for sustainability. Fraser Basin Management Program, Vancouver, British Columbia.

Goldsmith, E., R. Allen, M. Allaby, J. Davoll, and S. Lawrence. 1972. A blueprint for survival. Penguin Books Ltd., Harmondsworth, England.

Groot, C. 1982. Modifications on a theme—a perspective on migratory behavior of Pacific salmon. Pages 1–21 *in* E.L. Brannon and E.O. Salo, editors. Proceedings of the salmon and trout migratory behavior symposium. School of Fisheries, University of Washington, Seattle.

Groot, C., and L. Margolis, editors. 1991. Pacific salmon life histories. UBC Press, Vancouver, British Columbia.

Groot, C., L. Margolis, and W.C. Clarke, editors. 1995. Physiological ecology of Pacific salmon. UBC Press, Vancouver, British Columbia.

Hall, J.D., G.W. Brown, and R.L. Lantz. 1987. The Alsea watershed study: a retrospective. Pages 399–416 *in* E.O. Salo and T.W. Cundy, editors. Streamside management: forestry and fishery interactions. Contribution 57, University of Washington, Institute of Forest Resources, Seattle.

Hardin, G.J. 1993. Living within limits: ecology, economics and population taboos. Oxford University Press, Oxford.

Hartman, G.F. 1996. Impacts of growth in resource use and human population on the Nechako River: a major tributary of the Fraser River, British Columbia, Canada. GeoJournal 40(12):147–164.

Hartman, G.F., and J.C. Scrivener. 1990. Impacts of forestry practices on a coastal stream ecosystem, Carnation Creek, British Columbia. Canadian Bulletin of Fisheries and Aquatic Sciences 223.

Henderson, M.A., D.A. Levy, and J.S. Stockner. 1992. Probable consequences of climate change on freshwater production of Adams River sockeye salmon (*Oncorhynchus nerka*). GeoJournal 28 (1):51–59.

Hengeveld, H. 1991. Understanding atmospheric change: a survey of the background science and implications of climate change and ozone depletion. State of Environment Report 91-2. Environment Canada, Ottawa.

Hoar, W.S. 1965. The endocrine system as a chemical link between the organism and its environment. Transactions of the Royal Society of Canada 4:175–200.

Holtby, L.B. 1988. Effects of logging on stream temperatures in Carnation Creek, British Columbia, and associated impacts on the coho salmon (*Oncorhynchus kisutch*). Canadian Journal of Fisheries and Aquatic Sciences 45:502–515.

Holtby, L.B., and J.C. Scrivener. 1989. Observed and simulated effects of climatic variability, clear-cut logging, and fishing on the numbers of chum salmon (*Oncorhynchus keta*) and coho salmon (*O. kisutch*) returning to Carnation Creek, British Columbia. Pages 62–81 *in* C.D. Levings, L.B. Holtby, and M.A. Henderson, editors. Proceedings of the national workshop on effects of habitat alteration on salmonid stocks. Canadian Special Publication Fisheries and Aquatic Sciences 105.

Huntington, C., W. Nehlsen, and J. Bowers. 1996. A survey of healthy native stocks of anadromous salmonids in the Pacific Northwest and California. Fisheries 21(3):6–14.

Hyatt, K.D., and B.E. Riddell. 2000. The importance of "stock" conservation definitions to the concept of sustainable fisheries. Pages 51–62 *in* E.E. Knudsen, C.R. Steward, D.D. MacDonald, J.E. Williams, and D.W. Reiser, editors. Sustainable fisheries management: Pacific salmon. Lewis Publishers, Boca Raton, Florida.

IPPC (Intergovernmental Panel on Climate Change). 1990. Climate change: The IPCC scientific assessment. World Meteorological Organization/United Nations Environment Programme. University Press, Cambridge.

Johnson, S.C., R.B. Blaylock, J. Elphick, and K. Hyatt. 1996. Disease caused by the salmon louse *Lepeophtheirus salmonis* (Copepoda: Caligidae) in wild sockeye salmon *(Oncorhynchus nerka)* stocks of Alberni Inlet, British Columbia. Canadian Journal of Fisheries and Aquatic Sciences 53:2888–2897.

Knudsen, E.E. 2000. Managing Pacific salmon escapements: The gaps between theory and reality. Pages 237–272 *in* E.E. Knudsen, C.R. Steward, D.D. MacDonald, J.E. Williams, and D.W. Reiser, editors. Sustainable fisheries management: Pacific salmon. Lewis Publishers, Boca Raton, Florida.

Langer, O., B. MacDonald, J. Patterson, and B. Schouwenburg. 1992. A strategic review of fisheries resources and management objectives: Stuart/Takla Habitat Management Area. Fraser River Action Plan, Department of Fisheries and Oceans, Vancouver, British Columbia.

Levy, D.A., L.U. Young, and L.W. Dwernychuk. 1996. Strait of Georgia fisheries sustainability review. Hatfield Consultants Ltd., West Vancouver, British Columbia.

Maser, C., and J.R. Sedell. 1994. From the forest to the sea: the ecology of wood in streams, rivers, estuaries, and oceans. St. Lucie Press, Delray Beach, Florida.

Miles, M. 1996. Hydrotechnical assessment: Louis Creek watershed. British Columbia Ministry of Environment, Lands and Parks, Kamloops.

Naiman, R.J. 1992. New perspectives for watershed management: Balancing long-term sustainability with cumulative environmental change. Pages 3–11 *in* R.J. Naiman, editor. Watershed management: Balancing sustainability and environmental change. Springer-Verlag, New York.

Nehlsen, W., J.E. Williams, and J.A. Lichatowitch. 1991. Pacific salmon at the crossroads: stocks at risk from California, Oregon, Idaho and Washington. Fisheries 6(2):4–21.

Northcote, T.G. 1992. Prediction and assessment of potential effects of global environmental change on freshwater sport fish habitat in British Columbia. GeoJournal 28(1):39–49.

Northcote, T.G. 1996. Effects of human population growth on the Fraser and Okanagan River systems, Canada: A comparative inquiry. GeoJournal 40(1–2):127–133.

Northcote, T.G., and D.Y. Atagi. 1996. Pacific salmon abundance trends in the Fraser River watershed compared with other British Columbia systems. Pages 199–219 *in* D.J. Stouder, P.A. Bisson, and R.J. Naiman, editors. Pacific salmon and their ecosystems: status and future options. Chapman & Hall, New York.

Pirages, D.C., and P.R. Ehrlich. 1974. Ark II: social response to environmental imperatives. W.H. Freeman and Company, San Francisco.

Rees, W.E. 1996. Revisiting carrying capacity: area-based indicators of sustainability. Population and Environment: A Journal of Interdisciplinary Studies 17(3):195–215.

Roos, J.F. 1986. Restoring Fraser River salmon: a history of the International Pacific Salmon Fisheries Commission 1937–1985. Pacific Salmon Commission, Vancouver, British Columbia.

Schouwenburg, W.J. 1990. A report of the Kemano Task Force. Review of the environmental studies report submitted by Alcan in support of the Kemano Completion proposal in relation to continued fish production from the rivers involved. Report of six Task Force members, edited by W.J. Schouwenburg, Department of Fisheries and Oceans, Vancouver, British Columbia.

Schreier, H., S.J. Brown, and K.J. Hall. 1991. The land-water interface in the Fraser River basin. Pages 77–116 *in* A.H. Dorcey and J.R. Griggs, editors. Water in sustainable development: exploring our common future in the Fraser River basin. Westwater Research Centre, University of British Columbia, Vancouver.

Slaney, T.L., K.D. Hyatt, T.G. Northcote, and R.J. Fielden. 1996. Status of anadromous salmon and trout in British Columbia and Yukon. Fisheries 21(10):20–35.

Smith, R.C., and twelve coauthors. 1992. Ozone depletion: ultraviolet radiation and phytoplankton biology in Antarctic waters. Science 255:952–959.

Steer, G.J., and K.D. Hyatt. 1987. Use of a run timing model to provide in-season estimates of sockeye salmon *(Oncorhynchus nerka)* returns to Barkley Sound, 1985. Canadian Technical Report of Fisheries and Aquatic Sciences 1557.

Welch, D.W., A.I. Chigirinsky, and Y. Ishida. 1995. Upper thermal limits on the oceanic distribution of Pacific salmon *(Oncorhynchus* spp.*)* in the spring. Canadian Journal of Fisheries and Aquatic Sciences 52:489–503.

Welch, D.W., Y. Ishida, and K. Nagasawa. 1998a. Thermal limits and ocean migrations of Pacific salmon: long-term consequences of global warming. Canadian Journal of Fisheries and Aquatic Sciences 55:937–948.

Welch, D.W., Y. Ishida, K. Nagasawa, and J.P. Eveson. 1998b. Thermal limits on the ocean distribution of steelhead trout *(Oncorhynchus mykiss)*. North Pacific Anadromous Fisheries Commission Bulletin 1:396–404.

Williams, I.V. 1987. Attempts to re-establish sockeye salmon *(Oncorhynchus nerka)* populations in the upper Adams River, British Columbia, 1949–84. Pages 235–241 *in* H.D. Smith, L. Margolis, and C.C. Wood, editors. Sockeye salmon *(Oncorhynchus nerka)* population biology and future management. Canadian Special Publication of Fisheries and Aquatic Sciences 96.

Williamson, C.E. 1995. What role does UV-B radiation play in freshwater ecosystems? Journal of Limnology and Oceanography 40:386–392.

4 The Importance of "Stock" Conservation Definitions to the Concept of Sustainable Fisheries

Kim D. Hyatt and Brian E. Riddell

Abstract.—Ambiguous definitions of the resource unit that constitutes a "stock" of fish and of what is meant by "conservation for sustainable fisheries" create confusion about the specific aims and objectives of fisheries management. Examination of two recent definitions of conservation, proposed for use in fisheries management, reveals how subtle differences in terminology may result in highly divergent operational objectives (e.g., management to sustain either harvest biomass or regional biodiversity) and procedures that have important consequences at levels from local fish populations to entire ecosystems. In recent years, fisheries "stakeholders" on Canada's west coast have been faced with accommodating a potent combination of events involving legal decisions, institutional policy changes, new agreements forged in response to interagency conflicts, and increasing public pressure for "sustainable" resource management within an ecosystem context. These events have facilitated rapid movement by resource agencies away from traditional definitions of fisheries conservation that embodied a single species management focus for maximum sustained yield, and toward new definitions that reflect a greater interest in maintenance of biodiversity at multispecies and biological population levels for fish. However, there is substantial uncertainty about the magnitude and rate of future changes we can expect in fisheries management regimes to accommodate broader conservation objectives for the maintenance of ecosystem linkages, the productive capacity of habitats, and general regional biodiversity.

INTRODUCTION

Canada's Department of Fisheries and Oceans (DFO), the British Columbia Ministry of Environment, Lands and Parks (MELP), and the British Columbia Ministry of Fisheries (BCF) are jointly responsible for managing fisheries resources on Canada's west coast. DFO is the lead agency involved in management of ocean fisheries for anadromous salmon, while provincial agencies have greater involvement in management of freshwater fisheries. Both levels of government share responsibilities for habitat protection and enforcement of related legislation under various acts (e.g., the federal Fisheries Act, the provincial Water Management Act).

At the federal level in Canada, two documents guide the conservation and management of fishery resources: (1) the 1967 Fisheries Act, which, in a 1995 amendment, included a statement of purpose "to provide for the conservation and protection of fish and waters frequented by fish" (Parsons 1993), and (2) the Policy for the Management of Fish Habitat (Anonymous 1986), which establishes the goal of no net loss (i.e., the conservation) of productive capacity of fish habitats. Taken together, the Fisheries Act and National Habitat Policy documents provide a mandate for DFO to define and implement a wide range of operational objectives pertaining to management of fish and fish habitats. In addition, Supreme Court of Canada decisions (e.g., *Regina vs. Sparrow* 1990) along with multiple rulings in the lower courts have established the primacy of conservation needs over all other uses of the fisheries resource.

1-56670-480-4/00/$0.00+$.50

Although conservation of fish (and fish habitat) clearly take precedence over other management activities (e.g., exploitation or enhancement) pertaining to fisheries resources, no formal definition of conservation for fisheries resources has been adopted to date by a federal agency for application to fisheries throughout Canada. Because conservation means different things to different people, the absence of a clear definition creates considerable confusion about the aims and objectives of various agencies (e.g., DFO, BCF, Canada Department of the Environment, MELP) charged with management of the resource. Here, we attempt to dispel some of this confusion by first summarizing two recent definitions of conservation suggested for application to fisheries resource management and then by expanding on how these definitions relate to operational principles or objectives that may be applied to fisheries conservation in the Pacific Region.

CONSERVATION DEFINITIONS

Olver et al. (1995) reviewed the history of the term conservation and offered the first definition considered here. They also provided a set of operating principles, applicable to the management of fish stocks, consistent with an ecological or ecosystemic view of conservation (see Appendix 4.3). They suggested that a new definition of conservation should embrace a conservation ethic based on ecological values such that humans derive social, economic, recreational, and cultural benefits in a sustainable manner. They proposed conservation be defined as "the protection, maintenance and rehabilitation of native biota, their habitats and life-support systems to ensure ecosystem sustainability and biodiversity."

A second definition considered here originated with the Canadian Atlantic Fisheries Scientific Advisory Council (CAFSAC), which initiated a process to formally define conservation principles for Atlantic salmon and provide an operational translation of conservation to be applied for management of the resource (Chaput, 1997). The formal definition of Atlantic salmon conservation was recorded in a 1991 CAFSAC Advisory Document (Anonymous 1992). Subsequently, the Atlantic Zone Fisheries Resource Conservation Council adopted this definition with minor changes in wording for general application to east coast fisheries resources. This definition was recommended for review prior to adoption in the Pacific Region by DFO's Pacific Stock Assessment Review Committee (Rice et al. 1995). The wording of the CAFSAC definition recommended for consideration in the Pacific Region was: "Fisheries conservation is that aspect of the management of the fisheries resource which ensures that its use is sustainable and which safeguards its ecological processes and genetic diversity for the maintenance of the resource. Fisheries conservation ensures that the fullest sustainable advantage is derived from the resource and that the resource base is maintained."

The above definitions contain both similarities and differences, reflecting their origins and the thinking behind them. Both definitions identify sustainability of "the resource" as their prime objective and both stress that maintenance of ecosystem integrity and resource biodiversity are essential to this objective. The definitions differ in that the CAFSAC definition explicitly recognizes maximum sustainable use of the resource while the Olver et al. (1995) definition leaves the issue of resource use as the implicit issue that sustainability must inevitably address. Similarly, the Olver et al. definition explicitly identifies "native biota" as the focus for fisheries conservation while the CAFSAC definition incorporates less specific language identifying only the generic fisheries "resource" as the focus. Finally, the Olver et al. definition identifies the maintenance of both ecosystem function and biodiversity as end points for conservation, while the CAFSAC definition revolves around maintenance of ecological processes and genetic diversity inherent in the fisheries resource itself without encompassing all ecosystem processes or biodiversity values. Although subtle, these differences may result in divergent operational objectives derived from translation of one definition or the other.

EVOLUTION OF THE CONSERVATION PERSPECTIVE AND OPERATIONAL OBJECTIVES

Historically, fisheries agencies have identified operational conservation objectives shaped by the intersection of management objectives and resource units serving as the focal point of interest. Given this perspective, it is possible to (1) visualize an abstract fisheries management domain bounded by the sum of the intersection points that lay along separate management-objective and resource-unit (e.g., salmon "stock") axes (Figure 4.1) and (2) identify that a diversity of conservation objectives or options occupy different portions of this domain (Hyatt 1996a). To elaborate on this perspective, consider that conservation and protection of anadromous salmon may involve choices, at one extreme to manage stocks aggregated by species to maximize or conserve harvestable biomass (i.e., management to maximize biomass), or at the other extreme to manage stocks separated into populations or sub-populations to sustain or conserve salmon genetic diversity, ecosystem linkages, and general biodiversity (i.e., management for biodiversity). Although management for either biomass or for biodiversity may both be described superficially as conservation of the "stocks" or of the fisheries "resource," it should be apparent that their extreme expressions constitute very different conservation objectives (e.g., see Geiger and Gharrett 1997) requiring radically different management approaches with major social, economic, and biological implications.

Fisheries resource agencies in our jurisdiction have historically attempted to achieve success in satisfying conservation objectives at multiple points within the management domain without articulating priorities for competing objectives. Consequently, over most of the past century, fisheries management regimes have been applied on Canada's west coast under general policy guidelines that, by default, have given greater emphasis to stock conservation options positioned in the biomass rather than the biodiversity portion of the management response domain outlined in Figure 4.1. Thus, the management focus has been biased toward harvest and protection of major commercial stock or species aggregates (e.g., large stocks of sockeye salmon *Oncorhynchus nerka* or Pacific herring *Clupea harengus*) rather than toward consideration of conservation to avoid the biological extinction of the myriad less productive populations of either the same (e.g., small, local populations of sockeye salmon) or different species (e.g., white sturgeon *Acipenser transmontanus* or eulachon *Thaleichthys pacificus*) that have played a lesser role in the commercial fishery. These efforts may be regarded as either highly successful or as a failure, depending on one's viewpoint. The biologically healthy state of most salmon stocks and the maintenance of many sustainable fisheries on the coast represent a considerable measure of success (e.g., see Roos' 1991 account of salmon management by the Pacific Salmon Commission). In contrast, evidence of failure may be found in the dozens to hundreds of local populations of salmon that have been extinguished in the last century and the large number of stocks that are at high risk of extinction as a consequence of both critical and cumulative anthropogenic impacts (Riddell 1993; Slaney et al. 1996). Similarly, Scott and Crossman (1973) provide notes on the commercial extinction of non-salmonid fishes, such as white sturgeon and eulachon, in B.C. fisheries.

The beginnings of a global trend in fisheries for movement toward the biodiversity end of the management continuum (Figure 4.1) is reflected by the proliferation of statements of principles and definition of operational objectives at regional, national, and international levels of resource management (Appendices 4.1–4.5) that recognize conservation within a multispecies, ecosystem context. This trend is highly advanced in Canada's Pacific region where a potent combination of biological and socioeconomic events (e.g., Figure 4.1a–e) have combined to facilitate a rapid movement of resource agencies away from traditional, and toward new, definitions of fisheries conservation and management objectives. Traditional operational perspectives embodied a preoccupation with management of single species, multistock aggregates for maximum sustained yield (MSY) or maximum economic return (Larkin 1977). These are rapidly giving way to fisheries conservation and management objectives that place a greater emphasis on simultaneous consideration of

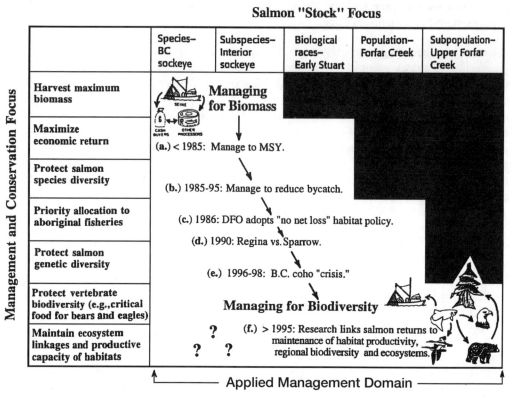

FIGURE 4.1   Summary of the applied management response domain defined by combinations of salmon
"stock" units and management and conservation objectives. (a) Maximum sustained yield (MSY) served as
the most common management focus prior to 1985 (e.g., Larkin 1977). (b) Between 1985 and 1995 federal-
provincial conflict over mixed stock fishery threats to steelhead salmon *O. mykiss* plus an emphasis on
rebuilding depressed chinook salmon *O. tshawytscha* populations under terms of the 1985 U.S.–Canada Pacific
Salmon Treaty (PST) shifted management away from MSY toward protection of salmon species diversity.
(c) See text for comments. (d) The Supreme Court of Canada considered Section 35 of the Canadian Constitution Act in *Regina vs. Sparrow* (1990) and confirmed the rights of aboriginal people to fish for food, societal,
and ceremonial needs. Subsequent management actions for priority allocation of salmon to "Section 35"
fisheries furthered reductions in offshore, mixed stock fisheries that are relatively unselective. (e) A 20-year
decline of coho salmon (Slaney et al. 1996) reached "crisis" levels such that in 1998 many local populations
(e.g., Upper Skeena and Thomson River coho) were threatened with extinction. Canada's fisheries minister
introduced blanket closures of major commercial fisheries throughout vast areas of the coast (Anderson 1998).
These actions signify the remarkable extent to which the DFO has moved away from managing for biomass
and toward managing for biodiversity of fisheries resources. (f) New research has highlighted the "keystone"
role that fish such as salmon play in maintenance of habitat productivity and regional biodiversity in coastal
ecosystems (see text for details). Future changes to fisheries management regimes to accommodate broader
conservation objectives for maintenance of ecosystem linkages, the productive capacity of habitats, and
regional biodiversity remain uncertain (figure modified after Hyatt, 1996a).

(1) priority allocation of fish to meet food, social and ceremonial needs of aboriginal groups;
(2) protection of species diversity; and (3) protection of genetic diversity within species.

   It is difficult to anticipate how far or how fast fisheries management regimes will develop to
accommodate objectives such as maintenance of regional biodiversity or ecosystem linkages (see
comments below). However, it is clear that the applied (as opposed to theoretical) management domain
(Figure 4.1) has expanded at an unprecedented rate over the past 15 years to deal with an increasingly
complex array of management and conservation objectives. Depending on one's perspective, the result

may be interpreted as either the end of sustainable fisheries that maximize for harvest biomass alone or the beginning of sustainable fisheries that manage for maintenance of biodiversity.

The dichotomy of managing for alternative conservation objectives such as harvestable biomass vs. biodiversity is not unique to just the fisheries sector. Recent moves by the Province of British Columbia to implement new policy guidelines governing forest harvest practices (i.e., the Forest Practices Code, Anonymous 1998a) reflect a major change in thinking about what forest resources consist of, as well as what management practices are necessary to sustain them. The new perspective recognizes forests are not just a source from which to maximize the biomass of harvestable fiber to the exclusion of other resource values (e.g., fish and wildlife production) but rather admits to requirements for maintenance of plant and animal biodiversity within a sustainable forest ecosystem context (Anonymous 1996a,b). Thus, there is a general trend, driven by widespread public support, for resource management jurisdictions, including fisheries and forestry, to move toward more multi-species and ecosystem level management approaches and objectives.

CONSEQUENCES OF AN ECOSYSTEMIC CONSERVATION PERSPECTIVE

Policy trends identified above will favor development of resource management and conservation practices not only to achieve sustainable development (originally defined as "development that meets the needs of the present without compromising the ability of future generations to meet their own needs," Brundtland 1987) but also to ensure protection, maintenance, rehabilitation, restoration, and enhancement of populations and ecosystems (Munro 1991). Thus, resource management definitions that stress single species or single resource sustainability are likely to be less desirable for application in future than general definitions such as those noted above (Olver et al. 1995 or Anonymous 1992) which embrace an ecosystemic view of conservation objectives.

The absence of a well-developed methodology that can be applied directly to manage resources at the ecosystem level (Chapin et al. 1996) means that population-based conservation is likely to remain the most practical way to implement management for whole system viability in the foreseeable future (Soule 1987). This limitation need not be a fatal flaw where significant protection for the entire ecosystem may be provided by actions designed to preserve the viability of populations that fill key functional roles in the system. This approach, founded on such familiar concepts as keystone predators, indicator organisms, and integrator species may be the most effective way currently available to achieve ecosystem sustainability (Rice et al. 1995).

ANADROMOUS SALMONIDS AS KEYSTONE OR INDICATOR SPECIES

In the Pacific Northwest, anadromous salmon fulfill all of the requirements of a keystone species in many freshwater and associated land-based ecosystems (Willson and Halupka 1995; Willson et al. 1998). In addition, several life history stages and traits of anadromous salmonids are highly sensitive to changes in the state of marine ecosystems (Hyatt 1996b). Therefore, the assessment, management, and conservation of anadromous salmonids may be an operationally effective means to diagnose and ensure the long-term maintenance of both biodiversity and function in many marine, freshwater, and associated terrestrial ecosystems. This is not to say that the historic perspective on management and conservation of anadromous salmonids requires no change to accommodate an ecosystemic conservation perspective. Indeed, the population focus of fisheries management has not been effective as an ecosystemic management proxy in the past when exploited populations have been treated as isolated entities, cut off from their past (i.e., evolutionary history and genetic uniqueness) and independent of the abiotic and biotic components of their supporting ecosystem (Olver et al. 1995). Rather, the approach as applied to date has favored management for the conservation of biomass and economic capital to the detriment of management for the conservation of biodiversity, genetic capital, and ecological linkages.

In the future, the benchmarks for success in management of fisheries resources such as anadromous salmon will focus not only on the conservation of biomass and economic opportunities pertaining directly to harvest activities but also on maintenance of both genetic diversity and important functional linkages in ecosystems. Maintenance of small, spatially isolated stocks will assume more importance because they embody thousands of years of evolutionary adaptations to local environments which make them important guarantors of the genetic diversity of a species (Riddell 1993). In particular, stocks that occur in marginal habitats that may be active sites of natural selection may be of adaptive significance to the species as a whole (Scudder 1989; Northcote 1992) such that these populations clearly qualify as evolutionarily significant units of the species (ESUs in the U.S., resulting from the 1973 Endangered Species Act).

Research and increased recognition of the roles that salmon or other keystone species populations play in maintenance of ecosystem linkages will favor future changes to management and conservation reference points for many populations regardless of their commercial or evolutionary importance. For example, it has been a long-standing principle of fisheries management that harvest must not exceed the regeneration rate of a population or its individual stocks (Appendix 4.3; Olver et al.1995). In natural populations, annual regeneration rates vary widely from year to year and cannot be accurately predicted. To provide a sufficient margin of safety, total allowable catch should be set well below maximum annual regeneration rates.

A wide variety of formulae have been developed in fisheries science with the aim of establishing defensible limits or reference points such as maximum sustainable yield (MSY) or optimal yield (OY) for exploitation (Parsons 1993). These approaches have generally been developed for single species stocks, ignoring the ecosystem contexts in which they exist. However, Trotter (1996) has pointed out that failure to explicitly recognize the feedback loop between salmon escapements to small streams and the subsequent productivity of stream ecosystems, as determined by the delivery of limiting nutrients from salmon carcasses (Bilby et al. 1996; Finney 1998), dooms harvest management plans based on MSY or OY concepts to long-term outcomes involving declines in ecosystem and subsequently stock productivity. This obviously violates DFO's stated goals, embodied in the National Habitat Policy, of conservation of productive capacity for "no net loss" of habitat and stocks (Anonymous 1986). Although this is clearly not the intent of DFO's species and stock focused management, failures to either understand or acknowledge ecosystem linkages and feedback loops may frustrate not only stock management objectives but also threaten the underlying integrity of entire ecosystems. Identification of such linkages will be accompanied by requirements to either (1) modify the reference points used for stock management and harvest (see Annex II of Appendix 4.5 for guidelines on the development and application of precautionary reference points) to accommodate maintenance of ecological linkages (e.g., salmon escapement objectives might be set to abundance levels that ensure adequate delivery of critical nutrients to streams) or (2) modify management activities to compensate for the threat that harvest poses to ecosystem linkages (e.g., by artificial supplementation of limiting nutrients in stream or lake ecosystems that have been chronically depleted; Hyatt and Stockner 1985; Bilby et al. 1996).

The examples above, as well as others, suggest that fisheries conservation policies that recognize intrinsic ecosystem dynamics are likely to be more stringent regarding definition of stock exploitation limits than those obtained by assessing each species in isolation (Olver et al. 1995). Further, definition of operational conservation objectives based on genetic, biodiversity, or ecosystem considerations will continue to be "a work in progress" because they have complex data requirements and will entail great uncertainty in their implementation because information sources are imperfect. The latter circumstance will frequently lead to application of conservation principles embodied in the "precautionary approach" to fisheries management and conservation (Appendix 4.5, Articles 5 and 6).

Finally, as a matter of practical concern, Rice et al. (1995) note that conservation objectives are usually defined within the context of human impacts on the probability of loss of species, maintenance of ecological processes, and ecosystem integrity. Human impacts are generally realized through (1) degradation of physical, chemical, or biological properties of the habitat; (2) introduction of exotic species; or (3) exploitation. These three proximate causes for species and ecosystem alteration usually act in concert. Consequently, guidelines on how to assess, manage, and monitor communities will continue to operate on three levels which include (1) physical protection of the habitats so that constituent populations will be able to survive over the long term; (2) constrained exploitation of resources at each of several levels (i.e., community, species, population) for long-term sustainability in an ecosystem context; and (3) effective rehabilitation of threatened populations.

ACKNOWLEDGMENTS

The idea for this chapter grew from initial discussions with Pete Bisson and Tim Slaney during work on an American Fisheries Society salmon stocks at risk project. The content benefited from informal reviews by Chris Wood, Max Stocker, and Mike Henderson, and formal reviews by Tom Northcote, Gordon Hartman, and Pete Bisson.

REFERENCES

Anderson, D. 1998. Announcement of Canada's coho recovery plan and federal response measures. Communications Directorate, Department of Fisheries and Oceans, Vancouver, British Columbia.

Anonymous. 1986. The Department of Fisheries and Oceans policy for the management of fish habitat. Communications Directorate, Department of Fisheries and Oceans, Ottawa, Ontario.

Anonymous. 1992. Definition of conservation for Atlantic salmon. CAFSAC Advisory Document 91/15. Pages 147–150 *in* the Canadian Atlantic Fisheries Scientific Advisory Committee Annual Report 14 (1991). Department of Fisheries and Oceans, Dartmouth, Nova Scotia.

Anonymous. 1995. The draft agreement for the implementation of the provisions of the United Nations Convention on the Law of the Sea of 10 December 1982 relating to the conservation and management of straddling fish stocks and highly migratory fish stocks. United Nations Conference. Fifth session, New York.

Anonymous. 1996a. 1996/97 Handbook for land-based programs. Forest Renewal B.C., Victoria, British Columbia.

Anonymous. 1996b. Research program funding application handbook. Forest Renewal B.C., British Columbia.

Anonymous. 1998a. Forest Practices Code of British Columbia Act. Information Management Group, British Columbia Ministry of Forests. URL:http://www.for.gov.bc.ca/tasb/legsregs/fpc/fpcact/contfpc.htm

Anonymous. 1998b. Consortium for International Earth Science Information Network (CIESIN). Environmental treaties and resource indicators (ENTRI). University Center, Michigan: CIESIN.URL: http://sedac.ciesin.org/entri/

Bilby, R. E., P. A. Bisson, and B. R. Fransen. 1996. Incorporation of nitrogen and carbon from spawning coho salmon in the trophic system of small streams. Canadian Journal of Fisheries and Aquatic Sciences 53:164–173.

Brundtland, G. H. 1987. Our common future: World Commission on Environment and Development. Oxford University Press, Oxford.

Chapin III, F. S., M. S. Torn, and M. Tateno. 1996. Principles of ecosystem sustainability. The American Naturalist 148:1016–1037.

Chaput, G. J. 1997. Proceedings of a workshop to review conservation principles for Atlantic salmon in eastern Canada: March 11 to 15, 1996. Halifax, Nova Scotia. Canadian stock assessment proceedings series; 97/15. Canada Department of Fisheries and Oceans. Maritimes Region, Science Branch, Moncton, New Brunswick.

Finney, B. P. 1998. Long-term variability of Alaskan sockeye salmon abundance determined by analysis of sediment cores. North Pacific Anadromous Fish Commission Bulletin 1:388–395.

Geiger, H. J., and A. J. Gharrett. 1997. Salmon stocks at risk: What's the stock and what's the risk. Alaska Fishery Research Bulletin 4:178–180.

Hyatt, K. D., and J. G. Stockner. 1985. Responses of sockeye salmon (*Oncorhynchus nerka*) to fertilization of British Columbia coastal lakes. Canadian Journal of Fisheries and Aquatic Sciences 42:320–331.

Hyatt, K. D., 1996a. Stewardship for biomass or biodiversity: A perennial issue for salmon management in Canada's Pacific Region. Fisheries 21(10):4–5.

Hyatt, K. D. 1996b. Salmon as barometers of the state of marine and aquatic environments on Canada's west coast. Pages 37–39 *in* D. E. Hay, R. D. Waters, and T. A. Boxwell, editors. Proceedings, marine ecosystem monitoring network workshop. Canadian Fisheries and Aquatic Sciences Technical Report 2108.

Larkin, P. A. 1977. An epitaph for the concept of maximum sustainable yield. Transactions of the American Fisheries Society. 106:1–11.

Munro, David A. 1991. Caring for the earth: A strategy for sustainable living. The World Conservation Union, United Nations Environment Programme and World Wide Fund for Nature. Gland, Switzerland.

Northcote, T. G. 1992. Migration and residency in stream salmonids: some ecological considerations and evolutionary consequences. Norwegian Journal of Freshwater Research 67:5–17.

Olver, C. H., B. J. Shuter, and C. K. Minns. 1995. Toward a definition of conservation principles for fisheries management. Canadian Journal of Fisheries and Aquatic Sciences 52:1584–1594.

Parsons, L. S. 1993. Management of marine fisheries in Canada. Canadian Bulletin of Fisheries and Aquatic Sciences 225.

Rice, J., R. D. Humphreys, L. Richards, R. Kadowaki, D. Welch, M. Stocker, B. Turris, G. A. McFarlane, F. Dickson, and D. Ware, editors. 1995. Pacific Stock Assessment Review Committee (PSARC) annual report for 1994. Appendix 3. Biological objectives working group report. Canadian Fisheries and Aquatic Sciences Manuscript Report 2318.

Riddell, B. 1993. Spatial organization of Pacific salmon: what to conserve? Pages 23–42 *in* J. G. Cloud and G. H. Thorgaard, editors. Genetic conservation of salmonid fishes. Plenum Press, New York.

Roos, J. F. 1991. Restoring Fraser River salmon: a history of the International Pacific Salmon Fisheries Commission, 1937–1985. The Pacific Salmon Commission, Vancouver, British Columbia.

Scott, W. B., and E. J. Crossman. 1973. Freshwater fishes of Canada. Fisheries Research Board of Canada Bulletin 184.

Scudder, G. G. E. 1989. The adaptive significance of marginal populations: a general perspective. Pages 180–185 *in* C. D. Levings, L. B. Holtby, and M. A. Henderson, editors. Proceedings of the national workshop on effects of habitat alteration on salmonid stocks. Canadian Fisheries and Aquatic Sciences Special Publication 105.

Slaney, T. L., K. D. Hyatt, T. G. Northcote, and R. J. Fielden. 1996. Status of anadromous salmon and trout in British Columbia and Yukon. Fisheries 21(10):20–35.

Soule, M. E. 1987. Viable populations for conservation. Cambridge University Press, Cambridge, England.

Trotter, P. 1996. Maximum sustained yield doesn't protect stocks or, "How to manage a fish stock to extinction while thinking you're getting a free lunch!" Washington Trout 6:4–6.

Willson, M. F., and K. C. Halupka. 1995. Anadromous fish as keystone species in vertebrate communities. Conservation Biology 9:489–497.

Willson, M. F., S. M. Gende, and B. H. Marston. 1998. Fishes and the forest. Bioscience 48:455–462.

APPENDICES

Appendix 4.1 Summary of principles and objectives associated with fisheries conservation as identified by the Biological Objectives Working Group of the Pacific Stock Assessment Review Committee (PSARC, Rice et al. 1995).

Operational principles implicit in the CAFSAC definition of fisheries conservation:

(i) the rationale for maintaining an area's natural biodiversity is continuation of the structure and function of the natural community, and its economic opportunities i.e., sustainable use will require that genetic diversity, community structure and ecological processes are conserved.

(ii) rational management of fish populations ought to prevent the loss of genetic diversity which in turn implies: the preservation of genetic variation, the maintenance of subpopulation structure, and the avoidance of artificial selection and hybridization.

Appendix 4.2 Operational conservation objectives suggested for application in the Pacific Region of DFO by the PSARC Biological Objectives Working Group (Rice et al. 1995):

1. Ensure that subpopulations over as broad a geographic and ecological range as possible do not become biologically threatened (in the COSEWIC sense of "threatened").
2. Operationally, management must allow enough spawners to survive, after accounting for all sources of mortality (including all fisheries and natural mortality), to ensure production of enough progeny that they will be able to replace themselves when mature.
3. Fisheries may have collateral effects on other species, mediated by the ecological relationships of the target species. Fisheries should be managed in ways that do not violate the above biological objectives for ecologically related species, as well as target species.

Appendix 4.3 Operational conservation principles to guide fisheries management (adapted from Olver et al. 1995):

Aquatic ecosystems should be managed to ensure long-term sustainability of native fish stocks and reciprocally, native fish stocks should be managed to ensure the long term sustainability of aquatic ecosystems where important feedbacks exist between fish stocks and the state of the aquatic ecosystem.

The sustainability of a fish stock requires protection of the specific physical and chemical habitats utilized by the individual members of that stock.

The sustainability of a fish stock requires maintenance of its supporting native community.

Vulnerable, threatened and endangered species must be rigidly protected from all anthropogenic stresses.

Exploitation of populations or stocks undergoing rehabilitation will delay, and may preclude, full rehabilitation.

Harvest must not exceed the regeneration rate of a population or its individual stocks.

Direct exploitation of spawning aggregations increases the risk to sustainability of fish stocks.

Appendix 4.4 Conservation Principles from Article II of the Convention for Antarctic Exploitation, Scientific Committee for the Conservation of Antarctic Marine Living Resources 1985 (Anonymous 1998b).

1. Conservation includes rational use;
2. The ecosystem is to be managed in a manner which maintains the ecological relationships among harvested, dependent and competing populations;
3. Management of the ecosystem will include the restoration of depleted populations (to a level defined in (4));
4. Management should prevent any harvested populations from falling below a level close to that which ensures the greatest net annual increment;
5. Management of the marine ecosystem should not only be concerned with the effects of harvesting, but also with the affects of the introduction of alien species and the impacts of other activities likely to affect environmental changes that are not potentially reversible over two or three decades.

Appendix 4.5   Excerpts from the draft agreement for the conservation and management of straddling fish stocks and highly migratory fish stocks (Anonymous 1995).

Article 5: General Principles

In order to conserve and manage straddling fish stocks and highly migratory fish stocks, coastal States and States fishing on the high seas shall, in giving effect to their duty to cooperate in accordance with the Convention:

(a) adopt measures to ensure long-term sustainability of straddling fish stocks and highly migratory fish stocks and promote the objective of their optimum utilization;

(b) ensure that such measures are based on the best scientific evidence available and are designed to maintain or restore stocks at levels capable of producing maximum sustainable yield, as qualified by relevant environmental and economic factors, including the special requirements of developing States, and taking into account fishing patterns, the interdependence of stocks and any generally recommended international minimum standards, whether subregional, regional or global;

(c) apply the precautionary approach in accordance with article 6;

(d) assess the impacts of fishing, other human activities and environmental factors on target stocks and species belonging to the same ecosystem or associated with or dependent upon the target stocks;

(e) adopt, where necessary, conservation and management measures for species belonging to the same ecosystem or associated with or dependent upon the target stocks, with a view to maintaining or restoring populations of such species above levels at which their reproduction may become seriously threatened;

(f) minimize pollution, waste, discards, catch by lost or abandoned gear, catch of non-target species, both fish and non-fish species, in particular endangered species, through measures including, to the extent practicable, the development and use of selective, environmentally safe and cost-effective fishing gear and techniques;

(g) protect biodiversity in the marine environment;

(h) take measures to prevent or eliminate overfishing and excess fishing capacity and to ensure that levels of fishing effort do not exceed those commensurate with the sustainable use of fisheries resources;

(i) take into account the interests of artisanal and subsistence fishers;

(j) collect and share in a timely manner, complete and accurate data concerning fishing activities on, *inter alia*, vessel position, catch of target and non-target species and fishing effort, as set out in Annex I, as well as information from national and international research programs;

(k) promote and conduct scientific research and develop appropriate technologies in support of fishery conservation and management; and

(l) implement and enforce conservation and management measures through effective monitoring, control and surveillance.

Article 6:   Application of the Precautionary Approach

1. States shall apply the precautionary approach widely to conservation, management and exploitation of straddling fish stocks and highly migratory fish stocks in order to protect the living marine resources and preserve the environment.

2. States shall be more cautious when information is uncertain, unreliable or inadequate. The absence of adequate scientific information shall not be used as a reason for postponing or failing to take conservation and management measures.

3. In implementing the precautionary approach, States shall:
    (a) improve decision making for fishery resource conservation and management by obtaining and sharing the best scientific information available and implementing improved techniques for dealing with risk and uncertainty;
    (b) apply the guidelines set out in Annex II and determine, on the basis of the best scientific information available, stock-specific reference points and the action to be taken if they are exceeded;
    (c) take into account *inter alia* uncertainties relating to the size and productivity of the stocks, reference points, stock condition in relation to such reference points, levels and distribution of fishing mortality and the impact of fishing activities on non-target and associated or dependent species, as well as existing and predicted oceanic, environmental and socio-economic conditions; and
    (d) develop data collection and research programs to assess the impact of fishing on non-target and associated or dependent species and their environment, and adopt plans which are necessary to ensure the conservation of such species and to protect habitats of special concern.
4. States shall take measures to ensure that, when reference points are approached, they will not be exceeded. In the event that they are exceeded, States shall, without delay, take the action determined under paragraph 3(b) to restore the stocks.
5. Where the status of target or non-target or associated or dependent species is of concern, States shall subject such stocks and species to enhanced monitoring in order to review their status and the efficacy of conservation and management measures. They shall revise those measures regularly in the light of new information.
6. For new or exploratory fisheries, States shall adopt as soon as possible cautious conservation and management measures, including, *inter alia*, catch limits and effort limits. Such measures shall remain in force until there are sufficient data to allow assessment of the impact of the fisheries on the long-term sustainability of the stocks, whereupon conservation and management measures based on that assessment shall be implemented. The latter measures shall, if appropriate, allow for the gradual development of the fisheries.
7. If a natural phenomenon has a significant adverse impact on the status of straddling fish stocks or highly migratory fish stocks, States shall adopt conservation and management measures on an emergency basis to ensure that fishing activity does not exacerbate such adverse impact. States shall also adopt such measures on an emergency basis where fishing activity presents a serious threat to the sustainability of such stocks. Measures taken on an emergency basis shall be temporary and shall be based on the best scientific evidence available.

Annex II. Guidelines for the application of Precautionary Reference Points in Conservation and Management of Straddling Fish Stocks and Highly Migratory Fish Stocks

A precautionary reference point is an estimated value derived through an agreed scientific procedure, which corresponds to the state of the resource and of the fishery, and which can be used as a guide for fisheries management.

Two types of precautionary reference points should be used: conservation, or limit, reference points and management, or target, reference points. Limit reference points set boundaries which are intended to constrain harvesting within safe biological limits within which the stocks can produce maximum sustainable yield. Target reference points are intended to meet management objectives.

Precautionary reference points should be stock-specific to account *inter alia* for the reproductive capacity, the resilience of each stock and the characteristics of the fisheries exploiting the stock, as well as other sources of mortality and major sources of uncertainty.

Management strategies shall seek to maintain or restore populations of harvested stocks, and where necessary associated or dependent species, at levels consistent with previously agreed precautionary reference points. Such reference points shall be used to trigger pre-agreed conservation and management action. Management strategies shall include measures which can be implemented when precautionary reference points are approached.

Fishery management strategies shall ensure that the risk of exceeding limit reference points is very low. If a stock falls below a limit reference point or is at risk of falling below such a reference point, conservation and management actions should be initiated to facilitate stock recovery. Fishery management strategies shall ensure that target reference points are not exceeded on average.

When information for determining reference points for a fishery is poor or absent, provisional reference points shall be set. Provisional reference points may be established by analogy to similar and better-known stocks. In such situations, the fishery shall be subject to enhanced monitoring so as to enable revision of provisional reference points as improved information becomes available.

The fishing mortality rate which generates maximum sustainable yield should be regarded as a minimum standard for limit reference points. For stocks which are not overfished, fishery management strategies shall ensure that fishing mortality does not exceed that which corresponds to maximum sustainable yield, and that the biomass does not fall below a predefined threshold. For overfished stocks, the biomass which would produce maximum sustainable yield can serve as a rebuilding target.

5 The Elements of Alaska's Sustainable Fisheries*

(The Honorable) Fran Ulmer

Abstract.—Prior to Alaska statehood in 1959, Alaska's salmon and steelhead stocks were severely depleted, primarily as a result of unsustainable fishing practices and inadequate federal fisheries management. Over the past 38 years, Alaska has focused considerable state resources on the conservation and management of its salmon stocks and associated habitats. Sustained yield management has been a cornerstone of the state's fisheries management program. Other key elements of the program include effective habitat protection, separation of allocation and conservation decisions through the establishment of the Alaska Board of Fisheries, and strong support from the fishing industry and the communities that depend on these resources. This chapter provides a brief description of Alaska's approach for achieving sustainable fisheries and outlines some of the recent initiatives that emphasize Alaska's commitment to this goal.

INTRODUCTION

Alaska is unique in many respects. First, Alaska is one fifth the size of the continental U.S. or about 586,412 square miles. It has more than 3,000 rivers, 3 million lakes, and 33,904 miles of shoreline. Importantly, Alaska has nearly 15,000 waterbodies that contain anadromous fish. This large land base and diverse physiography, combined with a small human population and relatively pristine conditions, provides Alaska with a unique perspective for managing its fisheries on a sustainable basis.

A HISTORICAL PERSPECTIVE ON FISHERIES MANAGEMENT IN ALASKA

One hundred and twenty years of fishing history have produced a complex mosaic of commercial fisheries in Alaska. Over this time, a great number of separate and distinct commercial fisheries have spread along Alaska's vast coastline. A diversity of sport, subsistence, and personal use fisheries have also operated in Alaska's waters for a protracted period. As a result of overfishing and inadequate regulation prior to statehood, Alaska's fisheries were on the brink of disaster. The status of salmon stocks at that time demonstrated the need for a more effective approach to fisheries management in the state.

One of the driving forces behind Alaska statehood was the desire to gain state control over the management of its fisheries. As a result, Alaska has focused considerable state resources on the conservation and protection of its salmon stocks and associated habitats in the 38 years since the state was established. The central element of the state's fisheries management strategy is a sustained yield model. This requirement for sustained yield management of its renewable resources is written into Alaska's constitution and is supported by nearly 40 years of legislation, regulation, and management experience.

---

* Excerpted from remarks at the "Toward Sustainable Fisheries" Conference, April 1996.

Implementation of this sustained yield management strategy has come at a cost, both to individual fishers and to the state. During years of lower returns, harvests had to be restricted to assure adequate escapements. However, these investments in escapements during periods of poor environmental conditions have resulted in the healthy salmon populations that currently exist in the state. In turn, these populations support a sustainable fishing industry, which now harvests up to eight times as many salmon as it did in 1959.

ELEMENTS OF ALASKA'S APPROACH TO SUSTAINABLE FISHERIES

The primary goal of Alaska's fisheries management program is conservation and to ensure that adequate numbers of fish return to spawn. This is the foundation of healthy fish stocks which, in turn, support a healthy fishing industry and healthy communities. The four main elements of Alaska's fisheries management program include:

- Habitat Protection;
- Sustained Yield Management;
- Open Allocation System; and
- Community and Industry Commitment.

Effective habitat protection is a central element of the overall fisheries management program in Alaska. Importantly, Alaska has made a conscious decision to forego the economic benefits associated with hydroelectric power developments where it would damage salmon resources. Alaska statutes now explicitly protect salmon runs by making it illegal to dam or in any way obstruct the free passage of anadromous fish into or out of stream systems. In addition, strict standards have been established for road building, coastal development, and mining to protect spawning and rearing habitats in freshwater, estuarine, and coastal areas. The Forest Practices Act, which is among the toughest in the country, restricts timber harvest within buffer zones along salmon streams to prevent erosion and associated impacts on freshwater habitats. Wastewater discharges are also tightly regulated to assure high water quality in receiving waters. Furthermore, the state works closely with federal government agencies to ensure that similar measures are applied to federally-managed lands. Together, these measures have ensured that freshwater, estuarine, and nearshore habitats throughout the state have been effectively protected and now support expanding populations of anadromous fish.

Sustained yield management forms a cornerstone of Alaska's fisheries management system. The key to the success of this system is localized management, which provides the flexibility to respond quickly to local conditions. Under the general sustained yield guidelines, in addition to more specific guidelines established by the Alaska Board of Fisheries, local biologists have full authority to open and close fisheries based on inseason assessments of run strength. As this approach is dependent on the provision of timely scientific and technical information, the state has devoted substantial resources to collecting and analyzing biological data. Research on the effects of human activities and natural phenomena on fisheries resources also provides essential information for managing the state's fisheries.

Progress toward full implementation of sustained yield management was further supported in 1972 through approval of a constitutional amendment to limit entry into Alaska's fisheries. This amendment provided a legislative basis for limiting the number of participants in the fishery to a manageable level. Since that time, the productivity of the fishing industry has increased due to advances in technology and management, rather than by increasing the number of vessels. This limited entry system guarantees that the fishing industry will have a vested interest in habitat protection and sustainable fisheries management. The industry has also demonstrated its commitment to the future by taxing itself to pay for fisheries enhancement.

The third element of Alaska's fisheries management model is the Alaska Board of Fisheries, which is a citizen-based organization that makes decisions on the allocation of fisheries resources. The Board members, who are appointed by the Governor and confirmed by the Legislature, represent a broad array of fishing groups and other interests. By taking on the task of resolving fishery disputes, the Board takes the politically-charged issue of allocation away from fishery managers and politicians. While this system is not without its flaws, it has dramatically increased the credibility of the management program by effectively separating decisions regarding allocation from those related to conservation. This separation of allocation and conservation decisions is critical for achieving sustainable fisheries in the state and elsewhere in the Northwest.

The fourth element of Alaska's approach is a strong commitment to sustainable communities, which depend on a healthy fishing industry. Alaska recognizes that much of the fishing fleet is comprised of small boat owners, both Native and non-Native, who regard fishing as their livelihood and their way of life. These fishers are strongly committed to keeping salmon stocks and associated habitats healthy and viable. The protection of stream habitats, statutory ban on fish farming, and emphasis on wild stock production that have been incorporated into the state's management program would not have been possible without the support of the industry. Furthermore, implementation of sustained yield management would not have been possible without the long-term participation and support of the fishing industry.

Alaska continues to ban finfish farming even though there has been tremendous economic and political pressure to allow it. While other countries have focused their efforts on producing farmed fish, often at the expense of natural runs, Alaska continues to concentrate its energy and its resources on the conservation of its wild stocks. Alaska also has strict laws on siting and managing its hatcheries to minimize genetic interference with wild stocks. Wild stocks have a statutorily mandated priority over hatchery stocks in Alaska's fisheries management.

ALASKA'S LEADERSHIP IN FISHERIES MANAGEMENT

While continued implementation of this integrated approach to fisheries management will help to assure the sustainability of Alaska's salmon resources, there are a number of threats to the sustainability of North Pacific salmon and steelhead stocks. These include ongoing international disputes, continuing degradation of habitats in British Columbia and the continental U.S., and expanding fish-farming activities, to name a few. Alaska recognizes that wild salmon fisheries can no longer be sustained through the efforts of any one political jurisdiction. For this reason, Alaska is committed to demonstrating leadership in the development of cooperative management regimes that are based first and foremost on conservation. Some of Alaska's most important initiatives in this area involve:

- North Pacific Anadromous Fish Commission;
- Columbia River Salmon Recovery;
- Sitka Salmon Summit; and
- Pacific Salmon Treaty.

The North Pacific Anadromous Fish Commission is a good example of a multijurisdictional effort to promote international cooperation in the conservation of anadromous fish stocks in the North Pacific. Alaska led the initiative to create the Commission, which was established in 1992 by the Convention for the Conservation of Anadromous Fish Stocks in the North Pacific Ocean. Notably, the Convention also spelled the end to fishing for salmon on the high seas. In October 1996, the Commission hosted a major international scientific forum on Pacific Rim salmon and their role in the ecosystem, which was attended by scientists from the U.S., Canada, Russia, and Japan. These cooperative efforts have resulted in significant progress toward the sustainable management of salmon in the North Pacific.

Alaska is also participating in the recovery of depleted Columbia River chinook salmon stocks. Most recently this initiative has involved evaluation of the impacts on these stocks of fisheries conducted within Alaska's Effective Economic Zone. In the longer term, Alaska will participate in a broader evaluation of the impacts of various fisheries on the sustainability of Pacific Northwest salmon and steelhead populations.

On May 20, 1996, the governors of Alaska, Washington, and Oregon met in Sitka, Alaska to discuss their shared interests in, and develop strategies for, assuring the health and vitality of Pacific Northwest salmon populations. As a result of these deliberations, the governors adopted a set of Principles of Cooperation for Salmon Conservation that reflected their concern for wild Pacific salmon stocks and associated habitats. In the spirit of cooperation, the governors challenged each state to increase the production of salmon for harvest, to carefully manage salmon harvests, and to reduce the impacts on shared salmon resources. The governors also set an ambitious goal to achieve a net increase in salmon habitat by identifying and protecting key habitats through appropriate designations (i.e., critical habitat areas, refuges, and wild and scenic rivers).

The governors agreed on several actions to implement the principles that were established at the Sitka Salmon Summit. First, they proposed the creation by Congress of a $250 million Pacific Salmon Conservation and Restoration Fund to restore wild salmon runs to sustainable levels. Second, the governors agreed to enhance communications and understanding among communities that depend on salmon. Third, the three governors called for the permanent protection of the salmon spawning grounds of the Hanford Reach of the Columbia River under the Wild and Scenic Rivers Act. The governors also called for a formal review of the Mitchell Act hatchery operations and for full and stable funding to implement the recommendations resulting from the review. It was also agreed to pursue reconciliation of the various approaches that have been used to address hydropower impacts in the Columbia River Basin. Finally, each governor directed their senior fisheries officials to work together toward the initiatives agreed to at the summit.

The Pacific Salmon Treaty remains one of the most challenging issues that faces everyone involved in the management of Pacific Northwest salmon resources. Nonetheless, Alaska is convinced that the issues associated with the Treaty can be resolved, if we agree to work together. The recent agreement with Washington, Oregon, and the Northwest Tribes on the conservation of chinook salmon provides a model that could be used to develop similar conservation regimes for the other chinook fisheries that are managed under the Treaty. In addition, Alaska has significantly reduced its harvest of chinook salmon in recent years to address conservation concerns in British Columbia and the other Pacific Northwest states. The equity issue of the allocation dispute is a particularly contentious and complex matter that needs to be addressed under the Treaty. Because the livelihoods of many people depend on fair and equitable sharing of the resource, we believe that a solution to the allocation dispute should include fishing groups from both sides of the border.

In summary, Alaska is committed to the conservation and wise use of salmon and steelhead throughout the Pacific Northwest. However, this goal cannot be achieved through the imposition of court orders or heavy-handed political interference in locally-based decisions. Rather, realization of this goal will require solid scientific and technical information, effective communication among all participants, and focused conservation and enhancement efforts. Alaska has made tremendous progress in this area by implementing sustained yield management as a keystone of its overall fisheries management program. This approach represents a relevant model that could be used to support sustainable fisheries by other jurisdictions in the Northwest. We owe healthy, sustainable fisheries to future generations of Alaskans, Washingtonians, Oregonians, and British Columbians. We must do everything in our power to make this a reality.

ACKNOWLEDGMENTS

I appreciate the helpful review comments of R. LeBrasseur, D. Levy, and M. Taylor.

6 Review of Salmon Management in British Columbia: What Has the Past Taught Us?

David W. Narver

Abstract.—Over the past 30 years our knowledge, experience, and ability in managing anadromous salmonids in British Columbia have increased significantly. However, the complexity of human influences on habitat and fish continues to frustrate attainment of fisheries management goals. There remains a huge gap between knowing how to manage to an ecologically sustainable maximum and actually achieving it, while allowing for beneficial uses of the resource. This chapter briefly reviews the lessons learned in managing Pacific salmonids. In public policy, we have learned the value of keeping hydroelectric dams off main salmon rivers and maintaining a current Pacific Salmon Treaty. In addition, we generally recognize the need for developing and maintaining strong habitat legislation, and keeping fisheries management separate from social welfare support, but progress in these areas is slow. In management policy we have learned that wild fish must be accorded the highest priority, that commercial fleets are overcapitalized and must be reduced in size, and that area licensing can improve manageability of commercial fisheries. Operational measures that have proven effective under certain circumstances include catch and release fisheries, selective and terminal fisheries, limited entry fisheries, stock-specific management, and planning for enhancement and recreation. Largely through trial and error, salmon management in British Columbia is evolving into a practical framework that aims to conserve the resource, is equitable and fair for all sectors, and is sustainable, both biologically and economically, over the long term. The process has been abetted by the elimination of policies that have proven ineffective or deleterious, by the publicity surrounding recent downward trends in salmon stock abundance, and by the emergence of operational measures that can be selectively applied to remedy specific problems. If we remain true to our commitment to place the resource ahead of special interests, then salmon and steelhead should regain some semblance of their former abundance.

INTRODUCTION

Over the past 30 years of managing anadromous salmonids in British Columbia, our knowledge, experience, and ability to manage have become much more sophisticated. At the same time the complexity of largely negative human influences on freshwater habitat and on fish stocks continues to frustrate most management goals. A huge gap remains between knowing how to manage to an ecologically sustainable maximum and actually achieving it. The objective of this chapter is to review what we have learned about anadromous salmonid management over the last three decades.

This 30-year span overlaps with my professional fisheries career in British Columbia: 9 years as a federal research scientist and 21 years as a provincial manager and agency Director. The review will focus on the management of anadromous salmonids, including salmon and steelhead, and freshwater habitat in the province. Note that there are only two fisheries management agencies in British Columbia: Fisheries Branch of the B.C. Ministry of Environment, Lands and Parks, responsible for freshwater fish and fisheries, and Department of Fisheries and Oceans (DFO), which oversees management of all tidal fish and fisheries as well as conservation of salmon and their habitat in freshwater. This chapter discusses the experience of fisheries management in British Columbia in three areas: public policy, management policy, and operational measures.

PUBLIC POLICY

The category of public policy comprises four areas of experience: hydroelectric development, Pacific Salmon Treaty, habitat legislation, and social welfare policy.

HYDROELECTRIC DEVELOPMENT

Among the many valuable and successful Canadian experiences in salmon management, none stands out more prominently than the public policy decision to maintain free of dams our major salmon rivers such as the Fraser and Skeena. It was the "Two Rivers" policy of Premier W.A.C. Bennett in the early 1960s that directed hydroelectric development to the Peace and upper Columbia Rivers. As a result the Skeena and Fraser remain truly great salmon rivers while the once magnificent salmon runs of the heavily dammed Columbia River have been decimated. When one compares the relative costs and benefits of a public policy that prevents hydroelectric development on rivers like the Fraser, it is evident that public interests were well served by Premier Bennett's decision.

PACIFIC SALMON TREATY

As an instrument of public policy, Canada strongly supports the Pacific Salmon Treaty, which it co-signed with the U.S. in 1985. By providing a forum for consultation and negotiation between the two countries on Pacific salmon issues, and by emphasizing conservation as the main problem to address, the Pacific Salmon Treaty improves upon previous arrangements. After several years of periodic and difficult negotiations, a new Pacific Salmon Treaty was signed in June, 1999. While not as encompassing as either country might wish, the Treaty will provide certainty to fishers and fisheries management for the next 10 years.

HABITAT LEGISLATION

Like the U.S., Canada has a great deal of public policy embedded in federal and provincial legislation that was designed to protect fisheries habitat. The strongest Canadian legislation is the Federal Fisheries Act, which encourages habitat protection and enables strong punitive actions for damages to fish habitat. Unfortunately, punitive action can be taken only after the damage is done, and damage often requires litigation to prove. Proposals to amend the Fisheries Act to make it more proactive and thus a better deterrent and planning tool have met with little success.

A significant feature of the Fisheries Act is the federal "No Net Loss" policy requiring that habitat lost due to approved developments be replaced by habitat of equal or greater quality and quantity. This policy, while well conceived and written, has fared poorly in practice. Habitat often degrades slowly, for example, following streamside logging on floodplains, and many decades may pass before the full impact is expressed. In situations where impacts are immediately apparent, they are usually so severe that preventative and mitigative measures are of limited benefit. As a consequence, habitat managers have had to overcompensate just to "break even" in maintaining habitat at current levels.

Provincial legislation has generally proven ineffective in protecting fish and their habitat. For example, management under the provincial Water Act has been weak due to little monitoring and enforcement of water licenses, with the result that many streams are greatly oversubscribed. The Municipal Act is an instrument that could protect urban waterways with green strips and setbacks, but the mechanism for doing so is voluntary and requires protracted and costly amendments to

Official Community Plans. Many communities have opted instead to do nothing to protect the streams within their jurisdiction.

Some new provincial legislation shows considerable promise for improving habitat protection in agricultural and forested areas. The Environmental Review Process is designed to ensure minimal environmental damage in the planning and approval of large projects. The Forest Practices Code is less than two years old but already appears to be reducing impacts of forest harvesting on Crown lands. However, it has yet to be applied to logging on private lands. Forest Renewal B.C. is a recent legislative initiative that seeks to stimulate major investments in reforestation and restoration of degraded watersheds. The initiative is based on the premise that watershed protection must be accompanied by restoration if fish stocks are to recover.

SOCIAL WELFARE POLICY

To some extent Canada has managed its commercial fisheries as a social net through Unemployment Insurance (now called Employment Insurance). Although change is in the wind, fisheries management for years has been influenced by the need for fishermen to make enough deliveries and fish enough days to qualify for full "U.I." benefits. This pressure has resulted in some fisheries for which there were few fish or that endangered co-migrating weak stocks.

MANAGEMENT POLICY

Three areas of experience in the category of management policy are conservation of wild fish, commercial fleet reduction, and area licensing.

CONSERVATION OF WILD FISH

Conservation of indigenous, naturally produced stocks of salmon and steelhead has been a goal of provincial and federal managers for decades, but conservation has often been subordinated to other management goals. Two recent developments have forced managers to reassert the primacy of wild stock conservation as a management objective. First, there has been the broad decline in stock abundances across much of the province, necessitating fishery closures and other stringent conservation measures. Second, the Supreme Court of Canada recently affirmed the priority right of First Nations to fish for food and ceremonial purposes, but only after conservation requirements have been achieved. Exactly how many fish are required for the conservation of a stock has been the subject of much debate.

Both salmon and steelhead have been affected by these developments. Steelhead have fared better than salmon, in part because managers have generally not used hatchery production to justify or mitigate for overfishing of wild stocks, as has been the case for salmon. Many of the less productive stocks of salmon have been overharvested in commercial fisheries where exploitation rates are traditionally indexed to the abundance of more productive stocks. For example, Fraser sockeye fisheries have long been managed for aggregates of the more abundant wild stocks, with little concern shown for individual weak stocks. Today, because poor ocean survival conditions exert additional pressure on all stocks, the smaller, less productive wild stocks face the greatest risk of extinction.

COMMERCIAL FLEET REDUCTION

It is well documented that the commercial fishing sector has been overcapitalized for most of the past 30 years. Fleet reduction programs have been planned, debated, and, in some cases, implemented by various federal governments with little lasting effect. In 1996, a new federal program was initiated to substantially reduce the size of the commercial fleet by as much as 50%. Although everyone agrees on the goal, there is no consensus on how to achieve it.

AREA LICENSING

Most of the commercial salmon fisheries of coastal British Columbia have historically been managed as a single license area. The result has been a chaotic rush of gill net, purse seine, and troll boats from up and down the coast to major commercial openings. Consequently, when large fleets suddenly appeared, fishing times were greatly restricted and everyone caught less and made less money. Because of the diversity of gear types allowed, stock-specific management was difficult to pursue. In 1996, for the first time, the federal government divided the coast into smaller license areas and assigned different gear types to those areas: north and south for purse seines; north, south, and Fraser River for gill nets; and north, west coast of Vancouver Island, and Strait of Georgia for troll.

OPERATIONAL MEASURES

In varying degrees, nine operational measures have been employed by fisheries managers in British Columbia during the last 20 years: catch and release; selective mark fisheries; limited entry; "weak" stock management; enhancement and restoration planning; salmon farming; economic value; partnerships; and enforcement.

CATCH AND RELEASE

Catch and release (nonretention) regulations were first implemented by the provincial Fisheries Branch in 1977 for the Coquihalla summer steelhead. By 1984 all wild steelhead on Vancouver Island and the Lower Mainland were put on catch and release. Presently all but a few steelhead-producing rivers on the north coast are covered by catch and release regulations. This management approach has been the savior of many British Columbia steelhead stocks, particularly in light of the recent prolonged "trough" in marine survival. As it is, returns of winter run fish to most streams are barely sufficient for escapement needs; the situation is even more precarious for commercially intercepted summer runs.

Extensive research on illegal harvest, delayed mortality, and other effects of catch and release has been conducted by DFO on sport-caught chinook and coho salmon of various sizes. In response to unprecedented conservation concerns over mackerel predation, nonretention was implemented by DFO for adult chinook salmon in many major 1996 ocean sport fisheries. Compliance and overall results met management objectives. However, the reality is that it takes some years and considerable work to get anglers to accept the notion of releasing adult fish. DFO has had nonretention restrictions for certain salmon species in various troll, net, and freshwater sport fisheries for a number of years.

SELECTIVE MARK FISHERIES

Many chinook and coho salmon and steelhead fisheries are "mixed stock" in that hatchery and wild fish are commingled and accessible to fishers. So that the more productive hatchery fish can be harvested and the relatively scarce wild fish can be released unharmed, all the hatchery fish must be marked. This has worked very well with steelhead since 1989, when all hatchery pre-smolts in British Columbia began receiving an adipose fin clip. Several sport fishing groups have urged DFO to implement a similar marking program with hatchery coho salmon. This would harmonize with a similar marking effort now underway in Washington State, but it would mean major changes in the coded wire tagging program for hatchery and wild fish since a missing adipose fin is the standard indicator that a fish has a coded wire nose tag. Fisheries managers are concerned that marking all hatchery coho salmon by removal of the adipose fin would undermine the coded wire tag program, which is essential for wild stock management. However, in anticipation of rapid improvement in coded wire detecting equipment designed to handle large numbers of adult salmon, the DFO committed in November 1997 to implementation of a selective mark fisheries for coho salmon starting in 1998.

Limited Entry

Limited entry is a lottery process by which recreational fishing opportunities can be provided on a stock of limited size or in a limited geographical area. Similar systems are utilized by most North American wildlife agencies for management of specific populations of big game. The provincial Fisheries Branch introduced limited entry on the Dean River in the mid-1980s for nonresident anglers as a means of controlling effort and maintaining a quality experience. It has worked well for that purpose and has been expanded to several other summer steelhead rivers. However, it can be administratively cumbersome.

"Weak" Stock Management

DFO has historically managed commercial fisheries on the most productive stocks. This focus, coupled with recent poor survival in the ocean, has been devastating to less productive stocks. The depressed condition of summer steelhead and summer coho salmon on the Skeena are good examples, but there are many others. In response, DFO's management of commercial salmon fisheries has shifted significantly to include directed stock management for less productive stocks, and reduction of mixed stock fisheries. This will be accompanied by a move toward more selective fishing, and more terminal and inriver fishing, particularly over the short term while abundances are low due to poor ocean survival.

In recent years, mainly at the urging of sport fishing groups, DFO has encouraged experimentation with selective fishing gear and techniques. The intent has been to eliminate the bycatch of scarce and valuable sport species, or stocks at risk, in mixed stock fisheries for more abundant sockeye, chum, and pink salmon. Pilot fisheries have employed weirs, beach seines, floating fish wheels, dip nets (rather than gaffs), modified gill nets, and modified purse seining techniques. Many of these gear types have a demonstrable utility in specific conditions or situations.

Enhancement and Restoration Planning

Over the last 20 years of large-scale salmon enhancement we have learned that careful planning is required to avoid overharvest of nonenhanced stocks when fishing on the newly created surplus. For example, the Babine spawning channels were developed on the faulty assumption that no additional fishing effort would be required to harvest the enhanced sockeye. In fact, because fishing effort did increase, some of the less productive Skeena sockeye stocks, as well as other species, were overfished and have been severely depressed.

Another focus of enhancement planning is to avoid deleterious genetic and ecological interactions between hatchery and wild fish. Where hatchery programs are necessary, they should use the wild, native broodstock whenever possible. While the public and the stakeholders have been persuaded, and in turn have supported salmon hatcheries in British Columbia, experience tells us that we must place wild stocks first. Thus, provincial and federal fisheries managers have shifted their focus away from enhancement and toward habitat restoration. With its emphasis on repairing past damages and restoring habitat and fish stocks to their normal condition, the Watershed Restoration Program serves as a counterpoint to the provincial Forest Practices Code, which emphasizes preventative measures.

Salmon Farming

Salmon farming in British Columbia has developed to the point where production is approaching half that of the catch of wild salmon. The fact that salmon farming has relieved commercial fishing pressure on some wild stocks was, from a salmon management standpoint, entirely fortuitous. The abundance of farmed salmon on the world market (primarily from Norway and Chile) has acted to severely reduce the price/market for wild British Columbia salmon. In spite of strong concerns about the industry's ecological impact on wild fish, including colonization by escaped farmed Atlantic salmon and introductions of non-native diseases, the provincial government has apparently endorsed the future of salmon farming. Farmed salmon likely will continue to find an expanding market relative to wild British Columbia chinook and coho.

Economic Value

The steadily rising economic value of chinook and coho salmon as sport fish in recent years has caused fisheries managers to reconsider their role in the management equation. The most recent federal/provincial study shows that in 1994 the net economic value of sport-caught chinook and coho was 12 and 5 times, respectively, greater than chinook and coho caught commercially. If the principle of "highest and best use" is adhered to, future commercial chinook and coho salmon fisheries likely will be reduced to fishing on occasional surpluses.

Partnerships

New government partnerships of various types have been or are being implemented and are certainly the wave of the future. Federal and provincial partnerships with the public, including commercial fishing groups, angling groups, school systems, and municipalities, have been extremely successful in raising public awareness and concern about salmon and salmon habitat.

Another form of partnership is exemplified by the Skeena Watershed Committee, which comprises the three fishing sectors (sport, commercial, and Native) plus the federal and provincial governments. The group is mandated to develop by consensus the annual management plan for the Skeena River, with the overarching goal of achieving sustainable fish populations and associated fisheries. Although the Skeena Watershed Committee's immediate future as a public advisory body is uncertain, most observers feel the Skeena model has worked and should be emulated elsewhere.

A third type of partnership is between government fisheries agencies and First Nations groups in the actual management of various fisheries. These arrangements have existed for several years but have not been completely successful. However, First Nations are expected to play an increasing role in the management of many salmon stocks.

Enforcement

Enforcement capability and performance are always a high priority within the various fishing sectors. The one issue on which all sectors agree is the need for more fisheries enforcement. Fisheries Officers in DFO have been reduced, reassigned, and retrained to the point that there are actually fewer now than in 1948. However, those remaining are focused exclusively on enforcement, whereas previously they had a much wider range of management responsibilities.

CONCLUSIONS

So what have we learned that will help achieve and maintain sustainability of Pacific salmon in British Columbia?

Public Policy Summary

1. *Hydroelectric development.*—We should continue the policy of avoiding construction of hydroelectric dams on salmon rivers.

2. *Pacific Salmon Treaty.*—International cooperation is essential to the sustainable management of widely ranging salmonids. Several years of acrimony and uncertainty preceded the signing by Canada and the U.S. of a new 10-year Treaty in June, 1999.
3. *Habitat legislation.*—Strong legislation is an important but, by itself, insufficient condition for meaningful habitat protection. Political and public will to implement protective legislation is needed and is becoming manifest now that some salmon populations are at risk.
4. *Social welfare.*—Social welfare support and fisheries management do not mix. Decisions based primarily on social welfare concerns tend to compromise management policies designed to achieve sustainable fisheries goals.

MANAGEMENT POLICY SUMMARY

1. *Conservation of wild fish.*—Our "wild fish first" policy gives the salmon resource priority and management agencies the authority they need to successfully fend off or modify many development proposals, from hatcheries and logging to gold mines and subdivisions.
2. *Commercial fleet reduction.*—We keep learning this lesson again every fishing season – we are many times overcapitalized for the amount of commercial fish available on a sustainable basis. Firm commitment and action to reduce the commercial fleet size is essential to long-term sustainability.
3. *Area licensing.*—We have learned the value of area licensing from Alaska, where it has been part of its management system for many years. In concert with fleet reduction, area licensing would significantly improve management of salmon fisheries.
4. *Catch and release/nonretention.*—This measure has a proven track record in maintaining wild steelhead stocks at viable levels while permitting a recreational benefit. Its use should be expanded to other salmon species, but it should not be used to compensate for poor management practices. A major outreach effort is needed to gain wide public acceptance.
5. *Selective mark fishery.*—This is a proven technique that enables the harvest of hatchery fish while protecting wild stocks. Already successfully applied to steelhead in conjunction with catch and release regulations, a selective mark could also be used in other large sport fisheries that are not dependent on harvesting wild fish.
6. *Limited entry.*—Effective and popular as a means of protecting special wild rivers and selected stocks, limited entry needs to be more widely applied to fisheries in British Columbia.
7. *"Weak" stock management.*—Clearly, management to protect less productive stocks must be the priority of fisheries managers. We have learned the hard way that harvest rates cannot be set solely by the abundance of the most productive stocks. Early application of selective fishing gear and techniques shows some promise of reducing the incidental catch of weak stocks in mixed stock fisheries.
8. *Enhancement and restoration planning.*—Careful, long-range planning of enhancement is vital to avoid overfishing or ecological impacts on wild stocks. Watershed restoration, with its focus on restoring wild stock productivity, is an essential accompaniment to habitat protection and the Forest Practices Code.
9. *Salmon farming.*—Two developments related to salmon farming will influence wild salmon stocks and fisheries management in British Columbia in the future. First, reported incidences of sexually mature farmed Atlantic salmon that have escaped from sea pens and entered freshwater to spawn are becoming increasingly common. Second, farmed salmon appear to be replacing the traditional troll-caught wild chinook and coho salmon in the fresh fish markets of the world. These developments have both negative and positive implications.

10. *Economic value.*—Sport-caught salmon have a very high economic value, compared to commercially caught salmon. There is expectation that commercial fishing on some stocks will be scaled back to maximize economic benefits.

11. *Partnerships.*—There was a time in fisheries when government did it all, but no longer. We are learning that partnerships work, that fish belong to the people, not to bureaucrats or biologists, and that there is some exceptional talent and commitment in the public with which to partner.

12. *Enforcement.*—There is a critical need for funding and training to enable adequate enforcement of existing laws and regulations.

A FINAL WORD

The bottom line of what we have learned in British Columbia is that no management program or initiative will work for long without strong public and political support. With freedom of information regulations and widespread public participation, it is impossible for agencies or governments to perpetuate poorly planned and ecologically unsound programs that benefit special interest groups at the expense of the resource. The salmon resource has provided sustenance and recreation to the people of British Columbia for centuries; it is now time for us to restore her salmon populations to levels that enable them to sustain themselves and a diversity of beneficial uses.

7 Aboriginal Fishing Rights and Salmon Management in British Columbia: Matching Historical Justice with the Public Interest

Parzival Copes

Abstract.—At the time of European contact, Aboriginal peoples inhabiting the Fraser and Skeena watersheds in what is now British Columbia pursued river fisheries on spawning runs of salmon, for both domestic use and trade. In support of a salmon canning industry established in the 1870s, the Canadian government severely restricted Native river fisheries, limiting them to fish for house-hold needs, and prohibiting the use of their most productive gears, such as weirs and traps. This contributed to the impoverishment of Aboriginal communities, with many descending into social dissolution and despair. Recent decisions by the Supreme Court of Canada have affirmed the constitutionally protected priority rights of Aboriginal peoples to fisheries integral to their culture at the time of European contact. An Aboriginal Fisheries Strategy, implemented in June 1992 in response to Supreme Court edicts, has modestly increased salmon available to some river tribes and has allowed parts of the catch to be sold. This has engendered strong opposition from commercial fishing interests, fearful of a decline in their catches. It has also added to current public concerns regarding alarming declines in some salmon stocks, particularly those of interest to recreational fishers. This chapter proposes a salmon management strategy for the dual purpose of strengthening the economies of First Nation river communities and substantially increasing the production of salmon from the Fraser and Skeena watersheds for the benefit of all stakeholders. This would involve an additional transfer of a modest share of the salmon harvest from the mixed-stock commercial sea fisheries to the Aboriginal river fisheries. The latter would be restructured to use selective fishing gears and terminal fisheries to improve stock-specific spawning escapements significantly. Spawning habitat restoration and improvement, undertaken by local Aboriginal com-munities, would further assist stock recovery. The objective would be to greatly increase sustainable salmon harvests, by rebuilding and enhancing weak stocks and allowing fuller utilization of surplus production from strong stocks. A program of experimental fisheries to develop and test appropriate gears and fishing techniques is proposed, in conjunction with the development of management models to establish optimum escapement targets, fishing locations, and harvest allocations. These would balance biological and economic considerations and observe the need for an equitable distribution of benefits in which all stakeholder groups would participate.

INTRODUCTION

For thousands of years prior to the arrival of Europeans, the abundant salmon of what is now British Columbia (B.C.) provided a major source of food for Aboriginal peoples (Mitchell 1925; Ware 1978; Carlson 1992). Surpluses from the catch, at times, served as important trading goods. Because Pacific salmon swim long distances up numerous tributaries from which they originated, they have

1-56670-480-4/00/$0.00+$.50
© 2000 by CRC Press LLC

been available not only to the coastal peoples of the region, but also to the tribes inhabiting the inland watersheds. Indeed, as coastal people had access to many alternative marine resources, it was often among inland river tribes that the greatest dependence on salmon was established. Coincidentally, salmon harvesting was exceptionally productive in the "gauntlet" fishery of the rivers, where fish were concentrated in dense spawning runs.

In the 1870s, Euro-Canadians established a salmon canning industry in B.C., with the raw material being supplied by a newly developed coastal commercial fishery. While the canning industry and associated harvesting operations provided employment for many Aboriginal people, primarily from coastal settlements, these operations adversely impacted the established salmon harvest of inland river tribes. Added to the existing Aboriginal catch, the expanding demand for cannery fish was leading to unsustainable harvest levels. Faced with the need to conserve salmon stocks, the federal government, in support of the canneries, severely curtailed the Aboriginal river catch under the 1888 Fisheries Act.

The 1888 regulations confined Native people to a "food fishery" for their domestic use and prohibited them from selling salmon, thus eliminating what for many was their primary source of cash and barter. They were also prohibited from using their efficient weirs and traps, and confined to less efficient fishing devices. These regulations above all impacted tribes inhabiting the watersheds of the two great rivers, the Fraser and Skeena (Figure 7.1), who mostly lived too far from the sea for them to join the commercial fishery in which many of the coastal Native people participated. This chapter primarily concerns the Fraser and Skeena, although the analysis is also applicable to a number of Aboriginal fisheries on smaller B.C. river systems.

In recent decades the Aboriginal peoples of B.C., now identifying themselves as "First Nations," have become much more assertive of their rights, for which they have received increasing recognition through appeals to the courts. The 1990 "Sparrow" decision by the Supreme Court of Canada has been particularly important in strengthening Aboriginal fishing rights (Binnie 1990; Helin 1994). Demands by First Nations for greater access to salmon catches have been coupled with extant land claims, pursued in part by court appeals and in part by negotiations in a treaty-making process that will take many years to complete. The most elaborate case heard by the courts thus far has been that of the Gitxsan and Wet'suwet'en inhabiting the upper Skeena region (Copes and Reid 1995).

The Sparrow decision assigned priority to the quantitatively undefined Native food fishery over all other fisheries. In response, Canada's federal government, which has jurisdiction over the salmon fisheries, implemented an Aboriginal Fisheries Strategy (AFS). In agreements with a number of First Nations, the federal Department of Fisheries and Oceans (DFO) started in 1992, on a modest scale, to allocate more fish to the food fishery. The sale of fish from these negotiated allocations was also permitted. This removed, for the time being, an important source of friction, legal uncertainty, and repeated court challenges. In return the Aboriginal groups concerned agreed to a negotiated cap on the amount of fish allocated to the food fishery, which helped to restore DFO control over aggregate catch levels. The expanded Aboriginal food fishery permitted under the AFS was initially plagued by problems of hasty implementation, including inadequate controls on the catch. This added to other management problems in the salmon fishery, provoking a major government-ordered enquiry in 1994 (Fraser 1995) that resulted in the application of stronger enforcement measures.

The AFS has been fiercely opposed by groups representing the commercial salmon fishery. This opposition has been predicated largely on the notion that reapportionment of the salmon catch is a "zero-sum game," with gains by First Nations on the rivers resulting in corresponding losses suffered individually by commercial operators. Yet the government has taken pains to demonstrate that this has not been the case and will not be so in the future. For example, additional allocations to First Nations are being matched by catches attributable to vessels withdrawn from the commercial fishery (through voluntary buyback from fishers retiring from the industry) and by increased production through stock and habitat enhancement. Therefore, average catches of individual operators remaining in the fishery should not be affected by the additional allocations made to First Nations.

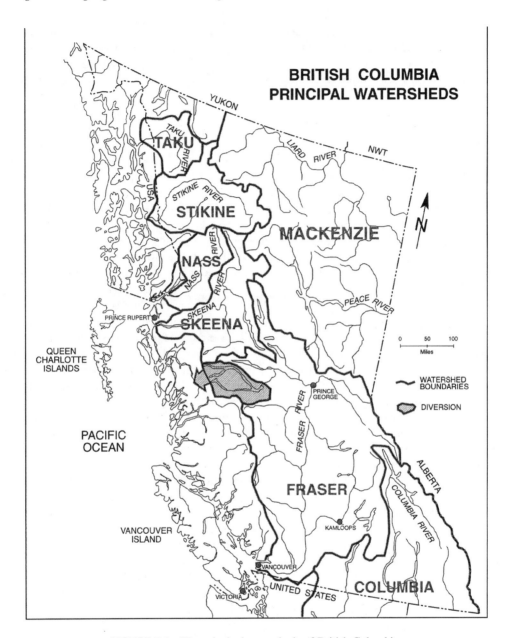

FIGURE 7.1 The principal watersheds of British Columbia.

The purpose of this chapter is twofold. It will present brief arguments to establish, on grounds of historical justice, a case for the allocation to river-based First Nations of significantly greater catches of salmon than they currently receive. The equally important second purpose is to make the case that an increase in the allocation of salmon to river fisheries makes it possible to introduce greater stock-specific selectivity in harvesting salmon, both through the increased use of selective inriver gears and through the use of terminal fisheries in appropriate locations and circumstances. This should allow for improved spawning escapement of weaker stocks and fuller harvest utilization of stronger stocks, both leading to increased total stock productivity. The end result should be much greater sustainable catches of salmon in which all stakeholder groups may share. To reconcile non-Aboriginal salmon fishery participants to this process, it should be emphasized that the proposal envisages maintaining their individual harvest shares in the short run and increasing them in the long run.

The case made in this chapter is speculative, based on *a priori* reasoning in relation to the known general circumstances of the salmon fisheries. It lacks hard data to demonstrate the feasibility of the management strategy proposed on a scale sufficient to generate large benefits. Therefore, a modeling exercise is also proposed, supported by experimental fisheries, to test feasibility and provide cost-benefit estimates.

THE SALMON RESOURCE AND ITS MANAGEMENT

Pacific salmon comprise the genus *Oncorhynchus*, of which six species are represented in the B.C. salmon catch: sockeye *Oncorhynchus nerka*, pink *O. gorbuscha*, coho *O. kisutch*, chinook *O. tshawytscha*, chum *O. keta*, and steelhead *O. mykiss*. See Burger (2000) in this volume for a description of their life history and habitat requirements. To demonstrate the possibilities for a major improvement in productivity of B.C. salmon stocks, it is necessary to refer to the basic features of the salmon life cycle and the principles by which the stocks may be effectively managed.

An important factor affecting the total size of the salmon resource is the extent to which available spawning and/or rearing habitats are used to full capacity. To obtain the best rate of reproduction (i.e., resulting in the largest number of surviving offspring) the number of spawners should be sufficient to fully utilize the available spawning or rearing habitat, depending on which is limiting for a given population. Returning salmon in excess of such numbers should be kept off the spawning grounds to prevent them from disturbing the redds of preceding spawners and thereby causing mortality of deposited eggs.

Fisheries scientists concerned with salmon reproduction management are particularly interested in identifying "stocks" consisting of groups of one or more distinguishable breeding populations (i.e., fish of the same species spawning at a particular location and time that as a consequence do not spawn with fish spawning at another location or time). With fish of the same stock managers attempt to secure optimal spawning by opening and closing the fishery in particular locations to allow for safe passage of the right "escapement" for each stock, while ensuring that the remainder of the stock is taken in the fishery as a harvestable surplus.

There are, however, many hundreds of distinct breeding stocks in the Fraser and Skeena systems. Thus, there often are several migrating stocks mixed in the river at the same time. Some of these are very strong, requiring only a short period of fisheries closure to secure sufficient escapement, while others are weak and require a longer period of closure—or even a complete closure—to guarantee adequate escapement. This creates a mixed-stock management problem with compromise closures that are not long enough to produce adequate escapements for weak stocks, but too short to allow full utilization of surpluses from strong stocks (Knudsen 2000).

The mixed-stock problem has been exacerbated in recent years by the successes of the Salmonid Enhancement Program (SEP), carried out by DFO in cooperation with provincial authorities and various local interests. The objective of this program has been to increase British Columbia's salmon resources by various artificial means. The largest undertaking was a 1960s project on Babine Lake in the Skeena River watershed, which greatly expanded the spawning capacity for sockeye salmon through the construction of a large number of artificial spawning channels, to supplement those naturally available (West and Mason 1987). This has resulted in runs of Babine sockeye that dwarf the runs of other Skeena salmon stocks. On the Fraser River, enhancement projects have also produced a number of very large sockeye stocks.

Heavy fishing on enhanced sockeye stocks has resulted in serious depletion of smaller wild stocks that are mixed in with them (Healey 1993) and has threatened many with extinction (Slaney et al. 1996). Reduction or extirpation of the weaker wild stocks is a serious matter because, not only will these stocks no longer contribute to the total salmon catch but, more importantly, the available gene pool will also be reduced, threatening loss or reduction of diversity and extent of many desirable traits. These relate, for instance, to disease resistance, adaptability to environmental change, ability to utilize diverse ecological niches, time and place of availability, and end product variety and quality. The practice of developing large stocks of enhanced salmon has been implicated in a variety of deleterious effects (Hindar et al. 1990). For example, approximately 60% of all Skeena sockeye are derived from the enhanced Babine Lake stocks, suggesting limited genetic diversity for the Skeena system sockeye (Sprout and Kadowaki 1987; Jakubowski 1990). One concern is that, if large stocks are depleted by disease and need rebuilding, it may be difficult to draw sufficient numbers of disease-resistant spawners from the few remaining wild stocks. Similar considerations apply to the large enhanced stocks of Fraser River sockeye.

Because many of the smaller and weaker stocks of the Fraser and Skeena systems are now threatened with extinction (Slaney et al. 1996), managers have imposed some fisheries closures to reduce the threat. In turn, this has resulted in large numbers of surplus sockeye being left to rot unharvested in some years (e.g., on the Babine River) because they could not be fished during the closures.

To solve or reduce the mixed-stock problem, two strategies are available. One is to reduce the fishing effort on mixed stocks as much as possible, by using "terminal fisheries" that target stocks separately at points where there is no mixing or little mixing with other stocks. A second strategy is to employ selective fishing techniques, using gear such as weirs and traps, fishwheels, beach and purse seines, fykes, reefnets, dragnets, bagnets, and dipnets. These allow for retention of fish from strong stocks and live release (with low mortality) of fish from weak stocks. This will be discussed further below.

If the mixed-stock problem is adequately contained, considerable advantage may be drawn from the remaining potential for an increase in B.C. salmon stocks, which appears to be substantial. Such a conclusion is supported by the evident availability of additional ecological capacity and by the larger size of B.C. stocks in former times. Large increases in salmon runs are considered possible (Ricker 1987, 1989). Sockeye constitute the most valuable component of the B.C. salmon fishery, accounting for about 60% of the commercial landed value. Henderson (1991) notes that the Fraser watershed, in the 1970s and 1980s, produced about 66% of the province's sockeye, and refers to estimates that Fraser sockeye stocks could be tripled to an average of about 30 million fish per year.

The launching of SEP is evidence of the opinion that the stocks could be brought back to larger size and the successes of the SEP so far confirm the practical feasibility of the techniques that have been developed. These SEP techniques, by and large, have concentrated on physical manipulation of the stocks and their environment in a harvest management setting that has otherwise remained constrained and little changed by institutional conditions.

This chapter suggests that progress toward larger salmon stocks and larger harvestable surpluses may be substantially advanced by strategic changes in institutional structures. This would be based on an increased role for Aboriginal river fisheries, with the proviso that these fisheries would be conducted in a highly selective manner to optimize stock-specific spawning escapements and minimize non-harvest mortality. A review of past and present patterns of salmon fishing will set the stage for the subject of stock-specific management.

THE TRADITIONAL ABORIGINAL FISHERY

There is clear prehistorical evidence of the utilization of fish resources by the people inhabiting the Fraser and Skeena watersheds (Carlson 1992). On the Skeena, for instance, numerous settlements were established in prime fishing areas (MacDonald et al. 1987). While relevant archaeological exploration has not been extensive, excavations have turned up bone fragments of salmon and other fish at four Skeena sites, namely at Hagwilget (Ames 1979), at Kitwanga (MacDonald 1989), and at two Kitselas Canyon sites (Allaire 1978; Coupland 1985). Some of the fragments may date back as far as 2000 B.C.

At the time of European contact in the late 18th century it was evident that the Northwest Coast First Nations had developed societies that had—in relation to time and place—a notable level of material sufficiency, cultural expressiveness, and artistic refinement. Fish resources played an especially important role in providing them with a plentiful supply of food, in response to which they developed a variety of ingenious and very efficacious fishing techniques. These have been described effectively, attractively, and artistically by Stewart (1977) in her book on Indian fishing. Among the fish resources salmon was evidently the most important. Garfield (1966) wrote: "Salmon was the decisive food resource of the Tsimshian, as it was of most other Northwest Coast tribes. Cohoes or spring salmon and sockeye salmon furnished the bulk of the fish dried for winter use."

When, in the early 19th century, the North West and Hudson's Bay (HBC) Companies brought the fur trade to the region, several tribes gained an opportunity for a significant extension of their salmon trade. Native-caught salmon—both fresh and smoke-dried—became an important staple for provisioning the HBC, and later other frontier groups, such as miners and construction workers (Ray 1984; McDonald 1985; Morrell 1985; Shepard and Argue 1989). Salmon was so important a commodity that it was accepted as a form of currency, with a well-known exchange rate (Ray 1984).

In their traditional fishery, the First Nations of the upper Fraser and Skeena rivers made great use of a variety of highly effective weir and trap systems (Stewart 1977; Morrell 1985), which intercepted salmon on their migration paths. These systems, operating under the authority of local chiefs, were eminently compatible with effective conservation. Fish were easily taken from the traps or along the weirs with dipnets, gaffs, and baskets. When enough fish was taken to occupy fully those engaged in processing the fish, the weirs and traps would be opened to let migrating fish pass through. Intermittently, the weirs and traps would be put into operation again to provide further raw material for processing, which mostly involved drying and smoking. This fishing system ensured bountiful harvests with escapement that was quite adequate to maintain the stocks in a healthy state, as evidenced by the prosperous condition of the tribes at the time of European contact and the healthy state of salmon stocks then observed (Morrell 1985, 1989).

The establishment of salmon canneries in the 1870s marked the beginning of a Euro-Canadian commercial salmon fishery. Initially it relied to a great extent on local Aboriginal labor, with the men employed in fishing and the women in processing. Over the ensuing decades, the commercial fishery expanded by attracting non-Aboriginals, often from outside the region. Competition for raw material with local tribal fisheries became more acute. Government fishery managers, evidently preoccupied with the interests of the Euro-Canadian dominated commercial fishery, were concerned that the effective weir and trap fisheries of the upriver Natives would take too much fish and endanger the stocks. In reality, of course, it was the additional pressure of an expanding commercial fishery that upset the pre-existing balance of catches and escapement, leading initially to large increases in harvests, but thereby subsequently depressing stock strength and threatening sustainability of catch levels.

As the commercial fishery expanded, progressively tighter restrictions were placed on upriver Aboriginal fishing (Lane and Lane 1978; Morrell 1985; Copes et al. 1994). In 1888 regulations were proclaimed that limited the authorized Aboriginal catch to the satisfaction of domestic needs, the so-called "food fishery," while prohibiting the Aboriginal salmon trade (Ray 1984). More productive Aboriginal fishing gears, such as weir-and-trap systems, referred to as "barricades" by fisheries officials, were banned by order-in-council in 1884.

For some years many Aboriginal groups managed to evade the restrictions. John T. Williams, Inspector of Fisheries, in 1905 reported to the Dominion Commissioner of Fisheries that conditions for salmon in the head waters of the Skeena River were "dangerous in the extreme" because of "illegal fishing by the Indians" which threatened "complete annihilation of this valuable fish and entire depletion of the river" (Williams 1906). Fishery Officer Hans Helgeson was sent out to enforce regulations and reported how he secured the destruction of all Aboriginal barricades throughout the area.

Helgeson's report (1906) incidentally gave impressive evidence of the size of the Aboriginal salmon industry in the area and of its importance to the Native population. Most striking was his description of two barricades on the Babine River, of "formidable and imposing appearance ...

constructed of an immense quantity of materials, and on scientific principles." He found "no less than 16 houses ... filled with salmon" and "an immense quantity of racks" for drying, which if stood close together would have covered "acres and acres of ground." He judged the catch "to be nearly three quarters of a million fish ... and though the whole tribe had been working for six weeks and a half it was a wonder that so much salmon could be massed together in that time."

Helgeson also reported the local chief's protest that

> said they have had an indisputable right for all time in the past, that if it was taken away the old people would starve, that by selling salmon they could always get "iktahs" [i.e., goods], and he wanted to know to what extent the government would support them, he thought it unfair to forbid them selling fish when the cannerymen sold all theirs, and I had to promise him to tell the government to compel the canners to let more fish to come up the rivers, as some years they did not get enough, that the canners destroyed more spawn than they, that formerly he could not see the water below his barricade for fish, that they were so plentiful that some of them were forced out on the beach, but latterly they had diminished, little by little every year.

In justification of their destruction of Aboriginal barricades, official reports described them as utterly incompatible with conservation. This ignored the survival of healthy salmon stocks over hundreds or thousands of years of Aboriginal fishing and the evidence that the barricades, as they were operated with alternating openings and closures at their accustomed rates of exploitation, demonstrated an understanding of, and adherence to, essential conservation practices. Nevertheless, there is good reason to believe that the British Columbia salmon stocks were seriously threatened. The Aboriginal fishery, by itself, had proven to be sustainable on the evidence of a long past. But the combination of a strongly developed salt-water fishery and continuation of the traditional Native fishery for domestic consumption and trade goods might well spell disaster.

While Native peoples on the upper Skeena were not the originators of the new pressures on the fish stocks, they were easy targets to blame for the results. Apart from having little influence with government, the media, or the general public, they were also in the unfortunate position of being the last user group in line along the migration path of salmon to their spawning grounds. By force of circumstance the ultimate onus of letting enough fish through to make up an adequate escapement was then placed on them. Any attempt by them to maintain their historical harvest levels, in the face of much greater catches downstream, except in very good years, might then have the "depensatory" effect of not leaving enough spawning escapement (Peterman 1980).

The foregoing account indicates that a large part of the salmon resources held and utilized by Native peoples in the Upper Skeena region (and elsewhere) about 100 years ago, was forcibly taken from them, without any significant compensation. It was handed over, essentially free of charge, to a new user group favored by the government and protected by the force of law. The Aboriginal population was compelled to use often inefficient and wasteful fishing practices with nets and gaffs, instead of weirs and traps. The supply of food fish left to them was at times inadequate, occasionally leading to starvation (e.g., in 1916; Sprout and Kadowaki 1987). Their salmon trade, which supplied them with many needed goods, was prohibited by law and greatly inhibited in practice.

The descent of several Skeena and Fraser River tribes from economically and culturally vibrant societies, by their standards of place and time in the mid-19th century, to socially and materially depressed communities 100 years later, is no doubt due in part to the severe reduction in access to the salmon resource that they suffered. The equity implications of the foregoing require no elucidation. Surely any attempt in that direction today would be quickly negated by the courts. Indeed, the courts are now in the process of reversing, to some degree, the effects of the injustices exemplified in the above account.

There is a concept in law, known as the "abstention principle," which is used both intra- and internationally. It holds that when a (fishery) resource is fully exploited by a user group, no new group is entitled to join in the exploitation of that resource. Canada and the U.S. have called on this principle in persuading Japan and other countries to refrain from fishing for salmon of Canadian or American origin in most of the North Pacific. The principle, in fact, is enshrined in Article 66 of the United Nations Convention on the Law of the Sea (United Nations 1983). Past actions of the Canadian government in suppressing Aboriginal fishing rights would appear to constitute a flagrant violation of the abstention principle.

After the removal of barricades, Aboriginal fishers in the B.C. interior were restricted largely to using nets and gaffs, which were relatively inefficient and damaging to the stocks. When gaffs are not used in conjunction with traps or weirs that inhibit the escape of fish, they become a wasteful technique. Many fish drop off gaffs gravely wounded and die. In an open-river fishery, they cannot be retrieved and are therefore lost for purposes of both consumption and spawning. Similar stock damage results when nets are lost in swift-flowing river waters and continue to entangle and kill fish ("ghost-fish"), unseen and unattended on the river bottom.

CURRENT USER GROUPS

The salmon fishery has been referred to as a gauntlet fishery. After spending much of their life on ocean feeding grounds, salmon return to the coast to assemble in concentrations that migrate upstream to spawn. In doing so, they are intercepted by various fishing groups along the way. The order of interception has significant implications. Fishing fleets that are furthest out to sea have the first opportunity to capture returning fish and thus have the most fish available for exploitation. Fishing groups on the river can exploit only what is left of a stock after other groups have taken their catch. However, they do have the advantage of fishing in confined waters where fish are concentrated and easy to capture.

During their return migration from the high seas some of the salmon originating in B.C. rivers first pass through Alaskan fisheries. When they reach Canadian coastal waters, returning salmon are subject to a commercial fishery conducted by trollers and seiners. Next in line is the commercial gillnet fleet, operating mostly near or in estuaries of major salmon rivers. It is only after all commercial fleets have taken their catches that the various First Nations on the rivers have an opportunity to exploit the remainder of the stocks for their food fishery. One further user group should be noted; sport fishermen angle for salmon in coastal waters and for salmon and steelhead on many rivers.

It is important to note that with current techniques and equipment, the various groups exploiting B.C. salmon stocks have an aggregate fishing capacity that is several times as large as necessary to take the entire harvestable surplus (Pearse 1982). With no constraints on fishing effort they could fish the salmon stocks to extinction in a few years, with each group blaming the others for the result. Only the restraint of government regulation and management has prevented this from happening.

Fisheries regulations imposed by DFO, particularly including time and area restrictions that impact diversely on the different components of the harvesting sector, are the prime determinant of how much fish each group is able to take. Thus the government in effect has responsibility for deciding the allocation of benefits from the fishery among the various user groups.

ECONOMIC EFFICIENCY

Fisheries regulation and catch allocation have bearing both on efficiency in resource use and on equity in the distribution of benefits among user groups. Efficiency comparisons may be made in terms of the total value of net benefits (i.e., the total value produced minus the costs of production) achieved under alternative regimes or processes. Value concepts underlying efficiency calculations are subject to conceptual controversy and practical ambiguity, which would be quite substantial in the case of the complex socioeconomic implications of this chapter. Moreover, given the extent and uncertainty of the material changes in the fishery that are being considered, calculation of meaningful efficiency estimates in absolute terms would require a major project, far beyond the scope of this chapter. Instead, the focus here will be on identifying major *improvements* in efficiency that may be achieved as a result of the strategic management changes proposed in this chapter. These changes will be linked to the allocation of more fish to river tribes and will refer to the resulting quantities and qualities of salmon harvested by different stakeholder groups in different locations.

A large body of knowledge has developed regarding the estimation of harvestable surplus (see Hilborn and Walters 1992 for a review). Suffice to say that, for each stock, it is theoretically possible to apportion the returning run into the harvest taken; the non-catch mortality of fish from natural causes and from harvesting-induced stress while en route to the spawning grounds; and the final escapement of fish onto the spawning grounds.

The primary test of efficiency for the proposed new management measures in the Fraser and Skeena systems will be (cost-effective) achievement of sustainable harvests that are much greater than those currently obtained. This is to be accomplished by pursuit of two goals: (1) maximizing production of fish (recruitment) through optimal stock-by-stock spawning escapement and (2) minimizing non-catch mortality losses from harvesting-induced stress. I abstract here from environmental questions, important as they are, to simplify the arguments advanced. Environmental objectives, such as habitat improvement, should be pursued for the net benefits they may bring, regardless of whether we are dealing with the current management system or the one proposed.

Stock-specific management is a requirement to achieve optimal spawning escapements in support of maximum production. There is considerable mixing of stocks in coastal waters where they are targeted by the largely indiscriminate fishing of commercial fleets, which is a major factor in the overfishing of weak stocks and underfishing of strong stocks. Once in the river, stocks are separated from those migrating to other rivers and also from many same-river stocks by the timing of their runs. As they move upriver the stocks separate further into different tributaries on the way to their respective spawning areas. Stock separation in time and space facilitates the culling of individual stocks to achieve precise escapement targets. Nevertheless, particularly in lower river sections and larger river systems, some mixing of migrating stocks remains, calling for selective fishing of the mixed aggregates.

A number of experimental Aboriginal fisheries, using selective harvesting techniques with live release of fish from weak stocks, have been authorized and supported by federal and provincial departments in conjunction with the AFS, notably in the Fraser and Skeena watersheds, on the Nass River (Figure 7.1), and on Vancouver Island. The gears used have included weirs and traps, fishwheels, beach seines, purse seines, reefnets, dragnets, bagnets, and dipnets. In the course of federal and provincial negotiations with B.C. First Nations in respect of land claim settlements, a first Agreement in Principle has been reached with the Nisga'a on the Nass River. It includes a proposed salmon allocation to the Nisga'a structured to mesh with management objectives including, notably, the use of fishwheels.

First Nations on the rivers have been demanding a greater allocation of salmon to allow them an income from fish sales in addition to their supply of fish for domestic consumption. Strategically located, large traps would be a particularly appropriate gear for an inriver fishery with a commercial component. Given the high concentration of fish during river migration, there is ample reason to believe that traps which are well designed and appropriately placed will be both harvest productive and cost effective. From an industry-wide standpoint, what is even more important is that traps have great utility, both in escapement management and in catch handling. Traps may be constructed to allow fish to be readily live-sorted by species. In well-designed and well-operated traps, stress on captured fish may be minimized. Fish from weak stocks may then be released with relatively low mortality. This contrasts with commercial net fisheries in which most fish are killed outright. The exception is that seiners may recover fish live by brailing them from the seine net, but that is a costly operation and not lacking in stress or mortality for fish meant to be released.

Commercial gillnet fisheries present an additional deleterious effect in that many fish are killed by entanglement but are not part of the catch because they drop out of the net dead or struggle free to die later from the stress endured. A report to the Minister of Fisheries and Oceans regarding salmon management problems on the Fraser (Larkin 1992; Pearse 1992) offered a rough estimate that in a phase of the (mostly) gillnet fishery that produced a catch of 583,000 fish, another 248,000 were lost to mortality, half of which was from natural causes while the other half was fishing-induced. A similar problem occurs in the troll fishery with "shaker mortality," wherein hooked fish struggle loose but die uncaught as a result of injury and stress.

In a well-designed trap fishery, fishing-induced, non-catch mortality could be greatly reduced or nearly eliminated. By using liftable culling platforms, from which fish from strong stocks could be harvested and the remainder returned to the water without handling, mortality in non-target stocks may be minimized. In the case of the Fraser and Skeena rivers, selective trap fisheries that released all steelhead, chinook, and coho—most stocks of which are weak—would greatly advance their conservation. Pink salmon that are often discarded (dead) by Aboriginal fishers, because of their poor quality when caught upriver, could also be released live when taken in upriver traps at locations where discarding is otherwise a problem. Released fish would add to spawning escapement and allow for a larger commercial pink catch in salt water.

There remains a problem, most notably with sockeye stocks on the Fraser and Skeena, when strong and weak stocks of the same species migrate upriver simultaneously. As the fish are indistinguishable, visual sorting is infeasible. However, intraspecific stock identification is possible using various laboratory techniques—depending on the species—including electrophoresis, scale pattern analysis, and parasite identification (James Woodey, Pacific Salmon Commission, personal communication). Test results may be secured within a few days (or in one day for some procedures), so that by sampling fish near the start of their river migration, separate breeding stocks may often be identified in time to determine appropriate schedules of fisheries openings and closures. In the important case of sockeye the current practice is to use scale analysis, for which next-day results may be obtained. Ongoing experiments with DNA analysis may lead to even greater precision in stock identification (Park and Moran 1994).

With migrating runs taking about a month to pass a location on the river, the overlap of strong and weak stocks can be severe. However, the peak of a run takes much less time to pass, so there is much more separation of the run peaks. By sampling the runs, peak migration assemblages for various weak stocks may be identified and appropriate closures set. This is already being done to the extent feasible in the current sockeye fishery.

The fact that salmon deteriorate in perceived quality as they enter freshwater and ascend their home rivers underlies a widely held prejudice against commercial inriver fisheries. This notion is subject to challenge. The extent to which change or deterioration takes place varies greatly by species, stock, and river system. Generally speaking, chum and pink salmon show a distinct deterioration in quality as fish move upriver, while sockeye, chinook, and coho are much less affected. Steelhead, which do not necessarily die after spawning, are least affected. Extensive quality testing of gillnet-caught sockeye on the Skeena was undertaken for DFO in 1982 (Slaney and Birch 1983). The results generally showed that sockeye caught in the lower river were of "number one" quality and those caught in the middle reaches of the river were of "number two" quality, apparently largely related to increased water- and netmarking. Fish taken near Hazelton, in the upper Skeena, were found suitable for export grade canned products and yielded smoked products of "acceptable quality."

Ocean-caught salmon, at its best, has qualities that cannot be matched—at least not in every respect—by salmon harvested upriver. Ocean fish may be superior particularly in producing commodities in the fresh and frozen product sector. However, there are some quality-based counter-considerations. River traps allow fish to be taken live and butchered fresh at an adjacent processing facility, whereas seiner- and gillnet-caught fish normally arrive at processing facilities dead, often after an extended trip for delivery. It appears also that both the quality and quantity of salmon roe (a by-product, now of considerable value) increase as the fish migrate upriver. In addition, upriver fish is considered more suitable for certain smoked products. It bears noting that upriver Aboriginal groups in British Columbia have used local salmon for ages as a staple food and as a valuable trading commodity. Their continuing trade (whether or not illegal) has demonstrated that the general population also finds their smoked salmon to be an attractive product. Therefore, it is evident that an upriver Aboriginal fishery should have little trouble in producing marketable commodities.

In judging the net economic benefits that may be generated by an enhanced First Nations river fishery, as proposed in this chapter, there are evidently trade-offs to be considered. There is the quantity–quality trade-off of generating potentially much larger fish stocks and harvests through selective inriver fishing, as against the lower quality of inriver fish for the end-product market. This trade-off loses significance if it is established that increased harvest production will be sufficient to achieve larger catches both in the commercial marine and Aboriginal river sectors. Moreover, the perceived lower quality of inriver fish from physiological change in freshwater may be offset, in whole or in part, by quality–quality trade-offs. In a well-designed trap system, fish may be freshly butchered and iced directly out of the trap, whereas boat-caught fish is often drowned and net-marked during harvesting and may have to travel a considerable distance for delivery.

Both from a macroeconomic efficiency viewpoint and that of social equity, there is one other important consideration. The lack of employment opportunities in Aboriginal communities is severe. Annual labor force surveys of Aboriginal communities in the northwest of British Columbia— covering the Skeena watershed and areas to the north and west (Figure 7.1)—have been conducted over the years 1994–1997 by Pacific Northwest Employment Training and Development (a First Nations organization). Their census data for 1997 showed unemployment for the 25 Aboriginal communities in the area ranged from 45 to 86%, with an average of 68% (Pacific Northwest Employment Training and Development 1997). The provision of a worthwhile amount of employment that would result from establishment of commercial inriver fisheries for inland tribes is likely to have a positive impact on the economy by reducing structural unemployment. While some employment downstream would likely be displaced initially, this might be absorbed by normal turnover of labor in the coastal commercial fishery. In any case, the coastal fishing labor force is drawn from a population that generally has much better prospects for alternative employment than are available to inland Aboriginal communities. It is also important that increased employment in an industry highly compatible with traditional activities of Native communities may help significantly to overcome the chronic conditions of economic depression, demoralization, and dissolution in many Aboriginal communities.

DISTRIBUTIONAL EFFECTS

Predictably, there has been a strongly negative response to the AFS and to claims for a greater share of salmon catches by river-based First Nations, coming from other stakeholder groups, including some Aboriginal fishers working in the commercial sector. These other stakeholders know that the immediate effect of more fish for Aboriginal river communities will mean less fish for them collectively, and fear (erroneously, so far) that it will mean less for them individually. Understandably, they are inclined to look at the equity issue in terms of maintaining their current share of the catch, with most of them probably ill-informed about the historical injustices suffered by Aboriginal river communities.

To allay the concerns of stakeholders other than the river-based First Nations, it is important to make two points at the outset. The first is that the diversion of fish from saltwater commercial fisheries to Aboriginal river fisheries need not be very large and should be undertaken with full compensation to those affected. Provided the commercial fleet is targeted as much as possible on stronger stocks with relatively low admixtures of weaker stocks, it should remain possible to leave the bulk of the catch to the commercial fishery. As the long-term allocation of food fish to both coastal and river First Nations has been about 4% of the salmon catch, the allocation to river communities only would be somewhat less than that. A tripling or quadrupling of that allocation would still leave the bulk of the harvest to the commercial fishery, while allowing for a multiple increase in weak stocks reaching the rivers.

This leads directly to the second important point. If Aboriginal river fisheries convert to selective gear and terminal fisheries with live-release of fish from weak stocks, then a three- or four-fold increase of fish in the river may be parlayed into a much larger multiple of fish from weak stocks reaching the spawning grounds, resulting in considerable restoration of those stocks. Given selective harvesting on the river it should then also be possible to engage in the strengthening of many other stocks with a potential for enhancement. Allocating a greater share of the catch to Aboriginal fishers on the river then is not just a matter of redistributing the harvest among stakeholders. It is also extremely important for improved management potentially leading to much larger aggregate catches.

To make inriver commercial fisheries for inland First Nations at all palatable to commercial fishermen, it is undoubtedly necessary to provide them with compensation for any reduction in the marine commercial harvest. This appears to be recognized by the federal government, which has justified the AFS allocations made to Aboriginal river fisheries by actions of two kinds. The government has bought out licensed vessels from some operators retiring from the commercial fishery, withdrawing those vessels from the fishery and transferring their estimated catch allocations to the river fisheries. Second, they have made allocations to the river fisheries from some of the salmon stock additions attributable to government-financed enhancement projects. Continuing such compensation measures for further salmon allocations to river fisheries appears in order, although this might not satisfy the processing companies who would fear reduced throughput for their plants. However, the larger total harvests in prospect, plus the likelihood that much of the river catch will flow through their establishments, could result in an outcome they would eventually find amenable.

The catching capacity of the salmon fleet is in excess of any current or prospective needs, resulting in excessive harvesting costs in relation to the value of the catch (Pearse 1982). Fleet capacity reduction should result in greater net benefits for the harvesting sector. Unfortunately, vessel buybacks by government so far have done little to reduce excess capacity overall (Pearse and Wilen 1979; Copes 1990, 1997). The federal government, however, appears to have a commitment to further rationalization, so a reduction of the coastal salmon fleet to accommodate more upriver fishing could be made part of any larger program of capacity management.

Sport fishers are likely to benefit significantly from stock-specific management. Their target species consist largely of steelhead, chinook, and coho, all of which are represented by stocks that have been quite vulnerable to depletion in mixed-stock fisheries. Initially there was much opposition to the allocation of more fish to Aboriginal communities among recreational fishers, who simply feared that it would leave less fish for them. This may change with the discovery that a shift of fishing effort, from the mixed-stock marine fisheries to a selective inriver Aboriginal fishery, could result in much better conservation, particularly of the vulnerable stocks of coho, chinook, and steelhead, with which they are primarily concerned.

MANAGEMENT AUTHORITY

An effective fisheries management agency should possess a competent administration, well-developed scientific capability, powerful regulatory capacity, and a correspondingly adequate budget. Most essentially, the management agency needs to have the legal power to structure, administer, and enforce a system-wide management plan. Effective use of the salmon resource requires that the fishery for each river system be carefully regulated and coordinated by a management authority able to follow a consistent plan and enforce regulations for all participants, so they will not exceed catch allocations or otherwise subvert the plan. This authority also needs the power to apply inseason management (i.e., to impose fishery closures and other strictures at short notice in any part of the system, based on information on stock conditions). In Canada, only the federal government has all of the requisite powers and resources to provide a management agency meeting the above criteria.

The Gitxsan-Wet'suwet'en land claims policy includes an important fisheries component. It incorporates a stock-specific management strategy for the tribal area and recommends "that Tribal Council take the necessary legal and political steps to establish a Gitxsan and Wet'suwet'en fishery agency with full authority over fishery management within the territory and with a mandate to negotiate with agencies from other jurisdictions regarding management of Skeena stocks while they are outside of Gitxsan and Wet'suwet'en territory" (Morrell 1985, 1989). This approach reflects the tribal group's position that it is unacceptable to be considered just another supplicant "user group" pleading for a favored share of the fishery resource from an omnipotent federal government.

In their current mood of self-assertion, First Nations are demanding recognition of what they consider to be unextinguished and inextinguishable rights to their traditional fishery resources. The establishment and legal recognition of a tribal fishery agency, such as proposed by the Gitxsan-Wet'suwet'en Tribal Council, undoubtedly would be of great political advantage in defending their fishery claims. However, the full range of powers envisioned—at least in their literal form—have little chance of being accepted *in toto* by the Canadian government. If they were taken literally and applied with full force, they might well risk unacceptable levels of conflict with other stake-holder groups, while complicating efforts and weakening authority of DFO in meeting its responsibilities for management of the Skeena River salmon stocks. Protracted conflict with Americans over interception of B.C.-origin fish already seriously debilitates DFO's salmon management capacity. To concede competing autonomous fisheries jurisdiction to any stakeholder group within Canada could well lead to unresolvable conflicts over allocation and other matters, with a consequent further erosion of effective management. There is notably strong and wide support in B.C.—publicly expressed—for the retention of final and effective authority by DFO in managing the salmon resource, notwithstanding constant criticism aimed at this department of government.

Despite these reservations, the Canadian government now appears well disposed to community-based Aboriginal fishery agencies with special and unique responsibilities in tribal areas, in the context of a developing system of comanagement (Cassidy and Dale 1988). This would involve the many tribes in addition to the Gitxsan-Wet'suwet'en that have shown an interest in assuming fisheries management responsibilities of some kind within their territories (Richardson and Green 1989; MacLeod 1989). Given the strong propensity for controversy and confrontation in the fishing industry, there is, in any case, merit in the establishment of a "comanagement" process involving the government management authority and all user groups, in an effort to foster cooperation, understanding, and mutual consultation (Pinkerton 1989). Elements of this process are already present in Canadian fisheries through various advisory councils but, overall, the arrangements made so far represent, at best, a low-level form of comanagement. There remains much room for further delegation of management authority to be exercised by user groups in agreement with the constitutionally empowered authority, provided the delegated responsibilities are not subject to serious conflicts of interest, are carried out competently, and are adequately monitored (Copes 1997).

In conjunction with the AFS some comanagement arrangements with Aboriginal agencies have already been made by DFO on the Fraser, Skeena, and Nass rivers, and at Port Alberni on Vancouver Island. The initial arrangements focused particularly on administering the "pilot sales" fisheries agreements that DFO concluded with a limited number of First Nations, which essentially converted their open-ended food fisheries into quantitatively capped fisheries with permission to sell any part of the catch. Comanagement here included delegated authority to Aboriginal agencies in the policing of the fisheries, as well as control and recording of landings. Other comanagement features, that have been started or may be included, are collaboration in establishing optimally located selective and terminal fisheries, the operation and control of such fisheries, the design of enhancement works, and the carrying out of enhancement activities. The complex of such activities could provide considerable economic benefits for some of the First Nations. There would be larger fish catches, much of which would be available for commercial sale, possibly with value-added benefits from processing. In addition, work carried out on enhancement, of benefit to salmon fisheries overall, presumably would be paid for from federal funds and provide much needed employment.

CONCLUSION

Historical evidence indicates that, at the time of first European contact, many of the First Nations inhabiting the Fraser and Skeena watersheds had economies strongly dependent on local salmon resources. These economies were suppressed in the interest of a Euro-Canadian commercial marine fishery and fish processing industry. A restoration of salmon allocations to Aboriginal river communities, sufficient to allow reestablishment of a commercial fishery component, would provide some redress for past injuries and provide much needed employment and income.

The salmon resources of B.C. have been significantly reduced by excessive mixed-stock fishing and a variety of environmental impacts. Scientific estimates suggest that prior to European contact salmon stocks in B.C. were considerably larger than now and that sustainable harvests could be at least doubled. Important factors in achieving this would include selective fishing practices for stock-specific management and reduction of non-harvest mortality, restoration of degraded habitats, and additional enhancement undertakings.

The strategy proposed in this chapter is for the simultaneous achievement of justice in restoring substantial salmon harvest benefits to Aboriginal river communities and of significantly increasing the salmon catch in B.C. for the ultimate benefit of all stakeholders. This dual strategy draws on the felicitous circumstance that river tribes are in a position to play a key role in selectively harvesting salmon on the final leg of their journey to the spawning grounds, when fine-tuning of target escapements for individual stocks is possible. Aboriginal river communities are also well situated to carry out needed habitat restoration projects.

There is at present no evidence of a clear policy, at either the federal or provincial government level, to address the question of salmon management consistently and forcefully in the light of combined considerations of Aboriginal entitlement and optimal salmon resource management. However, the time seems right for establishing a joint federal-provincial policy and comanagement plan of action in collaboration with First Nations and other stakeholder groups. Indeed, a few positive steps have already been taken consistent with the strategy proposed in this chapter.

What is now needed is recognition by the federal and provincial governments of the importance of linking the fisheries question in Aboriginal land claims with that of improved salmon management, and a decision to pursue an optimal joint solution as a matter of public policy. A key component of land claim settlements with river tribes should be that the amount of additional fish made available would be linked to the use of approved selective fishing techniques as part of an optimal salmon management strategy. The resulting prospect of greater salmon harvests would provide both the means and the justification for a more generous catch allocation to Aboriginal fisheries.

Full implementation of the suggested policy will take time. It should be supported by a vigorous program of technical experimentation, coupled with successive refinements of a management plan model. Among gears to be developed and tested should be various trap or trap-and-weir systems, as well as fishwheels, and a full range of live-capture net systems. Gear design considerations should include timing and location of their use, harvest success, low mortality from stress of fish intended for live release, and cost-effectiveness in construction and operation.

Agencies currently responsible for overall management of B.C. salmon stocks and runs, namely DFO and the Pacific Salmon Commission, use a variety of models in setting and executing their present management plans. Moving a larger part of the fishery into the river should allow for superior escapement management. However, it will require an elaborate and complex plan of harvest allocations to various sites and gears. Biological concerns for various escapement management scenarios will have to be balanced in new models with economic considerations, such as fish quality, and with equity considerations in terms of allocations to stakeholder groups. Moreover, new models will support better management under the plan by allowing for adjustments from preseason estimates to "real time" allocations in accordance with run sizes and patterns actually experienced inseason.

The current claims by river tribes for reestablishment of their entitlement to a greater share of the salmon resource may be settled by the courts, by a political process, or both. Whatever the process, it is important that the solution arrived at be compatible with sound use of the resource and with a high level of added benefits. Historical justice suggests that a priority share of these benefits be assigned to First Nations. I contend there is a credible "win-win" management strategy leading to a sufficiently substantial increase in the harvest of salmon that may both do justice to Aboriginal claims and provide a share of the benefits to all other stakeholders.

ACKNOWLEDGMENTS

The original research for this chapter was supported by grants awarded by the Science Subvention Program of DFO and by Simon Fraser University from its Special Research Project Fund. The most recent work was supported by a grant from the Social Sciences and Humanities Research Council of Canada. I was also informed by the research I undertook for the Royal Commission on Aboriginal Peoples (Copes et al. 1994). I wish to express my deep appreciation to the many individuals from Aboriginal communities, from the commercial and recreational fishing sectors, and from DFO, who gave of their time and knowledge to discuss the fishery issues with which this chapter is concerned. I am grateful to T. Carrothers and P. Panek for their skillful research assistance and to G. Taylor for his guidance in exploring the Skeena River fishery and in establishing contact with many local interests in that fishery. Thanks also go to my colleagues R. L. Carlson, M. N. Stark, and C. S. Wright for helpful discussions and provision of relevant background information. I am indebted to F. Fortier, Chair of the Shuswap Nation Fisheries Commission, for his invitation to make a joint presentation of our perspectives on Aboriginal river fisheries at the Victoria Sustainable Fisheries Conference. In preparing this chapter I have profited from technical information provided by B. White and J. Woodey of the Pacific Salmon Commission and from incisive comments of two referees, R.W. Stone and K. Wilson.

REFERENCES

Allaire, L. 1978. L'archéologie des Kitselas d'après le site stratifié de Gitaus GdTc: sur la rivière Skeena en Colombie Britannique. Mercury Series No. 72. National Museum of Man, Archaeological Survey of Canada, Ottawa.

Ames, K. 1979. Report of Excavations at GhSv 2, Hagwilget Canyon. Mercury Series No. 87. National Museum of Man, Archaeological Survey of Canada, Ottawa.

Binnie, W. C. 1990. The Sparrow doctrine: beginning of the end or end of the beginning? Queen's Law Journal 15:217–253.

Burger, C. V. 2000. The needs of salmon and steelhead in balancing their conservation and use. Pages 15–29 in E. E. Knudsen, C. R. Steward, D. D. MacDonald, J. E. Williams, and D. W. Reiser, editors. Sustainable fisheries management: Pacific salmon. Lewis Publishers, Boca Raton, Florida.

Carlson, R. L. 1992. The Native fishery in British Columbia: the archaeological evidence. Discussion Paper 92–3. Simon Fraser University, Institute of Fisheries Analysis, Burnaby, British Columbia.

Cassidy, F., and N. Dale. 1988. After Native claims? The implications of comprehensive claims settlements for natural resources in British Columbia. Oolichan Books, Lantzville, British Columbia.

Copes, P. 1990. The attempted rationalization of Canada's Pacific salmon fisheries: analysis of failure. Pages 1–18 in Papers of the Fifth International Conference of the International Institute of Fisheries Economics and Trade. Fundación Chile, Santiago.

Copes, P. 1997. Salmon fishery management post-Mifflin: where do we go from here? Pages 69–75 in P. Gallaugher, editor. British Columbia salmon: a fishery in transition. Pacific Fisheries Think Tank, Report No. 1. Simon Fraser University, Institute of Fisheries Analysis, Burnaby, British Columbia.

Copes, P., T. Glavin, M. Reid, and C. Wright. 1994. West Coast Fishing Sectoral Study: Aboriginal peoples and the fishery on Fraser River salmon. Royal Commission on Aboriginal Peoples, Ottawa. (Released on CD-Rom).

Copes, P., and M. Reid. 1995. An expanded salmon fishery for the Gitksan-Wet'suwet'en in the Upper Skeena Region: equity considerations and management implications. Institute of Fisheries Analysis, Discussion Paper 95-3. Simon Fraser University, Burnaby, British Columbia.

Coupland, G. 1985. Prehistoric cultural change at Kitselas Canyon. Doctoral dissertation. University of British Columbia, Vancouver.

Fraser, J. A. (Chairman). 1995. Fraser River sockeye 1994: problems and discrepancies. Report of the Fraser River Sockeye Public Review Board. Public Works and Government Services Canada, Ottawa.

Garfield, V. E. 1966. The Tsimshian and their arts: the Tsimshian and their neighbours. University of Washington Press, Seattle.

Healey, M. C. 1993. The management of Pacific salmon fisheries in British Columbia. Pages 243–266 *in* L. S. Parsons and W. H. Lear, editors. Perspectives on Canadian marine fisheries management. Canadian Bulletin of Fisheries and Aquatic Sciences 226.

Helgeson, H. 1906. Report by fishery officer, Hans Helgeson. Sessional Paper 22, 38th Annual Report of the Department of Marine and Fisheries, 1905. S. E. Dawson, Ottawa.

Helin, C. D. 1994. The fishing rights and privileges of B.C.'s First Nations. Simon Fraser University, Institute of Fisheries Analysis, Burnaby, British Columbia.

Henderson, M. A. 1991. Sustainable development of the Pacific salmon resources in the Fraser River Basin. Pages 133–154 *in* A. H. J. Dorcey, editor. Perspectives on sustainable development in water management: towards agreement in the Fraser River Basin. University of British Columbia, Westwater Research Centre, Vancouver.

Hilborn, R., and C. J. Walters. 1992. Quantitative fisheries stock assessment: choice, dynamics and uncertainty. Chapman & Hall, New York.

Hindar, K., N. Ryman, and F. Utter. 1990. Genetic effects of cultured fish on natural fish populations. Canadian Journal of Fisheries and Aquatic Sciences 48:945–957.

Jakubowski, M. J. 1990. Review of the Babine River counting fence biological program for 1989. Canadian Data Report of Fisheries and Aquatic Sciences 788.

Knudsen, E. E. 2000. Managing Pacific salmon escapements: the gaps between theory and reality. Pages 237–272 *in* E. E. Knudsen, C. R. Steward, D. D. MacDonald, J. E. Williams, and D. W. Reiser, editors. Sustainable fisheries management: Pacific salmon. Lewis Publishers, Boca Raton, Florida.

Lane, B., and R. Lane. 1978. Recognition of B.C. Indian fishing rights by federal-provincial commissions. Union of British Columbia Indian Chiefs fishing portfolio: fishing research. Union of British Columbia Indian Chiefs, Vancouver.

Larkin, P. A. 1992. Analysis of possible causes of the shortfall in sockeye spawners in the Fraser River. A technical appendix to "Managing salmon in the Fraser" by Peter H. Pearse. Department of Fisheries and Oceans, Vancouver, British Columbia.

MacDonald, G. 1989. Kitwanga Fort Report. Canadian Museum of Civilization, Ottawa.

MacDonald, G. F., G. Coupland, and D. Archer. 1987. The Coast Tsimshian, ca. 1750. Pl. 13 *in* R. C. Harris, editor. Historical Atlas of Canada, Vol. I. University of Toronto Press, Toronto.

MacLeod, J. R. 1989. Strategies and possibilities for Indian leadership in co-management initiatives in British Columbia. Pages 262–272 *in* E. Pinkerton, editor. Co-operative management of local fisheries: new directions for improved management and community development. UBC Press, Vancouver, British Columbia.

McDonald, J. A. 1985. Trying to make a life: the historical political economy of Kitsumkalum. Doctoral dissertation. University of British Columbia, Vancouver.

Mitchell, D. S. 1925. A story of the Fraser River's great sockeye runs and their loss: being part of a local history written for my neighbours of the Shuswaps. Unpublished manuscript cited in Copes et al. 1994.

Morrell, M. 1985. The Gitksan-Wet'suwet'en fishery in the Skeena River system. Gitksan-Wet'suwet'en Tribal Council, Hazelton, British Columbia.

Morrell, M. 1989. The struggle to integrate traditional Indian systems and state management in the salmon fisheries of the Skeena River, British Columbia. Pages 231–248 *in* E. Pinkerton, editor. Co-operative management of local fisheries: new directions for improved management and community development. UBC Press, Vancouver, British Columbia.

Pacific Northwest Employment Training and Development. 1997. Report on 1997 labour market census. Terrace, British Columbia.

Park, L. K., and P. Moran. 1994. Development in molecular genetic techniques in fisheries. Reviews in Fish Biology and Fisheries 4:272–299.

Pearse, P. H. (Commissioner). 1982. Turning the tide: a new policy for Canada's Pacific fisheries. The Commission on Pacific Fisheries Policy, Vancouver, British Columbia.

Pearse, P. H. 1992. Managing salmon in the Fraser: report to the Minister of Fisheries and Oceans on the Fraser River salmon investigation. Department of Fisheries and Oceans, Vancouver, British Columbia.

Pearse, P. H., and J. E. Wilen. 1979. Impact of Canada's Pacific salmon fleet control program. Journal of the Fisheries Research Board of Canada 36:764–789.

Peterman, R.M. 1980. Dynamics of Native Indian food fisheries on salmon in British Columbia. Canadian Journal of Fisheries and Aquatic Sciences 37:561–566.

Pinkerton, E. W., editor. 1989. Co-operative management of local fisheries: new directions for improved management and community development. UBC Press, Vancouver, British Columbia.

Ray, A. J. 1984. The early economic history of the Gitksan-Wet'suwet'en-Babine tribal territories, 1822–1915. Exhibit 960 in Supreme Court of British Columbia 0843, Smithers Registry in: Delgamuukw vs. Queen.

Richardson, M., and B. Green. 1989. The fisheries co-management initiative in Haida Gwaii. Pages 249–261 in E. Pinkerton, editor. Co-operative management of local fisheries: new directions for improved management and community development. UBC Press, Vancouver, British Columbia.

Ricker, W. E. 1987. Effects of the fishery and of obstacles to migration on the abundance of Fraser River sockeye salmon, *Oncorhynchus nerka*. Canadian Technical Report of Fisheries and Aquatic Sciences 1522.

Ricker, W. E. 1989. History and present state of the odd-year pink salmon runs of the Fraser River region. Canadian Technical Report of Fisheries and Aquatic Sciences 1702.

Shepard, M. P., and A. W. Argue. 1989. The commercial harvest of salmon in British Columbia, 1820–1877. Canadian Technical Report of Fisheries and Aquatic Sciences 1690.

Slaney, T., and G. Birch. 1983. Commercial quality of sockeye salmon collected from the Skeena River. Aquatic Resources Limited. Vancouver, British Columbia.

Slaney, T. K., K. D. Hyatt, T. G. Northcote, and R. J. Fielden. 1996. Status of anadromous salmon and trout in British Columbia and Yukon. Fisheries 21(10):20–35.

Sprout, P. E., and R. K. Kadowaki. 1987. Managing the Skeena River sockeye salmon fishery: the process and the problems. Pages 385–395 in H. D. Smith, L. Margolis, and C. C. Wood, editors. Sockeye salmon (*Oncorhynchus nerka*) population biology and future management. Canadian Special Publication of Fisheries and Aquatic Sciences 96.

Stewart, H. 1977. Indian fishing: early methods on the Northwest Coast. University of Washington Press, Seattle.

United Nations. 1983. The law of the sea: United Nations Convention on the Law of the Sea. United Nations, New York.

Ware, R. M. 1978. Five issues, five battlegrounds: an introduction to the history of Indian fishing in British Columbia, 1850–1930. Coqualeetza Education Training Centre, Sardis, British Columbia.

West, C. J., and J. C. Mason. 1987. Evaluation of sockeye salmon (*Oncorhynchus nerka*) production from the Babine Lake development project. Pages 176–190 in H. D. Smith, L. Margolis, and C. C. Wood, editors. Sockeye salmon (*Oncorhynchus nerka*) population biology and future management. Canadian Special Publication of Fisheries and Aquatic Sciences 96.

Williams, J. T. 1906. Report by Inspector of Fisheries, John T. Williams, from Port Essington, B.C., to the Dominion Commissioner of Fisheries. Sessional Paper 22, 38th Annual Report of the Department of Marine and Fisheries, Ottawa.

Section II

Stock Status

8 The Status of Anadromous Salmonids: Lessons in Our Search for Sustainability

Jack E. Williams

Abstract.—Recent review articles have indicated a widespread decline in the status of anadromous salmonid stocks from the U.S. West Coast. Many factors contribute to this decline and complicate attempts to generalize about the causes and solutions of the problem. For society to manage anadromous fish resources in a sustainable manner, it is important to carefully review and interpret available information on stock status and improve monitoring efforts. Several lessons emerge from the data and reports to date. First, each stock is unique and should be managed accordingly. At the same time, stocks are a product of the interactions among many freshwater, estuarine, and oceanic conditions. They cannot be managed successfully in a context that is separate from the watersheds and oceans they inhabit. Many of our technological and engineering approaches related to hatcheries and dam passage ignore this basic ecological principle. Finally, resource managers and scientists must strive to improve long-term monitoring of stock status data. A joint U.S./Canada coastwide tracking and reporting system is proposed to monitor stock status data and evaluate efforts to restore sustainable fisheries.

INTRODUCTION

In 1995, a book by Naiman et al. entitled *The Freshwater Imperative: A Research Agenda* was published that has timely messages for anyone concerned about the status of aquatic resources in North America. As a way of introduction, the authors made the following observation:

> Over the past fifty to two hundred years [our] fresh waters have undergone the most significant transformation they have experienced in nearly ten thousand years. In effect, a large human-induced experiment has been set into motion without the means to monitor its results or alter its outcome.

This chapter reports on some of the results of this vast experiment and on some of the messages from it that are relevant to the concept of sustainable fisheries in the Pacific Northwest.

REVIEW OF EXISTING BROADSCALE REPORTS OF STOCK STATUS

In *Pacific Salmon at the Crossroads* (Nehlsen et al. 1991), the authors reviewed the status of salmon *Oncorhynchus* spp., steelhead *O. mykiss*, and sea-run cutthroat trout *O. clarki clarki* stocks in California, Oregon, Idaho, and Washington. These investigators documented 106 stocks as extinct and widespread declines in other stocks throughout the 4-state area, including 101 stocks at high risk of extinction, 58 at moderate risk of extinction, 54 as special concern, and 1 stock—the Sacramento River winter chinook *O. tshawytscha*—already listed as threatened pursuant to the Endangered Species Act (ESA), for a total of 214 stocks at risk. Declines were widespread and affected nearly all areas in the survey. An exception was along the northern Washington coast, where chinook and certain other salmon stocks seemed to be holding their own.

The decline of each of the 214 stocks has its own unique history and complexity, which confounds attempts to generalize about causes of these declines. The coho salmon *O. kisutch* in Oregon's Grande Ronde River serves as but one example. Historically, the Grande Ronde supported 2,000 to 4,000 spawning adult coho salmon annually (Howell et al. 1985). These salmon initially declined during the early 1900s because excessive logging and livestock grazing degraded freshwater habitat. During the 1960s and 1970s, construction of Snake River dams further reduced the population. Management agencies then concluded that the stock was too weak to protect in mixed-stock fisheries. By 1980, only about 50 adults returned. By the time the Nehlsen et al. report was published in 1991, the Grande Ronde coho salmon was one of 106 stocks listed as extinct.

Local adaptations and unique life history traits have been lost as a result of the 106 extinctions. For the other 214 stocks, there is still an opportunity for survival, but the opportunity is fading fast. Broadscale declines continue despite additional ESA listings of the Snake River sockeye salmon *O. nerka*, Snake River fall chinook salmon, Snake River spring/summer chinook salmon, Umpqua River cutthroat trout, and numerous steelhead stocks.

Utilizing data from *Pacific Salmon at the Crossroads* and data from the Columbia Basin Fish and Wildlife Authority's (1990) integrated system plan, a review of Columbia River Basin stocks was made in 1992. Of 192 anadromous salmonid stocks historically known from the Columbia River Basin, 67 (35%) are extinct, 36 (19%) are at high risk of extinction, 14 (7%) are at moderate risk of extinction, 26 (13%) are of special concern, and 49 (26%) are considered to be secure (Williams et al. 1992). Nearly three quarters of the stocks in the Columbia River Basin have been extirpated or are at some level of risk.

In 1993, the Forest Ecosystem Management Assessment Team (FEMAT 1993) reviewed the status of anadromous salmonids within the range of the northern spotted owl *Strix occidentalis caurina* in western Washington, western Oregon, and northwestern California. Using data from Pacific Salmon at the Crossroads (Nehlsen et al. 1991) plus more detailed surveys of Higgins et al. (1992) for northern California, Nickelson et al. (1992) for coastal Oregon, and Washington Department of Fisheries et al. (1993) for coastal Washington, they found 384 stocks at risk within the coastal zone, including 314 stocks at high or moderate risk of extinction.

More recently, Huntington et al. (1996) published a survey of healthy native anadromous salmonid stocks from the same four-state area as reported in the 1991 survey: Idaho, Washington, Oregon, and California. Ninety-nine stocks are listed as healthy. Of the 99, 98 are in Oregon or Washington, one is in California, and none is in Idaho. "Healthy" was defined as those stocks having adult abundance that ranged from one third as great as would be expected without human impacts to those that are nearly as robust as occurred historically.

During October 1995, a conference was held in Reedsport, Oregon on the coast-wide status of sea-run cutthroat trout. The conference was impressive in that it was a community event, with participation from local citizens, angler groups, and the town mayor, as well as state, federal, and university scientists. Sea-run cutthroat trout are good indicators of the health and productivity of coastal ecosystems (Williams and Nehlsen 1997). Spawning migrations of this fish into freshwaters seldom reach far from the ocean, with many populations completing their life history in coastal streams less than 16 km long (Sumner 1962; Gerstung 1981). Behnke (1992) reported that sea-run cutthroat trout seldom occur in open oceans, preferring instead to rear in estuaries and bays.

According to data presented at the Reedsport conference the status of sea-run cutthroat is poor in the southern part of their range but improves northward. Populations from California, Oregon, and the Columbia River Basin clearly had declined from historic levels (Williams and Nehlsen 1997). Many of these declines were drastic. On Oregon's North Umpqua River, the numbers of adult fish passing over Winchester Dam averaged 950 from 1946 to 1956. However, counts have averaged just 23 adults between 1992 and 1995 (Hooton 1997), prompting a 1994 proposal from the National Marine Fisheries Service (NMFS) to list coastal cutthroat trout in the Umpqua Basin as threatened pursuant to the ESA. In Puget Sound of Washington, stocks of sea-run cutthroat trout mostly are healthy compared to those in more southern parts of the state (Leider 1997). In British Columbia, about 50% of the stocks were considered to be healthy, but little or no information was available for populations north of Vancouver Island (Slaney et al. 1997). In southeastern Alaska, the few populations for which data existed showed increasing population trends, including counts in 1994 and 1995 that were the highest on record (Schmidt 1997).

Relatively few reviews of stock status have been published for salmon, steelhead, and sea-run cutthroat trout in Canada and Alaska. In general, more positive reports are emerging from these regions because of reduced human impacts, less interference from hatchery production, and better ocean productivity in recent years. In southeastern Alaska, Baker et al. (1996) estimated that 9,296 spawning aggregates exist, but that sufficient data to determine status and population trends were available for only 928 (10%). They avoided using the term "stock" because they were uncertain whether the locations for spawning aggregates actually reflected stocks or other population units. Most of the 928 spawning aggregates were increasing or stable in numbers. Only about 1% were ranked at moderate or high risk of extinction. For anadromous salmonids in British Columbia and Yukon, Slaney et al. (1996) estimated that sufficient data were available to determine status for 5,487 (57%) of 9,662 identifiable stocks. Of the 5,487 stocks, 624 (11%) were ranked at a high risk of extinction, 78 (1%) at moderate risk, 230 (4%) at special concern, and 142 (3%) are extinct.

LESSONS LEARNED FROM BROADSCALE REVIEWS

Two facts become clear from these reviews. First, fisheries management should focus at the level of the stock, which is defined as a spawning population that has little interbreeding with other spawning populations and is uniquely adapted to its watershed (Ricker 1972). Many of these stocks are genetically distinct entities with unusual life history traits and requirements. As stocks are lumped together for management purposes, we run an increasing risk of masking these unique biological differences and therefore increasing the likelihood that management actions will not sustain stock-level distinctions. Ocean fisheries that target stock aggregates, for example, may overharvest those stocks with small or declining populations. The relative lack of data available for many Alaskan and Canadian stocks also argues for conservative management philosophies. In the face of limited understanding, management should focus on the smallest identifiable population units. Too often species and their ecosystems are manipulated by resource managers without a clear understanding of the complex relationships inherent among populations or the relationships between populations and their watersheds. The following quote from Stanford and Ward (1992) echoes a familiar complaint.

> Managers often want simplistic methodology that will explicitly satisfy an increasingly circumspect public. Unfortunately, in the absence of practical and conceptual understanding of ecosystem structure and function, management actions often produce results significantly different from what was predicted.

The second fact that is clear from these reviews is that stocks of salmon, steelhead, and sea-run cutthroat trout occur within the context of their ecosystems and cannot successfully be managed in isolation from their watersheds, estuaries, and oceans. Yet, perhaps because fish biologists prefer to work with the fish themselves, the underlying factors influencing fish status often are ignored. In the past, fish biologists too often have treated the symptoms of the problem without addressing the underlying cause of the problem itself (Frissell 1997). For example, logs have been anchored along stream banks to prevent erosion, while the primary cause of increased erosion—too many roads in the watershed or livestock overgrazing—was left untreated. In another example, hatchery supplementation has been increased to offset overfishing or habitat loss. Perhaps it is just "easier" to work with the fish rather than tackle the tougher problems. As Meffe (1992) described it, "a management strategy that has as a centerpiece artificial propagation and restocking of a species that has declined as a result of environmental degradation and overexploitation, without correcting the causes for decline, is not facing biological reality." Fish biologists and resource managers need to be certain to focus on understanding and correcting causes of our problems rather than affixing bandages and developing new technological silver bullets that only address symptoms.

Riverine ecosystems are dynamic and multidimensional. Stocks of anadromous salmonids that evolve in these ecosystems are no less dynamic and complex. Fish biologists are beginning to understand this complexity and the need to address the broader ecosystem context within which anadromous fishes occur. Ultimately, success will depend on a broadscale watershed approach that brings together efforts of a variety of biological, physical, and social scientists (Roper et al. 1997). Many of the problems faced by declining stocks are rooted in society's fundamental relationship with the planet's natural resources. Too many people believe that resources must be "used" to have served any useful purpose, and far too few people comprehend the basic ecological principles that guide all life on this planet, including the persistence of our own species (Cairns 1997). As Orr (1992; 1994) has repeatedly noted, we must increase the level of ecological literacy within society. For sustainability of fisheries to be achieved, the general public must comprehend fundamental ecological principles, including the ramifications of our daily decisions and lifestyles on the ability of salmon and steelhead to be sustained through time. Quoting from Orr (1994):

> It is widely assumed that environmental problems will be solved by technology of one sort or another. Better technology can certainly help, but the crisis is not first and foremost one of technology. Rather, it is a crisis within the minds that develop and use technology.

MYRIAD FACTORS AFFECTING STOCKS

Broadscale reviews of stock status, such as the report by Nehlsen et al. (1991), clearly demonstrate that many factors often are responsible for declines or extinctions of individual stocks. Based on the limited data available so far, the status of stocks in Canada and Alaska clearly seem to be more robust than are stocks further south. Why is this? Are freshwater habitats in better condition further north? Is fishing pressure less? Has artificial propagation and introductions confounded stock structure more in the south? Are ocean conditions more favorable further north? And, if the answer to some or all of these questions is yes, how long will these more favorable conditions in the north persist? What will happen if ocean productivity in northern latitudes declines significantly in the coming years? Will resource managers in Alaska, British Columbia, and the Yukon be able to reduce harvests and lessen impacts on freshwater spawning grounds? These are important questions that scientists, politicians, and the general public must be able to answer intelligently.

Multiple factors affect stock abundance. Many of these clearly are caused by human disturbance and fishing, while others are caused by natural cycles of drought, floods, and fire. There also appears to be a synergy among natural and human disturbances, such as occurred during the 1964 flood in northern California, and more recently, in spring 1996 flooding in western Oregon and Washington and northern Idaho. For example, long-term increases in peak stream flows occur in those watersheds affected by forest clear-cutting and road building (Jones and Grant 1996). These changes can lead to increased flood damage, such as occurred during a 100-year flood event in February 1996 in the Fish Creek watershed of Oregon's Mt. Hood National Forest. The Forest Service inventoried 236 landslides that resulted from the storm in the Fish Creek watershed and attributed 42% to timber harvest practices, 34% to roads, and 24% to nonmanagement-related causes (Reeves et al. 1997). Increased road construction, timber harvest, livestock grazing, or urbanization within a watershed can spell the difference between a moderate flood that improves stream habitat conditions and a catastrophic flood that adds tons of silt and debris to salmon spawning areas.

Fortunately, salmon are resilient. If given half a chance, they can recover from disturbances. In fact, pulsed or short-term natural disturbances may even rekindle natural recovery and productivity processes. How can such information be tracked to make accurate predictions on cause-and-effect relationships? How can broadscale trends be understood and interpreted? These questions are addressed in the next section on a coastwide tracking and reporting system for stock status.

NEED FOR COASTWIDE TRACKING AND REPORTING SYSTEM

There literally may be tons of paper containing data on the status of West Coast anadromous fish stocks. Unfortunately, the data are spread as far and wide as the fish themselves. As described in the reports mentioned earlier, there have been few attempts at regional synthesis of these data or their implications. Part of the reason so few broadscale syntheses have been completed is the difficulty in compiling and organizing all the data. A transboundary database to monitor and report status throughout the West Coast—a joint U.S./Canada program—is needed to interpret large-scale trends and report their findings to our respective governments and the public.

The recent proliferation of Pacific Northwest fisheries information on the Internet and World Wide Web provide timely models of how such a coastwide tracking and reporting system might be organized and operated. StreamNet is a Web site providing baseline data on fish distribution, production, habitat, and management of anadromous fishes in the Columbia River Basin. Data are geographically referenced on a 1:100,000-scale River Reach System developed by the U.S. Geological Survey. StreamNet is a cooperative effort of the Northwest Power Planning Council, Bonneville Power Administration, Pacific States Marine Fisheries Commission, Columbia River Inter-Tribal Fish Commission, federal agencies, state fish and wildlife agencies, and the Shoshone-Bannock Tribes. Another database, the Klamath Resource Information System, has been established to monitor fisheries and water quality information for the Klamath Basin of northwestern California and southern Oregon. Trends in spawner escapement are provided for various species and subbasins within the Klamath Basin. The system is funded by the U.S. Fish and Wildlife Service, U.S. Environmental Protection Agency, and California Water Resources Control Board and was developed by William M. Kier Associates.

At a minimum, a coastwide tracking and reporting system should include data on spawning escapements and total run size for each subbasin from southern California through Alaska. Where available, data should be referenced by 1:100,000-scale watershed maps. Stock status information from StreamNet and the Klamath Resource Information System should be included or hyperlinked to such a coastwide system. Databases developed for recent reviews of southeastern Alaska (Baker et al. 1996) and British Columbia and Yukon (Slaney et al. 1996) stocks can be incorporated to facilitate inclusion of stock information from northern latitudes where less information is available. Recent compilation of data for sea-run cutthroat trout (Hall et al. 1997) should facilitate inclusion of data for this coastal fish.

A coastwide tracking system would be invaluable for (1) providing an early warning system to detect stock declines; (2) interpreting the broadscale effects of climate change and changes in ocean currents and productivity cycles; (3) interpreting the multiple factors affecting stock status; (4) determining the success of existing management strategies; (5) providing a means to regularly report and update stock status; and (6) providing easier access to existing data on stock status to the scientific community and general public. Factors affecting stock status are multiple and complex. Without a coastwide monitoring system, our attempts to sustain the anadromous fish resource will continue to be hampered by our inability to adequately detect and track changes in the many hundreds of stocks spread along a five-state and two-province region.

NEED FOR INCREASED COMMUNICATION WITH PUBLIC

A coastwide tracking system implies the need for improved communication. This improved networking needs to extend to the public as well as among scientists and government agencies. The status of anadromous salmon stocks is a complex issue, which can be very confusing to scientists and the general public alike. As should be clear to all, many factors affect the status of individual stocks: changing ocean conditions, degradation of freshwater habitats, offshore and inriver fisheries, hatchery programs, hydroelectric dam operations, and other factors. The public and politicians often are baffled by a complex finger-pointing game of who is to blame for the demise of salmon in a particular river system. The public needs to understand—clearly *wants* to understand—the status of these resources, the causes behind stock collapse, and how to better manage these resources sustainably in the future.

Along the West Coast, community-based coalitions of landowners, private interests, and public agencies are being formed to facilitate common understanding in the search to restore sustainable fisheries. More than 300 watershed coalitions were active in California, Oregon, and Washington by 1996 (For the Sake of the Salmon 1996). These coalitions present social and scientific opportunities for fish biologists and resource advocates to foster sustainability of salmonid resources (Dombeck et al. 1997). They also provide excellent forums for communication among diverse interests and help to institutionalize community involvement and commitment to sustainable fisheries and watershed management.

Public interest and concern for salmon, steelhead, and sea-run cutthroat trout should not be underestimated. When the American Fisheries Society published the Pacific Salmon at the Crossroads report (Nehlsen et al. 1991), the Society issued a press release about the results and widely communicated the findings of large-scale stock declines and extirpations in California, Oregon, Washington, and Idaho. The response—from the conservation community, general public, and information media—was much greater than anticipated (Cone and Ridlington 1996). Findings from the *Crossroads* paper were reported in dozens of newspapers, major articles were published in national conservation magazines such as *Audubon* and *Sierra*, the article was reprinted and expanded in *Trout* magazine, and the authors were besieged with interview requests from CNN and Northwest radio stations: all for reporting something that the authors and many other fish biologists took for granted, that salmon, steelhead, and sea-run cutthroat trout were subject to broadscale declines in Idaho, Oregon, Washington, and California.

Salmon biologists and natural resource managers should regularly inform the public about salmon declines and causative factors. The following elements are especially important to communicate to the public.

1. The ecological, social, and economic values of salmon;
2. The multiplicity of factors contributing to the declines and how each can be addressed;
3. The need to reverse the overall decline in the health and productivity of our rivers and watersheds;
4. The importance of ecological restoration that treats the primary causes of declines, rather than symptoms;
5. The fallacy of depending on hatcheries, dam bypass structures, and other technological methodologies that attempt to mitigate for declines in habitat quality;
6. The importance of long-term monitoring and understanding of habitat condition and stock status; and
7. The realization that salmon are resilient and that despite the broadscale declines there is still time to act to save these valuable resources.

CLOSING NOTE

The Salish Indians of Puget Sound believed that the lower-class members within a tribe—the second-class citizens—belonged to families who had lost their history (Lichatowich et al. 1993). Second-class members belonged to families who had failed to pass on the oral history of their family, failed to pass on the codes of proper behavior and the skills needed to maintain a high standard of life. Society must understand where we have been to know who we are and what our potential is.

Our history is slipping away as well. The genetic information and life histories contained in the individual stocks of salmon, steelhead, and sea-run cutthroat trout are being lost through extinctions. This is history that took thousands of years to establish. Diverse and abundant salmon stocks have been a fundamental part of the social, economic, and ecological fabric of life along the West Coast. Many stocks remain, but their fate is uncertain. This is history at its most fundamental level: the basis of life, sustainability, and productivity of rivers, watersheds, and oceans. If we let this history slip away, we will all become second-class citizens.

ACKNOWLEDGMENTS

Robin Lebasseur, David Levy, Margaret C. Taylor, and Christopher A. Wood provided thoughtful reviews on earlier drafts of this manuscript. Their comments are greatly appreciated.

REFERENCES

Baker, T. T., and eight coauthors. 1996. Status of Pacific salmon and steelhead escapements in southeastern Alaska. Fisheries 21(10):6–18.

Behnke, R. J. 1992. Native trout of western North America. American Fisheries Society Monograph 6, Bethesda, Maryland.

Cairns, J., Jr. 1997. Eco-societal restoration: creating a harmonious future between human society and natural systems. Pages 487–499 in J. E. Williams, C. A. Wood, and M. P. Dombeck, editors. Watershed restoration: principles and practices. American Fisheries Society, Bethesda, Maryland.

Columbia Basin Fish and Wildlife Authority. 1990. Integrated system plan for salmon and steelhead production in the Columbia River Basin. Public Review Draft. Columbia Basin Fish and Wildlife Authority, Portland, Oregon.

Cone, J., and S. Ridlington, editors. 1996. The Northwest salmon crisis: a documentary history. Oregon State University Press, Corvallis.

Dombeck, M. P., J. E. Williams, and C. A. Wood. 1997. Watershed restoration: social and scientific challenges for fish biologists. Fisheries 22(5):26–27.

For the Sake of the Salmon. 1996. Directory of watershed groups in the Pacific region. For the Sake of the Salmon, Gladstone, Oregon.

FEMAT (Forest Ecosystem Management Assessment Team). 1993. Forest ecosystem management: an ecological, economic, and social assessment. U.S. Forest Service, National Marine Fisheries Service, U.S. Bureau of Land Management, U.S. Fish and Wildlife Service, National Park Service, and U.S. Environmental Protection Agency. Portland, Oregon.

Frissell, C. A. 1997. Ecological principles. Pages 96–115 in J. E. Williams, C. A. Wood, and M. P. Dombeck, editors. Watershed restoration: principles and practices. American Fisheries Society, Bethesda, Maryland.

Gerstung, E. R. 1981. Status and management of coastal cutthroat trout (*Salmo clarki clarki*) in California. Cal-Neva Wildlife Transactions 1981:25–32.

Hall, J. D., P. A. Bisson, and R. E. Gresswell, editors. 1997. Sea-run cutthroat trout: biology, management, and future conservation. Oregon Chapter, American Fisheries Society, Corvallis, Oregon.

Higgins, P., S. Dobush, and D. Fuller. 1992. Factors in northern California threatening stocks with extinction. Humboldt Chapter, American Fisheries Society, Arcata, California.

Hooton, B. 1997. Status of coastal cutthroat trout in Oregon. Pages 57–67 in J. D. Hall, P. A. Bisson, and R. E. Gresswell, editors. Sea-run cutthroat trout: biology, management, and future conservation. Oregon Chapter, American Fisheries Society, Corvallis, Oregon.

Howell, P., K. Jones, D. Scarnecchia, L. LaVoy, W. Kendra, and D. Ortmann. 1985. Stock assessment of Columbia River anadromous salmonids. Report to Bonneville Power Administration, Portland, Oregon.

Huntington, C., W. Nehlsen, and J. Bowers. 1996. A survey of healthy native stocks of anadromous salmonids in the Pacific Northwest and California. Fisheries 21(3):6–14.

Jones, J. A., and G. E. Grant. 1996. Peak flow responses to clear-cutting and roads in small and large basins, western Cascades, Oregon. Water Resources Research 32:959–974.

Leider, S. A. 1997. Status of sea-run cutthroat trout in Washington. Pages 68–76 in J. D. Hall, P. A. Bisson, and R. E. Gresswell, editors. Sea-run cutthroat trout: biology, management, and future conservation. Oregon Chapter, American Fisheries Society, Corvallis, Oregon.

Lichatowich, J., W. Nehlsen, and J. Williams. 1993. Pacific salmon: resource at risk. Current 12(2):26–28.

Meffe, G. K. 1992. Techno-arrogance and halfway technologies: salmon hatcheries on the Pacific Coast of North America. Conservation Biology 6:350–354.

Naiman, R. J., J. J. Magnuson, D. M. McKnight, and J. A. Stanford. 1995. The freshwater imperative: a research agenda. Island Press, Washington, D.C.

Nehlsen, W., J. E. Williams, and J. A. Lichatowich. 1991. Pacific salmon at the crossroads: stocks at risk from California, Oregon, Idaho, and Washington. Fisheries 16(2):4–21.

Nickelson, T. E., J. W. Nicholas, A. M. McGie, R. B. Lindsay, D. L. Bottom, R. J. Kaiser, and S. E. Jacobs. 1992. Status of anadromous salmonids in Oregon coastal basins. Oregon Department of Fish and Wildlife, Portland.

Orr, D. W. 1992. Ecological literacy: education and the transition to a postmodern world. State University of New York Press, Albany.

Orr, D. W. 1994. Earth in mind: on education, environment, and the human prospect. Island Press, Washington, D.C.

Reeves, G. H., and six coauthors. 1997. Fish habitat restoration in the Pacific Northwest: Fish Creek of Oregon. Pages 335–359 *in* J. E. Williams, C. A. Wood, and M. P. Dombeck, editors. Watershed restoration: principles and practices. American Fisheries Society, Bethesda, Maryland.

Ricker, W. E. 1972. Hereditary and environmental factors affecting certain salmon populations. Pages 19–160 *in* R. C. Simon and P.A. Larkin, editors. The stock concept in Pacific salmon. UBC Press, Vancouver, British Columbia.

Roper, B. R., J. J. Dose, and J. E. Williams. 1997. Stream restoration: is fisheries biology enough? Fisheries 22(5):6–11.

Schmidt, A. E. 1997. Status of sea-run cutthroat trout stocks in Alaska. Pages 80–83 *in* J. D. Hall, P. A. Bisson, and R. E. Gresswell, editors. Sea-run cutthroat trout: biology, management, and future conservation. Oregon Chapter, American Fisheries Society, Corvallis, Oregon.

Slaney, T. L., K. D. Hyatt, T. G. Northcote, and R. J. Fielden. 1996. Status of anadromous salmon and trout in British Columbia and Yukon. Fisheries 21(10):20–35.

Slaney, T. L., K. D. Hyatt, T. G. Northcote, and R. J. Fielden. 1997. Status of anadromous cutthroat trout in British Columbia. Pages 77–79 *in* J. D. Hall, P. A. Bisson, and R. E. Gresswell, editors. Sea-run cutthroat trout: biology, management, and future conservation. Oregon Chapter, American Fisheries Society, Corvallis, Oregon.

Stanford, J. A., and J. V. Ward. 1992. Management of aquatic resources in large catchments: recognizing interactions between ecosystem connectivity and environmental disturbance. Pages 91–124 *in* R. J. Naiman, editor. Watershed management: balancing sustainability and environmental change. Springer-Verlag, New York.

Sumner, F. B. 1962. Migration and growth of the coastal cutthroat trout in Tillamook County, Oregon. Transactions of the American Fisheries Society 91:77–83.

Washington Department of Fisheries, Washington Department of Wildlife, and Western Washington Treaty Indian Tribes. 1993. 1992 Washington state salmon and steelhead stock inventory. Washington Department of Fisheries, Olympia.

Williams, J. E., J. A. Lichatowich, and W. Nehlsen. 1992. Declining salmon and steelhead populations: new endangered species concerns for the West. Endangered Species UPDATE 9(4):1–8.

Williams, J. E., and W. Nehlsen. 1997. Status and trends of anadromous salmonids in the coastal zone with special reference to sea-run cutthroat trout. Pages 37–42 *in* J. D. Hall, P. A. Bisson, and R. E. Gresswell, editors. Sea-run cutthroat trout: biology, management, and future conservation. Oregon Chapter, American Fisheries Society, Corvallis, Oregon.

9 Endangered Species Act Review of the Status of Pink Salmon from Washington, Oregon, and California

Jeffrey J. Hard, Robert G. Kope, and W. Stewart Grant

Abstract.—The National Marine Fisheries Service received a petition in March 1994 seeking protection under the U.S. Endangered Species Act for two populations of pink salmon *Oncorhynchus gorbuscha* in the state of Washington. In response to the petition, NMFS established a Biological Review Team (BRT) of scientists to review the status of pink salmon in Washington, Oregon, and California. Two evolutionarily significant units (ESUs) were identified during the review that correspond to the even- and odd-year broodlines of pink salmon spawning in Washington and southern British Columbia; no permanent pink salmon populations occur in Oregon or California, and no pink salmon are known to occur in Idaho. The BRT unanimously concluded that neither ESU as a whole is presently at risk of extinction. However, the BRT expressed concern about the status of individual populations within both ESUs and recommended that these populations be closely monitored.

INTRODUCTION

The Endangered Species Act (16 U.S.C. §§1531–1543) allows listing of "distinct population segments" of vertebrates as well as named species and subspecies. The policy of the National Marine Fisheries Service (NMFS) on this issue for anadromous Pacific salmonids is that a population will be considered "distinct" for purposes of the ESA if it represents an evolutionarily significant unit (ESU) of the species as a whole (Federal Register 56 [20 November 1991]:58612). To be considered an ESU, a population or group of populations must (1) be substantially reproductively isolated from other populations, and (2) contribute substantially to ecological/genetic diversity of the biological species (Waples 1991, 1995). Once an ESU is identified, a variety of factors related to population abundance are considered in determining whether a listing is warranted.

NMFS initiated a status review of pink salmon in Washington, Oregon, and California in March 1994, in response to a petition seeking protection under the Endangered Species Act for two populations of pink salmon *Oncorhynchus gorbuscha* in the state of Washington (PRO-Salmon 1994). A NMFS status review is a comprehensive analysis undertaken to identify conservation units for Pacific salmon under the ESA and determine their risk of extinction. The review incorporates information from a wide variety of sources developed during a public process to address two key questions: (1) is the entity in question a "species" as defined by the ESA? (2) If so, is the "species" threatened or endangered? Several factors are considered in evaluating the first question, including natural rates of straying and recolonization, natural barriers to migration, and measurements of genetic differences among populations. Factors considered in evaluating the second question include absolute abundance and distribution, trends in abundance, natural and human-influenced factors that cause variability in survival and abundance, and possible threats to genetic integrity. Upon completion of the review, its major conclusions and recommendations for listing determinations are published by NMFS in the Federal Register.

NMFS formed a Biological Review Team (BRT) of 12 scientists to conduct the review. Scientists were selected from NMFS' Northwest Fisheries Science Center, the Northwest Region, and the Alaska Fisheries Science Center on the basis of their expertise in salmon biology and ecology, population biology, and genetics. This chapter briefly summarizes the major conclusions reached by the BRT and updates abundance information for Washington streams since the review. The reader is referred to the status review (Hard et al. 1996) for a full evaluation of the biological and environmental information used to support these conclusions.

PROPOSED PINK SALMON ESUs

Pink salmon spawn around the Pacific Rim from 44°N to 65°N in Asia and from 48°N to 64°N in North America. Washington appears to be the southern limit of the spawning distribution of pink salmon in North America (Figure 9.1); no persistent populations have been documented in Oregon or California, and pink salmon are not known to occur in Idaho. The BRT examined genetic, life-history, biogeographic, physiographic, and environmental information to identify ESU boundaries for pink salmon. Patterns of genetic differentiation and life history variation (primarily adult run timing and body-size variation) were found to be the most informative in identifying pink salmon ESUs (Hard et al. 1996). Snohomish River even-year pink salmon are genetically distinct from all odd-year pink salmon in Washington and southern British Columbia; adult even-year pink salmon from the Snohomish River also tend to be smaller and spawn earlier than most odd-year pink salmon in northwestern Washington. Based largely on these differences, the BRT identified two pink salmon ESUs in Washington and southern British Columbia (Figure 9.2), which reflect the distinction between even- and odd-year broodlines that are characteristic of pink salmon throughout their natural range (Heard 1991).

EVEN-YEAR PINK SALMON

The only persistent population of even-year pink salmon in Washington occurs in the Snohomish River (Figure 9.1). Although several attempts were made in the 20th century to transplant even-year pink salmon from Alaska and British Columbia to the Puget Sound region, there is no indication that these attempts were successful. Furthermore, life-history and genetic information for Snohomish River even-year pink salmon is consistent with the hypothesis that this population resulted from a natural colonization event. The nearest even-year pink salmon populations occur in British Columbia, at least 130–150 km away. However, these populations are not well characterized, and their relationships with the Snohomish River even-year pink salmon population are unknown. Because of uncertainty about the origin of the Snohomish River population and its relationship to other even-year populations, the BRT could not resolve the extent of the ESU that contains the Snohomish River even-year pink salmon population. After considering all available information, about half of the BRT members concluded that the Snohomish River even-year population is in an ESU by itself, whereas half judged that the ESU also included populations from British Columbia, as shown in Figure 9.2. In any case, the BRT unanimously agreed that any conclusion about the extent of the even-year pink salmon ESU should be regarded as provisional and subject to revision should substantial new information become available.

ODD-YEAR PINK SALMON

The BRT considered several possible ESU scenarios for odd-year pink salmon. The majority of BRT members concluded that all odd-year pink salmon populations in Washington are part of a

FIGURE 9.1 Principal pink salmon spawning streams in northwestern Washington. Stream names are in boldface (figure modified from IPSFC 1958).

single ESU. This ESU includes populations in Washington as far west as the Dungeness River (or the Elwha River, if that population is not already extinct) and in southern British Columbia (including the Fraser River and eastern Vancouver Island) as far north as Johnstone Strait (Figure 9.2). A minority of BRT members concluded that populations from Washington rivers draining into the Strait of Juan de Fuca are members of a separate ESU. All members agreed that, collectively, odd-year pink salmon in Washington contain a considerable amount of genetic and life-history diversity, with populations from the Dungeness, Nooksack, and Nisqually Rivers being the most distinctive in this regard. Several small odd-year populations occur on southwestern Vancouver Island, but insufficient information is available to ascertain the relationship of these

Even-year

FIGURE 9.2 Proposed evolutionarily significant units (ESUs) for even- and odd-year west coast pink salmon. In the panel on this page, a heavy line identifies the Snohomish River, which supports the only persistent even-year run in the U.S. south of Canada and may be the sole representative of the even-year ESU. However, the even-year ESU may also include runs in southern British Columbia, perhaps as far north as Johnstone Strait (dashed boundary). In the panel on the facing page, three rivers that support distinctive odd-year pink salmon populations (Dungeness, Nisqually, and Nooksack) in the odd-year ESU are also identified with heavy lines. The Elwha River population (identified with a dotted line) is believed to be extinct. See Figure 9.1 for identification of the other principal pink salmon rivers in northwestern Washington that compose the odd-year ESU (adapted from Hard et al. 1996).

populations to the proposed odd-year ESU. Additional information on these populations is needed to resolve the question of whether odd-year pink salmon in Washington and southern British Columbia are in one or more ESUs.

ASSESSMENT OF EXTINCTION RISK

The Endangered Species Act (Section 3) defines the term "endangered species" as "any species which is in danger of extinction throughout all or a significant portion of its range." The term "threatened species" is defined as "any species which is likely to become an endangered species within the foreseeable future throughout all or a significant portion of its range." According to the Act, the determination whether a species is threatened or endangered should be made on the basis of the best scientific information available regarding its current status, after taking into consideration conservation measures that are proposed or are in place. This determination involves several steps.

Odd-year

FIGURE 9.2 (continued)

In its review, the BRT was charged only with determining the degree to which each ESU is at risk of extinction, but not the likely or possible effects of conservation measures. Rather, the BRT drew scientific conclusions about the risk of extinction faced by identified ESUs under the assumption that present conditions will continue (see Hard et al. 1996). Consideration of factors contributing to population declines and development of conservation measures to address these factors are taken into account after the scientific conclusions of the BRT are used by NMFS to support a proposed listing recommendation. This report considers only the scientific evidence used to identify ESUs and evaluate their risk of extinction absent conservation measures. The resulting conclusions for each ESU follow.

EVEN-YEAR PINK SALMON

The BRT was unanimous in concluding that this ESU is not presently at risk of extinction. This conclusion was based on escapement estimates from populations in Washington and southern British Columbia (see Hard et al. 1996). Nevertheless, nearly all BRT members expressed concern about the status of this ESU. Although available escapement data indicate that the even-year pink salmon population in the Snohomish River has been increasing since 1980 (Table 9.1), the BRT identified three factors that pose risk to the sustainability of this population: (1) its marked isolation from other even-year populations; (2) its low abundance; and (3) the lack of variable age structure in pink salmon, coupled with considerable inter-generational variability in abundance. Collectively, these factors indicate that the population is at some risk due to demographic or environmental

TABLE 9.1

Recent escapements of odd- and even-year pink salmon in U.S. streams south of Canada (all populations occur in Washington State). Escapements were calculated from spawner abundance surveys of Washington Department of Fish and Wildlife stream reaches.

	Odd-year runs									
Population	1979	1981	1983	1985	1987	1989	1991	1993	1995	1997
Puget Sound										
Nooksack	19,472	26,814	66,966	24,914	32,862	126,006	21,298	51,680	207,092	26,000
Skagit	300,384	100,268	470,128	710,300	593,535	401,300	351,000	530,000	857,000	60,000
Stillaguamish	135,000	18,000	126,100	200,000	119,000	70,400	86,000	110,000	208,000	88,572
Snohomish	180,724	90,096	198,283	302,192	152,418	80,149	174,000	100,000	101,600	103,537
Puyallup	29,481	12,115	11,600	34,100	34,400	49,850	14,000	10,069	17,586	2,732
Nisqually	1,300	500	500	500	7,700	12,300	1,900	500	579	
Hood Canal										
Dosewallips	13,900	1,700	4,300	30,800	20,300	11,900	8,200	17,316	19,034	1,954
Duckabush	11,900	2,300	8,700	21,000	28,500	34,600	72,000	4,578	6,659	3,376
Hamma Hamma	11,300	2,450	12,000	12,000	13,300	14,300	38,000	13,432	5,593	2,969
Olympic Peninsula										
Dungeness	50,100	2,900	4,888	4,730	1,906	10,902	9,895	1,695	8,252	4,935
Elwha	7,100	200	200	100	50	1	1	1	2	18

	Even-year runs									
Population	1980	1982	1984	1986	1988	1990	1992	1994	1996	1998
Snohomish	151		137	1,016	1,097	2,187	2,723	1,640	3,734	

fluctuations. All BRT members agreed that this population should be closely monitored, even if it is determined to be part of a larger ESU. Uncertainty about the status of other populations that may exist in this ESU makes it difficult to characterize the degree of risk posed by these factors to the ESU as a whole. Based on this information, NMFS proposed that this ESU does not currently warrant a listing as a threatened or endangered species under the Endangered Species Act (Federal Register 60 [4 October 1995]:51928). The estimate of escapement to the Snohomish River in 1996 indicates that this population increased substantially since 1994 (Table 9.1).

ODD-YEAR PINK SALMON

The BRT was unanimous in concluding that this ESU as a whole is not presently at risk of extinction. Most populations have relatively large escapements (Table 9.1), and overall abundance appears to be close to historical levels. The two most distinctive Puget Sound populations (from the Nooksack and Nisqually rivers) both show nonsignificant trends in recent abundance, and no other factors were found that would suggest that either of these populations is at immediate risk. However, all BRT members expressed concerns about the status of certain populations within this ESU, especially those along the Strait of Juan de Fuca. Pink salmon in the Dungeness River are declining in abundance in the lower river, and the Elwha River population may already be extinct (Table 9.1). Two of the three populations nearest to these have also shown recent declines. The petitioner, PRO-Salmon, identified water withdrawals, habitat degradation due to increased urbanization, diking for flood control, possible sewage contamination, logging, and predation by hatchery coho salmon as primary threats to the viability of lower Dungeness River pink salmon, and the petitioner suggested that some of these factors may have also contributed to the decline in Elwha River pink salmon (PRO-Salmon 1994). However, the steep decline in the Elwha River population shows a somewhat different pattern than that in the Dungeness River population, and may also have been affected by the presence of the Elwha and Glines Canyon Dams on this system, so the nature of the risks is likely to differ somewhat between these neighboring rivers (PRO-Salmon 1994).

However, the remaining odd-year populations in the U.S., as well as most of those in southern British Columbia, show no evidence of sustained declines, and several are increasing in abundance (Table 9.1). In addition, the other U.S. populations that are the most distinctive based on genetic and life history characteristics appear to be healthy. For the ESU as a whole, the greatest risks to sustainability may arise from natural variation such as large-scale climatic change or habitat degradation due to human encroachment on stream habitat.

The BRT members therefore concluded that the odd-year pink salmon ESU is not presently at risk of extinction or endangerment. Based on this conclusion, NMFS proposed that this ESU does not currently warrant a listing as a threatened or endangered species under the Endangered Species Act (Federal Register 60 [4 October 1995]:51928). However, the BRT members emphasized their concern that further erosion of the marginal populations along the Strait of Juan de Fuca might eventually pose risk to a significant portion of the ESU as a whole. Prominent risk factors for these populations include the quality of freshwater habitat and water withdrawal. In addition, evidence exists for a recent decline in body length of odd-year Washington pink salmon, which increases risk to these populations by limiting their reproductive potential. A similar decline has also been observed in pink salmon from southeastern Alaska.

Although the BRT determined that each of the two ESUs as a whole are not at risk of extinction, there were strong concerns for individual populations within each ESU. The declines in some of these pink salmon populations, and uncertainty about the status of many others, raise questions about the basis of population sustainability and how these pink salmon populations respond to natural as well as anthropogenic factors. These factors may have a broad array of effects on population productivity. The adverse consequences of some anthropogenic factors, such as excessive harvest, may be more easily addressed than others, such as habitat degradation. However, superimposed on the effects of these factors are the consequences of natural variability in habitat and population productivity. A degree of scientific uncertainty about the effects of these factors on population abundance and productivity compounds the difficulty in assessing the sustainability of populations and ESUs. Coping effectively with this problem will require innovative approaches to population and habitat management, supported by what we do know and what can be learned through dedicated research. Equally important is a coordinated effort toward implementing changes in management of fish and their habitat for individual populations clearly at risk of extinction. Although these changes can be risky themselves and may not always translate to increased abundance, the sustainability of pink salmon populations ultimately depends on the greater understanding that will grow out of these efforts.

ACKNOWLEDGMENTS

We thank all who contributed information for the status review, especially the Washington Department of Fish and Wildlife, Oregon Department of Fish and Wildlife, California Department of Fish and Game, U.S. Fish and Wildlife Service, and Northwest Indian Fisheries Commission. Additional information was provided under contract by Big Eagle and Associates and LGL, Ltd., and Natural Resource Consultants. We are grateful to several scientists and managers for discussions of this information and their analyses, as well as critical evaluations of the status review: J. Ames, J. Haymes, D. Hendrick, J. Shaklee, J. Uehara, and S. Young, Washington Department of Fish and Wildlife; G. Graves and K. Lutz, Northwest Indian Fisheries Commission; A. Gharrett and W. Smoker, University of Alaska; T. Beacham, L. Hop Wo, and W. Luedke, Canadian Department of Fisheries and Oceans; and F. Thrower, NMFS Alaska Fisheries Science Center. We are also grateful to T. Parker, W. Waknitz, and R. Waples for their contributions to the status review. Members of the Biological

Review Team for pink salmon were S. Grant, J. Hard, R. Iwamoto, O. Johnson, R. Kope, C. Mahnken, M. Schiewe, W. Waknitz, R. Waples, and J. Williams, NMFS Northwest Fisheries Science Center; P. Dygert, NMFS Northwest Region; and W. Heard, NMFS Alaska Fisheries Science Center. We thank A. Thomas, Bureau of Land Management, and an anonymous reviewer for their constructive criticism of an earlier version of this chapter.

REFERENCES

Hard, J. J., R. G. Kope, W. S. Grant, F. W. Waknitz, L. T. Parker, and R. S. Waples. 1996. Status review of pink salmon from Washington, Oregon, and California. U.S. Department of Commerce, NOAA Technical Memorandum NMFS-NWFSC-25.

Heard, W. R. 1991. Life history of pink salmon (*Oncorhynchus gorbuscha*). Pages 121–230 *in* C. Groot and L. Margolis, editors. Pacific salmon life histories. UBC Press, Vancouver, British Columbia.

IPSFC (International Pacific Salmon Fisheries Commission). 1958. A preliminary review of pertinent past tagging investigations on pink salmon and proposal for a co-ordinated research program for 1959. Pink Salmon Co-ordinating Committee, Report. Number 1.

PRO-Salmon (Professional Resources Organization-Salmon). 1994. Petition for a rule to list nine Puget Sound salmon populations as threatened or endangered under the Endangered Species Act and to designate critical habitat. Document submitted to the USDOC NOAA NMFS Northwest Region, Seattle, Washington, September 1993. (Available from PRO-Salmon, Washington Public Employees Association, 124 10th Ave. S.W., Olympia, WA 98501.)

Waples, R. S. 1991. Pacific salmon, *Oncorhynchus* spp., and the definition of "species" under the Endangered Species Act. Marine Fisheries Review 53(3):11–22.

Waples, R. S. 1995. Evolutionarily significant units and the conservation of biological diversity under the Endangered Species Act. Pages 8–27 *in* J. L. Nielsen, editor. Evolution and the aquatic ecosystem: defining unique units in population conservation. American Fisheries Society, Symposium 17, Bethesda, Maryland.

10 Review of the Status of Coho Salmon from Washington, Oregon, and California

Laurie Weitkamp, Thomas C. Wainwright, Gregory J. Bryant, David J. Teel, and Robert G. Kope

Abstract.—In response to petitions to list coho salmon *Oncorhynchus kisutch* under the U.S. Endangered Species Act (ESA), the National Marine Fisheries Service (NMFS) initiated a status review of coho salmon in the Pacific Northwest and formed a Biological Review Team (BRT) to conduct the review. The BRT identified six evolutionarily significant units (ESUs), or "species" under the ESA, for coho salmon and evaluated the risk of extinction for these ESUs. Identification of these ESUs was based on genetic, life history, biogeographic, geologic, and environmental information. The six ESUs encompass coho salmon habitats from the southern limit of their range in Monterey Bay to tributaries flowing into the north end of the Strait of Georgia. Both qualitative and quantitative information were used to assess extinction risks faced by these ESUs. This information indicated a latitudinal trend in stock health, with coho salmon in the southernmost ESUs facing the greatest risks and those in the northern ESUs facing fewer risks. Based on this latitudinal trend in stock health, and on present conservation efforts, NMFS proposed listing three coho salmon ESUs south of the Columbia River as threatened under the ESA. Of the three remaining ESUs, only one was not proposed for listing. The other two were identified as "candidate" species, for which listing may be warranted in the future. Since the NMFS proposed listings, the three southernmost ESUs have been listed as "threatened" under the ESA and the fate of the remaining candidate ESUs are being determined as of this printing.

INTRODUCTION

The Endangered Species Act (ESA; 16 U.S.C. §§1531–1543) allows listing of "distinct population segments" of vertebrates as well as named species and subspecies. The policy of the National Marine Fisheries Service (NMFS) for anadromous Pacific salmonids *Oncorhynchus* spp. on this issue is that a population will be considered distinct for purposes of the ESA if it represents an evolutionarily significant unit (ESU) of the species as a whole (Federal Register 56 [20 November 1991]:58612). To be considered an ESU, a population or group of populations must 1) be substantially reproductively isolated from other populations and 2) contribute substantially to the ecological/genetic diversity of the biological species (Waples 1991, 1995). Once an ESU is identified, a variety of factors related to population abundance are considered in determining whether it is at risk of extinction or endangerment.

Identifying conservation units, or ESUs, and determining their extinction risks forms the basis of the status review. Such status reviews for Pacific salmon are conducted by a Biological Review Team (BRT), consisting of scientists with expertise in Pacific salmon biology. Team members are appointed largely from staff of the NMFS Northwest Fisheries Science Center and Southwest

Region. Upon completion of a status review, the BRT makes a recommendation to NMFS Northwest and Southwest Regional directors regarding its findings. The U.S. Department of Commerce, which is the NMFS parent agency, must then consider both the BRT recommendations *and* existing or proposed conservation measures to determine whether an ESU should be listed under the ESA and whether it should be listed as threatened or endangered.

The NMFS initiated a status review of coho salmon *O. kisutch* in Washington, Oregon, and California in October 1993, in response to three petitions seeking protection for coho salmon under the ESA (Oregon Trout et al. 1993; Pacific Rivers Council et al. 1993; SCCPD 1993). This document summarizes the BRT conclusions regarding ESU boundaries and risks of extinction, as well as the NMFS proposal for listing and the eventual listing status of coho salmon. The biological and environmental information considered in the review process is described in detail in the status review document (Weitkamp et al. 1995) and various status reports (e.g., Nehlsen et al. 1991).

PROPOSED COHO SALMON ESUs

The BRT examined genetic, life history, biogeographic, geologic, and environmental information to identify ESU boundaries for coho salmon populations from southern British Columbia to Monterey Bay, the southern limit of their spawning distribution. The best information for this process indicated clear differences between areas. This information consisted of data on the physical environment and ocean conditions/upwelling patterns, estuarine and freshwater fish distributions, coho salmon genetics, river entry and spawn timing, and marine coded-wire-tag (CWT) recovery patterns. Regional precipitation patterns are one example. The central California coast (roughly between Monterey Bay and Cape Mendocino/Punta Gorda) has relatively low precipitation levels, while northern California and the Oregon coast receive moderate precipitation, and the Olympic Peninsula is extremely wet with large accumulations of snow at higher elevations. Puget Sound, Strait of Georgia, and the Willamette Valley all lie in a rainshadow, yet have large accumulations of snow at higher elevations. River flow patterns also show similar groupings. For example, in central California many rivers have relatively low flows throughout the year punctuated by a single, short peak of flow, whereas in Washington, many rivers have relatively high flows throughout the year with two peaks of flow—one due to rain in the winter, the other to melting snow in the spring. River flow patterns are important for coho salmon because they affect the accessibility to, and amount of, available freshwater rearing habitat. Terrestrial vegetation also changes along the coast: most low-elevation forests in California and southern Oregon are dominated by redwoods (*Sequoia sempervirens*), coastal areas from southern Oregon to Alaska are typified by Sitka spruce (*Picea sitchensis*) forests, in Puget Sound western hemlock (*Tsuga heterophylla*) forms the dominant forest type, and in Oregon white oak (*Quercus garryana*) dominates Willamette Valley forests. Coho salmon themselves also show fairly discrete and similar groupings based on both genetics and marine CWT recovery patterns. River entry and spawn timing of coho salmon also exhibited distinctive patterns, with fish inhabiting central California streams spawning within days of entering freshwater, while those farther north spend 1 to 2 months in freshwater before spawning.

Many of these patterns changed in unison at distinct geographic locations or over fairly narrow ranges. For example, ocean upwelling, terrestrial vegetation, CWT recovery patterns, coho salmon genetics, and environmental features exhibited marked changes north and south of Cape Blanco. Other areas where similar changes occurred included Punta Gorda/Cape Mendocino, the mouth of the Columbia River, and along the Strait of Juan de Fuca.

Based on these patterns, the BRT identified six coho salmon ESUs in California, Oregon, Washington, and southern British Columbia, as shown in Figure 10.1. These were identified as (1) Central California Coast ESU; (2) Southern Oregon/Northern California Coasts ESU; (3) Oregon Coast ESU; (4) Lower Columbia River/Southwest Washington Coast ESU; (5) Olympic Peninsula ESU; and (6) Puget Sound/Strait of Georgia ESU.

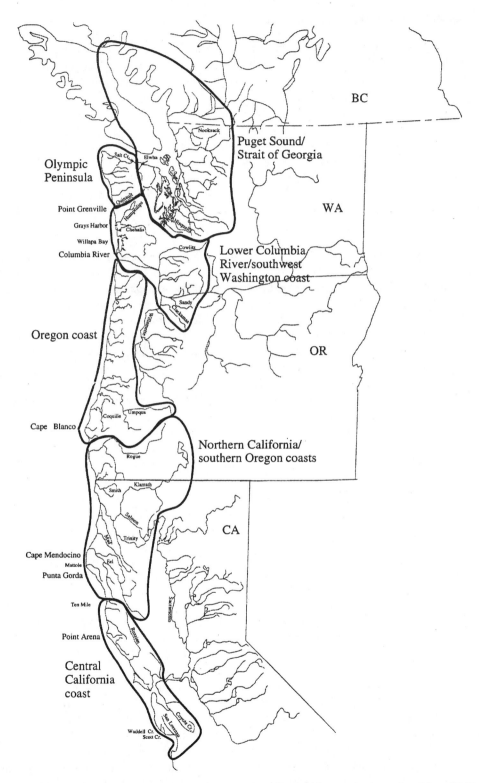

FIGURE 10.1 Proposed west coast coho salmon evolutionarily significant units (reproduced from Weitkamp et al. 1995).

ASSESSMENT OF EXTINCTION RISK

The ESA (Section 3) defines the term "endangered species" as "any species which is in danger of extinction throughout all or a significant portion of its range." The term "threatened species" is defined as "any species which is likely to become an endangered species within the foreseeable future throughout all or a significant portion of its range." According to the ESA, determination of whether a species is threatened or endangered should be based on the best scientific information available regarding its current status, after taking into consideration conservation measures that are proposed or in place. In assessing extinction risks, the BRT did not evaluate likely or possible effects of conservation measures, and therefore did not make recommendations as to whether identified ESUs should be listed as threatened or endangered species. Rather, the BRT drew scientific conclusions about the risk of extinction faced by identified ESUs under the assumption that present conditions would continue. Summaries of risk considerations for the six coho salmon ESUs are provided in Table 10.1, and the resulting conclusions for each ESU follow.

1. *Central California Coast.*—There was unanimous agreement among the BRT that natural populations of coho salmon in this ESU are presently in danger of extinction. The chief reasons for this assessment were extremely low current abundance (especially compared to historical abundance), widespread local extinctions, clear downward trends in abundance, extensive habitat degradation and associated decreased carrying capacity, and a long history of artificial propagation with the use of non-native stocks. In addition, recent droughts and current ocean conditions may have further reduced run sizes.

2. *Southern Oregon/Northern California Coasts.*—There was unanimous agreement among the BRT that coho salmon in this ESU are not in danger of extinction but are likely to become endangered in the foreseeable future if present trends continue. Current run size, the severe decline from historical run size, the frequency of local extinctions, long-term trends that are clearly downward, degraded habitat and associated reduction in carrying capacity, and widespread hatchery production using exotic stocks are all factors that contributed to the assessment. As for the central California ESU, recent droughts and current ocean conditions may have further reduced run sizes for this ESU.

3. *Oregon Coast.*—The BRT concluded that coho salmon in this ESU are not in danger of extinction but are likely to become endangered in the future if present trends continue. The BRT reached this conclusion based on low recent abundance estimates that are 5 to 10% of historical abundance estimates, clear downward long-term trends, recent natural spawner-to-spawner ratios that are below replacement levels, extensive habitat degradation, and widespread hatchery production of coho salmon. Drought and current ocean conditions may have also contributed to a reduction of run sizes.

4. *Lower Columbia River/Southwest Washington Coast.*—Previously, NMFS concluded that it could not identify any remaining natural populations of coho salmon in the lower Columbia River (excluding the Clackamas River) that warranted protection under the ESA (Johnson et al. 1991; Federal Register 56 [27 June 1991]:29553). The Clackamas River produces moderate numbers of natural coho salmon. The BRT could not reach a definite conclusion regarding the relationship of Clackamas River late-run coho salmon to the historic lower Columbia River ESU. However, the BRT did conclude that *if* the Clackamas River late-run coho salmon is a native run that represents a remnant of a lower Columbia River ESU, it is not presently in danger of extinction but is likely to become so in the foreseeable future if present conditions continue.

   For southwest Washington coho salmon, there is considerable uncertainty about the ancestry of coho salmon runs, given the high historical and current levels of artificial production. This uncertainty prevented the BRT from reaching a definite conclusion regarding the relationship between coho salmon runs in that area and those in the historical lower Columbia River/Southwest Washington ESU. If new information becomes available, the relationship and status of the ESU will be reexamined.

5. *Olympic Peninsula.*—There is continuing cause for concern about habitat destruction and hatchery practices within this ESU. However, the BRT concluded that there is sufficient native, natural, self-sustaining production of coho salmon in this ESU that it is not in danger of extinction and is not likely to become endangered in the foreseeable future unless conditions change substantially.

6. *Puget Sound/Strait of Georgia.*—The BRT was concerned that, if present trends continue, this ESU is likely to become endangered in the foreseeable future. Although current population abundance is near historical levels, and recent trends in overall population abundance have not been downward, there is substantial uncertainty relating to several of the risk factors considered. These include widespread and intensive artificial propagation, high harvest rates, extensive habitat degradation, a recent dramatic decline in adult size, and unfavorable ocean conditions. Further consideration of this ESU is warranted to attempt to clarify these uncertainties.

PROPOSED LISTING STATUS

Following the BRT conclusions regarding extinction risks, NMFS considered likely effects of conservation measures that were proposed or in place. Based on these considerations, in July 1995, NMFS proposed that the three southernmost ESUs—Central California, Southern Oregon/Northern California Coasts, and Oregon Coast—be listed as "threatened" under the ESA. The Lower Columbia River/Southwest Washington Coast and Puget Sound/Strait of Georgia ESUs were proposed as "candidate" species because of considerable uncertainty regarding their extinction risk. The NMFS concluded that listing the Olympic Peninsula ESU under the ESA was not warranted (Federal Register 60 [25 July 1995]:38011).

After considering public comments, new and updated information about the status of coho salmon, and existing and proposed conservation measures, the NMFS finalized the listing of the Central California Coast and Southern Oregon/Northern California Coasts ESUs as threatened in October 1996 and May 1997, respectively (Federal Register 61 [31 October 1996]:56138; 62 [6 May 1997]:245811). Also in May 1997, NMFS announced that, because of recent increases in spawning escapement and substantial conservation measures that are part of the State of Oregon's Oregon Coast Salmon Restoration Initiative, listing the Oregon coast ESU was not warranted. This finding was challenged and subsequently overturned in court, and the Oregon coast ESU was listed as a threatened species in August, 1998 (Federal Register 63 [10 August 1998]:42587). As of this printing, the status of the candidate ESUs is currently being reviewed. This review process is based on new information about factors identified during the review process as posing uncertain degrees of risk, updated trend information (e.g., spawner abundance, harvest rates, hatchery releases), and new conservation measures. Any proposed listings resulting from this review are expected to be announced in early 2000, with final listings occurring in 2001.

In this century, we have seen West Coast coho salmon decline from an abundant, commercially important species to one with widespread fisheries closures and ESA listings or proposed listings coastwide. The causes of this decline come from numerous natural and anthropogenic factors. Although some of these causes, such as overfishing and hatchery practices, are relatively easy to address in theory, and prompt actions may have immediate effects on populations, other causes, namely freshwater habitat degradation, are more difficult to address and it may take years before population-scale improvements are seen. Ocean conditions may also contribute to declines in coho salmon populations, but they are outside of human control. Only by implementing reforms, with the cooperation and efforts of management agencies, watershed councils, landowners, and concerned citizens, among others, can we hope to have sustainable coho salmon populations, let alone sustainable fisheries, in the future.

TABLE 10.1

Biological Review Team's summary and conclusions of risk considerations for six coho salmon evolutionarily significant units (ESUs). As described in the text, these conclusions do not consider conservation measures that are proposed or in place. Based on data through 1998 (modified from Weitkamp et al. 1995).

Risk category	Central California	Southern Oregon/Northern California Coasts	Oregon Coast
Absolute numbers (Recent average)	Escapement ca. 6,000, ca. 160 "native" with no history of hatchery influence.	Run size ca. 10,000 natural, 20,000 hatchery. Current production largely in the Rogue and Klamath basins.	Escapement ca. 45,000 natural, unknown hatchery.
Numbers relative to historical abundance and carrying capacity	Abundance substantially below historical levels. More than 50% of coho streams no longer have spawning runs. Widespread habitat degradation.	Substantially below historical levels. In California portion of ESU, ca. 48% of coho streams no longer have spawning runs. Widespread habitat degradation.	Natural production ca. 5–10% of historical levels, near 50% of current capacity. Widespread habitat degradation.
Trends in abundance and production	Long-term trends clearly downward. No data to estimate recent trends.	Long-term trends clearly downward. Main data are for Rogue River basin, where runs declined to very low levels in 1960s and 1970s, then increased with start of hatchery production.	Long-term trends clearly downward. Escapement declined substantially since early 1950s, but majority of decline was in early 1970s. Recent average spawner-to-spawner ratios below replacement. Recruits-per-spawner show a continuous decline up to present. Southern portion of ESU recently increasing.
Variability factors	Low abundance or degraded habitat may increase variability.	Low abundance or degraded habitat may increase variability.	Low abundance or degraded habitat may increase variability.
Threats to genetic integrity	Most existing populations have history of hatchery plantings, with many out-of-state stock transfers.	Most existing populations have hatchery plantings, with many out-of-state stock transfers in California portion of the ESU.	Most existing populations have hatchery plantings, with many out-of-basin (but largely within-ESU) stock transfers. Magnitude of hatchery influence declines from north to south.
Recent events	Recent droughts and change in ocean production have probably reduced run sizes.	Recent droughts and change in ocean production have probably reduced run sizes.	Recent droughts and change in ocean production have probably reduced run sizes.
Other factors	None identified.	None identified.	None identified.
Conclusion	Presently in danger of extinction.	Not presently in danger of extinction, but likely to become so.	Not presently in danger of extinction, but likely to become so.

TABLE 10.1 (continued)
Biological Review Team's summary and conclusions of risk considerations for six coho salmon evolutionarily significant units (ESUs). As described in the text, these conclusions do not consider conservation measures that are proposed or in place. Based on data through 1998 (modified from Weitkamp et al. 1995).

Risk category	Lower Columbia/ Southwest Washington Coast	Olympic Peninsula	Puget Sound/ Strait of Georgia
Absolute numbers (Recent average)	Total natural production unknown. Late Clackamas River escapement is less than 5,000.	Escapement ca. 19,000 natural, 18,000 hatchery.	Escapement ca. 216,000 191,000 hatchery in U.S. portion
Numbers relative to historical abundance and carrying capacity	Native, natural production near zero in much of the geographic area. Unable to identify extant natural populations, except possibly in Clackamas River. Widespread habitat degradation.	Substantially below historical levels. Widespread habitat degradation in most of geographic range, but headwater areas within Olympic National Park protected.	Total run is near historical levels, natural run is substantially below historical levels. Widespread habitat degradation.
Trends in abundance and production	Long-term trend in natural production is clearly downward. No substantial recent upward or downward trend in Clackamas River.	No substantial upward or downward trends were detected in terminal run size or in ocean exploitation rates.	Long-term trends in total run size relatively flat in WA portion of ESU, downward in BC portion. Recent escapement trends are mixed upward and downward, majority of stocks show no substantial trend.
Variability factors	Low abundance or degraded habitat may increase variability.	Low abundance or degraded habitat may increase variability.	Degraded habitat may increase variability.
Threats to genetic integrity	Widespread hatchery production far exceeds that for any other ESU, with many out-of-basin (but largely within-ESU) stock transfers.	Some populations have continuing hatchery plantings, largely within-basin although numerous small out-of-ESU transfers have occurred. Hatchery influence restricted to a few major rivers; several stocks have little or no hatchery influence.	Most existing populations have continuing hatchery plantings, with many out-of-basin (but largely within-ESU) stock transfers. Average hatchery contribution rate to runs is 62%, with largest effect on the Nooksack-Samish and South Puget Sound stock complexes.
Recent events	Recent droughts and change in ocean production have probably reduced run sizes.	Recent droughts and change in ocean production have probably reduced run sizes.	Recent droughts and change in ocean production have probably reduced run sizes.
Other factors	Harvest rates have been very high, but declining in recent years.	None identified.	Sharp decline in adult coho body size. Recent harvest rates have been high.
Conclusion	If ESU still exists, it is not presently in danger of extinction, but is likely to become so.	Not presently in danger of extinction, nor likely to become so.	Not presently in danger of extinction, but likely to become so.

ACKNOWLEDGMENTS

We thank all who contributed information for the status review, especially the Washington Department of Fish and Wildlife, Oregon Department of Fish and Wildlife, California Department of Fish and Game, U.S. Fish and Wildlife Service, and Northwest Indian Fisheries Commission. The biological review team for this status review included: Peggy Busby, Dr. David Damkaer, Robert Emmett, Dr. Jeffrey Hard, Dr. Orlay Johnson, Dr. Robert Kope, Dr. Conrad Mahnken, Gene Matthews, George Milner, Dr. Michael Schiewe, David Teel, Dr. Thomas Wainwright, William Waknitz, Dr. Robin Waples, Laurie Weitkamp, Dr. John Williams, and Dr. Gary Winans, all from the Northwest Fisheries Science Center (NWFSC); and Gregory Bryant from the NMFS Southwest Region. Craig Wingert, from the NMFS Southwest Region, and Steven Stone, from the NMFS Northwest Regional Office, also participated in the discussions and provided information on coho salmon life history and abundance. Jason Griffith and Megan Ferguson (University of Washington), and Don Vandoornik, Dave Kuligowski, and Kathleen Neely (NWFSC) provided considerable assistance in the completion of the status review.

REFERENCES

Johnson, O. W., T. A. Flagg, D. J. Maynard, G. B. Milner, and F. W. Waknitz. 1991. Status review for lower Columbia River coho salmon. U.S. Department of Commerce, NOAA Technical Memorandum NMFS F/NWC-202.

Nehlsen, W., J. E. Williams, and J. A. Lichatowich. 1991. Pacific salmon at a crossroads: Stocks at risk from California, Oregon, Idaho, and Washington. Fisheries 16(2):4–21.

Oregon Trout, Portland Audubon Society, and Siskiyou Regional Education Project. 1993. Petitions for listing species of Pacific coast *Oncorhynchus kisutch* pursuant to the Endangered Species Act of 1973 as amended. Petition to U.S. Department of Commerce, National Marine Fisheries Service, Northwest Region, Seattle. (Available from Oregon Trout, 5331 S.W. Macadam Ave., No. 228, Portland, OR 97201.)

Pacific Rivers Council and 22 coauthors. 1993. Petition for a rule to list, for designation of critical habitat, and for a status review of coho salmon throughout its range under the Endangered Species Act. Petition to U.S. Department of Commerce, National Marine Fisheries Service, Northwest Region, Seattle. (Available from Pacific Rivers Council, P.O. Box 309, Eugene, OR 97440.)

SCCPD (Santa Cruz County Planning Department). 1993. Petition to list central California coast coho salmon as an endangered species. Petition to U.S. Department of Commerce, National Marine Fisheries Service, Southwest Region. (Available from National Marine Fisheries Service, Southwest Region, 501 W. Ocean Blvd., Suite 4200, Long Beach, 90802.)

Waples, R. S. 1991. Pacific salmon, *Oncorhynchus* spp., and the definition of "species" under the Endangered Species Act. U.S. National Marine Fisheries Service Marine Fisheries Review 53(3):11–22.

Waples, R. S. 1995. Evolutionarily significant units and the conservation of biological diversity under the Endangered Species Act. Pages 8–27 *in* J. L. Nielsen, editor. Evolution and the aquatic ecosystem: defining unique units in population conservation. American Fisheries Society Symposium 17, Bethesda, Maryland.

Weitkamp, L. A., and six coauthors. 1995. Status review of coho salmon from Washington, Oregon, and California. U.S. Department of Commerce, NOAA Technical Memorandum NMFS-NWFSC-24.

11 Status Review of Steelhead from Washington, Idaho, Oregon, and California

Peggy J. Busby, Thomas C. Wainwright, and Gregory J. Bryant

Abstract.—In response to petitions to list populations of steelhead (anadromous *Oncorhynchus mykiss*) under the U.S. Endangered Species Act (ESA), the National Marine Fisheries Service (NMFS) initiated a status review of steelhead from the states of Washington, Idaho, Oregon, and California. A Biological Review Team (BRT) of 12 scientists from NMFS and the Biological Resources Division of U.S. Geological Survey identified 15 evolutionarily significant units (ESUs), or "species" under the ESA, for steelhead and evaluated the risk of extinction for these ESUs. Identification of these ESUs was based on genetic, life history, biogeographic, geologic, and environmental information. These ESUs include steelhead populations from the U.S.-Canada border south to the southern limit of their range in southern California. The BRT considered qualitative and quantitative information to assess extinction risks faced by the steelhead ESUs. Available information indicated a pattern of greater risk for ESUs approaching the edges of the species' range, to the south in central and southern California, and to the east in the Columbia and Snake River basins. Based largely on the findings of the BRT, and after considering proposed and implemented conservation measures, NMFS has taken several listing actions for steelhead. As of July 1999, nine ESUs have been listed (two as endangered and seven as threatened) and three ESUs have been designated as candidate species for which listing may be warranted in the future. Three ESUs have been determined not to be warranted for listing at this time.

INTRODUCTION

In February 1994, the National Marine Fisheries Service (NMFS) received a petition seeking protection under the Endangered Species Act (ESA) for steelhead (anadromous *Oncorhynchus mykiss*) in Washington, Idaho, Oregon, and California (ONRC et al. 1994). A Biological Review Team (BRT) of 12 scientists from the Northwest and Southwest Regions of NMFS and the Biological Resources Division of U.S. Geological Survey considered published biological and environmental information as well as information submitted by interested parties relative to assessing steelhead populations under the ESA. This document summarizes the findings of the BRT; the specific information that was used by the BRT to reach their conclusions is presented more completely in Busby et al. (1996).

The ESA (16 U.S.C. §§1531–1543) allows listing of "distinct population segments" of vertebrates as well as named species and subspecies. The policy of NMFS on this issue for anadromous Pacific salmonids is that a population will be considered "distinct" for purposes of the ESA if it represents an evolutionarily significant unit (ESU) of the species as a whole (Federal Register 56 [20 November 1991]:58612). To be considered an ESU, a population or group of populations must

(1) be substantially reproductively isolated from other populations and (2) contribute substantially to the ecological/genetic diversity of the biological species (Waples 1991). Once an ESU is identified, a variety of factors related to population abundance are considered in determining whether a listing is warranted.

WEST COAST STEELHEAD ESUs

Within the range of steelhead considered in this review, there are two recognized phylogenetic groups that have been proposed as subspecies (Behnke 1992). Inland steelhead *O. m. gairdneri* occur in the Fraser and Columbia River basins east of the Cascade Mountains; coastal steelhead *O. m. irideus* occur in river basins west of the Cascade Mountains. In each subspecies there is a nonanadromous form, referred to as inland redband trout and coastal rainbow trout, respectively. Inland and coastal *O. mykiss*, whether anadromous or nonanadromous, are genetically distinct groups, indicating that substantial reproductive isolation has occurred between the two forms. A comparable degree of reproductive isolation does not appear to occur between anadromous and nonanadromous life history forms within a geographic area (Busby et al. 1996).

Biologically, steelhead can be divided into two basic reproductive ecotypes, based on the state of sexual maturity at the time of river entry and duration of spawning migration (Burgner et al. 1992). The stream-maturing ecotype (commonly known as summer steelhead in the Pacific Northwest and northern California) enters fresh water in a sexually immature condition and requires several months to mature and spawn. The ocean-maturing ecotype (commonly known as winter steelhead) enters fresh water with well-developed gonads and spawns shortly thereafter. Although there is potential for reproductive isolation between summer and winter steelhead, there actually appears to be substantial temporal, if not spatial, overlap in spawning between the ecotypes where they co-occur. Inland steelhead are predominately summer steelhead, while coastal steelhead are largely winter steelhead with several geographically distinct populations of summer steelhead. Inland summer steelhead are genetically dissimilar to coastal summer steelhead (Busby et al. 1996).

After considering available information on steelhead genetics, phylogeny and life history, freshwater ichthyogeography, and environmental features that may affect steelhead, the BRT identified 15 ESUs: 12 for coastal steelhead and 3 for the inland form (Figure 11.1). The BRT reviewed population abundance data and other risk factors for these steelhead ESUs. The risk factors that were considered for each ESU, and the BRT's conclusions are summarized in Table 11.1.

The BRT concluded that, in general, the steelhead ESUs described below include both anadromous and resident *O. mykiss* in cases where the two forms have the opportunity to interbreed. Resident populations above long-standing natural barriers, and those that have resulted from the introduction of non-native rainbow trout, would not be considered part of the ESUs. Resident populations that inhabit areas upstream from human-caused migration barriers (e.g., Grand Coulee Dam, the Hells Canyon Dam complex, and numerous smaller barriers in California) may contain genetic resources similar to those of anadromous fish in the ESU, but little information is available on these fish or the role they might play in conserving natural populations of steelhead. The status, with respect to steelhead ESUs, of resident fish upstream from human-caused migration barriers must be evaluated on a case-by-case basis as more information becomes available. Regardless of any decision by the BRT to include resident trout within a steelhead ESU, NMFS' listing authority is restricted to anadromous salmonid species; therefore, any actual listing of a resident trout population would fall to the U.S. Fish and Wildlife Service.

Coastal Steelhead ESUs

1. *Puget Sound.*—This ESU occupies river basins of the Strait of Juan de Fuca, Puget Sound, and Hood Canal, Washington. Included are river basins as far west as the Elwha River and as far north as the Nooksack River. This ESU is primarily composed of winter

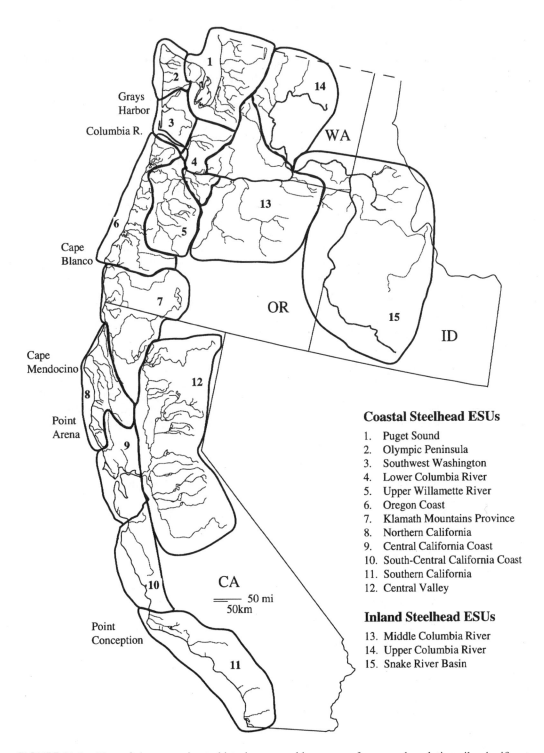

Grays
Harbor
Columbia R.
Cape
Blanco
Cape
Mendocino
Point
Arena
Point
Conception

Coastal Steelhead ESUs

1. Puget Sound
2. Olympic Peninsula
3. Southwest Washington
4. Lower Columbia River
5. Upper Willamette River
6. Oregon Coast
7. Klamath Mountains Province
8. Northern California
9. Central California Coast
10. South-Central California Coast
11. Southern California
12. Central Valley

Inland Steelhead ESUs

13. Middle Columbia River
14. Upper Columbia River
15. Snake River Basin

FIGURE 11.1 Map of the approximate historic geographic ranges of proposed evolutionarily significant units (ESUs) for west coast steelhead. Present distribution may be less than that shown. See text for more details.

TABLE 11.1

Summary of risk considerations and initial BRT conclusions for 15 steelhead Evolutionarily Significant Units (ESUs). See Busby et al. (1996) for more details.

	Coastal Steelhead ESUs		
Risk Category	**ESU 1** **Puget Sound**	**ESU 2** **Olympic Peninsula**	**ESU 3** **Southwest Washington**
Absolute numbers (recent average)	Total run size at least 45,000, natural escapement at least 22,000 (based on sum of major stocks only).	Total run size at least 54,000, natural escapement at least 20,000 (based on sum of major stocks only).	Total run size at least 20,000, natural escapement at least 13,000 (based on sum of major stocks only).
Numbers relative to historical abundance and carrying capacity	Total historical abundance and capacity unknown, but abundance certainly below historical levels. Widespread habitat degradation from logging, urbanization, and agriculture.	Total historical abundance and capacity unknown, but abundance certainly below historical levels. Widespread habitat degradation.	Total historical abundance and capacity unknown, but abundance certainly below historical levels. Widespread habitat degradation.
Trends in abundance and production	Long term trends downward. Recent trends all downward except in Skagit and Snohomish river basins.	Long-term trends downward. Recent trends predominantly upward, some downward.	Long-term trends downward. Recent trends almost entirely downward.
Variability factors	Low abundance and/or degraded habitat may increase variability.	Low abundance and/or degraded habitat may increase variability.	Low abundance and/or degraded habitat may increase variability.
Threats to genetic integrity	Most existing populations have history of hatchery plantings, largely from a single stock. Management efforts have reduced interactions between hatchery and natural fish.	Most existing populations have history of hatchery plantings, largely from a few stocks. Management efforts have reduced interactions between hatchery and natural fish.	Most existing populations have history of hatchery plantings, largely from out-of-ESU stocks. Management efforts have reduced interactions between hatchery and natural fish.
Recent events	Recent droughts and change in ocean production have probably reduced run sizes and exacerbated downward trends.	Recent droughts and change in ocean production have probably reduced run sizes and exacerbated downward trends.	Recent droughts and change in ocean production have probably reduced run sizes and exacerbated downward trends.
Other factors	None identified.	None identified.	None identified.
Original BRT conclusion	Not presently in danger of extinction. Not likely to become so.	Not presently in danger of extinction. Not likely to become so.	Not presently in danger of extinction. Not likely to become so.
ESA status as of July 1999	Not warranted for listing.	Not warranted for listing.	Not warranted for listing.

	Coastal Steelhead ESUs		
Risk Category	**ESU 4** **Lower Columbia River**	**ESU 5** **Upper Willamette River**	**ESU 6** **Oregon Coast**
Absolute numbers (Recent average)	Total run size at least 16,000, natural escapement at least 4,000 (based on sum of major stocks only).	Total run size about 16,000 of which about 4,000 are native. Natural escapement unknown.	Total run size about 130,000, total escapement about 100,000. Natural escapement unknown.
Numbers relative to historical abundance and carrying capacity	Total historical abundance and capacity unknown, but abundance certainly below historical levels. Widespread habitat degradation.	Total historical abundance and capacity unknown, but abundance certainly below historical levels. Widespread habitat degradation.	Total historical abundance and capacity unknown, but abundance certainly below historical levels. Widespread habitat degradation.

TABLE 11.1 (continued)
Summary of risk considerations and initial BRT conclusions for 15 steelhead Evolutionarily Significant Units (ESUs). See Busby et al. (1996) for more details.

Coastal Steelhead ESUs (continued)

Risk Category	ESU 4 Lower Columbia River	ESU 5 Upper Willamette River	ESU 6 Oregon Coast
Trends in abundance and production	Long-term trends downward. Recent trends predominantly downward, some strongly upward.	Long-term trends downward. Recent trend in native (late winter) run downward, introduced summer run upward.	Long-term trends downward. Recent trends downward except in southernmost portion of ESU.
Variability factors	Low abundance and/or degraded habitat may increase variability.	Low abundance and/or degraded habitat may increase variability.	Low abundance and/or degraded habitat may increase variability.
Threats to genetic integrity	Half of existing populations have history of hatchery plantings, largely from a few stocks. Current hatchery production is widespread. Most streams with estimates have at least 50% hatchery fish in natural spawning escapement.	Widespread hatchery production, mostly of non-native summer and early winter runs. Degree of interaction with native runs is unknown. Most native production is in one basin with substantial hatchery supplementation.	Most existing populations have history of hatchery plantings, largely from a few stocks. Current hatchery production is widespread. Most streams have at least 50% hatchery fish in natural spawning escapement.
Recent events	Recent droughts and change in ocean production have probably reduced run sizes and exacerbated downward trends. Eruption of Mount St. Helens strongly affected some stocks.	Recent droughts and change in ocean production have probably reduced run sizes and exacerbated downward trends.	Recent droughts and change in ocean production have probably reduced run sizes and exacerbated downward trends.
Other factors	None identified.	None identified.	None identified.
Original BRT conclusion	Not presently in danger of extinction. Likely to become so.	Not presently in danger of extinction. Not likely to become so.	Not presently in danger of extinction. Likely to become so.
ESA status as of July 1999	Listed as threatened.	Listed as threatened.	Candidate species.

Coastal Steelhead ESUs

Risk Category	ESU 7 Klamath Mountains Province	ESU 8 Northern California	ESU 9 Central California Coast
Absolute numbers (Recent average)	Total run size about 170,000, natural escapement unknown.	Total run size about 100,000 in early 1980s, no current estimate available.	Total run size (in streams with information) less than 9,000, natural escapement unknown.
Numbers relative to historical abundance and carrying capacity	Total historical abundance and capacity unknown, but abundance certainly below historical levels. Widespread habitat degradation.	Total historical abundance and capacity unknown, but abundance certainly below historical levels. Widespread habitat degradation.	Total historical abundance and capacity unknown. Major habitat blockages on two main rivers. Several stock extirpations in San Francisco Bay.
Trends in abundance and production	Long-term trends downward. Recent trends predominantly downward except for small summer run stocks and stocks with substantial hatchery production.	Long-term trends downward. Recent trends downward except for summer stocks in Eel River Basin. Special concern for summer steelhead stocks.	Long-term trends downward. Recent trends severely downward for main rivers, some small streams stable at low levels.

TABLE 11.1 (continued)
Summary of risk considerations and initial BRT conclusions for 15 steelhead Evolutionarily Significant Units (ESUs). See Busby et al. (1996) for more details.

	Coastal Steelhead ESUs (continued)		
Risk Category	ESU 7 Klamath Mountains Province	ESU 8 Northern California	ESU 9 Central California Coast
Variability factors	Low abundance and/or degraded habitat may increase variability.	Low abundance and/or degraded habitat may increase variability.	Low abundance and/or degraded habitat may increase variability.
Threats to genetic integrity	Most existing populations have history of hatchery plantings, largely from a few stocks. Current hatchery production is widespread.	Most existing populations have history of hatchery plantings, largely from a few stocks. Current hatchery production is localized.	Small population size and stock fragmentation may be problems.
Recent events	Recent droughts and change in ocean production have probably reduced run sizes and exacerbated downward trends.	Recent droughts and change in ocean production have probably reduced run sizes and exacerbated downward trends.	Recent droughts and change in ocean production have probably reduced run sizes and exacerbated downward trends.
Other factors	Poaching of adult summer steelhead.	Poaching of adult summer steelhead, predation by non-native squawfish in Eel River.	None identified.
Original BRT conclusion	Not presently in danger of extinction. Likely to become so.	Not presently in danger of extinction. Likely to become so.	Presently in danger of extinction.
ESA status as of July 1999	Candidate species.	Candidate species.	Listed as threatened.

	Coastal Steelhead ESUs		
Risk Category	ESU 10 South-Central California Coast	ESU 11 Southern California	ESU 12 Central Valley
Absolute numbers (Recent average)	Total run size in streams with information less than 500.	Total run size in streams with information less than 500.	Total run size probably less than 10,000.
Numbers relative to historical abundance and carrying capacity	Total historical abundance and capacity unknown. Widespread blockage and loss of habitat, extirpation of local stocks.	Total historical abundance and capacity unknown. Widespread blockage and loss of habitat, extirpation of local stocks.	Total historical abundance and capacity unknown. Widespread blockage and loss of habitat, extirpation of local stocks.
Trends in abundance and production	Long-term trends downward. Recent trends predominantly downward.	Long-term trends downward. Recent trends almost all downward.	Long-term trends downward. Recent trends unknown.
Variability factors	Low abundance and/or degraded habitat may increase variability.	Low abundance and/or degraded habitat may increase variability.	Low abundance and/or degraded habitat may increase variability.
Threats to genetic integrity	Small population size and stock fragmentation may be problems. Widespread stocking of hatchery rainbow trout.	Widespread stocking and transfers of hatchery steelhead. Small population size and stock fragmentation may be problems. Widespread stocking of hatchery rainbow trout.	Large-scale hatchery production, with widespread stock transfers. Small population size and stock fragmentation may be problems.

TABLE 11.1 (continued)
Summary of risk considerations and initial BRT conclusions for 15 steelhead Evolutionarily Significant Units (ESUs). See Busby et al. (1996) for more details.

	Coastal Steelhead ESUs (continued)		
Risk Category	**ESU 10** **South-Central California Coast**	**ESU 11** **Southern California**	**ESU 12** **Central Valley**
Recent events	Recent droughts and change in ocean production have probably reduced run sizes and exacerbated downward trends.	Recent droughts and change in ocean production have probably reduced run sizes and exacerbated downward trends.	Recent droughts and change in ocean production have probably reduced run sizes and exacerbated downward trends.
Other factors	None identified.	None identified.	None identified.
Original BRT conclusion	Presently in danger of extinction.	Presently in danger of extinction.	Presently in danger of extinction.
ESA status as of July 1999	Listed as threatened.	Listed as endangered.	Listed as threatened.

	Inland Steelhead ESUs		
Risk Category	**ESU 13** **Middle Columbia River**	**ESU 14** **Upper Columbia River**	**ESU 15** **Snake River**
Absolute numbers (Recent average)	Total run size about 140,000, natural run size about 40,000 (based on subtraction of mainstem dam passage data).	Total run size about 7,500, natural run size about 1,200.	Total run size about 70,000, natural run size about 9,500.
Numbers relative to historical abundance and carrying capacity	Historic abundance and capacity unknown. Major habitat areas blocked by dams. Habitat degradation from grazing and water diversions.	Historic abundance and capacity unknown. Major habitat areas blocked by dams. Widespread habitat degradation from grazing and water diversions.	Major habitat areas blocked by dams. Recent parr densities much below capacity (especially "B" group). Local areas of habitat degradation.
Trends in abundance and production	Total (natural + hatchery) trend increasing, natural stocks largely decreasing recently.	Total (natural + hatchery) trend stable, but natural production not self-sustaining.	Almost all trends for natural production are downward, severe overall recent decline.
Variability factors	Low abundance and/or degraded habitat may increase variability.	Low abundance and/or degraded habitat may increase variability.	Low abundance and/or degraded habitat may increase variability.
Threats to genetic integrity	Moderate hatchery production, largely using within-basin stocks.	Large-scale homogenized hatchery production in all populations.	Widespread large-scale hatchery production affecting most of basin.
Recent events	Recent droughts and change in ocean production have probably reduced run sizes and exacerbated downward trends.	Recent droughts and change in ocean production have probably reduced run sizes and exacerbated downward trends.	Recent droughts and change in ocean production have probably reduced run sizes and exacerbated downward trends.
Other factors	Migration corridor survival affected by irrigation diversions and hydropower development.	Migration corridor survival affected by irrigation diversions and hydropower development.	Migration corridor survival affected by irrigation diversions and hydropower development.
Original BRT conclusion	Not presently in danger of extinction. No decision whether it is likely to become so.	Presently in danger of extinction.	Not presently in danger of extinction. Likely to become so.
ESA status as of July 1999	Listed as threatened.	Listed as endangered.	Listed as threatened.

steelhead but includes several populations of summer steelhead. The steelhead in this ESU generally smolt at age 2 years, whereas most steelhead in British Columbia smolt at age 3. Steelhead from this area are genetically distinct from those in other areas of Washington, both chromosomally and electrophoretically. Habitat in the Puget Sound region is dominated by glacial effects, including extensive alluvial floodplains, and the fjord-like structure of Puget Sound may promote distinctive steelhead migration patterns. Recent population trends within the Puget Sound ESU are predominantly downward; however, trends in the two largest stocks (Skagit and Snohomish rivers) have been upward. The BRT was concerned about the large proportion of hatchery steelhead in Puget Sound and their origination primarily from a single stock; however, most hatchery fish appear to have earlier run timing and to be harvested prior to spawning, thus limiting their interactions with naturally spawning steelhead. Another concern of the BRT was the lack of information on the abundance and status of summer steelhead in this ESU.

2. *Olympic Peninsula.*—This ESU occupies river basins of the Olympic Peninsula, Washington, west of the Elwha River and south to, but not including, the rivers that flow into Grays Harbor on the Washington coast. The Olympic Peninsula ESU is primarily composed of winter steelhead but includes several populations of summer steelhead in the larger rivers. Olympic Peninsula steelhead are genetically distinct from other steelhead ESUs; this isolation is also supported by zoogeographic patterns of other species of fish and amphibians, indicating a faunal shift in the vicinity of the Chehalis River Basin. Population trends within this ESU are generally upward, but some stocks are declining. As was the case with the Puget Sound ESU, there is very little information regarding the abundance and status of summer steelhead in this region, and there is also uncertainty regarding the degree of interaction between hatchery and natural stocks.

3. *Southwest Washington.*—This ESU occupies the tributaries to Grays Harbor, Willapa Bay, and the Columbia River below the Cowlitz River in Washington and below the Willamette River in Oregon. This ESU is primarily composed of winter steelhead but includes summer steelhead in the Humptulips and Chehalis River basins. Genetic data show differentiation between steelhead of this ESU and those of adjacent regions. The ecological connectivity of the region occupied by the Southwest Washington ESU is demonstrated by similarities in riverine and estuarine ichthyofauna and current-driven sediment transfer from the Columbia River to Grays Harbor and Willapa Bay. Most population trends within this ESU have been declining in the recent past. There is very little information regarding the abundance and status of summer steelhead in this region, and there is also uncertainty regarding the degree of interaction between hatchery and natural stocks.

4. *Lower Columbia River.*—This ESU occupies tributaries to the Columbia River between the Cowlitz and Wind rivers in Washington and the Willamette and Hood rivers in Oregon, inclusive. Excluded are steelhead in the upper Willamette River Basin above Willamette Falls (see ESU 5-Upper Willamette River), and steelhead from the Little and Big White Salmon rivers, Washington (see ESU 13-Middle Columbia River ESU). This ESU is composed of both winter and summer steelhead. Genetic data show distinction between steelhead of this ESU and adjacent regions, with a particularly strong difference between coastal and inland steelhead in the vicinity of the Cascade Crest. The majority of stocks for which we have data within this ESU have been declining in the recent past, but some have been increasing strongly. However, the strongest upward trends are either non-native stocks (Lower Willamette River and Clackamas River summer steelhead) or stocks that are recovering from major habitat disruption and are still at low abundance (mainstem and North Fork Toutle River). The data series for most stocks is quite short, so the preponderance of downward trends may reflect the recent general coastwide decline in steelhead.

5. *Upper Willamette River.*—This ESU occupies the Willamette River and its tributaries upstream from Willamette Falls. The native steelhead of this basin are late-migrating winter steelhead, entering fresh water primarily in March and April. This unusual run timing appears to be an adaptation for ascending Willamette Falls, which functions as an isolating mechanism for upper Willamette River steelhead. Early migrating winter steelhead and summer steelhead have been introduced to the Upper Willamette River Basin; however, these non-native populations are not components of this ESU. Native winter steelhead within this ESU have been declining on average since 1971 and have exhibited large fluctuations in abundance. The main production of native (late-run) winter steelhead is in the North Fork Santiam River, where estimates of hatchery proportion in natural spawning range from 14 to 54%.

6. *Oregon Coast.*—This ESU occupies river basins on the Oregon coast north of Cape Blanco; excluded are rivers and streams that are tributaries of the Columbia River (see ESU 3-Southwest Washington). Native Oregon Coast steelhead are primarily winter steelhead; native summer steelhead occur only in the Siletz and Umpqua River basins. Recent genetic data for steelhead in this ESU show population differences from those in Washington, the Columbia River Basin, and coastal areas south of Cape Blanco. Ocean migration patterns also suggest a distinction between steelhead populations north and south of Cape Blanco. Steelhead, as well as chinook *O. tshawytscha* and coho *O. kisutch* salmon, from streams south of Cape Blanco tend to be south-migrating rather than north-migrating.

Most steelhead populations within this ESU have been declining in the recent past, with increasing trends restricted to the southernmost portion (south of Siuslaw Bay). There is widespread production of hatchery steelhead within this ESU, largely based on out-of-basin stocks, and approximately half of the streams (including the majority of those with upward trends) are estimated to have more than 50% hatchery fish in natural spawning escapements. Given the substantial contribution of hatchery fish to natural spawning throughout the ESU and the generally declining or slightly increasing trends, it is likely that natural stocks throughout the ESU are not replacing themselves.

7. *Klamath Mountains Province.*—This ESU occupies river basins from the Elk River in Oregon to the Klamath and Trinity rivers in California, inclusive. This ESU includes both winter and summer steelhead. Steelhead from this region are genetically distinct from populations to the north and south. The "half-pounder" life history* is reported only from this region. The Klamath Mountains Province is a unique geographical area with unusual geology and plant communities. Although absolute abundance of steelhead within the ESU remains fairly high, since about 1970 trends in abundance have been downward in most steelhead populations for which data are available, and a number of populations are considered by various agencies and groups to be at some risk of extinction. Declines in summer steelhead populations are of particular concern. This ESU was previously studied under a separate status review that was completed in December 1994 (Busby et al. 1994).

8. *Northern California.*—This ESU occupies river basins from Redwood Creek in Humboldt County, California south to the Gualala River, inclusive, and includes winter and summer steelhead. Allozyme and mitochondrial DNA (mtDNA) data indicate genetic discontinuities between steelhead of this region and those to the north and south. Freshwater fish species assemblages in this region are derived from the Sacramento River

---

* The *half-pounder* (Snyder 1925) is a life history trait of steelhead that is found only in the rivers of southern Oregon and northern California. Following smoltification, half-pounders spend only 2-4 months in the ocean, then return to fresh water. They overwinter in fresh water and emigrate to salt water again the following spring. This is often termed a false spawning migration, as few half-pounders are sexually mature.

Basin, whereas streams to the north include fishes representative of the Klamath-Rogue ichthyofaunal province. Population abundances are very low relative to historical estimates, and recent trends are downward in stocks for which we have data, except for two small summer steelhead stocks. Summer steelhead abundance is very low. Risk factors identified for this ESU include freshwater habitat deterioration due to sedimentation and flooding related to land management practices and introduced Sacramento squawfish *Ptychocheilus grandis* as a predator in the Eel River. For certain rivers (particularly the Mad River), the BRT was concerned about the influence of hatchery stocks, both in terms of genetic introgression and potential ecological interactions between introduced stocks and native stocks.

9. *Central California Coast.*—This ESU occupies river basins from the Russian River to Soquel Creek, Santa Cruz County (inclusive) and the drainages of San Francisco and San Pablo bays; excluded is the Sacramento-San Joaquin River Basin of the Central Valley of California (see ESU 12-Central Valley ESU). Allozyme and mtDNA data indicate genetic differences between the steelhead from this region and those from adjacent areas. Environmental features (e.g., precipitation patterns, vegetation, and soils) show a transition in this region from the northern redwood forest ecosystem to the more xeric southern chaparral and coastal scrub ecosystems. Steelhead populations within the major streams occupied by this ESU appear to be greatly reduced from historical levels; for example, steelhead abundance in the Russian River has been reduced roughly sevenfold since the mid-1960s, but abundance in smaller streams appears to be stable at low levels. The primary risk factor for this ESU is deteriorated habitat due to sedimentation and flooding related to land management practices. Uncertainty regarding the genetic heritage of the natural populations in tributaries to San Francisco and San Pablo bays makes it difficult to determine which of these populations should be considered part of the ESU.

10. *South-Central California Coast.*—This ESU occupies rivers from the Pajaro River, Santa Cruz County to (but not including) the Santa Maria River. Data from mtDNA provide evidence for a genetic transition in the vicinity of Monterey Bay. Both mtDNA and allozyme data show large genetic differences among populations in this area but do not provide a clear picture of population structure. The climate in this region is drier and warmer than it is to the north, resulting in chaparral and coastal scrub vegetation and stream mouths that are closed seasonally by sand berms. In addition to vegetation transitions, the northern end of this region is the southern limit of coho salmon distribution. The southern boundary of this ESU is near Point Conception, a well-recognized transition area for the distribution and abundance of marine flora and fauna. Total abundance of steelhead in this ESU is extremely low and declining. Risk factors for this ESU are habitat deterioration due to sedimentation and flooding related to land management practices and potential genetic interaction with hatchery rainbow trout.

11. *Southern California.*—This ESU occupies rivers from the Santa Maria River to the southern extent of the species range. Steelhead occur at least as far south as Malibu Creek, Los Angeles County, and may have historically occurred as far south as the U.S.-Mexico border. Genetic data show large differences among steelhead populations within this ESU as well as between these and populations to the north. Average rainfall is substantially lower and more variable in southern California than in regions to the north, resulting in increased duration of sand berms across the mouths of streams and rivers and, in some cases, complete dewatering of the lower reaches of these streams from late spring through fall. This affects steelhead migration patterns, as well as apparent adaptations to residualize and survive elevated water temperatures. Steelhead have already been extirpated from much of their historical range in this region. The BRT had a strong concern about the widespread degradation, destruction, and blockage of freshwater habitats

within the region, and the potential results of continuing habitat destruction and water allocation problems. There was also concern about the genetic effects of widespread stocking of rainbow trout.

12. *Central Valley.*—This ESU occupies the Sacramento and San Joaquin rivers and their tributaries. Recent allozyme data show that samples of steelhead from Deer and Mill creeks and Coleman National Fish Hatchery on the Sacramento River are well differentiated from all other samples of steelhead from California. The Sacramento and San Joaquin rivers offer the only migration route for anadromous fish to drainages of the Sierra Nevada and southern Cascade mountain ranges. The distance from the ocean to spawning streams can exceed 300 km, providing unique potential for reproductive isolation among steelhead in California. Steelhead have already been extirpated from most of their historical range in this region. Habitat concerns in this ESU focus on the widespread degradation, destruction, and blockage of freshwater habitats within the region, and the potential results of continuing habitat destruction and water allocation problems. The BRT also had a strong concern about the pervasive opportunity for genetic introgression from hatchery stocks within the ESU, and a strong concern for potential ecological interactions between introduced stocks and native stocks.

Inland Steelhead ESUs

13. *Middle Columbia River.*—This ESU occupies the Columbia River Basin upstream of the Wind River in Washington and the Hood River in Oregon to, and including, the Yakima River, Washington. Snake River Basin steelhead are not included (see ESU 15-Snake River Basin). This ESU includes the only winter-run populations of inland steelhead in the U.S., in the Klickitat River and Fifteenmile Creek. Some uncertainty exists about the exact boundary between coastal and inland steelhead, and the western margin of this ESU reflects currently available genetic data. There is good genetic and meristic evidence to separate this ESU from steelhead of the Snake River Basin. The boundary upstream of the Yakima River is based on limited genetic information and environmental differences including physiographic regions, climate, topography, and vegetation. All BRT members expressed special concern for the status of this ESU, particularly Yakima River and winter steelhead stocks. Total steelhead abundance in the ESU appears to have been increasing recently, but the majority of natural stocks for which we have data within this ESU have been declining, including those in the John Day River, which is the largest producer of wild, natural steelhead. There is widespread production of hatchery steelhead within this ESU. Habitat degradation due to grazing and water diversions has been documented throughout the range of the ESU.

14. *Upper Columbia River.*—This ESU occupies the Columbia River Basin upstream from the Yakima River. All upper Columbia River steelhead are summer steelhead. The streams of this region utilized by steelhead drain primarily the northern Cascade Mountains of Washington State. Streamflow is supplied by snowmelt, groundwater, and glacial runoff, often resulting in extremely cold water temperatures that retard the growth and maturation of steelhead juveniles, causing some of the oldest smolt ages reported for steelhead as well as residualization of juvenile steelhead that fail to smolt (McMichael et al. 2000). All anadromous fish in this region were affected by the Grand Coulee Fish Maintenance Project (1939 through 1943), wherein anadromous fish returning to spawn in the upper Columbia River were trapped at Rock Island Dam, downstream of the Wenatchee River. Some of these fish were then released to spawn in river basins above Rock Island Dam, while others were spawned in hatcheries and the offspring were released into various upper Columbia River tributaries; in both cases, no attempt was made to return these fish to their natal streams, resulting in an undetermined level of stock mixing within the

upper Columbia River fish. While total abundance of populations within this ESU has been relatively stable or increasing, this appears to be true only because of major hatchery supplementation programs. Estimates of the proportion of hatchery fish in spawning escapements are 65% for the Wenatchee River and 81% for the Methow and Okanogan Rivers. The major concern for this ESU is the clear failure of natural stocks to replace themselves. The BRT also had a strong concern about problems of genetic homogenization due to hatchery supplementation within the ESU. There was also concern about the apparent high harvest rates on steelhead smolts in rainbow trout fisheries and the degradation of freshwater habitats within the region, especially the effects of grazing, irrigation diversions, and hydroelectric dams.

15. *Snake River Basin.*—This ESU occupies the Snake River Basin of southeastern Washington, northeastern Oregon, and Idaho. This region is ecologically complex and supports a diversity of steelhead populations; however, genetic and meristic data suggest that these populations are more similar to each other than they are to steelhead populations occurring outside of the Snake River Basin. Snake River Basin steelhead spawning areas are well isolated from other populations and include the highest elevations for spawning (up to 2,000 m) as well as the longest migration distance from the ocean (up to 1,500 km) (Thurow et al. 2000).

Snake River steelhead are often classified into two groups, A- and B-run, based on migration timing, ocean age, and adult size. The A-run enters fresh water from June to August and passes Bonneville Dam before 25 August; the B-run enters fresh water from late August to October, passing Bonneville Dam after 25 August. A-run steelhead are thought to be predominately age-1-ocean, while B-run steelhead are defined as age-2-ocean. Adult B-run steelhead are also thought to be on average 75–100 mm larger than A-run steelhead of the same age; this is attributed to their longer average residence in salt water (see Busby et al. 1996 for more discussion).

While total (hatchery + natural) run size for Snake River steelhead has increased since the mid-1970s, the increase has resulted from increased production of hatchery fish, and there has been a severe recent decline in natural run size. Most natural stocks for which we have data within this ESU have been declining. Parr densities in natural production areas have been substantially below estimated capacity in recent years. Downward trends and low parr densities indicate a particularly severe problem for B-run steelhead, the loss of which would substantially reduce life history diversity within this ESU. The BRT had a strong concern about the pervasive opportunity for genetic introgression from hatchery stocks within the ESU. There was also concern about the degradation of freshwater habitats within the region, especially the effects of grazing, irrigation diversions, and hydroelectric dams.

CONCLUSIONS

Based largely on the recommendations of the BRT for west coast steelhead, NMFS proposed listing several steelhead ESUs under the Endangered Species Act (Federal Register 61 [9 August 1996]:41541). After public review and comment, receipt of additional information, and considering proposed and enacted conservation measures, NMFS has made listing determinations for seven ESUs. Two ESUs (Southern California and Upper Columbia River) have been listed as endangered (Federal Register 62 [18 August 1997]:43937). Seven ESUs (Lower Columbia River, Upper Willamette River, Central California Coast, South-Central California Coast, Central Valley, Middle Columbia River, and Snake River Basin) have been listed as threatened (Federal Register 62 [18 August 1997]:43937, Federal Register 63 [19 March 1998]:13347, and Federal Register 64 [25 March 1999]:14517). Additionally, three ESUs (Oregon Coast, Klamath Mountains Province, and Northern California) have been designated candidate species (Federal Register 63 [19 March 1998]:13347), and their status will be reevaluated within four years of this designation to determine whether listing is warranted.

The stated goal of the Endangered Species Act is to conserve endangered and threatened species and the ecosystems on which they depend. The listing of a species constitutes recognition that current management of the species and/or its ecosystem are not consistent with sustaining the species. As with other Pacific salmonids, steelhead populations have declined across the species' range and local extirpations of some runs have occurred. Many factors were identified during the status review process as driving the decline of steelhead. These include lost access to habitat due to dams and habitat degradation caused by land management practices. Anthropogenic factors may have synergistic effects superimposed on natural variation in climate and ocean conditions that collectively could jeopardize the sustainability of steelhead populations.

Fortunately, steelhead have a remarkable degree of natural plasticity in their life history strategies. Their ability to residualize may have allowed populations to persist upstream of barriers, and their iteroparous capability allows at least some members of a population multiple opportunities at spawning, enhancing the likelihood of passing on advantageous inherited characteristics. Also, their variability in run timing allows adaptation to changing hydrologic regimes. Clearly, if any species of salmonid has inherent sustainability, it is the steelhead. This is at once reassuring in terms of optimism for restoring steelhead populations, and disturbing in terms of the hope we can hold out for less resilient species.

ACKNOWLEDGMENTS

The west coast steelhead status review represents the combined efforts of dozens of people who submitted information on steelhead directly to NMFS, attended Biological and Technical Committee meetings in Washington, Idaho, Oregon, and California, and answered seemingly endless telephone questions from the authors and BRT members. The authors particularly wish to acknowledge Stevan Phelps and Steve Leider of the Washington Department of Fish and Wildlife for generous sharing of newly emerging genetic data, and Jennifer Nielsen of Hopkins Marine Laboratory for sharing her DNA studies in progress. Significant contributions in the compilation and analyses of data were made by L. Lierheimer, K. Neely, T. Parker, D. Teel, R. Waples, and F. Waknitz, all from NMFS Northwest Fisheries Science Center, and I. Lagomarsino of NMFS Southwest Region.

The biological review team for this status review included P. Busby, S. Grabowski, R. Iwamoto, C. Mahnken, G. Matthews, M. Schiewe, T. Wainwright, R. Waples, and J. Williams, from NMFS Northwest Fisheries Science Center; G. Bryant and C. Wingert from NMFS Southwest Region; and R. Reisenbichler from the U.S. Geological Survey, Biological Resources Division, Seattle.

REFERENCES

Behnke, R. J. 1992. Native trout of western North America. American Fisheries Society Monograph 6. Bethesda, Maryland.

Burgner, R. L., J. T. Light, L. Margolis, T. Okazaki, A. Tautz, and S. Ito. 1992. Distribution and origins of steelhead trout (*Oncorhynchus mykiss*) in offshore waters of the North Pacific Ocean. International North Pacific Fisheries Commission Bulletin 51. Vancouver, British Columbia.

Busby, P. J., T. C. Wainwright, and R. S. Waples. 1994. Status review for Klamath Mountains Province steelhead. U.S. Department of Commerce, NOAA Technical Memorandum NMFS-NWFSC-19. Seattle, Washington.

Busby, P. J., and six coauthors. 1996. Status review of west coast steelhead from Washington, Idaho, Oregon, and California. U.S. Department of Commerce, NOAA Technical Memorandum NMFS-NWFSC-27. Seattle, Washington.

McMichael, G. A., T. N. Pearsons, and S. A. Leider. 2000. Minimizing ecological impacts of hatchery reared juvenile steelhead trout on wild salmonids in a Yakima Basin watershed. Pages 365–380 *in* E. E. Knudsen, C. R. Steward, D. D. MacDonald, J. E. Williams, and D. W. Reiser, editors. Sustainable fisheries management: Pacific salmon. Lewis Publishers, Boca Raton, Florida.

ONRC (Oregon Natural Resources Council), and 15 coauthors. 1994. Petition for a rule to list steelhead trout as threatened or endangered under the Endangered Species Act and to designate critical habitat. Unpublished document submitted to the USDOC NOAA NMFS Northwest Region, Seattle, Washington, February 1994 (available from Oregon Natural Resources Council, 522 S.W. 5th, Number 1050, Portland, OR 97204).

Snyder, J. O. 1925. The half-pounder of Eel River, a steelhead trout. Calif. Fish Game 11(2):49–55.

Thurow, R. F., D. C. Lee, and B. E. Rieman. 2000. Status and distribution of chinook salmon and steelhead in the interior Columbia River basin and portions of the Klamath River basin. Pages 133–160 *in* E. E. Knudsen, C. R. Steward, D. D. MacDonald, J. E. Williams, and D. W. Reiser, editors. Sustainable fisheries management: Pacific salmon. Lewis Publishers, Boca Raton, Florida.

Waples, R. S. 1991. Pacific salmon, *Oncorhynchus* spp., and the definition of "species" under the Endangered Species Act. Marine Fisheries Review 53(3):11–22.

12 Status and Distribution of Chinook Salmon and Steelhead in the Interior Columbia River Basin and Portions of the Klamath River Basin

Russell F. Thurow, Danny C. Lee, and Bruce E. Rieman

Abstract.—This chapter summarizes information on presence, absence, current status, and probable historical distribution of steelhead *Oncorhynchus mykiss* and stream-type (age-1 migrant) and ocean type (age-0 migrant) chinook salmon *O. tshawytscha* in the interior Columbia River basin and portions of the Klamath River basin. Data were compiled from existing sources and via surveys completed by more than 150 biologists working in the region. We developed models to quantitatively explore relationships among fish status and distribution, the biophysical environment, and land management. Biophysical setting was an important determinant of species distributions and habitat suitability. We applied model results to predict fish presence in unsampled areas and mapped expected distributions in more than 3,700 subwatersheds. Chinook salmon and steelhead are extirpated from more than 50% of their potential historical ranges. Most remaining populations are severely depressed; less than 2% of the watersheds in the current range were classified as supporting strong populations of steelhead or stream-type chinook salmon. Wild, indigenous fish are rare; 22% of remaining steelhead stocks and less than 17% of chinook salmon stocks were judged to be genetically unaltered by hatchery-reared fish. Much of the historical production has been eliminated. However, a core for maintaining and rebuilding functional areas remains. Protection of core areas critical to stock persistence and restoration of a broader matrix of productive habitats will be necessary for productive and sustainable fisheries. This effort will require conservation and restoration of sufficient habitats to ensure the full expression of phenotypic and genotypic diversity in chinook salmon and steelhead.

INTRODUCTION

Historically, the Columbia River basin supported immense runs of anadromous salmonids. Diverse stocks included spring, summer, and fall chinook *Oncorhynchus tshawytscha*, sockeye *O. nerka*, coho *O. kisutch*, and chum salmon *O. keta*, and steelhead *O. mykiss*. Chapman (1986) estimated peak runs of Pacific salmon and steelhead in the Columbia River in the late 1800s were about 7.5 million fish. Estimates of annual chinook salmon returns prior to 1850 range from 3.4 to 6.4 million fish (NWPPC 1986). Commercial harvest of chinook salmon in the mainstem Columbia River peaked in 1883 at 2.3 million fish and was about 1.3 million fish annually from 1890–1920 (Mullan et al. 1992). Steelhead have been reported in the commercial Columbia River catch since 1889 and 2.23 million kg of canned steelhead were produced in 1892 (Fulton 1970). Estimates of steelhead runs were derived after Bonneville Dam was constructed, and, in 1940, 423,000 summer-run steelhead passed the dam (NWPPC 1986). Annual steelhead sport harvests averaged 117,000 summer-run and 62,000 winter-run fish from 1962 to 1966 (Fulton 1970). Many native people in the Columbia River basin shared a significant dependence on anadromous salmonids as a subsistence and ceremonial resource (NWPPC 1986; Mullan et al. 1992). Since European settlement, anadromous salmonids have continued to influence social and economic systems.

Many native stocks of Columbia River anadromous salmonids are now considered imperiled (Williams et al. 1989; Moyle and Williams 1990; Nehlsen et al. 1991; Frissell et al. 1993, 1995). Concern for the persistence of steelhead in the study area culminated in a final rule (Office of the Federal Register 62[August 18, 1997]:43937) listing one evolutionarily significant unit (ESU) as endangered (Upper Columbia River) and one as threatened (Snake River Basin) under the Endangered Species Act of 1973 (ESA). In 1998, a third stock (Lower Columbia River) was listed as threatened (Office of the Federal Register 63[March 19, 1998]:13347). Returns of wild steelhead to the uppermost Snake River dam have declined from more than 80,000 in the 1960s to an estimated 7,900 in 1995 (C. Petrosky, Idaho Department of Fish and Game, personal communication). Snake River spring, summer, and fall chinook and sockeye salmon are listed as threatened or endangered, and Columbia River coho salmon and upper Columbia River spring chinook salmon are candidate species under ESA. In the Snake River, an estimated 1,882 naturally produced stream-type chinook salmon reached Lower Granite Dam in 1994 (NMFS 1995) compared to an estimated production of 1.5 million fish in the late 1880s (Bevan et al. 1994). During 1985–1993 an average of 387 naturally produced ocean-type chinook salmon annually reached Lower Granite Dam (NMFS 1995). Redd counts in four tributaries to the Middle Fork Salmon River in central Idaho have declined from more than 2,000 redds in the 1960s (Hassemer 1993) to 11 in 1995.

Several recent status reviews of Columbia River basin anadromous salmonids exist (Howell et al. 1985a, 1985b; NWPPC 1986; CBFWA 1990; IDFG et al. 1990; WDF et al. 1993; Chapman et al. 1994a, 1994b; Kostow et al. 1994; Anderson et al. 1996; Busby et al. 1996). Different methods, lack of spatially explicit information, and focus on either declining stocks (Nehlsen et al. 1991) or healthy stocks (Huntington et al. 1996), have prevented a synthesis across the Columbia River basin. Frissell (1993) completed an extensive analysis of native fish extinctions within the Pacific Northwest, but provided little resolution below the scale of major river subbasins.

This chapter summarizes information collected by the Aquatic Science Team of the Interior Columbia River Basin Ecosystem Management Project (ICBEMP) (Lee et al. 1997). In response to the President's directive, the Chief of the Forest Service and the Director of the Bureau of Land Management (BLM) established the ICBEMP, which includes a scientific assessment of ecological, social, cultural, and economic systems; two environmental impact statements; and an evaluation of impact statement alternatives. One goal of the assessment was a consistent evaluation of the status and distribution of fishes throughout the interior Columbia River basin. Here we describe the potential historical range and current status and distribution of chinook salmon and steelhead. We consider the factors that have influenced status and distribution, and opportunities for conservation and restoration.

STUDY AREA

The complete study area included all U.S. waters entering the Columbia River basin east of the Cascade Mountains and those portions of the Klamath River basin and Great Basin in Oregon (Figure 12.1). A hierarchical system of subbasins, watersheds, and subwatersheds was defined by topography. Within the 58.3×10^6 ha study area, 164 large subbasins were defined (Figure 12.2). The subbasins were further divided into watersheds which average about 22,820 ha in surface area. The watersheds were ultimately divided into 7,498 subwatersheds averaging 7,800 ha. The hydrologic divisions follow the hierarchical framework of aquatic ecological units described by Maxwell et al. (1995). The delineations and map coverages were provided by the Columbia Basin Project

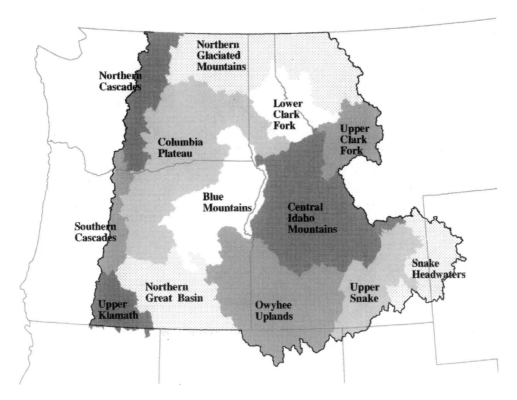

FIGURE 12.1   The interior Columbia River basin in the U.S. and portions of the Klamath River basin and Great Basin represented in this study. Ecological reporting units (ERUs) used to summarize information across broad regions of similar biophysical characteristics are also shown.

(Quigley and Arbelbide 1997). We used the subwatersheds as our basic unit of summary and prediction. Because anadromous fish were not native to the Great Basin those subwatersheds were excluded from further analysis.

   We also considered 13 larger landscape classifications defined as ecological reporting units (ERUs) (Figure 12.1) representing distinct land areas of broadly homogeneous biophysical characteristics (Jensen et al. 1997). Chinook salmon and steelhead are distributed over a broad geographic range and have likely evolved under a broad range of environmental conditions. If gene flow has been limited by distance, there is potential for local and unique adaptations to those environments (Lesica and Allendorf 1995). We summarized our results by ERU to consider potentially important differences in the representation of distinct populations and environments.

METHODS

Steelhead, the anadromous form of rainbow/redband trout, are distributed within the study area as two genetically distinct subspecies, coastal (*O. m. irideus*) and inland (*O. m. gairdneri*) (Utter et al. 1980). Coastal steelhead are found primarily below Bonneville Dam in the lower Columbia River and inland steelhead occur primarily from the Deschutes River, Oregon upstream. The two subspecies overlap near the crest of the Cascade Range (K. Kostow, Oregon Department of Fish and Wildlife, personal communication). Coastal steelhead are predominately winter-run and inland steelhead summer-run fish. Within the study area, winter steelhead are found in two tributaries to the lower Columbia River in Oregon, 15-mile Creek and Hood River (Kostow et al. 1994). We summarized information for summer steelhead only.

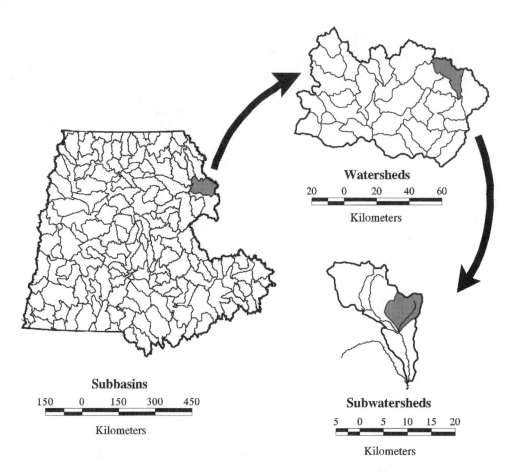

FIGURE 12.2 Hydrologic divisions of subbasins, watersheds, and subwatersheds used to characterize land-scape information and classify the status and distribution of salmon and steelhead across the interior Columbia River basin in the U.S. and portions of the Klamath River basin and Great Basin. The hierarchical framework of drainage boundaries was defined from topography as described by Maxwell et al. 1995. Detailed descriptions and derivations are available in Quigley and Arbelbide (1997).

Columbia River basin chinook salmon have traditionally been described as spring, summer, and fall races—separated primarily by their time of passage over Bonneville Dam (Matthews and Waples 1991). Spring chinook salmon cross Bonneville Dam from March to May, summers from June to July, and falls from August to September (Burner 1951). This nomenclature has led to some confusion across the study area because stocks of similar run timing may differ considerably between the Snake and Columbia rivers in their spawning areas, life histories, behavior, and genetic characteristics. Healey (1991) categorized juvenile chinook salmon that migrate seaward after one or more years as stream-type and those that migrate as subyearlings as ocean-type. We adopted these definitions to characterize chinook salmon stocks in the study area. Within the Snake River basin and tributaries to the Columbia River downstream from the Snake River, stream-type chinook salmon include spring- and summer-run fish and ocean-type chinook salmon include fall-run fish (Parkhurst 1950; Gebhards 1959; Fulton 1968; IDFG 1992). Within the Columbia River upstream from its confluence with the Snake River stream-type chinook salmon include spring-run fish and ocean-type chinook salmon include summer- and fall-run fish (Matthews and Waples 1991; Mullan et al. 1992).

Known Status and Distribution

Status was summarized by more than 150 private, agency, and tribal fish biologists working across the study area (Lee et al. 1997). Through a series of workshops in 1995, biologists were asked to characterize the status of naturally reproducing populations in each subwatershed within their jurisdiction. If populations were supported solely by hatchery-reared fish, naturally spawning fish were considered absent. Biologists classified subwatersheds where fish were present as either spawning and rearing habitat, overwintering and migratory-corridor habitat, or as supporting populations of unknown status. Subwatersheds containing spawning and rearing habitat were further classified as strong or depressed, based on population characteristics. We asked biologists to rely on target species' biological characteristics and not to infer status from habitat or landscape information or presence of introduced fishes.

The following criteria guided classification.

- *Present strong.*—Included those subwatersheds where 1) spawning and rearing occur; 2) all major life history forms that historically occurred are present; 3) numbers are stable or increasing, and the local population is likely to be at half or more of its historical size or density; and 4) the population within the subwatershed, or metapopulation within a larger region of which the subwatershed is a part, contains at least 5,000 individuals or 500 adults.
- *Present depressed.*—Included those subwatersheds where fish are present and use the area for spawning and rearing, but the population is not judged to be strong using the criteria listed above.
- *Absent.*—The species does not inhabit the subwatershed at any life history stage.
- *Present-unknown.*—The species is present, but there is no reliable information to determine current status.
- *Present, migration corridor.*—Migration corridors are habitats that do not support spawning, do not support initial rearing prior to juvenile migration, and function primarily as routes or staging and wintering areas for migrating fish. The mainstem migration routes of the Columbia River and major tributaries are examples of corridors. Some areas currently classified as migration corridors may have been important juvenile rearing areas prior to alteration by human activities (Lichatowich and Mobrand 1995).
- *Unknown.*—No information exists regarding the current presence or absence of the species.

To map areas of potential importance as wild, indigenous gene pools, we asked biologists to identify chinook salmon and steelhead spawning and rearing areas they judged to be unaltered by hatchery stocking, regardless of whether the populations were strong or depressed. Hatchery programs may erode genetic diversity and alter co-adapted gene complexes characteristic of locally adapted stocks (Waples and Do 1994; Reisenbichler 1997). The effects may include a loss of fitness or performance (such as growth, survival, and reproduction) and a loss of genetic variability important to long term stability and adaptation in varying environments. Because data describing genetic purity of populations was not available across the study area, we chose not to rely solely on genetic analysis. Instead, we defined wild, indigenous areas as subwatersheds with a low probability of non-indigenous strays spawning with indigenous fish *and* as

1. subwatersheds that had no history of hatchery-reared or non-indigenous introductions; or
2. subwatersheds that had been stocked rarely with hatchery-reared or non-indigenous fish, but where evidence suggested poor survival of stocked fish and a low probability of introgression; or
3. subwatersheds that had been stocked regularly in the past, but genetic analyses found existing wild fish identical to the original wild gene pool.

We submitted status classifications for review by others familiar with the area in question and attempted to use only the most current information. We summarized fish distribution information from existing state databases to validate and augment presence or absence classifications. We restricted our use to databases that had been created or updated after 1993. In general, biologists provided information in addition to and directly supporting that already available from the databases. When there was no survey response, no information in electronic databases, or conflicting responses between the survey and electronic databases that could not be resolved, unknown was the default classification.

Despite the criteria provided for classification, an element of subjectivity remains in the data (Rieman et al. 1997). Most of the information represented by the classifications is not published or peer-reviewed. Inconsistencies probably occur in classifications and the quality of available information may vary. Recognizing the potential for errors, we believe these data represent as complete a summary of the current (or recent), collective knowledge of these species as is possible. The resolution of our data may also produce more optimistic estimates of current distributions than work based on stream reaches. That is, if chinook salmon or steelhead occurred anywhere in a subwatershed they were considered present in the entire subwatershed. An analysis at the subwatershed level of resolution was necessary simply because of logistical and computational constraints imposed by the scale of our study. We believe, however, because of the potential for extended movement and dispersal and the structuring of populations within tributaries, subwatersheds are actually the most appropriate level for our analysis. We believe that subwatersheds will best approximate the distributions of potentially discrete groups or local populations and thus represent a better summary unit than stream reaches. To minimize any potential bias all estimates of the distributions were based on the number of subwatersheds and not area.

POTENTIAL HISTORICAL RANGE

Potential historical ranges, hereafter referred to as potential ranges, were defined as the likely distributions in the study area prior to European settlement. Potential ranges were characterized from historical distributions in prior databases and augmented through published and anecdotal accounts. In some cases, the potential ranges remain speculative; few records are available to ascertain the historical range of chinook salmon and steelhead in the Bruneau River subbasin, for example, because a 1906 dam blocked access prior to detailed population surveys (IDFG 1992). For a complete description of the data sources used to infer potential ranges for chinook salmon and steelhead, see Lee et al. (1997). We included all subwatersheds that were accessible as potential range based on the known current and historical occurrences because chinook salmon and steelhead are highly mobile, moving seasonally through subwatersheds, watersheds, subbasins, and basins at different life stages (Withler 1966; Healey 1991, 1994). Subwatersheds that were known to be historically isolated by barriers were excluded from potential ranges. We recognize that, within subwatersheds, the potential range may be further restricted by elevation, temperature, and local channel features but did not attempt to define potential ranges at a finer scale.

PREDICTIVE MODELS

Our knowledge of the current status of chinook salmon and steelhead was limited to areas that have been sampled. We attempted to generate the most complete picture of the current distribution of these fishes for unsampled areas by quantitatively exploring relationships among fish status and distribution, the biophysical environment, and land management. We produced a set of predictions that reflect the likelihood of a species' presence, or alternatively, the likely status of the population within a subwatershed. The predictions arose from statistical models, called classification trees (Breiman et al. 1984), which elucidate the relationship between a set of predictor variables and a single response variable. Tree-based models represent a nonparametric alternative to conventional linear models that have more constraining assumptions about data structure (see Breiman et al.

1984; Clark and Pregibon 1992; Crawford and Fung 1992; Taylor and Silverman 1993). For more information on the advantages of tree-based models, see Lee et al. (1997).

Classification trees consist of a dichotomous rule set that is generated through a process of recursive partitioning. Recursive partitioning involves sequentially splitting the data set into more homogeneous units, relative to the response variable, until a predefined measure of homogeneity is reached or no further subdivision is desired or feasible. Data are split at each juncture based on a single predictor variable that produces the greatest differences between the two resultant groups of observations. Predictor variables can be reused at subsequent splits. The objective of the classification algorithm is to derive a terminal set of nodes, each containing a subset of the original data, where the distribution of the response variable is independent of the predictor variables to the greatest extent possible. Details of the algorithm used to build the classification trees can be found in Clark and Pregibon (1992) and Statistical Sciences (1993).

All analyses involving classification trees were performed using the Splus^2 programming language, following the procedures outlined in Clark and Pregibon (1992). We used ERU as a variable and summarized 29 additional predictor variables (Table 12.1) from more than 200 coverages representing landscape characteristics across the study area. We limited our variables to those with potential influence on aquatic ecosystems and generally eliminated many of those that were strongly correlated with or directly derived from others (Lee et al. 1997). The variables were both categorical and continuous, representing vegetative communities, climate, geology, landform and erosive potential, and past management or the relative intensity of human disturbance (Table 12.1). Much of the information used in developing the landscape variables was developed from continuous 90-meter digital elevation data, 1-km vegetation data (from satellite imagery), 1:100,000 hydrography, state geologic maps, land ownership and activity maps, and extrapolated landscape characteristics derived from aerial photographs. Detailed descriptions of the complete landscape coverages, variables and their derivations can be found in Quigley and Arbelbide (1997) and Lee et al. (1997).

We present the classification tree for stream-type chinook salmon (Table 12.2) to illustrate the general procedure that we followed for stream-type and ocean-type chinook salmon and steelhead. Status of stream-type chinook salmon (*stc*), was used as the response. Four possible values indicated whether the species was absent (A), present in spawning and rearing areas at depressed levels (D), present in spawning and rearing areas at strong levels (S), or transient in migration corridors (T). The tree-fitting routine produced a full model that used 22 of the 29 potential predictor variables. The model had 64 terminal nodes and many of the nodes near the termini of the full tree added little to change the predictions or reduce the deviance. The next step was to prune the tree by removing the least important nodes. Breiman et al. (1984) developed a process of cross-validation which uses the full set of observations to build the full model, and mutually exclusive subsets of the data to iteratively prune and test the tree. For stream-type chinook salmon, a tree with 26 nodes was selected (Table 12.2). A smaller tree may predict responses equally well, but we chose to keep the additional nodes for the insight they provide.

Part A of Table 12.2 lists the variables used in the reduced tree and summarizes overall performance. Part B of Table 12.2 provides a complete listing of the tree structure and identifies each node, the variables and ranges defining each split, the number of observations, the deviance (a measure of within-node heterogeneity), the modal response (most frequent response), and the relative frequencies of each response level at that node. When used in a predictive fashion, these frequencies are equivalent to a probability of a given response. Normally, the modal response is equivalent to the predicted value. With the status data, however, three of the responses indicate different levels of presence (D, S, or T) in contrast to a single absent value (A). Thus, we chose to predict absence only when the relative frequency was greater than or equal to 0.5. The probability of presence was defined as one minus the probability of absence.

Classification trees were used to estimate the probability of chinook salmon or steelhead presence or absence in subwatersheds classified as unknown and to predict status in subwatersheds classified as unknown or in subwatersheds with fish of unknown status. All estimates and predictions were limited

TABLE 12.1
Descriptions of landscape variables used to build classification trees in the analysis of chinook salmon and steelhead distribution and status. All values for physiographic, geophysical, and vegetation variables were summarized for subwatersheds from complete coverages with a pixel resolution of 1 km². All values expressed as percents refer to the percent area of the subwatershed. Detailed descriptions and derivations are described in Quigley and Arbelbide (1997).

Variable Name	Description
ERU	Ecological Reporting Unit (see text and Figure 12.2)

Physiographic and Geophysical

Variable Name	Description
slope	area weighted average midslope based on 90 m digital elevation maps
slope2	percent of area in slope class 2 (slopes >10%, <30%)
con1	percent weakly-consolidated lithologies
con2	percent moderately-consolidated lithologies
con3	percent strongly-consolidated lithologies
sdt1	percent lithologies that produce coarse-textured sediments
sdt2	percent lithologies that produce medium-textured sediments
sdt3	percent lithologies that produce fine-textured sediments
alsi1	percent felsic lithologies
alsi2	percent intermediate aluminosilicated lithologies
alsi3	percent mafic lithologies
alsi4	percent carbonate lithologies
pprecip	mean annual precipitation extrapolated to complete coverage using a regional climate simulation model
elev	mean elevation
mtemp	mean annual air temperature extrapolated to complete coverage using a regional climate simulation model
solar	mean annual solar radiation based on topographic shading, latitude, and aspect
streams	total length of streams within subwatershed based on 1:100,000 scale hydrography
drnden	drainage density (stream length/watershed area)
anadac	access for anadromous fish (0 = no, 1 = yes)
dampass	number of intervening dams
hucorder	number of upstream subwatersheds tributary to the watershed of interest (0 designates a headwater subwatershed)
hk	soil texture coefficient
baseero	base erosion index representing relative surface erodability without vegetation
ero	surface erosion hazard, a derived variable based on *baseero* and other modifying factors
bank	streambank erosion hazard, a derived variable representing relative bank stabilities

Vegetation

Variable Name	Description
vmf	vegetation amelioration factor, a derived variable representing the relative amount of ground cover in vegetation
vegclus	current vegetation class (12 possible types) based on structure and composition

Ownership and Management

Variable Name	Description
roaddn	estimated road density class (5 classes: none to very high)
mngclus	management class (10 possible types, intense commodity extraction to wilderness) based on ownership and management emphasis

TABLE 12.2
The pruned classification tree for stream-type chinook salmon status (absent (A), depressed (D), transient (T), strong (S)) showing discriminating variables, sample sizes, splitting criteria, and frequency distributions within spawning and rearing areas in 1,262 subwatersheds used to develop the model. Nodes and accompanying data are hierarchical and represent the structure of a tree. The root node represents the complete distribution. The first split occurs at hucorder (number of upstream watersheds tributary to the watershed of interest), node 2 is <30 compared to node 3 that is >30. These two nodes are further independently subdivided. See text and Table 12.1 for a description of other variables and Lee et al. (1997) for further information.

A. Summary

Variables actually used in tree construction:					
hucorder	pprecip	dampass	streams	mtemp	eru
mngclus	alsi1	solar	alsi3	con3	

Number of terminal nodes: 26
Residual mean deviance: 0.8282   Misclassification error rate: $0.1672 = 211/1262$

B. Tree structure

Node) Split criterion	Sample size	Deviance	Modal response	Relative frequencies			
				Absent	Depressed	Transient	Strong
1) root	1262	2718.00	A	0.43900	0.36450	0.190200	0.006339
2) hucorder < 30	991	1652.00	A	0.54990	0.41370	0.028250	0.008073
4) hucorder = 0	500	622.40	A	0.75800	0.22200	0.020000	0.000000
8) pprecip < 601	145	93.04	A	0.92410	0.03448	0.041380	0.000000 *
9) pprecip > 601	355	473.80	A	0.69010	0.29860	0.011270	0.000000
18) dampass < 4	23	13.59	D	0.08696	0.91300	0.000000	0.000000 *
19) dampass > 3	332	418.60	A	0.73190	0.25600	0.012050	0.000000
38) streams < 23	145	112.40	A	0.88280	0.11030	0.006897	0.000000 *
39) streams > 23	187	274.20	A	0.61500	0.36900	0.016040	0.000000
78) mtemp < 6.9	158	242.70	A	0.54430	0.43670	0.018990	0.000000
156) streams < 43.6	122	177.00	A	0.63110	0.35250	0.016390	0.000000
312) pprecip < 960.7	52	50.68	A	0.84620	0.13460	0.019230	0.000000 *
313) pprecip > 960.7	70	106.00	D	0.47140	0.51430	0.014290	0.000000 *
157) streams > 43.6	36	49.04	D	0.25000	0.72220	0.027780	0.000000
314) streams < 64.1	31	32.40	D	0.12900	0.83870	0.032260	0.000000 *
315) streams > 64.1	5	0.00	A	1.00000	0.00000	0.000000	0.000000 *
79) mtemp > 6.9	29	0.00	A	1.00000	0.00000	0.000000	0.000000 *
5) hucorder > 0	491	841.50	D	0.33810	0.60900	0.036660	0.016290
10) eru: 5,7	116	117.80	A	0.81900	0.17240	0.008621	0.000000
20) mngclus: BR,PA,PF,PR	106	85.95	A	0.87740	0.11320	0.009434	0.000000
40) alsi1 < 1.56	77	18.55	A	0.97400	0.02597	0.000000	0.000000 *
41) alsi1 > 1.56	29	45.20	A	0.62070	0.34480	0.034480	0.000000 *
21) mngclus: FG,FH,FM,TL	10	10.01	D	0.20000	0.80000	0.000000	0.000000 *
11) eru: 1,2,6,13	375	568.10	D	0.18930	0.74400	0.045330	0.021330
22) mngclus:BR,PA,PR	68	129.60	A	0.45590	0.44120	0.102900	0.000000
44) mtemp < 10.54	62	94.79	A	0.50000	0.48390	0.016130	0.000000 *
45) mtemp > 10.54	6	0.00	T	0.00000	0.00000	1.000000	0.000000 *

TABLE 12.2 (continued)
The pruned classification tree for stream-type chinook salmon status (absent (A), depressed (D), transient (T), strong (S)) showing discriminating variables, sample sizes, splitting criteria, and frequency distributions within spawning and rearing areas in 1,262 subwatersheds used to develop the model. Nodes and accompanying data are hierarchical and represent the structure of a tree. The root node represents the complete distribution. The first split occurs at hucorder (number of upstream watersheds tributary to the watershed of interest), node 2 is <30 compared to node 3 that is >30. These two nodes are further independently subdivided. See text and Table 12.1 for a description of other variables and Lee et al. (1997) for further information.

B. Tree structure (continued)

Node) Split criterion	Sample size	Deviance	Modal response	Relative frequencies			
				Absent	Depressed	Transient	Strong
23) mngclus: FG,FH,FM,FW,PF,TL	307	394.20	D	0.13030	0.81110	0.032570	0.026060
46) dampass < 4	56	82.10	D	0.10710	0.75000	0.000000	0.142900
92) solar< 330.6	30	8.77	D	0.03333	0.96670	0.000000	0.000000 *
93) solar > 330.6	26	53.37	D	0.19230	0.50000	0.000000	0.307700
186) pprecip < 535.2	10	13.86	A	0.50000	0.50000	0.000000	0.000000 *
187) pprecip > 535.2	16	22.18	D	0.00000	0.50000	0.000000	0.500000 *
47) dampass > 4	251	280.20	D	0.13550	0.82470	0.039840	0.000000
94) hucorder <4	130	149.20	D	0.22310	0.76920	0.007692	0.000000 *
95) hucorder > 3	121	105.00	D	0.04132	0.88430	0.074380	0.000000
190) mngclus: FG,FM	51	0.00	D	0.00000	1.00000	0.000000	0.000000*
191) mngclus: FH,FW,PF	70	88.31	D	0.07143	0.80000	0.128600	0.000000*
3) hucorder > 30	271	334.40	T	0.03321	0.18450	0.782300	0.000000
6) hucorder < 58	74	138.00	T	0.09459	0.39190	0.513500	0.000000
12) eru: 5,7	10	12.22	A	0.70000	0.00000	0.300000	0.000000 *
13) eru: 1,6,13	64	88.16	T	0.00000	0.45310	0.546900	0.000000
26) alsi3 < 59.7	52	65.73	T	0.00000	0.32690	0.673100	0.000000
52) alsi1 < 0.37	9	6.28	D	0.00000	0.88890	0.111100	0.000000 *
53) alsi1 > 0.37	43	44.12	T	0.00000	0.20930	0.790700	0.000000 *
27) alsi3 > 59.7	12	0.00	D	0.00000	1.00000	0.000000	0.000000 *
7) hucorder > 57	197	155.60	T	0.01015	0.10660	0.883200	0.000000
14) dampass < 3	16	19.87	D	0.00000	0.68750	0.312500	0.000000*
15) dampass > 2	181	99.13	T	0.01105	0.05525	0.933700	0.000000
30) con3 < 0.49	90	19.18	T	0.02222	0.00000	0.977800	0.000000*
31) con3 > 0.49	91	63.00	T	0.00000	0.10990	0.890100	0.000000*

* denotes terminal node

to the potential range. We summarized both the known and predicted information by ERU and across the study area. Known and predicted information was summed to estimate current status and distribution. To estimate how much of the current distribution was protected by special land-use designations, we summarized the number of occupied subwatersheds that were found within designated wilderness or National Park Service lands. We also summarized occupied subwatersheds on federally administered lands. Chinook salmon and steelhead distributions were mapped using a geographic information system (GIS). The GIS coverages depicting the subwatersheds, land ownership, and management status were developed as part of the landscape information noted above (Quigley and Arbelbide 1997).

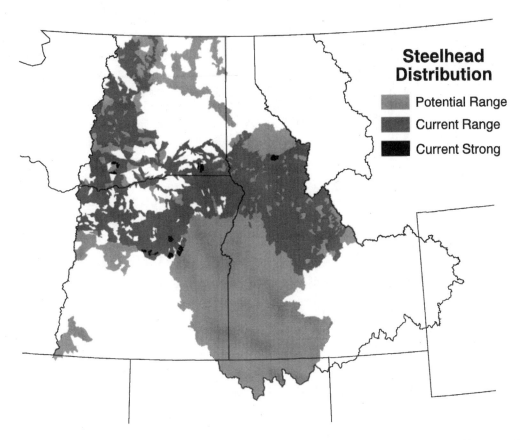

FIGURE 12.3 Map of the potential historical range, the known and predicted current range, and known and predicted strong populations of steelhead within the interior Columbia River basin in the U.S. and portions of the Klamath River basin.

RESULTS

STEELHEAD

Potential Historical Range.—The broad potential range of steelhead is well documented (Howell et al. 1985b; NWPPC 1986). Steelhead were present in most streams—both perennial and intermittent—that were accessible to anadromous fish including all accessible tributaries to the Snake (Evermann 1896; Parkhurst 1950) and Columbia rivers (Fulton 1970; Howell et al. 1985b; Bakke and Felstner 1990) (Figure 12.3). Steelhead formerly ascended the Snake River and spawned in reaches of Salmon Falls Creek, Nevada, more than 1,450 km from the ocean. Approximately 16,935 km of stream were accessible to steelhead in the Columbia River basin, including Canada (NWPPC 1986). Steelhead occupied about 50% of the subwatersheds and nine of the 13 ERUs in the study area (Table 12.3). The largest areas of unoccupied habitat were waters above Spokane Falls on the Spokane River and above Shoshone Falls on the Snake River.

Known Status and Distribution.—Steelhead were reported as present in about 33% of the subwatersheds within the potential range, with strong populations present in 1.9% of the current known range and <1% of the potential range (Table 12.3). The distribution of steelhead was unknown or unclassified in 7% of the potential range (279 subwatersheds) and reported as present but of unknown status in another 289 subwatersheds.

TABLE 12.3

Summary of the potential historical range and known (as reported by biologists) and predicted classifications of current range and status (number of subwatersheds) for summer steelhead. The numbers predicted are based on the classification trees and are shown in parentheses. Known status is the remainder of known and predicted minus predicted. Twenty-two subwatersheds classified as unknown had incomplete information and were omitted from analysis. Transient refers to transient in migration corridors. Wild indigenous subwatersheds were those judged to be unaltered by hatchery programs.

Ecological Reporting Unit	Total sub-watersheds	Potential historical range	Current known range	Current known and predicted range	Known and predicted status				
						Where present			Wild indigenous
					Strong	Depressed	Transient	Absent	
Northern Cascades	340	274	123	201 (78)	1 (0)	192 (77)	8 (1)	65 (7)	3
Southern Cascades	141	67	38	42 (4)	0	37 (4)	5 (0)	25 (0)	0
Upper Klamath	175	35	0	0	0	0	0	35 (0)	0
Columbia Plateau	1,089	554	295	392 (97)	6 (0)	289 (89)	97 (8)	160 (22)	56
Blue Mountains	695	584	364	378 (14)	13 (0)	326 (14)	39 (0)	206 (2)	169
Northern Glaciated Mountains	955	193	21	24 (3)	0	9 (3)	15 (0)	169 (10)	0
Lower Clark Fork	415	98	0	0	0	0	0	98 (0)	0
Owyhee Uplands	956	898	0	0	0	0	0	887 (0)	0
Central Idaho Mountains	1,232	1,051	392	674 (282)	3 (0)	578 (275)	93 (7)	376 (27)	150
Entire Assessment Area	7,498*	3,754	1,233	1,711 (478)	23 (0)	1,431 (462)	257 (16)	2,021 (68)	378

*This total includes 4 ERUs with no potential range

TABLE 12.4
Cross classification comparison of predicted status with observed status for chinook salmon and steelhead, and list of leading predictor variables. Correct classifications match on a diagonal from left to right. Leading predictor variables accounted for the largest amount of variance in the models and are listed in order of the relative proportion of variance accounted for (see Table 12.1 for variable descriptions).

| | | | Observed status | | | |
Species	Predicted status	Absent (A)	Present- depressed (D)	Present- strong (S)	Transient in migration corridor (T)	Leading predictor variables
Steelhead	A	124	30	0	2	hucorder, ERU, drnden, mngclus
	D	53	870	9	28	
	S	0	2	11	0	
	T	1	29	0	196	
Ocean-type chinook	A	89	5	0	2	hucorder, ERU, mngclus, dampass
salmon	D	2	47	2	1	
	S	1	0	17	2	
	T	1	0	1	54	
Stream-type chinook	A	440	40	0	12	hucorder, dampass, slope, elev, ERU
salmon	D	112	401	8	19	
	S	0	0	0	0	
	T	2	19	0	209	

Predictive Models.—A total of 1,355 subwatersheds within the current range, with both complete landscape information and known status, were used to develop the predictive models. We excluded all subwatersheds in the potential range that were no longer accessible. The overall model predictive success rate was 88.6% (Table 12.4). The model was most successful in classifying present-depressed (93%) and migration corridors (87%) and least successful in discriminating strong populations (55%). Variables including hucorder, ERU, slope, solar radiation, vegetation, management clusters, and variables related to lithology and sensitivity to erosion were useful in discriminating subwatersheds supporting steelhead (Table 12.4). Our results suggest that spawning and rearing areas for steelhead were likely to be found within specific ERUs, in small to mid-size streams, in erosive land types and in steeper, higher elevation subwatersheds. Spawning and rearing occurred primarily on Forest Service lands. Migration corridors represented probable steelhead occurrence in larger mainstem streams at lower elevations. Disturbance-related variables had a relatively minor influence in the model classifications. Management cluster and the number of dams steelhead must pass enroute to the ocean were useful but not dominant predictors.

Current Status and Distribution.—Steelhead were known or predicted to presently occur in 46% of their potential range and 23% of the subwatersheds in the study area (Table 12.3; Figure 12.3). Steelhead have been extirpated from large portions of their potential range, however, and are no longer present in the Upper Klamath, Lower Clark Fork, and Owyhee Uplands ERUs. In ERUs that still support steelhead, much of the potential range is no longer accessible.

Despite their relatively broad distribution, very few strong steelhead populations exist and many populations have been influenced by non-indigenous forms. We estimated about 39% of the potential range still supports spawning and rearing, but only 2% of those subwatersheds were classified as supporting strong populations (Table 12.3; Figure 12.3). None of the unknown or unclassified subwatersheds were predicted to support strong populations, suggesting that we know the distribution of most strong populations. Wild, indigenous steelhead, unaltered by hatchery stocks, are

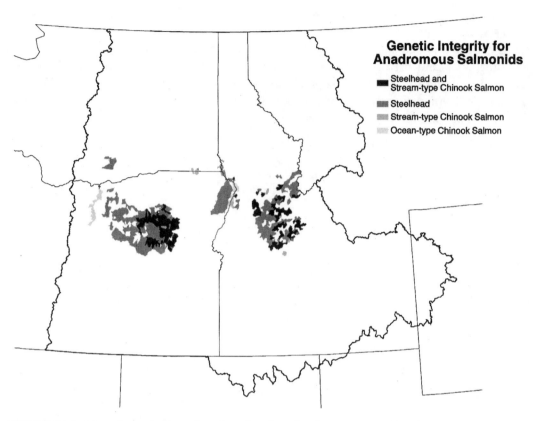

FIGURE 12.4 Map of genetically unaltered stocks of steelhead, stream-type chinook salmon, and ocean-type chinook salmon within the interior Columbia River basin in the U.S.

rare (Figure 12.4) and present in 10% of the potential range and 22% of the current distribution. Remaining wild stocks are concentrated in reaches of the Salmon River in central Idaho and the John Day River basin in Oregon. Although few wild stocks were classified as strong, the only subwatersheds classified as strong were those sustaining wild stocks (6%). About 9% of strong populations were found in designated wilderness and 70% were found on Forest Service or BLM lands.

CHINOOK SALMON

Potential Historical Range.—Chinook salmon were historically found in all accessible areas of the Snake River downstream from Shoshone Falls and in all accessible areas of the Columbia River downstream from Windermere Lake, British Columbia (Evermann 1896; Parkhurst 1950; Fulton 1970; Howell et al. 1985a; NWPPC 1986) (Figures 12.5 and 12.6). Prior to overfishing and habitat alterations, migrating chinook salmon in the Columbia River formed a continuum from March to October with the largest part of the run likely consisting of summer chinook salmon (Thompson 1951). Like steelhead, chinook salmon formerly ascended the Snake River and spawned in reaches of Salmon Falls Creek, Nevada, more than 1,450 km from the ocean. An estimated 16,935 km of stream were accessible to chinook salmon in the Columbia River basin in the U.S. and Canada (NWPPC 1986).

Stream-type chinook salmon were the most widely distributed form, occupying about 46% of the subwatersheds and nine ERUs in the study area (Table 12.5). Ocean-type chinook salmon were much less widely distributed, occupying about 7% of the subwatersheds and six ERUs (Table 12.6).

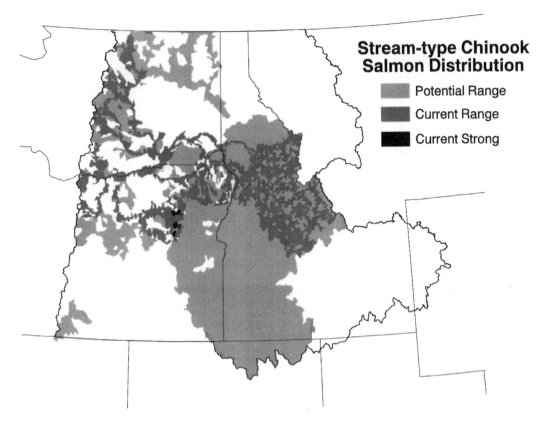

FIGURE 12.5 Map of the potential historical range, the known and predicted current range, and known and predicted strong populations of stream-type chinook salmon within the interior Columbia River basin in the U.S. and portions of the Klamath River basin.

The Snake River was considered the most important production area for ocean-type chinook salmon in the Columbia River basin (Fulton 1968).

Known Status and Distribution.—Stream-type chinook salmon were reported as present in about 21% and ocean-type chinook salmon in about 25% of the subwatersheds within the potential range (Tables 12.5 and 12.6). Strong populations were judged to be present in <1% of the potential range of stream-type chinook salmon and in 4% of the potential range of ocean-type chinook salmon. The distribution of stream-type chinook salmon was unknown or unclassified in 8% of the potential range (281 subwatersheds) and reported as present of unknown status in an additional 141 subwatersheds. The distribution of ocean-type chinook salmon was unknown or unclassified in 10% of the potential range (50 subwatersheds) and reported as present of unknown status in an additional 20 subwatersheds.

Predictive Models.—For stream-type chinook salmon, a total of 1,262 subwatersheds within the current range, with complete landscape information and known chinook status, were used to develop the predictive models. For ocean-type chinook salmon, 224 subwatersheds were used in the model. We excluded all subwatersheds in the potential range that were no longer accessible. The overall model predictive success rate was 83% for stream-type chinook salmon and 92% for ocean-type chinook salmon (Table 12.4). Variables including ERU, number of dams, hucorder, slope, elevation, and management cluster were useful in discriminating subwatersheds supporting chinook salmon (Table 12.4). Stream-type chinook salmon spawning and rearing areas were more likely to be found within certain ERUs in mid-size streams, above fewer mainstem dams, in steeper landscapes, and on Forest Service lands. Larger mainstem streams at lower elevations were more

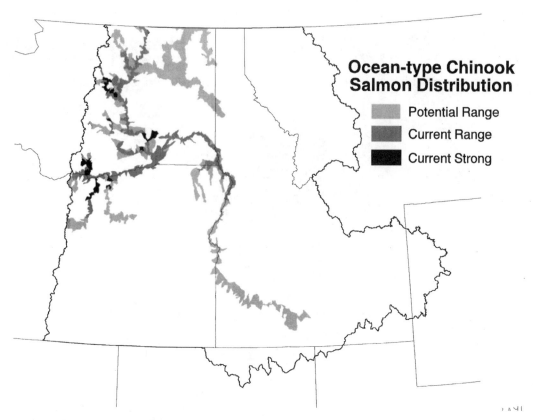

FIGURE 12.6 Map of the potential historical range, the known and predicted current range, and known and predicted strong populations of ocean-type chinook salmon within the interior Columbia River basin in the U.S. and portions of the Klamath River basin.

likely to represent migratory corridors. In contrast, the distribution of ocean-type chinook salmon spawning areas were associated with mainstem segments of the Columbia and Snake rivers.

Current Status and Distribution.—Chinook salmon populations have been extirpated in large portions of their potential range. Current known and predicted distributions of stream-type and ocean-type chinook salmon include 28% and 29%, respectively, of the potential range (Tables 12.5; 12.6). Stream-type chinook salmon are no longer present in the Upper Klamath, Lower Clark Fork, and Owyhee Uplands ERUs; and they have been extirpated from portions of their potential range in other ERUs (Figure 12.5). Ocean-type chinook salmon remain in all ERUs in the potential range but are extirpated from more than 70% of the potential subwatersheds (Figure 12.6).

Most chinook salmon populations are depressed and have been influenced by non-indigenous forms. We estimated that about 20% of the potential range still supports spawning and rearing areas for stream-type chinook salmon and 1.2% of those subwatersheds (0.2% of the potential range) were classified as supporting strong populations (Table 12.5; Figure 12.5). The North Fork of the John Day River contains the only strong population of stream-type chinook salmon. About 17% of the potential range still supports spawning and rearing areas for ocean-type chinook salmon and 26% of those subwatersheds (4.5% of the potential range) were classified as supporting strong populations (Table 12.6; Figure 12.6). The Northern Cascades, especially the Hanford reach of the Columbia River, and the Columbia Plateau ERUs support the remaining core of strong ocean-type chinook salmon populations. Like steelhead, many remaining chinook salmon populations have been influenced by hatchery-reared fish. Wild populations unaltered by hatchery stocks are rare

TABLE 12.5

Summary of the potential historical range and known (as reported by biologists) and predicted classifications of current range and status (number of subwatersheds) for stream-type chinook salmon. The numbers predicted are based on the classification trees and are shown in parentheses. Known status is the remainder of known and predicted minus predicted. Eight subwatersheds classified as unknown had incomplete information and were omitted from analysis. Transient refers to transient in migration corridors. Wild indigenous subwatersheds were those judged to be unaltered by hatchery programs.

Ecological Reporting Unit	Total subwatersheds	Potential historical range	Current known range	Current known and predicted range	Known and predicted status		Where present	Transient	Absent	Wild indigenous
					Strong	Depressed				
Northern Cascades	340	219	89	123 (34)	0	104 (33)	19 (1)	94 (20)	0	
Southern Cascades	141	51	18	26 (8)	0	21 (8)	5 (0)	25 (0)	0	
Upper Klamath	175	35	0	0	0	0	0	35 (0)	0	
Columbia Plateau	1,089	465	114	129 (15)	0	39 (5)	90 (10)	331 (46)	0	
Blue Mountains	695	504	192	208 (16)	8 (0)	152 (14)	48 (2)	295 (22)	68	
Northern Glaciated Mountains	955	172	7	7 (0)	0	1 (0)	6 (0)	165 (5)	0	
Lower Clark Fork	415	98	0	0	0	0	0	98 (0)	0	
Owyhee Uplands	956	866	0	0	0	0	0	866 (0)	0	
Central Idaho Mountains	1,232	1,051	314	465 (151)	0	351 (138)	114 (13)	586 (97)	60	
Entire Assessment Area	7,498*	3,461	734	958 (224)	8 (0)	668 (198)	282 (26)	2,495 (190)	128	

*This total includes 4 ERUs with no potential range.

TABLE 12.6

Summary of the potential historical range and known (as reported by biologists) and predicted classifications of current range and status (number of subwatersheds) for ocean-type chinook salmon. The numbers predicted are based on the classification trees and are shown in parentheses. Known status is the remainder of known and predicted minus predicted. Five subwatersheds classified as unknown had incomplete information and were omitted from analysis. Transient refers to transient in migration corridors. Wild indigenous subwatersheds were those judged to be unaltered by hatchery programs.

Ecological Reporting Unit	Total sub-watersheds	Potential historical range	Current known range	Current known and predicted range	Known and predicted status				
					Where present				Wild indigenous
					Strong	Depressed	Transient	Absent	
Northern Cascades	340	79	25	39 (14)	13 (1)	21 (9)	5 (4)	39 (11)	0
Southern Cascades	141	31	9	9 (0)	0	0	9 (0)	22 (3)	0
Columbia Plateau	1,089	147	51	57 (6)	12 (3)	14 (3)	31 (0)	86 (21)	11
Blue Mountains	695	99	38	41 (3)	0	27 (2)	14 (1)	58 (7)	16
Northern Glaciated Mountains	955	149	15	15 (0)	0	9 (0)	6 (0)	134 (0)	0
Owyhee Uplands	956	48	0	0	0	0	0	48 (0)	0
Entire Assessment Area	7,498*	553	138	161 (23)	25 (4)	71 (14)	65 (5)	387 (42)	27

*This total includes 7 ERUs with no potential range.

and present in 4% of the potential range and 15% of the current range for stream-type chinook salmon and 5% of the potential range and 17% of the current range for ocean-type chinook salmon (Figure 12.4). With the exception of strong populations in the Hanford reach, the only subwatersheds classified as strong were those sustaining wild stocks. About 50% of stream-type chinook salmon strong populations were found in designated wilderness or National Park Service lands and 88% were found on Forest Service or BLM lands. No strong populations of ocean-type chinook salmon were found in wilderness and 20% were found on Forest Service or BLM lands.

DISCUSSION

CURRENT STATUS AND DISTRIBUTION

The status and distribution of chinook salmon and steelhead within the study area is very different than it was historically; we estimated that stream-type and ocean-type chinook salmon have been extirpated from more than 70% of their potential range and steelhead from 54% of their potential range. About 12,452 km of streams in the potential range in the Columbia River basin in the U.S. and Canada are no longer accessible to anadromous fish (NWPPC 1986). Construction of numerous dams without passage facilities, such as Swan Falls in 1906, Grand Coulee in 1941, and the Hells Canyon complex in 1967, blocked fish access (IDFG 1992). Where accessible habitats remain, most wild chinook salmon and steelhead are declining. Populations in 99% of remaining spawning and rearing areas for stream-type chinook salmon, 76% of those for ocean-type chinook salmon, and 98% of those for steelhead were classified as depressed. Because our potential ranges remain speculative and because the resolution of our potential and current range data was not sufficient to map upper and lower distributional boundaries within subwatersheds, we cannot quantify extirpations at finer scales. Quantification of extirpations are further complicated by the likelihood that the distribution of steelhead and both chinook salmon forms was naturally restricted by physical characteristics of streams (for example, see Burner 1951; Platts 1974; Mullan et al. 1992). Spawning areas for ocean-type chinook salmon in Idaho, for example, appear to have been restricted to mainstem reaches where at least 960 temperature units accumulate from about November 15 to a late April-early May emergence (B. Connor, U.S. Fish and Wildlife Service, personal communication).

Extinctions and declining populations have resulted in lower diversity and total abundance of chinook salmon and steelhead. Cumulative habitat changes that eliminate or isolate segments of populations may increase both demographic and environmental stochasticity (thereby increasing the risk of extinction of remaining populations) because of reduced population structure or distribution (Rieman et al. 1993). Local extinction and subsequent refounding may be a natural and perhaps common element in the dynamics of many species (Rieman and McIntyre 1995; Schlosser and Angermeier 1995). Large scale processes such as climatic variation associated with fluctuating ocean conditions, drought and flood patterns, and local storm frequency and intensity may influence abundance and distributions in shorter time scales. The recent declines in Columbia River salmon and steelhead, for example, have likely been influenced by ocean conditions (Lichatowich and Mobrand 1995) and a period of extended drought (Anderson 2000). Although environmental variability is a factor and may be the proximate cause of decline or extinction in some cases, the effects of human-caused disturbance appear to be far more important to the declines in abundance of chinook salmon and steelhead.

FACTORS INFLUENCING STATUS

The association between status of chinook salmon and steelhead and disturbance variables (*dampass*, *mngclus*) in our models suggests these species have responded negatively to environmental disruption (Table 12.4). Because the models were designed to develop a complete database and not test hypotheses, there are several possible reasons for the lack of strong influence of these and other disturbance variables (such as *roaddn*) in discriminating anadromous fish status (e.g., Table 12.2). First, the status of anadromous fish is influenced by environmental factors within both the freshwater environment and in areas outside the study area. Our disturbance variables may not adequately reflect factors such as passage mortality, harvest, and ocean survival that occur outside the subwatersheds. Second, the model did not differentiate between wild and natural fish and did not address the influence of introduced stocks. In some heavily disturbed areas, remnant numbers of fish may be present only because of annual hatchery supplementation so disturbance may not be a good predictor of fish presence. Third, too few strong populations were represented in the model to relate their presence to disturbance. The number of subwatersheds where strong populations were known to be present included 8, 21, and 23 for stream-type chinook salmon, ocean-type chinook salmon, and steelhead, respectively. For these reasons, although the model included disturbance variables, it does not provide an adequate prediction of risk associated with different management scenarios. Salmon and steelhead declines can be associated with a variety of human-caused factors that include habitat disruption linked to land management; watershed development for hydropower and irrigation; competition, hybridization, and predation linked to the introduction of non-native species or stocks of fish; and harvest.

Freshwater Habitat Degradation.—Work at finer scales has confirmed that degradation of freshwater habitats is a consistent and pervasive problem facing the productivity and persistence of aquatic animals in the study area and throughout much of the western U.S. (Williams et al. 1989; Nehlsen et al. 1991). Numerous studies describe the negative effects of land-use activities on habitat conditions, and link habitat conditions to survival and productivity of anadromous fish (Meehan 1991; Murphy 1995; NRC 1996). More than 95% of the healthy native stocks of anadromous fish identified by Huntington et al. (1996) were judged to be threatened by habitat degradation. Nehlsen et al. (1991) identified habitat loss or degradation as a major problem for 90% of the 195 at-risk salmon and steelhead stocks they identified. Alterations of anadromous fish habitat can be attributed to mining (Meehan 1991), timber harvest (Meehan 1991), agriculture (NWPPC 1986), industrial development, and urbanization (NWPPC 1986).

Hydropower Development.—Construction and operation of mainstem dams on the Columbia and Snake rivers is considered to be the major cause of recent declines of anadromous fish (CBFWA 1990). Hydroelectric development changed Columbia and Snake river migration routes from mostly free-flowing in 1938 to a series of dams and impoundments by 1975. Reservoirs reduce flows in most years by about 50% during smolt migration (Raymond 1979). Anadromous fish must navigate as many as nine mainstem dams. At each dam, adults are delayed during upstream migrations. Smolts may be killed by turbines; become disoriented or injured, making them more susceptible to predation; or become delayed in impoundments behind dams (IDFG et al. 1990). Smolt-to-adult return rates declined from more than 4% in 1968 to less than 1.5% from 1970–1974. In 1973 and 1977, low flows resulted in 95% of migrating smolts never reaching the ocean (Raymond 1979). Losses of mid- and upper-Columbia ocean-type chinook salmon were estimated to be about 5% per dam for adults and 18 to 23% per dam for juveniles (Chapman et al. 1994b). The influence of passage mortality is illustrated by Huntington et al. (1996), who concluded that, although much of the Pacific Northwest's best remaining spawning and rearing habitat is in central Idaho, no healthy stocks of anadromous fish were found there.

Introduced Species.—Predation, competition, and genetic introgression from introduced species or stocks has influenced the status of anadromous salmonids. More than 55 introduced fishes occur within the current range of salmon and steelhead. In many reaches of the Snake and Columbia rivers, introduced species outnumber indigenous fish (Li et al. 1987). Hobbs and Huenneke (1992) suggested that non-natives may pose a greater risk to native species where habitat has been disturbed. Dams have created habitat that is suitable for a variety of native and non-native predators and potential competitors (e.g., Beamesderfer and Rieman 1991).

With few exceptions, most watersheds supporting chinook salmon and steelhead are also likely to be influenced by hatchery stocks. Existing populations are comprised of four types: wild, natural (non-indigenous progeny spawning naturally), hatchery, and mixes of natural and hatchery fish. Most existing chinook salmon and steelhead production is supported by hatchery and natural fish as a result of large-scale hatchery mitigation programs. By the late 1960s hatchery production surpassed natural production in the Columbia River basin (NWPPC 1986). Meanwhile, production of wild anadromous fish in the Columbia River basin has declined by about 95% from historical levels (Huntington et al. 1996). Hatcheries may affect salmon and steelhead populations through genetic introgression and loss of fitness, creation of mixed-stock fisheries, competition for food and space, and disease organisms (Reisenbichler 1977). Studies of the interaction between wild and hatchery fish illustrate that survival of progeny from hatchery or hybrid (wild × hatchery) parentage is less than for progeny of wild fish pairings (Reisenbichler and McIntyre 1977; Chilcote et al. 1986). Rainbow trout have been introduced throughout the current steelhead range (Lee et al. 1997) and have the potential to hybridize with steelhead (Chapman et al. 1994b). Byrne et al. (1992) suggested that supplementation of native stocks with hatchery fish have typically resulted in replacement, not enhancement of native fish. Most (107/121) of the healthy anadromous salmonid stocks identified by Huntington et al. (1996) either have had no fish culture activities in the home watershed or have been exposed to little risk from stock transfers or interaction with hatchery fish. Adult collection and egg-taking may also be detrimental to individual populations (Chapman et al. 1994b). Introductions of large numbers of hatchery-reared parr may cause localized decreases in the density of juveniles (Pollard and Bjornn 1973; McMichaels et al. 2000) and induce early migration of wild fish (Hillman and Mullan 1989). While information is not available across the study area to judge the effects of hatchery releases on genetic structure of steelhead (Busby et al. 1996) and chinook salmon, wild stocks appear to be rare. Biologists judged wild stocks of steelhead and stream-type and ocean-type chinook salmon unaltered by hatchery releases to be present in 10, 4, and 5% of the potential range, respectively.

Harvest.—Salmon and steelhead stocks have historically provided harvest opportunities for tribal, commercial, and sport fisheries. Harvest has contributed to the decline of spring and summer chinook salmon in the study area since the late 1800s (Fulton 1970) and to the decline of fall chinook salmon after 1920 (Lichatowich and Mobrand 1995). Historical ocean and river harvest rates exceeded 80% (Ricker 1959). Thompson (1951) reported that by 1919, as a result of excessive harvest, the characteristics of the Columbia River chinook salmon run had changed. Formerly large portions of the run were reduced making smaller portions of the run more important to the fishery. The once nearly continuous run of salmon became segregated into more discrete groups. Lichatowich and Mobrand (1995) divided the fishery into four phases: initial development (1866–1888), sustained production (1889–1922), resource decline (1923–1958), and maintenance at a depressed level (post-1958). Wild populations have declined as numbers of hatchery steelhead have increased, creating harvest management problems. Hatchery fish that are surplus to egg-taking needs can be harvested while declining runs of wild fish cannot. In response, sport and commercial harvest of most wild adult salmon and steelhead has been closed (NWPPC 1986; IDFG 1992; Chapman et al. 1994b), harvest of wild juvenile steelhead has been restricted (IDFG 1992), and tribal fisheries are regulated (NWPPC 1986; Chapman et al. 1994b). Although the harvest of wild stocks has been reduced, wild salmon and steelhead are killed during tribal fisheries, commercial salmon fisheries in the Columbia River and coastal marine waters, and in high seas driftnet fisheries (Cooper and Johnson 1992; Chapman et al. 1994b).

Emphasis Areas

With the exception of the Central Idaho Mountains and perhaps the Northern Cascades ERUs, most of the areas supporting naturally reproducing salmon and steelhead exist as patches of scattered watersheds. Many are not well connected or are restricted to much smaller areas than was the case historically. Many are associated with high-elevation, steep, and more erosive landscapes. These may be more extreme or variable environments contributing to higher variability in the associated populations, and higher sensitivity to watershed disturbances. Risks could be aggravated with further development. While we may be able to reestablish some populations in portions of the range, reopening large tracts of the former range to anadromous salmonids is unlikely in the short term. Identification and conservation of the most important populations and habitats will likely be critical to broad-scale persistence of these fish species.

Refugia or emphasis areas providing high quality habitat and populations of special significance are cornerstones of most species conservation strategies (Lee et al. 1997). Emphasis areas can be defined as watersheds and habitats that support remaining areas of high intrinsic value for aquatic species. These include areas supporting strong populations for one or multiple species, areas of high genetic integrity, areas located on the fringe of a species range, areas that support listed species, and areas with excellent habitat but too few fish. Within the study area, concern for the persistence of native anadromous fish has led to several efforts to identify emphasis areas. These include the concept of "key watersheds" (Reeves and Sedell 1992; FEMAT 1993), Section 7 and High Priority Watersheds in the Snake River basin (Lee et al. 1997), and PACFISH watersheds (U.S.D.A. Forest Service and U.S.D.I. Bureau of Land Management 1995).

We focused on identifying emphasis areas of two types: subwatersheds with designated strong populations and those retaining naturally reproducing populations of salmon and steelhead, including genetically intact populations. We used consistent rules to spatially locate these populations. As a result of the diverse life history and extensive migrations of salmon and steelhead, factors outside spawning and rearing areas influenced the distribution of strong populations. Consequently, strong populations were rare or absent even in relatively undisturbed habitats in the Central Idaho Mountains ERU. Because salmon and steelhead have very few strongholds, subwatersheds supporting naturally reproducing populations and populations with high genetic integrity may represent the only areas available from which to anchor a conservation strategy.

Although much of the native ecosystem has been altered, emphasis areas remain for rebuilding and maintaining functional native aquatic systems. Watersheds that support strong populations likely represent a fortuitous balance of habitat quality, climatic and geologic constraint, and geographic location, which effectively minimizes cumulative threats to these species (Lee et al. 1997). The most productive, abundant, and diverse populations are likely to be most resilient. Thus, they are more likely to serve as sources for the support of weak or at-risk populations, refounding of locally extinct populations, or repopulating habitats made available through restoration (Schlosser and Angermeier 1995). Because native gene complexes likely offer the best resources for refounding extinct populations in similar environments, conservation of locally adapted and marginal populations will also be critical (Scudder 1989). Populations that historically were distributed over broad geographic areas have likely evolved under relatively distinct environments with little gene flow across the species range (Lesica and Allendorf 1995). Conservation of the genetic diversity in these species then implies sustaining populations over the broad geographic area (Allendorf and Leary 1988; Leary et al. 1993).

Federal land management will be crucial to the establishment of emphasis areas. Both steelhead (70%) and stream-type chinook salmon (88%) have most of their strong populations on federal land. The recovery of depressed populations will also depend on management of federal lands because 61% and 77% of depressed stream-type chinook salmon and steelhead populations, respectively, occupy federal land. Only ocean-type chinook salmon are less influenced by federal land management since they tend to occupy larger mainstem systems which are more influenced by private lands.

Any changes in the environment that increase survival and productivity of remaining chinook salmon and steelhead, including improvements in rearing habitats, harvest, predation, and main stem passage, will improve chances for persistence (Emlen 1995; NRC 1996). Under current migrant survival conditions, many stocks are at serious risk. The differences between those that persist and those that are extirpated depends on chance events combined with stock survival and productivity, which are largely influenced by freshwater habitat quality, quantity, and accessibility. Without substantial improvement in migrant survival, securing and restoring the quality of freshwater habitats may make the critical difference in persistence for many of the remaining populations. In the short term, conservation and rehabilitation of habitats available to remaining populations will be key. In the long term, main stem passage conditions must be improved, followed by conservation and restoration of broader habitat networks to support the full expression of life histories and species (Lichatowich and Mobrand 1995; NRC 1996). Rehabilitation of depressed populations cannot rely on habitat improvement alone, but requires a concerted effort to address causes of mortality in all life stages including freshwater spawning, rearing, and overwintering, juvenile migration, ocean survival, and adult migration.

IMPLICATIONS FOR SUSTAINABLE FISHERIES

To conserve fisheries resources and ensure sustainability, agencies will need a multilevel understanding of spatial variation in the demographic attributes of fish populations (Schlosser and Angermeier 1995). This understanding can help to identify, protect, and restore key habitats and the ecosystem processes generating them. In the case of Pacific salmon, Healey and Prince (1995) report that genotypic variation maps mainly on the population scale whereas phenotypic variation maps strongly on the ecosystem and landscape scales. This suggests that local populations may have a much larger effect on phenotypic diversity than on overall genotypic diversity. The authors suggest that, because phenotypic diversity is a consequence of the genotype interacting with a particular environment, it is critical to preserve unique habitats and ensure their accessibility to Pacific salmon. The critical conservation unit, therefore, is the population within its habitat.

Conservation of salmon and steelhead populations will require the maintenance or rehabilitation of a network of well-connected, high-quality habitats that support the full expression of potential life histories and dispersal mechanisms, and the genotypic and phenotypic diversity necessary for long-term persistence and adaptation in a variable environment (Lichatowich and Mobrand 1995; Healey and Prince 1995). Protection of stronghold emphasis areas will not be sufficient. Such reserves will never be large enough or sufficiently distributed to maintain biological diversity (Franklin 1993). Watershed restoration and the development of more ecologically compatible land-use policies are required to ensure the long-term productivity of many systems (Thurow et al. 1997).

Protection and maintenance of system integrity and functioning will require innovative approaches. Simple solutions such as setting aside small, scattered watersheds probably will not be adequate for the persistence of even current distributions and diversity. The problems are too complex and too pervasive. However, there are several actions which could be taken to maintain or restore the integrity of larger aquatic ecosystems: first, definition and conservation of emphasis areas; second, reconnection and expansion of the mosaic of habitats for widely distributed species to enhance the integrity of larger systems. For wide-ranging fishes such as salmon and steelhead, this includes conservation and restoration of migratory corridors, as well as spawning and rearing areas. Third, an approach is necessary to provide for multiple species or complete communities (Gresswell et al. 1994; Frissell et al. 1995). Conservation and restoration of important habitats should provide for associated species (e.g., cutthroat trout and bull trout) and will sustain important processes that influence structure and function within these systems.

Restoration and management of watersheds on federal lands is critical but will not be sufficient. River corridors surrounded largely by private lands are a particularly important part of the habitat networks. Much of the overlap in species distributions occurs in the larger river corridors because many of the species range widely. Anadromous species use the entire system of rivers as migratory corridors and for overwintering and short-term rearing. The connections and habitat provided by larger river systems are thus critical to the maintenance of anadromous populations. Although much of the highest-quality habitat for anadromous fish probably remains in the Central Idaho Mountains

ERU, no strong populations persist there, due largely to passage mortality in migration corridors. These corridors provide a critical link maintaining the complex life histories of other species as well.

All remaining populations and habitats within the Central Idaho Mountains, Blue Mountains, Northern Cascades, Columbia Plateau, Northern Glaciated Mountains, and Southern Cascades ERUs are critical to the persistence of chinook salmon and steelhead within the study area. Protection of those core areas critical to stock persistence and restoration of a broader matrix of productive habitats will be necessary for full expression of phenotypic and genotypic diversity and establishment of sustainable fisheries.

ACKNOWLEDGMENTS

We thank the more than 150 agency, tribal, university, and private biologists who participated in preparation of the databases (listed in Lee et al. 1997) and especially P. Howell, K. MacDonald, T. Shuda, and D. Heller. G. Chandler and D. Myers assisted with creating, correcting, merging, and managing many of the databases with assistance from B. Butterfield, S. Gebhards, J. Gebhards, J. Gott, J. Guzevich, J. Hall-Griswold, L. Leatherbury, M. Radko, and M. Stafford. Development of landscape information and map coverages was supported by J. Clayton, K. Geier-Hayes, B. Gravenmeier, W. Hann, M. Hotz, C. Lorimar, S. McKinney, P. Newman, and G. Stoddard. P. Howell and R. House provided constructive reviews of the manuscript. The use of trade or firm names in this chapter does not imply endorsement of any product or service by the U.S. Department of Agriculture.

REFERENCES

Allendorf, F. W., and R. F. Leary. 1988. Conservation and distribution of genetic variation in a polytypic species, the cutthroat trout. Conservation Biology 2:170–184.

Anderson, D. A., G. Christofferson, R. Beamesderfer, B. Woodward, M. Rowe, and J. Hansen. 1996. Streamnet: the northwest aquatic resource information network. Report on the status of salmon and steelhead in the Columbia River basin, 1995. U.S. Department of Energy, Bonneville Power Administration, Environment, Fish and Wildlife, Portland, Oregon.

Anderson, J. J. 2000. Decadal climate cycles and declining Columbia River salmon. Pages 467–484 in E. E. Knudsen, C. R. Steward, D. D. MacDonald, J. E. Williams, and D. W. Reiser, editors. Sustainable fisheries management: Pacific salmon. Lewis Publishers, Boca Raton, Florida.

Bakke, B., and P. Felstner. 1990. Columbia Basin anadromous fish extinction record. Oregon Trout, Portland.

Beamesderfer, R. C., and B. E. Rieman. 1991. Abundance and distribution of northern squawfish, walleye, and smallmouth bass in John Day Reservoir, Columbia River. Transactions of the American Fisheries Society 120:439–447.

Bevan, D., and six coauthors. 1994. Snake River salmon recovery team: Final recommendations to the National Marine Fisheries Service—Summary. Portland, Oregon.

Breiman, L., J. H. Friedman, R. Olshen, and C. J. Stone. 1984. Classification and regression trees. Wadsworth International Group, Belmont, California.

Burner, C. J. 1951. Characteristics of spawning nests of Columbia River salmon. U.S. Fish and Wildlife Service Fishery Bulletin 52:97–110.

Busby, P. J., and six coauthors. 1996. Status review of west coast steelhead from Washington, Idaho, Oregon, and California. U.S. Department of Commerce. NOAA Technical Memorandum NMFS-NWFSC-27.

Byrne, A., T. C. Bjornn, and J. D. McIntyre. 1992. Modeling the response of native steelhead to hatchery supplementation programs in an Idaho River. North American Journal of Fisheries Management 12:62–78.

Chapman, D. W. 1986. Salmon and steelhead abundance in the Columbia River in the nineteenth century. Transactions of the American Fisheries Society 115:662–670.

Chapman, D., and eight coauthors. 1994a. Status of summer/fall chinook salmon in the mid-Columbia Region. Don Chapman Consultants, Boise, Idaho.

Chapman, D., C. Pevan, T. Hillman, A. Giorgi, and F. Utter. 1994b. Status of steelhead in the mid-Columbia River. Don Chapman Consultants, Boise, Idaho.

Chilcote, M. W., S. A. Leider, and J. J. Loch. 1986. Differential reproductive success of hatchery and wild summer-run steelhead under natural conditions. Transactions of the American Fisheries Society 115:726–735.

Clark, L. A., and D. Pregibon. 1992. Tree based models. Pages 377–419 *in* J. M. Chambers and T. J. Hastie, editors. Statistical models. Wadsworth and Brooks/Cole Advanced Books and Software, Pacific Grove, California.

CBFWA (Columbia Basin Fish and Wildlife Authority). 1990. Integrated system plan for salmon and steelhead production in the Columbia Basin. Northwest Power Planning Council, Portland, Oregon.

Cooper, R., and T. Johnson. 1992. Trends in steelhead (*Oncorhynchus mykiss*) abundance in Washington and along the Pacific Coast of North America. Washington Department of Wildlife Report No. 92–20. Olympia, Washington.

Crawford, S. L., and R. M. Fung. 1992. An analysis of two probabilistic model induction techniques. Statistics and Computing 2:83–90.

Emlen, J. M. 1995. Population viability of the Snake River chinook salmon (*Oncorhynchus tshawytscha*). Canadian Journal of Aquatic Sciences 52:1442–1448.

Evermann, B. W. 1896. A report upon salmon investigations in the headwaters of the Columbia River, in the State of Idaho, in 1895, together with notes upon the fishes observed in that state in 1894 and 1895. U.S. Fish Commission Bulletin 16:149–202.

FEMAT (Forest Ecosystem Management Assessment Team). 1993. Forest ecosystem management: an ecological, economic, and social assessment. U.S. Forest Service, Portland, Oregon.

Franklin, J. F. 1993. Preserving biodiversity:species, ecosystems, or landscapes? Ecological Applications 3:202–205.

Frissell, C. A. 1993. Topology of extinction and endangerment of native fishes in the Pacific Northwest and California (USA). Conservation Biology 7:342–354.

Frissell, C. A., W. J. Liss, and D. Bales. 1993. An integrated, biophysical strategy for ecological restoration of large watersheds. Pages 449–456 *in* D. Potts, editor. Proceedings of the symposium on changing roles in water resources management and policy. American Water Resources Association, Herndon, Virginia.

Frissell, C. A., J. Doskocil, J. Gangemi, and J. A. Stanford. 1995. Identifying priority areas for protection and restoration of aquatic biodiversity; A case study in the Swan River basin, Montana, USA. Open File Report. Flathead Lake Biological Station, Polson, Montana.

Fulton, L. A. 1968. Spawning areas and abundance of chinook salmon, *Oncorhynchus tshawytscha*, in the Columbia River Basin—past and present. U.S. Fish and Wildlife Service Special Scientific Report, Fisheries 571, Washington, D.C.

Fulton, L. A. 1970. Spawning areas and abundance of steelhead trout and coho salmon, sockeye, and chum salmon in the Columbia River Basin — past and present. U.S. Department of Commerce, National Oceanic-Atmospheric Administration, National Marine Fisheries Service Report, NOAA SSRF-618, Washington, D.C.

Gebhards, S. V. 1959. Columbia River Fisheries Development Program. Salmon River Planning Report. Idaho Department of Fish and Game, Boise.

Gresswell, R. E., W. J. Liss, and G. L. Larson. 1994. Life history organization of Yellowstone cutthroat trout (*Oncorhynchus clarki bouvieri*) in Yellowstone Lake. Canadian Journal of Fisheries and Aquatic Sciences 51(Supplement 1):298–309.

Hassemer, P. F. 1993. Salmon spawning ground surveys. Project F-73-R-15, Pacific Salmon Treaty Program. Idaho Department of Fish and Game, Boise.

Healey, M. C. 1991. Life history of chinook salmon (*Oncorhynchus tshawytscha*). Pages 311–393 *in* C. Groot and L. Margolis, editors. Pacific salmon life histories. UBC Press, Vancouver, British Columbia.

Healey, M. C. 1994. Variation in the life history characteristics of chinook salmon and its relevance to conservation of the Sacramento winter run of chinook salmon. Conservation Biology 8:876–877.

Healey, M. C., and A. Prince. 1995. Scales of variation in life history tactics of Pacific salmon and the conservation of phenotype and genotype. Pages 176–184 in J. L. Nielson, editor. Evolution and the Aquatic ecosystem. American Fisheries Society Symposium 17.

Hillman, T. W., and J. W. Mullan. 1989. Effect of hatchery releases on the abundance and behavior of wild juvenile salmonids. Pages 265–285 *in* Summer and winter ecology of juvenile chinook salmon and steelhead trout in the Wenatchee River, Washington. Don Chapman Consultants, Boise, Idaho.

Hobbs, R. J., and L. F. Huenneke. 1992. Disturbance, diversity, and invasion: implications for conservation. Conservation Biology 6:324–337.

Howell, P., and eight coauthors. 1985a. Stock assessment of Columbia River anadromous salmonids. Volume I: Chinook, coho, chum, and sockeye salmon stock summaries. Final Report 1984. U.S. Department of Energy, Bonneville Power Administration, DOE/BP-12737-1, Portland, Oregon.

Howell, P., and eight coauthors. 1985b. Stock assessment of Columbia River anadromous salmonids. Volume II: Steelhead stock summaries, stock transfer guidelines — information needs. Final Report 1984. U.S. Department of Energy, Bonneville Power Administration, DOE/BP-12737-1, Portland, Oregon.

Huntington, C. W., W. Nehlsen, and J. Bowers. 1996. A survey of healthy native stocks of anadromous salmonids in the Pacific Northwest and California. Fisheries 21:6–15.

IDFG (Idaho Department of Fish and Game). 1992. Anadromous fish management plan 1992–1996. Idaho Department of Fish and Game, Boise.

IDFG (Idaho Department of Fish and Game), Nez Perce Tribe of Idaho, and Shoshone-Bannock Tribes of Fort Hall. 1990. Salmon River subbasin salmon and steelhead production plan. Idaho Department of Fish and Game, Boise.

Jensen, M., I. Goodman, K. Brewer, T. Frost, G. Ford, and J. Nesser. 1997. Biophysical environments of the basin. Pages 99–320 in T. M. Quigley and S. J. Arbelbide, editors. An assessment of ecosystem components in the Interior Columbia Basin and portions of the Klamath and Great Basins. U.S. Forest Service General Technical Report PNW-GTR-405.

Kostow, K., and seven coauthors. 1994. Biennial report on the status of wild fish in Oregon and the implementation of fish conservation policies. Oregon Department of Fish and Wildlife, Portland.

Leary, R. F., F. W. Allendorf, and S. H. Forbes. 1993. Conservation genetics of bull trout in the Columbia and Klamath River drainages. Conservation Biology 7:856–865.

Lee, D. C., J. R. Sedell, B. E. Rieman, R. F. Thurow, and J. E. Williams. 1997. Broadscale assessment of aquatic species and habitats. An assessment of ecosystem components in the interior Columbia Basin and portions of the Klamath and Great Basins. Volume 3, Chapter 4. U.S. Forest Service General Technical Report PNW-GTR-405.

Lesica, P., and F. W. Allendorf. 1995. When are peripheral populations valuable for conservation? Conservation Biology 94:753–760.

Li, H. W., C. B. Schreck, C. E. Bond, and E. Rexstad. 1987. Factors influencing changes in fish assemblages of Pacific Northwest streams. Pages 193–202 in W. J. Matthews and D. C. Heins, editors. Community and evolutionary ecology of North American stream fishes. University of Oklahoma Press, Norman.

Lichatowich, J. A. and L. E. Mobrand. 1995. Analysis of chinook salmon in the Columbia River from an ecosystem perspective. Report for U.S. Department of Energy, Bonneville Power Administration, Contract No. DE-Am79-92BP25105. Portland, Oregon.

Matthews, G. M., and R. S. Waples. 1991. Status review for Snake River spring and summer chinook salmon. NOAA Technical Memorandum NMFS F/NWC-200. National Marine Fisheries Service, Seattle, Washington.

Maxwell, J. R., C. J. Edwards, M. E. Jensen, S. E. Paustian, H. Parrott, and D. M. Hill. 1995. A hierarchical framework of aquatic ecological units in North America (Nearctic Zone). U.S. Forest Service General Technical Report NC-176. St. Paul, Minnesota.

McMichaels, G. A., T. N. Pearsons, and S. A. Leider. 2000. Minimizing ecological impacts of hatchery-reared juvenile steelhead trout on wild salmonids in a Yakima River watershed. Pages 365–380 in E. E. Knudsen, C. R. Steward, D. D. MacDonald, J. E. Williams, and D. W. Reiser, editors. Sustainable fisheries management: Pacific salmon. Lewis Publishers, Boca Raton, Florida.

Meehan, W. R., editor. 1991. Influences of forest and rangeland management on salmonid fishes and their habitats. American Fisheries Society, Special Publication 19, Bethesda, Maryland.

Moyle, P. B., and J. E. Williams. 1990. Biodiversity loss in the temperate zone: Decline of the native fish fauna of California. Conservation Biology 4:275–283.

Mullan, J. W., K. R. Williams, G. Rhodus, T. W. Hillman, and J. D. McIntyre. 1992. Production and habitat of salmonids in mid-Columbia River tributary streams. U.S. Department of Interior, Fish and Wildlife Service, Monograph 1, Portland, Oregon.

Murphy, M. L. 1995. Forestry impacts on freshwater habitats and anadromous salmonids in the Pacific Northwest and Alaska: Requirements for protection and restoration. NOAA Coastal Ocean Program. Decision Analysis Series Number 7. Silver Spring, Maryland.

NMFS (National Marine Fisheries Service). 1995. Reinitiation of consultation on 1994–1998 operation of the federal Columbia River power system and juvenile transportation program in 1995 and future years. Seattle, Washington.

NRC (National Research Council). 1996. Upstream: Salmon and society in the Pacific Northwest. Chapter 3:39–66. National Academy Press, Washington, D.C.

Nehlsen, W., J. E. Williams, and J. A. Lichatowich. 1991. Pacific salmon at the crossroads: Stocks at risk from California, Oregon, Idaho and Washington. Fisheries 16(2):4–21.

NWPPC (Northwest Power Planning Council). 1986. Compilation of information on salmon and steelhead losses in the Columbia River basin. Columbia River Basin Fish and Wildlife Program, Portland, Oregon.

Parkhurst, Z. E. 1950. Survey of the Columbia River and its tributaries. U.S. Fish and Wildlife Service, Special Scientific Report, Fisheries 39.

Platts, W. S. 1974. Geomorphic and aquatic conditions influencing salmonids and stream classification. U.S. Department of Agriculture, Surface Environment and Mining Program. Boise, Idaho.

Pollard, H. A., and T. C. Bjornn. 1973. The effects of angling and hatchery trout on the abundance of juvenile steehead trout. Transactions of the American Fisheries Society 102:745–752.

Quigley, T. M., and S. J. Arbelbide, editors. 1997. An assessment of ecosystem components in the Interior Columbia Basin and portions of the Klamath and Great Basins. U.S. Forest Service, General Technical Report PNW-GTR-405. Portland, Oregon.

Raymond, H. L. 1979. Effects of dams and impoundments on migrations of juvenile chinook salmon and steelhead from the Snake River, 1966 to 1975. Transactions of the American Fisheries Society 108:505–529.

Reeves, G. H., and J. R. Sedell. 1992. An ecosystem approach to the conservation and management of freshwater habitat for anadromous salmonids in the Pacific Northwest. Transactions of the 57th North American Wildlife and Natural Resources Conference 57:408–415.

Reisenbichler, R. R. 1977. Effects of artificial propagation of anadromous salmonids on wild populations. Pages 2–3 in T. J. Hassler and R. R. Vankirk, editors. Genetic implications of steelhead management. Special Report 77-1. California Cooperative Fishery Research Unit, Arcata.

Reisenbichler, R. R. 1997. Genetic factors contributing to declines. Pages 223–244 in D. Stouder, P. Bisson, and R. Naiman, editors. Pacific salmon and their ecosystems. Chapman & Hall, New York.

Reisenbichler, R. R., and J. D. McIntyre. 1977. Genetic differences in growth and survival of juvenile hatchery and wild steelhead trout, *Salmo gairdneri*. Journal of the Fisheries Research Board of Canada 34:123–128.

Ricker, W. E. 1959. Evidence for environmental and genetic influence on certain characters which distinguish stocks of the Pacific salmons and steelhead trout. Fisheries Research Board of Canada, Biological Station, Nanaimo.

Rieman, B., D. Lee, J. McIntyre, K. Overton, and R. Thurow. 1993. Consideration of extinction risks for salmonids. U.S. Department of Agriculture, Forest Service. Fish Habitat Relationships Technical Bulletin Number 14. Ogden, Utah.

Rieman, B. E., and J. D. McIntyre. 1995. Occurrence of bull trout in naturally fragmented habitat patches of varied size. Transactions of the American Fisheries Society 124:285–296.

Rieman, B. E., D. C. Lee, and R. F. Thurow. 1997. Distribution, status, and likely trends in bull trout within the Columbia River and Klamath River basins. North American Journal of Fisheries Management 17:1111–1125.

Schlosser, I. J., and P. L. Angermeier. 1995. Spatial variation in demographic processes of lotic fishes: conceptual models, empirical evidence, and implications for conservation. Pages 392–401 in J. L. Nielson, editor. Evolution and the aquatic ecosystem: defining unique units in population conservation. American Fisheries Society Symposium 17, Bethesda, Maryland.

Scudder, G. G. E. 1989. The adaptive significance of marginal populations: A general perspective. Pages 180–185 in C. D. Levings, L. B. Holtby, and M. A. Henderson, editors. Proceedings of the national workshop on effects of habitat alteration on salmonid stocks. Canadian Fisheries and Aquatic Sciences Special Publication 105.

Statistical Sciences. 1993. S-PLUS guide to statistical and mathematical analysis, Version 3.2. Statistical Sciences, a division of Mathsoft, Inc., Seattle, Washington.

Taylor, P. C., and B. W. Silverman. 1993. Block diagrams and splitting criteria for classification trees. Statistics and Computing 3:147–161.

Thompson, W. F. 1951. An outline for salmon research in Alaska. Paper presented at the Meeting of the International Council for the Exploration of the Sea. Amsterdam, October 1–9, 1951. University of Washington, Fisheries Research Institute Circular No. 18, Seattle.

Thurow, R. F., D. C. Lee, and B. E. Rieman. 1997. Distribution and status of seven native salmonids in the interior Columbia River basin and portions of the Klamath River and Great basins. North American Journal of Fisheries Management 17:1094–1110.

U.S.D.A. Forest Service and U.S.D.I. Bureau of Land Management. 1995. Decision Notice/Decision Record, Appendices for the Interim Strategies for Managing Anadromous Fish-Producing Watersheds in Eastern Oregon and Washington, Idaho, and Portions of California. U.S.D.A. Forest Service and U.S.D.I. Bureau of Land Management, Washington, D.C.

Utter, F. M., D. Campton, S. Grant, G. Milner, J. Seeb, and L. Wishard. 1980. Population structures in indigenous salmonid species of the Pacific Northwest. Pages 285–304 *in* W. J. Neil and D. C. Himsworth, editors. Salmonid ecosystems of the North Pacific. Oregon State University Press, Corvallis.

Waples, R. S., and C. Do. 1994. Genetic risk associated with supplementation of Pacific salmonids: Captive broodstock programs. Canadian Journal of Fisheries and Aquatic Sciences 51(Supplement 1):310–329.

WDF (Washington Department of Fisheries), Washington Department of Wildlife, and the Western Washington Treaty Indian Tribes. 1993. 1992 Washington state salmon and steelhead inventory, Olympia.

Williams, J. E., and seven coauthors. 1989. Fishes of North America endangered, threatened, or of special concern: 1989. Fisheries 14(6):2–20.

Withler, I. L. 1966.Variability in life history characteristics of steelhead trout (*Salmo gairdneri*) along the Pacific Coast of North America. Journal of the Fisheries Research Board of Canada 23:365–392.

13 Status and Stewardship of Salmon Stocks in Southeast Alaska

Benjamin W. Van Alen

Abstract.—Wild stocks of coho *Oncorhynchus kisutch*, chinook *O. tshawytscha*, sockeye *O. nerka*, pink *O. gorbuscha*, and chum *O. keta* salmon in Southeast Alaska were generally at historically high levels in the 1980s and 1990s, well above the weak runs that occurred in the 1950s, 1960s, and early 1970s. These healthy runs reflect the success of efforts to (1) protect the habitat; (2) rebuild and attain escapements that are appropriate in spatial and temporal distribution and abundance; and (3) appropriately enhance stocks. Good survival due to favorable ocean and climatic conditions also have contributed. Current stock assessment and management programs generally are adequate for recognizing and responding to conservation concerns. Fishery management actions are routinely taken to attain escapements needed to maximize yields, i.e., escapement levels that sustain healthy stocks far above levels risking collapse. A management strategy has evolved that harvests the thousands of island and coastal stocks in mixed stock fisheries; this effectively moderates exploitation rates and reduces the risk of overexploiting individual runs or temporal segments of runs, as occurred historically. The ability to reduce exploitation rates in years with weak runs is an important element for these mixed stock fisheries. Fisheries targeting local stocks are effectively managed for desired escapements, as facilitated by stock-specific estimates of runs and escapements. Habitat protection regulations are generally adequate to promote a healthy environment for salmon in the future, although the cumulative impacts of timber harvesting and localized mining and urbanization are of concern. If current escapements and habitat are maintained, future salmon production should remain healthy, fluctuating primarily with natural environmental variability.

INTRODUCTION

The Southeast Alaska management region, extending from Cape Suckling to Dixon Entrance, is one of the most productive areas in the world for five species of Pacific salmon—coho *Oncorhynchus kisutch*, chinook *O. tshawytscha*, sockeye *O. nerka*, pink *O. gorbuscha*, and chum *O. keta*. Salmon production in this area supports highly valued commercial, sport, and subsistence/personal use fisheries. Salmon have always been important in the diet and culture of local residents. Commercial fisheries for salmon began in Southeast Alaska in the late 1870s, and the salmon industry is currently the largest private employer in the region. Commercial harvests between 1987 and 1996 averaged 43.8 million pink, 6.4 million chum, 3.0 million coho, 2.2 million sockeye, and 0.26 million chinook salmon, and the annual combined exvessel value in 1996 dollars averaged $108 million. During this 10-year period, an average of 390 seine, 466 drift gillnet, 157 set gillnet, 827 power troll, and 638 hand troll vessels participated annually in commercial openings (ADF&G 1997). Limited entry regulations since the mid-1970s have capped the number of vessels that can participate in commercial openings. Approximately 3,100 salmon subsistence/personal use permits were issued in

1996, and sockeye salmon composed 89% of the harvest. Sport fisheries, particularly for chinook and coho salmon, are important throughout the region (Bentz et al. 1996); 114,000 anglers fished 510,000 angler-days in Southeast Alaska in 1996 (Howe et al. 1997). In addition, 642 sport fishing businesses with 1,165 employees registered 1,028 sport charter vessels in 1996 (P. Suchanek, Alaska Department of Fish and Game, personal communication).

The Southeast Alaska region consists of a relatively narrow strip of mainland on the west side of the Coast Range Mountains, the Alexander Archipelago comprising six large islands, and over 750 smaller islands within an area about 200 km wide and 800 km long (Figure 13.1). The mountain and glacial valley topography creates many small watersheds, and streams are usually less than 24 km long (USFS 1995). The more than 17,700 km of shoreline exceeds that of the West Coast of the continental U.S. The region is heavily forested with hemlock, spruce, and cedar trees below the 760–900-m treeline. Approximately 73,000 residents are distributed throughout the region among many small coastal communities. Few communities are linked to each other by road and only Haines, Skagway, and Hyder are accessible by road from mainland Canada and the lower 48 states. Government, fishing, tourism, forest products, and mining support the basic economy in the region.

The status of salmon in Southeast Alaska was widely questioned after the publication of some salmon stock declines in the Pacific Northwest (Nehlsen et al. 1991; NRC 1996) and British Columbia (Slaney et al. 1996), and biologists wondered whether salmon stocks in Southeast Alaska were in similar decline. Although managers of Alaska's fisheries and habitat resources have learned from past mistakes, wise conservation decisions in the future will depend on improved knowledge if they are to assure the long-term sustainability of salmon stocks and the fisheries dependent on them. To that end, this chapter assesses the current status of wild Pacific salmon in Southeast Alaska based on historical comparisons of catch and escapement data and a review of the efficacy of fisheries management, enhancement, and habitat protection. The future sustainability of salmon stocks in the region and the biological, environmental, and social factors affecting production are also discussed. I also further evaluate stock status as a complement to a prior assessment of Southeast Alaska salmon (Baker et al. 1996), which compared the relative magnitude of short- and long-term escapement count averages, as in Nehlsen et al. (1991), to categorize the escapement trend and risk of extinction for "spawning aggregates" (i.e., escapement survey sites).

DATA SOURCES AND DATA LIMITATIONS

Several information sources were selected to examine the past, present, and future status of salmon in Southeast Alaska. Trends in abundance primarily were assessed by comparing current and historical catch and escapement data. These data primarily are collected and maintained by Alaska Department of Fish and Game (ADF&G). Salmon catch and escapement data from 1960 to 1996 were obtained from the Division of Commercial Fisheries, Integrated Fisheries Database (L. Talley, ADF&G, personal communication). Historical catch estimates for Yakutat, northern Southeast, and southern Southeast Alaska, are documented back to the inception of commercial fisheries in 1878 (Rigby et al. 1991). These historical catch estimates probably are the best assessment of abundances prior to 1960 because harvests generally reflect run sizes. Estimates of hatchery contributions come from ADF&G Coded Wire Tag database (Johnson 1990). Bean (1889), Moser (1899), Pennoyer (1979), Royce (1989), and Meacham and Clark (1994) review and assess the historical development of the region's fisheries. Estimates of escapements and catches of sockeye salmon originating from the Nass and Skeena rivers were obtained from NBTC (1997) and joint NBTC documents (G. Oliver, ADF&G, personal communication). Sea surface and air temperature data were obtained from the National Oceanic and Atmospheric Administration (NOAA).

The Department's salmon escapement computer database was developed principally to record escapement survey counts for pink salmon, but it also includes survey counts for other species and weir and mark–recapture estimates of escapement. Most survey counts were made by management

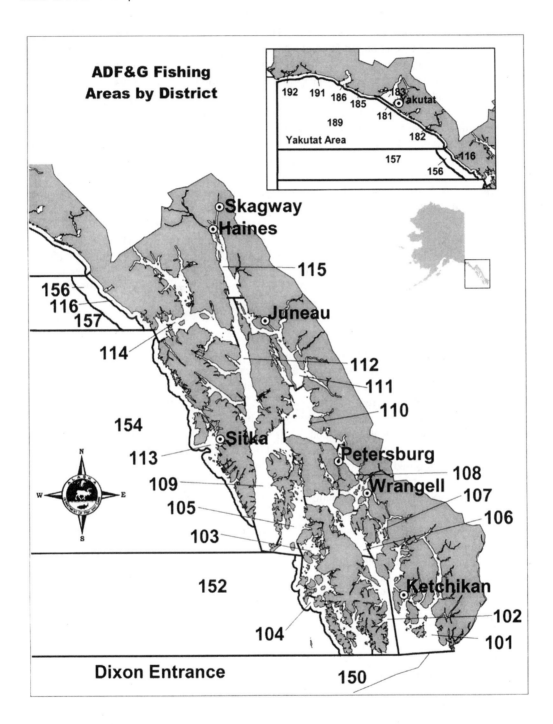

FIGURE 13.1 Map of Southeast Alaska and Alaska Department of Fish and Game's (ADF&G) commercial salmon fishing districts.

personnel for subjective inseason assessments of pink salmon timing, magnitude, and distribution in the streams surveyed. Most surveys are conducted from small, two-seat, fixed-wing aircraft or by foot surveys. Aerial or foot counts of sockeye, chum, and coho salmon are often obtained coincidentally to surveys for pink salmon but most of these survey counts are only useful for indexing escapements of the targeted species. The survey counts in this database underestimate total escapements. Observers do not attempt to correct for poor visibility, fish in the stream before or after the survey, unsurveyed parts of a stream, poor timing of surveys, or unsurveyed streams. Escapement survey counts are simply an estimate of the number of fish observed. Consistency among streams and years and obtaining precise counts has only been a primary objective in certain helicopter surveys for chinook (Pahlke 1995a) and coho salmon in specific index streams and sockeye salmon foot surveys in Hatchery Creek (McDonald Lake). The problems and limitations of using this qualitative data for quantitative indices of total escapement must be recognized. The "peak" escapement counts reported in this chapter underestimate the true escapement, are subject to among-observer bias, and have an unquantified level of precision (Bue 1998; Jones et al. 1998) and therefore should only be considered a relative indicator of escapement magnitude.

There is also a long regional history of counting salmon escapements through weirs. The midsummer timing of sockeye, pink, and chum salmon runs into the relatively small island and coastal streams with low midsummer flows is conducive to this method. However, there is an undercounting bias inherent in weir counts. Experience has shown that weir counts obtained at nonpermanent structures are not considered accurate unless validated by associated mark-recovery studies (e.g., McGregor and Bergander 1993; Shaul 1994; Kelley and Josephson 1997).

DEFINITIONS

Stock.—The term "stock" is used in a biological context similar to Ricker's (1972) definition of stock: "the fish spawning in a particular lake or stream (or portion of it) at a particular season, which fish to a substantial degree do not interbreed with any group spawning in a different place, or in the same place at a different season." Generally, stocks are characterized by the survey sites from which their individual escapement counts are obtained. These stocks are believed to have a high degree of reproductive isolation although infrequent interbreeding among stocks (i.e., straying) occurs naturally. Straying rates may be exacerbated by overcrowding or impediments to migration. Stocks might be composed of one or more "distinct populations" or "demes" (Geiger and Gharrett 1998); however, without specific knowledge of the geographic perimeters defining those individual demes, stocks have become the pragmatic units of real-world management.

Stock Group.—A term originating with salmon management that refers to geographic groupings of two or more stocks that experience similar environmental influences and have similar migration routes and timing. This enables stock groups to be managed as discrete units; thus, stock groups share common patterns of exploitation because management actions similarly harvest or protect fish in a stock group. Stocks in a stock group presumably have similar levels of productivity. A stock group might include a few stocks that are targeted in a local-stock fishery or include hundreds of stocks targeted in a passing-stock fishery (see definitions below).

Mixed Stock Fishery.—A fishery that targets two or more stocks, but generally many stocks or stock groups.

Passing-Stock Fishery.—A mixed stock fishery that targets stocks or stock groups that are actively migrating toward their spawning streams but are removed in space and time from their terminal areas.

Local-Stock Fishery.—A mixed-stock fishery that targets stocks returning to streams in a geographically confined area (often a bay, inlet, or canal) where there is a reasonable degree of spatial separation from stocks bound for other areas.

Terminal Fishery.—A fishery that occurs in a terminal area, i.e., an area immediately adjacent to the natal stream or within the natal stream of a targeted stock, where there is a significant degree of spatial separation from stocks bound for other streams. A terminal fishery therefore targets a single stock of fish although in reality the spatial and temporal separation from other stocks are seldom great enough to avoid harvesting at least some other stocks/species.

Escapement Distribution.—This refers to the spatial (geographic) and temporal distribution of salmon escapement within and among watersheds. Ideal escapement distribution results when fisheries are managed to achieve escapements that reflect the natural spatial and temporal distribution of each run so that the number, age, and sex composition of spawners maximizes the likelihood of their sustainability.

ABUNDANCE, ASSESSMENT, AND MANAGEMENT

The production of salmon is widely distributed throughout Southeast Alaska and Yakutat. There are currently 5,432 cataloged water bodies in the region producing anadromous fishes, including 2,993 principal salmon-producing streams (ADF&G 1994). Coho and pink salmon are found in virtually all of these, chum salmon in most, and sockeye salmon in those with accessible lakes and often in streams without lakes; chinook salmon are primarily found in mainland, glacially fed rivers. Some mainland rivers originate in Canada and flow to the sea through the U.S. These transboundary rivers include the Alsek, Taku, and Stikine rivers; their salmon production and management have been addressed by Jensen (2000) and will not be treated in detail here. Many of these Southeast Alaska streams are small producers, but collectively, they contribute a substantial portion of the region's annual salmon production. A brief review of the biology, abundance trends, and current stock assessment and management programs for each species is presented below.

Coho Salmon

Southeast Alaska provides excellent habitat for coho salmon, a species often associated with small coastal streams. High precipitation and a vast, convoluted shoreline provide flow and drainage for the almost 3,000 principal salmon-producing streams currently known in the region. Coho salmon spawn or rear in most of these streams. Side-channel, beaver pond, slough, and tributary areas are preferred rearing habitat for coho, especially in the larger mainland rivers (Murphy et al. 1989).

The number of spawners in the many small to medium streams often is well below 1,000 fish, but collectively, they represent a substantial portion of the overall production. Lake systems also are important, and each lake typically produces runs of 1,000 to 8,000 fish. Large stocks occur in the larger mainland rivers, including the Taku, Chilkat, Berners, Stikine, Unuk, and Chickamin, and in most systems along the Yakutat forelands. The productivity (return per spawner) of coho salmon appears to be less for inland spawning stocks in the large transboundary rivers than in the smaller island and coastal streams (COHOTC 1986, 1991; Shaul 1990). However, much of the regional coho salmon production comes from these large, mainland rivers.

Most coho salmon rear in freshwater for 1, 2, or 3 years and in the ocean for 16–18 months. Their homeward migration from the Gulf of Alaska takes a northwest to southeast routing. Thus, southern inside stocks are primarily harvested in northern outside and southern inside fisheries, whereas stocks returning to northern inside waters are harvested primarily in northern areas. Maturing coho salmon primarily return to their natal streams from late August to early October; spawning usually occurs in October and November.

Commercial harvests of wild coho salmon during the 1990s exceeded harvest levels in the 1940s and were well above the lows experienced in the late 1950s, 1960s, and early 1970s (Figure 13.2). Annual harvests have been 1.0–5.7 million during 1980–1996. Eight of the 11 highest harvests in the 117-year history of the region's fishery have occurred since 1984.

The majority of coho salmon harvested in Southeast Alaska are wild stocks. However, hatchery contributions have been almost 1 million fish in some recent years (Figure 13.2). Hatchery contributions averaged almost 794,000 (21%) during 1992–1996 and have remained relatively constant.

Roughly 80% of the coho salmon harvested in Southeast Alaska originate from Alaskan streams, the transboundary rivers, and Alaskan hatcheries; the remainder originate primarily from coastal streams in northern British Columbia (B.C.) (COHOTC 1994). Negligible numbers originate from

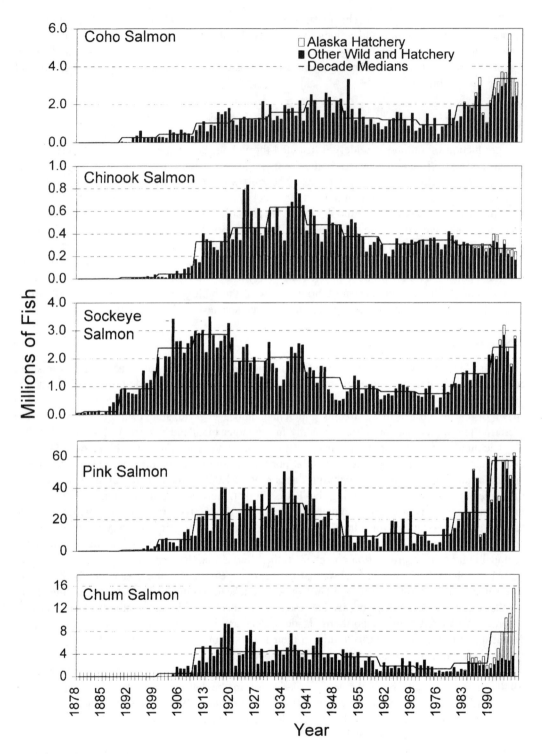

FIGURE 13.2 Annual harvest of coho, chinook, sockeye, pink, and chum salmon in Southeast Alaska, 1878–1996.

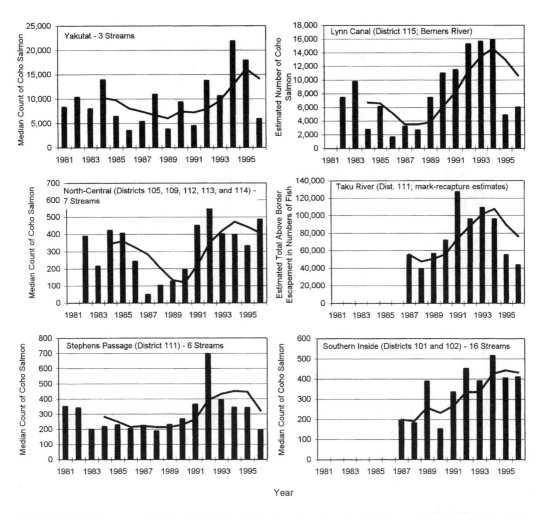

FIGURE 13.3   Escapement indices for Southeast Alaska coho salmon by stock group, 1981–1996. The solid line represents the 4-year moving average.

southern B.C., Washington, and Oregon. Northern B.C. coho salmon stocks are most concentrated in southern inside waters in July and early August but can occur anywhere in the region throughout the summer. Coho salmon harvest levels in northern B.C. in the 1980s and 1990s have not increased as they have in Southeast Alaska, indicating that survival and production have been lower for northern B.C. stocks (COHOTC 1991).

Current escapement trends, by stream and stock group, can be assessed from yearly "peak" survey counts. I enhanced the reliability of these indices by restricting streams to those consistently surveyed for coho salmon. The E-M algorithm (McLachlan and Krishnan 1997) was used to fill in missing values under the assumption that the expected count is determined by the individual stream and the individual year in a multiplicative way (i.e., counts across years for a stream are multiples of counts in other streams, and that counts across streams for a year are multiples of counts in other years). The estimated expected count for a given stream in a given year is then equal to the sum of all counts for the stream times the sum of all counts for the year divided by the sum of all counts over all streams and years and is iterated to convergence.

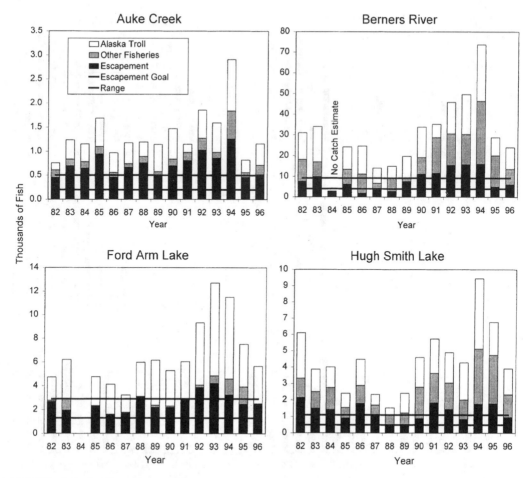

FIGURE 13.4 Total run size, catch, escapement, and escapement goal range for four wild Southeast Alaska coho salmon indicator stocks, 1982–1996.

None of the 34 streams in the six stock groups for which there is sufficient data (Figure 13.3) had significant declining trends in escapement (Spearman's rho rank correlation trend test, at α = 0.05; Conover 1980). At the α = 0.20 level, escapement trends were positive in 12 of the 34 streams and positive in the Southern Inside and North Central stock groups. No streams or stock groups had declining trends in escapement at the α = 0.20 level.

Estimates of escapement goals, escapements, returns, exploitation rates, and survivals during 1982–1996 are available for four wild indicator stocks: Ford Arm Lake (District 113), Berners River (District 115), Auke Creek (District 111), and Hugh Smith Lake (District 101) (Figure 13.1) (Shaul 1994; Shaul and Crabtree 1996; L. Shaul, ADF&G, personal communication). Escapement goal ranges for maximizing the sustained yields of these four stocks are based on spawner-recruit analysis (Clark et al. 1994). Total runs show a relatively stable production trend from the early 1980s to the early 1990s with a rise in production in the mid-1990s. Annual escapements have consistently been within or above goal ranges in each of these indicator stocks (Figure 13.4). There is no indication of a pattern of overexploitation during this 13-year period. Given the strong escapements to these indicator stocks, and the relatively stable smolt production (Shaul and Crabtree 1996), it is likely that rearing areas have been fully utilized in recent years.

Given adequate distribution and abundance of escapements, coho salmon smolt production appears to be limited more by rearing than spawning habitat. Competition for space and food among coho salmon juveniles limits the number that rear in the preferred off-channel, beaver pond, and slough/tributary habitats. Coho salmon juveniles unsuccessful in competing for space in a preferred rearing area are forced into less productive habitat, or are forced out of the drainage altogether (Mason and Chapman 1965; Crone and Bond 1976). Survival of displaced coho salmon is much lower, on average, and more dependent on favorable environmental conditions. Such a rearing limitation would explain the relatively stable smolt production observed in these four indicator stocks over the years (Shaul and Crabtree 1996).

Most harvest presently occurs in highly mixed stock or gauntlet fisheries distant from the streams of origin. The region's commercial troll fishery takes about 60–65% of the coho salmon catch, mostly in offshore and corridor areas where stocks are highly mixed and well in advance of when maturing fish enter their natal streams. Coho salmon are targeted during certain weeks in the Lynn Canal, Taku/Snettisham, Prince of Wales, and Stikine drift gillnet fisheries and in all Yakutat setnet fisheries. Coho salmon are caught incidentally to the harvest of other species in the purse seine fisheries. Recreational fisheries target mixed stocks in marine waters near each community. Subsistence fisheries for coho salmon occur near Yakutat, Haines, and Angoon. All production releases of coho salmon from hatcheries in the region are represented by a coded wire tag code, and tag recovery programs are in place for estimating hatchery contributions inseason.

The wide distribution of coho salmon production among thousands of streams in the region necessitates that the majority of the harvest occur in mixed stock areas where the effort can be moderated and spread over as many stocks as possible. This reduces the risk of overexploiting individual stocks. Escapements are only monitored in a small subset of these streams because of practical and financial limitations.

The primary management objective for stocks that are harvested predominately in highly mixed stock fisheries is to regulate average exploitation rates to a sustainable level and to reduce these rates in years of poor wild stock runs. ADF&G staff monitors harvest rates through indicator stock coded wire tagging programs on the four wild stocks described above. Catch rates in the District 106 gillnet fishery have also proven to be a reliable inseason indicator of returns to southern inside waters. The troll fishery is managed with the aid of an early season forecast of the wild coho salmon regionwide harvest based on troll catch per boat day and coded wire tag information collected in mid-July (Figure 13.5). If the run appears weak, harvest rates are reduced by decreasing troll fishing time and area and by one or more areawide closures. Since 1980 (except 1994), an annual mid-August closure of about 10 d has provided for conservation and allocation to net fisheries, as mandated by the Alaska Board of Fisheries in 1989. Otherwise, the troll fishery has a relatively continuous season and harvests a relatively constant fraction of the return. Average troll fishery harvest rates for the four indicator stocks are moderate, averaging 42% since 1982, but fishery restrictions have dropped harvest rates below 35% in years of low returns, such as 1988 (Shaul and Crabtree 1996). This ability to take early, fishery-wide action to reduce troll harvest rates is most beneficial for the southern inside stocks, which contribute heavily to the regionwide troll fishery in July and August and have the highest average exploitation rate because they pass through more fisheries than do the more northerly stocks.

Local-stock gillnet fisheries in Lynn Canal, Taku River, Stikine River, and Yakutat systems are managed to achieve levels of escapement that are thought to be near optimum or at least sustainable. In late August drift gillnet fisheries usually begin to target returns to southern inside waters in Districts 106, Stikine River in District 108, Taku River in District 111, and Lynn Canal systems in District 115 (Figure 13.1). These fisheries are managed primarily by comparing current and his-torical catch and effort statistics for wild stocks. Feedback from escapement assessment programs is generally not timely enough for inseason management needs with two exceptions: escapement surveys of Yakutat area streams (Clark and Clark 1994) and a fishwheel mark–recapture project on the lower Taku River.

Coho salmon escapements in the monitored systems have been at levels that have sustained, if not maximized, production. This also is likely to be the case for the many unmonitored systems given the wide distribution of production, the highly mixed stock nature of the harvest, and probable

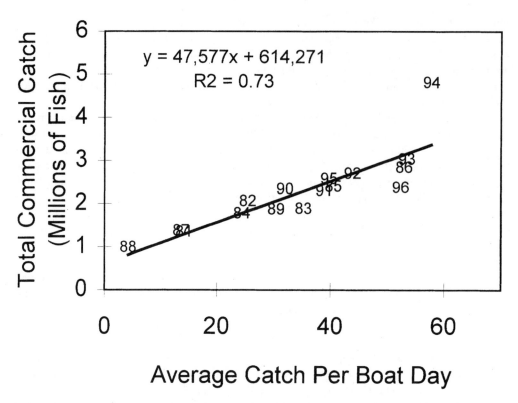

FIGURE 13.5 All-gear commercial catch of wild coho salmon in Southeast Alaska predicted from the mean average troll catch per boat day during statistical weeks 28 and 29, 1981–1996.

similarity in productivity among streams within stock groups. Annual coho salmon returns likely vary primarily from natural fluctuations in survival, assuming production is rearing limited and escapements are generally adequate to fully seed available freshwater spawning and rearing habitat. Since 1982, for the three indicator stocks with complete run reconstruction estimates, marine survival has accounted for 62% (range 49–80%) of the observed variation in abundance, and smolt production accounts for the remainder (L. Shaul, ADF&G, personal communication). Marine survivals vary annually but increased from an average of about 10% in the early 1980s to about 20% in the early 1990s. Thus, the record runs in the early 1990s were largely driven by favorable marine survivals. However, these strong runs would not have occurred if escapements had not been well distributed spatially, temporally, and numerically or if the freshwater habitat was broadly damaged.

CHINOOK SALMON

Wild chinook salmon stocks are known to spawn in at least 33 transboundary and mainland rivers and one island river in Southeast Alaska (Pahlke 1995a). The five large, glacial, transboundary rivers (Taku, Stikine, Alsek, Chilkat, and Unuk) and the Chickamin River are the largest producers. Joint Canadian and U.S. management and stock assessment of the five transboundary stocks are addressed by the Pacific Salmon Commission (PSC) under the terms of the Pacific Salmon Treaty. Details of Taku, Stikine, and Alsek chinook management and production are covered in Jensen (2000).

Southeast Alaska and transboundary chinook salmon are all "spring runs"; i.e., spawners usually enter rivers from late April through July and spawn from July through September. Most stocks produce yearling smolt (stream-type). The Situk River on the Yakutat Foreland is an exception, smolt are primarily age 0 (ocean-type; Johnson et al. 1992). Ocean residency is usually 3–4 years for females and 1–4 years for males (Olsen 1995). In general, the more inland a stock spawns, the more offshore its ocean distribution. Stocks that spawn east of the coast range mountains tend to rear in the Gulf of Alaska, whereas stocks that spawn in coastal and island rivers tend to rear in waters of the Alexander Archipelago for all, or most, of their ocean residency (Van Alen 1988; Pahlke 1995b).

Some chinook salmon from Pacific Northwest and British Columbia streams also migrate north and rear in waters of Southeast Alaska for at least a portion of their ocean life (Orsi and Jaenicke 1996). The majority of the traditional troll, net, and sport fisheries harvest these highly mixed stocks of chinook salmon. From 1985 to 1996, approximately 18% of the commercial chinook harvest originated from Alaskan or transboundary rivers, streams, and hatcheries, 51% from British Columbia, and 31% from Washington/Oregon (D. Gaudet, ADF&G, personal communication). Hatchery contributions to traditional fisheries are significant, both from Alaskan and non-Alaskan hatcheries. For example, all hatcheries contributed an average of 37% and Alaskan hatcheries 13% of the chinook salmon harvested by the troll fishery from 1983 to 1996.

Chinook salmon harvests are below historical levels (Figure 13.2); harvest ceilings have been in effect since 1981. Harvest peaks as high as 600,000 to nearly 900,000 occurred in the 1920s, 1930s, and early 1940s, prior to construction of dams on the Columbia River, whereas recent catches have averaged 270,000. However, annual harvests below 350,000 also occurred several times during this historical period.

It was apparent in the mid-1970s that wild chinook salmon production in the region was well below historical levels (Kissner 1974). In response, the ADF&G initiated closures of directed drift gillnet, setnet, and recreational fisheries starting in 1976, and a formal 15-year rebuilding plan was established in 1981 (ADF&G 1981). This plan used restrictions in directed commercial and recreational fisheries, spring closures of the regionwide troll fishery, and restrictive, all-gear catch ceilings to reduce exploitation rates and increase escapements. In 1985 this Alaskan rebuilding program was incorporated into the comprehensive PSC coastwide rebuilding program. The slow and steady decline in annual harvests in the 1980s and 1990s (Figure 13.2) reflects management under restrictive catch ceilings.

A program to index escapements of Southeast Alaska and transboundary river chinook salmon was started in the mid-1970s (Kissner 1976; Pahlke 1995a) and annual escapements are currently estimated from standardized escapement indices conducted by ADF&G and the Canadian Department of Fisheries and Oceans (Pahlke 1996). Helicopter surveys and weirs are used to count chinook escapements at 27 locations in 11 systems. ADF&G is conducting studies to convert escapement indices to total escapement estimates and to use coded wire tagging information to estimate fishery contributions; escapements are also sampled for age, sex, and size data (USCTC 1997). This information has been used to develop spawner–recruit-based escapement goals for six of these 11 systems: the Situk River near Yakutat (ADF&G 1991); the Alsek River (Jensen 2000); and the Unuk, Chickamin, Blossom, and Keta rivers in the Behm Canal area (McPherson and Carlile 1997). These escapement goal ranges are the escapements needed to maximize sustained yields (MSY). These spawner-recruit (S-R) escapement goals will replace the goals established prior to 1985, which were simply the largest escapement goals seen prior to 1981. New S-R-based goals so far have generally been lower than the original targets. This is to be expected because the original targets were set high under conditions of uncertainty and to obtain data on extreme escapements needed for S-R modeling. The appropriateness of these escapement goal ranges will be assessed periodically as additional years of spawner-recruit data become available.

Escapements of large (age-.3 and older) chinook salmon have been trending upward since the rebuilding program began in 1981 (Figure 13.6). Escapements to the six systems with S-R escapement goals have generally been within or above goal ranges in most years (ADF&G 1997). The five systems without quantitatively determined MSY escapement goals show increasing escapements during this rebuilding period. In fact, escapements had been well above goals in some systems

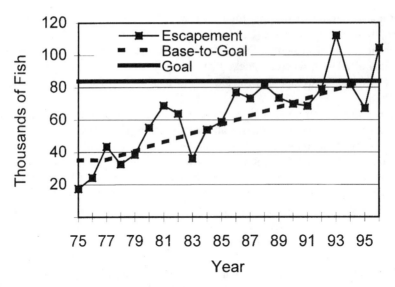

FIGURE 13.6   Escapements of Southeast Alaska/transboundary river chinook salmon relative to the rebuilding plan and escapement goal, 1975–1996.

in the late 1980s, and subsequent returns may have suffered somewhat as a result (McPherson and Carlile 1997). Efforts to increase escapements have worked well, and Alaska and transboundary river chinook stocks are now considered rebuilt and healthy by ADF&G.

Management and stock assessment programs currently focus on (1) maximizing the harvest of Alaska hatchery-produced chinook salmon; (2) achieving wild stock escapements at levels to maximize returns; and (3) regulating commercial and recreational fisheries to minimize incidental mortalities and comply with provisions of the Pacific Salmon Treaty and Endangered Species Act (McGee et al. 1996). Directed fisheries for Alaskan/transboundary chinook salmon must be actively managed inseason to achieve escapements within desired ranges. Unfortunately, escapement estimates have not been available or used to manage fisheries inseason (except Situk River), and this has necessitated a conservative management strategy. Nevertheless, stock assessment and management programs have been generally adequate for recognizing and responding to annual variations in run size of Alaska and transboundary river stocks. ADF&G has initiated new studies of the Alsek, Chilkat, Taku, Stikine, Unuk, and Chickamin stocks that will help assess run strengths and improve escapement management. Postseason evaluation of the consequences of management actions on wild stock escapements and exploitation rates on hatchery runs is also integral to management. All hatchery releases are represented with coded wire tags, and the tag-recovery data are used to estimate migration timing, routes, harvest rates, and fishery contributions for both hatchery and wild stocks. The catch and harvest rates of non-Alaskan chinook salmon stocks are also estimated using analysis of coded wire tag data.

SOCKEYE SALMON

Wild sockeye salmon originate from over 200 systems in Southeast Alaska and Yakutat. A large share of the production is from the transboundary Alsek, Taku, Stikine, and Chilkat rivers and other mainland systems such as the Situk and Chilkoot rivers and McDonald Lake. Production from the numerous island and coastal stocks is poorly monitored but substantial. Most production comes from systems with lakes accessible to rearing juveniles but some systems without lakes also produce sockeye salmon. Onshore migration from their Gulf of Alaska rearing area follows a general northeasterly direction circumnavigating the islands they encounter enroute to their natal streams (Hoffman et al. 1984; Burgner 1991). The migration timing of sockeye salmon through inside waters varies widely among stocks; abundances in the region's fisheries are highest in July, and the majority spawn from late August to early October.

Regionwide harvests of sockeye salmon in the 1980s and 1990s have been much higher than they were in the late 1940s through the late 1970s. However, harvests have not consistently been up to the 2.0–3.5 million range of 1900 to 1920 (Figure 13.2). Historical harvests reveal a 30-year trend of declining harvests between 1920 and 1950, followed by another 30 years of low harvests before the dramatic increases in the 1980s and 1990s.

Reliable indices, or estimates, of annual escapements are available for just a handful of the over 200 streams and rivers that produce sockeye salmon in the region and estimates of total runs and biological escapement goals, or goal ranges, are available for only a subset of these stocks. ADF&G's database contains counts of sockeye salmon in 206 different Southeast Alaska and transboundary streams from 1960 to 1996, but I consider escapement estimates sufficiently reliable to assess escapement trends in only 16 systems. Total runs are estimated for nine of these systems. These data are not without error, and all systems have years where data are missing or unsuitable for this purpose. Weir counts provide the basis for escapement estimates in most of these systems. Mark-recapture studies are now used to estimate escapements or serve as a backup to the weir estimates in five of these systems. With few exceptions, peak aerial, foot, or boat counts of sockeye salmon are poor indices of annual escapement. Usually only a small and unknown fraction of the run is present in the area(s) surveyed, and fish are difficult to see due to tannin-stained water that is often deep with a dark substrate and swift current.

Estimates of escapements and returns of sockeye salmon to the East Alsek, Italio, Situk, and Lost rivers on the Yakutat foreland are available for most years since the mid-1970s. Escapement goal ranges for MSY have been established for each of these systems (Clark et al. 1995a, b). Management of the set gillnet fisheries that target returns to each of these systems has effectively provided escapements within goal ranges in recent years. Sockeye salmon of the transboundary Stikine, Taku, and Alsek rivers are described in detail in Jensen (2000).

Chilkat and Chilkoot rivers are the principal contributors of sockeye salmon to the Lynn Canal gillnet fishery. Programs have been in place for estimating escapements, harvests, and S-R-based escapement goals for early and late runs in both systems since the 1970s (McPherson 1990).

Reliable estimates of escapement have only been available in recent years for two sockeye salmon systems in southern Southeast Alaska: Hugh Smith and McDonald lakes. A weir has been used to count adult returns of sockeye salmon into Hugh Smith Lake since 1982, and escapement estimates have been as low as 1.3 thousand and as high as 68.6 thousand during 1980–1996 (NBTC 1997).

McDonald Lake is the largest producer of sockeye salmon in southern Southeast Alaska. Production has been enhanced by lake fertilization since 1982. Annual returns have generally been in the 200,000 to 400,000 fish range in recent years and annual escapements averaged 90,400 from 1979–1996 and were in excess of ADF&G's informal 70,000–85,000 goal in 9 of the 18 years that estimates are available.

Using Spearman's rank correlation (rho) trend test (Conover 1980), I found a significant ($\alpha =$ 0.05) downward trend in sockeye salmon escapement counts over the 1980–1996 period in the Chilkoot and Italio rivers. Baker et al. (1996) also classified Italio River sockeye escapements in decline, the only one of 26 sockeye stocks they found to be decreasing. The drop in escapements to the Chilkoot River reflects escapement-based management of the Lynn Canal gillnet fishery and a lowering of the escapement goal from 80,000 to 70,000 in 1982 and then to 62,000 (range of 50,500 to 91,500) in 1989 (McPherson 1990). Total runs and escapements in 1994, 1995, and 1996 were down, probably because of overescapements in the parent years in this rearing-limited system. Densities of copepods *Cyclops columbianus*, the dominant macro-zooplankter, decreased 17-fold between 1987 and 1991 and biomass decreased 9-fold (Barto 1996). Zooplankton densities trended upward in 1995 and 1996 but were still well below 1987. If these trends continue, sockeye salmon production is expected to increase with returns from the 1995 and 1996 broods (D. Barto, ADF&G, personal communication).

In the Italio River, total escapements and runs dropped dramatically in 1988 after the river changed course and began flowing into the Akwe River Lagoon in 1988. Clark et al. (1995a) suspected that this hindered the homing ability of the sockeye adults returning to the Italio River over the next cycle of returns. Runs appear to be rebuilding in recent years.

Management of sockeye salmon runs differs substantially between areas and stocks. The drift gillnet fisheries in Districts 108, 111, and 115 (Figure 13.1) are intensively managed to achieve sockeye escapements within ranges estimated to maximize sustained yields of Stikine, Taku/Snettisham, and Lynn Canal stocks, respectively. Inseason programs are in place to estimate escapements, catches, and runs for individual stock groups. Differences in scale patterns and parasite prevalence are used for postseason stock identification. Run reconstruction is used to evaluate the success of management actions, set and reevaluate escapement goals, and serve as the basis for detailed management plans. A similarly intensive management program is in place for the set gillnet fishery targeting Situk River sockeye salmon, and Alsek River runs are also intensively managed with aid of an historical catch per unit effort model.

Sockeye salmon stocks in other areas of the region are managed less intensively. In the Yakutat area, local stocks are targeted in some terminal set gillnet fisheries. Terminal, or near-terminal, seine fisheries in Yes Bay target the McDonald Lake stock and likewise the Whale, Necker, and Red Fish Bay stocks on Baranof Island. Management of these local-stock fisheries relies on qualitative assessments of run strength to achieve escapements within desired ranges. Mixed stocks of sockeye salmon compose a substantial portion of the salmon harvested in June and July openings of the District 101 and 106 drift gillnet fisheries and the July seine openings in Districts 101, 102, 104, 105, 109, and 112 (Figure 13.1). Many other sockeye salmon are harvested incidentally in pink salmon purse seine fisheries.

Subsistence and personal use effort have traditionally been directed at sockeye salmon returning to streams located near each community. Harvests are relatively small in total but could be large in relation to the total run or escapement in some locations.

Sockeye salmon from the Nass, Skeena, and Fraser rivers in British Columbia contribute to fisheries in southern Southeast Alaska. This has occurred since commercial fishing began in the region in the late 1800s. Annual scale pattern analysis to identify stocks (Farrington et al. 1996; Jensen and Bloomquist 1994) is used to make postseason estimates of the contributions of Alaska, Nass, and Skeena sockeye salmon by gear, area, week, and year. Nass and Skeena sockeye salmon compose a 1987–1996 average of 60% of the 1.2 million sockeye salmon harvested in Districts 101 and 106 gillnet fisheries and District 101–104 seine fisheries (Figure 13.1). Sockeye escapements in the Nass and Skeena rivers have been at or above escapement targets in most years (Figure 13.7). Total returns have been quite variable between 1980 and 1996. The Alaskan exploitation rates ranged from 21 to 46% with a mean of 31% on Nass sockeye salmon and from 5 to 22% with a mean of 12% on Skeena sockeye salmon. Looking back further, the Nass and Skeena sockeye salmon return is trending upward and is now 2–3 times greater than in the late 1960s and early 1970s (Figure 13.7). A large share of the Skeena River production now originates from spawning channels and adults airlifted above a barrier falls on tributaries of Babine Lake.

Fraser sockeye salmon are caught in some years in a purse seine fishery targeting returns of Alaskan pink salmon along the west shore of Noyes and Dall islands in southern outside waters (District 104). The catch of Fraser sockeye salmon is relatively small; 1980–1996 averages are 0.5% of the District 104 catch and 0.6% of the total Fraser return (using estimates provided by the PSC). Annual harvests are extremely variable and depend on a poorly understood relationship between the size of the Fraser run, oceanic conditions that influence their migration patterns, and the seine effort in District 104, which depends on the strength of the pink salmon run to southern Southeast Alaska.

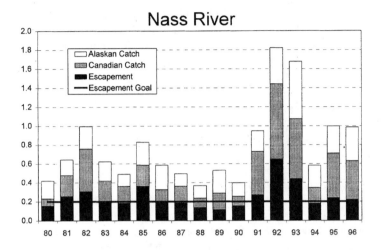

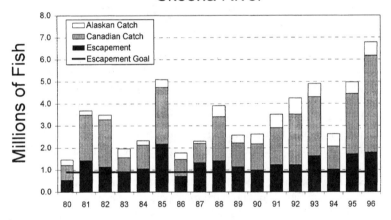

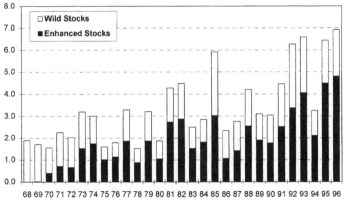

FIGURE 13.7 Nass and Skeena river sockeye run sizes, 1980–1996, and total wild and enhanced run, 1968–1996.

Enhancement programs for sockeye salmon have existed in the region almost since the start of commercial exploitation (Roppel 1982). These early hatchery programs were mostly failures, and total adult production was minor relative to the wild stock production. At present, Beaver Falls, Klawock, and Snettisham hatcheries incubate sockeye salmon. Enhancement efforts currently include hatchery releases of age-1 smolts, planting of age-0 fry in lakes, and lake fertilization programs. Total annual harvests of enhanced fish peaked at an estimated 358,000 in 1993 (McNair 1995) but have been 220,000 in recent years. Hatchery enhancement in the Stikine and Taku rivers are building up to runs of over 100,000 annually (Jensen 2000).

Pink Salmon

Pink salmon production originates from over 2,500 streams in Southeast Alaska, and these native fish predominate throughout the region's harvests. Northern British Columbia fish are also harvested in southern Southeast Alaska. Adult tagging studies have found a distinct separation in migration routes between pink salmon stocks in the northern (Districts 109–114) and southern (Districts 101–108) portions of the region (Nakatani et al. 1975); hence, management actions are independent. Management of runs to the outer coast of Chichagof and Baranof islands (District 113) is also independent from the management of other areas. Adults are most abundant in the fishing areas from late July through late August.

Pink salmon returns in the region do not show persistent trends of odd- or even-year dominance. I do not treat odd- and even-year classes independently in this review because the factors that influence the sustainability of these stocks are independent of year. Both year classes spawn and rear in the same habitat and are subjected to similar environmental conditions and patterns of exploitation.

Pink salmon harvests have been historically large since the mid-1980s (Figure 13.2). Eight of the 11 highest harvests in history have occurred from 1986 to 1996. Annual harvests currently range from 30 to 60 million fish. Hatchery-produced pink salmon compose a small fraction (<3%) of the harvest.

Local ADF&G staff intensively monitor escapements of pink salmon to Southeast Alaska streams from aircraft and by foot throughout the main part of the run, mid-July to early September. Counts are occasionally made from helicopters, at weirs, or by mark-recapture projects. Aerial surveys, flown several times each week during the peak weeks of the run, are used for inseason assessment of run strength and run timing, and managers try to survey all major streams and a consistent representation of smaller streams. Since 1960, managers have counted pink salmon in 1,588 different streams, but usually fewer than half are surveyed in any given year. However, most of the major producers and many of the minor producers are surveyed three or more times each season.

Escapement trends were estimated using "peak" aerial (fixed wing) survey counts from 652 streams that were surveyed by ADF&G area management personnel in 12 or more years from 1960–1996. The inherent variability in counting rates among observers was controlled in two ways. First, counts were limited to those made by observers who had flown more than 100 surveys a year in 5 or more years. Second, within each management area, each observer's counts were standardized to the counting rate of the current area management biologist by regressing instances that past and present pairs of observers had surveyed the same stream within 3 days of one another (K. Hofmeister, personal communication). The region was then divided into 50 stock groups; observer-standardized escapement data were available in two or more "index" streams in 43 of these stock groups. Missing data in each stock group were interpolated using the E-M algorithm (McLachlan and Krishnan 1997) applied across streams and years as described for coho salmon above. Missing data were not interpolated in any stock group and year when fewer than three streams were surveyed or if less than half of the streams were surveyed in any stock groups with four or fewer index streams. Escapement indices presented for northern and southern areas of Southeast Alaska were calculated using ADF&G's traditional method of summing the peak escapement counts for all streams surveyed in a year and expanding for unsurveyed streams (ADF&G 1997; K. Hofmeister, ADF&G, personal communication).

Overall, escapement indices show an upward trend for both northern ($\alpha = 0.0001$) and southern ($\alpha = 0.0001$) Southeast Alaska between 1960 and 1996 (Figure 13.8). This upward trend is also evident for individual stocks in each district (ADF&G unpublished data). There is a strong relationship between runs and subsequent escapements during this period. This relationship is most evident in southern Southeast Alaska for escapements of up to about 25 million (the escapement is approximated here by multiplying the index count by 2.5; Figure 13.9). The poor returns in 1987 and 1988 from the record parental escapements in 1985 and 1986, might be attributed to a combination of overescapement, an extremely harsh winter affecting the 1985 brood, daggertooth *Anotopterus pharao* predation affecting the 1986 brood, and anomalous lows in the upwelling indices measured off Dixon Entrance and Sitka during 1987 and 1988 (Hofmeister 1994; K. Hofmeister, ADF&G, personal communication).

Baker et al. (1996) reported declining trends in escapement in 8 of 389 odd-year and 15 of 363 even-year pink salmon stocks they evaluated from 1960–1994. Using Spearman's rank correlation (rho) trend test (Conover 1980), I found a significant ($\alpha < 0.05$) downward trend in escapement counts for two of these stocks – Snake Creek (District 107, Olive Cove) and Florence Creek (District 112, SW Admiralty Island). The decline in Snake Creek pink salmon escapement counts is probably an artifact of changing observers and survey methods. Survey counts for Snake Creek and the other 19 streams in this stock group all have positive Spearman's rho values when only observer bias-corrected escapement counts are used. The significant decline in pink escapement counts to Florence Creek may have also resulted from a change in survey methods. Some total escapement counts in the 1960s were inflated by high counts of fish off the stream mouth. Escapement surveys in this area in recent years have focused on instream counts because fish that school off the mouth of a stream often spawn in other streams (Pella et al. 1993). Nevertheless, Florence Creek was the only one of 652 index streams that had a significant downward trend (Spearman's rho, $\alpha < 0.05$) in escapement counts for years 1960 to 1996.

Management of pink salmon stocks is generally based on preseason and inseason forecasts of run size to southern and northern Southeast Alaska, inseason escapement counts, and catch, effort, and sex composition (run timing) data. Preseason forecasts have performed reasonably well in recent years (Geiger and Simpson 1995). The success of new projects for inseason run size forecasting still need to prove themselves over time (McKinstry 1993; Mathisen and Van Alen 1995). Management depends heavily on each area manager's subjective assessment of the run strength and migration timing. The overall management objective is to achieve escapements that are well distributed and at MSY levels. The consequences of management actions are evaluated postseason by a comparison of each district's observed escapement index with the minimum escapement goal (Northern Southeast) or goal range (Southern Southeast; ADF&G 1997).

At present, most harvest occurs in highly mixed stock fisheries where harvest rates are moderate and the risk of overexploiting individual stocks is minimized. Purse seine gear harvests most (90%) of the pink salmon caught in the region. Pink salmon are targeted by the seine fleet in most areas and openings and management actions focus on inseason assessment of this species' abundance. Purse seine fishing is allowed in all districts except 108, 115, and 116, the Yakutat districts, and the offshore districts (Figure 13.1). Pink salmon are also targeted in some weeks in late July and August in the Districts 101, 106, and 108 drift gillnet fisheries and in the Cross Sound experimental troll fishery in June.

The fishing effort (boat-days) in passing stock fisheries is directly related to the size of the pink salmon runs at low to moderate run sizes but is relatively stable at higher run sizes (Figure 13.10). Seine and gillnet fisheries throughout the region are opened concurrently, which promotes dispersion of the approximately 375 seine and 470 drift gillnet vessels that fish annually.

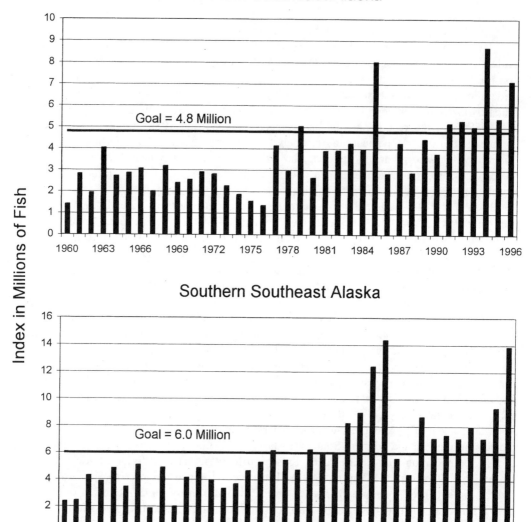

FIGURE 13.8   Escapement indices for pink salmon in northern and southern Southeast Alaska, 1960–1996.

It also reduces the risk of overharvesting fish in any one area. Seine and gillnet boats are allowed to fish in any open area in the region. These seine and gillnet areas do not overlap.

CHUM SALMON

Over 1,500 streams produce chum salmon in the region, both summer and fall runs being widely distributed. Summer runs peak in inside waters from mid-July to mid-August, whereas fall runs peak in September. Ages 0.3, 0.4, and 0.5 predominate in both runs.

Harvests of wild chum salmon in the 1980s and 1990s exceed those in the 1960s and 1970s but are well below the harvest levels observed in the first half of the 20th century (Figure 13.2).

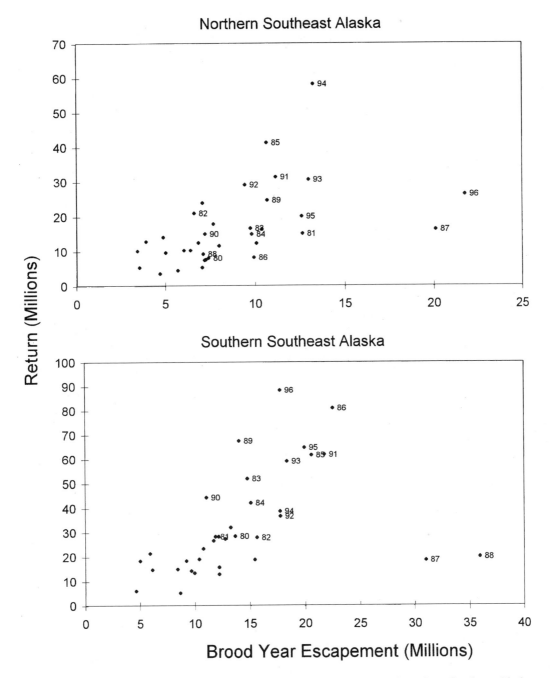

FIGURE 13.9 Spawner-recruit relationships for pink salmon in southern and northern Southeast Alaska, plotted for return years 1962–1996.

The lows in the 1960s and 1970s were preceded by a 40-year trend of declining chum harvests dating back to 1920.

Abundances of hatchery-origin chum salmon have increased dramatically since 1992, providing over half the chum harvest in 1994, 1995, and 1996 (Figure 13.2). Returns of Southern Southeast Regional Aquaculture Association releases to Naket, Neets, and Kendrick Bay; Northern Southeast Regional Aquaculture Association releases to Hidden Falls, Deep Inlet, and Boat Harbor; and

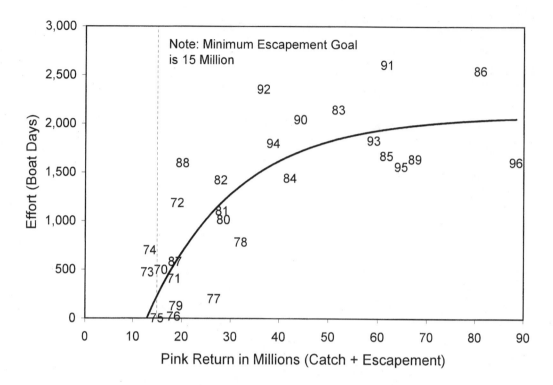

FIGURE 13.10 Fishing effort (number of boat-days fished) related to the pink salmon run. Example from the District 104 purse seine fishery, post statistical week 30, 1970–1996. Line is fit of asymptotic regression.

Douglas Island Pink and Chum, Inc. releases to Amalga Harbor and Limestone Inlet have contributed the bulk of the common property and hatchery cost-recovery harvests of enhanced chum salmon in recent years (McNair and McGee 1994).

ADF&G does not have a standardized program for indexing the escapement of chum salmon in the region although aerial and foot escapement survey counts dating back to 1960 are available in ADF&G's database. Most of these counts of chum salmon were obtained opportunistically in surveys directed at counting pink salmon. Summer chum runs tend to be earlier than pink salmon but there is a broad overlap in their escapement timing. Observers often record separate counts for pink and chum salmon in early surveys of a stream and record counts as "mostly pink" when pink salmon predominate and obscure counts for other species. Therefore, annual variability in pink salmon escapements has probably compromised indexing chum salmon escapements, and high pink salmon escapements may have masked high chum salmon escapements.

With this in mind, I present a rough index of chum salmon escapements to six stock groups in the region by plotting the median peak aerial count for the 180 streams in the region that have been surveyed in at least 10 years from 1960 to 1996. Missing data in each stock group were interpolated using the E-M algorithm (McLachlan and Krishnan 1997) applied across all streams and years as described for coho salmon above. I pooled districts geographically to construct the six stock groups.

The median counts tended to be above 1,000 fish per stock group in most of these 37 years (Figure 13.11). Counts tended to be cyclical and were highest in the 1960s, mid-1970s, mid-1980s, and mid-1990s. An examination of these rough indices does not indicate any persistent decline in chum salmon escapements in any stock group. However, counts tended to be higher in the 1960s and early 1970s.

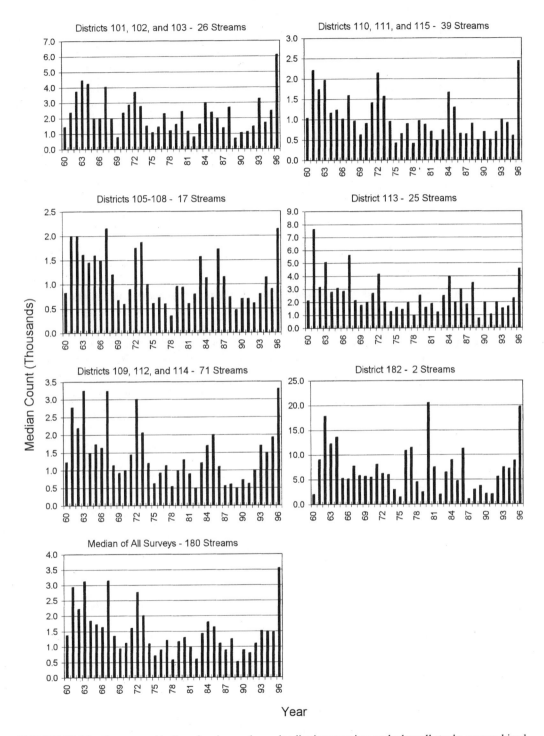

FIGURE 13.11 Escapement indices for chum salmon, by district grouping, and when all stocks are combined, 1960–1996.

Fall chum salmon production from the Taku and Chilkat rivers has declined in the 1990s, provided reduced terminal area gillnet harvests and escapement counts are a valid indicator. The reasons for this decline are unknown. The production drop might be related to changes in the rivers' course across the flood plain, to loss of productive upwelling areas where chum salmon have spawned in the past, to increased exploitation rates in fisheries directed at strong returns of coho salmon, or to poor early marine survivals.

Baker et al. (1996) evaluated escapement trends for 45 chum salmon stocks in the region and found declining escapements in 10. Escapement counts had a significant decline in 12 of the 180 stocks I evaluated (Spearman's rho, $\alpha = 0.05$; Conover 1980), but escapements were not declining in any of the six stock groups. Interestingly, the Chilkat River was the only stock classified as declining in both studies. This decline in escapements and production of Chilkat River chum salmon has been an ADF&G management concern since the mid-1980s.

Summer-run chum salmon are generally caught incidentally in fisheries targeting pink salmon and other salmon species. Chum salmon abundances factor into the management of several early-season seine and gillnet openings. In September and early October fall-run chum salmon returning to Cholmondeley Sound, Chaik Bay, and Excursion Inlet are often targeted by seine fisheries. Chum and coho salmon are targeted in fall gillnet openings in Lynn Canal. Management of these fisheries is passive; only catch, catch per effort, and non-indexed escapement counts are used to assess stock status. Efforts to reduce exploitation on chum salmon stocks are usually based on a comparison of inseason and historical catch and fishery performance.

Factors that limit chum salmon production in the region are poorly understood. Runs in successive years are often driven by a strong year class. S-R models have performed poorly and been hampered by imprecise estimates of escapements and returns by age and stock. Although chum salmon production in the region is clearly sustainable at present levels of escapement, management needs to be more responsive to annual variations in abundance to keep escapements within ranges likely to maximize production. More reliable and timely estimates of wild stock escapements and run sizes are needed to direct management of the mixed species fisheries which harvest chum salmon.

ADEQUACY OF PROGRAMS

The sustainability of wild salmon production is dependent, first, on the quality and quantity of the spawning, rearing, and migration habitat and second, on the quality and quantity of the escapement (Baker et al. 1996; Gregory and Bisson 1997). Salmon production and sustainability are maximized when the natural environment is consistent, the habitat pristine, and when escapements are naturally dispersed in space and time and at optimal numbers. If land-use activities destroy salmon habitat, then salmon production will be lost regardless of the quality and quantity of the escapement. Disturbing spawning, rearing, or migration habitats will almost always lower survival rates and production of salmon stocks that have evolved under the original environmental conditions. Over-fishing results in less than adequate escapement which also reduces total salmon production.

HABITAT PROTECTION

Extensive clearcut logging of old growth timber and associated roading, particularly on Prince of Wales, Baranof, and Chichagof islands, and localized urbanization, have altered salmon habitat and production potentials; nevertheless, vast areas of pristine salmon habitat still remain and must be protected. The actual impact that logging has had on salmon production is an issue of considerable discussion, but ongoing investigations should reveal information that may more clearly resolve the long-term, cumulative impact clearcut logging has on salmon production. The Tongass Timber Reform Act (Public Law 101-626) and the Alaska Forest Resources and Practices Act (AS 41.17) have required streamside buffers along anadromous fish-producing streams on federal, state, and private lands since 1990. These 100-ft buffer strips on federal and state lands and 66-ft strips on private lands provide minimum protection from impacts that we know most about—the loss of large woody debris and streamside canopy (Murphy and Koski 1989). The USDA Forest Service

(USFS) appropriately recommends additional protection of headwater areas and flood plains from logging (USFS 1995).

Alternatives to clearcut logging are also receiving serious consideration. The May 1997 revision of the Tongass Land Management Plan seeks to better balance the interests of timber harvesters with the public's desire to maintain fish and wildlife habitat. The plan provides additional protection for salmon habitat by reducing the maximum allowable timber harvest; placing more old-growth forest in nontimber reserves; and increasing nontimber buffers adjacent to streams, river mouths, and beaches. The plan also recommends 32 rivers for National Wild, Scenic, or Recreational River classification. However, the practice of clearcutting some old-growth forests will continue for the foreseeable future.

State and federal regulations appear to be effectively protecting most of the region's habitat from other land use activities. The National Environmental Policy Act and Alaska Coastal Management Plan mandate analysis of environmental impacts and air, water, and habitat quality standards. ADF&G updates the catalog of anadromous fish-producing streams annually, and permits are needed to do any work on, or over, such streams; stiff fines may be imposed if work is done without a permit, or in violation of the permit. Most importantly, a strong public interest in maintaining good salmon habitat and awareness of detrimental forest practices elsewhere have deterred those activities in Southeast Alaska. Dams have been built on only a few small anadromous streams in the region, and Alaska law prohibits hindering the free passage of salmon without compensation. The cost of complying with the current environmental regulations is not trivial, and do become important debits in the benefit–cost feasibilities of mining and other resource development activities.

FISHERIES MANAGEMENT

Alaska's constitution provides an important separation in responsibility of regulatory decisions affecting allocation and those affecting conservation. The Alaska Board of Fisheries, a citizen panel composed of commercial, recreational, and subsistence fishery representatives working in an open public process, establishes management plans and regulations for allocating salmon among user groups. The Board of Fisheries has the authority to establish optimum escapement goals or goal ranges usually based on recommendations of ADF&G. These escapement goals might be set below what biologists estimate is needed to maximize sustained yield but ADF&G policy requires that the goals be high enough to sustain the resource. ADF&G is responsible for management of salmon harvest on a sustained yield basis in a manner that meets the Board's conservation and allocation objectives. Commercial, subsistence, and personal use fisheries are managed by the Division of Commercial Fisheries Management and Development. Recreational fisheries are managed by the Division of Sport Fish. Allocations among gear groups are usually independent of annual abundances and expressed as long-term average percent of fish harvested by each gear type or by setting fishing time for one gear type in relation to fishing time afforded another gear type that is targeting the same stocks of fish. Allocations of transboundary Taku and Stikine River sockeye salmon between Canadian and U.S. fishers are expressed as percentages of the total allowable catch. Hatchery brood stock and cost recovery needs are also considered in the management, but priority is given to management of commingled wild stocks. To protect the manageability and economic vitality of the commercial fleet, the number of troll, gillnet, and seine permits was set by limited entry regulations in the mid-1970s. The Board of Fisheries fixed vessel and gear dimensions and the areas that each gear type can fish to help constrain growth in fishing power. Enforcement of fisheries regulations is the responsibility of state troopers with the Department of Public Safety, Division of Fish and Wildlife Protection, with assistance from deputized ADF&G biologists.

Area management biologists and their support staff, stationed in the principal ports of landing, Ketchikan, Petersburg, Sitka, Juneau, Wrangell, Haines, and Yakutat, closely monitor returns and escapements and the open fisheries for specific areas and times by "emergency order." ADF&G's

management of fisheries is intended to take advantage of the surplus production potential inherent in salmon stocks by managing for escapements that fall within optimal ranges well above the minimum number needed to sustain the stock. Management's primary goals are to achieve the distribution and abundance of spawners needed to (1) sustain, if not maximize, production, and (2) provide for traditional subsistence harvests. Secondary objectives are to facilitate an orderly harvest of salmon of the highest quality and value in commercial fisheries and of the greatest benefit to recreational and personal use fishers, consistent with user group allocations established by the Board of Fisheries. Transboundary and boundary area fisheries are managed to comply with terms of the Pacific Salmon Treaty as is the regionwide harvest of chinook salmon.

A mixture of terminal, local-, and passing-stock fisheries have evolved in Southeast Alaska in response to an ever-increasing mix of requisites: the physical geography of the area; the distribution, abundance, and migratory patterns of the wild and hatchery salmon; the capabilities of the boats and gear; markets for the fish; and constantly changing regulations (ADF&G 1997). Terminal fisheries accounted for an average of 15% of the sockeye and 5% of the pink salmon harvested in the region from 1993 to 1996. Terminal fisheries include most of the set gillnet fisheries in the Yakutat area, that target discrete stocks by fishing in the mouth and estuary of the streams, and a few purse seine and troll fishing openings in the region that target salmon returning to specific rivers or hatchery release sites.

Local-stock fisheries include the drift gillnet fisheries that target, depending on the season, sockeye, chum, and coho salmon returning to Chilkat and Chilkoot rivers in Lynn Canal (District 115), Taku River in Taku Inlet (District 111), and the Stikine River in upper Frederick Sound and Stikine Flats (District 108). Management of these local-stock fisheries are usually supported by inseason programs to assess returns and the adequacy of escapements relative to quantitatively determined escapement goals. Escapement monitoring programs are generally adequate to recognize and respond to conservation concerns for nontargeted species. Where needed, inseason or postseason programs estimate hatchery contributions or contributions of principal wild stocks.

The majority of the salmon harvested in Southeast Alaska are harvested in mixed stock fisheries—some in local-stock fisheries and some in passing-stock fisheries. To prevent overfishing of individual stocks immigrating to the numerous and widely distributed streams, the majority of the effort is directed into mixed stock areas, held to conservative levels, and spread over as many stocks as possible. These passing-stock fisheries are managed inseason based on the abundance of the targeted stocks and conservation concerns for nontargeted stocks. Harvest rates are adjusted inseason to accommodate low or high wild stock runs. Managers of passing-stock fisheries must have timely assessments of run strengths and be able to reduce fishing effort when runs are weak. In years with larger runs, the terminal and local-stock fisheries are able to harvest a relatively larger share of the run because harvest rates in passing-stock fisheries are constrained by boat, gear, area, and time limitations to err on the side of conservation (Figure 13.10). Likewise, in years with small runs, effort is reduced in both passing-stock and terminal/local-stock fisheries, but because terminal/local-stock fisheries occur at the end of the run, and have the greatest direct impact and dependence on a stock, they bear the greatest responsibility for ensuring that escapements are acceptable. Foregoing harvests of healthy stocks is an inevitable consequence of management of mixed stock fisheries that exploit multiple stock groups.

Directing fisheries toward mixed stock areas has proven to be an effective management tool for some stocks. Because returning spawners often school at the mouth of their home stream awaiting a freshet before ascending to spawn, a single purse seine boat could sometimes overfish a stock with one set of the net. Furthermore, regulating harvests to avoid overexploiting individual runs or run timing segments was impossible given the broad distribution of production from hundreds of small- and medium-sized streams and limited stock assessment and enforcement capabilities. Terminal/local-stock fisheries are often difficult to regulate because removal rates are often quite high and inseason programs for estimating returns and escapements by stock are often inadequate, impractical, unproven, or too costly. In many locations escapements needed to sustain stocks and fisheries are poorly understood, and habitat and S-R information for establishing optimum escapement goals are usually lacking. Therefore, confining the fisheries to the more terminal bays and inlets, as practiced in the early 1900s resulted in systematic overharvests and a gradual deterioration of the resource. Thus, it is risk-averse to direct harvests outside of terminal areas where exploitation rates can be moderated and the risk of overexploiting individual runs is minimized. Managers now rely on surveys of these terminal areas and adjacent streams to assess abundances and adequacy of the escapements, and when appropriate allow local-stock fisheries to harvest the surpluses. This passing-stock fishing strategy also provides for an orderly harvest of bright, high quality fish with a low-cost management program.

The importance of reducing harvest rates in mixed stock areas in years of poor returns cannot be overemphasized. Management must be based on inseason assessments of run strength. Preseason forecasts might be available for specific stocks, but they have not proven reliable enough for effective abundance-based management. Similarly, harvest and harvest rate ceilings have no place in abundance-based management given the natural variability in survivals and returns and the need to reduce harvest rates for maintaining escapements when runs are smaller than expected.

With exception of chinook salmon managed under the Pacific Salmon Treaty, management is not based on preseason expectations of run strength but rather on comparisons of fishery performance data between current and past years as an indicator of run strengths and timing. Catch and catch per effort data from early season openings are used to predict harvests of chinook, coho, and pink salmon (Mathisen and Van Alen 1995; ADF&G 1996) and, combined with escapement monitoring programs, to predict total returns of most major sockeye salmon stocks in the region.

ADF&G lacks reliable information for assessing escapements of salmon to hundreds of small- and medium-production streams. There is presently no evidence that these smaller producers are being overexploited by fisheries targeted at large producers for three reasons. First, the larger producers are not specifically targeted because the fisheries are directed into areas where the stocks are highly mixed, making it impossible to target on specific large stocks. Second, much of Southeast Alaska's salmon production comes primarily from numerous small and medium producers, rather than large producers. Third, there is no evidence indicating that smaller producers are inherently less productive than larger producers, especially among stocks from the same geographic area. All evidence suggests a strong correlation in productivity (survival) between small and large producers within stock groups. That is, when runs are strong to the larger streams they are also strong in the smaller streams. When returns are quite large, as has been the case for most years in the 1980s and 1990s, then escapements to all streams, regardless of size or productive capacity, are also up. The runs have approached or exceeded the capacity of the fleet to take all the harvestable production (Figure 13.10). The number of commercial fishing permits is fixed by Limited Entry and boat and gear specifications (restrictions) set by Board of Fisheries regulations have constrained the fishing power of the fleet. The pattern of large escapements in years with large returns is related to both limitations in effort and processing capacity. Salmon stocks having similar migration routes and timing show a good correlation between escapements and runs as a result. Any future shift in management emphasis from quantity to quality (and value) will further reduce exploitation rates on pink salmon and increase escapements of co-migrating species.

ENHANCEMENT

Management of hatchery activities is important to the health of wild salmon stocks. Less productive wild salmon could be overharvested in fisheries flooded with hatchery fish. Without precise knowledge of the abundance of wild and hatchery stocks, managing mixed stock fisheries to ensure the escapement of wild salmon becomes less precise (Wilbur and Frohne 1989). The genetic "fitness" of a stock may be compromised by introgression such as when hatchery returns stray into non-natal systems and interbreed with indigenous runs. Last, hatchery-produced salmon could reduce survivals of wild fish by competing for space and food in both the freshwater and marine habitats.

Many of the details of Southeast Alaska salmon hatchery programs are described in Smoker et al. (2000).

Wild stocks spawning and rearing adjacent to release sites are at greatest risk from enhancement activities, and overharvesting less productive wild stocks in fisheries targeted on more productive hatchery fish also is possible. The impact that hatchery strays have on the fitness and productivity of wild stocks is less understood but of concern (Reisenbichler 1997). Although monitoring programs are generally adequate to track wild stock escapements, when enhanced fish stray and spawn with wild runs, some stock identification is needed (e.g., thermal mark recovery) to estimate escapements of wild fish. There are several hatchery programs in the region that release large numbers of fish in "remote" marine release sites. In some cases the hatchery fish are not provided a freshwater source to imprint in, or escape to. Thermal mark recovery studies have shown that a high proportion of the pink and chum salmon spawning in Juneau-area streams in recent years originated from Gastineau Hatchery releases (D. Steinman, Douglas Island Pink and Chum, Inc., personal communication). Experience has generally shown that wild salmon escapements will fluctuate with abundance of hatchery returns and that these runs will persist even if enhancement programs are curtailed.

Another potential interaction between production releases of salmon and local wild stocks is competition for space and food in shared habitats. High numbers of wild and hatchery salmon could exceed carrying capacities, particularly in freshwater and early-marine environments, and reduce overall survival, especially if environmental conditions are poor. The poor survival of the 1991 brood of pink and chum salmon in the Juneau area might have been related to the combined abundance of hatchery (54 million pink salmon and 74 million chum salmon fry released) and wild fry (parental escapements were good) in an early-marine environment lacking enough food to support them. However, brood year failures have occurred when no hatchery fish were involved (e.g., poor survivals of pink salmon in the early 1970s), and harsh environmental conditions almost always noticeably reduce subsequent runs. Large hatchery releases might also attract predators that could reduce survivals of both wild and hatchery fish, although there is no convincing evidence that this has occurred in Southeast Alaska.

We must recognize that large-scale production releases of hatchery fish will impact management and production of wild stocks regardless of the thoroughness of catch and escapement monitoring programs. However, most enhancement programs now in place have indeed made substantial contributions to commercial, sport, personal use, and subsistence harvests without any yet identified effect on the sustainability of local wild stocks. An important benefit of several enhancement programs has been to draw fishing effort onto hatchery returns and away from areas where wild salmon are targeted.

ASSESSING THE FUTURE BY REVIEWING THE PAST

TRENDS OF THE PAST

Overfishing caused the long downward trend in salmon production from the peak years in the early 1900s through the low years in the 1950s, 1960s, and early 1970s (Bean 1889; Moser 1899; Royce 1989). Overfishing began with the first commercial salmon fisheries when barricades, traps, and nets were used in a wide-open melee that captured nearly all salmon returning to many streams (Bean 1889). Moser (1899) pointed out that the pack was dominated by sockeye salmon in 1889 and that, only 10 years later, it was dominated by pink salmon because "all the streams within 70 to 80 miles of the canneries have been scoured [of sockeye salmon] with all the gear that could be devised or used."

Intensive, competitive, and poorly regulated seine, trap, gillnet, and troll fisheries overharvested whole runs or temporal segments over the 60-year period prior to statehood in 1959. Sockeye stocks were the first to decline from overfishing followed by chum salmon and pink salmon; have never

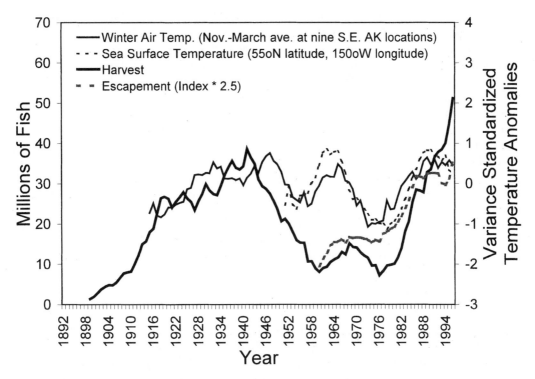

FIGURE 13.12 Winter air temperatures and sea surface temperatures and harvests and escapements of pink salmon in Southeast Alaska, 8-year moving averages, 1892–1996.

rebuilt to historical levels. The White Act, in effect from 1924 to 1959, required fisheries to harvest no more than half the run. This act was routinely implemented during years 1924 to 1945 by simply closing fisheries after mid-July (Thorsteinson 1950). Early-run segments were certainly overfished in these years (Alexandersdottir 1987). Coho salmon were the last species to decline in production, probably because their later run timing subjected them to lower harvest pressure during the White Act years. The decline in coho salmon abundance, which began in the early 1950s, coincides with the expansion of the commercial troll, seine, and gillnet fleets on top of a fully developed trap fishery (L. Shaul, ADF&G, personal communication) and a probable decline in natural survivals in the 1960s that affected all species.

Similarly, the increase in wild salmon production observed in the 1980s and 1990s resulted largely from the success of local fishery managers improving the distribution and abundance of escapements at a time of generally favorable environmental conditions. Enhancement (e.g., reducing harvest pressure on wild stocks) and advances in stock assessment and management have played important roles in the rebuilding of these stocks.

Oceanwide, interdecadal shifts in environmental conditions might help explain some of the variability in Alaskan salmon catches (Hare and Francis 1995). However, the influence that escapements have on the production of individual stocks was not considered in these analyses. Marshall's (1992) modeling of Alaskan pink salmon suggested that escapements were more useful in developing accurate forecasts of run size than were indices of local and regional environmental conditions. Escapements explain over 80% of the variability in ADF&G's 1967–1997 forecasts of pink salmon runs, whereas indices of local environmental factors (e.g., winter air temperature) explain only 5% (Figure 13.12; Hofmeister 1994; K. Hofmeister, ADF&G, personal communication). Farley and Murphy (1997) concluded that changes in escapement policies and short-term environmental fluctuations had more of an influence on sockeye salmon production than did large-scale atmospheric-oceanic regime shifts.

Increased returns of pink salmon in the 1980s and 1990s were correlated with increases in sea surface temperatures measured at 55°N latitude and 150°W longitude and November-February air temperatures averaged across nine Southeast Alaska NOAA weather stations (Figure 13.12). However, escapements were also trending upward during these years. If favorable environmental conditions were primarily responsible for the increased returns in the 1980s and 1990s, then why was there a long period of declining returns of pink and other salmon species in the 1940s and early 1960s when these environmental parameters were also trending upward? I suspect that favorable environmental conditions in these years did not compensate for the cumulative effects of past overfishing.

Environmental influences on runs is most evident when salmon populations are within MSY ranges because MSY escapement goals are, by definition, the escapements that maximize yield under average environmental conditions. Trends in coho salmon production are probably a good example. Over the 1980s and 1990s an average of 62% of three indicator stocks' run size can be explained by marine survivals (L. Shaul, ADF&G, personal communication). Coho salmon production was positively correlated with sea surface temperatures because (1) escapements were at, or above, MSY levels; (2) smolt production was limited by available rearing habitat; and (3) annual returns were strongly influenced by ocean survivals.

FUTURE OF STOCKS

At present, salmon stocks throughout Southeast Alaska are generally healthy and at little or no risk of collapse due to human actions. Only routine management actions are necessary to maintain escapements needed to sustain, if not maximize, production. The cumulative impacts of urbanization have put salmon in a small Juneau creek, Duck Creek, at risk and triggered a community-based restoration effort (Koski 1996). The health of our region's salmon stocks is ultimately dependent on this type of public awareness and involvement in protecting the spawning, rearing, and migration habitat. The need for a strong public interest, and the political will to maintain clean water, healthy watersheds, and abundant salmon is paramount to the long-term sustainability of the resource. The cumulative effect of urbanization and other land/water use activities pose the greatest anthropogenic threat to the sustainability of the region's salmon resource (Tarbox and Bendock 1996). The conversion of salmon habitat to other land and water uses is the primary reason for the decline in production, and production potential, of salmon in the Pacific Northwest (NRC 1996; Stouder et al. 1997). Therefore, the principal risk to the long-term sustainability of Alaska salmon is dependent upon compromises between habitat protection and resource development that face our elected officials in the years to come.

In many ways, the future looks good for sustaining healthy salmon stocks. Southeast Alaska is near the central part of the natural range of Pacific salmonids; environmental perturbations would have to be extreme to put these stocks at risk. The habitat is relatively intact. Habitat protection measures have been relatively effective and have been backed by a strong public interest in abundant salmon. The 1997 revision of the Tongass Land Management Plan provides salmon additional protection from the impacts of logging and other land use activities. The two large pulp mills in the region are now shut down and their long-term (50-year) timber contracts have been canceled. Fishery managers are strongly oriented to conservation; stock assessment programs are generally adequate for abundance-based management of mixed stock fisheries and escapement-based management of local-stock fisheries. Most importantly, local fishery managers closely monitor escapements and runs inseason and have full authority to regulate harvests to meet escapement needs. Continuing efforts by local management biologists to maintain the natural spatial and temporal distribution of wild salmon escapements within and among watersheds will be integral to the current health of Southeast Alaska salmon stocks.

We must be vigilant to human factors that can compromise the sustainability of wild salmon stocks. These include erosion of habitat protections, erosion of funding for basic stock assessment and management, management succumbing to industry or public pressure to overharvest in years with weak returns, and detrimental interactions between hatchery and wild fish if production is not moderated. On a larger and longer scale, global warming resulting from greenhouse gas emissions could lower survival and production of salmon throughout the Pacific Rim. If this happens, it will become more important than ever for managers to ensure optimal escapements and protect freshwater habitats.

At present, viability of the region's commercial fisheries is somewhat threatened because funding for basic stock assessment and management is eroding. Also, supplies of wild and hatchery salmon are exceeding market demand. The quantity of the harvest, albeit at record highs, is unable to compensate for the low prices paid to fishermen and processors. An inevitable shift in emphasis from harvest volume to harvest value in future years may result in less emphasis on hatchery production, reduced exploitation rates on wild stocks, and increased escapements. The rapidly expanding sport-charter industry will need to be better regulated to assure a fair allocation of salmon among the user groups. Finally, there is a real threat that future allocation arrangements under the Pacific Salmon Treaty will compromise continuation of the cooperative management of our shared international salmon resource for maximum sustained yield. Fisheries traditionally managed based on the abundance of local stocks might be disrupted to achieve a politically mandated numerical balance in interceptions.

ACKNOWLEDGMENTS

The information used in this status review came from many past and present salmon stock assessments projects in the region. In particular, I thank Leon Shaul for his input on coho salmon; Ron Josephson for transboundary river salmon; Scott McPherson for chinook, coho, and sockeye; Karl Hofmeister for pink and chum salmon; Keith Pahlke for chinook salmon; Glen Oliver for Boundary Area sockeye salmon; Fred Bergander and Dave Barto for Lynn Canal sockeye salmon; and Ben Kirkpatrick for habitat matters. Tim Baker provided computer files summarizing his review of Southeast Alaska salmon escapements. Jim Blick provided invaluable biometric support and Cori Cashen produced the maps. I am grateful for the technical and editorial reviews by Mason Bryant, John Burke, Dave Cantillon, John H. Clark, Phil Doherty, Doug Eggers, Kathleen Jensen, Robert Marshall, Andrew McGregor, Scott McPherson, Dave Peacock, Norma Jean Sands, and Leon Shaul. The statements and opinions in this chapter are those of the author and not necessarily those of the reviewers or ADF&G. This work was supported in part by NOAA Awards NA46FA0344 and NA57FP0425. This chapter is contribution PP-169 of the Alaska Department of Fish and Game, Division of Commercial Fisheries, Juneau.

REFERENCES

ADF&G (Alaska Department of Fish and Game). 1981. Proposed management plan for Southeast Alaska chinook salmon runs in 1981. Alaska Department of Fish and Game, Commercial Fisheries Management and Development Division, Unpublished Report RUR 1J81-3, Juneau.

ADF&G (Alaska Department of Fish and Game). 1991. Regulations of the Alaska Board of Fisheries for commercial fishing in Alaska, 1991–1993 Southeast-Yakutat commercial fishing regulations, Juneau.

ADF&G (Alaska Department of Fish and Game). 1994. Catalog of waters important for spawning, rearing or migration of anadromous fishes. As revised October 24, 1994. Alaska Department of Fish and Game, Division of Habitat and Restoration, Juneau.

ADF&G (Alaska Department of Fish and Game). 1996. Management plan for chinook and coho salmon in the Southeast Alaska/Yakutat summer troll fishery, 1996. Alaska Department of Fish and Game, Commercial Fisheries Management and Development Division, Regional Information Report 1J96-18, Juneau.

ADF&G (Alaska Department of Fish and Game). 1997. Commercial, subsistence, & personal use salmon fisheries Southeast Alaska-Yakutat Region 1996. Alaska Department of Fish and Game, Commercial Fisheries Management and Development Division, Regional Information Report 1J96-32, Juneau.

Alexandersdottir, M. 1987. Life history of pink salmon (*Oncorhynchus gorbuscha*) and implications for management in Southeast Alaska. Doctoral dissertation, University of Washington, Seattle.

Baker, T. T., and eight coauthors. 1996. Status of Pacific salmon and steelhead escapements in southeastern Alaska. Fisheries 21(10):6–18.

Barto, D. L. 1996. Summary of limnology and fisheries investigations of Chilkat and Chilkoot Lakes, 1987–1991. Alaska Department of Fish and Game, Commercial Fisheries Management and Development Division, Regional Information Report 5J96-07, Juneau.

Bean, T. H. 1889. Report of the salmon and salmon rivers of Alaska, with notes on the conditions, methods, and needs of the salmon fisheries. U.S. Fish Commission Bulletin, Volume 9.

Bentz, R., P. Suchanek, M. Bethers, S. Hoffman, A. Schmidt, M. Dean, and R. Johnson. 1996. Area management report for the sport fisheries of Southeast Alaska, 1994. Alaska Department of Fish and Game, Division of Sport Fish, Fishery Management Report 96-1, Anchorage.

Bue, B. G., S. M. Fried, S. Sharr, D. G. Sharpe, J. A. Wilcock, and H. J. Geiger. 1998. Estimating salmon escapements using area-under-the-curve, aerial observer efficiency, and stream-life estimates: The Prince William Sound pink salmon example. Proceedings of the International Symposium on Assessment and Status of Pacific Rim Salmonids. North Pacific Anadromous Fisheries Commission, Scientific Bulletin 1: Vancouver, British Columbia.

Burgner, R. L. 1991. Life history of sockeye salmon. Pages 1–118 *in* C. Groot and L. Margolis, editors. Pacific salmon life histories. UBC Press, Vancouver, British Columbia.

Clark, J. H., A. Burkholder, and J. E. Clark. 1995a. Biological escapement goals for five sockeye salmon stocks returning to streams in the Yakutat Area of Alaska. Alaska Department of Fish and Game, Commercial Fisheries Management and Development Division, Regional Information Report 1J95-16, Juneau.

Clark, J. H., S. A. McPherson, and A. Burkholder. 1995b. Biological escapement goals for Situk River sockeye salmon. Alaska Department of Fish and Game, Commercial Fisheries Management and Development Division, Regional Information Report 1J95-22, Juneau.

Clark, J. H., and J. E. Clark. 1994. Escapement goals for Yakutat area coho salmon stocks. Alaska Department of Fish and Game, Commercial Fisheries Management and Development Division, Regional Information Report 1J94-14, Juneau.

Clark, J. E., J. H. Clark, and L. D. Shaul. 1994. Escapement goals for coho salmon stocks returning to Berners River, Auke Creek, Ford Arm Lake and Hugh Smith Lake in Southeast Alaska. Alaska Department of Fish and Game, Commercial Fisheries Management and Development Division, Regional Information Report 1J94-26, Juneau.

COHOTC (Coho Technical Committee). 1986. Report of the Joint Coho Technical Committee to the Pacific Salmon Commission. Pacific Salmon Commission TCCOHO (86)-1, Vancouver, British Columbia.

COHOTC (Coho Technical Committee). 1991. Northern Panel area coho salmon status report. Pacific Salmon Commission TCCOHO (91)-1, Vancouver, British Columbia.

COHOTC (Coho Technical Committee). 1994. Interim estimates of coho stock composition for 1984–1991 southern area fisheries and for 1987–1991 northern panel area fisheries. Pacific Salmon Commission TCCOHO (94)-1, Vancouver, British Columbia.

Conover, W. J. 1980. Practical nonparametric statistic, 2nd edition. Wiley and Sons, New York.

Crone, R. A., and C. E. Bond. 1976. Life history of coho salmon, *Oncorhynchus kisutch*, in Sashin Creek, southeastern Alaska. Fishery Bulletin 74(4):897–923.

Farley, E. V., and J. M. Murphy. 1997. Time series outlier analysis: evidence for management and environmental influences on sockeye salmon catches in Alaska and Northern British Columbia. Alaska Department of Fish and Game, Alaska Fishery Research Bulletin 4:36–53.

Farrington, C. W., G. T. Oliver, and R. A. Bloomquist. 1996. Contribution of Alaskan, Canadian, and transboundary sockeye stocks to 1992 catches in Southeast Alaska purse seine and gillnet fisheries, Districts 101–108, based on analysis of scale patterns. Alaska Department of Fish and Game, Commercial Fisheries Management and Development Division, Regional Information Report 1J96-11, Juneau.

Geiger, H. J., and A. J. Gharrett. 1998. Salmon stock at risk: what's the stock, and what's the risk? Alaska Fishery Research Bulletin 4:178–180.

Geiger, H. J., and E. Simpson. 1995. Preliminary run forecasts and harvest projections for 1995 Alaska salmon fisheries and review of the 1994 season. Alaska Department of Fish and Game, Commercial Fisheries Management and Development Division, Regional Information Report 5J95-01, Juneau.

Gregory, S. V., and P. A. Bisson. 1997. Degradation and loss of anadromous salmonid habitat in the Pacific Northwest. Pages 277–314 *in* D. J. Stouder, P. A. Bisson, and R. J. Naiman, editors. Pacific Salmon and Their Ecosystems: Status and Future Options. Chapman & Hall, New York.

Hagen, P., K. Munk, B. Van Alen, and B. White. 1995. Thermal mark technology for inseason fisheries management: a case study. Alaska Fishery Research Bulletin 2:143–155.

Hare, S. R., and R. C. Francis. 1995. Climate change and salmon production in the northeast Pacific Ocean. Pages 357–372 *in* R.J. Beamish, editor. Climate change and northern fish populations. Canadian Special Publication of Fisheries and Aquatic Sciences 121.

Hoffman, S. H., L. Talley, and M. C. Seibel. 1984. U.S./Canada cooperative pink and sockeye salmon tagging, interception rates, migration patterns, run timing, and stock intermingling in southern southeastern Alaska and northern British Columbia. Alaska Department of Fish and Game, Division of Commercial Fisheries, Technical Data Report 110, Juneau.

Hofmeister, K. A. 1994. Southeast Alaska winter air temperature cycle and its relationship to pink salmon harvest. Pages 111–122 *in* Proceedings of the 16th Northeast Pacific Pink and Chum Salmon Workshop, Juneau, Alaska 1993. Alaska Sea Grand College Program Report 94-02, Fairbanks.

Hofmeister, K. A. 1998. Standardization of aerial salmon escapement counts made by several observers in Southeast Alaska. Proceedings of the 18th Northeast Pacific Pink and Chum Salmon Workshop. Department of Fisheries and Oceans, Nanimo, British Columbia.

Howe, A. L., G. Fidler, C. Olnes, A. E. Bingham, and M. J. Mills. 1997. Harvest, catch, and participation in Alaska sport fisheries during 1996. Alaska Department of Fish and Game, Division of Sport Fish, Fishery Data Series 97-29, Anchorage.

Jensen, K. 2000. Research programs and stock status for salmon in three transboundary rivers: the Stikine, Taku, and Alsek. Pages 273–294 *in* E.E. Knudsen, C. R. Steward, D. D. MacDonald, J. E. Williams, and D. R. Reiser, Sustainable fisheries management: Pacific salmon. Lewis Publishers, Boca Raton, Florida.

Jensen, K. A., and R. Bloomquist. 1994. Stock compositions of sockeye salmon catches in Southeast Alaska District 111 and the Taku River, 1990, estimated with scale pattern analysis. Alaska Department of Fish and Game, Division of Commercial Fisheries, Regional Information Report 1J94-23, Juneau.

Johnson, K. J. 1990. Regional overview of coded wire tagging of anadromous salmon and steelhead in Northwest America. American Fisheries Society Symposium 7:127–133.

Johnson, S. W., J. F. Thedinga, and K. V. Koski. 1992. Life history of juvenile ocean-type chinook salmon (*Oncorhynchus tshawytscha*) in the Situk River, Alaska. Canadian Journal of Fisheries and Aquatic Sciences 49: 2621–2629.

Jones, E. L. III, T. J. Quin II, and B. W. Van Alen. 1998. Accuracy and precision of survey counts of pink salmon in a Southeast Alaska stream. North American Journal of Fisheries Management 18:832–846.

Kelley, M. S., and R. P. Josephson. 1997. Sitkoh Creek weir results, June 8 to September 7, 1996. Alaska Department of Fish and Game, Commercial Fisheries Management and Development Division, Regional Information Report 1J97-05, Juneau.

Kissner, P. D. 1974. A study of chinook salmon in Southeast Alaska. Alaska Department of Fish and Game, Division of Sport Fish, Annual Report 1973–1974, Project F-9-7, 16 (AFS-41), Juneau.

Kissner, P. D. 1976. A study of chinook salmon in Southeast Alaska. Alaska Department of Fish and Game, Division of Sport Fish, Annual Report 1975–1976, Project F-9-8, 17(AFS-41), Juneau.

Koski, K. 1996. Habitat restoration in Alaska. Oncorhynchus: newsletter of the Alaska Chapter of the American Fisheries Society 16(3):1.

Marshall, R. P. 1992. Forecasting catches of Pacific salmon in commercial fisheries of Southeast Alaska. Doctoral dissertation, University of Alaska, Fairbanks.

Mason, J. C., and D. W. Chapman. 1965. Significance of early emergence, environmental rearing capacity, and behavioral ecology of juvenile coho salmon in stream channels. Journal of Fisheries Research Board of Canada 22:173–190.

Mathisen, O. A., and B. W. Van Alen. 1995. Southeast Alaska pink salmon management. Final Report, Alaska Science and Technology Foundation, Grant Number 90-1-008. University of Alaska Fairbanks, Juneau Center for Fisheries and Ocean Sciences, JC-SFOS 95-02, Fairbanks.

McGee, S., and seven coauthors. 1996. 1996 Annex chinook salmon plan for Southeast Alaska. Alaska Department of Fish and Game, Commercial Fisheries Management and Development Division, Regional Information Report 1J96-24, Juneau.

McGregor, A. J., and F. Bergander. 1993. Crescent Lake sockeye salmon mark-recapture studies, 1991. Alaska Department of Fish and Game, Division of Commercial Fisheries, Regional Information Report 1J93-13, Juneau.

McKinstry, C. A. 1993. Forecasting migratory timing and abundance of pink salmon (*Oncorhynchus gorbuscha*) runs using sex-ration information. Master's thesis, University of Washington, Seattle.

McLachlan, G. J., and T. Krishnan. 1997. The EM algorithm and extensions. John Wiley & Sons.

McNair, M. 1995. Alaska fisheries enhancement program 1994 annual report. Alaska Department of Fish and Game, Commercial Fisheries Management and Development Division, Regional Information Report 5J95-06, Juneau.

McNair, M., and S. G. McGee. 1994. Contribution of enhanced fish to commercial common property harvests in Alaska, 1979–1993. Alaska Department of Fish and Game, Commercial Fisheries Management and Development Division, Regional Information Report 5J94-10, Juneau.

McPherson, S. A. 1990. An in-season management system for sockeye salmon returns to Lynn Canal, Southeast Alaska. Master's thesis, University of Alaska, Fairbanks.

McPherson, S. A., and J. K. Carlile. 1997. Spawner-recruit analysis of Behm Canal chinook salmon stocks. Alaska Department of Fish and Game, Commercial Fisheries Management and Development Division, Regional Information Report 1J97-06, Juneau.

Meacham, C. P., and J. H. Clark. 1994. Pacific salmon management—the view from Alaska. Alaska Fisheries Research Bulletin 1:76–80.

Moser, J. F. 1899. The salmon and salmon fisheries of Alaska. U.S. Fish Commission Bulletin for 1898, Volume 18. 179 pp.

Murphy, M. L., and K. V. Koski. 1989. Input and depletion of woody debris in Alaska streams and implications for streamside management. North American Journal of Fisheries Management 9(4):427–436.

Murphy, M. L., J. Heifetz, J. F. Thedinga, S. W. Johnson, and K. V. Koski. 1989. Habitat utilization by juvenile Pacific salmon (*Oncorhynchus*) in the glacial Taku River, Southeast Alaska. Canadian Journal of Fisheries and Aquatic Sciences 46:1677–1685.

Nakatani, R. E., G. J. Paulik, and R. van Cleve. 1975. Pink salmon, *Oncorhynchus gorbuscha*, tagging experiments in S. E. Alaska 1938–1942 and 1945. NOAA Technical Report NMFS-SSRF-686, Washington, D.C.

NBTC (Northern Boundary Technical Committee). 1997. U.S./Canada Northern Boundary Area 1996 salmon fisheries management report and 1997 preliminary expectations. Pacific Salmon Commission TCNB (97)-1, Vancouver, British Columbia.

NRC (National Research Council). 1996. Upstream: salmon and society in the Pacific Northwest. National Academy Press, Washington, D.C.

Nehlsen, W., J. E. Williams, and J. A. Lichatowich. 1991. Pacific salmon at the crossroads: stocks at risk from California, Oregon, Idaho, and Washington. Fisheries 16(2):4–21.

Olsen, M. A. 1995. Abundance, age, sex, and size of chinook salmon catches and escapements in Southeast Alaska in 1988. Alaska Department of Fish and Game, Commercial Fisheries Management and Development Division, Technical Fishery Report 95-02, Juneau.

Orsi, J. A., and H. W. Jaenicke. 1996. Marine distribution and origin of prerecruit chinook salmon, *Oncorhynchus tshawytscha*, in southeastern Alaska. Fishery Bulletin 94:482–497.

Pahlke, K. A. 1995a. Escapements of chinook salmon in Southeast Alaska and Transboundary Rivers in 1994. Alaska Department of Fish and Game, Division of Sport Fish, Fishery Data Series 95-35, Anchorage.

Pahlke, K. A. 1995b. Coded wire tagging studies of chinook salmon of the Unuk and Chickamin Rivers, Alaska, 1983–1993. Alaska Fishery Research Bulletin 2:93–113, Juneau.

Pahlke, K. A. 1996. Abundance of the chinook salmon escapement on the Chickamin River, 1995. Alaska Department of Fish and Game, Division of Sport Fish, Fishery Data Series 96-37, Anchorage.

Pella, J., M. Hoffman, M. Masuda, S. Nelson, and L. Talley. 1993. Adult sockeye and pink salmon tagging experiments for separating stocks in northern British Columbia and southern Southeast Alaska, 1982–1985. U.S. Department of Commerce, National Oceanographic and Atmospheric Administration Technical Memorandum NMFS-AFSC-18.

Pennoyer, S. A. 1979. Development of management of Alaska's fisheries. Pages 17–25 *in* B. R. Melteff, editor. Alaska fisheries: 200 years and 200 miles of change. Proceedings of the 29th Alaska Science Conference, Fairbanks, Alaska, 15–16 August 1978. University of Alaska, Fairbanks.

Reisenbichler, R. R. 1997. Genetic factors contributing to declines of anadromous salmonids in the Pacific Northwest. Pages 223–244 *in* Stouder, D. J., P. A. Bisson, and R. J. Naiman, editors. Pacific salmon and their ecosystems: status and future options. Chapman & Hall, New York.

Ricker, W. E. 1972. Hereditary and environmental factors affecting certain salmonid populations. Pages 19–160 *in* R. C. Simon and P. A. Larkin, editors. The stock concept in Pacific salmon. UBC Press, Vancouver, British Columbia.

Rigby, P., J. McConnaughey, and H. Savikko. 1991. Alaska commercial salmon fisheries, 1878–1991. Alaska Department of Fish and Game, Division of Commercial Fisheries, Regional Information Report 5J91-16, Juneau.

Roppel, P. 1982. Alaska's salmon hatcheries 1891–1959. Alaska Historical Commission Studies in History No. 20. National Marine Fisheries Service, Portland.

Royce, W. F. 1989. Managing Alaska's salmon fisheries for a prosperous future. Fisheries 14(2):8–13.

Shaul, L. D. 1990. Taku River coho salmon investigations, 1989. Alaska Department of Fish and Game, Commercial Fisheries Management and Development Division, Regional Information Report 1J90-19, Juneau.

Shaul, L. D. 1994. A summary of 1982–1991 harvests, escapements, migratory patterns, and marine survival rates of coho salmon stocks in Southeast Alaska. Alaska Fishery Research Bulletin 1:10–34, Juneau.

Shaul, L. D., and K. F. Crabtree. 1996. Harvests, escapements, migratory patterns, smolt migrations, and survival of coho salmon in Southeast Alaska based on coded wire tagging. Alaska Department of Fish and Game, Commercial Fisheries, Management and Development Division, Regional Information Report 1J96-14, Juneau.

Slaney, T. L., K. D. Hyatt, T. G. Northcote, and R. J. Fielden. 1996. Status of anadromous salmon and trout in British Columbia and Yukon. Fisheries 21(10):20–34.

Smoker, W. W., B. A. Bachen, G. Freitag, H. J. Geiger, and T. J. Linley. 2000. Alaska ocean ranching contributions to sustainable salmon fisheries. Pages 407–420 *in* E.E. Knudsen, C. R. Steward, D. D. MacDonald, J. E. Williams, and D. R. Reiser, Sustainable fisheries management: Pacific salmon. Lewis Publishers, Boca Raton, Florida.

Stouder, D. J., P. A. Bisson, and R. J. Naiman, editors. 1997. Pacific salmon and their ecosystems: status and future options. Chapman & Hall, New York.

Tarbox, K. E. and T. Bendock. 1996. Can Alaska balance economic growth with fish habitat protection? A biologist's perspective. Alaska Fishery Research Bulletin 3:49–53.

Thorsteinson, F. V. 1950. Statistics of the Southeastern Alaska salmon fishery. University of Washington Fisheries Research Institute Circular 3, Seattle.

USCTC (U.S. Chinook Technical Committee). 1997. A review of stock assessment data and procedures for U.S. chinook salmon stocks. Pacific Salmon Commission USTCCHINOOK (97)-1, Vancouver, British Columbia.

USFS (U.S. Forest Service). 1995. Report to Congress on anadromous fish habitat assessment. R10-MB-279. Juneau, Alaska.

Van Alen, B. W. 1988. Feasibility of using scale and tag data to estimate origins of chinook salmon harvested in Southeast Alaska fisheries in 1982. Alaska Department of Fish and Game, Division of Commercial Fisheries, Fishery Research Bulletin 88-04, Juneau.

Wilbur, R. L., and I. Frohne. 1989. Management implications and planning for salmon enhancement in mixed wild and enhanced fisheries. Alaska Department of Fish and Game, Division of Commercial Fisheries, Regional Information Report 5J89-13, Juneau.

14 Kamchatka Steelhead: Population Trends and Life History Variation

Ksenia A. Savvaitova, Kirill V. Kuzishchin, and Sergei V. Maximov

Abstract.—Kamchatka steelhead *Oncorhynchus mykiss* are the last of their race where the natural diversity of population structure still exists and all stocks remain free of genetic interactions from fish-culture operations. In recent years, the number of Kamchatka steelhead populations, especially anadromous forms, has declined sharply. The present study is based on previous data (1971–72) and on results of joint Russian-American expeditions conducted during 1994 and 1995. In Asia, *O. mykiss* primarily occurs in small tundra rivers of the Kamchatka Peninsula and Bolshoy Shantar Island. Throughout its range, *O. mykiss* is represented by anadromous and resident forms. The typically anadromous form of steelhead is of the stream-maturing type, which enters the rivers with immature gonads in fall, stays there in winter, and spawns in early spring. The life strategy and structure of steelhead populations can change over time. In the Utkholok River, the correlation of the intrapopulation groups has changed in favor of coastal and river forms as a result of declines in the typically anadromous form. Understanding and protecting the natural diversity of population structure and life history forms are critical to sustaining the biological diversity inherent to *O. mykiss*.

INTRODUCTION

Kamchatka steelhead *Oncorhynchus mykiss* are the last of their race where the natural diversity of life history variation still exists in its natural habit and where all stocks remain free of genetic interactions with fish-culture operations. Kamchatka steelhead represent, therefore, an extremely valuable scientific resource for understanding the genetic diversity, local adaptations, and habitat-driven life strategy variations of this species. This understanding has important implications for recovery strategies of North American steelhead populations.

In recent years, however, Kamchatka steelhead, especially the anadromous form on the west coast of the Peninsula, have declined sharply. As a result, Kamchatka steelhead was recently entered into the Russian Red Book of Rare and Endangered Species. The reasons for this decline may be uncontrolled fishing by local residents, inshore commercial fishing, and the intense Japanese offshore fishery. Environmental and climatic changes may also be factors.

Freshwater trout from the Kamchatka Peninsula were first described as an independent species *Salmo mykiss* by Walbaum in Petri Artedi Genera Piscium in 1792. In 1811, Pallas split the anadromous form of Kamchatka steelhead and described it as the species *S. purpuratus*. In subsequent years the status of these species changed several times depending upon their being viewed either as synonyms in the framework of one species or as independent species (Berg 1948). More

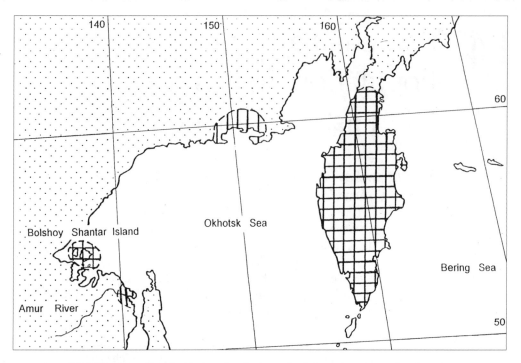

FIGURE 14.1   The distribution (hatched areas) of *Oncorhynchus mykiss* in Asia.

recent research conducted by scientists of the Ichthyology Department of Moscow State University and R. J. Behnke showed that only one species of the genus *Oncorhynchus, O. mykiss,* exists in Kamchatka (Savvaitova and Lebedev 1966; Savvaitova et al. 1973; Behnke 1966, 1992).

In the course of the past 100 years, the relationship of Kamchatka *O. mykiss* to American salmonids of the genus *Oncorhynchus* has been questioned repeatedly. Presently there is convincing evidence of the close relationships between the Kamchatka and American steelhead on the basis of genetic and morpho-ecological data (Savvaitova et al. 1973; Okazaki 1983, 1986; Behnke 1992; Osinov and Pavlov 1993).

On the Asian coast, *O. mykiss* is known primarily from the Kamchatka Peninsula, but it is found in single streams of the continental coast of the Okhotsk Sea and Amur River delta (Berg 1948). Lately it has also been discovered on Bolshoy Shantar Island (Figure 14.1; Alekseev and Sviridenko 1985).

Throughout its range, *O. mykiss* is represented by anadromous steelhead trout and resident rainbow trout forms, the local populations of which are each distinguished by unique ecological and morphological characteristics. The anadromous form is found in the rivers of the western coast of Kamchatka Peninsula. However, the extent of their distribution throughout the remainder of Kamchatka is not well documented (Savvaitova et al. 1973). The resident form inhabits many rivers on Kamchatka but reaches its greatest numbers along the eastern coast (Savvaitova 1975).

Against this background, Moscow State University, The Wild Salmon Center, University of Washington, and National Marine Fisheries Service have agreed to cooperate in a joint 20-year program to more fully understand Kamchatka steelhead. The objectives of the program are to (1) evaluate the status of populations of Kamchatka steelhead, including their phenotypic and genetic diversity throughout the range; (2) develop habitat and environmental protection strategies; and (3) develop fisheries management systems to restore and preserve Kamchatka steelhead populations. The project will also create a database to support the management and conservation of Kamchatka steelhead.

STUDY SITE

Kamchatka steelhead are distributed in small rivers of 100 to 200 km in length that originate in the high tundra and glaciers of the Central Kamchatka range. The tidal water moves upstream 5 to 40 km from the estuary mouths in these rivers. The rivers are fed by water originating from both mountain springs and tundra areas, with the latter predominating. Habitat characteristics along the entire length of the rivers alternates among shallow riffles, pools, and deep holes (up to 5–7 m deep). It appears that the Kamchatka steelhead reaches its highest abundance precisely in this type of river (Savvaitova et al. 1973).

METHODS

The present study is based on previous data (Savvaitova et al. 1973; Savvaitova 1989) and on results of joint Russian-American expeditions in 1994 and 1995. During the 1994 and 1995 expeditions specimens were collected from three rivers in northwest Kamchatka (Figure 14.2), including the Snatolvayam and Kvachina rivers (20 September to 10 October 1994 and 10 September to 17 October 1995) and the Utkholok River (16 September to 17 October 1995). Two hundred and seventy-six steelhead were studied: 101 from Utkholok River, 104 from Kvachina River, and 71 from Snatolvayam River. Twelve rainbow trout from the Utkholok River also were investigated.

Body length was measured from the top of the snout to the fork of the caudal fin (fork length). Sex was determined by direct observations during dissection of the fish, and also (in the case of released fish) by the shape of the head and by the relative length of the maxillary bone to the rear edge of the eye (Figure 14.3). Scale samples were collected from the body above the lateral line, between the dorsal fin and insertion of the anal fin.

Local populations of steelhead consist of a series of intrapopulation groups: typically anadromous, coastal-anadromous, and freshwater riverine (consisting largely of males). The freshwater form is represented by riverine populations, lake-riverine populations, estuarine groups (Maximov 1972; Savvaitova 1975). Following Maximov (1972), we used scale morphology to differentiate three groups of steelhead that differ drastically from one another in life history as: (1) typically anadromous fish with scales exhibiting wide ocean growth zones; (2) coastal fish that do not go far into the sea and whose scales show small marine growth zones; and (3) river fish that mature in the river without migrating to the sea and with scales showing only narrow growth zones.

RESULTS

Kamchatka steelhead belong to the autumn seasonal race, which enter rivers with immature gonads, stay during winter, and continue the migration to spawning grounds in the spring. The life history of Kamchatka steelhead, according to Savvaitova et al. (1973) and recently collected data, is summarized in Table 14.1.

In the three rivers that we explored, the Utkholok, Kvachina, and Snatolvayam, the *O. mykiss* life strategies and population structures differ (Figure 14.4, Table 14.2). In the Utkholok River, the local population is represented by typically anadromous, coastal, and riverine groups that have formed a single system of mutually connected groups, probably belonging to one gene pool. In the neighboring Kvachina and Snatolvayam rivers, where the numbers of anadromous fish have not diminished, the local population is represented mainly by anadromous groups: the typically anadromous group and the coastal group (Figure 14.4). Moreover, the typically anadromous group predominates, whereas the resident freshwater form is virtually absent (a total of only three fish were found, all males).

The structure of Kamchatka steelhead populations in the rivers of northwest Kamchatka was first studied in 1971 and 1972 (Maximov 1972; Savvaitova et al. 1973; Savvaitova 1975). In the intervening period some of the population characteristics have changed (Table 14.2). In the

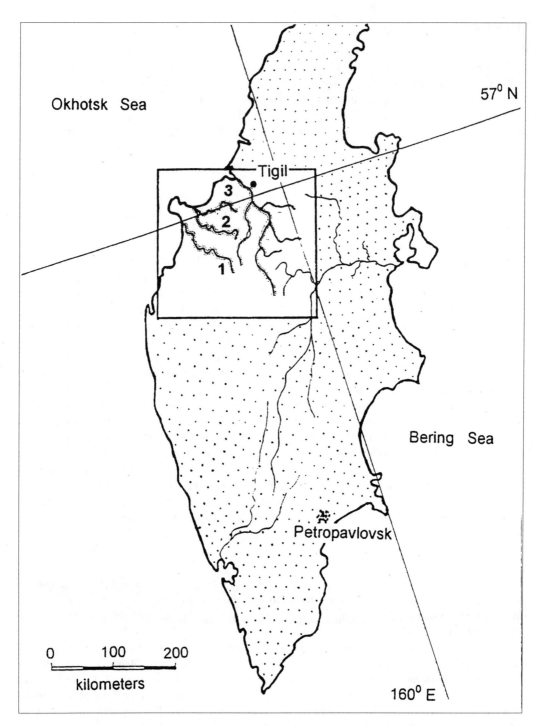

FIGURE 14.2   Location of river study areas in northwestern Kamchatka Peninsula:
1—Utkholok; 2—Kvachina; 3—Snatolvayam.

Utkholok River population, the ratio of intrapopulation groups has changed in favor of the coastal
and riverine forms. The relative numbers of the typically anadromous fish in the Utkholok River
have decreased during the monitoring period, whereas the numbers of the groups more closely tied
to freshwater, the coastal and purely riverine fish, have increased. Several fish of different sizes

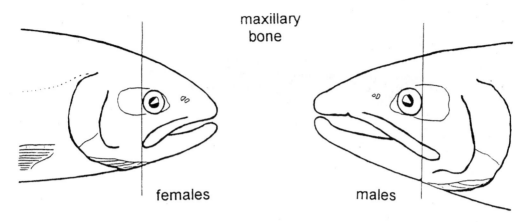

FIGURE 14.3 The method of sex determination in steelhead. In females the maxillary bone is shorter and reaches to about the back edge of the eye; in males the maxillary is longer and extends back beyond the back edge of the eye.

TABLE 14.1
The life history of Kamchatka steelhead.

Run timing	Beginning late August or early September; peaking middle of October, when the temperature decreased to 2–5°C; and ending in November, under the ice. In May, after overwintering, sexually mature fish move to spawning grounds.
Time of spawning	Spring (mid-May to early June), coinciding with flood waters and with increasing water clarity and temperature
Location of spawning grounds	Upper and middle river reaches
Substrate at spawning grounds	Sand and small pebbles
Spawning behavior	Digging of redds
Duration of freshwater period	1–4 years, mostly 3
Time of juvenile's seaward migration	June to July
Duration of sea life to first migration	2–6 years, mostly 2–3
Number of spawnings per life	1–6, mostly 2–3

(220–461 mm fork length; 300–900 g total weight) were collected, whose residence period in salt water was limited to a few months. These fish return to the river sexually immature, evidently to overwinter and fatten. It is possible that such fish can repeatedly go to sea from the rivers, returning before they have reached maturity.

During the monitoring period, changes in the sexual structure of the population, probably the result of poaching, were discovered. In the typically anadromous form from the Utkholok River, females continue to be in the majority, and their numerical domination over males increased. In the coastal group, the tendency of males to predominate remains, but the relative number of females increased. Most striking, females have appeared in the river group, where they formerly were completely absent (Table 14.2). The sex ratio also changed in the Kvachina and Snatolvayam rivers (Table 14.2).

The size–weight structure of the population has not changed substantially (Table 14.2), except in the coastal grouping in the Utkholok River. Longevity of the typically anadromous forms increased, although the coastal and riverine forms remained at previous levels. The age of first maturation did not change. At the same time, the percentage of fish that spawn repeatedly has increased. Previously, most fish in the typically anadromous group died after the first spawning, but now, the majority of fish in this group spawn more than once (Table 14.2).

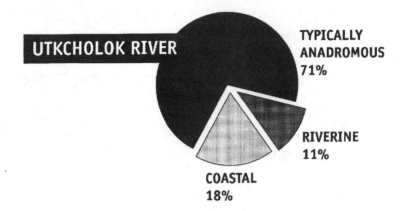

FIGURE 14.4 The structure of steelhead stocks from different rivers of northwestern Kamchatka Peninsula during 1995.

TABLE 14.2
Life history characteristics of the steelhead population during the monitoring periods.

| Parameters | Years | Utkholok River | | | Kvachina River | Snatolvayam River |
		Typical Anadromous	Coastal	Riverine		
Ratio of intrapopulational	1971–72	82%	12.4%	2.6%		
groupings in catch	1995	71%	18%	11%		
Sex ratio	1971–72	1:1.8	1.8:1	only males	1:1	3.81:1
males: females	1995	1:2.1	1.6:1	females appear	1.21:1	1:1
Size limits	1971–72	550–870 mm	610–930 mm	380–545 mm	540–960 mm	610–998 mm
		1,900–7,700 g	1,800–8,000 g	620–640 g	1,800–10,500 g	2,500–10,000 g
	1995	560–950 mm	220–630 mm	310–532 mm	620–960 mm	647–1,193 mm
		2,300–8,400 g	230–2,900 g	400–1,950 g	4,530–7,000 g	3,850–10,000 g
Longevity (years)	1971–72	7	9	8	to 8+	to 6+
	1995	10	9	9	to 10+	to 10+
Age of maturation	1971–72	4+–5+	—	—	3–4	3
	1995	4+	5+	6+	3–4	4
Number of spawnings in	1971–72	to 4, more 1	to 5	to 4	—	—
life	1995	to 6, more 2–3	to 5, more 1	to 4	to 4, more 2–3	to 6, more 1–2
Post-spawning mortality	1971–72	more than 65%	—	—	—	—
	1995	up to 20%	up to 35%	up to 65%	up to 30%	up to 30%
Periodicity of spawning	1971–72	annually	annually	annually	annually	annually
	1995	annually	annually	annually	annually	annually

DISCUSSION

O. mykiss exhibits a great diversity of life history patterns, including varying degrees of anadromy, differences in reproductive biology, and plasticity of life history between generations (Busby et al. 1996). Along the west coast of North America, steelhead can be divided into two basic reproductive ecotypes, based on the state of sexual maturity at the time of river entry and duration of spawning migration (Busby et al. 1996). The stream-maturing type enters freshwater in a sexually immature condition and requires several months to mature and spawn. Fall steelhead in Alaska and summer steelhead in the Pacific Northwest and northern California are of this type. The ocean-maturing type enters rivers with well-developed gonads and spawns shortly thereafter. Spring steelhead in Alaska and winter steelhead elsewhere represent this type. In North America, there is considerable overlap in migration and spawning time among populations of the same run. There is a high degree of overlap in spawning time among populations regardless of run type (Busby et al. 1996).

The "half-pounder" steelhead known from the Rogue, Klamath, Mad and Eel rivers of southern Oregon and northern California are immature fish which, following smoltification, spend only 2 to 4 months in the ocean and then return to freshwater. They overwinter in rivers and descend to salt water again in spring. Half-pounders are generally less than 400 mm and only a few half-pounders are sexually mature (Busby et al. 1996).

Along the west coast of the Kamchatka Peninsula, the autumn steelhead seasonal race exhibits a life history similar to that of the stream-maturing type in North American steelhead and enters rivers in fall with immature gonads and takes several months to mature for spawning in spring. The ocean-maturing type is unknown on the west coast of Kamchatka, but probably exists in separate rivers on the east coast of the Peninsula. Half-pounders have not been observed migrating into Kamchatka rivers, although separate specimens, similar to half-pounders, were caught during the run of stream-maturing steelhead in the Utkholok and Kvachina rivers.

The exact relationships between anadromous and nonanadromous forms in any given area of North America is not well understood (Busby et al. 1996). In coastal populations, it is rare for the two forms to coexist. In inland populations, coexistence of two forms appears to be more frequent. In this case it is possible that offspring of resident fish may migrate to the sea and offspring of steelhead may remain in streams as resident fish (Busby et al. 1996).

Previously it was determined that the local populations of the anadromous form of *O. mykiss* in Kamchatka consist of a series of intrapopulation groups that are tied to each other to a varying degree (Maximov 1972; Savvaitova 1975). The correlation of anadromous and resident forms and the smaller intra-population groups was not the same in various parts of their range, nor was the degree of their reproductive isolation from each other. As one moved to the south of the range of the anadromous form, there was a noticeable increase in the numbers of groups more closely tied to freshwater (Savvaitova 1975, 1989). In the northern rivers along the western coast of Kamchatka Peninsula, the riverine form was represented by small numbers of males, who spawned with the females of the anadromous steelhead. In the more southerly rivers, riverine females were also present. In the south of the range, there was also an increase in the diversity of intrapopulation groups within one local population. The greatest diversity of life history strategies was displayed by the Bolshaya River population (western Kamchatka), where all known types were found: typically anadromous; coastal; permanently resident river fish; and estuarine river fish, who venture into the saline estuary (Maximov 1974). The freshwater groups here were represented by individuals of both sexes in approximately equal proportions (Savvaitova 1975). In the Kamchatka River basin on the eastern coast of the Peninsula, relatively isolated local populations were tied to flows of small tributaries, not moving significantly, and were represented by males and females (Savvaitova 1975).

Information gathered during the 1994 and 1995 expeditions, while not altering the essence of the earlier findings, nevertheless supplemented them substantially. For example, intraspecific diversity is not necessarily tied to the size and geography of the rivers, as previously thought (Savvaitova 1975). The same intraspecific life history forms that exist in larger rivers can occur in small rivers, such as the Utkholok, providing that similar environmental conditions are present. Unfortunately, there are no exact abundance data on the stock size of Kamchatka steelhead. In the Utkholok River, the population structure has changed over a rather short time (22–23 years), probably as a result of the decrease in abundance of the typically anadromous group by poaching. At the same time, in neighboring Kvachina and Snatolvayam rivers, where the influence of poaching is not so strong, and where the abundance of typically anadromous steelhead has not changed noticeably, populations are represented by mainly anadromous groups. This change cannot be attributed to a response in natural environmental fluctuations, because all these rivers are situated close to each other (approximately 20 km). It is possible that the diversity and kinship of intrapopulational forms within local populations of steelhead are connected with environmental conditions. One can suppose that, under specific conditions, environment acts as a switch that determines the path of development according to one of several possible alternative routes (Schwartz 1980).

The present study shows the natural diversity and variation of life history forms inherent to *O. mykiss* throughout their range. Primary attention should be paid to differentiation of steelhead life history forms, to the study of their reproductive relations, and the possibility of transitions among them. These local populations and intrapopulational groupings comprise the biological diversity of *O. mykiss*, which must be protected or restored if steelhead are to be sustained into the future.

ACKNOWLEDGMENTS

We thank Pete Soverel, the Chairman of the Wild Salmon Center, and Serge Karpovich, J. Aalto, J. Sager, G. M. Inozemtsev, and T. Bevan for their diligent efforts in organizing these expeditions. Thanks to H. Li and two anonymous reviewers for their thoughtful contributions to the manuscript.

REFERENCES

Alekseev, S. S., and M. A. Sviridenko. 1985. Mikizha *Salmo mykiss* Walbaum (*Salmonidae*) from Shantar Islands. Journal of Ichthyology 25:68–73.

Behnke, R. J. 1966. Relationships of the far eastern trout, *Salmo mykiss* Walbaum. Copeia 1966:346–348

Behnke, R. J. 1992. Native trout of western North America. America Fisheries Society, Monograph 6. Bethesda, Maryland.

Berg, L. S. 1948. Freshwater fishes of the USSR and contiguous countries. 1. Moscow-Leningrad.

Busby, P. G., T. C. Wainwright, G. J. Bryant, L. Lierhaimer, R. S. Waples, F. W. Waknitz, and V. Lagomarsino. 1996. Status review of west coast steelhead from Washington, Idaho, Oregon and California. U.S. Department of Commerce, NOAA Technical Memorandum NMFS-NWFSC 27.

Maximov, V. A. 1972. Some data on the ecology of the Kamchatkan *Salmo mykiss* Walb. of the Utkholok River. Journal of Ichthyology 12:827–834.

Okazaki, J. 1983. Distribution and seasonal abundance of *Salmo gairdneri* and *Salmo mykiss* in the North Pacific Ocean. Japanese Journal of Ichthyology 30:235–246.

Okazaki, J. 1986. Studies on closely related species *Salmo gairdneri* and *Salmo mykiss*: their distribution and migration in the North Pacific and systematics. Far Seas Fisheries Research Laboratory Bulletin 23:1–68.

Osinov, A. G., and S. D. Pavlov. 1993. On the genetic similarity of Kamchatka noble trouts and American rainbow trout. Journal of Ichthyology 33:626–630.

Savvaitova, K. A. 1975. The structure of populations of species *Salmo mykiss* Walbaum in its natural range. Journal of Ichthyology 15:984–997.

Savvaitova, K. A. 1989. Arctic charrs (The structure of population systems and the future of economical exploration). Moscow. Translated by Canada Institute for Scientific and Technical Information National Research Council. Ottawa, Ontario. Translated in Canadian Transaction of Fisheries and Aquatic Science. 1993. N 5607.

Savvaitova, K. A., and V. D. Lebedev. 1966. On the systematic status of the Kamchatka steelhead *Salmo penshinensis* Pallas and Mikizha *Salmo mykiss* Walb., and their interrelationship with the American representatives of the genus *Salmo*. Journal of Ichthyology 6:593–608.

Savvaitova, K. A., V. A. Maximov, M. V. Mina, G. G. Novikov, L. V. Kokhmenko, and V. E. Matsuk. 1973. The noble trouts of Kamchatka (systematics, ecology, and the possibilities of using them as the object of trout-culture and acclimatization). The Voronezh University Publishers. Voronezh. Translated by University of Washington. 1987. Seattle.

Schwartz, S. S. 1980. The ecological rules of evolution. Moscow. "Nauka."

Section III

Existing Management

15 International Management of Fraser River Sockeye Salmon

James C. Woodey

Abstract.—International management of Fraser River sockeye salmon fisheries from 1946 to the present has evolved into a complex, week-to-week, in-season, decision-making process featuring a comprehensive data collection and analysis program designed to produce the information required for management action. Intensive catch monitoring, test fishing, racial analysis, and hydroacoustic programs are implemented each season. Analysis of catch, escapement, and biological information generates in-season updates of run size, arrival timing, and migration route by stock or stock group. The International Pacific Salmon Fisheries Commission regulated fisheries from 1946 to 1985 and, under the Pacific Salmon Treaty, the Fraser River Panel of the Pacific Salmon Commission (PSC) has regulated fisheries in Panel Area waters since 1986 (except 1992 and 1994) to harvest annual runs in a fashion that meets escapement objectives and achieves the required international allocation and domestic allocations of the available catch. The PSC's monitoring programs provide the Panel with the ability to track the progress of escapement and catch objectives and to modify regulations in-season to achieve these objectives. The cooperative efforts of Canada and the U.S. in the rational management of the Fraser River sockeye salmon resource over a period of 50 years have resulted in the rebuilding of Fraser River sockeye stocks. The yield to all fisheries over the 8 years from 1987–1994 totals 81 million fish or just over 10 million per year.

BACKGROUND

Fraser River sockeye salmon *Oncorhynchus nerka* stocks form the largest run of this species other than the combined runs to rivers in Bristol Bay, Alaska. Approximately 20 stocks produce the large majority of fish returning annually to the Fraser River. Over 100 individual spawning populations throughout the watershed are monitored for escapement and biological data. The southern migration route of Fraser adult sockeye takes these fish through Juan de Fuca Strait, then U.S. waters of the San Juan Islands and Point Roberts, before re-entering Canadian waters in the Strait of Georgia and the Fraser River (Figure 15.1). An annually variable proportion of the run also approaches the Fraser River from the north through Johnstone Strait. Fraser sockeye stock abundance exhibited a distinctive pattern of quadrennial dominance on the 1901–05–09–13 cycle and this had been observed as early as the 1820s (Ward and Larkin 1964). Early in this century, the dominant cycle years produced from 4 to 20 times (average run ≥35 million fish) the off-cycle year returns (average run = 4.2 million, Figure 15.2). Most Upper Fraser River sockeye stocks were dominant on the 1901 cycle.

Intensive commercial fishing began in both countries in the late 1800s. By the early 1900s the stocks on non-dominant cycle lines had begun to show the impacts of over-fishing by the concentration of fishing gear on both sides of the border that built up to harvest the large numbers of fish returning on the dominant cycle. Then in 1913, rock dumped into the river during railroad construction activities in the Fraser Canyon at Hells Gate resulted in the blockage of the river to sockeye migration for much of that year's dominant cycle return. Millions of fish died in the Canyon, reducing the dominant cycle spawning abundance to a fraction of the pre-1913 abundance

1-56670-480-4/00/$0.00+$.50
© 2000 by CRC Press LLC

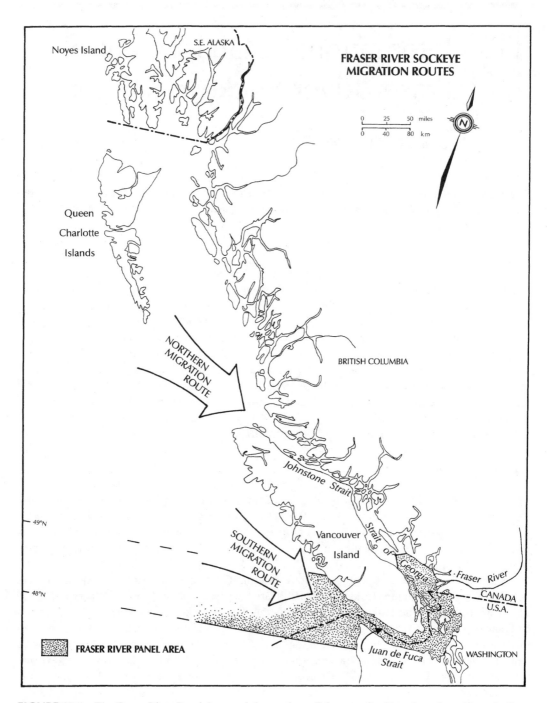

FIGURE 15.1 The Fraser River Panel Area and the northern (Johnstone Strait) and southern (Juan de Fuca Strait) routes for sockeye migrating to the Fraser River.

(Thompson 1945). The reduced numbers of fish returning in 1917 from the diminished spawning in 1913 were greatly over-fished by the large commercial gear concentration in both countries. This sequence of events, along with the construction of dams on the Quesnel River (1897) and the Lower Adams River (1907), led to the severe depletion of the resource by the late 1910s and early 1920s (1918–1925 average run = 1.5 million).

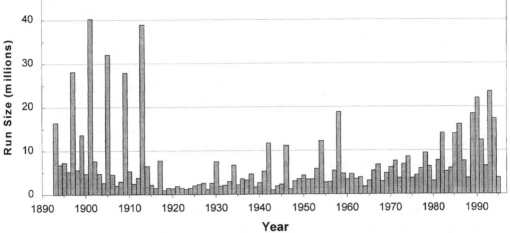

FIGURE 15.2   Annual returns (catch plus escapement) of Fraser River sockeye salmon, 1893–1995.

Negotiations between officials from Canada and the U.S. regarding the conduct of fisheries on Fraser River sockeye salmon go back over 100 years, beginning in 1892 (Roos 1991). However, it took the near-destruction of the resource for the two countries to reach agreement in 1930 on international management of the fisheries. This agreement was implemented in 1937 with the formation of the International Pacific Salmon Fisheries Commission (IPSFC) for "the protection, preservation and extension of the sockeye salmon fisheries in the Fraser River system." In 1938, the IPSFC commenced field programs designed to assess the status of the stocks and to determine the factors limiting production. Construction of fishways at Hells Gate to remedy fish passage problems in the Fraser Canyon began in 1944. However, under the Treaty, the IPSFC did not begin to actively manage the sockeye salmon fisheries in Convention (Treaty) Waters until 1946.

When IPSFC assumed management responsibility in 1946, sockeye returns had increased to an average of 3.9 million fish annually (1938–1945), largely due to the strong recovery of the late-timed Lower Adams River run. However, most early-timed sockeye stocks had shown little recovery. Noteworthy in the rebuilding of the Lower Adams River was the switch of the dominant cycle from the 1901–1905 line to the 1902–1906 line.

The IPSFC managed the stocks for 40 years, before being terminated on December 31, 1985, through agreement of the Parties. Responsibility for management of the commercial fisheries in Panel Area (formerly Convention) waters (Figure 15.1) was transferred to the new Pacific Salmon Commission (PSC) on January 1, 1986. Since then, Canada and the bilateral Fraser River Panel of the PSC have shared the management responsibility (except for 1992 and 1994 when catch-sharing agreements were not achieved). Canada has the responsibility to set escapement goals, to enumerate escapement, and to manage fisheries for Fraser River sockeye salmon in Canadian waters outside the Panel Area. The Panel, through the PSC, has the mandate to regulate the commercial fisheries for Fraser River sockeye salmon in the Panel Area to achieve a bilaterally agreed-to set of objectives. The international staff of the PSC is charged with the responsibility to interpret data and other information provided by the Parties, to collect catch and biological data, and to conduct the analyses on which the Fraser River Panel bases its in-season fishery management. A bilateral Fraser River Panel Technical Committee ensures that both sides of the Fraser Panel are fully informed of technical aspects of the advice from PSC staff, and provides technical review of the PSC's data and interpretations.

Detailed studies of the biology and management of Fraser River sockeye were conducted by the IPSFC and, more recently, by the PSC and Canada Department of Fisheries and Oceans (DFO). These studies generated the data upon which the current management process is based. First, the production database consists of biological information on brood-year spawning abundances, fry or juvenile numbers, returns of progeny by age of maturity, and size and scale growth information. These data are essential for analysis of the productive capacities of the stocks. Early tagging studies defined the migration timing and delay behavior of the stocks (Verhoeven and Davidoff 1962) and research established the in-river behaviour patterns (Killick 1955). Studies of fish scale patterns (Clutter and Whitesel 1956) showed that freshwater growth in each rearing lake is characteristic and can be used to identify the racial composition of mixed-stock samples. Analyses of the catching capacity of the fishing fleets have provided essential data for structuring fisheries. Hydroacoustic methods of estimating daily in-river escapements were first developed in the 1970s, and have been refined in recent years (Banneheka et al. 1995). With the historical database and in-season analytical techniques, fisheries in the U.S. and Canada can be managed effectively to achieve the objectives of the countries through the Fraser River Panel.

The actual fishery management decisions are made by the Fraser River Panel in consultation with the respective national management agencies and the user groups. The Panel normally adopts a pre-season fishing plan based on options developed by PSC staff in consultation with the Fraser River Panel Technical Committee. In preparing the fishing plan, a pre-season fishery simulation model (Cave and Gazey 1994) is used to achieve, to the extent practicable, the objectives of the Panel. These objectives are, in order of priority: (1) to achieve gross escapement requirements (adult sockeye for spawning escapements and in-river Indian fishery catches) as set by Canada for each stock or stock group (stocks with similar arrival timing and behavior are grouped into four stock groups: early Stuart, early summer, summer, and late runs); (2) to achieve the international division of catch as agreed by the PSC and the Parties; and (3) to allocate the catch among the user groups in each country in accordance with national objectives. In carrying out these objectives the Panel must consider the conservation requirements of other stocks and species mingling with Fraser River sockeye salmon. The simulation model is initiated with DFO forecasts of return abundance and arrival timing by stock group and Johnstone Strait diversion rates.

IN-SEASON DATA COLLECTION

The staff of the PSC is responsible for the design of an annual data collection plan focusing on estimating commercial fishery catches, conducting test fishing operations, estimating daily escapement by means of a hydroacoustic program at Mission, B.C., and estimating the racial composition of catches wherever Fraser sockeye salmon are harvested (Woodey 1987). PSC staff implements these programs in Panel Area waters and the PSC annually requests the appropriate agencies to do so in non-Panel waters.

PSC staff estimates daily catches in Panel Area fisheries by area and gear. Canadian and U.S. agencies are also integrally involved in catch estimation work, partly because many of the fisheries of concern are outside the Panel Area, but also due to the PSC's limited ability to obtain on-water estimates. On-water catch and effort estimates provided by Canadian and U.S. agencies are updated later with more complete delivery information obtained from fish buyers. Catch data from fish sales slips are substituted into the database as they become available.

Prior to the commercial fishing season and during closed periods, the PSC conducts, or requests that Canada conduct, test fishing operations along the sockeye migration routes. Test fishing operations are carried out annually for two purposes. First, they monitor the fluctuations in abundance of fish in the migration routes, principally Juan de Fuca Strait, Johnstone Strait, and the Fraser River. Second, scales and biological data collected from catches are used to estimate the stock composition of catches. Catch and effort data from daily test fishing operations are collated and faxed to Panel Members, Technical Committee members, and fishery agencies along the coast each morning.

The PSC conducts an extensive hydroacoustic program at Mission, B.C., some 80 km upstream of the river mouth, to estimate the daily passage of sockeye salmon escaping through the lower river. Meeting gross escapement goals is the first priority of the Panel, so accurate estimation of escapements is crucial to the management process. Estimates are produced around 1000 hours each day, 5 hours after data for each 24-hour period are collected and sent to the PSC office. The program uses species composition data from two lower Fraser River variable mesh gillnet test fishery catches and racial composition estimates from sockeye taken in these test fisheries. Having daily and cumulative escapement data by stock or stock group enables the Panel to make adjustments to the fisheries in the Panel Area (Canadian and U.S. areas) to achieve gross escapement goals. In addition, the PSC provides these data to the Parties so that Canada can adjust its outside troll fishery and Johnstone Strait fisheries, and Canada and First Nations can regulate in-river Indian fisheries.

Intensive racial data collection and analyses provide the basic estimates for management of stock-specific catch and escapement requirements. Fish scales provide the principal data for racial analyses. Freshwater circuli counts and circuli spacing data are used in discriminant function analysis (DFA) methods of stock composition estimation (Gable and Cox-Rogers 1993). The turnaround time on scale samples is often 24 hours or less, which ensures that up-to-date information on stock composition is available from a variety of commercial and test fishing locations along the coast. In-season standards for the DFA are developed using sibling samples (e.g., from last year's jack returns) or adult scales from prior years. Fishery samples are reanalysed post-season using DFA standards developed from spawning ground scales collected by DFO. Scale data are augmented with fish size data and information collected on the presence or absence of the body cavity parasite, *Philonema oncorhynchi* (PSC 1996a, 1996b), and the brain parasite, *Myxobolus articus*.

IN-SEASON DATA ANALYSIS

Data analysis focuses on three areas: estimation of run size by stock or stock group, arrival timing of each stock or stock group in coastal areas, and the Johnstone Strait diversion rate (i.e., the proportion of the stock returning to the Fraser River by the northern route). Management of fisheries directed toward Fraser sockeye requires that these three parameters be estimated in-season.

Pre-season forecasts of return abundance are provided to the PSC by DFO. However, since pre-season forecasts are usually less accurate than in-season estimates, the Panel requires in-season run-size estimates to estimate the total allowable catch (TAC) for setting catch objectives. PSC in-season run-size estimation models take several forms, utilizing catch data collected from commercial and test fisheries and escapement estimates at Mission. Regression methods are applied to historical data on run size to: (1) peak week purse seine catch and catch per unit of effort (CPUE); (2) test fishing CPUE data at peak abundance; and (3) cumulative catch and escapement. A fourth model (cumulative-normal model) does not rely on historical data but tests the fit of observed cumulative abundances for each day of the migration to the equivalent cumulative abundances from a suite of normally distributed pre-set curves of varying abundance, duration, and peak timing. The theoretical abundance/timing curve that best approximates the observed data simultaneously provides estimates of both abundance and timing. The identification of peak time of arrival is essential in all these methods.

Results of empirical models and nonquantitative data are summarized by PSC staff and provided to the Panel along with recommendations for adopting in-season run-size estimates. Recent introduction of Bayesian statistical methods has led to the potential of objectively combining independent estimates. However, the judgment of experienced staff is critical to accurate estimation of run sizes for management purposes.

DFO provides forecasts of the Johnstone Strait diversion rate to the PSC just prior to the fishing season, when the oceanographic data that generate such forecasts become available. The PSC makes in-season estimates of Johnstone Strait diversion from the estimates of absolute abundance or the weekly relative abundance on the two routes, plus catches taken seaward of the assessment points that reasonably can be attributed to one route or the other. Good diversion rate estimates are critical to the management process, since they relate to the availability of sockeye to the fleets in each major area.

PANEL DECISION-MAKING

The Fraser River Panel is composed of members and alternate members appointed by Canada and the U.S. These appointees, up to 12 from Canada and 8 from the U.S., consist of both agency and user-group representatives. Panel agreements are by consensus, with each national section having one vote.

The Panel normally announces pre-season fishing plans that include an agreement that all Panel Area fisheries remain closed until opened by the Panel. Thus, decisions to allow fishing at any location within the Panel Area require both sides of the Panel to agree. This process places in-season decision-making squarely on the Panel, which maintains a close watch on the progress of escapement and catch allocation. The membership of the Panel and the form of the decision-making process provides balance to Panel deliberations, and normally works well. Canada retains the authority to make decisions regarding escapement policy and used this authority in 1985 to 1989 to accelerate rebuilding by placing more sockeye (than called for in the Treaty) on the spawning grounds from the Canadian share. Canada also has the responsibility for regulating major Canadian fisheries outside the Panel Area, upriver management of spawning escapement and, with First Nations, management of Indian fisheries.

In-season management is a hectic weekly cycle of: (1) PSC staff collection of commercial and test fishing catch data, estimation of daily escapements, and racial identification work; (2) melding the resulting catch and biological data into estimates of abundance, timing, and diversion rate; (3) staff analysis of potential fishery strategies and development of recommendations to the Panel; and (4) Panel review of these analyses and recommendations prior to decisions on the conduct of fisheries in the following week. The Panel also makes mid-week adjustments to the initial regulations, which often requires several meetings each week during the peak of the sockeye season. The Panel has met up to six times in a week to achieve the hierarchy of objectives set before it by fine-tuning regulations. Many of these meetings are by conference call, but Panel members often find that in-person meetings provide better insight into the data and the views of the other section.

DOES THE PROCESS WORK?

For the objectives of the Panel to be met, both the results of PSC staff analyses and the decisions of the Panel must be correct. This is challenging given the difficulties in accurately estimating catch, escapement, and racial composition in-season. Staff estimates of run size provide the crucial pieces of information (i.e., total abundance by stock group) from which the TAC is calculated. The run-size estimation process involves making sequential estimates based on steadily improving data and models from early in each run to near the completion of the migration. These periodic adjustments usually become smaller because the models give converging estimates as the migration progresses and an increasing proportion of the run is "seen" in catch and escapement. Some in-season adjustments occur simply because of revisions to the catch and racial composition estimates over the course of the season (i.e., the quality of catch and racial estimates improve over the season).

Over- or underestimation of run sizes directly impact the achievement of catch and escapement goals. Likewise, decisions of the Panel or the Parties to pursue a fishery plan may turn out to be too aggressive or too conservative to accomplish all objectives of the Panel. How well then have the objectives of the Panel and countries been met? There are three areas in which this can be addressed: gross escapement, international allocation, and domestic allocation.

The first objective of the Panel is to achieve gross escapement goals by stock group. A comparison of the estimates of gross escapement obtained by the Mission hydroacoustic program with the goals set by Canada (Figure 15.3a) shows that in 5 of the 7 years between 1986 and 1994 during which the Panel managed the fishery, in-season estimates of summer run gross escapement exceeded the targets (i.e., above the 1:1 line). The immediate river entry of escaping summer-run fish and hydroacoustic estimation of daily abundances at Mission facilitated this result. The rapid analysis of catch and corresponding escapement data provides evaluation of harvest rates and allows the Panel to achieve goals through adjustments to the fisheries. Late-run sockeye gross escapement goals have been less well achieved (Figure 15.3b); in 3 of 7 years, in-season estimates were less than the goals. Late-run fish delay in the Strait of Georgia after escaping marine area fisheries, and, hence, do not provide timely enough feedback, in the form of escapement data from Mission, to corroborate harvest rate assessments and allow management actions to be modified before the run has passed through Juan de Fuca and Johnstone Straits. The in-season, late-run escapement to the Strait of Georgia is calculated by subtracting catch from estimated in-season run size. Unavoidably, differences between the run-size assessments and the actual abundances are transferred directly to the escapements.

Estimates of gross escapement at Mission can also be compared to upstream estimates of Indian fishery catches and spawning escapements. For summer-run sockeye stocks as a group, Mission hydroacoustic estimates were very close to upstream estimates for years prior to 1992 (Figure 15.4). Since that time, estimates have been at greater variance. These differences between Mission and upriver estimates have been the subject of intense discussion and review by the Pearse/Larkin investigation after the 1992 season (Pearse 1992), and by the Fraser River Sockeye Public Review Board inquiry after the 1994 season (Fraser River Sockeye Public Review Board 1995). The Pearse/Larkin review provided estimates of the unreported catches and natural and fishery-induced mortalities in 1992. However, the Fraser Board could not pinpoint the cause of the difference in 1994 but made extensive recommendations for better management and examination of potential errors and biases in all estimates. In 1995, errors were found in both Mission hydroacoustic estimates and Indian catch estimates. After these errors were corrected, the differences between the two estimates were small.

The second Panel objective is to achieve agreed-to international allocations to user groups of the two countries. The Panel regulates the majority of U.S. fisheries, but in Canada, DFO regulates major fisheries outside the Panel Area. The flexibility of the Panel in regulating Panel Area fisheries has resulted in catches that were reasonably close to the desired allocations. For example, Figure 15.5 shows a comparison of U.S. catches and calculated allocations for the years of inter-national management (1985–1991, 1993, 1995). A similar graph for Canada would show reciprocal deviations but because Canada allowed a portion of its share to escape in 1985 to 1989, the calculation is slightly different.

Allocation of catches among user groups within the two countries is the third priority of the Panel. In the U.S., the yearly balance of catches between Treaty Indian and non-Indian fishers varies somewhat, although payback provisions in the agreements between these user groups allows overages and underages to be balanced in subsequent years. Cumulative catches by each user group have been very close over the years of international management. Domestic goals also include allocation of the Treaty Indian catch between Juan de Fuca Strait and Puget Sound Tribal groups and allocation of the non-Indian catch between purse seine, gillnet, and reef net fleets.

Since 1986, Canada's Minister of Fisheries has provided annual gear allocation objectives for the catch of Fraser River sockeye salmon based on recommendations of the Minister's Advisory Council (MAC) or later, the Commercial Fishing Industry Council (CFIC). Since the Panel regulates only some Canadian fisheries in Panel Area waters, and none of the fisheries outside the Panel Area, comparison of catches by gear type in Canada to the Minister's allocation objectives do not represent Panel achievements, per se. However, catches by user group have been reasonably well

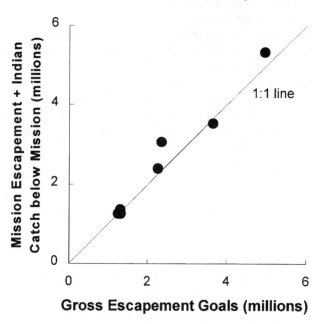

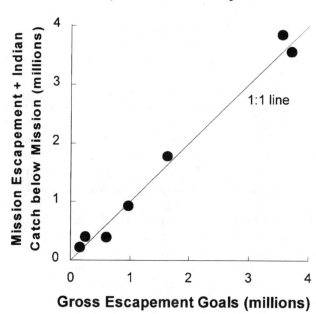

FIGURE 15.3   (a) Comparison of in-season estimates of gross escapement (hydroacoustic estimates of daily passage at Mission plus catch in Indian fisheries below Mission) to Canada's gross escapement goals for summer-run sockeye stocks, 1986–1991 and 1993. (b) Comparison of in-season estimates of gross escapement (hydroacoustic estimates or test fishing index estimates of daily passage at Mission plus catch in Indian fisheries below Mission) to Canada's gross escapement goals for late-run sockeye stocks, 1986–1991 and 1993.

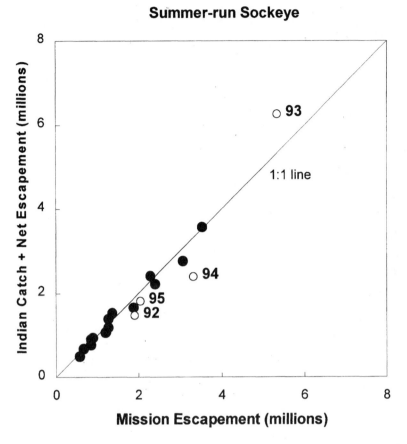

FIGURE 15.4 Comparison of post-season estimates of catches and spawning escapements above Mission to in-season estimates of gross escapement obtained at Mission (hydroacoustic estimates of daily passage) for summer-run sockeye stocks, 1977–1995.

achieved and, since 1990, internal agreements have allowed paybacks among the gear types in subsequent years to balance annual variations from the allocations.

In achieving catch allocation objectives, the Panel has adopted quite rigid fishing plans for Panel Area fisheries to limit the incidental harvest of nontarget species and stocks. For example, fisheries prior to or after certain dates have been constrained to conserve non-Fraser sockeye and other species of salmon. In recent years, Canada and the U.S. have instituted programs to encourage or mandate the release of non-target species. Overall, in Panel-regulated fisheries in Canada and the U.S., Fraser sockeye and pink salmon have amounted to approximately 92% of the total harvest between 1988 and 1995. An additional 5% were non-Fraser pink salmon, while 3% were fish of other species.

STOCK STATUS

The important question from a sustainable fisheries standpoint is: are the stocks healthy? One measure is the trend in spawning escapements. Spawner numbers have been at their highest levels since abundances were first measured in 1939, although 1992, 1994, and 1995 escapements were below the goals for those years. The largest populations to reach the spawning grounds since 1909 occurred in 1990 (6.1 million) and 1993 (5.8 million). As well, the diversity of stock production

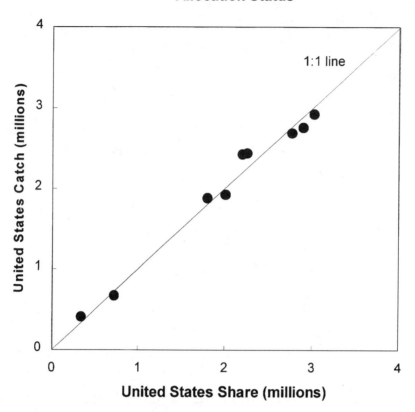

FIGURE 15.5 Comparison of the actual U.S. catch to the post-season estimate of catch share following Pacific Salmon Treaty agreements, 1985–1995 (excluding 1992 and 1994 when there was no bilateral agreement on catch sharing).

and escapement has increased. The proportion of the annual escapement to the principal stock each year has dropped from 59% (1948–1967) to 38% (1986–1995). Less reliance on individual stocks has benefited the fishery by stablizing production.

The other measure of stock health is the trend in recent years' production. Fraser River sockeye returns (catch plus escapement) in the 8-year period from 1987 to 1994 averaged 13.8 million fish (Figure 15.2), the highest 8-year average since the large runs in the early 1900s (1899–1906; 13.8 million; IPSFC, unpublished data). Total catch for this 8-year period 1987–1994 was 81 million fish or approximately 10.1 million per year.

Individual stocks have shown strong growth in recent years. The Quesnel watershed stocks, in particular, have expanded greatly on the dominant (1993) and sub-dominant (1994) cycles. In addition, the 1995 "off" cycle escapement reached 214,000 spawners, which approximated escapement on the dominant cycle in the 1950s. Modern day record productions have been recorded for several stocks in recent years, e.g., Quesnel, Lower Shuswap, and Late Stuart (PSC 1991, 1996b). Minor stocks have expanded, as well.

In overview, the international cooperation on management of Fraser River sockeye salmon through the IPSFC and the PSC over the past 50 years can be viewed as strongly positive. Such management has resulted in substantial success in rebuilding Fraser sockeye stocks and providing large, sustained catches for fishers from both countries for many years.

REFERENCES

Banneheka, S. G., R. D. Routledge, I. C. Guthrie, and J. C. Woodey. 1995. Estimation of in-river fish passage using a combination of transect and stationary acoustic sampling. Canadian Journal of Fisheries and Aquatic Sciences 52:335–343.

Cave, J.D., and W.J. Gazey. 1994. A pre-season simulation model for fisheries on Fraser River sockeye salmon (*Oncorhynchus nerka*). Canadian Journal Fisheries Aquatic Sciences 51:1535–1549.

Clutter, R.I., and L.E. Whitesel. 1956. Collection and interpretation of sockeye salmon scales. International Pacific Salmon Fisheries Commission Bulletin IX, New Westminster, British Columbia.

Fraser River Sockeye Public Review Board (Canada). 1995. Fraser River sockeye 1994: problems and discrepancies. Ottawa, Ontario.

Gable, J.H., and S.F. Cox-Rogers. 1993. Stock identification of Fraser River sockeye salmon: methodology and management application. Pacific Salmon Commission Technical Report 5, Vancouver, British Columbia.

Killick, S.R. 1955. The chronological order of Fraser River sockeye salmon during migration, spawning and death. International Pacific Salmon Fisheries Commission Bulletin VII, New Westminster, British Columbia.

PSC (Pacific Salmon Commission). 1991. Report of the Fraser River Panel to the Pacific Salmon Commission on the 1990 Fraser River sockeye salmon fishing season. Vancouver, British Columbia.

PSC (Pacific Salmon Commission). 1996a. Report of the Fraser River Panel to the Pacific Salmon Commission on the 1992 Fraser River sockeye salmon fishing season. Vancouver, British Columbia.

PSC (Pacific Salmon Commission). 1996b. Report of the Fraser River Panel to the Pacific Salmon Commission on the 1993 Fraser River sockeye and pink salmon fishing season. Vancouver, British Columbia.

Pearse, P. H. 1992. Managing salmon in the Fraser: report to the Minister of Fisheries and Oceans on the Fraser River salmon investigation with scientific and technical advice from Peter A. Larkin. Department of Fisheries and Oceans, Vancouver, British Columbia.

Roos, J.F. 1991. Restoring Fraser River Salmon. Pacific Salmon Commission. Vancouver, British Columbia.

Thompson, W.F. 1945. Effect of the obstruction at Hells Gate on the sockeye salmon of the Fraser River. International Pacific Salmon Fisheries Commission Bulletin I, New Westminster, British Columbia.

Verhoeven, L.A., and E.B. Davidoff. 1962. Marine tagging of Fraser River sockeye salmon. International Pacific Salmon Fisheries Commission Bulletin XIII, New Westminster, British Columbia.

Ward, F.J., and P.A. Larkin. 1964. Cyclic dominance in Adams River sockeye salmon. International Pacific Salmon Fisheries Commission Progress Report 11, New Westminster, British Columbia.

Woodey, J.C. 1987. In-season management of Fraser River sockeye salmon (*Oncorhynchus nerka*): Meeting multiple objectives. Pages 367–374 in H.D. Smith, L. Margolis, and C.C. Wood, editors. Sockeye salmon (*Oncorhynchus nerka*) population biology and future management. Canadian Special Publication in Fisheries and Aquatic Sciences 96.

16 The History and Status of Pacific Northwest Chinook and Coho Salmon Ocean Fisheries and Prospects for Sustainability

Gary S. Morishima and Kenneth A. Henry

Abstract.—Since the turn of the century, Pacific Northwest ocean fisheries for salmon have undergone remarkable changes. From their beginnings as the last frontier to escape the jurisdictional reach of state and federal management, these fisheries are now being subjected to intensive and extensive regulation. Today there is increasing societal concern for impacts on the salmon resource base caused by ocean salmon fisheries that harvest extraordinarily complex mixtures of stocks with different productivities. Computer simulation models have become important tools for managing ocean salmon fisheries, as demands for more detailed and precise information have escalated. Regulations are becoming ever more complex and intricate with various interests vying to squeeze out the last drop of fishing opportunity that political processes will permit. Yet, at the same time, funding for resource management is declining and public trust in government and regulation is rapidly disappearing. Within this social climate, sustainability of ocean fisheries for Pacific salmon will require a new management paradigm rooted in better information about the salmon populations and principles of risk aversion.

THE DEVELOPMENT OF OCEAN SALMON FISHERIES

For thousands of years, salmon supported the cultures and economies of Indian communities in the Pacific Northwest. Fishing methods varied, depending on location and species. While a few tribes harvested salmon to a limited extent in ocean waters, the predominant fishing strategy consisted of patiently waiting to catch the fish when they were easiest to harvest—in the rivers when the salmon returned on their spawning migrations. This pattern of limited use provided a successful management regime for countless generations.

In the mid-1800s, the U.S. entered into treaties with many Indian tribes of the Pacific Northwest as a peaceful means of opening the way for settlement by non-Indians. Through these treaties, Indian tribes ceded their claims of aboriginal title to millions of acres and received in return promises by the U.S. that their rights to fish, hunt, and gather would be protected. For the first few decades after the treaties were signed and ratified, Indian people continued to provide fish to sustain both their own tribes and a growing population of non-Indians (AFSC 1970). However, as more and more settlers flooded in to homestead, changes in the environment and utilization of resources began to take their toll on the salmon resource. Major improvements in fish canning and processing spurred the development of fisheries as business enterprises to process huge quantities of salmon for worldwide markets.

Under the open access, common property principle of fisheries management that prevailed in non-Indian society, harvesters began to leapfrog over one another in a race to catch the fish before their competitors. When state legislatures turned their attention to regulation of harvesting activity in rivers and estuaries, fishermen developed new methods and places of harvest further from the rivers. From fish wheels, to gillnets, to traps, to purse seines, salmon exploitation moved from the rivers to nearby marine waters. In the early 1900s, a highly mobile troll fishery developed to catch salmon on offshore banks, several months, even years, before the fish would return to their natal rivers (Cobb 1921). By moving the fishery out of the rivers, the trollers escaped the seasonality of river fisheries, the reach of the regulatory and enforcement agencies, and the pervasive influence of the canning industry. The pattern of exploitation that emerged to harvest salmon was complex, costly to administer, economically inefficient, and nearly impossible to control (Crutchfield and Pontecorvo 1969; Pearse 1988).

THE SALMON RESOURCE

Six species of salmon originate in river systems along the eastern Pacific coast: chinook *Oncorhynchus tshawytscha*, coho *O. kisutch*, chum *O. keta*, sockeye *O. nerka*, pink *O. gorbuscha*, and steelhead *O. mykiss*. This chapter focuses on the Pacific Northwest ocean troll and recreational fisheries for chinook and coho salmon. Pink, sockeye, and chum, the mainstays of commercial net fisheries, are harvested to a much lesser extent by these fisheries. These species are anadromous— they hatch and rear in freshwater, migrate to sea to grow 1 to 6 years, and finally return to their rivers of origin to spawn. Except for a small proportion of steelhead, all salmon species originating on the eastern Pacific coast die after spawning (Groot and Margolis 1991).

During their ocean residence, salmon are so similar in physical appearance that, for many years, agencies involved in the harvest and management of salmon believed that one fish was just like any other. Very little scientific knowledge was available about Pacific salmon as ocean salmon fisheries began to develop. For instance, Samuel Wilmot, whom Canada called upon to address problems with the developing Fraser River fishery in the late 1800s, knew surprising little about the biology of Pacific salmon. Hailed by Canada as the father of artificial propagation who had demonstrated the feasibility of producing "unlimited quantities of any species of fish" in hatcheries, Wilmot reportedly believed that all Pacific salmon were the same species and were capable of spawning several times, just like Atlantic salmon (Meggs 1995).

Scientific knowledge of salmon biology has advanced considerably since that time. The concept that salmon occur as locally-adapted, self-sustaining reproductive populations (i.e., stocks) began to emerge in the late 1920s, when the results of high-seas tagging studies off the coasts of Canada and southeast Alaska started to become available. However, the stock concept was still not fully accepted in the scientific community even into the late 1930s. For example, Henry Ward (Ward 1939), commenting on a conclusion reached by one of his colleagues, W.H. Rich, that the parent stream theory had been conclusively proven, stated:

> ...Under joint auspices extensive tagging of adult salmon was carried on off the west coast of Vancouver Island, and off the Queen Charlotte Islands. These fish scattered widely, reaching many different spawning grounds. But at no point in the chapter is any evidence furnished to show that the author who was in charge of the operation knew from what streams these fish had come as young or by what route they had reached the points at which they were captured.... this comment by Rich is purely an assumption based on the fact that these adults after being tagged were recaptured in the Columbia.

Information as to the stocks being exploited by ocean fisheries was scant into the late 1960s. Managers expressed great concern and frustration over the inability to constrain harvest impacts. The manager of the Washington troll fishery (DiDonato 1965), for instance, stated:

The obvious lack of information regarding our offshore chinook stocks becomes apparent to all. The offshore troll fishery seems to defy the conditions which effective biological management demands, that is, regulation by stock. The foremost question then is not how best to manage the offshore troll fishery, but simply if this fishery can be managed on any biological, social, or economic basis. Until the marine life history of the various chinook stocks is uncovered, this question will remain unanswered.

Today, we know that the salmon resource is comprised of several hundred distinct reproductive populations, which have evolved over millennia to adapt to environmental conditions that are characteristic of their parent streams. Each stock has its own specific pattern of survival — a pattern of growth, maturity, and migration that have proved successful in sustaining its cycle of life. A sustainable fishery depends upon limiting the harvest to production that is surplus to the needs of reproduction. When overfishing, or other sources of mortality, exceeds these limits, a vicious downward spiral is set in motion: a progressively diminishing resource, the loss of some stocks and lower productivity of others, reduced catches, escalating pressures to harvest at excessive rates to try to save a declining fishing industry, and ever increasing regulatory restrictions (PFMC 1997a). Donald Hodel, in a May 4, 1982 letter to U.S. Secretary of Commerce Malcolm Baldrige, nicely summarized the basic problem:

> Once a depleted stock becomes a minor component of a fishery, socioeconomic factors prohibit their protection because of benefits foregone from not harvesting "healthy stock." With this concept, depleted runs contributing to mixed-stock fisheries will not be permitted to recover.

IMPACTS OF OCEAN SALMON FISHERIES

Although the remarkable ability of the salmon to return to their rivers of origin provided an obvious, convenient, and efficient means of harvest, state management agencies have favored fisheries that exploited several stocks simultaneously. While such a strategy may have simplified tasks of regulation and enforcement, it ignored the complexity of the salmon resource and greatly increased the difficulty of protecting the individual stocks that were the basis for production. As the predominant harvest pattern moved away from rivers to marine areas, fishing on complex mixtures of mature and immature fish from several stocks became the norm.

This pattern of exploitation had two major repercussions. First, the biological yield attained from chinook and coho by ocean fisheries is considerably less than the maximum potential. Parker (1960) concluded that "Critical size is not attained prior to maturity" (critical size is the age at which rates of growth and mortality are equivalent). The Pacific Fishery Management Council (PFMC 1978) environmental impact statement prepared for consideration of alternative ocean fisheries management strategies echoed this conclusion:

> Achieving maximum yield levels in pounds would require elimination of ocean troll and sport fisheries and the taking of fish at or near river mouths. This action would be required because rate of growth exceeds rates of natural mortality in the ocean.

Ocean fisheries are also inefficient because immature fish, which are too small to market profitably, are often discarded. This practice results in substantial mortality losses and loss of yield. It has been estimated that between 63% and 98% of the potential yield for Columbia River chinook was lost due to harvest and incidental mortalities of immature fish (Ricker 1976).

Second, the resource base deteriorated as less productive stocks became progressively overfished. In Washington, Oregon, California, and British Columbia, whole stocks of salmon have been eliminated (Walters and Cahoon 1985; Nehlsen et al. 1991; Slaney et al. 1996). Unfortunately, the lack of protection for individual stocks by fisheries managers was coupled with massive degradation of the habitat, putting additional pressures on the resource (Palmisano et al. 1993;

Meggs 1995; Walters 1995). In many Pacific Northwest areas, the salmon resource now consists of an assortment of declining stocks, some of which have been perpetuated by management intent, many others of which exist only as a result of the vagaries of chance.

HATCHERY PANACEA

As the abundance of wild stocks declined, managers turned to the technology of fish hatcheries in an attempt to increase the supply of salmon. Hatcheries churned out millions of salmon while managers avoided politically unpopular restrictions on harvest and society refused to impose constraints on economic activity that was degrading the streams, rivers, and estuaries essential for healthy wild stocks. From 1933 through the early 1970s, major investments in hatchery propagation were made as mitigation for damage to fish runs due to the construction of hydropower projects. Early attempts at hatchery production proved unsuccessful. Improvements in hatchery practices in the 1960s, involving diet, disease control, and rearing conditions, greatly increased survival rates (ISG 1996). Hatchery production of chinook salmon increased from 6 million smolts in the early 1900s to over 375 million smolts by the early 1990s; for coho, production has increased from 2 million smolts in the early 1900s to nearly 160 million smolts in the 1990s (Ruggerone et al. 1995).

The availability of large numbers of hatchery fish spurred increased fishing pressure and improved the efficiency of ocean troll and sport fleets. Since hook and line fisheries are density dependent (Shardlow 1993), exploitation rates on stocks can be elevated by the availability of large numbers of fish produced by hatchery facilities. In addition, hatchery stocks can withstand higher exploitation rates than their wild counterparts. Since some fisheries were regulated to achieve hatchery production objectives with little regard for impacts on wild stocks, less productive wild stocks were gradually eliminated from the resource base as the mixed-stock fishing patterns harvested both hatchery and wild fish simultaneously (Kaczynski and Palmisano 1993). The indiscriminate introduction of large numbers of hatchery fish into the environment has also been implicated in the proliferation of disease, attraction of predators, and reduced productivity of naturally spawning populations (Weitkamp et al. 1995).

For a few years, increasing hatchery production supported the fisheries and masked the continuing decline of natural production from overharvest and habitat deterioration. By the late 1970s, however, the failure of the "put and take" strategy of hatchery production as a panacea to resource depletion problems had become apparent. Survivals of hatchery fish decreased, the fish were getting smaller, and ever fewer salmon were returning to spawn (Ricker 1980; Ricker and Wickett 1980; Healey 1986; CTC 1996).

REGULATION OF OCEAN SALMON FISHERIES

The need to regulate ocean salmon fisheries has long been recognized. The earliest expressions of concern centered on the harvest of small, immature fish by trollers.

> The taking of immature salmon in the Puget Sound and on the banks along the coast of Oregon, Washington and Vancouver Island is responsible for a great loss in one of the important food products of the region; not only is the loss great but much of the food is of inferior quality. The most inexcusable waste is caused by the spring fishing in the Puget Sound near the south end of Whidbey Island. Here many tons of young silver and spring salmon are taken whose weight average not more than one and one-half to two pounds. The silver salmon would mature in the fall of the same year and produce 5 to 7 tons of fish for every ton taken in the spring. The chinook salmon, if left until mature, which would be in two to three years, would yield even greater returns.*

---

* Smith, V.E. The taking of immature salmon in the waters of the State of Washington. State of Washington, Department of Fisheries, Olympia, 1920.

When the troll fishery moved to ocean waters, state fisheries managers were powerless to do much about the problem because of the lack of jurisdictional authority. Management of ocean fisheries was not really possible until federal regulatory authority was enacted and treaty obligations, both domestic and international, were firmly established.

REGULATION OF OCEAN TROLL FISHERIES

The ocean troll fishery began as a fleet of row and sail-powered boats using hand-pulled gear, which operated around the mouths of harbors and rivers. Over time, the entry of large, powered vessels with freezers, sophisticated navigational equipment, improved fishing gear, and the capability to make trips lasting several days drastically increased the range and efficiency of the fleet. Trollers have historically located salmon by searching, although fishing technology is becoming increasingly reliant on fish finders, geographic positioning systems, computers, and communications equipment. As trolling developed, the hook and line nature of the fishery made it more "sporting" than commercial net fisheries and a segment of the fleet historically was comprised of "com-sport" vessels operated on a part-time basis using rod and reel gear.

The early history of regulation of ocean troll fisheries has been reported by Rathbun (1899); CDFG (1949); Fry and Hughes (1951); Kauffman (1951); Van Hyning (1951); Parker and Kirkness (1956); ODFW (1976); NRC (1981); and Argue et al. (1987). Regulation of Pacific Northwest ocean troll fisheries has changed dramatically over the last few decades as jurisdictional authorities of coastal states have expanded to contend with a highly mobile fleet that had the capacity to move freely along the coast. Until the late 1970s, participation during liberal seasons enjoyed by ocean fisheries was constrained principally by weather and market-driven minimum size limits. Both foreign and domestic vessels commonly fished outside coastal 3-mile territorial limits. Canadian vessels harvested significant quantities of Washington coastal, Puget Sound, and Columbia River salmon off the U.S. coast, and U.S. vessels, principally from Washington, harvested significant quantities of those same stocks off the coast of Canada.

During the development of the ocean troll fisheries, the variety and complexity of regulations led to difficulty in enforcing state landing laws, such as minimum size limits. This in turn led to attempts to coordinate management among jurisdictions. The Pacific Marine Fisheries Commission (PMFC) was established in 1947 through a tri-state compact by the States of Washington, Oregon, and California with the consent of Congress (Idaho and Alaska later became members). The objectives of this compact were set forth in Article I as:

> The purposes of this compact are and shall be to promote the better utilization of fisheries, marine, shell, and anadromous, which are of mutual concern, and to develop a joint program of protection and prevention of physical waste of such fisheries in all of those areas of the Pacific Ocean over which the states of California, Oregon and Washington jointly or separately now have or may hereafter acquire jurisdiction.

During the late 1940s and early 1950s, there were efforts to constrain the effects of increased fishing pressure by controlling fishing seasons and establishing minimum size limits to reduce the loss in potential yield resulting from the harvest of small, immature fish. Enforcement, however, was difficult due to frequent changes in these regulations.

Managers were learning that coordination was also a biological necessity because salmon did not respect political jurisdictions. For example, the 1948 recommendations of the PMFC were made not only to the member states of Washington, Oregon, and California, but also to Canada and the U.S. Fish and Wildlife Service (since that agency managed Alaskan salmon fisheries within 3 miles of the Alaskan coast and all Alaskan inshore waters prior to statehood in 1959). Efforts were made throughout the ensuing years to coordinate regulations, but coastwide consistency was not achieved.

During the mid-1950s, the harvest by the Canadian ocean troll fleet off the west coast of Vancouver Island and the Washington coast rapidly increased, taking large numbers of U.S.-origin chinook and coho. This fishery was part of a larger and more serious international problem of salmon interceptions; that is, the harvesting of salmon originating in one country by fishermen of another. Interceptions were a matter of growing concern, especially in the northeast Pacific, where high seas fisheries conducted principally by Japanese fleets were catching large numbers of U.S. salmon. In 1953, the U.S., Canada, and Japan entered into the International Convention for the High Seas Fisheries of the North Pacific Ocean that banned Japanese fisheries from operating east of 175°W. This boundary was meant to avoid exposing the vast majority of North American-origin salmon to Japanese nets. The so-called *abstention principle* of the North Pacific Fisheries Convention holds that fully exploited, scientifically managed stocks should not be exposed to effort from other countries. Japan has continued to abide by this principle since 1963, even though it is entitled to terminate the agreement with a year's notice (Van Cleve and Johnson 1963).

At this same time, increasing competition was pushing Washington and Canadian net fisheries seaward at the mouth of the Strait of Juan de Fuca. This complicated domestic management of fisheries under their respective jurisdictions and raised questions about the inconsistency of these ocean fisheries with positions taken by the U.S. and Canada over high seas net fishing by Japan. In 1957, an agreement to ban gillnet fishing outside the "surfline" was reached between the U.S. and Canada at an informal Conference on Coordination of Fisheries Regulations. As part of that agreement, Canada agreed to adjust its minimum size and season regulations and the U.S. agreed to bring management of Fraser River pink salmon under the jurisdiction of the International Pacific Salmon Fisheries Commission.

In the late 1940s and 1950s, both Canada and the U.S. were becoming increasingly alarmed over the inability to exercise jurisdiction over their own fishing vessels and the intrusion of foreign vessels operating off their coasts. President Truman issued a proclamation on September 28, 1945 that asserted the U.S. had the right to "establish conservation zones in those areas of the high seas contiguous to the coasts of the United States wherein fishing activities have or in the future may be developed and maintained on a substantial scale." The primary motivation for the declaration was the incursion of Japanese fishing fleets into the sockeye fishery in Alaska's Bristol Bay. During the late 1950s, initiatives by Canada in the United Nations Conferences on the Law of the Sea were aimed at the extension of jurisdictional authority. Canada proposed that an exclusive economic zone (EEZ) be established within 12 miles of the coast while the U.S. proposed a 6-mile territorial sea and an additional 6-mile exclusive fishing zone, subject to recognition of historic fisheries conducted by foreign nationals. The U.S. proposal was influenced by the desire to protect the vested interests of Washington State trollers who had been fishing off the coast of Canada for several years. Concerns of Washington fishermen over the prospect of being prohibited from fishing grounds off the Canadian coast were reflected in a July 1958 memo from the Washington Department of Fisheries:

> If the 12-mile limit were ratified, it would not only deny our fishermen the right to historical fishing areas off Vancouver Island but the right to harvest and participate in conservation measures of stocks of salmon from United States rivers. In other words, we would be responsible for maintaining the runs of salmon in our rivers and streams, but have no means of regulating the fishing pressure placed on them once they leave the ocean waters adjacent to our coast.

As a result of shared concerns over salmon, the U.S. and Canada initiated a series of periodic "Conferences on Coordination of Fisheries Regulations." These discussions led to the establishment of a number of technical committees that began to focus on the information available for management and the implications of salmon interceptions for the ability to conserve the resource.

In the late 1960s and early 1970s, the Seabed Committee of the United Nations supported provisions for the establishment of 200-mile fishing zones by coastal states during its Conference on the Law of the Sea Treaty. When U.S. and Canadian proposals failed to be ratified at the U.N.'s Law of the Sea Conference, Canada proposed that a multilateral treaty be negotiated to provide for a 6-mile territorial sea and an additional 6-mile contiguous zone for restricting fisheries. When this proposal was rejected by the U.S., Canada unilaterally extended its jurisdiction to 12 miles offshore. The U.S. followed by establishing a 12-mile zone for fisheries regulation off its coast in 1966.

Since both countries had already established troll fisheries within 3 miles of the other's coastline, the extension of territorial seas and the desire to continue to provide for established fisheries prompted the U.S. and Canada to negotiate bilateral reciprocal fishing agreements. These agreements began in 1970 to restore the fisheries that were threatened by the 12-mile limit. The first agreement, signed on April 24, 1970 in Ottawa, Canada, lasted for 3 years and specified that salmon trolling by the Canadians would be permitted in the 3- to 12-mile limit of the U.S. only along the Washington coast north of the Columbia River. U.S. salmon trollers would be permitted to fish within 3 to 12 miles off Vancouver Island. The U.S. and Canada revised the agreement in 1973, reducing reciprocal fishing areas for both countries. The amended agreement was extended annually until 1977.

Prior to the mid-1970s, management of ocean salmon fisheries was virtually nonexistent. Catch limits were not in effect and even seasons and minimum size limits could not be enforced because different jurisdictions had different requirements. Regulation, to the extent that it existed at all, was principally through the landing laws and licensing requirements of coastal states. There was no capability to limit participation by ocean fisheries outside of 12 miles from shore and limited regulation of the fishery within 12 miles did not effectively restrict harvest. There were no limitations on the number of salmon that could be harvested by ocean fisheries under the jurisdiction of the states, U.S., or Canada. Management for individual stocks was virtually impossible.

In 1975, the United Nations Conference on the Law of the Sea reached a consensus over the right of coastal states to establish an EEZ for the harvest and management of resources found within 200 miles of shore. By 1977, both the U.S. and Canada had enacted EEZs and fishery conservation zones that extended 200 miles offshore. In the U.S., it was only after the Fishery Conservation and Management Act (FCMA) was enacted, establishing the 200-mile EEZ, that efforts were undertaken to directly constrain the number of fish harvested by U.S. ocean fisheries. When the Pacific Fishery Management Council (PFMC) began to manage ocean salmon fisheries off the coasts of Washington, Oregon, and California within the 200-mile zone, Canada raised considerable objections over reduced Canadian fishing opportunities.

The U.S. and Canada suspended the reciprocal fishing agreement in the late 1970s. On August 1, 1977, the Governments of Canada and the U.S. appointed Special Negotiators to reach a comprehensive agreement covering their maritime boundaries and related marine resource issues. The negotiators recommended to the two governments the terms of a reciprocal fishing agreement that provided new mechanisms for consultation and resolution of differences.

An important aspect of this new arrangement was Canada's agreement to consult with the U.S. about the conservation need to close the Swiftsure Bank Area of Canada (Area 21) to all salmon fishing from April 15, 1978 through June 14, 1978, because of the presence of a large number of immature salmon originating in U.S. rivers. Incidental mortality resulting from the capture and release of fish smaller than the minimum size limits, or of species that could not be legally retained, became a matter of increasing management concern. Canada's refusal to completely close Swiftsure Banks, along with disagreements over east coast fisheries, led both countries to close their reciprocal fishing areas in June 1978, denying fishermen from each country access to fish off the other's coast. The 1978 accord was the last annual reciprocal salmon fishing agreement between the U.S. and Canada. However, the two countries agreed to accelerate negotiations on other fishery problems of mutual interest, particularly salmon interception issues.

The debate by the U.S. and Canada over the implications of salmon interceptions had begun with the 1959 Conference on Coordinated Fisheries Regulations and would continue for the next

25 years. Discussions eventually turned into full-fledged treaty negotiations to establish principles for conserving the salmon resource and sharing available harvests. During this period, Canada escalated its ocean fisheries off the west coast of Vancouver Island, an area where coho and chinook from Washington and Oregon predominated, to exert pressure on the U.S. to agree to a treaty to constrain salmon interceptions. During the 1970s, there were extensive negotiations between the U.S. and Canada. Numerous bilateral committees were established to consolidate information and address various technical issues.

REGULATION OF OCEAN SPORT FISHERIES

The ability to catch chinook and coho salmon by hook and line methods attracted sportsmen to the ocean fishery in search of recreation. Combined, the ocean troll and sport fisheries collectively can impart intensive harvest pressures on salmon stocks. Troll and sport fisheries have been estimated to be capable of removing 15% of the available population of coho salmon per week during seasons lasting several months (CoTC 1984).

Ocean sport fisheries for salmon developed later than troll fisheries, after World War II, and were of relatively minor concern to fishery managers before catches, spurred by the rapid expansion of charter fishing businesses, began to soar. Harvest by the ocean sport fishery in Oregon peaked in 1976 at 79,000 chinook salmon and 500,000 coho salmon. The ocean sport fishery in Washington also peaked in 1976 with 538,000 angler trips and a catch of 170,800 chinook salmon and 942,800 coho salmon. Since the mid-1970s, the Washington ocean sport fishery has been severely restricted (PFMC 1993; PFMC 1997a).

Prior to the early 1980s, ocean sport fisheries enjoyed liberal bag limits and were subjected to only minimal time/area restrictions. Since then, bag limits have been reduced and more direct regulatory measures such as catch ceilings for individual ports have been implemented to control harvests in several areas off the U.S. and Canadian coasts.

Increasing competition between sport and commercial troll fisheries has led to the development of specific allocation schedules between the two fisheries off the Washington and Oregon coasts. Generally, sport fisheries in these areas receive a larger share of available harvests at low levels of abundance. Inter-port allocation for sport fisheries off the Washington coast has been incorporated into the PFMC's Framework Plan and has become a subject of increasing interest off Oregon. Historical data on the ocean fisheries off the coasts of Washington, Oregon, and California are documented by the PFMC's Salmon Technical Team (PFMC 1993).

However, even as agencies began to regulate the ocean fisheries within their respective jurisdictions, escapements and catches were plummeting. Although it was clear that something had to be done, it was not clear what needed to be done or how to go about doing it on a comprehensive basis.

DEVELOPMENT OF THE CAPACITY TO MANAGE OCEAN FISHERIES

Prior to 1970, fisheries agencies had neither the scientific nor the legal basis to effectively manage ocean fisheries in the Pacific Northwest. However, the need, the authority, and technical capacity to manage these fisheries evolved virtually simultaneously in the early and mid-1970s. As such, the convergence of three factors, including social circumstance, information availability, and technological advancement, led to the development of the capacity to manage ocean salmon fisheries in Canada and the U.S.

While dwindling stocks and declining catches emphasized the need for management of ocean salmon fisheries, several social developments were needed before fisheries managers could effectively address the impending conservation and economic concerns. Among the most important of these was the entrenchment of Indian fishing rights in 1974. In a landmark decision, the Honorable George H. Boldt ruled that the tribes have enforceable property rights to take up to 50% of the harvestable surplus of each run of fish that originates in or passes through their usual and accustomed fishing places. The concept of harvestable surplus of individual stocks necessitated a change in management focus by providing protection for the portion of the run required for escapement. In addition, this decision recognized the need to manage fisheries so as to assure that a substantial portion of each run's harvestable surplus of fish was allocated to more terminally-located fisheries. In this way, the individual stock was recognized as the basic unit of fisheries production and management.

Within the last decade, the stock concept has been further refined to include evolutionary significant units (ESUs), as defined under the Endangered Species Act (ESA) of 1973. A number of salmon and steelhead ESUs have now been listed as "threatened" or "endangered" under the ESA (Knudsen et al. 2000). Considering these new legal obligations, it was clear that fisheries managers needed novel scientific, administrative, and legislative tools to generate the information required to provide for specific harvest entitlements and ESU protection.

While the need for advancements in the management of ocean fisheries was apparent, the legislative basis for such reforms was lacking prior to 1976. Washington Senator Warren Magnuson described that situation succinctly when he indicated that fisheries management in the Pacific Northwest was characterized by weak, divided authority and inadequate enforcement (Magnuson 1977). As a result, the federal governments of Canada and the U.S. were virtually powerless to control fishing outside their territorial waters, where ocean troll fisheries often operated. Establishment of EEZs within 200 miles of their respective coasts improved the situation by making it possible to regulate the ocean fisheries that operated outside state and provincial jurisdictions. However, establishment of EEZs did not obviate the need for international cooperation because the fish still crossed international boundaries. Hence, the actions of one management entity could have profound effects on the stocks that were being managed by another.

In recognition of the need for international cooperation, the U.S. and Canada initiated negotiations to establish a comprehensive salmon treaty in the late 1970s. In 1981, interim agreements for a limited set of fisheries were reached. Two years later, a general framework for a salmon treaty was established. However, the preliminary agreement was not ratified by the U.S. due, in part, to the omission of the west coast Vancouver Island troll fisheries, which could prevent the U.S. from achieving its fisheries management objectives. Therefore, further negotiations focused on developing a coastwide approach, eventually resulting in the ratification of the Pacific Salmon Treaty in 1985. With the establishment of the treaty, the social circumstances needed to affect fisheries management reform had been established. Yet, advancements in the available scientific information and technology were still needed to fully achieve fisheries management objectives.

The scientific information that was available until the 1980s was rudimentary by today's standards. These data were largely limited to trends in spawning escapements and catches in and near the rivers of origin. The lack of an effective means of sharing information further constrained the ability of fisheries managers to conserve the salmon resource base. However, the development of coded wire tag (CWT) technology in the early 1970s made it possible to obtain previously unavailable information on the distribution, abundance, survival, and recovery of salmon and steelhead populations throughout their range. Data collected using this technology showed that many stocks were being harvested at several stages of maturity over an extensive geographic area. Moreover, many populations were being harvested in mixed stock fisheries at rates that jeopardized their sustainability.

The emergence of personal computer technology in the early 1980s provided fisheries scientists with an essential tool for managing ocean salmon fisheries. This technology enabled scientists to quickly process large quantities of data to develop a better understanding of salmon exploitation on a coastwide basis. This technology was rapidly incorporated into fisheries management, particularly through the development of computer models that made it possible to analyze the copious CWT data that had been gathered over the previous decade. Importantly, the computer models provided a means of utilizing data on stock distribution, fishing mortality rates, and other biological variables to evaluate the effects of various regulatory regimes on individual stocks and stock aggregates. As such, the models provided much-needed tools for establishing catch quotas for each species by geographic area for commercial and recreational fisheries. Since their inception, such models have been further developed and refined to meet the specific needs of a variety of regulatory agencies throughout the Pacific Northwest.

Experience with successful application of computer simulation models in salmon management on the Pacific coast reveals four major lessons. First, successful models have been developed through an open, multi-jurisdictional effort; there must be a shared sense of ownership of the data and analysis underlying the workings of the model. Second, successful models force resolution of technical issues so policy implications can be evaluated in terms meaningful to decision makers; technical debates over data, assumptions, analysis procedures, and interpretation are eliminated or at least contained. Third, successful models are readily modified to meet changing demands for information and they must be capable of producing the information needed within an acceptable decision time frame. Last, successful models provide a consistent, convenient framework for evaluating options for complex conservation problems; they are easily accessible to allow positions to be developed and evaluated independently within each affected jurisdiction.

IMPACTS ON THE BIOLOGICAL POPULATIONS

The reluctance of managers to constrain fisheries under their jurisdictions, unless others do so first, has led to the collapse of some salmon stocks. During the last decade, the mandate of law has appeared on the scene to force U.S. fishery managers to consider impacts of harvest activity on individual, imperiled stocks. The Endangered Species Act (ESA) of 1973 has come to wield a powerful influence on the management of U.S. ocean salmon fisheries. Under the ESA, ESUs are protected and must meet two requirements: (1) reproductive isolation; and (2) be deemed to represent an important component of the evolutionary legacy of the species. For salmon ESUs listed as "endangered" or "threatened" under the ESA, biological opinions are required to permit fisheries to incidentally take listed fish. Stocks that have been listed include Sacramento winter run chinook, Snake River spring, summer, and fall chinook, Snake River sockeye salmon, and California coho (PFMC 1997b). In addition, several coho salmon ESUs are considered to be depressed and are possible candidates for listing under the ESA (Weitkamp et al. 1995).

Today, ocean salmon fisheries along the entire west coast are being constrained to limit stock-specific impacts. Some of these constraints are long term, such as those resulting from the need to protect species listed as endangered under the ESA. Other constraints address short-term survival or annual abundance problems. Central California ocean fisheries are regulated for concerns for impacts on the endangered Sacramento winter chinook salmon run. Fisheries off northern California and Central Oregon are managed to constrain impacts on Klamath River fall chinook salmon. Fisheries from Washington to California are evaluated based on their impacts on the endangered Snake River fall chinook salmon stock. Chinook and coho salmon fisheries in the Strait of Georgia are constrained due to concerns for local stocks. West coast Vancouver Island and North/Central B.C. fisheries have been curtailed to address serious conservation problems for west coast Vancouver Island chinook salmon. Fisheries in Puget Sound, the Columbia River, and along the Washington coast are restricted to meet stock-specific escapement objectives. The ESA has affected Alaskan fisheries to a lesser degree than several other U.S. fisheries, but Alaskan impacts on listed stocks are still taken into consideration and harvests are still subject to incidental take permits under biological opinions issued by the National Marine Fisheries Service.

Confronted with the continuing decline of certain salmon stocks, primarily caused by habitat degradation and low ocean survival, managers are faced with increasing requirements to address the needs of individual stocks by regulating ocean fisheries. Managers have turned to a variety of tools to constrain impacts, most commonly seasons, catch ceilings, and size limits, all directed toward achieving spawner escapement objectives. Regardless of the methods, the ultimate result has been a quantum leap in regulatory complexity and demands for more information.

THE SITUATION TODAY

Unprecedented challenges currently confront ocean trollers, sportsmen, and fishery managers. For example, fisheries exploiting highly complex stock mixtures must be managed on a stock-by-stock basis to meet legal obligations stemming from reserved tribal rights and the ESA. The salmon resource has declined in response to continuing degradation of freshwater and estuarine habitats and poor marine survival conditions. Large sums of public funds are expended to support hatcheries and to maintain the management and information systems required to contend with difficult technical challenges. Ready supplies of fresh salmon from aquaculture operations in Norway, Chile, New Zealand, and Canada are making major inroads into markets once dominated by the troll industry, calling into question the wisdom and need to continue hunting for salmon on the high seas. Societal attitudes are rapidly changing from consumption and commercialization of naturally produced salmon toward recreation and preservation.

Today, managers of ocean salmon fisheries are in a quandary. They are struggling to make do with diminishing fiscal resources while trying to develop strategies for making scientifically supportable decisions in the face of uncertainty, to devise strategies for avoiding mistakes that can have far-reaching consequences for the salmon resource base, and to find defensible and effective means to meet their legal obligations. Three basic alternatives are currently under political and management consideration: (1) approaches that evade accountability by delaying implementation of unpopular management decisions affecting fisheries and habitat; (2) approaches that are less information-demanding, less risky, and more conservative, but which still do not address, in detail, problems of declining salmon abundance on a stock-by-stock basis; or (3) improved technology coupled with significantly increased critical stock data to do a better job.

Delaying implementation of necessary corrective management measures appears to be popular on several fronts. Some of this is understandable because the limited political tenure of elected officials and top fisheries administrators allows them to escape accountability. By the time the consequences of decisions become evident, either the cause has been forgotten or the decision-makers have moved on. A plethora of interests seek to evade responsibility by accusing each other of being the real cause of the salmon problem. Politicians, administrators, loggers, farmers, ranchers, irrigators, developers, power companies, and competing groups of fishermen, among many, have managed to entangle issues in copious rhetoric and blame. Few solutions have emerged. Indeed, conflict and controversy have forced fishery managers to divert resources from their clear responsibility: to manage fisheries and hatcheries, and to try to halt degradation of the habitat necessary for the salmon to exist.

The option to implement management strategies that are less information-demanding provides a means of managing fisheries in recognition of uncertainty. This option employs heuristic decision rules, based on general stock status indicators, to establish allowable harvest levels for broad geographic regions. Costs of implementation would be lower and would result in more realistic expectations by taking into account the uncertainty surrounding abundance forecasts and projected fishing patterns. This type of approach contrasts markedly with the current management practice in the Pacific Northwest, where complex time-area fishing regimes are devised to find every fishing opportunity.

The search for management approaches that are less information-demanding is also leading to consideration of other strategies. Managers of ocean fisheries have not given up the search for an elusive magic bullet that will resolve their problems quickly and easily. Presently, this magic bullet appears to be manifesting itself as a push for selective fisheries (e.g., Lawson and Comstock 2000). The superficial appeal of fisheries that selectively harvest hatchery fish while releasing wild fish makes this approach a modern-day Siren's song. Instead of basing decisions on scientifically supportable analyses, managers are turning to assumption and presumption that selective fisheries will resolve resource conservation and utilization problems. Selective fisheries strive to improve economic benefits from hatchery fish, particularly for certain recreational fisheries, but they are also likely to diminish stock assessment and management capabilities because they may obscure the true mortality imparted on wild stocks, increase the uncertainty as to impacts of fisheries, require the expenditure of scarce funds to mark fish and modify CWT sampling programs, and undermine the data collection and analysis systems required for management. Fears have also been expressed that selective fisheries for hatchery fish may undermine the political support needed to protect habitats for production of wild fish. There are many operational problems, including impacts on other stocks and species, which need to be resolved before the role of selective fisheries in protecting and restoring salmon stocks can be scientifically evaluated.

New technology has allowed management of Pacific Northwest ocean salmon fisheries to become information-intensive. Ocean fishery managers have invested considerable funds to collect the necessary stock data and develop the needed computer models to more effectively manage salmon. In response to increasing demands for stock-specific information on impacts of ocean fisheries, affected constituents have raised numerous questions regarding the validity of management models while at the same time insisting on ever more complex and sophisticated analyses. The thirst for more and perfect information seems insatiable, driving decision-makers to seek answers that go far beyond the limits supportable by available scientific information. Then, when results based on scientific data and information are unwelcome, some managers choose to ignore or discount them as being inconclusive. Although data are adequate to document the declines of some salmon stocks, some management entities dispute the scientific data in order to delay measures needed to successfully rebuild depleted stocks. As a consequence, substantial funds are being diverted from important management programs to support research, studies, and analysis before decisions can be made through political processes of fisheries regulation. The consequence of course is that the resource bears the burden of uncertainty. Impacts continue unabated and the inexorable loss of less productive stocks further depletes the diversity and resiliency of the resource base. Such problems are by no means unique to salmon fisheries. Similar experiences prompted the United Nations to adopt a draft agreement in 1995 containing the precautionary management principle that: "States shall be more cautious when information is uncertain, unreliable or inadequate. The absence of adequate scientific information shall not be used as a reason for postponing or failing to take conservation and management measures."

The public may not continue to support funding to obtain, analyze, and utilize the data necessary for information-intensive management of ocean fisheries. The challenge is to utilize existing information for decision-making within the context of uncertainty and implications for error while studies continue to improve knowledge for future decisions.

CURRENT APPROACHES SUPPORTING OCEAN FISHERIES SUSTAINABILITY

Other alternative approaches for management of ocean salmon fisheries are now coming into focus as deliberations over sustainability continue. Chan and Fujita (1994), among others, have made suggestions for reforming ocean salmon fisheries. In general, alternatives that appear to be gaining support can be characterized as being adaptive, collaborative, or risk averse.

Adaptive management is in vogue (Lee 1993). The idea is relatively appealing because it does not lock decision-makers into a course of action. It is not new and exotic, but is simply muddling, trial and error. It can produce needed data in a relatively short time, data that might be years in obtaining under normal conditions. Since the ecological and economic system for salmon management is extremely complex, this management strategy perturbs the system and evaluates its response. This approach will likely falter because: 1) the political environment staunchly resists the perturbation of ocean fisheries to the degree necessary to conduct scientific and sociological experiments on a scale required to cover the migratory range of the salmon; and 2) it is unlikely that the mood of fiscal austerity will provide the resources necessary to monitor and evaluate results.

Collaboration, or the voluntary imposition of management measures to protect depressed salmon stocks, has been attempted in the past. This approach is still being pursued, in part because of its simplicity, by assuming that the managers who impact stocks throughout their migratory ranges will "do the right thing." The stark reality, however, is that there is an overriding temptation to shift responsibility to someone else. For example, Washington and Oregon's repeated cries for help from Canada to address growing concerns for chinook and coho salmon generally have gone unheeded. Now Canada has its turn to ask Alaska to reduce harvests to help conserve its chinook and coho salmon stocks. Unless managers can agree on common goals and their respective responsibilities for resource conservation, collaboration will likely fall by the wayside.

Another disturbing development that will interfere with a collaborative approach is the tendency to question the minutiae of the technical basis for management — in some forums, every assumption, goal, data value, and analysis method comes under intense scrutiny. Management models, while providing the capacity to conveniently evaluate the impacts of alternative regulatory regimes, make fertile ground for conflict and controversy. In the Columbia River for example, policy issues are debated and buttressed by battling models that rely upon different assumptions and data. While we recognize the value of a good dose of skepticism and the need for scrutiny of models and data, the tactic too often has not been to improve the scientific basis for management, but rather to delay and increase uncertainty so that important decisions are not made. Thus, the status quo is preserved and only minor perturbations that can easily be accommodated are possible. The problem is that time is of the essence; delayed decisions to conserve the resource can have serious long-term consequences.

Even where competition between models is not a factor, the issue of compatibility between models surfaces. In evaluating impacts of regulations affecting chinook salmon, for instance, the Salmon Technical Team of the PFMC must interpret, consolidate, and reconcile the results of four distinct models that involve different stocks, effort measures, and time frames. Greater collaboration on modeling is needed to alleviate confusion, conflict, and redundancy.

Of the three strategies for ocean salmon fisheries management looming on the horizon, risk aversion holds the most promise (e.g., Hilborn et al. 1994). Increasing attention has recently been paid to the explicit consideration of uncertainty and incorporation of risk-averse strategies into harvest regulation. Risk aversion can take a variety of forms. For ocean fisheries, risk-averse strategies could employ conservative estimates of abundance projections and/or consideration of impacts at the upper end of an anticipated range or could involve less information-intensive approaches such as those described earlier. Risk-averse strategies have already been implemented for some ocean salmon fisheries. For example, the allowable harvest of chinook salmon produced by Southeast Alaskan hatchery facilities reflected a policy decision that there would not be more than a 5% chance that the actual harvest of hatchery fish would exceed the inseason projection. This approach was risk averse in that it attempted to avoid pushing the limits of resource sustainability and was acceptable to certain constituencies.

Risk-averse strategies, however, can have adverse consequences, principal among them being reduction in the short-term yield from fisheries. Because risk aversion would err on the side of the resource, fewer fish would be available for harvest and fishing opportunity would be reduced in anticipation of larger catches in the future. To minimize undesirable adverse economic impacts, there is likely to be a strong tendency to push decision-making beyond the limits supportable by science, to demand more precision in management, particularly in the areas of stock identification and abundance estimation. Political pressures will likely keep the economic cost of risk aversion to a minimum and ignore the variability in the data and the analyses.

CAN OCEAN SALMON FISHERIES BE SUCCESSFULLY MANAGED FOR SUSTAINABILITY?

Changes in the political climate have set the stage for deliberations over the sustainability of ocean salmon fisheries. Institutions and boundaries are forming with the establishment of regional fishery management councils, court rulings regarding joint tribal and state management authorities and specific tribal entitlements, the ratification of the Pacific Salmon Treaty, and the implementation of the ESA. In the face of the ESA, adverse consequences of management error are becoming increasingly severe and unacceptable. New management tools, like computerized simulation models, have emerged to help regulators contend with technological challenges. However, the fundamental, daunting complexity that characterizes ocean salmon fisheries still remains. Ocean salmon fisheries exploit dynamic mixtures of natural and hatchery stocks with vastly different productivities, migration routes, and exploitation patterns. These stocks originate over an extensive geographic range and have high interannual variability in survival, growth, and distribution.

Ocean salmon fisheries can be managed for sustainability, despite this complexity, if that is what society desires. Sustainability can take many different faces — from protecting productivity and diversity of the various salmon runs to stability of the fisheries in terms of season structure, catch, or opportunity. In the years ahead, management of ocean salmon fisheries will certainly undergo changes, some of them likely to be drastic. Even now, commercial trolling along the Washington coast is rapidly waning and ocean sport fishing is barely holding on. There is no sense in longing for the return of unfettered fisheries that gave trollers and sportsmen the freedom of less than two decades ago to hunt the high seas for as many salmon as they could catch. The days of the last frontier have come and gone. Ultimately, the form of sustainability that will evolve for ocean fisheries will depend upon the needs and objectives spawned and dictated by societal values.

In many ways, the sustainability of ocean fisheries will depend upon the willingness and ability of managers to maintain a viable information base to evaluate impacts on the resource. Managers must be willing to make substantial investments in the development of common data systems, methodologies, and sophisticated computer models. The viability of the coastwide system for collecting CWT data must be preserved. Abundance, age, and maturity data for both catch and escapement are vitally important for stock assessments and the establishment of scientifically supported spawning escapement goals. The degree to which exploitation patterns of hatchery stocks are representative of wild stocks must be determined. Finally, the relationship between production, fishery impacts, and habitat must be demonstrated and quantified. Proper management of ocean salmon fisheries will necessitate collecting detailed, stock-specific data and substantial investments in predictive information, such as effort response, migration routes, and timing, so that models can accurately estimate the expected impacts of various management options and reflect uncertainty. At the same time, a new management paradigm is needed to overcome the variability in the data and the environment. Therefore, the development of approaches that are less information-demanding and more risk averse will be critical to future decisions regarding sustainability of ocean salmon fisheries and the stocks upon which those fisheries depend.

Despite these technical challenges, the foremost hurdles that must be overcome to sustain ocean fisheries are fundamentally political (Netboy 1973). Because coho and chinook salmon, the primary species harvested by ocean fisheries, can be caught several months before maturity in places far distant from their natal rivers, extraordinary efforts will be required among interjurisdictional managers to achieve the degree of coordination necessary to protect the resource. The involvement of multiple political jurisdictions that affect harvest and habitat create the necessity to integrate disparate types and qualities of information on abundance, spawning escapements, and biological characteristics and the need to accommodate uncertain impacts by fisheries. Managers must, therefore, be willing to forego some of the flexibility they have become accustomed to exercising. Instead, multi-jurisdictional approaches to planning and management must become the norm. Each management entity must recognize the necessity of regulating its own fisheries in full consideration and appreciation of other entities that affect the resource throughout its migratory range. Sustainability of ocean salmon fisheries ultimately rests upon sustainability of the salmon populations. That can only be assured if cumulative impacts of all fisheries are considered and accommodated within the populations' productive capacity.

With ocean salmon fisheries, future decisions will be less concerned with salmon biology or coordination of harvest management than with the resolution of issues that pit diverse economic interests against strongly held moral and ethical values. Already the flavor of the debate raging among politicians and their constituencies over salmon and salmon fisheries has involved profound questions of social policy. Can society afford to spend hundreds of millions of taxpayer dollars to pay for complex management systems and hatcheries to subsidize fishing for salmon? To preserve inefficient fisheries to catch wild salmon when aquaculture operations can provide high quality fresh fish for consumption? To curtail or forego economic development and growth to protect and maintain salmon habitats? To protect every remnant run of salmon?

Managing ocean salmon fisheries for sustainability requires that political leaders confront all these issues and more in a way that assures the viability of the salmon populations. This will be a formidable task. Given the reluctance of decision-makers to take responsibility for divisive decisions that affect their constituencies, there is considerable doubt as to whether many natural salmon stocks can be sufficiently productive to support viable ocean fisheries, or even to simply survive. The focus of the debate over sustainability needs to change from the fisheries to the fish. If the resource succumbs, the fisheries will too.

ACKNOWLEDGMENTS

We thank P. Dygert, L. Rutter, and several anonymous reviewers for their thorough reviews of the manuscript.

REFERENCES

AFSC (American Friends Service Committee). 1970. Uncommon controversy: fishing rights of the Muckleshoot, Puyallup, and Nisqually Indians. University of Washington Press, Seattle.

Argue, A. W., M. P. Shepard, T. F. Shardlow, and A. D. Anderson. 1987. Review of the salmon troll fisheries in southern British Columbia. Canadian Technical Report of Fisheries and Aquatic Sciences 1502.

CDFG (California Department of Fish & Game). 1949. The commercial fish catch of California in the year 1947 with an historical overview 1916–1947. Bureau of Commercial Fisheries, Fish Bulletin 74, Sacramento.

Chan, F., and R.M. Fujita. 1994. Ocean salmon fishery management: reforms for recovering our biological and economic heritage. Environmental Defense Fund, Oakland, California.

Cobb, J. N. 1921. Pacific Salmon Fisheries. U.S. Bureau of Fisheries. Document 902, Washington, D.C.

Cohen, F. G. 1986. Treaties on trial: the continuing controversy over Northwest Indian fishing rights. University of Washington Press, Seattle.

Crutchfield, J. S., and G. Pontecorvo. 1969. The Pacific salmon fisheries: a study of irrational conservation. The Johns Hopkins University Press, Baltimore, Maryland.

CTC (Chinook Technical Committee). 1983. Report of the U.S./Canada Chinook Technical Committee. Prepared for the Advisors to the U.S./Canada Negotiations on the Limitation of Salmon Interceptions. Vancouver, British Columbia.

CTC (Chinook Technical Committee). 1987. Chinook Technical Committee Report to the November 1987 Meeting of the Pacific Salmon Commission. Report TCCHINOOK (87)-5. Vancouver, British Columbia.

CTC (Chinook Technical Committee). 1996. Chinook Technical Committee Report TCCHINOOK (96)-1. Vancouver, British Columbia.

CoTC (Coho Technical Committee). 1984. Preliminary report of the Canada/U.S. Technical Committee on coho salmon. Prepared for the advisors to the U.S./Canada negotiations on the limitations of salmon interceptions. Vancouver, British Columbia.

DiDonato, G. S. 1965. Evolution of seasonal closures on the offshore troll chinook fishery, their effect on Columbia River fall chinook stocks, and a justification for the present chinook season. Supplemental Progress Report, Washington Department of Fisheries, Olympia.

Fry, D. H., and E P. Hughes. 1951. The California salmon troll fishery. Pacific Marine Fisheries Commission, Bulletin 2:7–42.

Groot, C., and L. Margolis, editors. 1991. Pacific salmon life histories. UBC Press, Vancouver, British Columbia.

Healey, M.C. 1986. Optimum size and age at maturity in Pacific salmon and effect of size-selective fisheries. Pages 39–52 in D.J. Meerburg, editor. Salmonid age at maturity. Canada Special Publication of Fisheries and Aquatic Sciences 89.

Hilborn, R., E.K. Pikitch, and M.K. McAllister. 1994. A Bayesian estimation and decision analysis for an age-structured model using biomass survey data. Fisheries Research 19:17–30.

ISG (Independent Science Group). 1996. Return to the river: restoration of salmonid fishes in the Columbia River ecosystem. Bonneville Power Administration, Portland, Oregon.

Jensen, T. C. 1986. The United States-Canada Pacific Salmon Treaty: an historical and legal overview. 16 Environmental Law 365.

Johnson, F. C. 1977. The Washington Department of Fisheries-National Bureau of Standards catch/regulation analysis model for salmon. National Bureau of Standards (mimeo). Olympia, Washington.

Johnson, K. J. 1990. Regional Overview of Coded Wire Tagging of Anadromous Salmon and Steelhead in North America. American Fisheries Society Symposium 7:782–816.

Kaczynski, V. W., and J. F. Palmisano. 1993. Oregon's wild salmon and steelhead trout: a review of the impact of management and environmental factors. Oregon Forest Industries Council, Salem.

Kauffman, D. E. 1951. Research report on the Washington State offshore troll fishery. Pacific Marine Fisheries Commission, Bulletin 2:77–91.

Knudsen, E. E., D. D. MacDonald, and C. R. Steward. 2000. Setting the stage for a sustainable Pacific salmon fisheries strategy. Pages 3–13 in E. E. Knudsen, C. R. Steward, D. D. MacDonald, J. E. Williams, and D.W. Reiser, editors. Sustainable fisheries management: Pacific salmon. Lewis Publishers, Boca Raton, Florida.

Lawson, P. W., and R. M. Comstock. 2000. The proportional migration selective fishery model. Pages 423–433 in E. E. Knudsen, C. R. Steward, D. D. MacDonald, J. E. Williams, and D. W. Reiser, editors. Sustainable fisheries management: Pacific salmon. Lewis Publishers, Boca Raton, Florida.

Lee, K.N. 1993. Compass and gyroscope: integrating science and politics for the environment. Island Press, Washington, D.C.

Magnuson, W. G. 1977. The Fishery Conservation and Management Act of 1976: first step toward improved management of marine fisheries. 52 Washington Law Review 427, Seattle.

Meggs, G. 1995. Salmon—the decline of the British Columbia fishery. Douglas & McIntyre. Vancouver/Toronto.

Nehlsen, W., J .E. Williams, and J. A. Lichatowich. 1991. Pacific salmon at the crossroads: stocks at risk from California, Oregon, Idaho, and Washington. Fisheries 16(2):4–21.

Netboy, A. 1973. The salmon: their fight for survival. Houghton Mifflin, Boston.

NRC (Natural Resources Consultants). 1981. A study of the offshore chinook and coho salmon fishery off Alaska. North Pacific Fishery Management Council, Document 15, Seattle.

O'Brien, S. 1986. Undercurrents in international law: a tale of two treaties. Canada-United States Law Journal 9:1–57.

ODFW(Oregon Department of Fish and Wildlife). 1976. A history and current status of Oregon ocean salmon fisheries. Oregon Department of Fish and Wildlife, Fisheries Division.

Palmisano, J.F., R.H. Ellis, and V.W. Kaczynski. 1993. The impact of environmental and management factors on Washington wild anadromous salmon and trout. Washington Forest Protection Association and Washington State Department of Natural Resources, Olympia.

Parker, R. R., and W. Kirkness. 1956. King salmon and the ocean troll fishery of Southeastern Alaska. Research Report l, Alaska Department of Fisheries, Juneau.

Parker, R.R. 1960. Critical size and maximum yield for chinook salmon. Journal of Fisheries Research Board of Canada. 17:199–210.

Pearse, P. 1988. Rising to the Challenge. Canadian Wildlife Federation, Ottawa, Ontario.

PFMC (Pacific Fishery Management Council). 1978. Final environmental impact statement and fishery management plan for commercial and recreational salmon fisheries off the coasts of Washington, Oregon, and California. Pacific Fishery Management Council, Portland, Oregon.

PFMC (Pacific Fishery Management Council). 1993. Historical ocean salmon fishery data for Washington, Oregon, and California. Pacific Fishery Management Council, Portland, Oregon.

PFMC (Pacific Fishery Management Council). 1997a. Review of 1996 ocean salmon fisheries. Pacific Fishery Management Council, Portland, Oregon.

PFMC (Pacific Fishery Management Council). 1997b. Preseason Report III. Analysis of Council Adopted Management Measures for 1997 Ocean Salmon Fisheries. Pacific Fishery Management Council, Portland, Oregon.

Rathbun, R. 1899. A review of the fisheries in the contiguous waters of the State of Washington and British Columbia. U.S. Bureau of Fisheries. Report to the Commissioner of Fish and Fisheries.

Ricker, W. E. 1976. Review of the rates of growth and mortality of Pacific salmon in salt water, and noncatch mortality caused by fishing. Journal of the Fisheries Research Board of Canada 33: 1483–1524.

Ricker, W.E. 1980. Causes of the decrease in age and size of chinook salmon. Canadian Technical Report on Aquatic and Fisheries Sciences 944.

Ricker, W.E., and W.P. Wickett. 1980. Causes of the decrease in size of coho salmon. Canadian Technical Report on Aquatic and Fisheries Sciences 971.

Ruggerone, G.T., F. Kutcha, D. Bregar, H. Senn, and G.S. Morishima. 1995. Database of propagation of Pacific salmon. Prepared for the Northwest Fishery Science Center, National Marine Fisheries Service. Natural Resources Consultants, Seattle, Washington.

Sabella, J. 1983. The wild king salmon: peril and promise. Pacific Fishing IV(3):34–39.

Shardlow, T.F. 1993. Components analysis of a density dependent catchability coefficient in a salmon hook and line fishery. Canadian Journal of Fisheries and Aquatic Sciences 50:513–520.

Slaney, T.L., K.D. Hyatt, T.G. Northcote, and R.J. Fielden. 1996. Status of anadromous salmon and trout in British Columbia and Yukon. Fisheries 21(10):20–35.

Smith, V. E. 1920. The taking of immature salmon in the waters of the State of Washington. State of Washington, Department of Fisheries, Olympia.

Stone, L. 1896. The artificial propagation of salmon on the Pacific Coast of the United States. Bulletin of the U.S. Fish Commission for 1896. Washington, D.C.

United Nations. 1995. Draft Agreement for the Implementation of the Provisions of the United Nations Convention on the Law of the Sea of 10 December 1982 Relating to the Conservation and Management of Straddling Fish Stocks and Highly Migratory Fish Stocks. August 4, 1995 (not yet in force).

Van Cleve, R., and R. W. Johnson. 1963. Management of the high seas fisheries of the northeastern Pacific. Publications in Fisheries, New Series. Volume 2, Contribution 160, College of Fisheries, University of Washington, Seattle.

Van Hyning, J. M. 1951. The ocean salmon troll fishery of Oregon. Pacific Marine Fisheries Commission. Fisheries Bulletin 2:43–76.

Walters, C.J. 1995. Fish on the line. A report to the David Suzuki Foundation. Fisheries Project Phase I. Vancouver, British Columbia.

Walters, C.J., and P. Cahoon. 1985. Evidence of decreasing spatial diversity in B.C. salmon stocks. Canadian Journal of Fisheries and Aquatic Sciences. 42:1033–1037.

Ward, H. B. 1939. Factors controlling salmon migration. Pages 60–70 in F.R. Moulton, editor. The migration and conservation of salmon. AAAS Publication 8. Science Press, Washington, D.C.

Weitkamp, L.A., T.C. Wainwright, G. J. Bryant, G. B. Milner, D. J. Teel, R. G. Kope, and R.S. Waples. 1995. Status review of coho salmon from Washington, Oregon, and California. U.S. Department of Commerce. NOAA Technical Memorandum, NMFS-NWFSC-24, Seattle, Washington.

Wilkinson, C. F., and D. K. Conner. 1983. The law of the Pacific salmon fishery: conservation and allocation of a transboundary common property resource. 32 U. Kansas Law Review 17.

17 Managing Pacific Salmon Escapements: The Gaps Between Theory and Reality

E. Eric Knudsen

Abstract.—There are myriad challenges to estimating intrinsic production capacity for Pacific salmon populations that are heavily exploited and/or suffering from habitat alteration. Likewise, it is difficult to determine whether perceived decreases in production are due to harvest, habitat, or hatchery influences, natural variation, or some combination of all four. There are dramatic gaps between the true nature of the salmon spawner/recruit relationship and the theoretical basis for describing and understanding the relationship. Importantly, there are also extensive practical diffi- culties associated with gathering and interpreting accurate escapement and run-size information and applying it to population management. Paradoxically, certain aspects of salmon management may well be contributing to losses in abundance and biodiversity, including harvesting salmon in mixed population fisheries, grouping populations into management units subject to a common harvest rate, and fully exploiting all available hatchery fish at the expense of wild fish escapements. Information on U.S. Pacific salmon escapement goal-setting methods, escapement data collection methods and estimation types, and the degree to which stocks are subjected to mixed stock fisheries was summarized and categorized for 1,025 known management units consisting of 9,430 known populations. Using criteria developed in this study, only 1% of U.S. escapement goals are set by methods rated as excellent. Escapement goals for 16% of management units were rated as good. Over 60% of escapement goals have been set by methods rated as either fair or poor and 22% of management units have no escapement goals at all. Of the 9,430 populations for which any information was available, 6,614 (70%) had sufficient information to categorize the method by which escapement data are collected. Of those, data collection methods were rated as excellent for 1%, good for 1%, fair for 2%, and poor for 52%. Escapement estimates are not made for 44% of populations. Escapement estimation type (quality of the data resulting from survey methods) was rated as excellent for <1%, good for 30%, fair for 3%, poor for 22%, and nonexistent for 45%. Numerous recommendations for improvements in escapement mangement are made in this chapter. In general, improvements are needed on theoretical escapement management techniques, escape- ment goal setting methods, and escapement and run size data quality. There is also a need to change managers' and harvesters' expectations to coincide with the natural variation and uncertainty in the abundance of salmon populations. All the recommendations are aimed at optimizing the number of spawners—healthy escapements ensure salmon sustainability by providing eggs for future production, nutrients to the system, and genetic diversity.

INTRODUCTION

The recently documented declines in Pacific Northwest salmon populations (Nehlsen et al. 1991; Slaney et al. 1996; NMFS 1997) indicate a breakdown in the west coast salmon management paradigm. Salmon managers, harvesters, and the public face some challenging questions. Why are many Pacific Northwest salmon populations declining even though salmon managers use the best available management techniques? Are the current concepts of salmon management flawed? Is the gap between theoretical, scientific salmon management and the practical application of those theories so large that it often renders management ineffective? Why have these conditions persisted in salmon management despite clear and repeated warnings from scientists, most notably Larkin (1977), Wright (1981), Fraidenburg and Lincoln (1985), and Ludwig et al. (1993)? This review and analysis of the existing escapement management process may point to some solutions to these apparent dilemmas.

Although overharvest, dams, habitat degradation, hatcheries, and natural environmental fluctuations all have contributed to declines in Pacific salmon, the populations will not be sustainable without proper spawning escapements. *Sustainable* means providing the best possible economic and social benefits while maintaining natural biodiversity. Achieving sustainability will certainly require optimal escapements. *Optimal escapements* are those sufficient to fully realize the biological potential of the freshwater habitat, thereby maximizing smolt production.

Any meaningful discussion of Pacific salmon management must also be prefaced by definition of terms relevant to the stock concept. In this chapter, a *population* is defined as a spawning aggregation, having little interbreeding with other spawning aggregations other than the natural background stray rate, uniquely adapted to a spawning habitat, and inherently unique attributes (Ricker 1958) resulting in different productivity rates (Pearcy 1992; NRC 1996). A population is analogous to the spawning aggregations described by Baker et al. (1996) and the demes of NRC (1996). *Management units* are groups of one or more populations treated together for management purposes, such as executing fisheries, setting escapement goals, and estimating harvest rates.

In heavily exploited salmon populations and/or those suffering from habitat alteration, there is no adequate method to estimate the intrinsic production capacity or determine whether perceived decreases in production are due to harvest, habitat, hatchery influences, natural variation, or some combination of all four. (Even in an unexploited population, there is no known way to assess production capacity because there is no way to account for compensatory survival once harvest begins.) Until recently, the commonly accepted solution to this dilemma was to employ a spawner/recruit model (e.g., Ricker 1954), which expressed the relationship between the number of spawners and the resultant production of adult progeny (harvest plus returning spawners), while accounting for limitations in the productive capacity of the spawning environment. This model has been used to estimate population parameters such as optimum population size at maximum sustainable yield (MSY), optimum harvest rate at MSY, and the population size that produces maximum recruitment (Ricker 1975; Hilborn and Walters 1992). With these estimates, managers try to determine the harvestable surplus and the number of spawners required to perpetuate the population, referred to as the *escapement goal*.

In reality, however, estimates resulting from the spawner/recruit model can and do result in overharvest (NRC 1996). Deficiencies in the spawner/recruit approach fall into two broad categories: theoretical and practical. In the theoretical realm there are two major types of weaknesses. The first is that mathematical attempts to describe the actual productivity of the population have significant limitations (summarized by Hilborn and Walters 1992). The second is our inability to reasonably integrate the many effects of the salmon's environment into the spawner/recruit relationship (e.g., Drinkwater and Myers 1987; Walters and Collie 1988).

In the practical realm of salmon escapement management, deficiencies arise from a variety of interrelated sources. These include lack of or inaccurate catch and escapement data; institutional, political, and fiscal barriers preventing application of the most advanced run forecasting and inseason modeling; imprecise run-size predictions; inability of regulatory processes to keep abreast of changes in run abundance and harvesting efficiency as the run materializes in the various fishing areas; inability to account for varying freshwater and marine survival; and the effects of applying harvest management decisions to composite populations, where each population has a production capacity determined by its fitness and the environment (Wright 1981; Hilborn and Walters 1992; NRC 1996).

Because salmon management units often have been managed with some form of spawner/recruit model—typically only having population data from the exploited condition—and considering the deficiencies described above, escapement goals sometimes may have been set too low. Furthermore, in some populations, lower escapements have resulted in less nutrients (salmon carcasses) being transported into the system to support freshwater productivity and subsequent juvenile salmon production (Kline et al.1993; Bilby et al. 1996). This means there is a relationship between escapement and carrying capacity (i.e., escapement alone can influence the shape of the spawner/recruit curve).

The problems described above sometimes are further exacerbated when basic Ricker model concepts are applied in salmon management only loosely, i.e., without ensuring a rigorous modeling approach is used, but rather assuming that observed escapements are indicative of escapements at MSY if they appear stable (see, for example, Ames and Phinney 1977; Fried 1994). Since application of the spawner/recruit approach is limited, even under ideal circumstances, one can see why lax application of the concept can lead to inappropriate escapement goal setting.

Salmon harvest managers have used escapement goals as a method for setting the optimum or MSY population size and then determining the harvestable surplus each year. In some cases, particularly in abundant and stable populations, fixed escapement goals or a range of target escapements have worked well (e.g., Brennan et al. 1997). However, there are a number of management units for which escapement goals are not being met (e.g., Palmisano et al. 1993; WDFW and WWTIT 1994) and other management units in which the goals are being met but the habitat appears to have the capacity to produce larger runs (PFMC 1978; Hiss and Knudsen 1993; NRC 1996). Lack of salmon in salmon habitat cannot be completely explained by habitat destruction alone because some undamaged, unobstructed habitats are relatively void of salmon. Since salmon have a natural tendency to stray and colonize vacant habitat, as evidenced by the rapid colonization of habitats recently exposed by retreating glaciers in Glacier Bay, Alaska (Milner and Bailey 1989), one would expect to see fish straying into and colonizing unused habitats if they were available to do so.

There may also have been a tendency for escapement goals in some management units to evolve downward. Historic accounts of run sizes often indicate escapements and harvests were substantially greater than they are today. For one example, Fraser River sockeye runs were 25–35 million fish at the turn of the century but are presently managed for about 1–9 million fish (Collie and Peterman 1990). In another example, Chehalis River (Washington) chum salmon average run size was about 140,000 at the turn of the century but has averaged about 54,000 in recent years (Hiss and Knudsen 1993). Neither basin has had sufficient recent habitat degradation to fully explain the declines.

Escapement goals have sometimes been modified based on the spawner/recruit relationship of the most recent few years (e.g., Ames and Phinney 1977; Fried 1994). In some ways this can be an attempt to account for changing population structure. However, it may also be a response to a relatively short-term environmental influence combined with the socio-economic need for continued fishing (even though at low population levels, maintaining adequate escapement is most critical for continued population productivity). A spawner/recruit model of such a population, based on that same recent data only (i.e., without historic data at various population levels) would yield a lower escapement goal (Hilborn and Walters 1992). This example illustrates the misuse of spawner/recruit models.

Mixed population fisheries also complicate the attainment of adequate escapements in several ways. First, depending on how populations are grouped, a common harvest rate may result in overfishing less productive populations, since each population has an inherently different productivity rate (Pearcy 1992; NRC 1996). Second, when composite populations are harvested together, it is often difficult to apportion the catch to the various populations, making population-specific catch and run-size estimates difficult. The third problem resides in uncertainties about the influence of mixed populations, namely: difficulties in defining the uniquely adapted populations; the degree to which populations are grouped into management units; and difficulties in assessing productivity rates of individual populations. When populations are harvested together, it is more difficult to apportion the catch to the various populations, making population-specific catch and run-size estimates difficult or impossible (Mundy 1996). The most extreme impacts of mixed population fisheries occur where escapements for natural production have been severely reduced through harvest rates set so that hatchery fish can be fully harvested. This usually results in continual overharvest of wild fish (Hilborn 1992; NRC 1996).

The purpose of this chapter is to lay the groundwork for rebuilding and maintaining healthy Pacific salmon populations by identifying and describing the aspects of salmon escapement management presently precluding sustainability. I addressed the theoretical deficiencies of escapement management and developed a coastwide (but not exhaustive) collation and general analysis of information on the status of escapement management based on data provided by management agencies. Specific objectives of this study were to (1) identify and describe weaknesses in theoretical salmon management models and related weaknesses in setting escapement goals and estimating escapements; (2) determine whether some escapement goals have been gradually reduced over time; (3) summarize some effects of mixed population fisheries on escapements; and (4) make recommendations for improving escapements and escapement management.

METHODS

Weaknesses in theoretical escapement management were identified by reviewing the literature and developing a simple graphic illustration of the general concepts of why standard theoretical approaches to escapement management have been inadequate. For the remainder of the evaluation, information on escapement goal-setting methods and annual escapement estimation techniques was assembled and tabulated from the literature and through personal communications with agency management biologists. The general hypothesis that, for a significant proportion of management units and/or populations, there are serious deficiencies in the methods for setting escapement goals and estimating annual escapements was investigated by summarizing and categorizing escapement management information. A large spreadsheet table was prepared listing as many management units and populations as could be identified over a broad survey area covering much of Alaska, Washington, Oregon, and California. Escapement management information pertinent at the population or management unit level were incorporated into the table. The evaluation did not include data for hatchery populations or populations predominated by hatchery production, populations upstream of Bonneville Dam on the Columbia River, British Columbia, or the Arctic-Yukon-Kuskokwim area of Alaska. Every attempt was made to identify populations at their smallest, although biologically meaningful, level. However, this was not always possible, either because some populations were lumped by managers in their reports to me or because specific populations are yet to be differentiated. This means that my estimates of the number of populations were conservative.

Information on escapement goal-setting methods, escapement data collection methods, and escapement estimation types were subjectively and cursorily assigned to one of six categories, based on the availability and quality of the underlying methods and data. The classifications used in this study are described below.

ESCAPEMENT MANAGEMENT

Escapement Goal-Setting Methods.—The methods used to set escapement goals for each management unit (groups of one or more populations) were categorized according to their degree of technical sophistication and likelihood of accurate representation of productivity, as follows.

Excellent method:

    Combined-strong—a combination of methods, such as spawner/recruit modeling based on a comprehensive data record, with consideration for habitat production potential and allowance for annual variability, resulting in an accurate escapement goal.

Good methods:

    Habitat-advanced—the escapement goal is based on a relatively sophisticated application of accurate habitat-based production potential;

    Spawner/recruit—a spawner/recruit model is formally used to estimate the escapement goal; and

    Historic—escapement goal based on some notion of what the escapements were prior to heavy exploitation.

Fair methods:

    Combined-weak—escapement goal is based on a combination of methods, such as weak past escapement data combined with a general sense of habitat production potential;

    Habitat—goal based on a generalized or somewhat outdated estimate of the watershed's carrying capacity; and

    Recent escapements—escapement goal is based loosely on observed escapement data (e.g., average) or indices in the recent past (e.g., up to past 25 years).

Poor methods:

    Index—escapement goal is set for one or more key populations within a management unit with an assumption that performance of the key population reflects performance of other populations within the unit.

No method:

    No goal—no escapement goal set for the management unit.

No information:

    No information available—Information available identifying management unit and sometimes the escapement goal but no information available as to how the goal is set.

For some management units, the escapement goal-setting method could have been classified into more than one of the methods defined above. If enough information was available, a management unit was categorized into the category best describing the escapement goal-setting method for that unit. The proportions of management units that fell into each category of escapement goal setting were then calculated to examine the extent of the gap between the best salmon population theory and its actual application in escapement goal management.

Gradual Escapement Goal Reduction.—There may be cases where the escapement goal has gradually been reduced over time in response to lower observed returns. This may occur when analysis of spawner/recruit data using recent data from a depressed population indicates an apparent steady state population, but which is unknowingly below carrying capacity, thereby resulting in underestimation of the escapement goal. I reviewed the literature to find cases where management policy had perhaps inadvertently contributed to declines in productivity and/or sustainability.

Estimating Annual Escapements.—Regardless of the run management method employed, the most basic data for evaluating management unit and/or population performance is the escapement estimate (plus catch data, which is outside the realm of this study). For management to be most successful, escapement estimates should be accurate and information should be collected at the population level. Therefore, the quality of coastwide escapement estimates was evaluated at the population level in two ways. First, escapement estimation was characterized in terms of the recent data collection methods for each population and classified generally as to the quality of that method. These included:

Excellent:

> Trap or dam count—A complete enumeration of all fish of the species and race passing a trap, weir, or dam;

Good:

> Dam or trap estimate—Partial observations and/or extrapolations of the run at a dam or estimates when a trap or weir is at times overtopped by high water;
>
> Tower—Estimates or total counts from a tower, bridge, or other visual observation point;
>
> Sonar—Estimates using sonar to count upstream migrants;

Fair:

> Mark-recapture—Run estimated using mark and recapture methods;
>
> Combined—Estimates or indices based on a combination of survey methods;

Poor:

> Foot index—Estimates or indices based on foot surveys;
>
> Aerial index—Estimates or indices based on aerial surveys;
>
> Boat index—Estimates or indices based on boat or drift surveys;
>
> Snorkel survey—Estimates or indices based on snorkel surveys;

None:

> None—no escapement estimate made for this population.

No information:

> No information available—Information available identifying the population but no information on whether or how the escapement is estimated.

Second, the type of count, estimate, or index resulting from the escapement data, and the relative quality of those statistics, was characterized for each population. Categories included:

Excellent:

> Total—A complete count of all individuals; only practically possible from dam or trap counts;

Good:

> Total estimate—An estimate based on an enumeration technique, i.e., counting and expanding (possible sources include dam or trap partial counts, tower, sonar, or mark and recapture);
>
> Peak count—based on repetitive survey of adults over the duration of the run, where the peak count or maximum fish days is ascertained and reported;
>
> Good index—Repetitive surveys within a season, utilizing an estimation technique such as area under the curve, but not expected to estimate total run due to variable visibility, etc. Surveys done annually either for entire stream or some consistent index reach(es).
>
> Total redds—Annually consistent program by some standard and calibrated method to either count all redds in a river or in a consistent index area;

Fair:

> One count—An annual survey done sometime during spawning with no way of knowing whether it was at the peak or not.
>
> Fair index—Similar to a Good Index (above) but lacking in either annual consistency or reliable visibility.
>
> Redd survey—survey one or more times to estimate total redds or redds/mile over some stretch(es) of river, but without rigorous validation or annual consistency;

Poor:

> One count-sporadic—Same as One count above but not done every year;
>
> Poor index—Similar to a Good Index (above) but having significant deficiencies in either annual consistency, data consistency, or reliable visibility; or
>
> Carcass index—survey one or more times to estimate carcasses/mile over some stretch(es) of river.

None:

None—no escapement estimate made for this population.

No information:

No information available—Information available identifying the population but no information on the type of escapement estimate, if any.

To better understand the effects that various data collection methods and types of escapement estimation surveys have on salmon escapement management, the information collected on the populations was summarized and used to assess the overall quality of escapement data collected in the western U.S.

EFFECTS OF MIXED POPULATION FISHERIES

If every population was a single management unit and was harvested separately from all other management units, life would be much simpler for salmon harvest managers. Unfortunately, most populations, and many management units, are harvested together in mixed population fisheries. Complete assessment of the extent of this fisheries management dilemma would require a unique and extensive study unto itself. However, assembly of the data described above provided an opportunity to conduct a cursory evaluation of the effects of mixed population fisheries on escapement. To assess the degree to which intraspecific mixed population fisheries were occurring, the number of populations within each management unit was summarized, by species and geographical area. Several case histories where decisions to forego wild production for full utilization of hatchery fish (the most extreme case of a mixed population fishery hindering natural escapements) were also reviewed and presented.

Results of the various evaluations described above were used to identify the successes and deficiencies of present management schemes and to make general recommendations regarding improvement of escapement goals and escapement management approaches.

RESULTS AND DISCUSSION

THEORETICAL WEAKNESSES

The inability of Pacific salmon management programs to in some cases prevent dramatic declines in populations can at least partly be attributed to theoretical deficiencies in widely used spawner/recruit models (NRC 1996). Although the models do provide a conceptual framework for considering management alternatives, they are often insufficient to support quality management due to inherent weaknesses and biases as well as inaccurate or sparse data (NRC 1996). Specifically, frequent and substantial errors in the numbers upon which models are based (namely, the counts or estimates of spawners, returning run sizes, and catches) can lead to overestimates of optimum harvest rate and underestimates of optimum population size, especially in overexploited populations (Hilborn and Walters 1992). Time-series bias of parameters can also develop in the models because size of recruitment depends to some extent on size of recruitment in the parent year (i.e., the independent variable is not actually independent). This can result in underestimation of optimum population size (Hilborn and Walters 1992). In addition, the stock/recruitment model assumes the relationship between spawners and subsequent run sizes does not change over time, but it does. This is particularly influenced by temporal changes in the population structure which can lead to models indicating a healthy population when, in fact, it is overexploited (Hilborn and Walters 1992). Gradual temporal shifts in the degree to which each environmental variable influences salmon survival render a model developed over a series of years less meaningful (biased) in subsequent years. Extreme interannual variation due to environmental influences leads to imprecision (poor fit) in the models. Furthermore, in exploited populations there are rarely extremely high escapements

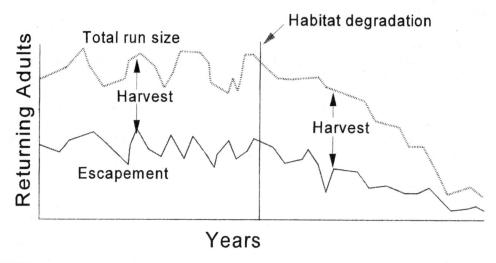

FIGURE 17.1   Hypothetical run size and escapement data from a salmon population under constant exploitation rate before (left of vertical line) and after (right of vertical line) habitat degradation begins in the watershed. Harvest is the difference between run size and escapement each year.

allowing assessment of the effects of large population on the stock/recruitment relationship (Hilborn and Walters 1992). This could also inadvertently lead to an underestimate of optimum population size and an escapement goal set too low.

Another serious challenge to the spawner/recruit model for managing Pacific salmon is that it does not adequately account for gradual habitat alterations nor can it discern whether changes in population abundance are due to habitat degradation or overfishing. A hypothetical data series is illustrated in Figure 17.1. During the years before habitat degradation (years for which there is usually no data), the population is in equilibrium, with steady-state exploitation that does not diminish the population. After habitat degradation begins, the population begins to decline. Assuming the harvestable surplus or harvest rate remains constant, which a spawner/recruit model based on previous years would support, the escapement and harvest begin to drop. Once the decline is underway, it is impossible to tell whether continuing losses are caused by habitat degradation or overfishing. It may have been, for example, that carrying capacity was only reduced somewhat, but that harvest rate either increased or, with decline in carrying capacity, the existing harvest level became too high for population perpetuation. Unfortunately, there is a time lag between when the carrying capacity is reduced and/or the harvest rate becomes too high and when the spawner/recruit model would indicate a recommended reduction in fishing effort. The biggest problem, however, is that once run sizes, escapements, and harvests have been reduced, the spawner/recruit model would indicate an escapement goal at some level below actual carrying capacity. For example, if a spawner/recruit model was developed based on the recent years of data (on the right in Figure 17.1), the escapement goal might be set lower than carrying capacity.

So one of the major flaws in salmon population models is that they assume a constant carrying capacity. To make matters worse, low escapements may be further diminished by the decreasing number of carcasses in the stream. Recent research demonstrates that a large proportion of productivity in healthy salmon streams is derived from nutrients in decaying salmon carcasses (Kline et al. 1994; Bilby et al. 1996). As the number of carcasses decreases, so does the biological carrying capacity, even when physical habitat is in good condition. If the physical habitat is simultaneously being degraded, it is impossible to discern whether the relative carrying-capacity reductions are due to habitat degradation, lack of carcass-transported nutrients, or both.

Oregon coastal coho historical catch and escapement data serve to illustrate that current "steady state" models do not reflect true production potential. Total run sizes have decreased to

10–20% of historical run sizes over the past 100 years or so (OFWC 1995, Figures 4a, 5a, and 7a), yet there is no information indicating the relative proportion of the long-term decline attributable to habitat degradation or overfishing (underescapement). What would production be like if escapements were of a historic magnitude, i.e., about ten or more times greater than today's? Of course, this unrealistically assumes pristine freshwater carrying capacity. However, since it is unlikely that freshwater habitat quality has been reduced 90% coastwide, as escapements have, ideal contemporary escapement goals probably lie somewhere between those observed today and ten times as much.

In summary, it seems that to pursue salmon management based solely on trying to improve the theoretical basis for standard spawner/recruit models and/or the accuracy and quality of data used in the models would be imprudent (NRC 1996). Because of significant uncertainty about factors influencing run sizes, even the best models simply will not perform to the degree that we can totally depend on them to accurately predict run sizes and, hence, catches. We will only make progress by finding new approaches that fully account for the influence of population abundance, habitat quality and carrying capacity, biological diversity, and variations in the marine and freshwater systems on the relationship between spawners and recruits. New models and approaches must also be addressed in an adaptive management framework (Walters 1986) and incorporate uncertainty in decision-making using some kind of decision theory (see, for example, Frederick and Peterman 1995; Adkison and Peterman 1996).

ESCAPEMENT MANAGEMENT

Escapement Goal-Setting Methods.—Using the information from published and unpublished sources, I identified 1,025 wild or mostly wild U.S. management units (see Appendix Table 17.1). Some management units are not represented, particularly for California, the Arctic-Yukon-Kuskokwim area of Alaska, and the Columbia River upstream of Bonneville, but I believe about 90% of the U.S. coastwide units have been accounted for. Of the 1,025 documented units, I could not find information on how escapement goals have been set for 171 management units (17%) (Table 17.1). Of the 854 for which I had information, escapement goals for only 8 management units (1%) are set by methods that were rated as excellent, i.e., using methods that combined information in a way that most effectively characterizes the management unit's production potential (Table 17.1). Escapement goals for 142 management units (16%) were rated as good; they are set by either advanced consideration of habitat potential, spawner/recruit models based on fairly accurate run size estimates, or some notion of the historic production potential (Table 17.1). Fifty-eight percent (499) of management units have their goals set by methods rated as fair; using relatively inaccurate, habitat-based production potential, recent escapements, or some combination of the two. Escapement goal-setting methods for 13 management units (2%) were rated as poor; these were developed by monitoring one or several index streams representing a large number of populations in the same geographical area (Table 17.1). Escapement goals have not been established at all for 192 U.S. management units (22%). Together, this assessment shows that far too many management units are being managed with inappropriate, weak, or no goals.

To achieve fisheries sustainability, the methods for setting escapement goals must be improved. With 80% of management units having goals developed by fair or poor methods or having no goal, there is clearly a serious problem with the salmon management system. Furthermore, it must be remembered that escapement goal setting is done at the management unit level and many units are composed of multiple populations (reviewed below). Since individual populations may have varying production capacity (Pearcy 1992), it is likely that some management unit escapement goals could be inappropriate for individual component populations, resulting in chronic overharvest.

Gradual Escapement Goal Reduction.—A downward trend in escapements may result when overfishing and/or habitat degradation causes decreased population productivity and managers, basing their predictions for future productivity on recent low productivity, lower the estimated

TABLE 17.1
Summary of escapement goal setting methods for U.S. salmon management units.

Quality	Method	Species						
		Chinook	Chum	Coho	Pink	Sockeye	Steelhead	All Species
Excellent	Combined-strong	0	0	4	0	4	0	8
Good	Habitat advanced	1	0	15	0	3	38	57
	Spawner/recruit	3	1	1	1	17	1	24
	Historic	14	12	1	26	8	0	61
Fair	Combined-weak	11	27	16	46	14	0	114
	Habitat	8	3	2	2	6	0	21
	Recent escapements	42	135	54	101	31	1	364
Poor	Index	2	0	4	1	6	0	13
No method	No goal	35	48	35	25	16	33	192
No information	No information	31	17	0	4	8	111	171

escapement goal. When, in the face of continued heavy harvest or habitat degradation, the population size is further reduced, managers are tempted to again lower the escapement goal, and so on. This occurs when, as is usually the case, the assumption of stable productivity is not met in spawner/recruit models (Hilborn and Walters 1992) and is exacerbated when analyses do not include an extensive historical record of population productivity. (Although revising the model to account for long-term changes in environment or habitat makes sense, differentiating between the sources of the changes, i.e., whether they are intractable or subject to human intervention, is impossible.)

An example of such a downward trend is the Klamath River chinook salmon, well documented in the beginning of its negative spiral by Fraidenburg and Lincoln (1985). As of 1978, the first year for which basin-wide escapement estimates were available, the escapement goal was 115,000 spawners, most of which were wild. In response to drought and overfishing, the Pacific Fishery Management Council (PFMC) adopted a 1980 "interim" escapement goal of 86,000 to prevent disruption of troll fisheries and cited a commitment to return to the original goal within 4 years. By 1983, PFMC had, in response to cries of economic hardship from user groups, adjusted the inriver run size target (escapement plus inriver catch) to 68,900 and the rebuilding schedule was lengthened to 16 years. Over the past few years, the escapement floor has been reduced to 35,000 with inriver run size (escapement plus inriver catch) targets set annually—in 1995 the target was 75,200 (PFMC 1996). The 1995 escapement exceeded the floor for the first time since 1989 (PFMC 1996). While this case provides an excellent example of how politics has influenced salmon management, it also illustrates how scientists and managers sometimes participate in regulating a fishery into overfishing.

ESTIMATING ANNUAL ESCAPEMENTS

Based on information collected from state biologists, I identified 9,430 discrete U.S. populations of Pacific salmon (see Appendix Tables 17.2 and 17.3). Based on my definition of a population, it was estimated that these 9,430 populations represent roughly two thirds of the wild or mostly wild U.S. Pacific salmon populations. There is little further information on the missing populations primarily because (1) populations are lumped together as reported by managers; (2) no work has been done to differentiate among populations; or (3) populations in some locations have not yet been documented.

Escapement Data Collection Methods.—Of the 9,430 identified populations, 6,614 (70%) had sufficient information to categorize the method by which escapement data have been collected (Table 17.2). Over 44% of populations (2,925) for which there is information are not monitored for escapement. Escapement data collection methods were rated as excellent for 79 U.S. populations (1%), where methods included total counts at dams, traps, or weirs. Methods were rated as good

TABLE 17.2
Number of U.S. populations for which each method of escapement estimation is used.

Quality	Method	Species						All Species
		Chinook	Chum	Coho	Pink	Sockeye	Steelhead	
Excellent	Dam count	9	0	2	0	0	0	11
	Trap count	5	6	21	20	16	0	68
Good	Dam or trap estimate	1	4	4	0	30	4	43
	Tower	1	0	0	0	5	0	6
	Sonar	0	0	1	0	4	0	5
Fair	Mark/recapture	0	0	0	0	1	0	1
	Combined	51	1	16	7	25	13	113
Poor	Foot survey	50	37	107	30	2	15	241
	Aerial survey	78	1230	744	928	197	18	3195
	Boat survey	3	0	0	0	0	0	3
	Snorkeling	0	0	0	0	0	2	2
None	No method	2	771	1733	0	84	335	2925
No information	No information	83	119	6	2418	7	184	2817

for 54 populations (1%); these methods included tower or sonar counts, or extrapolated estimates from trap, dam, or weir counts. Escapement data collection methods were fair for 114 populations (2%); these included mark and recapture methods and methods combining one or more survey types. Methods were poor for 3,441 populations (52%) and included individual survey types (foot, boat, aerial, snorkeling).

This summary was affected by several factors. First, the purpose was to generally characterize the quality of escapement data collection methods, not to be specific to locations. While recognizing that there are various possible criteria for categorizing and rating escapement data collection methods, the frame of reference for judging escapement data collection method quality in this study was a total count (i.e., the ideal escapement estimate). While some fishery managers may disagree with the categorizations, setting the quality categorizations relative to the ideal sets the tone for achieving sustainability. In regard to the outcome, even if all populations were upgraded by one category level, 96% of populations would still be rated as only fair or having no escapement estimation at all.

Second, the process of assigning the category for each run's escapement data collection method may have affected the summary. The quality of escapement data is strongly influenced by attributes like relative visibility, repetition, duration of annual survey records, and consistency of escapement estimation location and methods. Gathering information at that level of detail for each population would have been an insurmountable task, although more information was available for some populations than others, thereby allowing more accurate categorization. In most cases, however, my categorization decision was made based on limited information.

Third, the relative availability of information across populations affected the summary. It is possible that the information missing for 30% of the populations could skew the outcome of the summary one way or another. For example, the large number of Alaskan populations, mostly surveyed using aircraft (judged to be a poor method), tilted the overall outcome toward "poor" (see Appendix Table 17.2). On the other hand, of the Washington, Oregon, and California populations for which there was sufficient information, 93% were rated as having escapement estimation data collection methods rated as fair, poor, or non-existent (see Appendix Table 17.2). This suggests that the availability of information likely did not affect the summary outcome.

Even when these factors are considered, it remains obvious that both escapement data collection methods and the programs to collect high-quality escapement information are deficient. It is

TABLE 17.3
Number of U.S. populations for which each escapement estimation type is used.

Quality	Method	Species						
		Chinook	Chum	Coho	Pink	Sockeye	Steelhead	All Species
Excellent	Total	8	2	7	0	13	0	30
Good	Total estimate	13	258	27	514	52	5	869
	Peak count	88	252	106	462	109	0	1017
	Good index	12	0	29	0	0	6	47
	Total redds	7	0	7	0	0	0	14
Fair	One count	25	61	3	0	9	3	101
	Fair index	2	6	21	0	6	4	39
	Redd survey	17	0	21	0	0	15	53
Poor	One count—sporadic	1	680	644	0	90	17	1432
	Poor index	2	0	4	0	0	0	6
	Carcass index	9	0	0	0	0	0	9
None	No method	2	771	1733	0	84	335	2925
No information	No information	97	138	32	2427	8	186	2888

interesting to note that, while the NRC (1996) recommended adoption of a minimum sustainable escapement (MSE) approach and recommended more spawners in streams, they only indirectly alluded to the dearth of good quality escapement data essential for understanding the health of populations. Without improvements in the escapement monitoring system, achieving the laudable goals of MSE will be nearly impossible for many populations.

Escapement Estimate Type.—While the method of escapement data collection has important ramifications for the quality of the estimate, the type of estimate influences the quality of subsequent data analysis for run management. Of the 9,430 identified U.S. populations, 6,542 (69%) had sufficient information to categorize the type of escapement estimate (Table 17.3). Of the 6,542 populations, I rated 30 (>1%) as having excellent escapement estimation types; namely total counts. Escapement estimation types were rated as good for 1,947 populations (30%); they were based on total estimates extrapolated from dam, weir, sonar, or tower counts; peak counts derived from repetitive surveys; indices generated from reliable and consistent methods; or total redd counts based on some reliably calibrated method. Escapement estimate types for 193 populations (3%) were rated as fair; those based on one count per year, indices either not based on rigorous methodology or annually inconsistent, or surveys of redds not calibrated to estimate total production. Twenty-two percent of populations (1,447) have escapement estimate types considered to be poor; these estimates were based on one count per season but not in all years, loosely applied index areas, or based on carcass counts. No escapement estimation, therefore no estimation type, is generated for 2,925 populations (45%).

Since the ideal type of escapement estimate, a complete count, was set as the reference point for deciding how to categorize the quality of escapement estimation types, it could be argued that some types were underrated. However, even if some escapement estimation types were rated higher, most types would still be only fair. In terms of managing for sustainability, the goal should always be to acquire escapement counts or estimates that relate accurately to the harvest management processes used to set allowable catches.

The outcome of the escapement estimation type summary was strongly influenced by information availability. In particular, the large number of Alaskan populations, predominantly monitored by aerial surveys and rated as poor escapement estimation types (once per year, but sporadic, aerial indices), skewed the overall summary toward "poor." On the other hand, a large number of Alaskan populations are surveyed repetitively each year allowing a "peak count," rated as good. The lack

TABLE 17.4
Number of populations per management unit,
by U.S. state and species.

State	Species	Number of management units	Stocks per management unit
Alaska	Chinook	75	2.21
	Chum	159	12.67
	Coho	87	29.02
	Pink	193	17.56
	Sockeye	109	3.36
	Steelhead	7	48.00
Washington	Chinook	52	1.40
	Chum	82	1.77
	Coho	43	2.09
	Pink	13	1.08
	Sockeye	6	1.00
	Steelhead	92	1.25
Oregon	Chinook	12	2.92
	Chum	2	4.00
	Coho	2	9.50
	Steelhead	84	1.43
California	Chinook	7	1.00

of information about escapement estimation types may have also influenced the summary outcome. Perhaps additional information would have resulted in a somewhat different outcome.

EFFECTS OF MIXED POPULATION FISHERIES

The major effects of mixed population fisheries are experienced when fish from different populations, often having differing production capacities, are harvested simultaneously at the same harvest rate (NRC 1996). This can happen either with multiple wild populations or when hatchery populations are mixed with wild fish.

Number of Populations Within Management Units.—By summarizing the number of populations identified for each management unit (see Appendix Table 17.4), I was able to provide a minimal indication of the degree of mixed wild population fishery effects on populations. This is a minimal estimate because it does not account for interspecific mixed fisheries, nor fisheries where other management units are mixed with the management unit of interest, nor fisheries targeting hatchery populations.

In Alaska, chum, coho, and pink salmon and steelhead management units consist of a very large number of populations (Table 17.4). Much of this is largely influenced by Southeast Alaska and Prince William Sound where there are an extraordinary number of populations and where they are intentionally grouped together for management (Fried 1994; Baker et al. 1997; Van Alen 2000). Alaskan chinook and sockeye salmon management units averaged approximately three and four populations per management unit (Table 17.4).

In Washington, the average number of populations per management unit is less than in Alaska (Table 17.4). This is due partly to the smaller geographical area and easier access to most populations. However, the low numbers of populations per management unit may also be attributed to inadvertent or uninformed grouping of populations (as reported to me in the available information). For example, Hood Canal summer chum salmon are managed (secondarily to chinook and coho salmon) as four management units for which guideline escapement goals are set (PNPTC and WDFW 1995). Yet, spawning is known to occur in some 13 different locations which would result in a population to management unit ratio of 3.25 rather than the ratio of 1.77 based on published information or databases.

In Oregon chinook, coho, and chum salmon escapement goals are set for broad geographical groupings of populations (PFMC 1996), as indicated by high populations per management unit ratios (Table 17.4). The little information available for California chinook management units indicated they consisted of one population per management unit (Table 17.4).

The average number of populations per management unit varies widely among species and locations (Table 17.4). Much of this variation is likely due to differing management strategies for each species and the extent of the geographic area covered by the management agency. For example, many salmon and steelhead in Washington are managed on a population-by-population basis, whereas in Alaska, where there are tens of spawning streams within manageable geographical areas, very large numbers of populations are in some cases grouped together for management (Table 17.4). Alaskan managers reason that, because of the large numbers of spawning populations and the huge geographic areas, they have been successful at maintaining escapements by protecting near-terminal areas and moderating mixed population exploitation rates (e.g., Van Alen 2000).

My estimates of the numbers of populations per management unit are in many cases conservative for several reasons. First, I used only documented populations. There were cases where reports or databases indicated populations were grouped but there was no way to know which ones or how many. There were other cases where populations were grouped and I was not aware of it. I simply used the lowest published grouping level as the population.

Second, managers are only recently becoming aware of the appropriate spawning aggregation scale at which variability in productivity occurs (e.g., Varnavskaya et al. 1994). There is often a lack of definitive knowledge about population differences or, more critically, whether the populations have differing productivity rates. For one example, Alaska Department of Fish and Game sets one escapement goal for Kasilof River sockeye salmon and manages based on a spawner/recruit relationship (Fried 1994). However, recent studies are beginning to reveal that the Kasilof system sockeye run actually consists of several biologically unique populations (Burger et al. 1997; Woody 1998). A critical looming question in this and other similar cases is whether, having this new knowledge about smaller population units, harvest management could or should be changed to ensure abundance and biodiversity of all the populations.

Third, only intraspecific effects were considered. There are numerous cases where a secondary species is intercepted as by-catch during the prosecution of a fishery on a target species. This factor is not addressed in the foregoing summary but can have major impacts (e.g., Slaney et al. 1996).

In summary, many management units, upon which decisions are made, are composed of multiple populations (Table 17.4). The important consideration is that, when populations are grouped together for management, small and/or less productive populations may not be represented in the data and are at risk in light of high harvest rates set for larger and/or more productive populations (NRC 1996; Narver 2000). Small populations are at risk of being driven below a viable number of individuals and populations exhibiting low productivity are subject to chronic overharvest. As further testimony to this problem, it was clearly stated in recent status papers (Baker et al. 1996; Slaney et al. 1996) that many smaller populations were unaccounted for in their analyses. Trend analysis in those studies focused on the more abundant and productive populations, even though smaller or less productive populations are the most likely to be jeopardized first. Thus, those studies may have underestimated the risk to unmonitored populations. To achieve overall sustainability, it is always preferable to identify likely spawning aggregations as the smallest population unit, try to assess productivity at that level, and then develop management plans for protecting abundance and genetic diversity based on that knowledge. This does not necessarily rule out mixed population fisheries, but highlights the need for full understanding of the component populations being managed as a unit.

Harvest Priority for Hatchery Fish.—Extreme reductions in wild salmon escapements have occurred when intentional or inadvertent decisions have been made to prioritize harvest of hatchery fish over conservation of wild fish. Since management areas where this occurs can sometimes be quite extensive, the strategy can negatively affect a significant number of populations. Maximizing harvests of hatchery fish in mixed population fisheries often results in overharvest of wild fish and in large areas of potential salmon habitat being underutilized. Some noteworthy examples include Lower Columbia River coho salmon (WDF and WDW 1993), Willapa Bay salmon (PFMC 1994, NRC 1996) and Nooksack coho salmon (WDF et al. 1992) to name a few. The problem is not limited to the Pacific Northwest, but occurs in Alaska as well. For one example, increasing fishing effort on coho from Medvejie Hatchery in Southeast Alaska is resulting in a dramatic increase in harvest rate of wild Salmon Lake coho salmon which cannot be sustained (the harvest rate increased from 35% in 1985 to 72% in 1995, Schmidt 1996).

Effects of Run Management on Escapements.—While inseason run management is a complex topic unto itself and will not be treated in this study, it is important to recognize some of the ways that run management can influence escapements. These can include, but are not limited to: (1) socio-economic-political decisions affecting the setting of harvest rates, and, in turn, escapements; (2) managers being forced to provide proof there is overfishing rather than having to demonstrate there is a harvestable surplus above escapement needs; (3) allocation before escapements in many political jurisdictions; (4) scientists and managers becoming advocates of a fishery user group rather than advocates of the fish; and (5) uncertainty about population dynamics and harvest efficiency (Wright 1981; Fraidenburg and Lincoln 1985; Ludwig et al. 1993).

MANAGEMENT RECOMMENDATIONS

A variety of interrelated actions can be taken by fisheries managers, the public, and politicians to help achieve salmon sustainability through better escapement management, while additional remedial efforts are made in other areas (e.g., inseason run management, habitat protection and restoration, and hatchery program management). These recommended actions fall into three general categories: (1) improve the science of, and practical methods for, assessing escapements and setting goals; (2) ensure that escapements are sufficient to perpetuate maximum biomass production and biodiversity; or (3) change public attitudes and expectations about salmon production and fishing. The recommendations are summarized in Table 17.5; details follow.

IMPROVE ESCAPEMENT MANAGEMENT TECHNOLOGY

Theoretical salmon management must progress from being dependent upon the basic stock/recruitment model to new ways of expressing the natural cycles of salmon production, while accounting for human influences on salmon abundance (NRC 1996). The technology of escapement management can be improved by developing and testing new models for simulating and predicting salmon population dynamics, finding better ways to apply adaptive management, and incorporating decision theory and risk analysis. Research and development of escapement technology should be responsive to the needs of management agencies, perhaps coordinated by the Scientific Advisory Board recommended by NRC (1996).

Develop New Models.—There is an increasing call for salmon managers and scientists to move beyond the spawner/recruit modeling and rigid escapement goal setting used now, as these have not always provided the best management (NRC 1996) and have even led to erroneous conclusions and recommendations (Hilborn and Walters 1992). While present spawner/recruit modeling cannot yet be totally abandoned (because it is the best presently available and it takes time and money to replace the old models), scientists and managers should develop new models that better account for natural environmental variability and actual carrying capacity (when the freshwater habitat is fully seeded with carcasses). Several promising new models are worth investigation and further development; combinations of models may also warrant exploration.

TABLE 17.5
Summary of management actions which, taken in concert as necessary on a population-by-population basis, are likely to improve Pacific salmon escapements (see text for supporting explanations).

Improve Escapement Management Technology

Develop new models
    Habitat and environmental control models
    Time and space models
    Exploitation rate models
    Decision theory
Improve escapement estimation technology
Improve population discrimination
Increase research funding for population management

Ensure Healthy Escapements

Identify and achieve "safe" escapement levels
Collect accurate, consistent, and fully representative run size data
Avoid the use of temporary escapement goals
Reduce the number of populations per management unit
Improve escapement goal setting methods
Use smaller, more precise management areas
Guard against gradual escapement goal reduction
Improve harvest management
    Reduce harvest rates
        Reduce exploitation rates on all populations simultaneously in one fishery
        Increase specificity of fisheries
        Establish fishery refuges
        Use selective fisheries
        Invoke additional gear limitations
        Increase use of limited entry
        Buy back fishing boats and licenses
    Accept "overescapement" at hatcheries
Use adaptive management
Settle Pacific Salmon Treaty allocation issues
Separate allocation issues from biological process

Change Public Attitudes and Expectations

Improve public education
Increase public involvement in the process
Encourage harvesters to adapt to natural variation

- *Habitat and environmental control models.*—Production is normally limited by some combination of physical and biological carrying capacity, as influenced by climate and weather patterns. Models that incorporate such information and then examine the effects of human actions on production look promising for future salmon management. These have included Bayesian approaches (e.g., Hilborn et al. 1994; Geiger and Koenings 1991; Adkison and Peterman 1996), stochastic simulations (e.g., Cramer 2000), habitat-based approaches (e.g., Lestelle et al. 1996), and comprehensive planning approaches (e.g., Puget Sound Comprehensive Coho Program, CCW 1994). Whatever models are ultimately proven to be the best, they must incorporate terms to account for both marine and freshwater environmental variability. Walters and Parma (1996) have argued alternatively for fixed exploitation rate strategies wherein a percentage of the run is harvested regardless of run strength; this allows the escapement to track natural climatic variation. They postulated it may be more cost-effective to invest in research on how to implement fixed harvest rate strategies than on how to explain and predict climate effects.

- *Time and space models.*—Simulation models of fish populations as they migrate through time and space may help substantially in managing for better escapements because they will improve decisions about inseason management and/or help to evaluate consequences of alternative management scenarios. Several different approaches are being explored for quantifying the effects of migration, mortality, abundance, fishing effort, and gear types on fish abundance as they move through spatial cells or fishing districts (e.g., Walters et al. 1998; Lawson and Comstock 2000). "Nerkasim" is another promising technique, combining time and space attributes with environmental controls and individual fish bioenergetics (Rand et al. 1997).

- *Exploitation rate models.*—More work is needed on the effects of using exploitation rate models to set harvest rates, as described by CCW (1994), particularly regarding implementation of the same exploitation rate for mixed populations having varying productive capacities (i.e., variable spawner/recruit relationships). The degree to which spawners per recruit varies among neighboring populations is a central research question to be addressed before this approach can be fully implemented.

- *Decision theory.*—Researchers must continue investigating applications of decision theory and risk analysis (Walters 1986; Hilborn and Walters 1992) to salmon management. Subjecting the results of population/environmental modeling to decision analyses, as was done by Hilborn et al. (1994), in light of a variety of past management scenarios will allow testing of the predicted outcomes against reality and accounts for uncertainty in the management process.

Improve Escapement Estimation Technology.—As was seen in Tables 17.2 and 17.3, there is a great need for better escapement assessment and estimation techniques. Further research is needed in the areas of survey design, stratified sampling (e.g., Irvine et al. 1992), area under the curve techniques (e.g., English et al. 1992), remote sensing such as video and hydroacoustic techniques (e.g., Hatch et al. 1994), mark and recapture (e.g., Schwarz et al. 1993), or other as yet untried methods.

Improve Population Discrimination.—The ability to separate populations in fisheries is critical to managers wanting to direct or control harvest on specific populations. Salmon population composition is usually accomplished by genetic analysis (e.g., Carvalho and Hauser 1994), tag recoveries (e.g., Cormack and Skalski 1992), or by scale pattern recognition (e.g., Marshall et al. 1987). All of these processes currently have drawbacks in timeliness or cost that prevent managers from making expedient decisions to open or close fisheries. More technological research is needed to develop improved or new population discrimination techniques that can be rapidly applied inseason. Furthermore, much work is needed to establish baseline population differences, both in technology development and in practical research on population discrimination within management units. Debate will continue on the relevance of discriminating among populations at the finest scale, but I contend the critical question is whether biologically discernable populations have differing productivity rates; if so, they should either be managed separately from neighboring populations or as a group but with a conservative exploitation policy.

Increase Research Funding for Population Management.—There is a pressing need for improvements in technology for understanding, enumerating, predicting, and managing salmon runs. The NRC (1996) identified many critical gaps in knowledge, many of which are relevant to escapement management. The federal government presently supports most of the basic research that is conducted—that should continue and increase. However, it is also important that states and user groups provide contributions to research funding. Research on escapement management should be of particular relevance to states, since they are the primary escapement managers. While the costs for additional research may seem prohibitive, long-term recovery and sustainability may be worth the price, just as it has been for other depleted fisheries such as the Atlantic coast striped bass *Morone saxatilis* (Field 1997).

ENSURING HEALTHY ESCAPEMENTS

Increasing escapements of depleted populations and maintaining adequate escapements of healthy populations are the quickest ways to realize conservation goals (Riddell 1993) and should be the ultimate goal of fishery managers trying to achieve sustainability. As stated by NRC (1996), a shift must be made from focusing on catch to focusing on escapement. Salmon managers should be required to provide evidence that a population is healthy enough to allow a fishery rather than having to prove the population may be jeopardized by overfishing before curtailing fishing (Wright 1981). Several authors have demonstrated the concept that, in many cases, fishing less (increasing escapement) can result in larger catches in the long term (Hilborn and Walters 1992; NRC 1996). Optimal escapements are numbers that not only perpetuate the population and ensure biodiversity (Riddell 1993), but also provide enough carcasses to maximize the carrying capacity potential of the system. The goal should be to identify the appropriate harvest rate in light of each population's naturally varying mortality schedule. There are a number of specific fishery management recommendations that support sustainability through increasing and optimizing escapements.

Identify and Achieve "Safe" Escapement Levels.—As recommended by NRC (1996), the concept of MSY should be replaced with minimum sustainable escapements (MSE) for as many populations as possible. Rather than selecting a specific escapement goal, about which target escapements fluctuate, as has been done in the past, the MSE is an escapement level which should always be met. Most importantly, escapements should range well above the MSE. This will enhance productivity and biodiversity by allowing for years in which so-called excess escapement builds resiliency into the system, supplies abundant carcasses (nutrients), and allows for sufficient escapements of any smaller, weaker populations within the management unit. Further work will be required to estimate how much escapements should range above MSE.

Since salmon survival is intimately dependent on highly variable ocean conditions, it is critical that we ensure adequate escapements in years of poor ocean productivity (poor marine survival). It is important to remember, for example, that short-term upswings in apparent abundance may result from variation in marine production rather than improvements due to habitat changes or improvements in escapement management (see, for example, Lawson 1993).

Collect Accurate, Consistent, and Fully Representative Run Size Data.—Regardless of the theoretical modeling approach employed for data analysis and run prediction, basic data collection will always be a critical component of the salmon management process. Although this chapter is about escapement, it is essential that consistent data be collected on all aspects of the run size, including both catch and escapement. As can be seen in the summary of escapement estimation (Table 17.2), the majority of populations have poor or no escapement data. Management agencies need to increase emphasis on escapement assessments, as well as other critical population-specific data.

In large, remote areas, where it is impractical to survey escapement of every population, it is vitally important to routinely and accurately assess the status of populations of a range of sizes and productive capacities. Escapement assessment programs should be designed to include intensive monitoring of small and less productive populations in approximately the proportion that they occur naturally. In this way, early warnings can be raised when these important components of the population structure are thought to be jeopardized.

Avoid the Use of Temporary Escapement Goals.—Managers should also avoid the use of "interim," "target," "phased-in," "gradual," "eventual" or other short-term escapement objectives when dealing with depressed salmon populations because they tend to lead to deliberate overfishing of salmon runs (Wright 1981). For any salmon run returning at or below the level required for MSE, all target fisheries must be closed. There is no viable alternative (Wright 1981). In addition, management options to reduce incidental harvest should be invoked.

Reduce the Number of Populations per Management Unit.—Whenever possible, it is preferable that escapement goals be established for individual populations, i.e., in the context of this chapter, that each population become a management unit. This will help reduce the effects of mixed population fisheries on small or less productive populations. Whether the fisheries on these populations can actually be managed to harvest each population separately is a separate issue (addressed below); the salient point is that managers must understand how many populations occur within a management unit, the natural productivity of each population, and how fisheries are influencing their total production and viability.

Improve Escapement Goal-Setting Methods.—The results of this evaluation show that there are a large number of management units for which there are poor or no escapement goals, even when populations are combined (Table 17.1). I recommend that MSE (NRC 1996) be applied to all possible populations. This will require additional funding and personnel to implement but is essential for future sustainability. As new management technology develops and better information is collected for each unit, the goals should be refined. Escapement goal setting will also benefit from related improvements in escapement, stock discrimination, coded wire tag, age structure, smolt productivity, and habitat utilization data.

Use Smaller, More Precise Management Areas.—One way to gradually move salmon management toward sustainability is to decrease the size of some fishing management areas or districts. While this is now mostly limited by the inability to discern which populations of fish are being harvested in each area and the time required for processing information from the fishery to managers, I believe we should be striving in many fisheries for managing time and location of fishery openings on a smaller, more expedient scale. As the technical ability to rapidly process population discrimination information is further developed, fisheries can be opened in smaller areas or times to harvest any abundant populations and closed in areas to protect weak populations. More opportunities will be available for opening fisheries on discrete, abundant populations when management areas are smaller.

Guard against Gradual Escapement Goal Reduction.—Managers, decision-makers, and users should be vigilant against the temptation to reduce escapement goals. As described in detail above, standard spawner/recruit models can give the illusion that MSY will be attained with a lower escapement goal, particularly when based on recent population performance (Hilborn and Walters 1992). While there is often remarkable pressure from users, and the concomitant desire of fishery managers to satisfy constituents, decision-makers should require hard evidence that "excessive" escapements are actually reducing productivity before a goal or MSE is lowered.

Improve Harvest Management.—There is also a group of harvest management actions which can help to achieve healthy escapements, either by reducing the effects of mixed population fisheries or simply ensuring additional fish escape to the spawning grounds.

- *Reduce harvest rates.*—Reducing harvest rates where necessary will increase abundance (i.e., long-term catch) on strong populations (e.g., Cramer 2000), revitalize depleted populations, and protect weaker populations. Because abundant and depleted (or susceptible) populations are often mixed together in fisheries, it is important that management allow for separate harvest regimes for strong and weak populations (NRC 1996). Several recent cases demonstrate how reduced harvest rates have benefited escapement, particularly of smaller populations. In 1995, for example, Canada's Department of Fisheries and Oceans recognized the emergency nature of coho off the west coast of Vancouver Island and reduced the harvest rate from the previous 60 to 80% down to about 50%, increasing escapements of coho to Carnation and Clemens creeks at least tenfold (Tschaplinski 2000). Reduced harvest rates in many other locations will undoubtedly increase the size and diversity of spawning populations, as recommended by NRC (1996).

  The following management actions can be applied, in various combinations on a case by case basis, to reduce harvest rates and/or the effects of mixed population fisheries.

Reduce exploitation rates on all populations simultaneously in one fishery.—Closing or reducing effort in mixed population fisheries, as necessary, will protect weak populations and allow more productive populations to pass to the next fishery for either harvest or escapement. It is recognized this may result in short-term disruptions and complications to the economic and social infrastructure of salmon-based economies (NRC 1996), but will improve the chances of sustaining production of all populations for the long-term benefit of society.

Increase specificity of fisheries.—Some fisheries can and should be managed with more specific time and area openings and closings to control how they influence populations migrating through management areas. That way, weak populations can be protected when they are mixed with strong ones, but strong populations can be harvested as they separate from others during migration. This strategy will result in a larger emphasis on terminal fisheries, not only providing harvest opportunities and weak population protection, but with the added advantage of more accurate documentation of fishing mortality (Mundy 1997). It must be noted, however, that these shifts will have their own harvest management challenges and cause disruptions to the existing salmon fishery social and economic infrastructure.

Establish fishery refuges.—It may be preferable to close some harvest management areas for the long term. These may be areas where a large number of particularly sensitive populations congregate. This will also result in larger catches in terminal fishing areas.

Use selective fisheries.—Selective fisheries have been recommended as one method of effectively harvesting strong populations while allowing others to escape (Lincoln 1994). There are a number of gear and management options that can be combined to create selective fisheries. A most popular option being proposed and investigated is the fin-clipping of all hatchery-reared coho and chinook salmon (Lawson and Comstock 2000). Non-clipped, wild fish could be released from non-lethal fisheries, such as purse seines, trollers, sport, live traps, and fish wheels, while all fin-clipped fish could be retained. Fishers using those same gear types could also retain or release fish on a species-by-species basis as necessary.

Invoke gear limitations.—Use of less selective gears, such as gill-nets in certain fisheries, should be reduced or eliminated except in areas where it is demonstrated that they have no impact on weak populations. This again could have significant implications for existing salmon fisheries.

Increase use of limited entry.—Most salmon fisheries are already limited (NRC 1996). There is some hope that individual transferable quotas (ITQs) may provide incentive for harvesters to limit catches when run sizes are low (e.g., Fujita and Foran 2000). Since ITQs apply to specific runs, ITQ holders may recognize that an investment in future production (i.e., by sometimes reducing or eliminating fishing effort in the short term) will increase their catches in the longer term.

Buy back fishing boats and licenses.—Although buy-back programs have been implemented in certain fisheries in the past with mixed success, it is still a viable option to help reduce the potential effort in certain fisheries (NRC 1996) and the pressure on managers to open fisheries on populations that cannot withstand fishing mortality.

- *Accept "overescapement" at hatcheries.*—In areas currently managed for hatchery harvest rates, exploitation should be reduced to allow sufficient natural spawners to fully seed all available habitat. This may result in so-called overescapement of hatchery fish unless they can be harvested in a terminal area where they are separated from wild fish. In cases where too many hatchery fish might result in negative ecological or genetic impacts in the adjoining habitat, it might be preferable to reduce the hatchery program so that it simply augments wild production. If programs can be developed to market the excess hatchery salmon carcasses, then another plausible strategy might include fishing at the rate sustainable by the natural population while harvesting all excess at the hatchery rack. Some combination of these alternatives should allow hatchery production beyond what would be produced from wild production alone while protecting and maximizing wild production.

- *Use adaptive management.*—The principal of adaptive management (Walters 1986; Hilborn and Walters 1992) should be applied to as many management units as possible. This is because, regardless of the methods presently used or those to be used in the future, managers need to evaluate the success or failure of the variety of management alternatives that are intentionally or inadvertently invoked. Managers should follow the six steps of adaptive management (Walters 1986; Hilborn and Walters 1992), making new decisions each year using decision theory and evaluating the consequences of those decisions.

- *Settle Pacific Salmon Treaty allocation issues.*—Although an updated Pacific Salmon Treaty (PST) was recently signed, it would be naive not to recognize problems caused in past U.S. and Canadian escapement management by the inability to resolve international allocation issues. Many of the other recommendations in this chapter need to be implemented by one country to benefit populations originating in the other country and vice versa. The challenges of the PST have been discussed in detail by other authors (e.g., NRC 1996). Suffice to say that resolution of these international issues is essential to the future of all salmon populations originating in one country and migrating through the other country's fisheries.

- *Separate allocation issues from biological process.*—Fishery biologists charged with determining whether there is a harvestable surplus should not also be involved in allocation decisions. Biologists should be free to make recommendations of escapement levels or harvest rates necessary to maintain abundant populations and biological diversity. They should also make recommendations about whether there is a harvestable surplus and when and where the surplus will be available with the least impact on other populations. This information should then be provided to the political process for final allocation decisions. The Alaskan management process has generally worked well locally and serves as a good model (Holmes and Burkett 1996).

CHANGE PUBLIC ATTITUDES AND EXPECTATIONS

Until recently, the general goal of fisheries management was to stabilize fisheries so that user groups could count on a certain level of harvest and stable income. While there may be some viable strategies to reduce the likelihood of closed fisheries (such as fishing regimes based on steady, but most likely lowered, harvest rate) salmon managers, harvesters, and the public may ultimately benefit by accepting that salmon abundance follows natural, often extreme, cycles (Cramer 2000). This means that user groups should be encouraged to adjust to fluctuations in fish availability and income. There are several ways salmon managers can assist in disseminating this message, thereby helping to ease the negative ramifications of the natural downswings in salmon abundance.

Improve Public Education.—Salmon managers and scientists should help people understand the concepts of (1) variable productivity; (2) less fishing can mean more fish over the long term; (3) the importance of large escapements to long-term productivity; (4) the connections between human population growth (and associated impacts) and salmonid populations; and (5) the importance of genetic and population biodiversity. This can be accomplished through public forums and workshops and by incorporating these concepts into high school curricula.

Public education of salmon harvesters and recreational users will help to support increased funding for research and management. As a negative example of how this feedback loop functions, notice how, as soon as fish become unavailable, the users tend to blame government managers for ineptness. Yet, agency funding is continually being reduced in state legislatures, preventing scientists and managers from conducting the research and basic data collection so desperately needed to support quality run size predictions and escapement management. An informed public will pressure legislators to support and fund the necessary programs.

Increase Public Involvement in the Process.—There has been much discussion and progress toward an ecosystem-based, community approach to watershed management and salmon restoration (e.g., Lichatowich et al. 1995; Bingham 2000; Fields 2000; MacDonald et al. 2000). Yet these new public processes have usually failed to incorporate salmon production, escapement, and harvest management, primarily because harvest management remains the realm of agency and tribal fisheries managers. When salmon user groups and watershed landowners and citizens have the opportunity to hear all the evidence presented by harvest management biologists, and have the chance to voice their opinions about decisions, they may become more invested in the outcome of decisions and the status of the resource upon which they vitally depend (e.g., Riddell 1993). The salmon ecosystem extends from the ridgetops to the high seas. Watershed-oriented discussions designed to benefit salmon should include all stakeholders, cover all portions of the salmon ecosystem and all impacts along the way, and particularly include the effects of harvest and harvesters.

Encourage Harvesters to Adapt to Natural Variation.—A major public paradigm shift is particularly required, wherein all users' and managers' expectations are modified to coincide with the variable and unpredictable nature of salmon populations. Protection of the spawning escapement (the investment principle) must be given the highest priority (NRC 1996), rather than maximizing the catch. This may require significant economic and social adjustments because fishing patterns will necessarily be variable from year to year, resulting in disruptive and unpredictable employment patterns. However, if coastal communities can adapt to the variation, the pay-offs in improved long-term productivity will be substantial.

In closing, although it is obvious that invoking all these escapement management recommendations will be very expensive, the long-term economic, social, and cultural costs of not doing so (i.e., further depleting salmon populations and/or production) will be greater. Furthermore, voluntary, proactive implementation of these measures will forestall the otherwise inevitable, involuntary restrictions resulting from further Endangered Species Act listings or, worse, the eventual loss of additional populations. To truly achieve Pacific salmon sustainability depends on a public commitment to invest in expanded salmon research, management, and public education. We cannot count on repairing only one damaged aspect of salmon runs (e.g., degraded habitat) to fix the problem, but must work on all fronts simultaneously. Ultimately, though, both productivity and biodiversity depend on sufficient escapement of spawners to fully utilize the available freshwater habitat, fertilize the systems with carcasses, and optimize genetic diversity.

ACKNOWLEDGMENTS

I am particularly indebted to Claribel Coronado and Jerry Berg for their assistance in collecting and compiling management information. I thank A. Baracco, R. Brix, S. Fried, R. Leland, C. Smith, C. Swanton, W. Tweit, B. Van Alen, R. Williams, and R. Woodard for their assistance in supplying management information, consultation, and manuscript reviews. Thanks also to L. Buklis, C. Walters, and a particularly thorough anonymous reviewer for their helpful reviews of the manuscript.

REFERENCES

Adkison, M. D., and R.M. Peterman. 1996. Results of bayesian methods depend on details of implementation: an example of estimating salmon escapement goals. Fisheries Research 25:155–170.

Ames, J., and D. E. Phinney. 1977. 1977 Puget Sound summer-fall chinook methodology: escapement estimates and goals, run size forecasts, and in-season run size updates. Washington Department of Fisheries Technical Report 29. Olympia, Washington.

Baker, T. T., and eight coauthors. 1996. Status of Pacific salmon and steelhead in Southeastern Alaska. Fisheries 21:6–18.

Bilby, R. E, B. R. Fransen, and P. A. Bisson. 1996. Incorporation of nitrogen and carbon from spawning coho salmon into the trophic system of small streams. Canadian Journal of Fisheries and Aquatic Sciences 53:164–173.

Bingham, N., and A. Harthorn. 2000. Spring run chinook salmon workgroup: a cooperative approach to watershed management in California. Pages 647–654 in E. E. Knudsen, C.R. Steward, D. D. MacDonald, J. E. Williams, and D. W. Reiser, editors. Sustainable fisheries management: Pacific salmon. Lewis Publishers, Boca Raton, Florida.

Brennan, K., D. L. Prokopowich, and D. Gretsch. 1997. Kodiak Management Area commercial salmon annual management report, 1995. Alaska Department of Fish and Game, Regional Information Report 4K97-30, Kodiak.

Burger, C. V., W. J. Spearman, and M. A. Cronin. 1997. Genetic differentiation of sockeye salmon subpopulations from a geologically young Alaskan lake system. Transactions of the American Fisheries Society 126:926–938.

Carvalho, G. R., and L. Hauser. 1994. Molecular genetics and the stock concept in fisheries. Reviews in Fish Biology and Fisheries 4:326–350.

CCW (Comprehensive Coho Workgroup). 1994. Comprehensive coho management plan. Interim Report. Western Washington Treaty Tribes and Washington Department of Fish and Wildlife, Olympia.

Collie, J. S., and R. M. Peterman. 1990. Experimental harvest policies for a mixed-stock fishery: Fraser River sockeye salmon, *Oncorhynchus nerka*. Canadian Journal of Fisheries and Aquatic Sciences 47:145–155.

Cormack, R. M., and J. R. Skalski. 1992. Analysis of coded wire tag returns from commercial catches. Canadian Journal of Fisheries and Aquatic Sciences 49:1816–1825.

Cramer, S. P. 2000. The effect of environmentally driven recruitment variation on sustainable yield from salmon populations. Pages 485–503 in E. E. Knudsen, C.R. Steward, D. D. MacDonald, J. E. Williams, and D. W. Reiser, editors. Sustainable fisheries management: Pacific salmon. Lewis Publishers, Boca Raton, Florida.

Drinkwater, K. F., and R. A. Myers. 1987. Testing predictions of marine fish and shellfish landings from environmental variables. Canadian Journal of Fisheries and Aquatic Science 44:1568–1573.

English, K. K., R. C. Bocking, and J. R. Irvine. 1992. A robust procedure for estimating salmon escapement based on the area-under-the-curve method. Canadian Journal of Fisheries and Aquatic Sciences 49:1982–1989.

Field, J. D. 1997. Atlantic striped bass management: where did we go right? Fisheries 22(7):6–8.

Fraidenburg, M. E., and R. H. Lincoln. 1985. Wild chinook salmon management: an international conservation challenge. North American Journal of Fisheries Management 5:311–329.

Frederick, S. W., and R. M. Peterman. 1995. Choosing fisheries harvest policies: when does uncertainty matter? Canadian Journal of Fisheries and Aquatic Sciences 52:291–306.

Fried, S. M. 1994. Pacific salmon spawning escapement goals for the Prince William Sound, Cook Inlet, and Bristol Bay areas of Alaska. Alaska Department of Fish and Game, Commercial Fisheries Management and Development Division, Special Publication 8, Juneau.

Fujita, R. M., and T. Foran. 2000. Creating incentives for salmon conservation. Pages 655–666 in E. E. Knudsen, C.R. Steward, D. D. MacDonald, J. E. Williams, and D. W. Reiser, editors. Sustainable fisheries management: Pacific salmon. Lewis Publishers, Boca Raton, Florida.

Geiger, H. J., and J. P. Koenings. 1991. Escapement goals for sockeye salmon with informative prior probabilities based on habitat considerations. Fisheries Research 11:239–256.

Hatch, D. R., M. Schwartzberg, and P. R. Mundy. 1994. Estimation of Pacific salmon escapement with a time-lapse video recording technique. North American Journal of Fisheries Management 14:626–635.

Hilborn, R. 1992. Hatcheries and the future of salmon in the Northwest. Fisheries 17(1):5–8.

Hilborn, R. C., and C. J. Walters. 1992. Quantitative fisheries stock assessment. Chapman & Hall, New York.

Hilborn, R., E. K. Pikitch, and M. K. McAllister. 1994. A Bayesian estimation and decision analysis for an age-structured model using biomass survey data. Fisheries Research 19:17–30.

Hiss, J. M., and E. E. Knudsen. 1993. Chehalis Basin fishery resources: status, trends, and restoration. U.S. Fish and Wildlife Service, Fishery Resource Office, Olympia, Washington.

Holmes, R. A., and R. D. Burkett. 1996. Salmon stewardship: Alaska's perspective. Fisheries 21(10):30–32.

Irvine, J. R., R. C. Bocking, K. K. English, and M. Labelle. 1992. Estimating coho salmon (*Oncorhynchus kisutch*) spawning escapements by conducting visual surveys in areas selected using stratified random and stratified index sampling designs. Canadian Journal of Fisheries and Aquatic Sciences 49:1972–1981.

Kline, T. C., J. J. Goering, O. A. Mathisen, and P. H. Poe. 1993. Recycling of elements transported upstream by runs of Pacific salmon: II. 15N and 13C evidence in the Kivchak River watershed, Bristol Bay, Southwestern Alaska. Canadian Journal of Fisheries and Aquatic Sciences 50:2350–2365.

Larkin, P. A. 1977. An epitaph for the concept of maximum sustained yield. Transactions of the American Fisheries Society 106:1–10.

Lawson, P. W. 1993. Cycles in ocean productivity, trends in habitat quality, and the restoration of salmon runs in Oregon. Fisheries 18(8):6–10.

Lawson, P. W., and R. M. Comstock. 2000. The proportional migration selective fishery model. Pages 423–433 *in* E. E. Knudsen, C.R. Steward, D. D. MacDonald, J. E. Williams, and D. W. Reiser, editors. Sustainable fisheries management: Pacific salmon. Lewis Publishers, Boca Raton, Florida.

Lestelle, L. C., L. E. Mobrand, J. E. Lichatowich, and T.S.Vogel. 1996. Applied ecosystem analysis: a primer. BPA Project Number 9404600. Bonneville Power Administration, Portland, Oregon.

Lichatowich, J., L. Mobrand, L. Lestelle, and T. Vogel. 1995. An approach to the diagnosis and treament of depleted pacific salmon populations in pacific northwest watersheads. Fisheries 20(1):10–18.

Lincoln, R. 1994. Molecular genetics applications in fisheries: snake oil or restorative? Reviews in Fish Biology and Fisheries 4:389–392.

Ludwig, D., R. Hiborn, and C. Walters. 1993. Uncertainty, resource exploitation, and conservation: Lessons from history. Science 260:17, 36.

MacDonald, D. D., C. R. Steward, and E. E. Knudsen. 2000. One Northwest community — people, salmon, rivers, and sea: toward sustainable salmon fisheries. Page 687–701 *in* E. E. Knudsen, C. R. Steward, D. D. MacDonald, J. E. Williams, and D. W. Reiser, editors. Sustainable Fisheries Management: Pacific salmon. Lewis Publishers, Boca Raton, Florida.

Marshall, S., and nine coauthors. 1987. Application of scale pattern analysis to Alaska's sockeye salmon (Oncorhynchus nerka) fisheries. Pages 307–326 *in* H. D. Smith, L. Margolis, and C. C. Wood, editors. Sockeye salmon (*Oncorhynchus nerka*) population biology and future management. Canadian Special Publications in Fisheries and Aquatic Sciences 96.

Milner, A. M., and R. G. Bailey. 1989. Salmonid colonization of new streams in Glacier Bay, Alaska. Aquaculture and Fisheries Management 20:179–192.

Mundy, P. R. 1997. The role of harvest management in the future of Pacific salmon populations: shaping human behavior to enable the persistence of salmon. Pages 315–329 *in* D. J. Stouder, P. A. Bisson, and R.J. Naiman, editors. Pacific salmon and their ecosystems: status and future options. Chapman & Hall, New York.

Narver, D.W. 2000. Review of salmon management in British Columbia: What has the past taught us? Pages 67–74 *in* E. E. Knudsen, C.R. Steward, D. D. MacDonald, J. E. Williams, and D. W. Reiser, editors. Sustainable fisheries management: Pacific salmon. Lewis Publishers, Boca Raton, Florida.

Nehlsen, W., J. E. Williams, and J. A. Lichatowich. 1991. Pacific salmon at the crossroads: stocks at risk from California, Oregon, Idaho, and Washington. Fisheries 2(2):4–21.

NMFS (National Marine Fisheries Service). 1997. Final rule: endangered and threatened species; threatened status for southern Oregon/northern California coast coho evolutionarily significant unit (ESU) of coho salmon. 62-FR-24588. Washington, D.C.

NRC (National Research Council). 1996. Upstream: Salmon and Society in the Pacific Northwest. Committee on Protection and Management of Pacific Northwest Salmonids. National Academy Press, Washington, D.C.

OFWC (Oregon Fish and Wildlife Commission). 1995. Oregon coho salmon biological status assessment and staff conclusion for listing under the Oregon Endangered Species Act. Oregon Fish and Wildlife Commission, Portland, Oregon.

Palmisano, J. F., R. H. Ellis, and V. W. Kaczynski. 1993. The impact of environmental and management factors on Washington's wild anadromous salmon and trout. Washington Forest Protection Association and Washington Department of Natural Resources, Olympia.

Pearcy, W. 1992. Ocean ecology of north pacific salmonids. University of Washington Press, Seattle.

PFMC (Pacific Fishery Management Council). 1978. Freshwater habitat, salmon produced, and escapements for natural spawning along the Pacific coast of the United States. Pacific Fishery Management Council, Portland, Oregon.

PFMC. 1994. Review of 1993 ocean salmon fisheries. Pacific Fishery Management Council, Portland, Oregon.

PFMC. 1996. Review of 1995 ocean salmon fisheries. Pacific Fishery Management Council, Portland, Oregon.

PNPTC, and WDFW (Point No Point Treaty Council and Washington Department of Fish and Wildlife). 1995. 1995 Management framework plan and salmon runs' status for the Hood Canal region. Point No Point Treaty Council and Washington Department of Fish and Wildlife, Olympia.

Rand, P. S., J. P. Scandol, and E. E. Walter. 1997. Nerkasim: a research and educational tool to simulate the marine life history of Pacific salmon in a dynamic environment. Fisheries 22(10):6–13.

Ricker, W. E. 1954. Stock and recruitment. Journal of the Fisheries Research Board of Canada 11:559–623.

Ricker, W. E. 1958. Maximum sustained yields from fluctuating environments and mixed stocks. Journal of the Fisheries Research Board of Canada 15:991–1006.

Ricker, W. E. 1975. Computation and interpretation of biological statistics of fish populations. Fisheries Research Board of Canada, Bulletin 191, Ottawa.

Riddell, B. E. 1993. Spatial organization of Pacific salmon: what to conserve? Pages 23–41 in J. G. Cloud and G. H. Thorgaard, editors. Genetic conservation of salmonid fishes. Plenum Press, New York.

Schmidt, A. E. 1996. Interceptions of wild Salmon Lake coho salmon by hatchery supported fisheries. Alaska Department of Fish and Game, Fishery Data Series 96–26. Anchorage.

Schwarz, C. J., R. E. Bailey, J. R. Irvine, and F. C. Dalziel. 1993. Estimating salmon spawning escapement using capture-recapture methods. Canadian Journal of Fisheries and Aquatic Sciences 50:1181–1197.

Slaney, T. L., K. D. Hyatt, T. G. Northcote, and R. J. Fielden. 1996. Status of anadromous salmon and trout in British Columbia and Yukon. Fisheries 21(10):20–35.

Tschaplinski, P. J. 2000. The effects of forest harvesting, climate variation, and ocean conditions on salmonid populations of Carnation Creek, Vancouver Island, British Columbia. Pages 297–327 in E. E. Knudsen, C.R. Steward, D. D. MacDonald, J. E. Williams, and D. W. Reiser, editors. Sustainable fisheries management: Pacific salmon. Lewis Publishers, Boca Raton, Florida.

Van Alen, B. W. 2000. Status and stewardship of salmon stocks in Southeast Alaska. Pages XXX–XXX in E. E. Knudsen, C.R. Steward, D. D. MacDonald, J. E. Williams, and D. W. Reiser, editors. Sustainable fisheries management: Pacific salmon. Lewis Publishers, Boca Raton, Florida.

Varnavskaya, N. V., and six coauthors. 1994. Genetic differentiation of subpopulations of sockeye salmon (*Oncorhynchus nerka*) within lakes of Alaska, British Columbia, and Kamchatka, Russia. Canadian Journal of Fisheries and Aquatic Sciences 51(Supplement 1):147–157.

Walters, C. J. 1986. Adapative management of renewable resources. McMillan Press, New York.

Walters, C. J., and J. S. Collie. 1988. Is research on environmental factors useful to fisheries management? Canadian Journal of Fisheries and Aquatic Sciences 45:1848–1854.

Walters, C., and A. M. Parma. 1996. Fixed exploitation rate strategies for coping with effects of climate change. Canadian Journal of Fisheries and Aquatic Sciences 53:148–158.

Walters, C., D. Pauly, and V. Christensen. 1998. Ecospace: prediction of mesoscale spatial patterns in trophic relationships of exploited ecosystems, with emphasis on impacts of marine protected areas. International Council for the Exploration of the Sea, CM 1998/S:4.

WDF and WDW (Washington Department of Fisheries and Washington Department of Wildlife). 1993. 1992 Washington State salmon and steelhead stock inventory. Appendix Three: Columbia River Stocks. Washington Department of Fisheries and Washington Department of Wildlife, Olympia.

WDF et al. (Washington Department of Fisheries, Puget Sound Treaty Indian Tribes, and Northwest Indian Fisheries Commission). 1992. 1992 Puget Sound coho salmon forecasts and management recommendations. Washington Department of Fisheries and Northwest Indian Fisheries Commission, Olympia.

WDFW and WWTIT (Washington Department of Fisheries and Western Washington Treaty Indian Tribes). 1994. 1992 Washington State salmon and steelhead stock inventory. Appendix One, Puget Sound stocks, North Puget Sound Volume. Washington Department of Fish and Wildlife and Western Washington Treaty Indian Tribes, Olympia.

Woody, C. A. 1998. Ecologic and genetic variation of two sockeye salmon populations of Tustumena Lake, Alaska. Doctoral dissertation. University of Washington, Seattle.

Wright, S. 1981. Contemporary Pacific salmon fisheries management. North American Journal of Fisheries Management 1:29–40.

APPENDIX TABLE 17.1

Summary of escapement goal-setting methods for U.S. salmon management units, by species and state (or Alaskan region).

Quality	Method	Chinook	Chum	Coho	Pink	Sockeye	Steelhead	All Species
Alaska (Alaska Peninsula/Aleutian Islands)								
Excellent	Combined-strong							0
Good	Habitat-advanced					2		2
	Spawner/recruit							0
	Historic	1				3		4
Fair	Combined-weak		9		16			25
	Habitat					4		4
	Recent escapements	4		5		3		12
Poor	Index							0
No method	No goal	1	3	4	4	5		17
No information	No information							0
Alaska (Bristol Bay)								
Excellent	Combined-strong							0
Good	Habitat-advanced							0
	Spawner/recruit	1			1	9		11
	Historic							0
Fair	Combined-weak	1		3				4
	Habitat							0
	Recent escapements	1						1
Poor	Index							0
No method	No goal	25	24	10	9	9		77
No information	No information							0
Alaska (Chignik)								
Excellent	Combined-strong							0
Good	Habitat-advanced							0
	Spawner/recruit					2		2
	Historic		5		5			10
Fair	Combined-weak							0
	Habitat							0
	Recent escapements	1						1
Poor	Index							0
No method	No goal			2				2
No information	No information							0
Alaska (Cook Inlet)								
Excellent	Combined-strong					3		3
Good	Habitat-advanced							0
	Spawner/recruit							0
	Historic	3		1				4
Fair	Combined-weak	2	14		30	12		58
	Habitat							0
	Recent escapements	20		6		1		27
Poor	Index							0
No method	No goal							0
No information	No information							0

APPENDIX TABLE 17.1 (continued)
Summary of escapement goal-setting methods for U.S. salmon management units, by species and state (or Alaskan region).

Quality	Method	Chinook	Chum	Coho	Pink	Sockeye	Steelhead	All Species
Alaska (Kodiak)								
Excellent	Combined-strong							0
Good	Habitat-advanced							0
	Spawner/recruit					5		5
	Historic					5		5
Fair	Combined-weak					1		1
	Habitat							0
	Recent escapements	3	77	41	100	26		247
Poor	Index							0
No method	No goal		7	6	6	1		20
No information	No information		1			4		5
Alaska (Prince William Sound)								
Excellent	Combined-strong					1		1
Good	Habitat-advanced					1		1
	Spawner/recruit					1		1
	Historic		7		16			23
Fair	Combined-weak	1						1
	Habitat							0
	Recent escapements	1		2		2		5
Poor	Index							0
No method	No goal							0
No information	No information							0
Alaska (Southeast)								
Excellent	Combined-strong							0
Good	Habitat-advanced							0
	Spawner/recruit	1						1
	Historic							0
Fair	Combined-weak	3		1				4
	Habitat		2	2	2	1		7
	Recent escapements	3						3
Poor	Index			4	1	6		11
No method	No goal		10					10
No information	No information	3			3	1	7	14
Washington								
Excellent	Combined-strong			4				4
Good	Habitat-advanced			16			38	54
	Spawner/recruit	1					1	2
	Historic	3			5			8
Fair	Combined-weak	1	4	12		1		18
	Habitat	8	1			1		10
	Recent escapements	9	59		1		1	70
Poor	Index	2						2
No method	No goal	2	3	12	6	1	33	57
No information	No information	28	16	1	1	3	20	69

APPENDIX TABLE 17.1 (continued)
Summary of escapement goal-setting methods for U.S. salmon management units, by species and state (or Alaskan region).

Quality	Method	Species						All Species
		Chinook	Chum	Coho	Pink	Sockeye	Steelhead	
		Oregon						
Excellent	Combined-strong							0
Good	Habitat-advanced							0
	Spawner/recruit		1	1				2
	Historic	7						7
Fair	Combined-weak	3						3
	Habitat							0
	Recent escapements							0
Poor	Index							0
No method	No goal	2	1	1				4
No information	No information						84	84
		California						
Excellent	Combined-strong							0
Good	Habitat-advanced	1						1
	Spawner/recruit							0
	Historic							0
Fair	Combined-weak							0
	Habitat							0
	Recent escapements	1						1
Poor	Index							0
No method	No goal	5						5
No information	No information							0

APPENDIX TABLE 17.2
Number of populations, by species and state (or Alaskan region), for which each method of escapement estimation is used.

Quality	Method	Species						All Species
		Chinook	Chum	Coho	Pink	Sockeye	Steelhead	
		Alaska (Alaska Peninsula/Aleutian Islands)						
Excellent	Dam count							0
	Trap count						2		2
Good	Dam or trap estimate							0
	Tower							0
	Sonar							0
Fair	Mark/recapture							0
	Combined							0
Poor	Foot survey							0
	Aerial survey	11	117	20	300	51		499
	Boat survey							0
	Snorkeling							0
	Test fishing							0
None	No method							0
No information	No information		1	3				4

APPENDIX TABLE 17.2 (continued)
Number of populations, by species and state (or Alaskan region), for which each method of escapement estimation is used.

Quality	Method	Species						All Species
		Chinook	Chum	Coho	Pink	Sockeye	Steelhead	
		Alaska (Bristol Bay)						
Excellent	Dam count		0					0
	Trap count							0
Good	Dam or trap estimate							0
	Tower					5		5
	Sonar				1			1
Fair	Mark/recapture							0
	Combined	2	1	1	3	4		11
Poor	Foot survey							0
	Aerial survey	29	24	9	7	9		78
	Boat survey							0
	Snorkeling							0
	Test fishing							0
None	No method			2				2
No information	No information							0
		Alaska (Chignik)						
Excellent	Dam count							0
	Trap count							0
Good	Dam or trap estimate	1				2		3
	Tower							0
	Sonar							0
Fair	Mark/recapture							0
	Combined			1				1
Poor	Foot survey							0
	Aerial survey		48	3	48			99
	Boat survey							0
	Snorkeling							0
	Test fishing							0
None	No method							0
No information	No information							0
		Alaska (Cook Inlet)						
Excellent	Dam count							0
	Trap count							0
Good	Dam or trap estimate			1		4		5
	Tower							0
	Sonar					3		3
Fair	Mark/recapture							0
	Combined	8		3	4	5		20
Poor	Foot survey	2	3	3	15			23
	Aerial survey	15	11		12	4		42
	Boat survey							0
	Snorkeling							0
	Test fishing							0
None	No method							0
No information	No information							0

APPENDIX TABLE 17.2 (continued)
Number of populations, by species and state (or Alaskan region), for which each method of escapement estimation is used.

| Quality | Method | Species | | | | | | All Species |
		Chinook	Chum	Coho	Pink	Sockeye	Steelhead	
		Alaska (Kodiak)						
Excellent	Dam count							0
	Trap count	3	4	14	20	14		55
Good	Dam or trap estimate							0
	Tower							0
	Sonar							0
Fair	Mark/recapture							0
	Combined							0
Poor	Foot survey		1	33	15	1		50
	Aerial survey		86	24	138	28		276
	Boat survey							0
	Snorkeling							0
	Test fishing							0
None	No method							0
No information	No information			3		1		4
		Alaska (Prince William Sound)						
Excellent	Dam count			0				0
	Trap count							0
Good	Dam or trap estimate					1		1
	Tower							0
	Sonar					1		1
Fair	Mark/recapture							0
	Combined	1				16		17
Poor	Foot survey							0
	Aerial survey	9	202	30	419	3		663
	Boat survey							0
	Snorkeling							0
	Test fishing							0
None	No method							0
No information	No information		1					1
		Alaska (Southeast)						
Excellent	Dam count							0
	Trap count	2		4				6
Good	Dam or trap estimate		4	2		23	2	31
	Tower							0
	Sonar							0
Fair	Mark/recapture					1		1
	Combined	19		3				22
Poor	Foot survey					1		1
	Aerial survey	12	742	658	3	102	18	1535
	Boat survey							0
	Snorkeling							0
	Test fishing							0
None	No method	1	770	1707		84	316	2878
No information	No information	51			2405			2456

APPENDIX TABLE 17.2 (continued)
Number of populations, by species and state (or Alaskan region), for which each method of escapement estimation is used.

Quality	Method	Chinook	Chum	Coho	Pink	Sockeye	Steelhead	All Species
				Species				**All Species**
		Washington						
Excellent	Dam count			2				2
	Trap count		2	3				5
Good	Dam or trap estimate			1			2	3
	Tower							0
	Sonar							0
Fair	Mark/recapture							0
	Combined	21		8			13	42
Poor	Foot survey	19	25	52			15	111
	Aerial survey	2			1		1	4
	Boat survey	3						3
	Snorkeling						2	2
	Test fishing							0
None	No method		1	24			19	44
No information	No information	30	117		13	6	64	230
		Oregon						
Excellent	Dam count	6						6
	Trap count							0
Good	Dam or trap estimate							0
	Tower							0
	Sonar							0
Fair	Mark/recapture							0
	Combined							0
Poor	Foot survey	29	8	19				56
	Aerial survey							0
	Boat survey							0
	Snorkeling							0
	Test fishing							0
None	No method							0
No information	No information						120	120
		California						
Excellent	Dam count	3						3
	Trap count							0
Good	Dam or trap estimate							0
	Tower	1						1
	Sonar							0
Fair	Mark/recapture							0
	Combined							0
Poor	Foot survey							0
	Aerial survey							0
	Boat survey							0
	Snorkeling							0
	Test fishing							0
None	No method	1						1
No information	No information	2						2

APPENDIX TABLE 17.3
Number of populations, by species and state (or Alaskan region), for which each escapement estimation type is used.

Quality	Method	Chinook	Chum	Coho	Pink	Sockeye	Steelhead	All Species
				Species				
Excellent	Total							0
Good	Total estimate					2		2
	Peak count	10	115	18	294	51		488
	Good index							0
	Total redds							0
Fair	One count							0
	Fair index							0
	Redd survey							0
Poor	One count—sporadic							0
	Poor index							0
	Carcass count							0
	Carcass index							0
None	No method							0
No information	No information	1	3	5	6			15

Alaska (Alaska Peninsula/Aleutian Islands)

Quality	Method	Chinook	Chum	Coho	Pink	Sockeye	Steelhead	All Species
Excellent	Total							0
Good	Total estimate	2		2	2	8		14
	Peak count	26	25	9	8	10		78
	Good index							0
	Total redds							0
Fair	One count	3						3
	Fair index							0
	Redd survey							0
Poor	One count—sporadic							0
	Poor index							0
	Carcass count							0
	Carcass index							0
None	No method			2				2
No information	No information							0

Alaska (Bristol Bay)

Quality	Method	Chinook	Chum	Coho	Pink	Sockeye	Steelhead	All Species
Excellent	Total							0
Good	Total estimate	1	48	1	48	2		100
	Peak count							0
	Good index							0
	Total redds							0
Fair	One count			3				3
	Fair index							0
	Redd survey							0
Poor	One count—sporadic							0
	Poor index							0
	Carcass count							0
	Carcass index							0
None	No method							0
No information	No information							0

Alaska (Chignik)

APPENDIX TABLE 17.3 (continued)
Number of populations, by species and state (or Alaskan region), for which each escapement estimation type is used.

Quality	Method	Chinook	Chum	Coho	Pink	Sockeye	Steelhead	All Species
				Species				
Alaska (Cook Inlet)								
Excellent	Total							0
Good	Total estimate	3		4	31	10		48
	Peak count		14					14
	Good index							0
	Total redds							0
Fair	One count	22						22
	Fair index			3		6		9
	Redd survey							0
Poor	One count—sporadic							0
	Poor index							0
	Carcass count							0
	Carcass index							0
None	No method							0
No information	No information							0
Alaska (Kodiak)								
Excellent	Total					13		13
Good	Total estimate	3	3	14	14			34
	Peak count		88	35	159	29		311
	Good index							0
	Total redds							0
Fair	One count					1		1
	Fair index							0
	Redd survey							0
Poor	One count—sporadic							0
	Poor index							0
	Carcass count							0
	Carcass index							0
None	No method							0
No information	No information				25	1		26
Alaska (Prince William Sound)								
Excellent	Total							0
Good	Total estimate		203		419	3		625
	Peak count	10		30		18		58
	Good index							0
	Total redds							0
Fair	One count							0
	Fair index							0
	Redd survey							0
Poor	One count—sporadic							0
	Poor index							0
	Carcass count							0
	Carcass index							0
None	No method							0
No information	No information							0

APPENDIX TABLE 17.3 (continued)
Number of populations, by species and state (or Alaskan region), for which each escapement estimation type is used.

Quality	Method	Chinook	Chum	Coho	Pink	Sockeye	Steelhead	All Species
		Species						

Quality	Method	Chinook	Chum	Coho	Pink	Sockeye	Steelhead	All Species
**Alaska (Southeast)**								
Excellent	Total	2		1				3
Good	Total estimate		4	6		27	2	39
	Peak count	19		14	1	1		35
	Good index							0
	Total redds			2				2
Fair	One count		61			8	1	70
	Fair index							0
	Redd survey	2						2
Poor	One count—sporadic		680	644		90	17	1431
	Poor index							0
	Carcass count							0
	Carcass index							0
None	No method	1	770	1707		84	316	2878
No information	No information	61	1		2407	1		2470
**Washington**								
Excellent	Total		2	6				8
Good	Total estimate						3	3
	Peak count	1	2					3
	Good index	12		10			6	28
	Total redds	7		5				12
Fair	One count						2	2
	Fair index	2	6	18			4	30
	Redd survey	14		21			15	50
Poor	One count—sporadic	1						1
	Poor index	2		4				6
	Carcass count							0
	Carcass index	3						3
None	No method		1	24			19	44
No information	No information	33	134	2	14	6	66	255
**Oregon**								
Excellent	Total	6						6
Good	Total estimate							0
	Peak count	22	8					30
	Good index			19				19
	Total redds							0
Fair	One count							0
	Fair index							0
	Redd survey	1						1
Poor	One count—sporadic							0
	Poor index							0
	Carcass count							0
	Carcass index	6						6
None	No method							0
No information	No information						120	123

APPENDIX TABLE 17.3 (continued)
Number of populations, by species and state (or Alaskan region), for which each escapement estimation type is used.

| Quality | Method | Species | | | | | | All Species |
		Chinook	Chum	Coho	Pink	Sockeye	Steelhead	
			California					
Excellent	Total							0
Good	Total estimate	4						4
	Peak count							0
	Good index							0
	Total redds							0
Fair	One count							0
	Fair index							0
	Redd survey							0
Poor	One count—sporadic							0
	Poor index							0
	Carcass count							0
	Carcass index							0
None	No method	1						1
No information	No information	2						2

APPENDIX TABLE 17.4
Number of management units and stocks per management unit for each U.S. state (or Alaskan region) and species.

State or region	Species	Number of management units	Number of populations	Stocks per management unit
Alaska (Alaska Peninsula and Aleutian Islands)	Chinook	6	11	1.8
	Chum	12	118	9.8
	Coho	9	23	2.6
	Pink	20	300	15.0
	Sockeye	17	53	3.1
Alaska (Bristol Bay)	Chinook	28	31	1.1
	Chum	24	25	1.0
	Coho	13	13	1.0
	Pink	10	10	1.0
	Sockeye	18	18	1.0
Alaska (Chignik)	Chinook	1	1	1.0
	Chum	5	48	9.6
	Coho	2	4	2.0
	Pink	5	48	9.6
	Sockeye	2	2	1.0
Alaska (Cook Inlet)	Chinook	25	25	1.0
	Chum	14	14	1.0
	Coho	7	7	1.0
	Pink	30	31	1.0
	Sockeye	15	16	1.1
Alaska (Kodiak)	Chinook	3	3	1.0
	Chum	85	91	1.1
	Coho	47	74	1.6
	Pink	106	173	1.6
	Sockeye	43	45	1.0
Alaska (Prince William Sound)	Chinook	2	10	5.0
	Chum	7	203	29.0
	Coho	2	30	15.0
	Pink	16	419	26.2
	Sockeye	5	21	4.2
Alaska (Southeast)	Chinook	10	85	8.5
	Chum	12	1516	126.3
	Coho	7	2374	339.1
	Pink	6	2408	401.3
	Sockeye	8	211	26.4
	Steelhead	7	336	48.0
Washington	Chinook	52	75	1.4
	Chum	82	145	1.8
	Coho	43	90	2.1
	Pink	13	14	1.1
	Sockeye	6	6	1.0
	Steelhead	92	115	1.2
Oregon	Chinook	12	35	2.9
	Chum	2	8	4.0
	Coho	2	19	9.5
	Steelhead	84	120	1.4
California	Chinook	7	7	1.0

18 Research Programs and Stock Status for Salmon in Three Transboundary Rivers: The Stikine, Taku, and Alsek

Kathleen Jensen

Abstract.—Salmon from the transboundary Stikine, Taku, and Alsek rivers are harvested in U.S. and Canadian aboriginal, subsistence, personal use, recreational, and commercial fisheries. Cooperative research programs for inseason run strength assessment and postseason escapement evaluation on these rivers involve state, provincial, federal, and First Nations personnel. Stikine salmon researchers use biological data obtained from commercial catch samples, weirs, aerial surveys, and test fisheries to estimate marine harvest and inriver run size or index escapements. Stikine sockeye *Oncorhynchus nerka* are experiencing record high abundance and chinook *O. tshawytscha* populations are healthy. Throughout Southeast Alaska coho *O. kisutch*, chum *O. keta*, and pink *O. gorbuscha* runs are at healthy to record high abundances, but data are lacking to determine how closely Stikine stocks follow the regional trend. For the Taku River, marine harvest and inriver run size or index escapement are estimated using a wide range of biological data, including coded wire tagging of juveniles, analysis of biological markers, adult mark-recapture estimates, weirs, and aerial surveys. Taku sockeye, coho, and pink salmon are experiencing above average to record high abundances, chinook populations are healthy, and chum salmon are declining. A weir and aerial surveys provide index escapement estimates for Alsek salmon. These data indicate that sockeye and coho salmon remain near historical average abundance and the chinook indices appear stable. The joint U.S./Canada transboundary river research programs, although of recent origin, have provided an array of information critical for maintaining sustainable fisheries on these salmon stocks. The transboundary river programs provide a good model for international and inter-agency cooperation in management of shared fisheries resources.

INTRODUCTION

The Taku, Stikine, and Alsek rivers are transboundary rivers that originate in Canada and flow through the U.S. before entering the Pacific Ocean, near the Southeast Alaska communities of Petersburg, Juneau, and Yakutat, respectively. Anadromous stocks of chinook *Oncorhynchus tshawytscha*, sockeye *O. nerka*, coho *O. kisutch*, pink *O. gorbuscha*, and chum *O. keta* salmon from these rivers are harvested in commercial, recreational, aboriginal, subsistence, and personal use fisheries in both nations. The ability to manage these resources based on abundance and productivity is evolving as resource knowledge increases. The Pacific Salmon Commission (PSC) provides a forum in which U.S. and Canadian agencies run cooperative research programs.

1-56670-480-4/00/$0.00+$.50
© 2000 by CRC Press LLC

The U.S. and Canada have operated cooperative research programs on the Stikine, Taku, and Alsek rivers since the ratification of the U.S./Canada Pacific Salmon Treaty in 1985. Participants include the Alaska Department of Fish and Game (ADF&G) and the National Marine Fisheries Service (NMFS) for the U.S., and the Department of Fisheries and Oceans (DFO), First Nations, and the Province of British Columbia for Canada. Evaluations of the status of information on transboundary river stocks and existing research programs, development of new programs, and prioritization of specific programs were undertaken in 1989 and 1990 (Marshall et al. 1989; TTC 1990). Research and management plans have been jointly developed since 1983 and are detailed in reports of the Transboundary Technical Committee (TTC) (TTC 1999a). Catches in fisheries targeting transboundary river salmon stocks are monitored and the U.S. and Canada have jointly evaluated research and management programs annually since 1983 (TTC 1999b).

Concern about the status of Pacific salmon stocks has been increasing in recent years as stocks have become depleted or absent from historical ranges (Nehlsen et al. 1991). The NRC (1996) found that Pacific salmon were no longer present in 40% of their historical range from Washington, Idaho, Oregon, and California. Slaney et al. (1996) described the overall status of British Columbia and Yukon salmon as healthy but, where information was available, classified 9 to 10% of the stocks as of concern or at some level of risk. General overviews of the status of salmon stocks in Alaska (Wertheimer 1996) and Southeast Alaska (Baker et al. 1996; Van Alen 2000) indicated that most Alaska salmon stocks are healthy. This chapter documents salmon research programs in the Stikine, Taku, and Alsek rivers and reviews the status of salmon stocks in these transboundary rivers. Overall, salmon stocks in the transboundary rivers appear to be healthy and several have been at or near record high abundances in recent years. However, data are insufficient to assess the status of some stocks and a few stocks are below historical abundance despite conservative fisheries management. Monitoring programs are still being developed to gain better information on harvest rates, migratory characteristics, and fishery contributions for these salmon.

STUDY AREA

The Stikine, Taku, and Alsek rivers are large and glacially occluded but include a wide diversity of main-channel and off-channel habitats, such as lakes, ponds, side sloughs, and clear headwater tributaries. Much of the habitat in the three drainages has been relatively undisturbed by human activities with the exception of mine sites and associated access roads. The Stikine drainage encompasses a land area of over 52,000 km^2; more than 90% is inaccessible to salmon due to velocity blocks and other natural barriers (Figure 18.1). The upper Stikine drainage is accessible to humans via the Telegraph Creek Road and the Stewart Cassiar Highway. There is also restricted access via mining roads in the Tahltan, Iskut, and Chutine river valleys. The Taku River has a drainage of more than 17,000 km^2, with road access limited to the southeast portion of the watershed (Sheslay River) via a restricted use mining road (Figure 18.2). The Alsek drainage area is about 28,000 km^2, much of which is inaccessible to anadromous salmonids due to velocity barriers (Figure 18.3). Part of the upper drainage, the Klukshu and the upper Tatshenshini rivers are accessible to humans by the Dalton Post Road and the Haines Highway.

Several habitat studies undertaken prior to the cooperative research programs established under the PSC process have been used by both countries as a basis for salmon assessment in the Stikine River. Stream and habitat type and usage have been studied for most of the Stikine River accessible to salmon (McCart and Walder 1982; Edgington and Lynch 1984; Hancock and Marshall 1984). Studies in the Taku River documented the prey consumption (Brownlee 1991), rearing habitat use by juvenile salmon (Lorenz et al. 1991; Murphy et al. 1989) and spawning habitat use by adults (Lorenz and Eiler 1989). Additional Taku River studies focused on juvenile and smolt abundance (Thedinga et al. 1988; Murphy et al. 1991) and migrations (Meechan and Siniff 1962; Murphy et al. 1997) in the lower river and estuary.

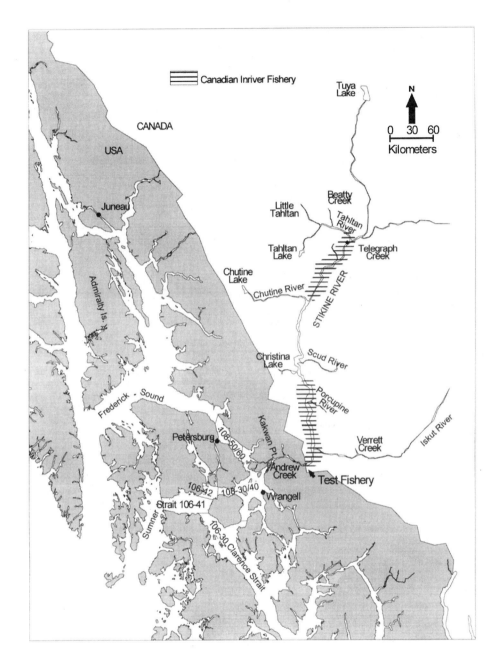

FIGURE 18.1   The Stikine River and U.S. and Canadian Fishing Areas.

FISHERIES

Salmon resources of the Stikine, Taku, and Alsek rivers have been important, both historically and currently, to aboriginal peoples for food, social, and ceremonial purposes. Commercial marine net fisheries were established in the late 1800s. In addition, fish traps were operated in straits throughout the southeast archipelago from around 1900 through 1958, with a limited number of traps still allowed near Annette Island. The troll fishery in Southeast Alaska was started in 1905 and had little regulation prior to statehood in 1959. Recreational fisheries have also become important on

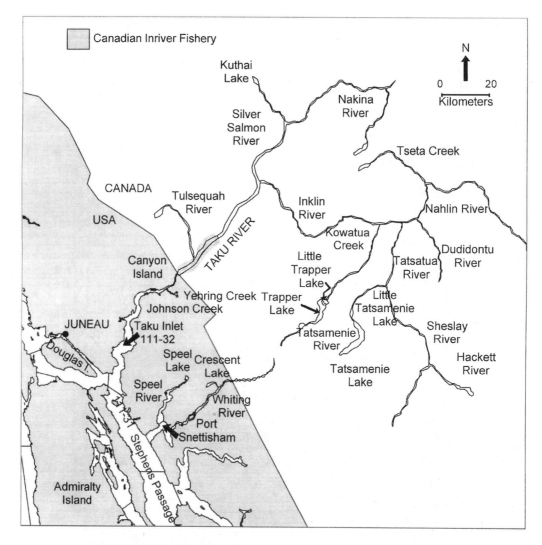

FIGURE 18.2 The Taku River and U.S. and Canadian Fishing Areas.

all three rivers. In the mid-1970s it became apparent that chinook salmon in Southeast Alaska were below historical abundance levels and closures of directed chinook fisheries were initiated. All Southeast Alaska catches of wild and non-Alaska hatchery chinook salmon were limited by harvest guidelines or catch ceilings under a 15-year rebuilding program established in 1981. The Alaska rebuilding program was incorporated into the PSC 15-year coastwide chinook rebuilding plan in 1984. Both U.S. and Canadian commercial fisheries directed at Stikine and Taku sockeye stocks have been managed under abundance-based harvest sharing agreements negotiated in the Pacific Salmon Treaty process since 1985.

The U.S. gillnet fisheries in Alaska Districts 106 and 108 have been in operation since at least 1895 (Rich and Ball 1933). These fisheries harvest a mixture of stocks, some of which are of Stikine origin. Sockeye salmon has been the principal target species, but all five Pacific salmon species have been of major commercial importance. The gillnet fishery directed on chinook salmon was suspended in 1977 because of conservation concerns. Restrictions on the spring troll fisheries directed at Stikine chinook salmon were initiated in the mid-1960s, with complete closure of the spring fishery observed since 1978. The current troll fishery harvests coho and some chinook salmon

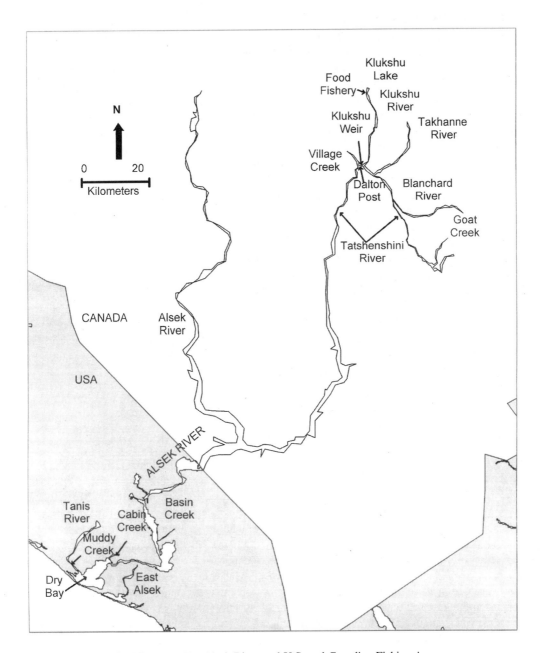

FIGURE 18.3   The Alsek River and U.S. and Canadian Fishing Areas.

of Stikine origin. Mixed stock seine fisheries in central Southeast Alaska also harvest salmon originating from the Stikine River. Canada has operated commercial gillnet fisheries near Telegraph Creek since 1975 and near the U.S./Canada border since 1979. The fisheries have been directed at sockeye salmon with incidental harvests of other salmonids, primarily chinook and coho salmon. A Canadian aboriginal fishery near the community of Telegraph Creek harvests primarily chinook and sockeye salmon. The Canadian coho harvest was under a treaty-imposed catch ceiling for most of the 1985–1992 period. Only a few permits have been fished in the U.S. personal use fishery in the lower river. There are U.S. marine sport fisheries near Wrangell and Petersburg while Canada has sport fisheries in the upper drainages of the Stikine River.

The U.S. gillnet fishery in Alaska District 111, in operation since at least 1897 (Rich and Ball 1933), was originally developed to harvest runs of chinook and sockeye salmon to the Taku River. Coho salmon and, in later years, pink and chum salmon have also contributed significantly to the total harvests. Similar to the Stikine, the U.S. gillnet fishery directed on chinook salmon was suspended in 1976 due to conservation concerns. The spring troll fishery directed on Taku chinook stocks was also suspended in 1976. The current troll fishery harvests coho and some chinook salmon of Taku origin. Mixed stock seine fisheries in northern Southeast Alaska also harvest salmon of Taku origin. Canada has operated an inriver commercial gillnet fishery upstream of the U.S./Canada border since 1979. This fishery, directed at sockeye salmon, incidentally harvests other salmonids, primarily chinook and coho salmon. As with the Stikine, the Canadian coho harvest was under a treaty-imposed catch ceiling for most of 1985 to 1992. The U.S. historical subsistence and current personal use fisheries are located in the lower river and Canada has an aboriginal fishery that usually occurs near the Canadian commercial fishery. Sport fisheries are active in marine waters of the U.S. and in upper river tributaries in Canada.

The U.S. commercial gillnet fishery in the Alsek River and Dry Bay area, which started between 1901 and 1908 (Rich and Ball 1933), harvests Alsek salmon stocks. Delayed openings of the U.S. set net fishery in and near the mouth of the Alsek River have occurred since 1983 to conserve chinook salmon. Sockeye salmon are the primary target species but chinook and coho salmon are also taken in the set net fishery. The Alaska troll fishery also harvests coho salmon and some chinook salmon of Alsek origin. Canada has no commercial fishery on the Alsek River. U.S. subsistence and sport fisheries occur in the lower river and Canadian sport and aboriginal fisheries occur in the upper drainage, principally on the Klukshu River near the settlements of Dalton Post and Klukshu Village, and in the Blanchard and Takhanne rivers and Village Creek.

STIKINE RIVER PROGRAMS AND STOCK STATUS

CHINOOK SALMON

Helicopter surveys of three upriver Stikine chinook spawning areas (Little Tahltan River, Tahltan River, and Beatty Creek) (Figure 18.1) have been conducted to provide escapement indices since 1975 (Pahlke 1996; TTC 1999b). A weir was operated on Andrew Creek in the lower river from 1979 to 1984 and aerial surveys of spawning grounds in the creek have been conducted since 1985 (TTC 1999b). These indices are used to evaluate trends in escapements. An adult enumeration weir on the Little Tahltan River has provided escapement counts since 1985 (TTC 1999b). However, except for the Little Tahltan, the fraction of the spawners represented by the survey counts of the Stikine index areas is unknown. The fraction of the total Stikine chinook run that uses the index areas is also unknown; therefore, total chinook escapement for the Stikine River cannot be accurately estimated. In 1995 an adult tagging program in the lower river was combined with recovery efforts at the spawning grounds to estimate the inriver run strength and the fraction of the chinook run accounted for by the indices (Pahlke and Etherton 1997). Coded wire tagging of chinook juveniles and smolts from 1978 to 1981 (Hubartt and Kissner 1987) successfully provided information on migratory routes and marine harvest areas, but recoveries were insufficient to estimate fishery contributions.

Stikine chinook populations appear to be healthy (Figure 18.4). Under conservation measures initiated during the mid-1970s the average Little Tahltan index counts have increased from 1,389 fish (1975–1984) to 2,540 fish for the last decade. Stikine chinook escapement goals were re-evaluated by the TTC in 1991 (TTC 1991), and a single, joint U.S./Canada goal of 5,300 fish for the Little Tahltan stock was established. The Little Tahltan weir count has exceeded this goal 4 of the last 10 years and was less than 80% of goal only once in the last decade. The lower river tagging study is providing information that will allow refinement of the Little Tahltan goal and development of a goal for the entire Stikine chinook escapement. Recent-year catches of all chinook salmon in the District 108 drift gillnet fishery are far lower than the unsustainable historical levels,

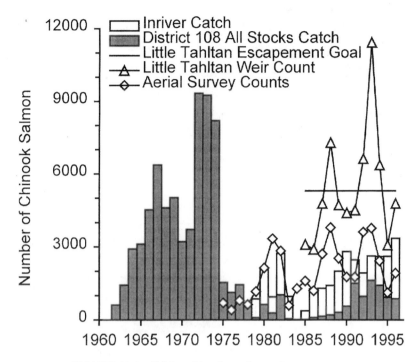

FIGURE 18.4   Stikine chinook catches and escapements.

due primarily to management directing the fisheries away from chinook stocks. The average annual harvest of 1,100 fish in District 108 since 1990 is at least partially due to the increased effort directed at strong runs of early migrating Stikine sockeye salmon.

SOCKEYE SALMON

Stikine sockeye populations are grouped into three stocks for management purposes. The Tahltan stock consists of fish that spawn and rear in Tahltan Lake, the mainstem stock is a conglomerate of fish that spawn and rear in the mainstem river, tributaries, sloughs, and other small lakes, and the Tuya stock is composed of fish planted into Tuya Lake. An adult enumeration weir, operated at Tahltan Lake since 1959, provides annual escapement counts for this sockeye stock (TTC 1999b). A smolt weir, operated at the outlet of Tahltan Lake since 1984, produces smolt abundance and age composition estimates (TTC 1999b). Sonar, fishwheels, and a test fishery were used in the lower river from 1982 to 1984 in conjunction with biological markers to estimate the total Stikine sockeye run (Mesiar 1984; Lynch and Edgington 1986). Data from these projects proved to be unreliable and the monitoring programs were discontinued. Test fishery catches at Kakwan Point on the lower river (1977–1985) were inadequate to assess run strength and the project was discontinued (Lynch et al. 1987). Joint U.S./Canada marine tagging of adult sockeye salmon in 1982 and 1983 provided information on the distribution of marine harvest of Stikine sockeye stocks (Hoffman et al. 1984). Scale pattern analysis (Jensen and Frank 1993) and analysis of egg diameters (Craig 1985) from inriver commercial and test fishery catches have been used to estimate the proportion of the inriver run composed of Tahltan fish. Egg diameter analysis has been used annually to estimate inriver stock-specific run strength, manage inriver fisheries, and reconstruct runs since 1989 (TTC 1999b). Electrophoretic data, population age structure, scale pattern characteristics, and parasite prevalence were studied to determine the separability of Stikine sockeye populations (Wood et al. 1988; Guthrie et al. 1994). Scale pattern analysis for sockeye stock identification in marine catches has been used annually since 1982. Contributions of transboundary Stikine River and other

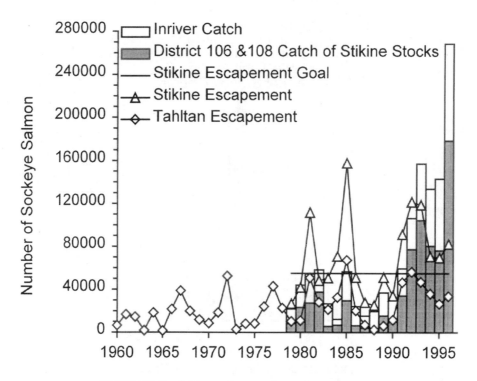

FIGURE 18.5   Stikine sockeye catches and escapements.

U.S. and Canadian stocks to Southeast Alaska District 106 and 108 gillnet catches are estimated for each week of the fisheries (Jensen and Frank 1995; TTC 1999b). Stock compositions of marine catches were also estimated based on prevalence of brain parasites (Moles et al. 1990) and a mixture of biological markers (Wood et al. 1989). Agreements among the stock composition estimates were generally high for the different methods where specific stocks could be identified. Scale pattern analysis with scales in current year catches, compared to scales sampled from spawning escapements the prior year, were used inseason in 1986 to 1989 to estimate stock contributions to District 106 and 108 catches. Inseason analysis was discontinued after 1990 because results became unreliable due to interannual variability of scale patterns (Jensen 1994).

   Since 1979 a program that combines stock composition data, catch per unit effort (CPUE) data from Canadian commercial and test fisheries, and the Tahltan escapement count has allowed calculation of the entire above-border run by estimating the fraction of the run comprised of the Tahltan stock (TTC 1999b). The marine contribution estimates and the inriver run strength assessment allow complete reconstruction of the Tahltan and mainstem Stikine sockeye runs, harvest rates in various fisheries, migratory timing, and escapements. Preseason forecasts of the Tahltan, mainstem, and Tuya runs based on the return of 4-year-old fish the prior year, on smolt outmigrations from Tahltan weir, and on spawner-recruit relationships have been made since 1987 (Jensen 1992; TTC 1999a). Historical stock composition estimates combined with migratory timing information and inriver test or commercial fishery CPUE provide inseason run strength assessment (TTC 1999a). Preseason and inseason run assessment allow intensive abundance-based management of fisheries by both the U.S. and Canada. Escapement goal ranges of 20,000 to 40,000 fish for each of the Tahltan and mainstem Stikine sockeye stocks were established in 1987 (TTC 1987). The status of Stikine sockeye runs was re-evaluated (Humphreys et al. 1994) and the escapement goal for the Tahltan stock was revised in 1993 to 24,000 fish (TTC 1993) while the mainstem goal remained 30,000 fish for a total escapement goal of 54,000 sockeye salmon (Figure 18.5).

In 1987 the TTC evaluated the feasibility of increasing Stikine sockeye production through hatchery programs (TTC 1988a). A goal of producing an additional 100,000 returning adults was set with the harvest to be shared equally between the U.S. and Canada. Broodstock is captured at Tahltan Lake, gametes are collected and flown to Snettisham Hatchery where they are incubated and thermally marked, and the resultant fry are planted into Tahltan and Tuya Lakes. This program has produced substantial numbers of returning adults that have benefited fishers from both countries. From 1994 through 1996 catches of thermally marked sockeye salmon have averaged 27,000 fish in U.S. fisheries and 11,000 fish in Canadian inriver fisheries (TTC 1999b). Otoliths are collected inseason from U.S. and Canadian fisheries and analyzed to estimate the number of thermally marked fish in the catches. This inseason analysis combined with the run strength assessment for the wild Stikine sockeye stocks aids managers in adjusting harvest levels to avoid overharvest of wild fish while maximizing harvest of thermally marked fish.

Wild Stikine sockeye stocks are at record high abundance levels. The five highest total run sizes have occurred since 1992, as have the five highest marine and inriver catches (Figure 18.5). Escapements of both the Tahltan and non-Tahltan stocks have been above their respective goals since 1991.

COHO SALMON

Research programs to provide management information for Stikine coho stocks were initiated in the late 1970s. Coho smolts and rearing juveniles were captured and tagged with coded wire tags (CWT) on four tributaries in U.S. sections of the Stikine River in 1978 and 1979 (Shaul et. al 1984). Tagged adults were recovered in 1979–1982. Coho juveniles were also captured and tagged with CWT at several sites within Canadian sections of the Stikine River in 1986 (P. Etherton, DFO, personal communication). These programs provided harvest distribution estimates and migratory timing information. An attempt to determine the fraction of the run tagged with CWTs and thus estimate harvest rates was unsuccessful when the Barnes Lake adult coho enumeration weir on the lower Stikine River was submerged due to extreme flooding in 1980 and 1981. In lieu of specific data for Stikine coho stocks, harvest rates in the U.S. troll fishery are assumed to be similar to rates estimated with CWT information for nearby wild and hatchery fish. Aerial survey estimates from up to eight Stikine tributaries have been made annually since 1984 (Shaul 1987, TTC 1999b). The proportions of the spawning populations counted for each index area and the fraction of the total Stikine coho run represented by the index areas are unknown.

Aerial survey index counts of Stikine coho salmon and U.S. District 108 gillnet catch of all coho salmon have been highly variable and exhibit no long-term trend (Figure 18.6). Some of the larger interannual catch variability is due to highly variable effort. The Canadian inriver commercial catch was under a PSC treaty limit for 1985, 1986, and 1988 to 1992. Therefore, inriver catches are not an indication of abundance. Catch and escapement data for other coho stocks in Southeast Alaska indicate that coho salmon throughout the region are at record high abundance (Shaul and Crabtree 1998), although there is insufficient data to determine whether Stikine stocks follow this trend.

PINK AND CHUM SALMON

Most known Stikine pink and chum salmon spawning grounds are located on the U.S. side of the border and are not of concern in the Pacific Salmon Treaty process. Aerial surveys of major spawning areas are made during the U.S. gillnet fishery to allow managers to assess the progress of the escapements inseason. There are no programs in place to estimate total escapement or harvest rates for either species.

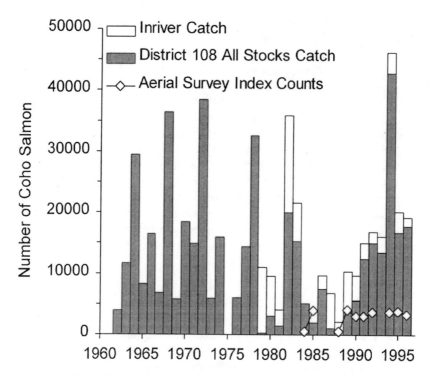

FIGURE 18.6   Stikine coho catches and escapement indices.

Aerial survey indices and catches of pink and chum salmon in the areas near to and within the Stikine River have ranged from average to record high levels (Van Alen 2000). Hatchery production contributed an unknown component of marine catches of these species in recent years.

TAKU RIVER PROGRAMS AND STOCK STATUS

CHINOOK SALMON

Aerial surveys to count spawning Taku chinook salmon have been flown in most years since the mid-1960s and have been standardized since 1974 (Pahlke 1996; TTC1999b). The enumerated fraction of chinook salmon on the spawning grounds is unknown, as is the fraction of the entire Taku chinook run that spawns in the index areas. In an attempt to estimate the fraction of the chinook run counted from aerial surveys, chinook juveniles were captured and tagged (CWT) in the Nahlin River in 1975 and 1977 and in upper Taku drainages and the Taku estuary from 1977 to 1983 (Kissner 1984). Data from these studies yielded harvest distributions but escapement recoveries were insufficient to estimate harvest rates. An adult enumeration weir was operated on the Nahlin River in 1993 through 1996 and a carcass weir has been used to collect age and sex composition on the Nakina River since 1973 (Pahlke 1995). In 1988 through 1990 spaghetti tags and radio transmitters were applied to chinook salmon captured in fish wheels located at Canyon Island to generate mark-recapture population estimates (Eiler et al. 1991; Pahlke and Bernard 1996). The adult tagging program resumed in 1995 and has continued annually (McPherson et al. 1997a). Marked-unmarked ratios are determined in the inriver fishery and at escapement weirs. These programs have provided spawning distribution information, inriver migratory timing data, and estimates for the entire above-border chinook run and can also be used to determine the run fraction counted in the aerial survey index counts. A program to capture and CWT outmigrant chinook and coho smolts in the lower river has been operated annually since 1991 (McPherson et al. 1997b).

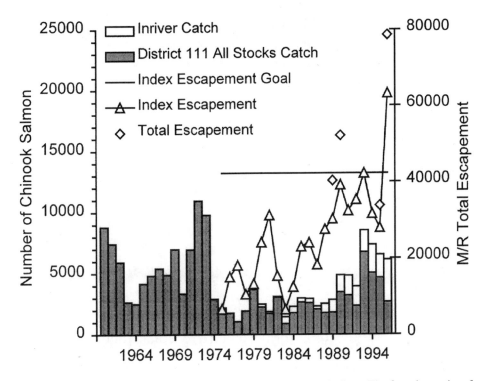

FIGURE 18.7   Taku chinook catches, escapements, and escapement indices. The four data points for total above-border escapements are based on mark-recapture estimates at Canyon Island.

Data from this program provide marine harvest distribution, harvest rates, and migratory information for Taku chinook stocks. Escapement goals for Taku chinook salmon were evaluated in 1991 and a goal of 13,200 fish in the index survey counts was established (TTC 1991).

Taku chinook populations appear to be healthy (Figure 18.7). Escapement indices have been trending upward since 1983 and were above the goal in 1993 and 1996. The current mark-recapture programs are providing information that will allow refinement of this goal and development of an escapement goal for the entire drainage. The marine catches of chinook salmon in recent years are not directly comparable to catches prior to 1975 when the U.S. gillnet fleet targeted chinook salmon. However, since 1990, the combined catch of chinook salmon in marine and inriver fisheries, which is taken incidentally during the sockeye fishery, is within the same order of magnitude as the directed fishery catches of the 1960s and early 1970s. Taku chinook stocks appear to have responded well to conservation measures inititated in the mid-1970s and the rebuilding program implemented in 1981.

SOCKEYE SALMON

Adult enumeration weirs have provided stock-specific escapement counts for some Taku sockeye stocks since the early 1980s: for Trapper Lake since 1983; Tatsamenie Lake since 1985; Hackett River between 1985 and 1988; Kuthai Lake during 1980, 1981, and 1992 to 1996; and Nahlin River during 1988, 1990, and 1992 to 1996 (Figure 18.2) (TTC 1999b). An adult tagging program, initiated at Canyon Island in 1983, and its associated inriver recovery effort are conducted annually (Kelley et al. 1997; TTC 1999b). Data from this program provide inseason estimates of the entire above-border run of Taku sockeye stocks and allow intensive abundance-based management of fisheries by the U.S. and Canada. An above-border escapement goal range of 71,000 to 80,000 sockeye salmon was established in 1987 (TTC 1987). In 1984 and 1986 sockeye were captured

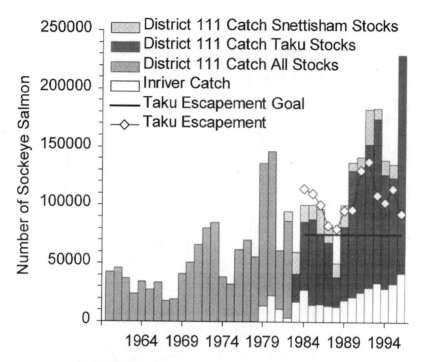

FIGURE 18.8   Taku sockeye catches and escapements.

and tagged with radio transmitters (Lorenz and Eiler 1989; Eiler et al. 1992). The study provided information on inriver migration patterns, timing, and spawning distributions. Scale pattern analysis was used from 1983 through 1985 to estimate the contributions of an all-Taku stock conglomerate and the all-Port Snettisham stock conglomerate to the District 111 gillnet catches (McGregor 1986). In 1986 the analysis was refined to include four individual stock groups from the Taku River and two groups from Port Snettisham and has been annually implemented since then (Jensen et al. 1993; TTC 1999b). Analyses of brain parasites (Moles et al. 1990) have also been used to estimate stock contributions to the marine harvests. Postseasonally, scale pattern analysis is combined with analysis of brain parasites to estimate stock compositions in marine catches and reconstruct Taku and Port Snettisham sockeye runs. However, inseason use of scale pattern analysis, where scales from current catches were compared to scales from escapements the prior year, was discontinued after 1990 because interannual variation in scale patterns resulted in unreliable stock composition estimates (Jensen and Bloomquist 1994).

As with the Stikine, the TTC evaluated the feasibility of supplementing Taku sockeye production through hatchery programs in 1987 (TTC 1988a). Broodstock were originally captured at Little Tatsamenie and Little Trapper lakes, gametes were collected and flown to Snettisham Hatchery, and the resultant fry were planted into Tatsamenie and Trapper lakes. This program has not produced substantial numbers of returning adults. Catches for the 1995 and 1996 fishing seasons averaged 4,000 thermally marked fish for the U.S. and 1,000 fish for Canada (TTC 1999b). Changes in the sites of brood stock collection and in fry planting techniques are being evaluated in an attempt to increase production (TTC 1999b).

Wild Taku sockeye salmon are experiencing record high abundance (Figure 18.8). Above-border escapements have been greater than the mid-goal range of 75,000 fish since 1984. The six largest inriver and marine catches of Taku sockeye stocks have occurred since 1990. The average catch of all sockeye salmon in the marine fishery for the last 10 years is 50% greater than the average catch from 1960 to 1987.

Coho Salmon

Aerial surveys of clear-water spawning areas provided index estimates of coho abundance for the upper Nahlin River and the Dudidontu River from 1986 to 1991 and for lower river tributaries of Flannigan Slough and Yehring, Johnson, and Fish creeks annually since 1984 (Shaul 1990; TTC 1999b). The aerial surveys of the Nahlin and Dudidontu spawning populations were discontinued because they proved to be unrepresentative of the main Taku coho run. The portions of the spawning populations enumerated at each site is unknown except for the Nahlin River in 1988 and Yehring Creek from 1986 to 1990 when weirs were also in place. The fraction of the entire Taku coho run using the index areas is also unknown. Adult enumeration weirs were operated on the Hackett River from 1985 to 1988, on the Nahlin River in 1988 and from 1992 to 1994, on Yehring Creek from 1986 to 1990 and annually on Tatsamenie since 1985 (Elliott and Sterritt 1991; TTC 1999b). As with the aerial surveys, the portion of the entire Taku coho run that passed the enumeration weirs is unknown. Some weir counts are partial estimates since the weirs were not operable throughout the entire coho migration. Most weirs were discontinued for economic reasons and the lack of an index stock that adequately represented the entire Taku coho run.

To provide harvest distribution and migratory timing information for marine fisheries, fluorescent pigment marking was applied to juvenile coho salmon in Moose, Johnson, and Yehring creeks in 1972–1974 (Gray et al. 1978). Rearing juveniles and smolts were captured and tagged with CWT at several locations in conjunction with the operation of adult enumeration weirs. Smolts were tagged at the U.S./Canada border in 1986 (Shaul 1987), on the Nahlin River in 1986 to 1988, on the Dudidontu in 1987, at Tatsamenie Lake in 1986 through 1990, at the Shesley River and the lower Taku mainstem in 1988 (Shaul 1990), and at Yehring Creek from 1987 to 1989 (Elliott and Sterritt 1990). These studies, in conjunction with adult enumeration weirs, provided harvest rate, distribution, and migratory timing estimates. In addition, the Yehring Creek program produced estimates of smolt abundance.

Returning adult coho salmon were captured in the lower river and tagged with radio transmitters in 1987, 1988, and 1992 (Eiler 1995). This study provided spawning distribution and inriver migratory timing information in addition to an estimate of the fraction of the Taku coho run that migrates past the U.S./Canada border. An adult tagging program at Canyon Island and associated recoveries from the inriver fishery, operated annually since 1987, provides estimates of the above-border run size (Kelley et al. 1997; TTC 1999b). In years when water levels force early termination of the mark-recapture program, the coho run estimate is expanded by the proportion of CPUE in District 111 that occurs during the remainder of the run. Emigrating coho smolts have been tagged with CWT in the lower river annually since 1991 (McPherson et al. 1997a) and recoveries of tagged adult coho salmon have been used to refine harvest, harvest rates, and distributions for above-border stocks of Taku coho salmon. These programs will allow re-evaluation of the above-border escapement goal range of 27,500 to 35,000 coho salmon.

Taku coho stocks appear to be healthy. Inriver escapements peaked in the early 1990s and have been above the goal since they were first estimated in 1987 (Figure 18.9). The catches of all coho salmon in the Southeast Alaska gillnet fishery have been at record levels in recent years and the 10-year average catch is more than twice the 1960–1987 average. The Canadian inriver coho harvest is not a reliable indicator of coho abundance due to a PSC-imposed catch ceiling in 1985, 1986, and 1988–1992.

Pink and Chum Salmon

Pink salmon are captured in fish wheels at Canyon Island and were tagged in 1984, 1985, 1987, 1989, and 1991 (Kelley et al. 1997; TTC 1999b). Data from this program provides escapement estimates for 5 years and indices of abundance for years when pink salmon are not tagged.

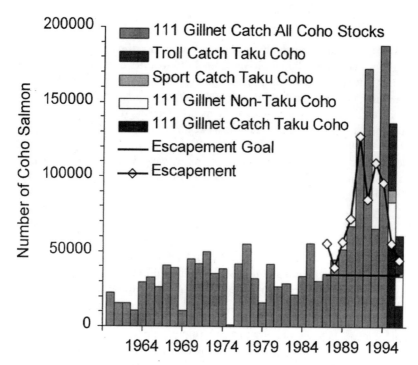

FIGURE 18.9   Taku coho catches and escapements. U.S. marine catches of Taku coho stocks are only available for 1995 and 1996.

There is currently no program to estimate the abundance of pink salmon in the Taku River. The average harvest in the District 111 fishery for the last 10 years is 50% higher than the average from 1960 to 1987 (TTC 1999b). Hatchery production contributed an unknown component of recent catches. The inriver fish wheel captures were well above average for 1994 and 1996 but were at record lows in 1993 and 1995.

The capture and tagging project at Canyon Island has provided escapement indices for chum salmon since 1984 (Kelley et al. 1997; TTC 1999b). Numbers of fish tagged and recovered have been insufficient to make population estimates. Aerial surveys are flown annually to assess the progress of migration onto the spawning grounds. However, the proportion of the run observed during the surveys is unknown.

Taku chum salmon are declining in abundance (Figure 18.10). Inriver fish wheel catches indicate that chum abundance has been far lower in the last 5 years than in the 1980s. Taku chum salmon are fall-run fish, as are chum salmon from the nearby Whiting River. There is no directed fishery on these stocks, but they are harvested incidentally during the fall coho fishery. U.S. catches of fall run chum salmon in District 111 have been similar to harvests in the 1960s but far less than observed in the 1970s and early 1980s. Other local wild and hatchery chum salmon are summer-run fish. The bulk of the District 111 chum catch occurs during the summer migration. The proportion of the District 111 fall chum catch composed of Taku fish vs. Whiting fish is unknown. Catches of fall chum in the marine and inriver fisheries and the fish wheels have continued to decline despite limited fishery openings during the peak of migration.

ALSEK RIVER PROGRAMS AND STOCK STATUS

CHINOOK SALMON

Aerial surveys of Alsek chinook salmon in the Klukshu, Blanchard, and Takhanne rivers and Goat Creek have been conducted since 1962 (Pahlke 1996). Aerial surveys of some major spawning

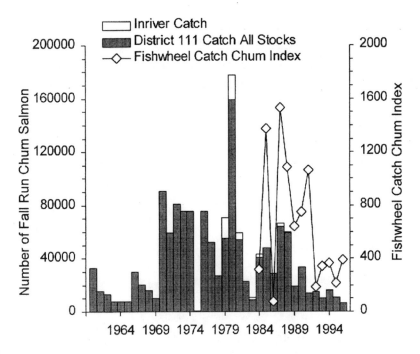

FIGURE 18.10   Taku chum catches and escapement indices.

grounds provide index escapement estimates, but the counted fraction of spawners is unknown, as is the portion of the entire Alsek salmon run that migrates to the index areas. An adult enumeration weir on the Klukshu River near the settlement of Dalton Post has been fished annually since 1976 (TTC 1999b) and provides escapement counts of chinook and sockeye salmon. The fraction of the Alsek chinook run which migrates up the Klukshu River is unknown. The weir count is used as an index for the entire chinook run.

There is no clear long-term trend in the Alsek chinook abundance indices other than an apparent increase in the last 4 years (Figure 18.11). Lower river and surf set net catches in recent years are not comparable to catches prior to 1983 when the U.S. directed a fishery on Alsek chinook stocks. An escapement goal for the Klukshu stock of 4,700 fish recommended by the TTC (1991) was met for the first time in 1995. This goal is currently being evaluated and will likely be lowered based on spawner-recruit analysis.

Sockeye Salmon

Aerial surveys of sockeye salmon on four lower and two upper Alsek spawning areas have produced escapement indices sporadically since 1985 (TTC 1999b). Sockeye salmon returning to Klukshu Lake have been enumerated annually at the weir on the Klukshu River since 1976 (TTC 1999b). An escapement goal range of 20,000 to 30,000 sockeye salmon for the Klukshu stock was established by the TTC (TTC 1988b). The Klukshu weir count is used as an index for the entire Alsek sockeye run. A tagging program for estimating the fraction of the Alsek sockeye run that migrated past the Klukshu weir in 1983 did not produce conclusive results (McBride and Bernard 1984). An electronic counter located on Village Creek has produced escapement estimates for that system since 1986 (TTC 1999b). The fraction of Alsek sockeye that migrate up the Klukshu River, into Village Creek, or to the aerial survey index sites is unknown.

The Alsek sockeye run appears to be stable (Figure 18.12). The average escapement of the Klukshu stock for the last 10 years is the same as for the 1976 to 1986 period. There is no noticeable trend in the lower river sockeye catches nor the upper river aboriginal and sport fishery catches.

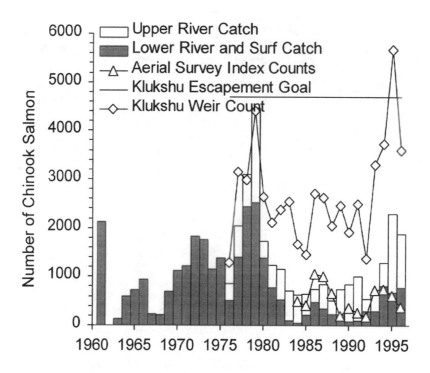

FIGURE 18.11   Alsek chinook catches and escapement indices.

Spawner-recruit data is being analyzed to determine the optimum escapement for Klukshu sockeye salmon.

Coho Salmon

Aerial surveys of lower river coho spawning areas have been conducted annually since 1985 (TTC 1999b). The enumerated proportions of the spawning populations are unknown, as is the fraction of the total Alsek coho run that uses the index areas. The weir at Klukshu provides an index of coho escapement to the Klukshu River but is usually not in operation through the entire coho run. The fraction of the run counted is unknown, as is the proportion of the Alsek coho run that uses this spawning area.

Little is known about the size of the Alsek coho run. There is no long-term trend in the counts through the weir at Klukshu (Figure 18.13). There is little correlation between lower river catches, Klukshu weir counts, and aerial survey counts. Catch and escapement data for other coho stocks in the Yakutat area indicate that coho salmon throughout the region are at record high abundance, although there is insufficient data to determine whether Alsek stocks follow this trend.

DISCUSSION

Few long-term data series have been collected on transboundary river salmon populations. Reliable catch statistics for U.S. gillnet fisheries near the river mouths are available only since 1960 and these might not be accurate indicators of run size for the river of interest because the catches are from a mixture of stocks. Most of the research programs to determine inriver abundance estimates or abundance indices were initiated in the early 1980s or later. Therefore, assessment of the status of transboundary river salmon stocks has only been possible in the last few decades. Catch data for U.S. fisheries from the late 1800s to 1927 are available (Rich and Ball 1933), but fishery areas

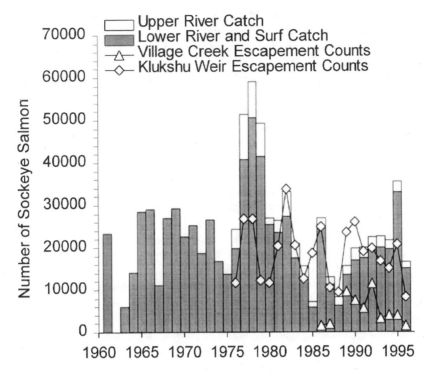

FIGURE 18.12   Alsek sockeye catches and escapement indices.

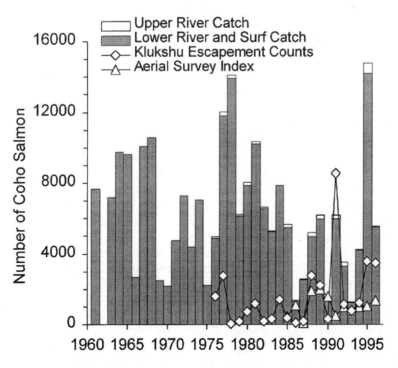

FIGURE 18.13   Alsek coho catches and escapement indices.

and gear are different enough from current conditions that catches are not directly comparable. In addition, much of the early historical information was derived from processors who bought fish from different areas and moved operations between areas. Area- and gear-specific catches for years between 1928 and 1960 are generally unavailable or poorly documented.

Biologically derived escapement goals for transboundary river salmon stocks are being refined as the database is expanded. Cooperative inseason run strength assessment allows U.S. and Canadian fishery managers to maximize harvest while assuring adequate escapement to the spawning grounds. Abundance-based management, with catches allowed to increase when run strength increases, is a goal for fisheries targeting transboundary river salmon stocks. Lack of data to sufficiently evaluate inseason run strength limits the ability of managers to fully utilize the resource and management must generally be conservative to ensure sustainable salmon runs and fisheries.

The Stikine chinook escapement estimation program will provide a basis for evaluating the Little Tahltan weir counts and aerial survey counts as indices of above-border run strength. However, a program is needed to estimate marine harvest of Stikine chinook stocks in the gillnet and troll fisheries. Earlier CWT studies indicated that the troll fishery harvested few chinook salmon of Stikine origin, but tag recoveries were insufficient to estimate actual harvests. Estimation of marine harvests of Stikine chinook salmon would be extremely valuable for fisheries management, particularly if fishery restrictions enacted to rebuild these stocks were relaxed. For Stikine sockeye salmon, the current programs for inseason run strength assessment and postseason run reconstruction allow intensive fisheries directed at these stocks. Programs to assess marine harvest and inriver escapement of Stikine coho salmon are not available because of difficulty in locating representative indicator stocks.

The current Taku research programs provide adequate inseason run strength assessment and escapement estimates for above-border runs of chinook, sockeye, and coho salmon. The aerial survey index counts for chinook salmon track with the 4 years of mark-recapture population estimates, but these counts appear to represent a smaller proportion of the total run than previously assumed. The smolt tagging program for emigrant chinook and coho salmon is beginning to provide harvest estimates for these stocks in marine fisheries. Catches and escapements are estimated for the Kuthai, Trapper, Tatsamenie, and mainstem sockeye stock groups. Of these, recent year escapements have been less than desired only for the Tatsamenie stock and managers have attempted to direct fishing effort on the more abundant stocks.

Information on Taku chum salmon is more difficult to obtain. The known chum spawning areas on the Taku River are subject to dramatic annual physical alteration due to natural processes. Substantial bed loading and shifting and channel changing occur during ice-out and floods. The water was clear in the principal known spawning area, Chum Salmon Slough, during aerial surveys in the 1970s and 1980s but has now become glacially occluded. Due to these observations, it is speculated that the decline in the chum run may be partially due to natural physical changes in the river. The low abundance of Taku chum salmon is particularly troubling in light of the apparent good survival conditions for other salmon stocks.

The Klukshu weir provides the best escapement index for Alsek chinook, sockeye, and coho stocks. Although it would be desirable to estimate the fraction of the Alsek runs utilizing this spawning area, and to assess the representativeness of these stocks, it is impossible with current funding limitations. It is not believed that these stocks are greatly impacted in fisheries other than the U.S. commercial set net fishery in the lower river and surf area and the Canadian upper river aboriginal and sport fisheries. Therefore, these stocks have not been given a high priority for research within the PSC process. Refinement of escapement goals for Klukshu chinook and sockeye salmon may allow more directed fishing on these species.

The transboundary river research programs, although of fairly recent origin, provide an array of information critical to successful management and sustainability of the shared salmon resources of these rivers. Many programs have been initiated and implemented jointly by the U.S. and Canada.

Cooperative research has allowed the development of otherwise unaffordable projects. Continual evaluation of research results has enabled fishery managers to prioritize research needs. There are data needs that have yet to be met, particularly for Stikine coho salmon, Taku chum salmon, and for Alsek chinook, sockeye, and coho salmon. As ongoing programs provide longer data sets and better understanding of the resources, it may be possible to address some of the unmet data needs with existing programs or to redirect resources if priorities change.

ACKNOWLEDGMENTS

Numerous people from many agencies were responsible for planning, developing, and implementing the research programs referred to in this report. I am indebted to the biologists and staff from ADF&G, NMFS, DFO, and First Nations that provided information used in this chapter. I wish to thank all authors who have documented their various projects, members of the Transboundary Technical Committee, and especially the people who participated in the actual field programs. I am grateful for editorial and technical reviews by Sandy Johnston, Norma Jean Sands, Leon Shaul, Ben Van Alen, Alex Wertheimer, Bob Wilbur, and two anonymous reviewers. This chapter is contribution PP-153 of the Alaska Department of Fish and Game, Commercial Fisheries Management and Development Division, Juneau. This work was supported in part by NOAA awards NA46FA0344 and NA57FP0425.

REFERENCES

Baker, T.T., and eight coauthors. 1996. Status of Pacific salmon and steelhead escapements in southeastern Alaska. Fisheries 21(10):6–18.

Brownlee, K.M. 1991. Prey consumption by juvenile salmonids on the Taku River, Southeast Alaska. Master's thesis. University of Alaska, Fairbanks.

Craig, P.C. 1985. Identification of sockeye salmon *Oncorhynchus nerka* stocks in the Stikine River based on egg size measurements. Canadian Journal of Fisheries and Aquatic Sciences 42:1696–1701.

Edgington, J., and B. Lynch. 1984. Revised anadromous stream catalog of Southeastern Alaska Volume I: District 108 the Stikine River and its tributaries. Alaska Department of Fish and Game, Division of Commercial Fisheries, Technical Data Report 125, Juneau.

Eiler, J.H. 1995. A remote satellite-linked tracking system for studying Pacific salmon with radio telemetry. Transactions of the American Fisheries Society 124:1–10.

Eiler, J.H., B.D. Nelson, and R.F. Bradshaw. 1991. Radio tracking chinook salmon *Oncorhynchus twhawytscha* in a large turbid river. Pages 202–206 *in* A. Uchiyama and C.J. Amlaner, Jr., editors. Proceedings of the Eleventh International Symposium of Biotelemetry. Waseda University Press, Tokyo.

Eiler, J.H., B.D. Nelson, and R.F. Bradshaw. 1992. Riverine spawning by sockeye salmon in the Taku River in Alaska and British Columbia. Transactions of the American Fisheries Society 121:701–708.

Elliott, S.T., and D. A. Sterritt. 1990. A study of coho salmon in Southeast Alaska, 1989: Chilkoot Lake, Yehring Creek, Auke Lake, Vallenar Creek. Alaska Department of Fish and Game, Fishery Data Series 90–53, Anchorage.

Elliott, S.T., and D. A. Sterritt. 1991. Coho salmon studies in Southeast Alaska, 1990: Auke Lake, Chilkoot Lake, Nahlin River, and Yehring Creek. Alaska Department of Fish and Game, Fishery Data Series 91–43, Anchorage.

Gray, P.L., K.R. Florey, J.F. Koerner, and R.A. Marriott. 1978. Coho salmon *Oncorhynchus kisutch* fluorescent pigment mark-recovery program for the Taku, Berners, and Chilkat Rivers in Southeastern Alaska (1972–1974). Alaska Department of Fish and Game, Division of Commercial Fisheries, Information Leaflet 176, Juneau.

Guthrie, C.M., III, J.H. Helle, P. Aebersold, G.A. Winans, and A.J. Gharrett. 1994. Preliminary report on the genetic diversity of sockeye salmon populations from Southeast Alaska and Northern British Columbia. U.S. Department of Commerce, National Marine Fisheries Service Processed Report AFSC 94-03, Juneau, Alaska.

Hancock, M.J., and D.E. Marshall. 1984. Catalogue of salmon streams and spawning escapements of sub-districts 120 and 130 (Alsek-Stikine-Taku watersheds). Canadian Data Report of Fisheries and Ocean Sciences 456.

Hoffman, S., L. Talley, and M. Seibel. 1984. U.S./Canada cooperative pink and sockeye salmon tagging, interception rates, migration patterns, run timing, and stock intermingling in southern Southeastern Alaska and northern British Columbia, 1982. Alaska Department of Fish and Game, Division of Commercial Fisheries, Technical Data Report 110, Juneau.

Hubartt, D.J., and P.D. Kissner. 1987. A study of chinook salmon in Southeast Alaska. Alaska Department of Fish and Game, Division of Sport Fish, Annual Report 1986–1987, Project F-10-2, Vol 28 (AFS-41-13), Juneau.

Humphreys, R.D., and six coeditors. 1994. Pacific Stock Assessment Review Committee (PSARC) annual report for 1993. Canadian Manuscript Report of Fisheries and Aquatic Science 2227:iv.

Jensen, K.A. 1992. Forecasts for the 1992 Stikine River sockeye salmon run. Alaska Department of Fish and Game, Division of Commercial Fisheries, Regional Information Report 1J92-14, Douglas.

Jensen, K.A. 1994. Differences between inseason and postseason stock composition estimates for sockeye salmon in gillnet catches in 2 districts in Southeast Alaska and in the Stikine River, 1986–1989. Alaska Fishery Research Bulletin 1:107–124.

Jensen, K.A., and R. Bloomquist. 1994. Stock compositions of sockeye salmon catches in Southeast Alaska District 111 and the Taku River, 1990, estimated with scale pattern analysis. Alaska Department of Fish and Game, Division of Commercial Fisheries, Regional Information Report 1J94-23, Douglas.

Jensen, K.A., and I.S. Frank. 1993. Stock compositions of sockeye salmon catches in Southeast Alaskan Districts 106 and 108 and in the Stikine River, 1989, estimated with scale pattern analysis. Alaska Department of Fish and Game, Division of Commercial Fisheries, Technical Fisheries Report 93-13, Juneau.

Jensen, K.A., and I.S. Frank. 1995. Stock compositions of sockeye salmon catches in Southeast Alaska Districts 106 and 108 gillnet fisheries, 1990, estimated with scale pattern analysis. Alaska Department of Fish and Game, Division of Commercial Fisheries, Technical Fisheries Report 1J95-13, Douglas.

Jensen, K.A., E.L. Jones, and A.J. McGregor. 1993. Stock compositions of sockeye salmon catches in Southeast Alaska's District 111 and the Taku River, 1989, estimated with scale pattern analysis. Alaska Department of Fish and Game, Division of Commercial Fisheries, Technical Fishery Report 93-15, Juneau.

Kelley, M.S., P.A. Milligan, and A.J McGregor, 1997. Mark-recapture studies of Taku River adult salmon stocks in 1996. Alaska Department of Fish and Game, Division of Commercial Fisheries, Regional Information Report 1J97-22, Douglas.

Kissner, P.D. 1984. A study of chinook salmon in southeast Alaska. Alaska Department of Fish and Game Project AFS 41-10, Vol. 25, Juneau.

Lorenz, J.M., and J.H. Eiler. 1989. Spawning habitat and redd characteristics of sockeye salmon in the glacial Taku River, British Columbia and Alaska. Transactions of the American Fisheries Society 118:495–502.

Lorenz, J.M., M.L. Murphy, J.F. Thedinga, and K.V. Koski. 1991. Distribution and abundance of juvenile salmon in two main channel habitats of the Taku River, Alaska and British Columbia. U.S. Department of Commerce. NOAA Technical Memorandum NMFS F/NWC-206, Juneau.

Lynch, B., and J. Edgington. 1986. Stikine River studies: adult salmon tagging, population investigations, and side scan sonar operations, 1983. Alaska Department of Fish and Game, Division of Commercial Fisheries, Technical Data Report 171, Juneau.

Lynch, B.L., N.C. Rattner, W.R. Bergmann, and R.L. Timothy. 1987. Kakwan Point test fishery and Stikine River Canadian gill net port sampling, 1985. Alaska Department of Fish and Game, Division of Commercial Fisheries, Technical Data Report 211, Juneau.

Marshall, S., N.J. Sands, A. McGregor, and K.A. Jensen. 1989. Data and programs for the transboundary Stikine, Taku, and Alsek Rivers needed to implement the Pacific Salmon Treaty. Alaska Department of Fish and Game, Division of Commercial Fisheries, Regional Informational Report 1J89-09, Douglas.

McBride, D.N., and D.R. Bernard. 1984. Estimation of the 1983 sockeye salmon (*Oncorhynchus nerka*) return to the Alsek River through analysis of tagging data. Alaska Department of Fish and Game, Division of Commercial Fisheries, Technical Data Report 115, Juneau.

McCart, P., and G. Walder. 1982. Fish populations associated with proposed hydroelectric dams on the Stikine and Iskut Rivers. Volume 1. Baseline studies. Report by Aquatic Environments Ltd. for B.C. Hydro and Power Authority, Vancouver, British Columbia.

McGregor, A.J. 1986. Origins of sockeye salmon (*Oncorhynchus nerka* Walbaum) in the Taku-Snettisham drift gillnet fishery of 1984 based on scale pattern analysis. Alaska Department of Fish and Game, Division of Commercial Fisheries, Technical Data Report 174, Juneau.

McPherson, S.A., D.R. Bernard, and M.S. Kelley. 1997a. Production of coho salmon from the Taku River, 1995–1996. Alaska Department of Fish and Game, Division of Sport Fish. Fishery Data Series 97-24, Anchorage.

McPherson, S.A., D.R. Bernard, M.S. Kelley, P.A. Milligan, and P. Timpany. 1997b. Spawning abundance of chinook salmon in the Taku River in 1996. Alaska Department of Fish and Game, Division of Sport Fish, Fishery Data Series 97-14, Anchorage.

Meechan, W.R., and D.B. Sinnif. 1962. A study of downstream migrations of anadromous fishes in the Taku River, Alaska. Transactions of the American Fisheries Society 91:399–407.

Mesiar, D. 1984. Vertical and horizontal in-river fish distribution, Stikine River, 1983. Alaska Department of Fish and Game, Division of Commercial Fisheries, Technical Data Report 118, Juneau.

Moles, A., P. Rounds, and C. Kondzela. 1990. Use of the brain parasite *Myxobolus neurobius* in separating mixed stocks of sockeye salmon. Pages 224–231 *in* R.C. Parker, and five coeditors. Fish Marking Techniques. American Fisheries Society, Symposium 7, Bethesda, Maryland.

Murphy, M.L., J. Heifetz, J.F. Thedinga, S.W. Johnson, and K V. Koski. 1989. Habitat utilization by juvenile Pacific salmon *Oncorhynchus* in the glacial Taku River, Southeast Alaska. Canadian Journal of Fisheries and Aquatic Sciences 46:1677–1685.

Murphy, M.L., K V. Koski, J.M. Lorenz, and J.F. Thedinga. 1997. Downstream migrations of juvenile Pacific (*Oncorhynchus* spp.) in a glacial transboundary river. Canadian Journal of Fisheries and Aquatic Sciences 54:2837–2846.

Murphy, M.L., J.M. Lorenz, and K.V. Koski. 1991. Population estimates of juvenile salmon downstream migrants in the Taku River, Alaska. U.S. Department of Commerce. NOAA Technical Memorandum NMFS F/NWC-203, Juneau.

NRC (National Research Council). 1996. Upstream: salmon and society in the Pacific Northwest. National Academy Press, Washington.

Nehlsen, W., J.E. Williams, and J.A. Lichatowich. 1991. Pacific salmon at the crossroads: stocks at risk from California, Oregon, Idaho, and Washington. Fisheries 16(2):4–21.

Pahlke, K.A. 1995. Escapements of chinook salmon in Southeast Alaska and transboundary rivers in 1994. Alaska Department of Fish and Game, Division of Sport Fish, Fishery Data Series 95-35, Anchorage.

Pahlke, K.A. 1996. Escapements of chinook salmon in Southeast Alaska and transboundary rivers in 1995. Alaska Department of Fish and Game, Division of Sport Fish, Fishery Data Series 96-35, Anchorage.

Pahlke, K.A., and D.R. Bernard. 1996. Abundance of the chinook salmon escapement in the Taku River, 1989 to 1990. Alaska Fishery Research Bulletin 3:9–20.

Pahlke, K.A. and P. Etherton. 1997. Chinook salmon research on the Stikine River, 1996. Alaska Department of Fish and Game, Fishery Data Series 97-37, Anchorage.

Rich, W.H., and E.M. Ball. 1933. Statistical review of the Alaska salmon fisheries, part IV: southeastern Alaska. U.S. Department of Commerce, Bureau of Fisheries, Volume XLVII, Bulletin 13.

Shaul, L.D. 1987. Taku and Stikine River coho salmon *Oncorhynchus kisutch* adult escapement and juvenile tagging investigations, 1986. Alaska Department of Fish and Game, Division of Commercial Fisheries, completion Report of National Marine Fisheries Service, Cooperative Agreement NA-85-ABH-00050, Juneau.

Shaul, L.D. 1990. Taku River coho salmon investigations, 1989. Alaska Department of Fish and Game, Division of Commercial Fisheries, Regional Information Report 1J90-19, Juneau.

Shaul, L.D., and K.F. Crabtree. 1998. Harvests, escapements, migratory patterns, smolt migrations, and survival of coho salmon in Southeast Alaska based on coded-wire-tagging, 1994–1996. Alaska Department of Fish and Game, Division of Commercial Fisheries Management and Development, Regional Information Report 1J98-02, Douglas.

Shaul, L.D., P. L. Gray, and J. F. Koerner. 1984. Migratory patterns and timing of Stikine River coho salmon, *Oncorhynchus kisutch* based on coded-wire tagging studies, 1978–1982. Alaska Department of Fish and Game, Informational Leaflet Number 232, Juneau.

Slaney, T.L., K.D. Hyatt, T.G. Northcote, and R.J. Fielden. 1996. Status of anadromous salmon and trout in British Columbia and Yukon. Fisheries 21(10):20–35.

Thedinga, J.F., K.V. Koski, M.L. Murphy, J. Heifetz, S.W. Johnson, and C.R. Hawks. 1988. Abundance and distribution of juvenile coho salmon *Oncorhynchus kisutch* in the lower Taku River, Alaska. *In* G. Gunstrom, editor. Southeast Alaska regional coho salmon program review. Alaska Department of Fish and Game, Division of Commercial Fisheries, Regional Information Report 1J88-1, Juneau.

TTC (Transboundary Technical Committee). 1987. Report of the Canada/U.S. Transboundary Technical Committee to the Pacific Salmon Commission, February 8, 1987. Vancouver, British Columbia.

TTC (Transboundary Technical Committee). 1988a. Sockeye salmon enhancement feasibility studies in the transboundary rivers. Pacific Salmon Commission, TCTR (88)-1, Vancouver, British Columbia.

TTC (Transboundary Technical Committee). 1988b. Salmon management plan for the transboundary rivers 1988. Pacific Salmon Commission, TCTR (88)-2, Vancouver, British Columbia.

TTC (Transboundary Technical Committee). 1990. Long-term research plans for the transboundary rivers. Pacific Salmon Commission, TCTR (90)-3, Vancouver, British Columbia.

TTC (Transboundary Technical Committee). 1991. Escapement goals for chinook salmon in the Alsek, Taku, and Stikine Rivers. Pacific Salmon Commission, TCTR (91)-4, Vancouver, British Columbia.

TTC (Transboundary Technical Committee). 1993. Salmon management plan for the Stikine, Taku, and Alsek Rivers, 1993. Pacific Salmon Commission, TCTR (93)-2, Vancouver, British Columbia.

TTC (Transboundary Technical Committee). 1999a. Salmon management plan for the Stikine, Taku, and Alsek Rivers, 1999. Pacific Salmon Commission, TCTR (99)-2, Vancouver, British Columbia.

TTC (Transboundary Technical Committee). 1999b. Pacific Salmon Commission Joint Transboundary Technical Committee estimates of transboundary river salmon production, harvest, and escapement, 1996. Pacific Salmon Commission, TCTR (99)-3, Vancouver, British Columbia.

Van Alen, B.W. 2000. Status and stewardship of salmon stocks in Southeast Alaska. Pages 161–193 *in* E.E. Knudsen, C.R. Steward, D.D. MacDonald, J.E. Williams, and D.W. Reiser, editors. Sustainable fisheries management: Pacific salmon. Lewis Publishers, Boca Raton, Florida.

Wertheimer, A. 1996. The status of Alaska Salmon. Pages 179–198 *in* D.J. Stouder, P.A. Bisson, and R.J. Naiman, editors. Pacific salmon and their ecosystems: status and future options. Chapman & Hall, New York.

Wood, C.C., G.T. Oliver, and D.T. Rutherford. 1988. Comparison of several biological markers used for stock identification of sockeye salmon *Oncorhynchus nerka* in Northern British Columbia and Southeast Alaska. Canadian Technical Report of Fisheries and Aquatic Sciences 1624.

Wood, C.C., D.T. Rutherford, and S. McKinnell. 1989. Identification of sockeye salmon *Oncorhynchus nerka* stocks in mixed-stock fisheries in British Columbia and Southeast Alaska using biological markers. Canadian Journal of Fisheries and Aquatic Sciences 46:2108–2120.

Section IV

Habitat Assessment

19 The Effects of Forest Harvesting, Fishing, Climate Variation, and Ocean Conditions on Salmonid Populations of Carnation Creek, Vancouver Island, British Columbia

Peter J. Tschaplinski

Abstract.—The Carnation Creek Fisheries-Forestry Interaction Project, initiated in 1970, is the longest, continuous study of the effects of forestry practices on biological and physical watershed processes in North America. This case study was initially designed to investigate the effects of different streamside forest-harvest treatments on stream channels, aquatic habitats, and fish. The salmonid populations of Carnation Creek have been monitored through 5 pre-logging, 6 during-logging, and 14 post-logging years as one component of this multidisciplinary study. Forest harvesting has had complex and often variable effects upon Carnation Creek fish species and life stages. Chum salmon *Oncorhynchus keta* have shown the sharpest decline. After logging, numbers of adults returning to the stream fell to about one third of the pre-logging average. This decline is due partly to reductions in egg-to-fry survival resulting from decreased quality of spawning and egg-incubation habitats in the lowermost stream reach. Reductions in summer rearing habitat appear to explain the roughly 50% post-logging decline in abundance of coho salmon *O. kisutch* fry inhabiting the stream. However, the fewer coho fry have produced >1.5-times more smolts after logging due to improved overwinter survival, which is in turn correlated with increased winter water temperatures and summer growth. Increased smolt abundance has not caused more adults to return. Coho returning to the system have declined after logging by 31%, due at least partly to both depressed marine survivals resulting from earlier timing of spring smolt migrations and ocean climate shifts. The production of salmonids from coastal streams clearly depends upon processes occurring both within watersheds and the marine environment. We cannot control natural shifts in marine ecosystems and climate. Therefore, to sustain our salmonid resources, we must always apply our best forest-harvest practices to ensure that adverse effects of natural variations are not compounded with those of inappropriate land use.

INTRODUCTION

The effects of forest harvesting on fish populations have been studied for over 25 years at Carnation Creek on the west coast of Vancouver Island, British Columbia. This single-watershed, intensive case study has generated the longest series of continuous data on fisheries-forestry interactions anywhere. The Carnation Creek Experimental Watershed Project was initiated in 1970 by the federal agency now known as Fisheries and Oceans Canada (DFO), and soon expanded into a multi-agency, multidisciplinary program on the effects of forest harvesting on a coastal watershed and its salmon and trout populations. In the 1960s, resource managers and planners used studies conducted elsewhere in North America, for example, in Oregon, Alaska, and as far away as New Hampshire, to make judgments on the effects of logging on B.C. fish populations. Both the forest industry and government resource agencies expressed concern that these extrapolations might not lead to the most appropriate planning decisions for areas on the west coast of the province. Therefore, the Carnation Creek study was initiated in order to provide fisheries-forestry information on at least one type of drainage basin in coastal B.C.

The objectives of the Carnation Creek study were to (a) provide an understanding of the physical and biological processes operating within a coastal watershed; (b) determine how the forest harvesting practices employed in the 1970s and early 1980s changed these processes; and (c) apply the results of the study to make informed decisions concerning land-use management, fish populations, and aquatic habitat protection. The project has achieved these goals despite limitations resulting from these intensive studies being conducted only in a single watershed. Over 180 publications have been produced from Carnation Creek research. The results from this project have made major contributions to the B.C. Coastal Fisheries-Forestry Guidelines (CFFG) implemented in 1987, and the legally binding provisions for aquatic habitat protection within the B.C. Forest Practices Code, which replaced the CFFG in 1995.

Fish populations have been studied at Carnation Creek virtually continuously since 1970. The objectives of this review are to illustrate (1) the changes in abundance, growth, and survival of coho *Oncorhynchus kisutch* and chum salmon *O. keta* in Carnation Creek between 1970 and 1995 through 5 pre-logging, 6 during-logging, and 14 post-logging years; (2) that the effects of forest harvesting are complex, and vary among species and among life stages within the same species; and (3) that salmonid production depends on biological and physical processes occurring not only within watersheds but also in marine environments (e.g., climate-associated changes, predation, and fishing). Long-term trends in the abundance of steelhead *O. mykiss* and cutthroat trout *O. clarki* are discussed briefly. I also demonstrate the value of long-term, multidisciplinary studies for clarifying complex interactions among land-use practices and natural processes occurring within watersheds which together determine salmonid abundance and growth in coastal streams.

METHODS

The design of the Carnation Creek Experimental Watershed Project and the methods employed for monitoring physical variables and biological processes before, during, and after forest harvesting are thoroughly described by Hartman and Scrivener (1990). They also summarized all aspects of the project and provided a comprehensive bibliography of the publications generated from Carnation Creek research current to 1990. The following summary is condensed from their detailed descriptions of the project.

Carnation Creek is located ~20 km northeast of Bamfield on the south shore of Barkley Sound in southwestern Vancouver Island (Figure 19.1; 49°N, 125°W). The watershed occurs within the Coastal Western Hemlock Biogeoclimatic Zone, which spans the west coast of North America from the Olympic Peninsula in Washington State to the Queen Charlotte Islands and southeast Alaska (Krajina 1969). The stream drains an area of 11 km² and contains rugged terrain between 0 and 800 m elevation. The valley walls have gradients up to 80%. The main stream is ~7.8 km long, but only the lowermost 3.1 km extending from the stream mouth to the base of a steep-gradient canyon is inhabited by anadromous salmonids including coho, chum, steelhead, and cutthroat. This lowest stream reach contains a valley bottom of about 55 ha that is 50 to 200 m wide. The coarse, well-drained soils, forest cover, hydrology, and heavy annual precipitation (varying from 210 to over 500 cm/yr) are typical of western Vancouver Island and many other areas of coastal B.C.

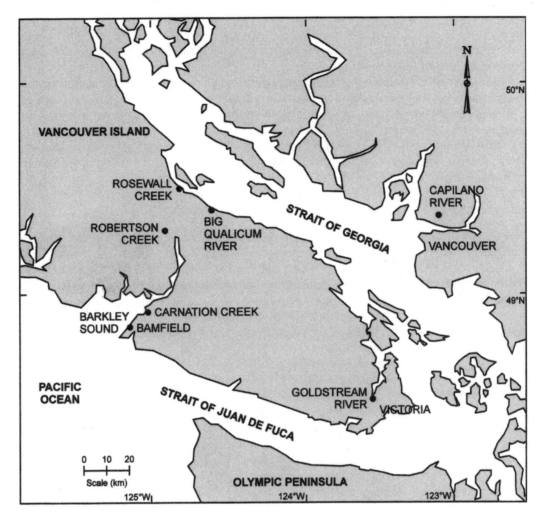

FIGURE 19.1 Location of the Carnation Creek study site in southwestern Vancouver Island.

About 95% of the annual precipitation falls as rain, primarily during autumn and winter. High variations in seasonal rainfall cause stream discharge to range from 0.03 m^3/s in summer to 64 m^3/s in winter. Stream flows may increase by 200-fold within 48 h because of rapid runoff from rainstorms that can produce up to 26 cm of precipitation within the same time.

The Carnation Creek study was designed initially to examine the effects of three different types of streamside forest-harvest treatments on stream channels and fish populations. These treatments were applied along the lowermost 3 km of the stream (Hartman and Scrivener 1990). A "leave-strip treatment" was applied from the estuary to 1,300 m upstream. This treatment was designed to buffer the effects of clearcut logging from the stream channel by leaving a riparian strip of trees that varied from 1 to 70 m wide. An "intensive treatment" was applied along 900 m of stream channel immediately upstream from the leave-strip treatment. In the intensive treatment, clearcut harvesting occurred simultaneously along both sides of the stream up to the channel margin. No riparian trees were left standing. Any activity within the channel that was considered operationally convenient, such as felling and yarding trees across the stream, was permitted. Economically valuable, wind-thrown trees lying within the stream channel were removed. Logging-associated debris was burned after harvesting was completed. The third treatment, called "careful clearcutting," was applied over the 900-m length of stream immediately upstream from the intensively treated area. No activity within the stream was permitted in this treatment with the exception that six trees leaning over the stream channel were felled across it and removed. Perennial vegetation on the streambanks, such as salmonberry *Rubus spectabilis,* was left alone; however, red alder trees *Alnus rubra* were removed from the streamside.

The responses of a comprehensive set of biological and physical variables within the Carnation Creek basin were determined relative to forest harvesting over (a) 5–6 pre-logging years spanning 1970–1975 (beginning in 1970 or 1971, depending upon the variable measured); (b) 6 years spanning 1976–1981 during which 41% of the watershed was harvested (including almost all of the valley bottom); and (c) 14 post-logging years, 1982–1995. From 1987 to 1995, an additional 21% of the basin was harvested. This later harvesting occurred in headwater areas remote from the main stream channel.

HISTORICAL DATA COLLECTIONS

Data collected historically have included comprehensive information on climate; stream temperatures and discharge; groundwater levels (by using piezometers); water chemistry; stream channel morphology and large woody debris abundance and distribution; streambed particle-size composition (by using frozen-core methods); suspended sediment transport during high flows (from automated sampling at one hydrological weir); streambed scour and deposition; ground disturbance, landslides, and post-logging revegetation; biomass of aquatic algae (periphyton); abundance and distribution of benthic macroinvertebrates; and fish populations. Details of all historic methods are given by Hartman and Scrivener (1990).

Fish population studies have included (1) abundance and distribution of adult salmonid spawners returning to the stream (autumn and winter); (2) numbers of juvenile fish, i.e., salmonid smolts and young-of-the-year (fry) migrating seaward in spring; (3) abundance, distribution, age structure, growth, and survival of juvenile salmonids rearing in freshwater and estuarine habitats during summer and early autumn; (4) seasonal movements of juvenile salmonids out of the main stream into "off-channel" overwinter habitats, and return movements in spring; (5) main-channel and "off-channel" abundance, distribution, and survival of juvenile salmonids in winter; (6) chum egg incubation, egg survival, and fry emergence (redds capped with trap nets); and (7) fecundity determinations for female chum and coho for estimates of annual egg-to-fry survival (Andersen 1983; Andersen and Scrivener 1992; Brown 1987; Brown and Hartman 1988; Brown and McMahon 1988; Bustard 1991; Bustard and Narver 1975; Tschaplinski 1982a, 1982b, 1988; Tschaplinski and Hartman 1983).

CURRENT DATA COLLECTIONS

Many variables and processes continue to be studied. Since 1990, data have been collected on fish populations and habitat, stream channel morphology, streambed movements (plus erosion and sedimentation), climate, hydrology, forest regeneration and growth, and hillslope processes. Water temperature, depth, and discharge are monitored at permanent hydrological weirs installed on the mainstream and on several principal tributaries. Climate stations are located in several sites at different elevations in the watershed. Some stations are co-located with the hydrological weirs. Climate variables monitored include air temperature, solar radiation, precipitation, relative humidity, and wind speed and direction. Climate and hydrology stations have been updated by the installation of continuous-operation, electronic data recorders (Tschaplinski et al. 1998).

Changes in channel morphology are determined annually in nine survey reaches of the stream (which incorporate the same sections used to determine seasonal fish population abundance and distribution). Standard survey and mapping techniques are employed (see Scrivener and Hartman 1990). Within each survey section (1) all pieces of large woody debris ("LWD," which includes tree trunks, root masses, and large limbs) are mapped and tagged to observe changes in distribution and abundance, and (2) textural distributions of surface sediments were described visually by using grid samplers (Hogan et al.1998). Each year, these ground-based surveys are supplemented with aerial photographic surveys of the entire creek channel. Stereo photographs (70-mm aperture) are used to determine changes in channel structure in areas between study reaches and to generate inventories of fish habitats throughout the stream. Aerial photographs are also employed to monitor the rates of canopy closure over the creek as the new forest grows. Canopy closure and forest growth will be studied relative to future water temperature changes in Carnation Creek.

Channel scour and deposition are studied in the same survey reaches by using scour-and-fill monitors (Haschenburger 1996). Estimates of sediment (bedload) transport are determined by following the annual movements (distances and depths) of painted, magnetic rocks placed onto the streambed. The rock samples represent the range in sizes and textural proportions within each study section (Haschenburger 1996).

Adult salmonids (coho, chum, steelhead, and cutthroat) returning to spawn in Carnation Creek are enumerated at the main fish weir ("fence") located at the tidewater limit. Spawners are identified to species and sex. Ages are determined from scale samples, and lengths are taken. Chum that spawn downstream of the fence are enumerated visually each day by observers on foot (Tschaplinski et al. 1998).

Juvenile salmonids (fry and smolts), prickly sculpins *Cottus asper*, and coastrange sculpins *C. aleuticus* migrating seaward in spring have also been enumerated and identified to species at the main fish weir. Large samples of salmonid fry and smolts (up to 50 individuals of each species per day) are measured for length and weighed. Scale samples are taken daily from up to 50 smolts of each species.

The abundance, habitat distribution, growth, and survival of juvenile salmonids and sculpins rearing in Carnation Creek from spring to autumn have been determined from two to three seasonal surveys conducted usually between 15 June and 30 September. During each survey, the two-catch removal method (Seber and LeCren 1967) is used to assess abundance within nine to ten representative study sections (95% confidence limits usually within 5 to 10% of the estimate). Fish are captured by electrofishing and seining in each of two fishing trials, identified to species, and measured for length (see Tschaplinski et al. 1998). Large samples of fish are weighed, and scales are taken to determine population age-size distributions and age-specific growth rates. The total abundance of fish in Carnation Creek is determined by extending the numbers of fish captured in the survey sections to the total length of stream inhabited by each species. Within each surveyed section, the total wetted surface area of the stream and its component pool, glide, and riffle areas are measured to determine population densities for specific habitat units. Fish habitat is classified and quantified according to methods adapted from Bisson et al. (1982) and Hankin and Reeves (1988).

Overwinter survival of juvenile salmonids is determined from the difference between population abundance estimated in late summer and the number of smolts migrating seaward from Carnation Creek in spring (plus any residual parr remaining in the stream as estimated from the population surveys). Seasonal changes in distribution and habitat use between summer (rearing) and winter (shelter) are determined each year from juvenile population surveys in the main channel in winter, and by monitoring the movements of salmon and trout between Carnation Creek and its valley-bottom tributaries through daily enumerations of fish at tributary weirs. Abundances of fish in specific off-channel sites are determined at intervals during winter by using removal methods and large numbers of Gee-type fish traps baited with salted fish roe. Seasonal use of these off-channel sites by fish, and changes in habitat characteristics, are determined annually.

RESULTS AND DISCUSSION

The principal trends in Carnation Creek fish population abundance, distribution, and survival over the past 25 years are discussed primarily for coho and chum (the numerically dominant species in the watershed).

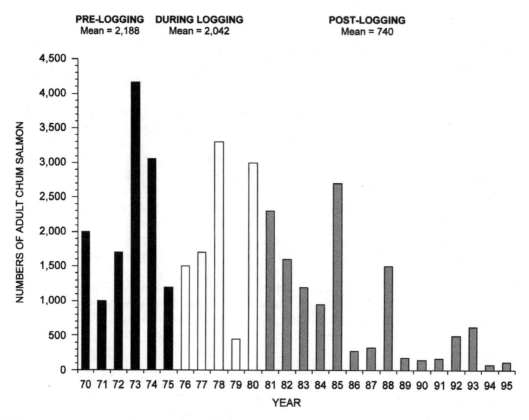

FIGURE 19.2 Numbers of adult chum salmon returning annually to Carnation Creek between 1970 and 1995.

ADULT CHUM AND COHO SALMON RETURNS

Adult chum return to spawn in Carnation Creek primarily as age 4 fish (usually >80%), mainly in October and November (Hartman and Scrivener 1990). In most years, chum have been the numerically dominant salmonid spawning in Carnation Creek (Figure 19.2). However, this species has shown the most drastic decline in abundance after forest harvesting. Prior to forest harvesting (Figure 19.2, 1970–1975), adult chum returns ranged from 1,000–4,168 and averaged 2,188 (95% confidence interval, ±1,272). During the 6 years of logging (1976–1981), chum returns were not significantly different from the pre-logging average. Mean numbers returning were 2,042 ± 1,102 (Figure 19.2). Spawner abundance varied between 450 and 3,300 during that period. Over the first 4 years of the post-logging period, average numbers returning exceeded 1,600; however, sharp declines have been observed in most years since 1986. Only 740 ± 444 chum have returned to spawn in Carnation Creek in the post-logging period (Figure 19.2, 1982–1995). Therefore, chum returns have averaged only about one third of their pre-logging levels (Student's t, $p < 0.05$), and were less than one sixth of the pre-logging mean in 6 of the 14 years for which post-logging data are currently available.

Fisheries on Barkley Sound chum are thought to have had little effect on the numbers of this species returning to Carnation Creek. Commercial harvesting in Barkley Sound has been restricted since 1962 (Lightly et al. 1985). Lightly et al. (1985) reported that the fishing rate (proportion) has been <0.01 in 15 of 24 years examined, and usually <0.15 in most years since 1951. In 1971, 1973, 1978, and 1980, fishing took an estimated 20 to 43% of Barkley Sound chum, but the extensive fishery in those years reflected exceptional adult returns to both the area and Carnation Creek (Lightly et al. 1985; Andersen 1983). The commercial gillnet and seine fisheries were concentrated in terminal areas far from Carnation Creek on the north side of Barkley Sound (Lightly et al. 1985), and suggest that few chum from Carnation Creek were caught.

Local aboriginal peoples have annually conducted a small food fishery for chum; however, this fishery often consisted of only one net-set during the peak of the adult return to the nearby Sarita River (1.8 km away). Holtby and Scrivener (1989) noted that up to 300 fish are taken annually, and some are probably Carnation Creek chum. These investigators believed that the long-term decline in chum returns to Carnation Creek coinciding with the post-logging period are not likely due to fishing mortality.

Two patterns in coho spawner abundance have been observed, and each is associated with one of two types of coho returning annually to Carnation Creek. In most years, the majority of coho spawners are age 3 or 4 adults that have spent roughly 18 months, including two summers, in the ocean (Hartman and Scrivener 1990). These fish called "large adult coho" are usually >44 cm long (fork length). Other coho called "jacks" are usually ≤44 cm long. They return to spawn after spending only about 5–6 months in the ocean (Hartman and Scrivener 1990). Most of these small fish are age 2, precocious males.

Returns of large adult coho have declined significantly after logging in Carnation Creek (Figure 19.3a; Student's t, $p < 0.05$) although the decrease has been less marked than that shown by chum. Before forest harvesting (1971–1975), 165 ± 17 large adult coho returned each year (Figure 19.3a). These returns decreased by about 31% to only 114 ± 36 in the post-logging period. Most of the decline is due to decreased returns of females. In pre-logging, during-logging, and post-logging periods, the mean numbers of females returning to spawn were 73 ± 6, 74 ± 53, and 48 ± 16, respectively. The abundance of adult females has declined on average by about 34% in the post-logging period compared with numbers in pre-logging years (Student's t, $p < 0.05$). In the pre-logging and during-logging periods, the male:female sex ratios were 1.25:1 and 1.20:1 on average. The mean proportion of males increased in the post-logging period to 1.42:1 and varied from 1:1.47 (i.e., females exceeding males) to 2.2:1 (ratios exclude the unusually low returns of 1994 when only one female and eight adult males were observed in the stream). The long-term decline in females has not been explained, but it appears to be widespread among coho stocks from streams in the Barkley Sound area and to the south (Simpson et al. 1996).

In contrast to large adults, the mean numbers of jacks returning show no statistically significant trend among pre-logging, during-logging, and post-logging periods (Figure 19.3b; Student's t, all $p > 0.05$). Jack returns actually exceeded the numbers of large adult coho in 1978, 1988, 1989, and 1994 when they made up, respectively, 69.6, 63.2, 50.5, and 92.0% of the total numbers of coho returning to spawn. Therefore, when large adults and jacks are combined, the significant decline in the total coho return to Carnation Creek between pre-logging (227 ± 26) and post-logging (182 ± 45) periods is due to the decline in numbers of large adult coho alone (Figure 19.3c; Student's t, $p < 0.05$).

The abundance of large adult coho was nearly invariant among years prior to logging (Figure 19.3a). However, the interannual variation in the abundance of both jacks and large adults has increased dramatically since 1976 when forest harvesting activities were initiated (Figures 19.3a and 19.3b). This increased variation included especially low adult returns observed in four of five years spanning 1984 and 1988 when large coho averaged only 60 at Carnation Creek (Figure 19.3a). These depressed numbers occurred at roughly the same time during the mid-1980s when chum returns also began to show sharp declines in some years although species-specific differences in annual patterns were observed (Figures 19.2 and 19.3). In contrast to chum, coho spawner abundance increased dramatically in the next 3 years (1989–1991) when the numbers of adults returning exceeded the pre-logging mean in each year (Figure 19.3a). However, these elevated returns completely reversed in the following 3 years when Carnation Creek coho (and chum) were subjected to the simultaneous effects of forest harvesting, fishing, and prolonged poor conditions for marine survival caused by "El Niño," the northward extension of warm, nutrient-poor waters with low

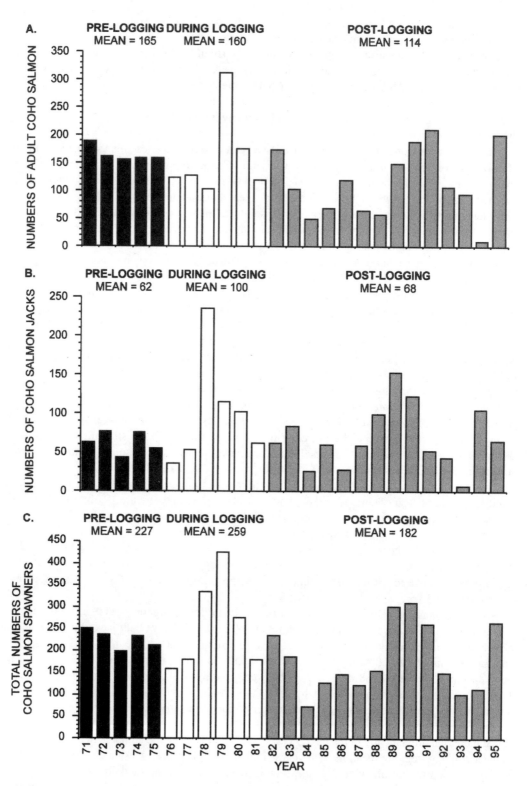

FIGURE 19.3 Numbers of coho salmon returning to Carnation Creek annually between 1971 and 1995, including large adults (A), jacks (B), and all spawners combined (C).

productivity and elevated predator abundance (Hargreaves and Hungar 1994; Rice et al. 1995). A similar combination of conditions coincided in 1983 (Holtby and Scrivener 1989; Karinen et al. 1985) and contributed to the declines in adult coho and chum observed during the mid-1980s (Holtby and Scrivener 1989); however, the El Niño phenomenon at that time persisted for only about 1 year. In the early and mid-1990s, El Niño-like conditions prevailed for perhaps 3 years (1992–1994; Hargreaves and Hungar 1994; Rice et al. 1995). The cumulative effects of forest harvesting and environmental shifts were associated with declining returns of both large adult coho and chum that reached historically low levels by 1994 (Figures 19.2 and 19.3).

In 1994, only nine large adult coho including just one female were enumerated (Figure 19.3a). Carnation Creek coho thus approached year-class extinction. Low salmon returns were not unique to Carnation Creek. Similar observations were made for coho and other salmon species returning to streams throughout the west coast of Vancouver Island and elsewhere in south coastal B.C. (Hargreaves and Hungar 1994; Heizer 1991; Nelson 1993; Rice et al. 1995). Recognizing that emergency action was required for population conservation, the DFO reduced the commercial catch of coho in the summer of 1995 (Rice et al. 1995) to allow more adults to return to their spawning grounds. The overall exploitation of coho stocks by the fishery in south coastal B.C. under this management regime may have been lowered to roughly 50% from levels that probably averaged ~67% (Holtby and Scrivener 1989).

Coho stocks appeared to respond immediately to this emergency conservation measure. Returns of large adult coho to Carnation Creek increased dramatically from only 9 in 1994 to 201 in 1995 (Figure 19.3a). This return exceeded the pre-logging average by nearly 22%. Coho stocks elsewhere responded similarly to reduced fishing pressure. For example, 2,300 coho were enumerated at Clemens Creek (tributary to Henderson Lake) in the Barkley Sound region, where coho returns to this system for the previous 10 years were generally fewer than 200 spawners (P. J. Tschaplinski, B.C. Ministry of Forests, and K. D. Hyatt, DFO, unpublished data).

The strong and immediate response shown by Carnation Creek coho to reduced commercial fishing demonstrates the vulnerability of numerically weak salmon stocks in mixed-stock commercial fisheries. This observation also begs the question of the historical effects of fishing pressure on adult coho returns to Carnation Creek and its influence on conclusions regarding the effects of forest harvesting. For example, the unusually high return of 312 large adult coho in 1979 was nearly twice the average return for the pre-logging and during-logging periods combined. Although the fishing rate on the aggregation of coho stocks from the west coast of Vancouver Island in 1979 reflected the long-term average (Simpson et al. 1996), one might speculate that the anomalously high adult returns for that year was due to the majority of Carnation Creek coho escaping the fishery in the summer of 1979 by chance.

In contrast with chum, coho originating from Carnation Creek, 281 other western Vancouver Island stocks, the Strait of Georgia, Fraser River, and the U.S. are subject to significant commercial and recreational fisheries each year off the west coast of Vancouver Island (Simpson et al. 1996). The commercial troll fishery in this region is the single largest harvester of coho in B.C. There is no direct measure of the number of Carnation Creek coho caught in various fisheries because smolts leaving Carnation Creek are not usually marked with coded-wire tags to study their patterns of ocean distribution and fishing mortality. The coho stock from the nearby Robertson Creek hatchery (Figure 19.1) is the only one from the west coast of Vancouver Island that is tagged annually and has the data required to calculate harvest rates. Since 1972, fisheries have taken 54.1 to 76.7% of Robertson Creek coho without upward or downward trend (Simpson et al. 1996). The great majority (>91%) are taken off the west coast of Vancouver Island by the troll fishery, and up to ~50% are caught in or near Barkley Sound.

Similar fishing rates and patterns have been presumed for Carnation Creek coho. Historically, fisheries scientists and managers have used Robertson Creek coho data to estimate fishing and natural mortalities for west coast Vancouver Island stocks including coho from Carnation Creek (Holtby and Scrivener 1989; Simpson et al. 1996). Despite some untested assumptions (e.g., that hatchery and wild-stock smolts behave similarly in the ocean), the application of Robertson Creek information to Carnation Creek coho has been justified because (1) the limited studies of coded-wire-tagged coho from Carnation Creek and other streams have shown that catch distributions are similar among west coast Vancouver Island stocks; therefore, coho from Carnation and Robertson creeks should be exposed to the same fisheries for similar periods of time; (2) temporal patterns of smolt-to-adult survival are significantly correlated between coho from Carnation and Robertson creeks ($r = 0.65$, $p < 0.001$), although survivals are significantly higher for Carnation Creek smolts (0.118 vs. 0.045, respectively; $p < 0.001$); and (3) temporal trends in adult escapements are generally similar between Carnation Creek coho and other western Vancouver Island stocks such as those of the Stamp River system (which includes Robertson Creek) and Gold River in northwest Vancouver Island, far from Carnation Creek ($r = 0.66$, $p < 0.01$; Holtby and Scrivener 1989; Simpson et al. 1996). From these correlations, Holtby and Scrivener (1989) estimated that annual fishing pressure on Carnation Creek coho varied between 65 and 70%. They concluded that fishing at these levels had little effect upon annual variations in adult coho returns to Carnation Creek.

Holtby and Scrivener (1989) used a series of sequentially-linked (and life-history–based) regression models to determine the relative effects of fishing, forest harvesting, and climate on the returns of adult chum and coho to Carnation Creek. The models predicted adult escapements based upon correlations between fish population responses (e.g., survival and growth) at different life stages with (a) climatic, hydrologic, and physical variables; (b) indices of freshwater habitats affected by logging; (c) realistic fishery exploitation rates ranging over 0–0.50 for chum and 0.59–0.80 for coho. Shifts in climate were determined from long-term trends in air and water temperatures from monitoring stations at Carnation Creek and other west coast Vancouver Island sites (to include data on ocean surface salinities and temperatures, and information prior to the start of the Carnation Creek study; Holtby 1988; Holtby and Scrivener 1989). Holtby (1988) used multiple regression analyses to partition the increases in stream temperatures observed after 1976 between the effects of forest cover removal and climate change. He determined that logging-associated increases in water temperatures varied from 0.7°C in December to ~3.3°C in August.

Simulations by Holtby and Scrivener (1989) based upon data available up to the late 1980s for Carnation Creek indicated most of the variation in observed and predicted adult spawner returns for both chum and coho resulted from climate variations in both freshwater and marine environments (in roughly equal measure). Fishing mortality generated little change in the inter-annual patterns of adult returns associated with climate variations; however, exploitation at the highest rates resulted in two- to threefold increases in interannual variation in adult numbers relative to moderate (i.e., observed) levels of fishing. Their model predicted the collapse of salmon stocks when the effects of habitat disturbance (forest harvesting), adverse oceanic conditions, and high fishing rates coincided. This prediction appears consistent with the collapse of adult coho returns to Carnation Creek and other Vancouver Island streams observed in 1994.

Although the analyses by Holtby and Scrivener (1989) and Scrivener (1991) indicated that forest harvesting alone reduced the numbers of chum adults returning to Carnation Creek after logging by ~26%, effects upon coho were relatively minor. Less than 10% of the decline in adult coho returns by the late 1980s was predicted from forest-harvest effects (Holtby and Scrivener 1989). Most of the effects upon adult returns were due to processes occurring early in the life history of salmon in both freshwater and marine environments (see following discussion on juveniles). The authors noted that their results were counter-intuitive, given that the relatively long time spent by coho in fresh water suggests that this species would be more strongly affected by forest harvesting than would chum. Conversely, chum spend more of their life cycle in marine environments and thus might be expected to be more strongly affected by marine climate shifts than by freshwater habitat changes. These results demonstrate the need for researchers and natural resource managers to be aware of biological and physical processes occurring within both watersheds and marine environments before interpreting observed patterns in salmonid production.

Other complex relationships are apparent. For example, in contrast with depressed returns of coho adults in 1994, the numbers of coho jacks returning in that year exceeded their pre-logging average (Figure 19.3b). The reason for this difference is unclear; however, returns of coho jacks were strongly depressed in 1993. The 1993 jacks and 1994 adults belong to the same brood year, indicating that this entire brood experienced low survival in the ocean. Simpson et al. (1996) speculated that the poor survivals of 1993 and 1994 were due to increased predation by piscivorous marine fishes which were abundant around Vancouver Island in both years.

EFFECTS UPON JUVENILE SALMONIDS

Chum Salmon

Forest harvesting is clearly one of several causes of the observed declines in chum abundance at Carnation Creek and elsewhere along the south coast of B.C. (Holtby and Scrivener 1989; Scrivener 1991). Two thirds of the post-logging decline of Carnation Creek chum that was attributed to forest harvesting by Holtby and Scrivener (1989) has been explained by reductions in egg survival due to sedimentation of spawning and egg incubation gravels (Scrivener 1991). Observed egg-to-fry survival for chum has declined by about one half from a mean of 20.3% in pre-logging years to 10.9% after logging (Hartman and Scrivener 1990).

Most chum at Carnation Creek spawn in the lowermost portion of the system located downstream of the main fish weir. From 68 to virtually 100% of all chum spawn in this area, all of which is under tidal influence, and most of which is regularly inundated with saline water (Tschaplinski 1982b, 1988). Most of the remaining spawners migrate usually only short distances (e.g., 100 m) upstream of the weir. Frozen-core gravel samples have shown that all of this area used by chum has been subjected to increases in fine sediment deposition after forest harvesting (Hartman and Scrivener 1990; Scrivener 1988a,b,c, 1991; Scrivener and Brownlee 1982, 1989).

Much of the sediment added to the stream came from eroding banks in areas upstream where both careful and intensive streamside forest-harvest treatments were applied (Hartman et al. 1987; Hartman and Scrivener 1990). During logging and after logging (1978–1985), bank erosion accelerated in these clearcut areas in association with increased frequencies and magnitudes of seasonal freshets (Hartman et al. 1987; Scrivener 1988b,c; Scrivener and Brownlee 1989). Freshets transported the eroded materials downstream into the leave-strip treatment including the sites used by chum (Hartman and Scrivener 1990; Scrivener 1988b). Analyses of frozen-core gravel samples showed that most of the material that reached the chum spawning sites and accumulated in the streambed consisted of sand which increased at depths where chum eggs would occur (in the middle and deep layers of the cores, representing streambed depths of 12 to 35 cm; Scrivener 1988b). Seasonal freshets cleared this material from the streambed in both the clearcut areas upstream throughout the study, and from the leave-strip sites downstream before logging (Scrivener and Brownlee 1989). However, after logging, the persistent source of fine sediments upstream caused the rate and depth of sand accumulation to increase in the chum spawning sites to the point where seasonal floods were no longer able to clean the streambed of these materials. Scrivener and Brownlee (1989) reported that volumes of sand and pea gravel (i.e., fines) increased significantly after logging. These investigators demonstrated that these changes in substrate composition explained 60% of the variation in chum egg survival between pre-logging and post-logging periods.

Accumulations of fine sediments causes increased egg mortality, principally by reducing intragravel water flow and dissolved oxygen concentrations around developing embryos (Everest et al. 1987). Fine sediments can also fill interstitial spaces and bury alevins, thus preventing fry emergence (Dill and Northcote 1970; Koski 1975; Sowden and Power 1985; Scrivener 1988c). The production of chum fry from Carnation Creek (i.e., the number of fry per spawner) has been reduced by one half after logging, principally by these mechanisms (Holtby and Scrivener 1989).

Coincident with the decline in adult chum returns to Carnation Creek, fewer chum have spawned upstream of the main fish weir after forest harvesting. Prior to logging, up to 32% of chum adults spawned upstream of the fish weir beyond any tidal (or saline-water) influence (Andersen 1983; Andersen and Scrivener 1992). Most chum now spawn downstream of the weir in areas that include some deep pools where brackish water ≥12‰ can remain in the streambed after high tides (Scrivener 1988a; Groot 1989). Groot (1989) demonstrated that chum eggs thrived at a salinity of 6‰, but 100% mortality occurred when eggs were continually exposed to salinities >12‰. Proportionally more of the total number of chum eggs deposited annually in Carnation Creek are now in areas influenced by moderately high estuarine salinities where increased risk of egg mortality occurs. For example, between 1990 and 1995, only 3.3% of all chum migrated upstream past the weir on average (range: 0 to 10.6%). With the exception of 1993, when eight females spawned upstream of the weir, less than three female chum spawned in the same areas in the 6 years since 1990. No chum spawned upstream of the weir in 1994, and no females spawned above tidal influence in 1990. This unexplained shift in spawner distribution might also have contributed to post-logging declines in chum fry production from the stream.

Stream habitat alterations have had little direct effect on chum fry after they emerge from the streambed in spring because these fry spend little time rearing in fresh water. They emigrate seaward shortly after they emerge (Andersen 1983; Andersen and Scrivener 1992). However, post-logging reductions in juvenile chum marine survival are indirectly linked to logging-associated effects in fresh water prior to fry emergence. First, the size (length) of chum fry has decreased after logging in association with reductions in the mean particle size of the spawning gravel (Scrivener 1988b,c; Scrivener and Brownlee 1989). Fine particles in the streambed are known to reduce interstitial spacing and thus selectively trap larger fry within redds (Dill and Northcote 1970; Koski 1975; Sowden and Power 1985). Second, beginning almost immediately after logging, seasonal increases in water temperatures were observed that were approximately proportional to the area of basin harvested (Holtby 1988). Increases occurring during winter were relatively subtle (i.e., 0.7°C mean weekly increase in December, and 1 to 2°C between February and April), but they allowed incubating eggs to develop more rapidly during autumn and winter, and consequently, fry to emerge and emigrate seaward earlier in spring (Hartman and Scrivener 1990; Holtby 1988).

Both reduced fry size and earlier seaward emigration were strongly correlated with reduced ocean survival (Hartman et al. 1987; Holtby and Scrivener 1989; Scrivener 1988c). Increased mortality of chum fry early in their ocean life history was attributed to increased susceptibility to predation (small size) and early-season entry into near-shore waters during winter-like conditions of relatively low salinity and biological productivity (Holtby and Scrivener 1989; Scrivener 1988c). Support for this explanation was provided by the observation that marine survival of chum was correlated positively with sea-surface salinities in spring (Scrivener 1988c). Very low survival was also observed in years when El Niño conditions prevailed (Fulton and Lebrasseur 1985; Hartman and Scrivener 1990). In El Niño years, chum fry entered the ocean in conditions that combined high predator abundances with low sea-surface salinity, warm water, and correspondingly reduced plankton productivity that resulted from suppression of upwelling of coastal waters (Fulton and Lebrasseur 1985; see following discussion for coho).

Coho Salmon: Stream-Resident Period

The decline in the numbers of adult coho returning to Carnation Creek after forest harvesting might be explained at least partly if the capacity of the stream to support populations of juvenile coho has decreased. Consistent with this notion, significant post-logging declines in the abundance of juvenile coho rearing in the stream during summer have been observed (Figure 19.4; Student's t, $p < 0.05$). Before forest harvesting, the freshwater habitats in Carnation Creek supported 11,944 ± 2,117 coho juveniles in late summer (late Sep–early Oct, fry and yearlings combined; Figure 19.4). Between 1976 and 1981, when most of the valley bottom was harvested, late-summer coho

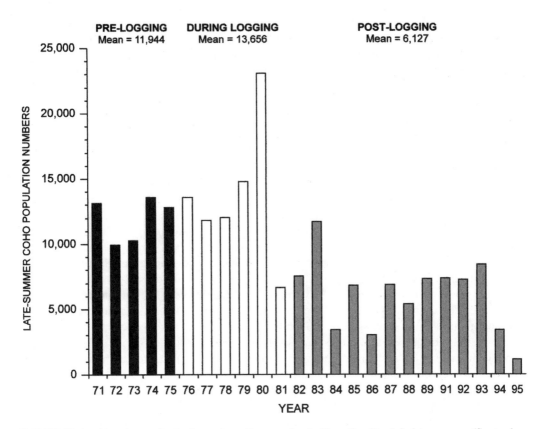

FIGURE 19.4   Abundance of coho fry and yearlings rearing in Carnation Creek in late summer (September–October). No data were available for 1990.

populations increased to 13,656 ± 5,661, but this increase was not statistically significant (Figure 19.4; Student's t, $p > 0.05$). However, the numbers of coho fry and yearlings (combined) rearing in Carnation Creek since 1982 have fallen by approximately one half to 6,127 ± 1,684 (Figure 19.4; Student's t, $p < 0.05$)).

This post-logging reduction in juvenile coho abundance is also due partly to marine survival variations that have reduced the numbers of adults returning to spawn (Hartman and Scrivener 1990; Holtby and Scrivener 1989; Scrivener and Andersen 1984). The decline in the numbers of adult females (~34%) has resulted in fewer eggs deposited into the streambed. Consequently, fewer fry have inhabited the system during summer since 1982. If there were no significant post-logging reductions in the amount of suitable rearing habitat, these reduced numbers of fry inhabiting the stream after logging would reflect a system well below rearing capacity due to insufficient spawners. However, several observations indicate that the post-logging reduction in the abundance of juvenile coho in Carnation Creek during summer is also the result of logging-caused reductions in egg-to-fry survival and the quantity and quality of summer rearing habitats (Hartman and Scrivener 1990; Hartman et al. 1996).

After forest harvesting, the survival of coho eggs to the time of fry emergence declined by about one half in Carnation Creek from 28.8 to 15.6% (Hartman and Scrivener 1990). This trend was similar to that shown by chum, and was again partly associated with increased percentages of streambed sand and pea-sized gravel (by 5.7 and 4.6%, respectively) in the leave-strip area downstream of the clearcut portions of the creek ($r = 0.81$, $p < 0.001$; Scrivener and Brownlee 1989). However, reductions in coho egg-to-fry survival were primarily associated with increased rates of streambank and channel erosion after logging in both the intensively-harvested and carefully-harvested clearcut treatments (Toews and

Moore 1982; Scrivener and Brownlee 1989). Clearcutting to the stream margins promoted the destabilization of LWD and mobilization of streambank and channel substrates through accelerated erosion (Hartman et al. 1996). The proximal cause of increased mortality of coho embryos was increased streambed scour and deposition in these sites during freshets after logging (Holtby and Scrivener 1989; McNeil 1966).

The capacity of the stream to support those fry that survived to emerge from the streambed in spring also declined after logging. Most of this decline can be attributed to stream morphology changes that have occurred both in the intensively harvested and carefully harvested clearcut areas. Prior to 1982, surveys of fish abundance showed that Carnation Creek was able to support as many as 23,095 juvenile coho at the end of summer (including 20,953 fry; Figure 19.4). The 1980 peak in juvenile abundance exceeded mean population sizes in the pre-logging and during-logging periods by >1.9- and 1.7-fold, respectively, and was produced from the 25-year peak spawner return in the autumn of 1979 (312 large adults, including 176 females). The greatly elevated number of juveniles rearing in Carnation Creek in 1980 suggests that habitat quantity in the stream did not limit the summer abundance of juveniles in most years prior to the post-logging period. A >1.9-fold increase in spawner abundance resulted in nearly doubling the rearing population during the following summer. Despite the elevated abundance of fry in 1980, strong density-dependent reductions in summer growth rates and mean size of fry were observed in that year, and indicate the upper limits of fry capacity for Carnation Creek may have been exceeded (Holtby 1988; Tschaplinski 1987). These juveniles produced 4,164 smolts, about 400 fewer than the 4,567 smolts produced from only 14,776 fry and yearlings of the previous brood year (see following discussion on smolts). The numbers of coho fry observed in the system before logging (~12,000 annually) may thus be a better indicator of the average rearing capacity of the stream.

In years during or after logging other than 1980, whenever relatively high numbers of adults returned, no corresponding increase in the abundance of juveniles rearing in the stream occurred the following summer. For example, 189 and 210 large adult coho returned to Carnation Creek in 1990 and 1991, respectively (Figure 19.3a), exceeding the pre-logging mean spawning escapements by 14.5 and 27.3%. Additionally, the number of females returning in 1990 exceeded the pre-logging mean by nearly 33% ($p < 0.05$), while the number of females returning in 1991 (76) was similar to the pre-logging mean ($p > 0.05$). Despite these relatively high spawner returns, the numbers of juveniles surviving in the stream by late summer throughout the late 1980s and early 1990s remained low and relatively invariant (Figure 19.4). This pattern indicates that the quantity (and quality) of rearing habitats in Carnation Creek (i.e., capacity) were reduced after logging and limited the numbers of juveniles the stream could support in summer.

Of the 3,070 m of Carnation Creek used by young coho upstream of the estuary, long-term post-logging reductions in the quantity and quality of available habitat have been observed, primarily in the 1,800-m portion subjected to clearcut harvesting (Hartman and Scrivener 1990). Habitat complexity in both the careful and intensive clearcut treatments decreased after logging due to reductions in the amount, size, and stability of LWD within the stream channel (Hartman and Scrivener 1990; Toews and Moore 1982). Volumes of LWD declined by at least 50% in sections of stream adjacent to the clearcut treatments (Hartman et al. 1996; Toews and Moore 1982). Stable pieces of LWD within the stream channel dissipate hydraulic energy, and in the fish-bearing reaches of Carnation Creek, are largely responsible for channel structure including the diverse sequence of riffles, pools, glides, meanders, and undercut banks which are important habitat features for stream salmonids. The decrease in stable LWD altered fluvial geomorphic processes by increasing stream velocities, bank erosion, and channel scour and deposition (Toews and Moore 1982). Habitat quantity and quality consequently declined. The channel became wider by at least twofold, and straighter (thus shorter) when meanders were cut. The proportion of the stream consisting of shallow, fast-flowing riffles increased (Hartman et al. 1987; Hogan et al. 1998). Favored coho-rearing habitat consisting of deep pools with cover in the form of LWD, undercut banks, and overhanging vegetation

(Tschaplinski 1987) became less abundant (Hartman et al. 1987). The remaining pools became shallower due to bedload deposition (Hartman et al. 1987). Long stretches of channel were filled with large deposits of gravel upstream of new logjams, thus creating ephemeral channels and reducing stream wetted area and salmonid rearing habitat (Tschaplinski et al. 1998).

Streamside clearcutting was partly responsible for some of these changes (see Hartman and Scrivener 1990). Logging activity around the streambanks resulted in killed or weakened tree roots in both the careful and intensive-treatment areas. Large amounts of small woody debris (pieces <3 m long) were added to the channel, especially in the intensive treatment site, which destabilized LWD within the stream. Additionally, LWD was directly broken or removed by machinery in the intensive-treatment area (Hartman and Scrivener 1990). In the first few years of the during-logging period, additions of woody debris initially increased habitat diversity and resulted in increases in fish density in the clearcut reaches of Carnation Creek (Hartman and Scrivener 1990). However, this habitat enhancement was short-lived. Major freshets after logging soon removed much of this material. Coho densities within the clearcut areas soon declined, both during summer and winter, relative to those observed in the leave-strip area downstream (Scrivener and Andersen 1984).

Although the Carnation Creek study was intended to investigate the effects of different stream-side harvesting treatments upon stream channels and fish, the most pronounced changes to the stream channel have been associated with increased frequencies of landslides and debris torrents that occurred after logging. Over 80 small landslides and three major debris torrents have occurred after logging in Carnation Creek, all in the logged portions of the watershed, and most have contributed sediment and debris into the creek channel. Overall, the volume of landslide material has increased by 12-fold after logging (Hartman et al. 1996).

No debris torrents occurred in the watershed in the pre-logging or during-logging periods; however, the three major torrents observed to date occurred early in the post-logging period in 1984. They occurred within the clearcut portions of three valley-wall tributaries (gullies) situated >1.5 km upstream of the portion of Carnation Creek containing anadromous fish (Hartman and Scrivener 1990; Hogan et al. 1998; Hogan and Millard 1998). These rainstorm-triggered torrents deposited large volumes of logging-associated woody debris and inorganic sediments into the stream channel where the materials were carried downstream into the carefully clearcut site inhabited by anadromous fish. Since 1984, the large logjams and associated sediments deposited by the torrents have moved progressively downstream into the intensively clearcut treatment and continue to cause major channel changes and fish habitat loss 12 years after their initiation. The post-logging widening of the channel, accelerated scour and deposition, and loss of stable LWD have been largely due to the stream moving around these logjams and sediment deposits, and redistributing materials downstream (Hogan et al. 1998; Hogan and Bird 1998).

As of 1996, most of the woody debris and large sediments associated with post-logging landslides and debris torrents had not reached the leave-strip area of Carnation Creek (Hogan et al.1998; Hogan and Bird 1998). The stream structure and fish habitat characteristics in that area remain much the same as observed in pre-logging years with the exception of fine sediment accumulations in the streambed (Hogan and Bird 1998; Scrivener 1988c). Surveys of juvenile salmon reveal that the majority of coho fry originate or rear in this lowermost portion of the stream or in off-channel habitats (Hartman et al. 1996; Tschaplinski et al. 1998). As excess sediment and debris move downstream into this area from the intensive and careful clearcut treatments, further reductions in the rearing capacity for coho will likely occur. Therefore, the full extent of the harmful effects of logging on the stream channel and fish habitats are yet to be observed at Carnation Creek.

Coho Smolt Abundance

If data on smolt abundance from Carnation Creek were unavailable, one might reasonably speculate that the numbers of coho smolts produced from this system would have declined after forest harvesting in parallel with the decreased numbers of fry and yearlings rearing in the watershed. However, the exact opposite has occurred. For several periods, both during logging and after, the numbers of coho smolts migrating seaward in spring increased (Figure 19.5a). Before logging, $2,213 \pm 424$ smolts were enumerated annually at the main fish weir. Their abundance increased early in the during-logging period; for example, it nearly doubled to 4,246 in 1978 compared with the pre-logging mean (Figure 19.5a). Annual smolt abundance increased by nearly 1.7-fold to 3,688 $\pm$ 813 in the 6 years during logging. Numbers have remained high to the present. A 25-year peak migration of 5,253 coho occurred in 1992, exceeding the pre-logging mean by nearly 2.4-fold. Mean annual smolt abundance in the post-logging period has been $3,441 \pm 495$ (Figure 19.5a), significantly greater than in the pre-logging period (Student's t, $p < 0.05$). Therefore, in the post-logging period, Carnation Creek fry, at roughly one half their pre-logging abundance, have produced 55% more smolts.

Smolt biomass has also increased after logging. The average weight of both age-1 and age-2 smolts has increased since 1977; for example, the mean weight of age-1 smolts increased by 1.5 g or one third of their mean weight prior to logging (Holtby 1988). This has occurred because coho fry are surviving the winter better after logging due to the temperature-related effects of forest harvesting and climatic warming which have resulted in larger fry due to earlier emergence and better seasonal growth (Hartman et al. 1990; Holtby 1988; Scrivener 1988c). Larger coho are apparently better able to survive winter conditions that include frequent scouring freshets (Tschaplinski and Hartman 1983; Brown and McMahon 1988).

Holtby (1988) demonstrated that the same logging-associated increases in winter water temperatures that allowed chum eggs to develop more rapidly, and chum fry to emerge earlier in spring, have had similar effects upon coho. During and after logging, coho fry were emerging from the streambed up to 6 weeks earlier in spring (Holtby 1988). Earlier emergence thus permitted these fish to experience a period for summer growth that was as much as 6 weeks longer than available to fry in pre-logging years. Additionally, the lower numbers of fry rearing in Carnation Creek after logging resulted in increased growth rates due to density-dependent reductions in competition for food (Holtby 1988; Scrivener and Andersen 1984). After logging, coho fry consequently grew 11 mm longer on average by the end of their first summer compared with growth in pre-logging years (trout fry also increased in mean length by 18 mm after logging; Hartman and Scrivener 1990). This larger body size was positively correlated with improved overwinter survival after logging ($r = 0.91$, $p < 0.001$; Holtby 1988).

The additional seasonal growth has also radically changed the age structure of coho smolt populations (Figure 19.5b). Prior to forest harvesting, nearly one half of all coho rearing in Carnation Creek required 2 years of growth before they were large enough to transform into smolts and migrate seaward (Figure 19.5b; 1971–1975). During and after logging, increased growth resulted in age-2 fish becoming relatively rare (Figure 19.5b).

Increases in water temperatures were observed almost immediately after clearcut harvesting removed significant portions of the streamside forest canopy (Hartman and Scrivener 1990; Holtby 1988). Coinciding with this rapid temperature shift, the numbers of coho fry able to grow to smolt size in about 1 year increased dramatically in 1976, the first year of forest harvesting (Figure 19.5b). The proportion of age-1 smolts increased significantly from $55.3 \pm 13.8\%$ in the pre-logging period to $76.7 \pm 11.0\%$ in the during-logging period (Student's t, $p < 0.05$). This proportion increased further to $91.4 \pm 3.2\%$ in the post-logging period (1982–1985; $p < 0.05$). Therefore, the temperature-related effects of forest harvesting upon juvenile coho growth and age structure, established soon after streamside harvesting began, persist in the watershed 20 years later. These effects will likely continue for several years until a new riparian forest canopy is established at Carnation Creek, and both water temperatures and fish growth decline toward pre-logging levels.

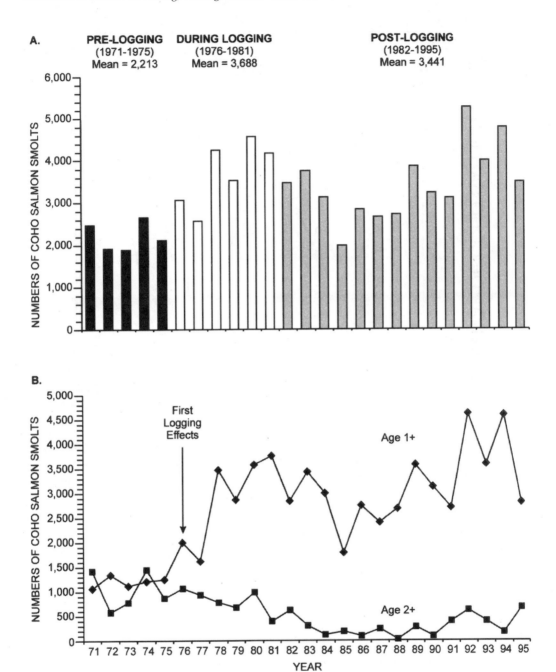

FIGURE 19.5 Total numbers of coho smolts (A) and numbers of age-1 and age-2 smolts (B) migrating seaward from Carnation Creek each spring between 1971 and 1995. Since 1982, >91% of coho smolts have been age-1 fish.

Marine Survival of Coho Juveniles

Although smolt numbers have increased after logging, reductions in their marine survival are implied from the declining numbers of adults returning to Carnation Creek since 1982 (Figure 19.3a). The marine survival of Carnation Creek coho smolts has decreased steadily and

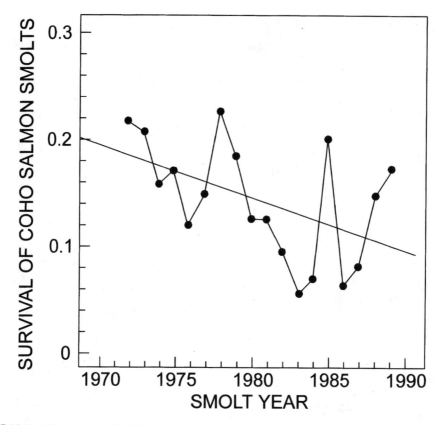

FIGURE 19.6 Marine survival of Carnation Creek coho smolts from the early 1970s to the late 1980s. Declining adult returns (Figure 19.3) suggest that this significant ($p < 0.05$) long-term decline in survival continued in more recent years. (Figure adapted from one provided by L. B. Holtby, Fisheries and Oceans Canada.)

significantly since the 1970s ($p < 0.05$), and the lowest survivals have occurred in the most recent years for which these data are available (Figure 19.6). Larger size and consequently better survival in fresh water have not translated to similar relationships in the ocean (Hartman and Scrivener 1990; Holtby et al. 1990). The marine survival of coho smolts appears unrelated to (1) the size of either age-1 or age-2 smolts, or (2) smolt age (Figure 19.7; paired t-tests for both age and size, all $p > 0.05$; Holtby et al. 1990). After logging, the mean size of coho smolts migrating seaward has actually declined because most smolts now are age-1 fish which are smaller on average than the now rare age-2 smolts (Hartman and Scrivener 1990).

The long-term decline in marine survival of smolts from Carnation Creek has been simultaneously reflected by coho from other streams in southwestern B.C. For example, declines in smolt survivals have been recorded for hatchery-reared coho released from Robertson Creek, the Big Qualicum River, and the Capilano River (Figures 19.1 and 19.8). These data indicate that long-term reductions in marine survival have been characteristic for the southwestern region of coastal B.C., and are largely associated with shifts in marine climate, decreased ocean productivity, and increased predator abundances (see discussion for adult coho and chum; Holtby and Scrivener 1989).

Although decreasing marine survival of coho smolts appears to be a regional phenomenon (Figure 19.8), forest harvesting appears to have made at least some contribution to this decline for smolts leaving Carnation Creek. This linkage is again ultimately associated with the temperature-related effects of forest harvesting. Seasonal increases in water temperatures after logging have shifted the timing of the seaward migration of coho smolts from the stream about 10 days earlier

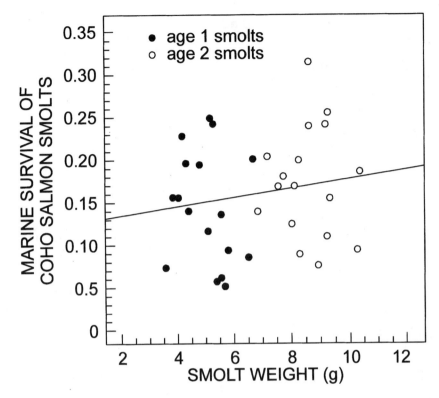

FIGURE 19.7 Marine survival vs. size and freshwater age of coho salmon smolts from Carnation Creek. (Figure adapted from one provided by L. B. Holtby, Fisheries and Oceans Canada.)

in the season compared with pre-logging migrations (Holtby et al. 1990). While this shift appears trivial, most mortality in salmonids in marine environments is known to occur soon after they enter the ocean (Mathews and Buckley 1976; Healey 1982; Holtby and Scrivener 1989; Holtby et al. 1990). Migration timing and ocean conditions in late winter and spring appear to be critical in determining whether chum fry or coho smolts survive to be adults (Healey 1982; Holtby et al. 1990). For example, Thedinga and Koski (1984) demonstrated that coho smolts leaving a small Alaskan stream 1–2 weeks earlier or later than the median day of the migration had only 45 to 60% of the survival of smolts that left in the middle part of the migration period. Walters et al. (1978) maintained that salmon fry and smolts entering the ocean must do so within a period that coincides with the vernal peak in estuarine and near-shore productivity. Timing is important because juvenile salmonids must maximize their access to food and thus their growth rates in order to spend the least amount of time as small fish vulnerable to predators (Holtby and Scrivener 1989).

Consistent with these observations, Holtby and Scrivener (1989) found that migration timing in coho smolts (and chum fry) was a significant correlate of survival. From analyses of the "early-ocean" growth patterns on the scales of adult coho returning to Carnation Creek (i.e., spacing of the first five marine circuli), Holtby et al. (1990) determined that a strong and significant correlation also occurred between marine survival and the growth rates of smolts soon after they enter the ocean (Figure 19.9, $p < 0.05$; relationship may be linear or stepped). Furthermore, strong and consistent correlations occurred between both early ocean growth and total marine survival and sea-surface salinities (Figure 19.10). Years with high-salinity (and low temperature) coastal waters were associated with the best years of coho survival and early-ocean growth. A threshold salinity of 31.5‰ appeared to separate years in which growth and survival were poor with years when both population statistics reached relatively high values (Figure 19.10; Holtby et al. 1990).

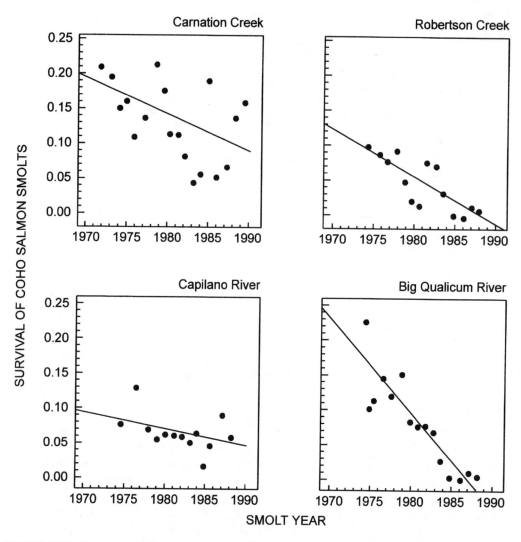

FIGURE 19.8 Long-term declines in the ocean-phase survival of coho salmon smolts from Carnation Creek and other streams in southwestern B.C. Dates represent brood years (figure adapted from one provided by L. B. Holtby, Fisheries and Oceans Canada).

The ocean conditions which promote the presence of cool, high-salinity water at the time that Carnation Creek smolts enter coastal waters are created by the seasonal upwelling of deep water off the northwest coast of Vancouver Island (Fulton and Lebrasseur 1985; Holtby et al. 1990). Winter conditions are characterized by relatively nutrient-poor, warm, and low-salinity water flowing to the northwest from the Washington coast and out from the Strait of Juan de Fuca (Figure 19.11). This pattern shifts to a counter-current flow during spring and summer that is associated with the upwelling of deep, cold, and nutrient-rich water that promotes plankton growth, and consequently high rates of salmonid growth and survival (Figure 19.11). The post-logging decline in the ocean survival of coho smolts from Carnation Creek may thus be explained partly by the shift to earlier smolt migrations which have increased the risk that these smolts enter the ocean during winter-like conditions less favorable for growth and survival.

Notwithstanding the effects of forest harvesting, coho smolt survival in the past several years has clearly been depressed further due to physical-regime shifts in the ocean. In years when ocean surface temperatures are high and salinities are low (e.g., El Niño years), the interface between the

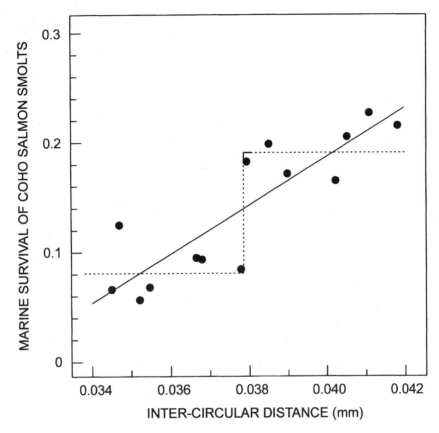

FIGURE 19.9   Marine survival as a function of coho smolt growth rates soon after these juveniles enter the ocean from Carnation Creek. Each point represents average inter-circular distance (an index of growth) for smolts leaving the stream in a given year. The dashed line represents a possible stepped model (figure adapted from one provided by L. B. Holtby, Fisheries and Oceans Canada).

Alaskan and California current systems (the "subarctic boundary") moves northward and away from the west coast of Vancouver Island. This zone is displaced because of a northward shift of warm, nutrient-poor water that suppresses seasonal upwelling, depresses ocean productivity, and is associated with low rates of survival for juvenile salmonids and other species such as herring *Clupea harrengus pallasi* (Holtby et al. 1990). At the same time, large numbers of predators, mainly chub mackerel *Scomber japonicus* and Pacific hake *Merluccius productus*, move northward into the coastal waters of B.C., and are believed to consume large numbers of juvenile salmonids including coho (Fulton and Lebrasseur 1985; Holtby et al. 1990).

Ware and McFarlane (1988) concluded that changes in the abundance of herring off Barkley Sound are due primarily to changes in the intensity of predation. They have noted that the biomass of piscivorous predators such as Pacific hake have been sufficiently high to account for all of the annual mortality within herring stocks in some years. Holtby et al. (1990) have shown that the survival of coho smolts and ages 1 and 2 herring covary (Figure 19.12; $r = 0.6$, $p < 0.01$). They noted that coho smolts and herring between 1 and 2 years old are similar in size, have overlapping diets, and occur together in both Barkley Sound and other rearing areas in the coastal waters of western Vancouver Island. Although Pacific hake and chub mackerel prey primarily upon herring which usually greatly outnumber coho, even incidental predation upon coho can cause substantial mortality in their populations. Predation probably increases in years when El Niño-like conditions reduce ocean productivity and consequently the abundance of food available for young salmon.

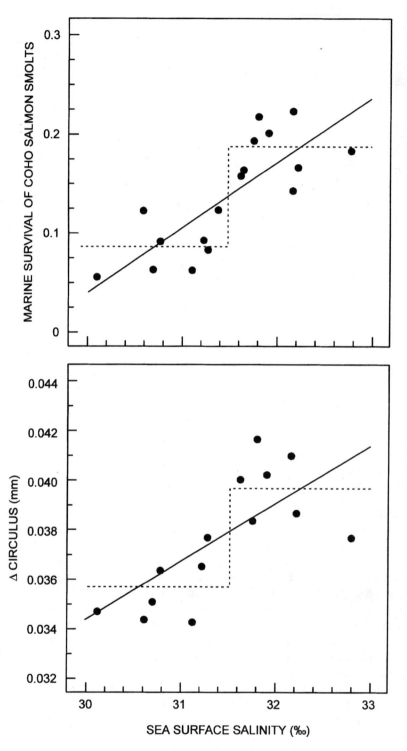

FIGURE 19.10 Growth (inter-circular distance from scale analyses) and marine survival of Carnation Creek smolts as functions of sea-surface salinities. Each point represents average inter-circular distance or survival for coho smolts leaving the stream in a given year. The dashed line represents a possible stepped model (figure adapted from one provided by L. B. Holtby, Fisheries and Oceans Canada).

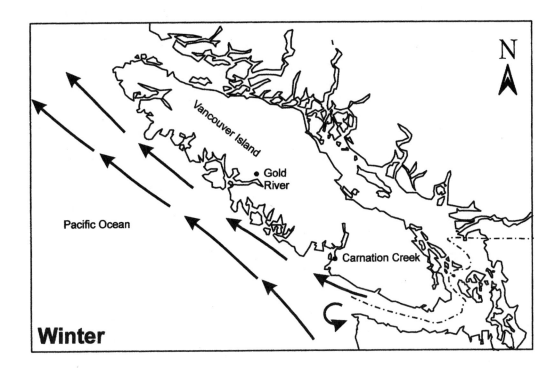

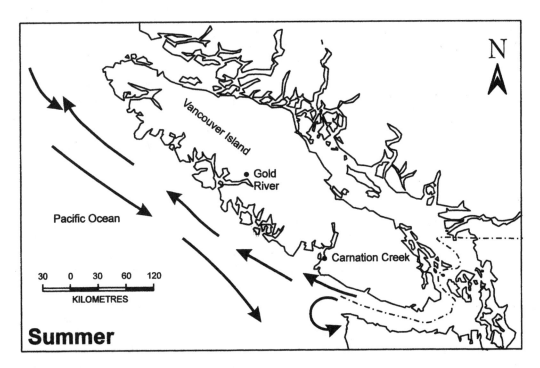

FIGURE 19.11 Seasonal shifts in the coastal currents off of the west coast of Vancouver Island (figure adapted from one provided by L. B. Holtby, Fisheries and Oceans Canada).

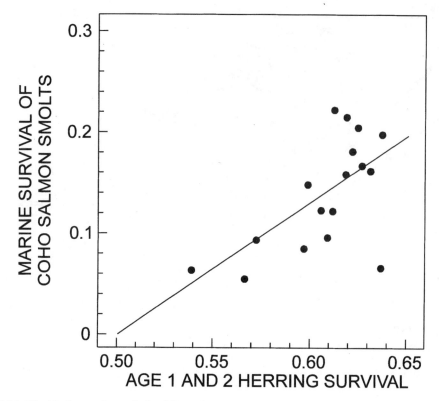

FIGURE 19.12 Marine smolt survivals of Carnation Creek coho covary with an index of survival for ages 1 and 2 herring off of the southwest coast of Vancouver Island from 1971–1987 (figure adapted from one provided by L. B. Holtby, Fisheries and Oceans Canada).

This results in slow growth that keeps juvenile coho small and susceptible to predators (Holtby et al. 1990). Therefore, Holtby et al. (1990) suggested that most of the reduction in the ocean survival of coho smolts in recent years was the result of predation as the ultimate consequence of low ocean productivity. Holtby and Scrivener (1989) concluded that this ocean process was the largest determinant of declines in survival of coho (and chum) from Carnation Creek and elsewhere from the west coast of Vancouver Island (Figure 19.8).

Steelhead and Anadromous Cutthroat Trout

Anadromous trout populations in Carnation Creek have always formed a relatively minor part of its fish fauna. Twelve or fewer adult steelhead and nine or fewer adult cutthroat are known to have returned to spawn in any year since the project was initiated (Hartman and Scrivener 1990). Because of these low numbers, interpretation of population trends relative to the effects of forest harvesting is difficult. Additionally, direct estimates of adult trout abundance are not available after 1990 because from that year onward adult counts were discontinued for the winter months (after 18 December) when many adult trout return. Prior to 1991, the accuracy of adult trout counts was often reduced due to floods that allowed spawners to migrate upstream over the fish weir without being enumerated (Hartman and Scrivener 1990). Despite these problems, the long-term trends in the abundance of trout smolts at Carnation Creek are briefly included in this review to demonstrate that both species continue to inhabit the lower 3 km of the stream even though (1) adult abundances have always been low and (2) juvenile trout might be especially sensitive to habitat alterations because they may reside for several years in fresh water (Figure 19.13).

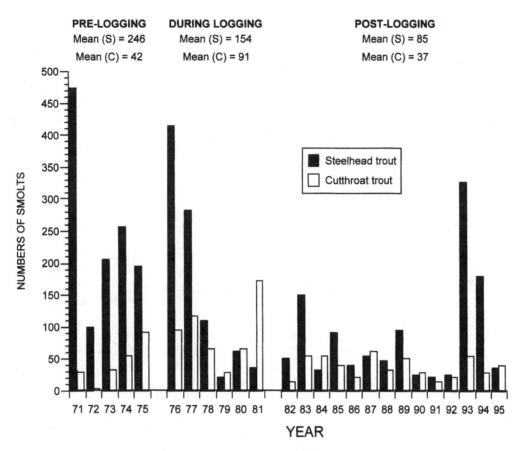

FIGURE 19.13 Numbers of steelhead trout (S) and cutthroat trout (C) smolts migrating seaward from Carnation Creek each spring between 1971 and 1995.

Very few cutthroat smolts have been produced annually from Carnation Creek. In no year have the numbers exceeded 174, and fewer than 95 smolts have been enumerated at the main weir in spring in most years (Figure 19.13). No statistically significant changes in the numbers of cutthroat smolts produced from Carnation Creek are apparent after logging. The pre-logging mean of 42 smolts is nearly identical to the post-logging mean of 37 (Student's t, $p > 0.05$; Figure 19.13). The trend to higher average numbers in the during-logging period is not significant due to high variability among years ($p > 0.05$).

The abundance of steelhead smolts has declined on average at Carnation Creek from 246 before logging to only 85 between 1982 and 1995 (Figure 19.13). Mean numbers of steelhead smolts have thus fallen to <35% of their pre-logging levels; however, variability among years has been so high that these trends are not statistically significant (Student's t, $p > 0.05$). Sharp reductions in some years are nevertheless apparent after 1978 when freshet-associated changes in the stream channel and consequent loss of rearing habitats were first observed in clearcut-logged areas of the stream (Hartman et al. 1987; Hartman and Scrivener 1990).

Steelhead may be more susceptible to main-channel habitat loss than either juvenile coho or cutthroat, especially in winter when freshets are common. Steelhead in Carnation Creek are restricted to main-channel habitats in contrast with coho and cutthroat which also occupy tributaries, especially in winter (Brown 1987; Tschaplinski and Hartman 1983). During winter, many juvenile coho and cutthroat (at least 20% of all age-0 fish) seek shelter from scouring freshets by inhabiting "off-channel" sites, including tributaries (Brown 1987; Hartman et al. 1996; Tschaplinski and

Hartman 1983). On the other hand, young steelhead must find shelter in main-channel pools and undercut banks associated with logs and tree roots (Bustard and Narver 1975).

In some years after logging (e.g., 1984), low abundance of steelhead smolts in spring occurred after winters with frequent severe freshets (Figure 19.13). With the loss of main-channel shelter habitats in clearcut sections of Carnation Creek (Hartman et al. 1987; Hartman and Scrivener 1990), the salmonid mortality associated with freshets was likely more pronounced in post-logging years. Winters without strong freshets were sometimes associated with high numbers of steelhead smolts the following spring, even after logging. For example, the abundance of steelhead smolts in the spring of 1993 is the third highest on record (Figure 19.13), and occurred after a relatively mild winter without strong freshets (unpublished project data). While there are some indications that steelhead have been affected by logging, patterns are not consistent among years and are obscured by generally low abundance of this species in Carnation Creek.

THE EFFECTS OF FOREST HARVESTING AND THE CARNATION CREEK STUDY DESIGN

When compared to the influence of climatic shifts in both freshwater and ocean environments, the effects of forest harvesting upon coho and chum at Carnation Creek have been relatively small (for each species, explaining, respectively, ~26% and <10% of the post-logging variability in the abundance of adult returns) (Holtby and Scrivener 1989; Holtby et al. 1990). Furthermore, the numbers of coho smolts leaving Carnation Creek have increased despite habitat degradation, not only because of the effects of temperature on seasonal growth and survival, but also because of the physical attributes of the basin. Unlike many other small coastal streams, Carnation Creek has a wide floodplain which contains important winter habitats consisting of seven ephemeral tributaries and side channels. These habitats remained largely intact after logging despite some deposition of debris and loss of aquatic vegetation (Brown 1987; Brown and Hartman 1988; Hartman et al. 1996). Consequently, the maintenance of good-quality "off-channel" habitats contributed to high rates of overwinter survival in juvenile coho after logging. Without these valley-bottom features, smolt abundance after logging may have been substantially lower than observed.

The effects of forestry practices upon both chum and coho would likely have been more severe had the network of forestry roads been more typical of those in most other coastal watersheds. Most of the forestry practices conducted at Carnation Creek, such as clearcut harvesting over progressively larger areas, were typical of those employed in the 1970s and 1980s. However, no roads crossed the main stream. Only one short section of road entered the floodplain in the lower portion of the watershed inhabited by anadromous salmonids, and it crossed only the main tributary to Carnation Creek. All other roads were located on relatively stable hillslopes on both sides of the basin. This atypical road network remains one of the limitations of the Carnation Creek study design. Reductions in the quality of streambed gravels and egg-to-fry survival were probably minimized in the absence of (1) the short-term peak of fine sediments from the construction of valley-bottom roads and stream crossings and (2) the chronic introduction of fines into the stream from road surfaces and stream crossings characteristic of many logged watersheds (Everest et al. 1987; Cederholm and Reid 1987).

The absence of an external, unlogged control watershed is another study design limitation. Comparing the long-term trends in numbers of smolts produced per spawner and the numbers of spawners produced per smolt between logged and control watersheds would likely have helped to clarify (1) the direct effects of forest harvesting upon salmonids in fresh water and (2) the indirect linkages to effects on growth and survival in marine environments. Despite the limitations of this project, much has been learned in general about the biological and physical processes operating within a small coastal watershed, and the effects of forestry practices on these processes. We will likely learn more as this project continues. The abundance of coho smolts leaving Carnation Creek remains high. However, the main-channel habitats at Carnation Creek continue to deteriorate: large portions of the stream will lack stable LWD for many decades. In a few years, the forest canopy will be re-established over the stream and begin to reduce water temperatures. Some of the increases in fish growth and survival associated with elevated stream temperatures after logging will disappear. With poorer habitat quality and lower water temperatures, smolt production from Carnation Creek may begin to decline within a few years.

SUMMARY

The 25 years of study on fish populations at Carnation Creek through 5 pre-logging, 6 during-logging, and 14 post-logging years has illustrated that forest harvesting has complex and often variable effects upon population processes at the different life stages of each fish species. The abundance of both adult chum and coho returning to Carnation Creek have declined significantly after logging. Chum populations have declined the most sharply. Adult returns are now only about one third of their pre-logging levels. Chum egg survival has declined by about one half after forest harvesting partly as the result of fine sediment deposition and other gravel quality changes in the lowest parts of the stream. The numbers of coho fry rearing in Carnation Creek are about one half of their pre-logging levels. The effects upon juvenile coho associated with forest harvesting include a reduction in egg-to-fry survival by about one half and decreased quantity and quality of the rearing habitat due to stream channel changes associated with streambank damage, loss of LWD, and increased erosion (scour and deposition). Loss of rearing habitat was associated with clearcut logging along streams and over steep-sloped terrain. Debris torrents and landslides from clearcut hillsides have introduced large volumes of sediment and woody debris into the stream channel that continue to cause pronounced changes to fish habitats 20 years after forest harvesting was initiated.

The Carnation Creek study has demonstrated the importance of managing forestry activities to minimize the risk of landslides and debris torrents. It has also demonstrated the importance of leaving forested riparian strips along streams to "buffer" channels and aquatic habitats from some forest-harvesting effects. Although the leave-strip treatment at Carnation Creek was located downstream of the clearcut treatments, and thus was partly impacted by activities upstream, the stream channel and the quality and amount of salmonid habitat in the leave-strip area remains largely the same as in pre-logging years. Major logging-associated changes will eventually be transmitted downstream into this area, and the main-channel habitats at Carnation Creek will thus deteriorate further. Therefore, the full extent of the harmful effects of logging on the stream channel and fish habitats is yet to be observed at Carnation Creek. The extended time-course of forestry effects demonstrates the value of long-term studies such as the Carnation Creek project.

After forest harvesting, roughly one half the number of coho fry have produced nearly 1.6-times more smolts due ultimately to increased water temperatures associated with forest canopy removal and climate change. Increased temperatures allowed coho fry to emerge earlier in spring, grow larger by autumn, survive the winter better because of larger size, and migrate seaward mainly as age-1 smolts. The post-logging increase in smolt abundance has not resulted in increased numbers of adult spawners returning to Carnation Creek because early-marine growth and survival has declined. Logging-associated temperature increases in fresh water have shifted seaward smolt migrations earlier in the spring, potentially into ocean conditions unfavorable for growth and survival. Additionally, marine climate changes have increased coho mortality in the ocean through reductions in biological productivity and simultaneous increases in predator abundance.

These results show that both freshwater and marine phases of salmonid life histories must be considered from the perspective of resource conservation and management. We must also note that our best forest-harvest practices might not ultimately result in more salmon returning to our coastal streams in every situation because of climate changes and marine processes. However, the application of the best practices for watershed stewardship is essential because the combination of forest harvesting, fishing, and climate change can together reduce salmon populations to levels so low that recovery is not possible (see Lawson 1993). This situation was approached for Carnation Creek coho in 1994. High-quality habitats for stream fishes after logging remain a high priority in the interests of resource sustainability.

ACKNOWLEDGMENTS

Many people should be thanked for their contributions to the Carnation Creek study. For contributions to this chapter, I thank L. B. Holtby for his data analyses and provision of several figures noted within the text, and D. Reiser, J. C. Scrivener, and B. Ward for their review of the manuscript and useful comments.

REFERENCES

Andersen, B. C. 1983. Fish populations of Carnation Creek and other Barkley Sound streams 1970–1980. Canadian Data Report of Fisheries and Aquatic Sciences 415.

Andersen, B. C., and J. C. Scrivener. 1992. Fish populations of Carnation Creek 1981–1990. Canadian Data Report of Fisheries and Aquatic Sciences 890.

Bisson, P. A., J. L. Neilson, R. A. Palmason, and L. E. Grove. 1982. A system of naming habitat types in small streams with examples of habitat utilization by salmonids during low stream flow. Pages 62–73 in N. B. Armantrout, editor. Acquisition and utilization of aquatic habitat inventory information. Western Division, American Fisheries Society, Portland, Oregon.

Brown, T. G. 1987. Characterization of salmonid over-wintering habitat within seasonally flooded land on the Carnation Creek flood-plain. British Columbia Ministry of Forests and Lands, Land Management Report 44.

Brown, T. G., and G. F. Hartman. 1988. Contribution of seasonally flooded lands and minor tributaries to coho (Oncorhynchus kisutch) salmon smolt production in Carnation Creek, a small coastal stream in British Columbia. Transactions of the American Fisheries Society 117:546–551.

Brown, T. G., and T. McMahon. 1988. Winter ecology of juvenile coho salmon in Carnation Creek: Summary of findings and management implications. Pages 108–117 in T. W. Chamberlin, editor. Proceedings of the workshop: Applying 15 years of Carnation Creek results. Carnation Creek Steering Committee, Pacific Biological Station, Nanaimo, British Columbia.

Bustard, D. R. 1991. Fish populations of Carnation Creek: 1991 data report. Forest Sciences Data Report, British Columbia Ministry of Forests, Prince Rupert Forest Region, Smithers, British Columbia.

Bustard, D. R., and D. W. Narver. 1975. Aspects of the winter ecology of juvenile coho salmon (Oncorhynchus kisutch) and steelhead trout (Salmo gairdneri). Journal of the Fisheries Research Board of Canada 32:667–680.

Cederholm, C. J., and L. M. Reid. 1987. Impact of forest management on coho salmon (Oncorhynchus kisutch) populations in the Clearwater River, Washington. Pages 373–398 in E. O. Salo and T. W. Cundy, editors. Streamside management: Forestry and fishery interactions. Institute of Forest Resources, Contribution 57, University of Washington, AR-10, Seattle.

Dill, L. M., and T. G. Northcote. 1970. Effects of gravel size, egg depth, and egg density on intragravel movement and emergence of coho alevins. Journal of the Fisheries Research Board of Canada 27:1191–1199.

Everest, F. H., R. L. Beschta, J. C. Scrivener, J. R. Sedell, and C. J. Cederholm. 1987. Fine sediment and salmon production: A paradox. Pages 98–143 in E. O. Salo and T. W. Cundy, editors. Streamside management: Forestry and fishery interactions. Institute of Forest Resources, Contribution 57, University of Washington, AR-10, Seattle.

Fulton, J. D., and R. J. Lebrasseur. 1985. Interannual shifts of the subarctic boundary and some of the biotic effects on juvenile salmonids. Pages 237–252 in W. S. Wooster and D. L. Fluharty, editors. El Niño North: El Niño effects in the eastern subarctic Pacific Ocean. Washington Sea Grant Program, University of Washington, Seattle.

Groot, E. P. 1989. Intertidal spawning of chum salmon: Saltwater tolerance of the early life stages to actual and simulated intertidal conditions. Master's thesis. Faculty of Graduate Studies, Department of Zoology, University of British Columbia, Vancouver.

Hankin, D. G., and G. M. Reeves. 1988. Estimating total fish abundance and total habitat area in small streams based on visual estimation methods. Canadian Journal of Fisheries and Aquatic Sciences 45:834–844.

Hargreaves, N. B., and R. M. Hungar. 1994. Robertson Creek chinook assessment and forecast for 1994. Part B: Early marine survival. Pacific Stock Assessment Review Committee Working Paper S94-01.

Hartman, G. F., and J. C. Scrivener. 1990. Impacts of forestry practices on a coastal stream ecosystem, Carnation Creek, British Columbia. Canadian Bulletin of Fisheries and Aquatic Sciences 223.

Hartman, G. F., J. C. Scrivener, L. B. Holtby, and L. Powell. 1987. Some effects of different streamside treatments on physical conditions and fish population processes in Carnation Creek, a coastal rainforest stream in British Columbia. Pages 330–372 *in* E. O. Salo and T. W. Cundy, editors. Streamside management: Forestry and fishery interactions. Institute of Forest Resources, Contribution 57, University of Washington, AR-10, Seattle.

Hartman, G. F., J. C. Scrivener, and M. J. Miles. 1996. Impacts of logging in Carnation Creek, a high-energy coastal stream in British Columbia, and their implication for restoring fish habitat. Canadian Journal of Fisheries and Aquatic Sciences 53 (Supplement 1): 237–251.

Haschenburger, J. K. 1996. Scour and fill in a gravel-bed channel: Observations and stochastic models. Doctoral thesis. Department of Geography, University of British Columbia, Vancouver.

Healey, M. C. 1982. Timing and relative intensity of size-selective mortality of juvenile chum salmon (*Oncorhynchus keta*) during early sea life. Canadian Journal of Fisheries and Aquatic Sciences 39:952–957.

Heizer, S. R. 1991. Escapement enumeration of salmon passing through Stamp Falls fishway on the Somass River system, 1986 through 1989. Canadian Manuscript Report of Fisheries and Aquatic Sciences 2067.

Hogan, D. L., and S. A. Bird. 1998. Stream channel assessment methods. Pages 189–200 *in* D. L. Hogan, P. J. Tschaplinski, and S. Chatwin, editors. Carnation Creek and Queen Charlotte Islands fish/forestry workshop: Applying 20 years of coastal research to management solutions. Land Management Workbook 41. British Columbia Ministry of Forests Research Branch, Victoria.

Hogan, D. L., S. A. Bird, and S. Rice. 1998. Stream channel morphology and recovery processes. Pages 77–96 *in* D. L. Hogan, P. J. Tschaplinski, and S. Chatwin, editors. Carnation Creek and Queen Charlotte Islands fish/forestry workshop: Applying 20 years of coastal research to management solutions. Land Management Handbook 41. British Columbia Ministry of Forests, Research Branch, Victoria.

Hogan, D. L. and T. H. Millard. 1998. Gully assessment methods. Pages 183–188 *in* D. L. Hogan, P. J. Tschaplinski, and S. Chatwin, editors. Carnation Creek and Queen Charlotte Islands fish/forestry workshop: Applying 20 years of coastal research to management solutions. Land Management Handbook 41. British Columbia Ministry of Forests, Research Branch, Victoria.

Holtby, L. B. 1988. Effects of logging on stream temperatures in Carnation Creek, British Columbia, and associated impacts on the coho salmon (*Oncorhynchus kisutch*). Canadian Journal of Fisheries and Aquatic Sciences 45:502–515.

Holtby, L. B., B. C. Andersen, and R. K. Kadowaki. 1990. Importance of smolt size and early ocean growth to interannual variability in marine survival of coho salmon (*Oncorhynchus kisutch*). Canadian Journal of Fisheries and Aquatic Sciences 47:2181–2194.

Holtby, L. B., and J. C. Scrivener. 1989. Observed and simulated effects of climatic variability, clearcut logging, and fishing on the numbers of chum salmon (*Oncorhynchus keta*) and coho salmon (*O. kisutch*) returning to Carnation Creek, British Columbia. Pages 61–81 *in* C. D. Levings, L. B. Holtby, and M. A. Henderson, editors. Proceedings of the national workshop on effects of habitat alteration on salmonid stocks. Canadian Special Publications of Fisheries and Aquatic Sciences 96.

Karinen, J. F., B. L. Wing, and R. Straty. 1985. Records of sightings of fish and invertebrates in the eastern Gulf of Alaska and ocean phenomena related to the 1983 El Niño event. Pages 253–267 *in* W. S. Wooster and D. L. Fluharty, editors. El Niño North: El Niño effects in the eastern subarctic Pacific Ocean. Washington Sea Grant Program, University of Washington, Seattle.

Koski, K. V. 1975. The survival and fitness of two stocks of chum salmon (*Oncorhynchus keta*) from egg deposition to emergence in a controlled stream environment at Big Beef Creek. Doctoral dissertation. University of Washington, Seattle.

Krajina, V. J. 1969. Ecology of forest trees in British Columbia. Pages 1–146 *in* V. J. Krajina and R. C. Brooke, editors. Ecology of western North America. Department of Botany, University of British Columbia, Vancouver.

Lawson, P. W. 1993. Cycles in ocean productivity, trends in habitat quality, and the restoration of salmon runs in Oregon. Fisheries 18:(8)6–10.

Lightly, D. T., M. J. Wood, and S. R. Heizer. 1985. The status of chum salmon stocks on the west coast of Vancouver Island, 1951–1982, Statistical Areas 22–27. Canadian Technical Report of Fisheries and Aquatic Sciences 1366.

Mathews, S. B. and R. Buckley. 1976. Marine mortality of Puget Sound coho salmon (*Oncorhynchus kisutch*). Journal of the Fisheries Research Board of Canada 33: 1677–1684.

McNeil, W. J. 1966. Effect of the spawning bed environment on reproduction of pink and chum salmon. U.S. Fish and Wildlife Service Fisheries Bulletin 65:495–523.

Nelson, T. C. 1993. Stamp Falls fishway counts, adipose clip/CWT recovery and biological sampling of chinook salmon escapements in Stamp River and Robertson Creek Hatchery, 1992. Canadian Manuscript Report of Fisheries and Aquatic Sciences 2213.

Rice, J., and nine coeditors. 1995. Pacific stock assessment review committee (PSARC) annual report for 1994. Canadian Manuscript Report of Fisheries and Aquatic Sciences 2318.

Scrivener, J. C. 1988a. Two devices to assess incubation survival and emergence of salmonid fry in an estuary streambed. North American Journal of Fisheries Management 8:248–258.

Scrivener, J. C. 1988b. Changes in the composition of the streambed between 1973 and 1985 and the impacts on salmonids in Carnation Creek. Pages 59–65 *in* T. W. Chamberlin, editor. Proceedings of the workshop: Applying 15 years of Carnation Creek results. Carnation Creek Steering Committee, Pacific Biological Station, Nanaimo, British Columbia.

Scrivener, J. C. 1988c. A summary of the population responses of chum salmon to logging in Carnation Creek, British Columbia, between 1971 and 1986. Pages 150–158 *in* T. W. Chamberlin, editor. Proceedings of the workshop: Applying 15 years of Carnation Creek results. Carnation Creek Steering Committee, Pacific Biological Station, Nanaimo, British Columbia.

Scrivener, J. C. 1991. An update and application of the production model for Carnation Creek chum salmon. Pages 210–219 *in* B. White and I. Guthrie, editors. Proceedings of the 15th Northeast Pacific pink and chum salmon workshop. Pacific Salmon Commission, Canada Department of Fisheries and Oceans, Vancouver, British Columbia.

Scrivener, J. C., and B. C. Andersen. 1984. Logging impacts and some mechanisms that determine the size of spring and summer populations of coho salmon fry (*Oncorhynchus kisutch*) in Carnation Creek, British Columbia. Canadian Journal of Fisheries and Aquatic Sciences 41:1097–1105.

Scrivener, J. C., and M. J. Brownlee. 1982. An analysis of Carnation Creek gravel-quality data, 1973–1981. Pages 154–176 *in* G. F. Hartman, editor. Proceedings of the Carnation Creek workshop, a 10-year review. Pacific Biological Station, Nanaimo, British Columbia.

Scrivener, J. C., and M. J. Brownlee. 1989. Effects of forest harvesting on spawning gravel and incubation survival of chum (*Oncorhynchus keta*) and coho salmon (*O. kisutch*) in Carnation Creek, British Columbia. Canadian Journal of Fisheries and Aquatic Sciences 46:681–696.

Seber, G. A. F., and E. D. LeCren. 1967. Estimating population parameters from catches large relative to the total population. Journal of Animal Ecology 36:631–643.

Simpson, K., B. Holtby, R. Kadowaki, and W. Luedke. 1996. Assessment of coho stocks on the west coast of Vancouver Island. Pacific Stock Assessment Review Committee. PSARC Working Paper S96-10.

Sowden, T. D., and G. Power. 1985. Prediction of steelhead trout survival in relation to groundwater seepage and particle size of spawning substrates. Transactions of the American Fisheries Society 114:804–812.

Thedinga, J. F., and K. V. Koski. 1984. A stream ecosystem in an old-growth forest in southeastern Alaska. Part IV: The production of coho salmon smolts and adults from Porcupine Creek. Pages 99–114 *in* W. R. Meehan, T. R. Merrell, Jr., and T. R. Hanley, editors. Fish and wildlife relationships in old-growth forests. American Institute of Fisheries Research Biology, Juneau, Alaska.

Toews, D. A., and M. K. Moore. 1982. The effects of streamside logging on large organic debris in Carnation Creek. Province of British Columbia, Ministry of Forests, Land Management Report 11, Victoria.

Tschaplinski, P. J. 1982a. Winter distribution of juvenile coho salmon (*Oncorhynchus kisutch*) in Carnation Creek and some implications to overwinter survival. Pages 273–288 *in* G. F. Hartman, editor. Proceedings of the Carnation Creek workshop, a 10-year review. 24–26 February 1982. Pacific Biological Station, Nanaimo, British Columbia.

Tschaplinski, P. J. 1982b. Aspects of the population biology of estuary-reared and stream-reared juvenile coho salmon in Carnation Creek: A summary of current research. Pages 289–307 *in* G. F. Hartman, editor. Proceedings of the Carnation Creek workshop, a 10-year review. Pacific Biological Station, Nanaimo, British Columbia.

Tschaplinski, P. J. 1987. Comparative ecology of stream-dwelling and estuarine juvenile coho salmon (*Oncorhynchus kisutch*) in Carnation Creek, Vancouver Island, British Columbia. Doctoral dissertation. University of Victoria, Victoria, British Columbia.

Tschaplinski, P. J. 1988. The use of estuaries as rearing habitats by juvenile coho salmon. Pages 123–142 *in* T. W. Chamberlin, editor. Proceedings of the workshop: Applying 15 years of Carnation Creek results. Carnation Creek Steering Committee, Pacific Biological Station, Nanaimo, British Columbia.

Tschaplinski, P. J., and G. F. Hartman. 1983. Winter distribution of juvenile coho salmon (*Oncorhynchus kisutch*) before and after logging in Carnation Creek, British Columbia, and some implications for overwinter survival. Canadian Journal of Fisheries and Aquatic Sciences 40:452–461.

Tschaplinski, P. J., J. C. Scrivener, and L. B. Holtby. 1998. Long-term patterns in the abundance of Carnation Creek salmon, and the effects of logging, climate variation, and fishing on adult returns. Pages 155–181 *in* D. L. Hogan, P. J. Tschaplinski, and S. Chatwin, editors. Carnation Creek and Queen Charlotte Islands fish/forestry workshop: Applying 20 years of coastal research to management solutions. Land Management Handbook 41. British Columbia Ministry of Forests, Research Branch, Victoria.

Walters, C. J., R. Hilborn, R. M. Peterman, and M. J. Staley. 1978. Model for examining early ocean limitation of Pacific salmon production. Journal of the Fisheries Research Board of Canada 35:1303–1315.

Ware, D. M., and G. A. McFarlane. 1988. Relative impact of Pacific hake, sablefish and Pacific cod on west coast Vancouver Island herring stocks. International North Pacific Fisheries Commission Bulletin 47.

20 Habitat Assessment in Coastal Basins in Oregon: Implications For Coho Salmon Production and Habitat Restoration

Kim K. Jones and Kelly M. S. Moore

Abstract.—Quantitative habitat surveys have been conducted in western Oregon streams since 1990. Over 950 streams, a total of 6,000 kilometers, have been surveyed in coastal basins with the results organized into over 3,100 reaches characterized by land use, channel morphology, and valley form. The data have been compiled into a comprehensive database that describes key attributes of instream habitat, riparian structure, and channel morphology. The information was used to describe current status of habitat throughout the coastal basins and the potential to support coho salmon *Oncorhynchus kisutch* populations. Example maps and evaluations were developed for the Yaquina River watershed to describe and compare coho salmon habitat. The datasets will support sustainability because they can be used to estimate potential survival and production of juvenile coho salmon in coastal basins, to identify core habitats, for designing and evaluating monitoring programs, and for developing restoration strategies.

INTRODUCTION

Stream surveys have been an integral part of fish management in coastal basins of Oregon for the past 100 years. In the late 1800s and early 1900s, surveys focused on single issues, such as siting capture facilities for broodstock, identifying potential hatchery sites, describing barriers to upstream passage, and estimating quantities of spawning gravel. In the late 1930s and 1940s the Oregon Game Commission surveyed biotic conditions in streams relative to the potential to support hatchery fish. The Fisheries Commission and the Game Commission of Oregon continued surveys to evaluate spawning gravel and rearing habitat, upstream passage problems, and overall watershed condition through the early 1970s. The physical and biological survey format (ODFW 1977) was accepted as the standard survey method after the Game and Fisheries Commissions were merged in 1975. Though broadly conceived, in practice these surveys were qualitative in nature and typically focused on selected fish populations or on one aspect of a species' life history.

During the past 20 years, many anadromous and resident fish populations in the coastal basins have declined in abundance or become fragmented and isolated. Habitat degradation was one of many factors contributing to these changes, but data that quantified the associated changes in habitat were largely absent. Hence, the inapplicability of old survey methodology and information to new questions and concerns became apparent.

The necessity to address a complex and broader set of issues required a new approach to stream surveys, one that considered all aspects of habitat throughout a watershed. The new approach needed to include both key characteristics of stream habitat and the processes that link those key characteristics

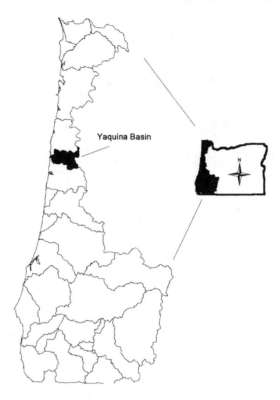

FIGURE 20.1   Major coastal watersheds in Oregon. Surveys described in this chapter were conducted in the Yaquina basin (highlighted in black).

within a stream and to the watershed. Managers needed to be able to relate the condition of aquatic habitat to salmonid production in streams and to develop effective stream restoration strategies, at both the local and regional levels. Results of new surveys can and must contribute to long-term sustainability of salmonid populations by helping to (1) identify critical or core habitat; (2) provide information for designing and evaluating monitoring programs; (3) develop effective restoration projects; (4) design sampling for estimating abundance and distribution of fish populations; and (5) refine production and survival models. In this chapter, however, we focus on the overall approach to contemporary habitat surveys, the habitat characteristics described from the survey data, and use of the survey data in habitat assessment. Equally important is the use of habitat characteristics to predict or estimate the potential of a stream to support juvenile salmon (Nickelson et al. 1992a; Nickelson 1998) and to identify core habitats (Nicholas 1997). Last, habitat assessments are useful in locating and designing restoration projects (Nicholas 1997). Examples from the Yaquina River basin (Figure 20.1), in Oregon's mid-coast, will be used to illustrate the analysis and application of the survey information.

METHODS

SURVEY DESIGN

An aquatic inventory protocol was designed by the Oregon Department of Fish and Wildlife (ODFW) in 1990 to facilitate the collection of quantitative information on stream habitat conditions throughout Oregon. The objectives of the habitat inventories were to provide technical information that could be used to:

- Develop habitat protection and restoration strategies
- Estimate juvenile fish production and survival
- Provide information for the aquatic component of watershed analyses
- Establish applicable and measurable monitoring standards

The inventory protocol was comprehensive enough to provide a standard methodology that could be applied across different ecoregions, and for different land-use or land-management regimes. The method was also designed to be integrated with other watershed monitoring activities, such as temperature monitoring, water quality sampling, and fish population surveys. Finally, the methodology provided flexibility of scale. Information was summarized at various levels, including microhabitat, associations of habitat, portions or reaches of streams, watersheds, and regions. Because ODFW was the only agency responsible for fish in all waters of the state, it was essential that the survey protocol provide a widely applicable base of information.

The conceptual background for this work came from the research experience of ODFW staff and from interactions with Oregon State University (OSU) scientists, the forest industry, and United States Forest Service (USFS) Pacific Northwest Research Station scientists (Bisson et al. 1982, Frissell et al. 1986; Everest et al. 1987, Grant 1988, Hankin and Reeves 1988, Moore and Gregory 1989, and Gregory et al. 1991). Annual review and modification of the methodology has been expedited with the assistance of ODFW and USFS biologists, who have applied the survey results or who are working on similar programs.

The survey design was based on a continuous walking survey of a stream from the mouth or confluence to the headwaters. The surveys were designed to be most appropriate for 1st through 5th order streams. Each stream was stratified into a series of long sections called reaches and into short habitat units within each reach. This survey approach uses a visual estimation technique described by Hankin and Reeves (1988) to allow expeditious estimates of habitat area and characteristics. Within a watershed, crews surveyed the major streams and a selection of small tributaries. Smaller tributaries were selected based on flow, fish use, land management, or randomly depending on the purpose of the survey. The streams usually represented about 30% of the streams on a 1:100,000 USGS stream coverage, and up to 50% of the stream miles inhabited by coho salmon *Oncorhynchus kisutch* within a watershed. Digital map layers in a geographic information system (GIS), such as digital elevation models, geology, and vegetation, were used to describe characteristics of unsurveyed streams and were verified by limited field observations. Accurate representation of unsurveyed streams for habitat monitoring at a basin scale was not the goal of this work but would require randomly selecting and surveying short reaches throughout a basin.

We inventoried all habitat units within a stream. Other survey approaches rely on surveying "representative" sections or on systematically surveying short sections of stream. However, it was considered that a continuous survey would provide more accurate estimates of habitat conditions throughout a stream (Dolloff et al. 1997), provide a complete inventory of barriers to up- or downstream migration of fish (such as falls or culverts), describe relationships of habitat conditions and hydrologic processes in one section of stream to another or to landscape features, and provide the sample template necessary to estimate distribution and abundance of fish populations throughout a stream (Hankin 1986; Hankin and Reeves 1988). A watershed-level survey also provides the sample template for designing monitoring programs at different scales (Sedell and Luchessa 1982; Hawkins et al. 1993; McIntosh et al. 1994). Continuous surveys also allow results to be integrated into GIS map layers for more powerful analyses, such as watershed analysis, and for display to managers and the public (McKinney et al. 1996).

Roper and Scarnecchia (1995) suggested that detailed studies of short stream sections may be more accurate than continuous surveys throughout a stream because of inconsistencies in surveyor observations. We minimized those by (1) requiring all surveyors and field supervisors, regardless of experience level, to attend a 3-day training session; (2) having field supervisors regularly work

TABLE 20.1
Description of key habitat features in stream survey.

Habitat feature	Description
Valley and channel morphology	Relationships of hillslopes, valley floor, terraces, floodplains to stream channel
Secondary channels	Amount of off-channel habitat
Gradient	Stream slope in percent
Pool characteristics	Type, size, depth, complexity
Fast water unit complexity	Slope, substrate, boulders, wood
Large woody debris	Size and type of wood within active channel
Substrate characteristics	Substrate type, percent of fine sediments in riffles
Bank erosion	Percent of bank erosion at active channel margin
Stream shade	Exposure of channel to sunlight
Riparian composition	Size and type of trees within 30 meters of channel margin

in the field with the surveyors and review their data during the field season; and (3) periodically conducting resurveys to better understand sources of error and amount of variation. We also improved consistency and efficiency by defining habitat units based on active channel width rather than wetted width.

Field Methods

Stream surveys began with a collection of general information from maps and other sources, followed by the direct observations of stream characteristics using Moore et al. (1997). The information was both collected and analyzed based on a hierarchical system of regions, basins, streams, reaches, and habitat units. The survey teams collected field data based on stream, reach, and habitat units. The field data focused on channel and valley morphology (stream and reach data), riparian characteristics and condition (reach data), and instream habitat (habitat unit data).

The stream surveys were organized from mouth to headwaters by reach and channel units. Reaches described the valley geomorphology, land use, riparian characteristics, and stream flow, and varied in length from as short as 1/2 km to more than 8 km. Valley and channel morphology defined the stream configuration and level of constraint that local landforms such as hillslopes or terraces imposed upon the stream channel (Moore and Gregory 1989; Gregory et al. 1991). The channel was described as hillslope constrained, terrace constrained, or unconstrained. These descriptions of channel morphology have equivalents within the Rosgen channel typing system (Rosgen 1985). The streamside vegetation within 30 m of the active channel was described by collecting information on diameter and type of tree every 0.50 to 1 km along the stream.

Within each reach, the stream was described as a sequence of channel or habitat units. Each unit was longer than one active channel width and was an area of relatively homogeneous slope, depth, and flow pattern representing different channel forming processes. The units were classified into 22 hierarchically organized types of pools, glides, riffles, rapids, steps, and cascades (Bisson et al. 1982; Hawkins et al. 1993; Moore et al. 1997). In each unit, the length and width were visually estimated, and adjusted by calibration (Hankin and Reeves 1988). At every unit, attributes were estimated or measured to describe water surface slope, depth, substrate, woody debris, shade, features of instream cover, and bank stability (Table 20.1). Substrate characteristics were visually estimated at every habitat unit. Estimates of percent silt, sand, and gravel in low gradient (1 to 2%) riffles were used to describe gravel quantity and quality.

The methodology and the characteristics used in surveys conducted by the ODFW were similar to surveys conducted by other public agencies such as the USFS and the U.S. Bureau of Land Management (BLM). This compatibility of survey methods, while allowing each agency or group to achieve its objectives, prevented a duplication of survey efforts.

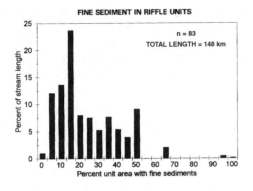

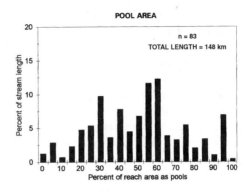

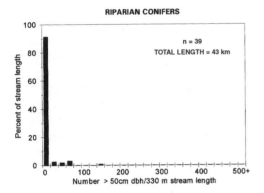

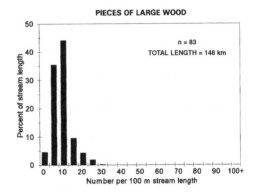

FIGURE 20.2   Frequency histograms of selected habitat characteristics in Yaquina basin streams expressed as a percent of stream length.

MANAGEMENT AND ANALYSES OF HABITAT DATA

The data from each stream were compiled by reach and by habitat unit type. A series of tables, longitudinal profiles, habitat graphs, maps, and photos were provided for each. These reports became the basis for summaries at a larger spatial scale: watersheds, basins, and ecoregions. The habitat survey results were also presented in a geographic format by dynamically segmenting the habitat units and reaches in an ArcInfo GIS onto a 1:100,000 or 1:24,000-scale digitized stream layer to display habitat features in map view.

The habitat characteristics summarized by reach were entered into a database that included watershed features such as ecoregion and basin size. Each record in the database included 80 fields that summarized information for each reach of stream. Summaries were developed that included all surveyed streams within a watershed, basin, or ecoregion. As an example, the streams we surveyed in the Yaquina basin totaled 100 km. Fields in the reach database summarized the key habitat features listed in Table 20.1 for streams in the Yaquina basin. Comparisons of important habitat features could be made among reaches, streams, watersheds, or ecoregions. Graphs using the data from the reach database were developed that provided a quick assessment of individual characteristics in the Yaquina basin (Figure 20.2). The data from each reach was adjusted for reach length so the height of the bars represent a true proportion of stream length in the watershed.

ESTIMATES OF HABITAT QUALITY

Two methods were used to assess the quality of habitat for juvenile coho salmon in the Yaquina River basin. The first was a qualitative description of individual characteristics or combination of characteristics that defined adult holding, spawning, and juvenile rearing habitat, while the second was described in terms of egg-to-smolt survival of coho salmon. In the first method, we compared the characteristics to a set of quantitative benchmarks (Nicholas 1997) that defined a range of conditions as desirable, intermediate, or undesirable. We mapped the location and quality of habitat features to graphically depict high quality stream reaches for different life history stages.

Benchmark values, derived from reference conditions, analysis of the distribution of observed values, and compiled from published values, provided the initial context for evaluating measures of habitat quality. Comparison of habitat measures to benchmark values, however, must be made with caution, taking into consideration both the geomorphic and biological template that defines the potential of the system and the combination of natural disturbance and management history that influence the expression of that potential. The benchmarks provided a context for interpretation and a starting point for more detailed and meaningful analysis. Benchmark values were also very useful for looking at overall conditions within a watershed, basin, or region. Under natural conditions some percentage of a watershed, basin, or region may always be classified as below desirable condition (Reeves et al. 1995).

The habitat benchmark values for desirable (good) and undesirable (poor) conditions were derived from a variety of sources. Habitat characteristics representative of conditions in stream reaches with high productive capacity for salmonid species were used as a starting point. Values from "reference" reaches were used to develop standards for large woody debris and riparian conditions. These reference values were then compared to the overall distribution of values for each habitat characteristic expressed as a frequency distribution within a basin or region. From this analysis, it was generally apparent that values from the 66th or higher percentile could represent desirable or good conditions and values from the 33rd or lower percentile represent undesirable or poor conditions. We recognize that these represent arbitrary cuttoffs, yet found the classification useful in subsequent analyses. This development of benchmarks from the frequency distributions of the habitat features data was made specific to appropriate stream gradient, regional, and geologic groupings of the reach data. Finally, values for habitat characteristics, such as pool frequency and amount of fine sediments, were developed by comparing the distributions to generally accepted or published values (Everest et al. 1987; Keller and Melhorn 1978).

In the second method of assessment, habitat quality was described in terms of coho salmon survival from egg to smolt using a regression model (Nickelson 1998). The model was an extrapolation of the habitat relationships presented in Nickelson et al. (1992b) and the habitat limiting factors model (Nickelson et al. 1992a) for streams in coastal basins of Oregon. Survival of coho salmon from egg to smolt was predicted as a function of active channel width (up to 20 m wide), gradient, the number of beaver dams per km, and the percent of pools in the reach. Survival was highest in reaches with a wide channel, low gradient, large number of beaver pools, and a high percentage of pools. The model weighted the variables most critical to overwinter survival of juvenile coho salmon (Nickelson 1998). Variables such as amounts of large woody debris and percent of fine sediment were not incorporated in the model because these variables did not vary much in the study streams used to develop the relationships. An important product of the model was the depiction of relative habitat quality as a function of measurable variables and the ability to compare habitat quality among reaches. Nickelson (1998) defined streams with egg to smolt survival greater than 3% as productive.

RESULTS AND DISCUSSION

HABITAT ASSESSMENT

Surveys conducted throughout the state represented all major ecoregions and many of the major basins in Oregon. Approximately 6,000 km of stream were surveyed in 950 streams located in 35 coastal watersheds which drained into 16 estuaries. The coastal basins varied by size and climatic

TABLE 20.2

Qualitative rating of key habitat characteristics relative to benchmarks for streams in the Yaquina basin expressed as percent of total length surveyed (n = 83 reaches).

Characteristic	Rating (Percent stream length)		
	Undesirable	Intermediate	Desirable
Pool Area (% total stream area)	4	27	69
Pool Frequency (distance between pools)	21	45	34
Gravel Availability (% gravel in riffles)	7	29	64
Gravel Quality (% fines in riffles)	50	37	13
Large Woody Debris (LWD) Pieces	40	58	2
LWD Volume	94	4	2
Riparian Conifers 50+ cm dbh/330 m*	98	2	0
Stream Shade (% canopy closure)**	0	—	100

* diameter at breast height

** streams <12 meters wide.

zone, and included three major ecoregions, which were defined based on climate, geology, and vegetation (Omernik et al. 1995).

The description of habitat characteristics in the Yaquina basin summarized in the stream reach database and frequency graphs provided an opportunity to assess the overall condition of aquatic habitat compared to benchmarks (Table 20.2). Surveyed streams in the Yaquina basin had channels that were often constrained by high terraces in the lowlands, and by hillslopes in the higher elevations. Very few streams had active floodplains. The streams had a high percentage of pools (70%) and an intermediate to desirable frequency of pools (79%) (Table 20.2). However, the number and volume of large wood debris in the channels were considered undesirable in more than 40% and 90% of the stream length, respectively. Very few pools deeper than 1 m and containing at least 3 pieces of large wood were observed, although numerous beaver ponds were present in the basin. Riparian areas were comprised almost entirely (98%) of hardwoods and small conifers (Table 20.2) which limits the recruitment of large wood from the riparian zone. The quantity of gravel was desirable in most of the streams, although the quality was undesirable in 50% of the stream length because of high amounts of silt and sand embedded in the gravel (Table 20.2).

Aquatic conditions in the Yaquina River basin have been influenced by urbanization, roadbuilding, agriculture, and timber harvest. These effects have been compounded by major floods. Many low gradient stream reaches formerly influenced by an adjacent flood plain have been isolated from the valley floor by high terraces along the stream. Riparian structure, large wood, complex pools, and off-channel habitats have been lost in many areas. Conifer-dominated riparian zones have largely been replaced by alder and small conifers. As a consequence, the quality of stream habitat in the Yaquina basin can now be characterized as undesirable to intermediate in terms of structural components, such as large woody debris and riparian structure. However, the number of beaver ponds and pools is considered to be desirable (Table 20.2). Changes in the structure and complexity of freshwater habitat have coincided with the decline of coho salmon in coastal basins (Nicholas 1997).

FISH PRODUCTIVITY

Habitat conditions influence the carrying capacity of the habitat for salmonid production and affect the survival of each freshwater life stage of salmonids. Everest et al. (1987), Nickelson et al. (1992a, b), Johnson and Solazzi (1995), Lestelle and Mobrand (1995), and Nickelson (1998) provided information on the relationship of habitat features to survival of salmonids and the carrying capacity

TABLE 20.3
Description of relationships between key habitat features and salmonid life history requirements.

Habitat feature	Importance to coho salmon
Valley and channel morphology	Unconstrained streams with adjacent floodplains have more opportunity for low gradient reaches with abundant off-channel and secondary channel habitat to benefit fry and juveniles
Secondary & off-channel areas	Provides quality low velocity rearing areas for fry and juveniles
Gradient	Low gradient reaches (1 to 4%) provide more low velocity areas for juveniles
Pool characteristics	Type, size, depth, complexity can improve rearing
Fast water unit complexity	Slope, substrate, boulders, wood can improve rearing by modifying current or reducing hydraulic forces
Large woody debris	Provides micro-habitats & cover for fry, juveniles, adults
Substrate characteristics	Influences fry emergence and rearing
Stream shade	Prevents lethal temperatures
Riparian composition	Maintains cooler temperatures, supplies large wood, and food

of habitat (Table 20.3). Our habitat assessment provided information on the amount and quality of habitat available in each stream relative to fish life history requirements.

The results of the Yaquina River basin habitat surveys were compiled into a digitized GIS stream layer to display habitat features and combinations of habitat features. The larger mainstem channels were migratory corridors from the juvenile rearing habitat to the estuary, while the smaller stream reaches were used for spawning and juvenile rearing. Figure 20.3 displays locations where some of the more important juvenile and adult coho habitat variables were rated as good, based on comparisons to benchmark values.

Spawning adults and incubating fry require proper size gravel that has low amounts of fine sediments embedded in the substrate (Everest et al. 1987). Distribution of these reaches is represented in Figure 20.3. Habitat attributes where rearing juvenile coho salmon survive best included low gradient, complex and deep pools or beaver ponds, and abundant large wood debris in the channel (Figure 20.3). We identified and depicted high quality juvenile rearing reaches as those stream reaches that flowed through a wide valley, were less than 10 m wide, had a gradient less than 4%, contained beaver pools, and had at least 30% of the surface area in pool habitat (Figure 20.3). Optimal rearing habitat for juvenile coho salmon also includes large wood debris, but very few of the reaches had intermediate or desirable amounts of large wood debris. Only three streams in the northeast portion of the basin were in excellent condition, containing abundant pools, more than 10 pieces of large wood debris per 100 m of stream length, and high quality substrate.

Using the numerical model of Nickelson (1998), the estimate of coho salmon survival from egg to smolt for the surveyed reaches ranged from less than 1% up to 4% (Figure 20.3, lower right panel). Most of the streams were predicted to have intermediate (1 to 3%) survival rates.

Estimates of the location of good habitat based on the benchmark method were different from that predicted by the survival model because the survival model weighted abundance and type of pools, whereas the benchmark method considered large wood debris and substrate as well. Both methods provided useful but slightly different information. Additional work to compare number and location of adult fish, fry densities, and summer and winter juvenile densities to habitat features is needed before the processes that influence productivity of the basin can be better understood.

Freshwater survival from egg to smolt in many of the surveyed Yaquina streams was predicted to be intermediate (1 to 3%), but there were some areas of potentially high productivity still present. In particular, the Upper Yaquina River in the northeast portion of the basin has good rearing potential because of the low gradient, pool-rich streams. Large wood debris provided some structural complexity

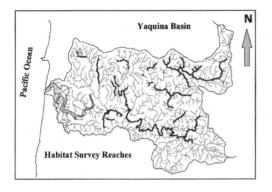
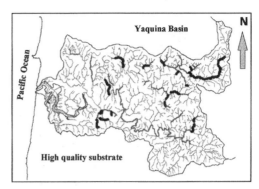
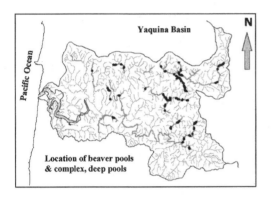
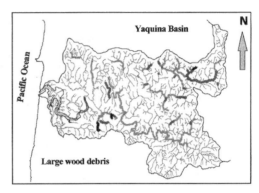
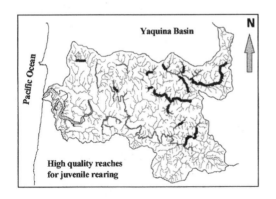
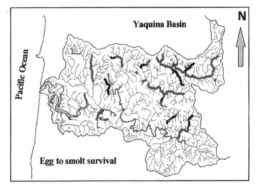

FIGURE 20.3 Maps depicting location of selected habitat features in the Yaquina basin. Top left: streams surveyed in the Yaquina basin. Top right: high quality spawning and incubation areas; reaches containing riffles with desirable amounts of gravel and low amounts of sand and silt. Center left: location of pools deeper than 1 meter with at least 3 pieces of large wood debris and beaver pools. Center right: amount of large wood in streams. Low (slim gray) = <10 pieces/100 meters of stream length; intermediate (dark gray) = 10–20 pieces/100 meters; high (black) = >20 pieces/100 meters. Bottom left: location of high quality rearing areas based on habitat characteristics (see text). Bottom right: estimates of freshwater survival of juvenile coho from egg to smolt. Low (slim gray) <1%; intermediate (dark gray) = 1 to 3%; high (black) >3%.

in those streams as well. These estimates of habitat quality have important implications for how we manage freshwater habitat and the fisheries that influence the number of returning adults (Nickelson 1998). The two habitat assessment methods were used in conjunction with known locations of high spawner abundance to map core habitats for coho salmon (Nicholas 1997)

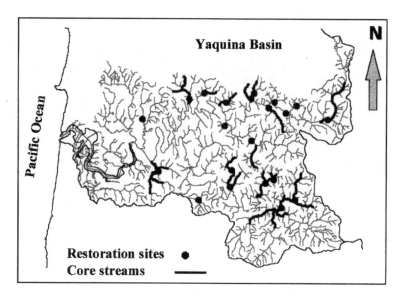

FIGURE 20.4   Location of habitat restoration sites and core streams for coho salmon in the Yaquina River system.

(Figure 20.4). The core habitats or streams will receive higher levels of habitat protection and will serve as primary sites for long-term monitoring.

RESTORATION STRATEGY DEVELOPMENT

The descriptions of habitat condition and production potential provided critical technical information for developing appropriate habitat restoration strategies in each basin. Mapping of important reach and unit characteristics was essential for identifying potential restoration sites that met specific hydrologic and biological criteria. Regional restoration plans were developed by considering both fish conservation strategies and reach-level habitat assessment information to identify candidate sites for instream and riparian restoration (Nicholas 1997) (Figure 20.4). In the absence of habitat surveys, GIS data layers were used to locate small to mid-sized stream reaches in low gradient, wide valleys. These tended to have the greatest potential for retaining instream structure because of their low energy and volume. Environments such as these also yielded the best opportunities for creating complex off-channel and secondary channel habitats. Site-specific project plans were based on the habitat unit-level information. For example, the habitat mapping data were used to identify locations that would benefit from large woody debris or other instream structure restoration or riparian composition improvement, depending on the existing stream characteristic or on fish life history requirements. At a watershed level, the habitat mapping information was used in the aquatic component of watershed analysis and was integrated into long-range management plans on public and private lands potentially influencing terrestrial, riparian, and aquatic ecosystems. By providing information at multiple scales, our habitat surveys have, and will continue to, support the sustainability of coho salmon in Oregon coastal basins.

ACKNOWLEDGMENTS

The stream surveys were conducted in cooperation with Oregon Department of Forestry, Oregon Forest Industry Council, private timber companies, BLM, USFS, and the USFWS. We appreciate thorough reviews by Ian Williams and one anonymous reviewer.

REFERENCES

Bisson, P. A., J. A. Nielsen, R. A. Palmason, and E. L. Grove. 1982. A system of naming habitat types in small streams, with examples of habitat utilization by salmonids during low stream flow. Pages 62–73 *in* N. B. Armantrout, editor. Acquisition and Utilization of Aquatic Habitat Inventory Information. Western Division, American Fisheries Society. Portland, Oregon.

Dolloff, C. A., H. E. Jennings, and M. D. Owen. 1997. A comparison of basinwide and representative reach habitat survey techniques in three southern Appalachian watersheds. North American Journal of Fisheries Management 17:339–347.

Everest, F. H., R. L. Beschta, J. C. Scrivener, K. V. Koski, J. R. Sedell, and C. J. Cederholm. 1987. Fine sediment and salmonid production: A paradox. Pages 98–142 *in* E. O. Salo and T. E. Cundy, editors. Streamside Management: Forestry and Fishery Interactions. Contribution 57, Institute of Forest Resources, University of Washington, Seattle.

Frissell, C. A., W. J. Liss, C. E. Warren, and M. D. Hurley. 1986. A hierarchical framework for stream habitat classification: viewing streams in a watershed context. Environmental Management. 10:199–214.

Grant, G. E. 1988. Morphology of high gradient streams at different spatial scales, western Cascades, Oregon. Pages 1–12 *in* Shizouka Symposium on Geomorphic Change and the Control of Sedimentary Load in Devastated Streams, Shizouka University, Shizouka, Japan.

Gregory, S. V., F. J. Swanson, and W. A. McKee. 1991. An ecosystem perspective of riparian zones. BioScience 40:540–551.

Hankin, D. G. 1986. Sampling designs for estimating the total number of fish in small streams. U.S. Forest Service Research Paper PNW-360, Portland, Oregon.

Hankin, D. G., and G. H. Reeves. 1988. Estimating total fish abundance and total habitat area in small streams based on visual estimation methods. Canadian Journal of Fisheries and Aquatic Sciences 45:834–844.

Hawkins, C. P., and ten co-authors. 1993. A hierarchical approach to classifying stream habitat features at the channel unit scale. Fisheries 18(6):3–12.

Johnson, S. L., and M. F. Solazzi. 1995. Development and evaluation of techniques to rehabilitate Oregon's wild salmonids. Oregon Department Fish and Wildlife, Fish Division Fish Research Project F-125-R, Annual Progress Report, Portland.

Keller, E. A., and W. N. Melhorn. 1978. Rhythmic spacing and origin of pools and riffles. Geological Society of America Bulletin 89:723–730.

Lestelle, L. C., and L. E. Mobrand. 1995. Application of an ecosystem analysis method to the Grande Ronde Watershed, using spring chinook salmon as a diagnostic species. Bonneville Power Administration. Portland, Oregon.

McIntosh, B. A., J. R. Sedell, N. E. Smith, R. C. Wissmar, S. E. Clarke, G. H. Reeves, and L. A. Brown. 1994. Historical Changes in fish habitat for select river basins of eastern Oregon and Washington. Northwest Science 68 (Special Issue):36–53.

McKinney, S. P., J. O'Conner, C. K. Overton, K. MacDonald, K. Tu, and S. Whitwell. 1996. A characterization of inventoried streams in the Columbia River basin. Aqua-Talk. R-6 Fish Habitat Relationship Technical Bulletin 11, Portland, Oregon.

Moore, K. M., and S. V. Gregory. 1989. Geomorphic and riparian influences on the distribution and abundance of salmonids in a Cascade Mountain Stream. Pages 256–261 *in* D. Abell, editor. Proceedings of the California Riparian Systems Conference. Pacific Southwest Forest and Range Experiment Station General Technical Report PSW-110. Berkeley, California.

Moore, K. M. S., K. K. Jones, and J. M. Dambacher. 1997. Methods for stream habitat surveys. Oregon Department of Fish and Wildlife, Information Report 97-4, Portland, Oregon.

Nicholas, J. W., editor. 1997. Coastal Salmon Restoration Initiative: The Oregon Plan. Report to the National Marine Fisheries Service. State of Oregon, Salem.

Nickelson, T. E. 1998. A habitat-based assessment of coho salmon production potential and spawner escapement needs for Oregon coastal streams. Oregon Department of Fish and Wildlife, Information Report 98-4, Portland.

Nickelson, T. E., M. F. Solazzi, S. L Johnson, and J. D. Rodgers. 1992a. An approach to determining stream carrying capacity and limiting habitat for coho salmon (*Oncorhynchus kisutch*). Pages 251–260 *in* L. Berg and P. W. Delaney, editors. Proceedings of the coho workshop. Nanaimo, British Columbia.

Nickelson, T. E., J. D. Rodgers, S. L Johnson, and M. F. Solazzi. 1992b. Seasonal changes in habitat use by juvenile coho salmon (*Oncorhynchus kisutch*) in Oregon coastal streams. Canadian Journal of Fisheries and Aquatic Sciences 49:783–789.

Omernik, J. M., S. A. Thieve, and D. Peter. 1995. Draft Level III and IV Ecoregions of Oregon and Washington. U.S. Environmental Protection Agency Research Lab, Corvallis, Oregon.

Oregon Department of Fish and Wildlife. 1977. Manual for Fish Management. Portland, Oregon.

Reeves, G. H., L. E. Benda, K. M. Burnett, P. A. Bisson, and J. R. Sedell. 1995. A disturbance-based ecosystem approach to maintaining and restoring freshwater habitats of evolutionarily significant units of anadromous salmonids in the Pacific Northwest. Pages 334–349 *in* J.L. Nielson, editor. Evolution and the aquatic ecosystem: defining unique units in population conservation. American Fisheries Society Symposium 17. Bethesda, Maryland.

Roper, B. B., and D. L. Scarnecchia. 1995. Observer variability in classifying habitat types in stream surveys. North American Journal of Fisheries Management 15:49–53.

Rosgen, D. L. 1985. A stream classification system. Pages 95–100 *in* Johnson, R. R., C. D. Zeibell, D. R. Patton, P. F. Folliott, and R. H. Hamre, editors. Riparian ecosystems and their management; reconciling conflicting uses. First American Riparian Conference, U.S. Forest Service General Technical Report RM-120. Fort Collins, Colorado.

Sedell, J. R., and K. J. Luchessa. 1982. Using the historic record as an aid to salmonid habitat enhancement. Pages 210–223 *in* N. B. Armantrout, editor. Acquisition and Utilization of Aquatic Habitat Inventory Information. American Fisheries Society, Western Division. Bethesda, Maryland.

21 An Overview Assessment of Compensation and Mitigation Techniques Used to Assist Fish Habitat Management in British Columbia Estuaries

Colin D. Levings

Abstract.—A comprehensive fisheries management strategy to ensure sustainability of salmon and steelhead needs to arrest and reverse the ongoing and piecemeal loss of estuarine fish habitat in the northeast Pacific. Prevention, mitigation, compensation, and restoration are techniques used by Canadian fisheries agencies when managing estuarine fish habitat for no net loss of productive capacity. In the context of the Department of Fisheries and Oceans habitat management policy, mitigation includes prevention, alternate siting, modification of proposals, ecological zoning, or other techniques used for avoiding site-specific damage. Mitigation by alternate siting and ecological zoning has been used extensively in the Fraser River estuary, where a comprehensive, large scale database was developed to determine the fisheries values of specific sites. In the Fraser River estuary and other British Columbia estuaries there have been more than 100 situations where compensation has been applied in efforts to recover productive capacity due to losses. Vegetation transplanting, channelization, construction of intertidal islands, creation of artificial reefs, and lowering the elevation of terrestrial habitat were some of the techniques used. The success of the compensation projects has been mixed. Some major shortcomings are the lack of performance criteria and remedial action where projects have failed. Habitat restoration is also being used as a strategy to achieve net gain in productive capacity, but few projects have been adequately evaluated. Better design criteria, long-term monitoring and use of the basic principles of adaptive management are needed to improve the success rate of both compensation and restoration projects. It will be very difficult to achieve sustainability of estuarine habitats because of the lack of comprehensive followup studies to assess performance of the projects and the ongoing disruption of small habitat areas.

INTRODUCTION

In response to industrial and urban development in coastal British Columbia (B.C.), habitat managers and scientists with the Canadian Department of Fisheries and Oceans (DFO) have developed methods and policies to minimize loss of productive fish habitat in estuaries. This chapter provides a brief summary of some of the contemporary methods used and problems encountered, with particular reference to the Fraser River estuary, near the City of Vancouver (population approximately 1.2 million).

Estuaries in B.C. are considered critical habitat for salmon as these fish are adapted to migrate through the mouth of the river twice, once as juveniles heading to sea and again as adults moving into the rivers to spawn. In addition, certain stocks of chinook salmon *Oncorhynchus tshawytscha* and chum salmon *O. keta* use the estuaries as nursery areas and are, therefore, dependent on estuarine habitat for feeding and as refuge from predation (e.g., Levings et al. 1989). Estuaries and their associated uplands are also critical for human settlements and transportation facilities. Resources from inland areas, such as coal and wood products, are moved from railways to ships in the estuaries. Fish habitats such as marshes, sand and mud flats, and seagrass beds can be adversely affected by such industrial activity.

POLICY CONTEXT

There are several aspects of the Canadian fish habitat policy that are particularly relevant to estuary management. The 1986 Policy on Fish Habitat Management (DFO 1986) gave the following working definitions for mitigation and compensation, which will be followed in this chapter:

> *mitigation**—actions taken during the planning, design, construction, and operation of works and undertakings to alleviate potential adverse effects on the productive capacity of fish habitats, including rejection of development proposals at a particular site.
> *compensation for loss*—the replacement of natural habitat, increase in the productivity of existing habitat, or maintenance of fish production by artificial means in circumstances dictated by social and economic conditions, where mitigation techniques and other measures are not adequate to maintain habitats.

More recent policy items include guidance pertinent to Section 35(2) under the Canadian Fisheries Act, which prohibits harmful alteration, disruption, and destruction of fish habitat unless authorized. If an authorization to alter or destroy fish habitat is permitted, then compensation must be provided (Metikosh 1997). The January 1997 passage of the Oceans Act calls for management of the coastal zone, including estuaries, on an ecosystem basis.

TYPES OF DISRUPTION

About 25 years ago, Odum (1970) described the insidious loss of estuarine habitat in North America. Loss was occurring in many small inconspicuous or unnoticeable areas through waste disposal and small scale disruption. This type of loss has also been prevalent in B.C. estuaries for the past two decades and is an ongoing problem that must be addressed. However, most fish habitat loss in B.C. estuaries has occurred as a result of diking and filling of intertidal and flood plain areas for agriculture and/or urban development, particularly in the late 19th century. In the Fraser River estuary, 70 to 90% of this type of habitat has been lost since European settlement began, with accompanying decreases in carbon supply to the ecosystem (Healey and Richardson 1996). There has also been substantial habitat loss from port construction, especially ferry terminals and bulk loading facilities (Levings and Thom 1994). Some lower reaches of rivers are prone to erosion from currents at high discharge. Riprap revetment has been placed at these areas to protect land, which leads to secondary loss of shoreline habitat. Changes in flow patterns in the estuaries have also occurred as training walls and causeways have been built to constrain flow in shipping channels (Levings 1980).

---

* Note the difference in meaning relative to U.S. policy language, where mitigation is similar to the Canadian terminology for compensation (see also Metikosh 1997)

METHODS FOR PREVENTION AND MITIGATION

In British Columbia, development proposals are reviewed through a referral system, which uses inference based on available information to make decisions. The referral method involves the exchange of information about a proposal among a variety of resource and planning agencies that are invited to provide comments. The shared information is then used to reach a joint consensus decision concerning the project in question. If permission is obtained for the development to proceed, the shared information is also used to specify the procedures needed to avoid unacceptable impacts.

Locating an industrial development on a shoreline characterized by habitat of lower value than other areas is a zoning technique that is routinely used in the Fraser River estuary. The Fraser estuary is perhaps the most intensively managed estuary in Canada because of its major fish stocks and rapidly growing human populations. To help decide where development will have less impact, an inventory of fish habitat along the entire shoreline (about 538 km) has been developed by the Fraser River Estuary Management Program (FREMP). The data are available on 1:2,500 scale maps and are currently being entered in a geographic information system. The shoreline is color coded, so that red zones are areas of high habitat value (including all compensation sites), yellow areas are intermediate value, and green areas are of low value. The data were originally gathered in the 1980s and have been updated recently (Kistritz et al. 1992). In other estuaries, where detailed data are not available or an estuary management plan not in place, a site-specific survey is conducted before a decision is made on habitat alteration.

There have also been some innovative methods developed to mitigate estuarine habitat loss. For example, a project was proposed where construction of a conveyor system was needed to move a product from a factory to a ship loading facility on the riverfront. To avoid disrupting the estuarine marsh between the factory and the loading facility, a suspended conveyor system was built on posts, with minimal impact on the marsh.

METHODS USED FOR COMPENSATION: SUCCESSES AND PROBLEMS

Several methods have been used to compensate for habitat loss in B.C. estuaries, but unfortunately very few follow-up studies have been conducted to evaluate their effectiveness. Therefore, in most cases, we do not know whether the compensation project was successful from a fisheries viewpoint or not. A brief review of some of the available evidence is given below.

VEGETATION TRANSPLANTING

The importance of estuarine marsh vegetation in fish food webs and as refuge from predation has been demonstrated in numerous studies (e.g., Sibert et al. 1977; Gregory and Levings 1996). Therefore, transplanting of marsh vegetation has been identified as an important tool for compensating habitat loss. Sedges, especially Lyngbyei's sedge *Carex lyngbyei*, have been used extensively in this application. In a long-term study at the Campbell River estuary, for example, sedges have survived 14 years after transplanting in 1982 on four intertidal artificial islands (Levings and Macdonald 1991; Levings, unpublished 1996 observations). Low wave energy, intermediate salinity, and stable substrates may be factors that have led to the relative success of this particular project. Other transplant projects elsewhere in the region have shown mixed success. For example, in the Fraser River estuary, a transplanted marsh developed on a dredged sand platform was buried by sand moving in from the adjacent river channel (Williams 1993). Another technique, used successfully in the Fraser estuary, is the construction of marsh benches. These structures are constructed at the appropriate elevation on the outside of the riprap revetment, which characterizes about 30% of the shoreline in the Fraser River estuary. Fine sediment is placed on the benches and then planted with cores of sedge rhizomes. Submergence and stability of all the transplant areas is a key factor affecting success since these physical variables influence whether or not fish can access the transplanted habitat. In a subsample of transplanted and natural marshes in the Fraser River estuary, some transplanted marshes showed more variability in submergence relative to natural sites (Levings and Nishimura 1996). In some instances higher elevations of the transplanted marshes resulted in reduced submergence time relative to natural sites, which meant the former locations would be flooded by water for relatively less time.

In view of the risk involved in developing successful compensation marshes, and the time required for them to approach the productive capacity of natural marshes, habitat managers apply safety factors to the particular areas involved. For example, a 2:1 factor is required for new marshes—that is, for each unit lost, two need to be transplanted (Levings 1991).

There have been eight compensation projects where the eelgrass *Zostera marina* has been transplanted in the Strait of Georgia. There are few published data on survival of transplanted eelgrass in British Columbia. Kistritz and Gollner (1995) reviewed seven transplant projects and found it difficult to draw conclusions about the success of the projects because monitoring data were sparse. In Puget Sound, techniques for transplanting this species have generally been unsuccessful (Thom 1990). Light availability, carbon reserves, and timing may be some of the determining factors for survival of the transplants (Zimmerman et al. 1995).

Pickleweed *Salicornia virginica* is a salt marsh plant which has been transplanted in an experiment at the Fraser River estuary (Pomeroy, et al. 1981) and a compensation project at an estuary on the east coast of Vancouver Island. Preliminary results at the latter site showed this species may be suitable for transplanting if conditions at the compensation site match those where the species occurs naturally.

CHANNELIZATION AND LOWERING OF TERRESTRIAL HABITAT

Because of extensive loss of intertidal and shallow subtidal habitat due to filling in estuaries, compensation projects have often involved lowering of land to create embayments or channels. There is an obvious and immediate advantage for fish as newly created productive capacity or living space becomes available for fish. An example is a project in the Fraser estuary where compensation was provided for habitat lost when supports for a highway bridge were built in the intertidal zone. An embayment of approximately 0.5 ha with a depth of about −2 m below chart datum was dug out of forested habitat. Sedge was then planted around the perimeter of the embayment. Unfortunately, the created shoreline was steep and unstable and the sedge transplants did not survive—ultimately colonization did occur from adjacent natural areas (Levings and Nishimura 1996). However, the habitat was definitely useable by fish even before marsh colonization was complete. Investigations of fish behavior showed that juvenile chinook salmon resided in the embayment about the same length of time relative to natural habitats (Hvidsten et al. 1996).

Lowering of terrestrial habitat to create an intertidal marine foreshore was used as a compensation measure in eight projects on the south coast of B.C., including Victoria and Nanaimo harbors, Menzies Bay, and sites in Knight Inlet (DFO, unpublished data). A study conducted on the northeast coast of Vancouver Island showed creation of intertidal lagoons could provide habitat for some stocks of juvenile coho salmon *O. kisutch* (Atagi 1994).

Fish are often adapted to feed on the edges or perimeters of habitats and, therefore, habitats that maximize shoreline length can also maximize productive capacity. The intertidal islands constructed at the Campbell River estuary, for example, were provided with channels and bays to increase their shape complexity (Levings and Macdonald 1991). Since they were constructed 14 years ago, some infilling of sediment has occurred, showing that maintenance of some aspects of the islands is necessary. In the Fraser River estuary, a system of freshwater tidal creeks about 2 km long was dug through riparian habitat as a habitat improvement measure. This technique was also used as a restoration measure at the Squamish River estuary, when tidal creeks were reconnected after a planned port development blocked flow (Ryall and Levings 1987).

ARTIFICIAL REEFS

Development of subtidal artificial reefs was initially attempted in our region in 1983 as a measure to compensate for sand and mudflat fish habitat lost to port construction at Roberts Bank on the Fraser River estuary. Since then, the technique has been applied at about eight other sites, including Nanaimo (Armstrong 1993) and Sidney harbors, Discovery Passage, Port Hardy, and in Seymour Inlet (DFO, unpublished data). Natural rock was usually used as a substrate but in some cases concrete blocks or large diameter concrete pipes were used. However, it should be noted there are difficulties determining whether fish use of these structures actually represents an increase in productive capacity or whether fish observed near the reefs are being attracted to it from adjacent habitats (Levings 1995).

HABITAT BANKS

Another emerging strategy is the construction of habitat banks, which are habitats created in anticipation of loss. In the Fraser River estuary, habitat managers and harbor authorities have developed about 3.5 ha of marsh habitat for use as a bank from which withdrawals can be made if industrial activity requires the loss of habitat in the future. The artificial reef in Nanaimo harbor was created to bank habitat toward future fill projects (DFO, unpublished). Some authors are of the opinion that mitigation banks, as they are called in the U.S., could be a very useful technique in estuarine management (e.g., Etchart 1995). However, this is more or less an unproven technique that tends to focus on the creation of a large amount of habitat in a particular reach of the estuary. In natural estuaries, habitat is more evenly distributed or has a pattern of patchiness that may be important for fish and wildlife. Research in estuaries from a landscape ecology perspective would help answer some of these questions and could perhaps be conducted by habitat fragmentation analyses, as used in forestry studies.

SUMMARY AND RECOMMENDATIONS

Because of the lack of comprehensive monitoring and follow-up studies to assess the results of habitat management in the estuaries, it is difficult to make general statements about how effective the programs described above have been in achieving sustainability of salmon habitat in British Columbia estuaries. Fiscal constraints have prevented assessment in the dozens of B.C. estuaries where habitat managers have practiced their trade. At the Fraser River estuary, it appears that estuarine habitat loss may have slowed. Kistritz (1996) determined that, since about 1985, there had been a "net gain" of 6 ha of brackish marsh habitat, at the expense of sand and mud flat habitat that had been planted with marsh vegetation. On the other hand, as Levings and Nishimura (1996) have shown, it is not clear whether all the created marshes will perform as planned. In addition, Hutchinson (1982) and Hutchinson et al. (1989) showed that expansion of brackish marsh was occurring fairly rapidly at certain areas on the Fraser estuary owing to natural sedimentation and colonization. Unfortunately, these gains have never been quantified, but need to be added to the area of created habitat identified by Kistritz (1996).

It is also unclear whether sufficient critical fish habitat remains to provide functions for the current anadromous fish populations that use the estuary. At any rate, estuarine fish habitat function should not be viewed in isolation from freshwater and marine influences (Bradford and Levings 1997). Major fish populations which are thought to be reliant on the Fraser River estuary as rearing habitat, such as chinook in the Harrison River, a lower Fraser tributary, have recently shown significant interannual variability in survival (Schubert et al. 1994), likely owing to the interactive levels of survival in their three primary habitats (freshwater, estuary, and ocean). On the other hand, non-anadromous fish, such as cyprinids, show relatively stable populations in the lower river and estuary (Healey and Richardson 1996) when data from recent surveys were compared with those obtained two decades ago.

Many estuaries around the world are in poor health and require rejuvenation. Many B.C. estuaries could also benefit from restoration, another strategy which may help. When humans suffer heart problems, we often can be revived using cardiopulmonary resuscitation (CPR). Estuarine ecosystems need ecological CPR: Conservation, Preservation, and Restoration. Because of the uncertain success of several of the compensation measures discussed above, fish habitat managers should strive to manage for total prevention of habitat loss, and ideally, net gain of habitat through restoration. Mitigation and prevention should be used to avoid habitat loss, in keeping with the precautionary principle. If compensation is to be used, as a last resort, performance criteria need to be established before the compensatory habitat is built. The basic principles of adaptive management should be followed. While safety factors are routinely applied in compensation projects, there are very few projects where fisheries-oriented performance criteria have been applied. Performance bonds may be needed at high risk areas to ensure that funds are in place to repair compensation works that may degrade over time or simply do not work as anticipated.

The cumulative loss of numerous small areas (range 10 to 1000 m^2), especially in small estuaries, will continue to be a major problem unless overall management plans are developed and implemented. These estuary plans require the involvement and commitment of the many community stakeholders in addition to fisheries agencies.

As estuaries begin to be managed on an ecosystem basis, as called for in the new Oceans Act, estuary management plans should be developed for all the major estuaries on the B.C. coast, providing an ideal opportunity for integrating the restoration assessments with other ecological projects in the coastal zone.

ACKNOWLEDGMENTS

I am grateful to Mike Flynn, Habitat and Enhancement Branch, DFO, for supplying unpublished information on habitat compensation projects on the coast of British Columbia. I also thank T. Brown, I. Williams, and an anonymous reviewer for helpful comments on the manuscript.

REFERENCES

Armstrong, J. W. 1993. A biological survey of an artificial reef at Newcastle Island, British Columbia. Canadian Manuscript Report of Fisheries and Aquatic Sciences 2212.

Atagi, D. Y. 1994. Estuarine use by juvenile coho salmon *Oncorhynchus kisutch*: is it a viable life history strategy? Master's thesis, University of British Columbia, Vancouver.

Bradford, M. J., and C. D. Levings. 1997. Report on the Workgroup "Partitioning Survival." Pages 279–282 in R. L. Emmett and M. H. Schiewe, editors. Proceedings of the workshop on estuarine and ocean survival of Northeastern Pacific salmon. NOAA Technical Memorandum NMFS-NWFSC-29.

DFO (Department of Fisheries and Oceans). 1986. The Department of Fisheries and Oceans Policy for the Management of Fish Habitat. DFO, Ottawa, Ontario.

Etchart, G. 1995. Mitigation banks: a strategy for sustainable development. Coastal Management 23:223–237.

Gregory, R. S., and C. D. Levings. 1996. The effects of turbidity and vegetation on the risk of juvenile salmon (*Oncorhynchus* spp.) to predation by adult trout (*O. clarkii*). Environmental Biology of Fishes 47:279–288.

Healey, M. C., and J. S. Richardson, 1996. Changes in the productivity base and fish populations of the lower Fraser River (Canada) associated with historical changes in human occupation. Archive Hydrobiologie (Supplement) 113 (Large Rivers 10) (1–4):279–290.

Hutchinson, I. 1982. Vegetation-environment relations in a brackish marsh, Lulu Island. Canadian Journal of Botany 60:452–462.

Hutchinson, I., A. C. Prentice, and G. Bradfield. 1989. Aquatic plant resources of the Strait of Georgia. Pages 50–60 in K. Vermeer and R. Butler, editors. The ecology and status of marine and shoreline birds in the Strait of Georgia, British Columbia. Special Publication, Canadian Wildlife Service. Catalogue Number CW66-98/1987E.

Hvidsten, N. A., C. D. Levings, and J. Grout. 1996. A preliminary study of hatchery chinook salmon smolts migrating in the lower Fraser River, determined by radiotagging. Canadian Technical Report of Fisheries and Aquatic Sciences 2085.

Kistritz, R. U. 1996. Habitat compensation, restoration, and creation in the Fraser River estuary: are we achieving a no net loss of fish habitat? Canadian Manuscript Report of Fisheries and Aquatic Sciences 2349.

Kistritz, R., G. Williams, and J. Scott. 1992. Inspection of red-coded habitat, Fraser River estuary, summer of 1992. Prepared for Fraser River Estuary Management Program. Burnaby, British Columbia.

Kistritz, R., and M. Gollner. 1995. Review and assessment of eelgrass *Zostera marina* transplanting projects in coastal British Columbia. Report prepared for Department of Fisheries and Oceans, West Vancouver Laboratory. West Vancouver, British Columbia.

Levings, C. D. 1980. Consequences of training walls and jetties for aquatic habitats at two British Columbia estuaries. Coastal Engineering 4:111–136.

Levings, C. D. 1991. Strategies for restoration and development of fish habitat in the Strait of Georgia—Puget Sound Inland Sea, northeast Pacific Ocean. Marine Pollution Bulletin 23:417–422.

Levings, C. D. 1995. Natural factors to be considered in restoration of marine and estuarine ecosystems supporting fisheries. Pages 47–50 *in* Murata, B., editor. Proceedings of the Osaka-Wan Symposium. Osaka, Japan.

Levings, C. D., and five coauthors. 1989. Chinook salmon and estuarine habitat: A transfer experiment can help evaluate estuary dependency. Pages 116–122 *in* C. D. Levings, L. B. Holtby, and M. A. Henderson, editors. Proceedings of the national workshop on the effects of habitat alteration on salmonid stocks. Canadian Special Publication of Fisheries and Aquatic Sciences 105.

Levings, C. D., and J. S. Macdonald. 1991. Rehabilitation of estuarine fish habitat at Campbell River, British Columbia. Pages 176–190 *in* J. Colt and R. J. White, editors. Proceedings of the fisheries bioengineering symposium. American Fisheries Society Symposium 10. Bethesda, Maryland.

Levings, C. D., and R. M. Thom. 1994. Habitat changes in Georgia Basin: implications for resource management and restoration. Pages 330–351 *in* R. C. H. Wilson, R. J. Beamish, F. Aitkens, and J. Bell, editors. Review of the marine environment and biota of Strait of Georgia, Puget Sound, and Juan de Fuca Strait. Canadian Technical Report of Fisheries and Aquatic Sciences 1948.

Levings, C. D., and D. J. H. Nishimura, editors. 1996. Created and restored sedge marshes in the lower Fraser River and estuary: An evaluation of their functioning as fish habitat. Canadian Technical Report of Fisheries and Aquatic Sciences 2126.

Metikosh, S. 1997. No net loss in the "real" world. Pages 9–17 *in* C. D. Levings, C. K Minns, and F. Aitkens, editors. Proceedings of the DFO workshop on research priorities to improve methods for assessing productive capacity for fish habitat management and impact assessment. Canadian Technical Report of Fisheries and Aquatic Sciences 2147.

Odum, W. E. 1970. Insidious alteration of the estuarine environment. Transactions of the American Fisheries Society 99:836–847.

Pomeroy, W. M., D. K. Gordon, and C. D. Levings. 1981. Experimental transplants of brackish and salt marsh species in the Fraser River estuary. Canadian Technical Report of Fisheries and Aquatic Sciences 1067.

Ryall, R., and C. D. Levings. 1987. Juvenile salmon utilization of rejuvenated tidal channels in the Squamish estuary, British Columbia. Canadian Manuscript Report of Fisheries and Aquatic Sciences 1904.

Schubert, N., Farwell, M. K., and L. W. Kalnin, 1994. Enumeration of the 1993 Harrison River chinook salmon escapement. Canadian Manuscript Report of Fisheries and Aquatic Sciences 2242.

Sibert, J., T. J. Brown, M. C. Healey, B. A. Kask, and R. J. Naiman. 1977. Detritus-based food webs: exploitation by juvenile chum salmon *Oncorhynchus keta*. Science 196:649–650.

Thom, R. M. 1990. A review of eelgrass *Zostera marina* transplant projects in the Pacific Northwest. Northwest Environmental Journal 6:121–137.

Williams, G. L. 1993. Mitchell Island marsh compensation project: Monitoring results and implications for estuarine management. Pages 415–425 *in* Proceedings of the 1993 Canadian coastal conference. National Research Council of Canada, Ottawa, Ontario.

Zimmerman, R. C., J. L. Reguzzoni, and R. S. Alberte. 1995. Eelgrass *Zostera marina* transplants in San Francisco Bay: Role of light availability on metabolism, growth, and survival. Aquatic Botany 51:67–86.

22 Human Population Growth and the Sustainability of Urban Salmonid Streams in the Lower Fraser Valley

Otto E. Langer, Fern Hietkamp, and Melody Farrell

Abstract.—Of the approximately 300 significant salmon spawning streams in the Fraser River system, about half are located in, or flow through, the urbanized area of the Lower Fraser Valley. These streams support spawning for all species of salmon, cutthroat trout, and many other fish species. They account for about 65% of the basin's coho production and almost all of its chum production. Between 1880 and 1960, diking, dredging and filling, dam construction, and land clearing had a devastating impact on the valley's fish habitat. Two million people now live in this rapidly growing area. Since the early 1970s, efforts have been made to protect salmon habitat in the face of continued urban population increase and sprawl. In 1978, land development guidelines were published in an attempt to mitigate the impacts of urban development on streams. These were inadequate, and new guidelines were implemented in 1992. In the face of unprecedented growth, these present guidelines are inadequately complied with and require upgrading. Even with the introduction of a "no net loss" policy in 1983, which requires that habitats lost to development be replaced with compensation habitat, a recent assessment of the pilot area in the Fraser Estuary shows that significant losses of fish habitat are still occurring. A radical societal rethinking of urban growth and sprawl is required to ensure that healthy urban streams survive the population growth projected for the Lower Fraser Valley. A new conservation strategy must be developed and implemented to protect streams and fish habitat. Without a new approach, the concept of sustainability of streams in such urban growth areas will be shown to be little more than what can be expected in a fool's paradise.

INTRODUCTION

THE LOWER FRASER VALLEY

The Fraser River originates in the Canadian Rocky Mountains and flows over 1,300 km to join the Pacific Ocean at Vancouver, British Columbia. It is British Columbia's largest river and drains about 230,400 km², about one quarter of the land area of the province. Detailed descriptions of the river and basin can be found in Birtwell et al. (1988), Northcote and Larkin (1989), and Dorcey and Griggs (1991).

The Fraser River Basin is characterized by a complex of mountain streams and lakes. A high energy system, the Fraser River is hemmed in along most of its length by mountains, but at Hope, British Columbia, approximately 150 km from the Pacific Ocean, it leaves its canyon environment to flow across a large, post-glacial floodplain. A large, active delta and associated wetland complex exists in the tidal reaches of the river and its extensive estuary. This area of the Fraser River Basin

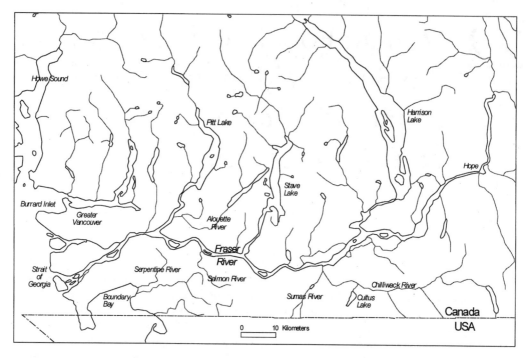

FIGURE 22.1   View of the Lower Fraser Valley.

is called the Lower Fraser Valley (LFV) and is the focus of this chapter. The area extends from Hope west to the Pacific Ocean and from the mountains on the north side of the floodplain to the Canada–U.S. border on the south side (Figure 22.1). This floodplain and the limited higher elevation areas within it cover 2.8% of the Fraser Basin's total area.

The Fishery Resource

The Fraser River is the most productive system on Canada's west coast. Unique in that it is the single largest producer of salmon in the world, it supports large runs of chinook, chum, coho, pink, and sockeye salmon that contribute to Canadian and U.S. commercial, sports, and aboriginal fisheries. Over 80 species of finfish, including the salmonids, frequent the Lower Fraser River, its tributary streams, and its estuary (Birtwell et al. 1988; Northcote and Larkin 1989).

The Fraser River produces over 50% of all salmon in British Columbia. In peak years up to 10 million adult salmon escape the fishery and spawn in the Fraser River system. Up to 1 billion juveniles migrate downstream during the spring to begin the ocean phase of their life cycle. About 150 of the river's 300 significant streams that support runs of spawning salmon originate in or flow directly through the LFV.

Although the LFV is a small part of the entire Fraser Basin, LFV streams support a disproportionately large part of the river's salmon production; specifically, about 80% of the chinook salmon, 90% of the chum salmon, 50% of the coho salmon, 80% of the pink salmon, and significant stocks of sockeye salmon (Fishery Management Group 1995a,b,c; 1996a,b).

Early Development of the Lower Fraser Valley

In the mid-1800s, non-aboriginal immigrants began to settle and develop the Fraser Basin. From the Cariboo Gold Rush to the present, tributaries of the river have been mined, dammed, cleared for agriculture, logged and encroached upon by roads and railways and other linear developments. As a result, extensive modifications have been made to the estuary and to the approximately 150 rivers and streams that flow into the Lower Fraser River and its estuarine area. This development has had a devastating impact on fish habitat. Present rates of human population growth and associated infrastructure development continue to exert pressure on remaining habitat.

This chapter reviews human development trends in the LFV, comments on past impacts of urban and industrial development, and examines the potential for sustaining salmon habitat into the next century. We did not examine the impacts of specific types of urban development on salmonid habitats nor consider water quality and related air quality concerns. Rather, we only reviewed the status of physical salmonid habitat in the LFV. When water and air quality concerns are taken into account, sustaining the salmonid resource will be an even greater challenge.

POPULATION GROWTH AND LAND USE IN THE LOWER FRASER VALLEY

Early exploitation of resources in the Lower Fraser focused on the fur and fishery resources of the basin. However, the discovery of gold in the Cariboo region in 1858 precipitated rapid development in the LFV. Roads and navigation infrastructures quickly appeared on the landscape. By the 1860s logging and local land clearing and settlements were occurring on the Burrard Peninsula and the first farms were being developed on diked lands in Richmond. The transcontinental railroad reached Burrard Inlet in 1885 and, coupled with shipping made possible by the Panama Canal, the Vancouver area became a center for the export of primary resources and a major growth area.

Private Property System.—The land settlement patterns that began 130 years ago have set the stage for today's land use conflicts in the LFV. When the British Royal Engineers began land surveys in the Fraser Valley in 1859, their surveys were notoriously inaccurate and often included wetlands and waterbodies. A letter from a disgruntled settler in the May 9, 1861 edition of the *British Columbian* records that, at a land sale, the Chief Commissioner of Lands said that he "would not guarantee whether what he was offering for sale should be land or water, or that it should contain 160 acres or that the spot could be found at all."

Since lands could not be surveyed quickly enough for settlers and speculators, the Governor passed the PreEmption Act (1860, 1861), which allowed any settler a preemptive right to claim 160 acres to be surveyed at a later date (Ross 1953). This attempt to give away land at any cost caused a land rush in the floodplains of the colony. As a result, property boundaries were often set in disregard for the existence of aquatic environments.

Development Trends.—By 1950 the development pattern of the LFV was set. The major urban centres were concentrated on the Burrard Peninsula (Vancouver, Burnaby, and New Westminster). By this time much of the remainder of the LFV had been converted to agricultural and rural uses. During the 1950s and 1960s, the private automobile had a pronounced impact on urban development in the LFV, encouraging urban sprawl and necessitating the construction of significant transportation infrastructures. These factors, in addition to the early settlement practices, established the contemporary growth patterns of the LFV. These settlement patterns and growth trends are responsible for extensive degradation of fish habitat in the streams of the LFV. They have also created the major habitat protection issues that we have today, which still must be resolved.

Population Growth Trends.—Continued population growth in the LFV is intensifying the pressure on fish habitat. Since the 1950s, the regional human population has been increasing at a high rate. During the past 30 years, the number of human migrants settling in the LFV have set records, making this one of the three fastest growing areas in North America (Cernetig 1996). Record high growth rates of 3.0% in the metropolitan areas of the LFV occurred in the 1980s. Between 1981 and 1986 the LFV had one of the fastest growth rates in the country at 9.1% (Moore 1990). The current overall growth rate for the LFV is 2.6% per annum (Baxter 1992). By 1985 the Fraser Valley was home to 1.5 million people. In 1995, the population in the LFV was 2.0 million and it is projected to reach between 3.2 million (GVRD 1993) to 4.1 million people by the year 2031 (Elliott et al., in press) (Figure 22.2). With up to 70,000 people moving to the LFV each year,

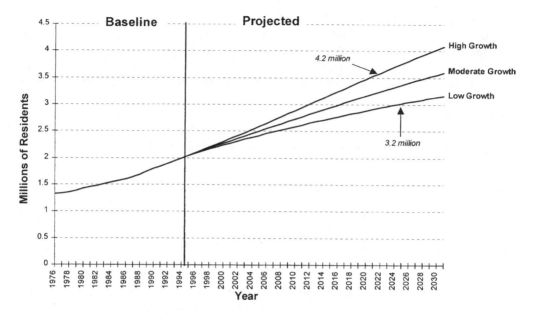

FIGURE 22.2   Population growth in the Lower Fraser Valley from 1976 to 1994 and high, medium and low growth projections to 2031 (Elliott et al., in press).

and with the settlement patterns that are currently in place, the sprawl of development will continue (Figure 22.2).

   Growth Management.—The results of early efforts at growth management are not encouraging. Each municipality appears to demand a fair share of growth. Rural communities in the eastern LFV, such as Abbotsford and Chilliwack, are experiencing significant growth due to the affordability of land and the opportunity of a rural lifestyle. The continued growth and demand for single-family dwellings will put additional pressure on LFV streams and wetlands in the future. Although many people are concerned about urban sprawl, attempts to reduce the demand for single-family dwellings and encourage densification appear to be largely ineffective (Munro 1996).

   With the high growth rate in the LFV, conversion of land from its natural state to agricultural and urban uses has been dramatic and almost complete in some areas (Figure 22.3). One of the few obstacles in the way of greater land conversion and sprawl is the Agricultural Land Reserve. As most floodplain streams flow through agricultural reserve lands, agriculture is, ironically, an important buffer, keeping urban development at bay and minimizing further urban impacts on valley streams. At the same time, it must be realized that agricultural practices have had a significant impact on most streams in farmed areas (due to water quality degradation, channelizing, diking, dredging, etc.).

IMPACTS ON FISH HABITAT

From the 1880s until 1970 major modifications were made to the landscape that caused great impacts on salmonid streams and on marine and estuarine habitats in the Vancouver area. Although three quarters of the population was confined to the Burrard Peninsula in 1920, this early population made extensive demands on the valley landscape with consequent impacts on all habitat types in most valley areas.

   Early development (i.e., 1870–1930) probably had a disproportionately significant impact on fish habitat and set the tone for future urban and associated industrial development in the Fraser Valley. Concern for the natural values of wetlands and streams was nonexistent. If anyone saw a connection between habitat and fish, a valued domestic and export food commodity, it was overruled

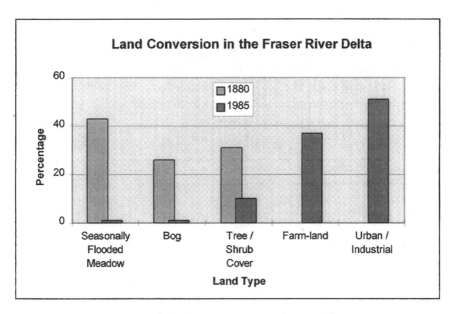

FIGURE 22.3   Conversion of natural land types in the LFV between 1880 and 1985 (Butler and Campbell 1987).

by the desire to profit by clearing and developing the land. While commenting on the future of lands along the Lower Fraser, Governor Douglas in 1860 noted that he wished that "those gorgeous forests might soon be swept away by the efforts of human industry and give place to cultivated fields and other accessories of civilization" (Ross 1953). The fisheries resources were taken for granted and, if any local stocks were impacted, there was always plenty of fish produced elsewhere.

ESTUARINE HABITATS

By 1890 major alienation of wetlands had taken place in Richmond, and most of Sea and Lulu islands and South Delta areas had been diked and drained. An examination of habitat along the North Arm of the Fraser River has indicated the significant loss of wetland habitat associated with the riverine tidal wetland complex. In 1880 the river flooded 2,896 ha of wetland habitat accessible to fish. By 1940 this area had been reduced to 109 ha (Levings et al. 1996). By 1915 the extensive wetlands in the eastern basin of False Creek had been filled in for industrial purposes (Roy 1988). This area contained the estuary of China Creek, Vancouver's most productive salmon stream.

Although significant efforts have been made to protect estuarine habitats since 1970, it is difficult to balance the social, economic, and environmental aspirations of a rapidly growing city located in an environmentally sensitive area. As part of an early pilot, the concept of "no net loss" of habitat was applied to the Fraser Estuary beginning in 1983. An assessment of that initiative has recently been conducted by Langer et al. (1994) and Kistritz (1996). This study shows that, even when a policy requires that habitats lost to development be replaced with compensation habitat, most types of habitat are still suffering significant losses (Table 22.1).

IMPACTS ON LOWER FRASER STREAMS AND RIVERS

Loss and Modification of Streams.—When settlement began in the Vancouver area, streams were greatly impacted. Many of them interrupted orderly grid development and were simply culverted and covered over. Larger streams became open sewers and were subsequently converted into underground storm sewers. This was the fate of China Creek, which became a storm sewer in 1950. Most of the streams in the earlier developed areas of Burrard Peninsula also suffered this

TABLE 22.1
Summary of total losses and gains from "no net loss" habitat compensation projects in the Fraser Estuary (1983–1993). The numbers in the table are not additive within each habitat type due to unlike habitat compensation and varying compensation formulae (Langer et al. 1994).

Habitat Type	Total Losses	Total Gains	Balance
Sub Tidal (m²)	−947,100	16,000	−931,000
Mud/Sandflat (m²)	−268,200	145,800	−98,200
Marsh (m²)	−75,900	157,800	61,200
Riparian (m)	−3,400	3,300	−100

TABLE 22.2
Effective impervious area of four watersheds in the Lower Fraser Valley (Rood and Hamilton 1994).

Stream	Municipality	Drainage	E.I.A.
Byrne	Burnaby	9 sq. km	27%
Mahood	Surrey	38 sq. km	23%
Schoolhouse	Coquitlam	7 sq. km	33%
Willband	Abbotsford	13.5 sq. km	26%

fate. Figure 22.4 shows that the great majority of presettlement streams in the Vancouver area have been buried or culverted, and are effectively lost (reproduced from Lost Streams of the Lower Fraser River map poster, 1995). Most of these streams would have been frequented by salmonids. By 1950 only one of 40 salmon spawning streams still survived in Vancouver (Harris 1989). The Fraser River Action Plan assessment of lost stream lengths indicated that, of 2,000 km of historic streams in the urban area of the LFV, about 588 km (approximately 30%) have been culverted and covered.

Although most of the streams in rural areas of the LFV did not become underground storm sewers, they did not escape the impacts of human development. Lower reaches of streams were channelized and diked to create and protect farmland. Sediment discharges, associated with many types of development, clogged the clean gravels required for spawning salmon. The headwaters of many streams originating in the floodplain often disappeared. Dams were also built on many streams (e.g., Alouette, Coquitlam, and Stave rivers) to supply water and generate electrical power for the growing population.

Impacts of Urbanization.—By the 1970s the physical integrity of most streams was better protected by improved legislation and increased diligence by fisheries agencies. However, indirect impacts from urbanization of watersheds continued to increase sediment loadings to streams, alter stream hydrology, and degrade water quality and riparian areas. Most of the activities causing these impacts have yet to be satisfactorily addressed.

An example of the extent of urbanization impacts on streams is the effect of increasing impervious surface areas in urban watersheds. The impervious area of many LFV watersheds exceeds 20% (Table 22.2). Studies in Washington State's King County have shown that, when impervious surfaces (i.e., through urbanization) cover more than 10% of a watershed, significant hydrological changes occur in the local streams (Gino Lucchetti, King County, personal communication). These impacts can induce adverse and irreversible changes in the quality of streams.

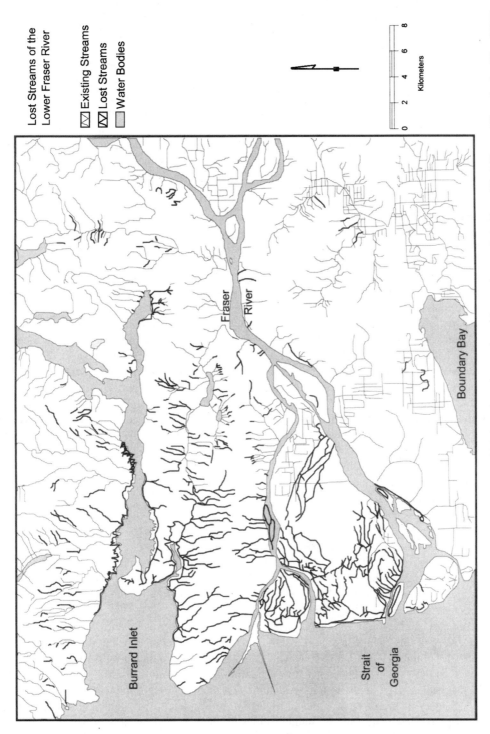

Lost Streams of the Lower Fraser River

⊠ Existing Streams
⊠ Lost Streams
☐ Water Bodies

Kilometers
0 2 4 6 8

Burrard Inlet

Fraser River

Strait of Georgia

Boundary Bay

FIGURE 22.4   Streams in the Greater Vancouver area which have been lost: culverted, paved over, drained, or filled (Fraser River Action Plan 1995).

Research has also shown that many streams become hydrologically and biologically simplified once watershed impervious cover exceeds 25% (Schueler 1996). Increases in impervious cover in the LFV continue to impact stream hydrology and associated fish habitat. Most LFV municipalities have ineffective or nonexistent programs to deal with the quantity and quality of stormwater in even moderately developed watersheds (Quadra Planning Consultants 1995).

Floodplain Modifications.—All large tributaries flowing into the lower Fraser River have been physically altered to the detriment of fish habitat. Diking to protect property and farmland has been extensive along all of the tributary rivers flowing through the floodplain. After the large flood of 1949, major dike improvements almost completely and permanently separated the river from its floodplain. Unfortunately, most dikes were not set back, which effectively eliminated the riparian zone and restricted access to many productive off-channel areas. As investment in property increases in certain areas, such as the Chilliwack Valley, more extensive dike and flood control works are being constructed at the expense of fish habitat (Brent Lister, Lister and Associates Ltd., personnal communication; Gordon Hartman, Fisheries and Research Education Services, personal communication).

THE PROTECTION OF FISH HABITAT

With few exceptions, habitat loss and water quality degradation continued virtually unabated until habitat conservation became an issue in the 1960s. Initially, the Canada Fisheries Act and early conservation efforts focused largely on water quality and fish safety (e.g., intake screening, provision of fishways). Legislation to protect fish habitat (e.g., spawning, food production, and rearing areas) was not promulgated until 1977.

Legislation Protecting Fish Habitat.—In Canada, the federal government has the constitutional power to manage and conserve the fishery and fish habitat. These powers are outlined in the Canada Fisheries Act. The modern provisions of this 130-year-old act make it the strongest piece of environmental legislation in Canada. It is a criminal offense to cause harmful alteration to, or destruction of, fish habitat unless authorized by the Department of Fisheries and Oceans (DFO). However, the Fisheries Act is reactive in nature, and cannot prevent activities from taking place unless they can be shown to be harmful to fish or fish habitat. The Fisheries Act was designed this way in order not to encroach on the constitutional right of the provinces to control land management. The management of water, land, and waste is a provincial responsibility. Yet what happens to the water and land fundamentally affects, and is directly linked to, fish habitat.

The provincial government has delegated land use decision-making powers in settlement areas to the municipalities (local government) under the provisions of the Municipal Act. The Municipal Act is designed to be enabling legislation, allowing each municipality to meet its own needs and desires. The content of community plans and bylaws created pursuant to the act is extremely flexible and is at the discretion of the local government. Thus, municipalities are under no obligation to use any provisions with their power to protect the environment or fish habitat. During the past few years, the Municipal Act has been upgraded to better address environmental concerns. However, in most cases a municipality must have a strong will, and a consistent vision over time, to protect environmental features, including aquatic habitats (Webb 1996).

To conserve fish habitat in settlement areas, and especially on private lands, regulatory change is needed. Most of this change is required at the jurisdictional level responsible for land and water use in British Columbia. An effective provincial habitat protection initiative would need to integrate and strengthen the various acts dedicated to regulating activities often at odds with protection of fish and fish habitat (e.g., lands, water, agriculture, and municipal legislation).

The voluntary nature of environmental protection provisions in the Municipal Act leads to a variety of approaches being used by municipalities to regulate land use around aquatic habitats. Since most streams and salmonids require similar protection, a more consistent approach is needed. For example, minimum stream protection standards within municipal, agricultural, and forest land legislation are required. An approach similar to that recently taken in the British Columbia Forest Practices Code (1996) to protect streams in public forest lands could be adopted.

In 1997 the Province of British Columbia introduced the B.C. Fish Protection Act which contained provisions that amended the Municipal Act. The amendment requires local governments to provide a level of streamside protection that is comparable to or exceeds that specified in a policy directive which is still under development. Unfortunately this legislation allows the local government to determine whether or not they have achieved a comparable level of protection to that in the directive. The directive could therefore be open to more of a political than a scientific determination. It remains to be seen how effective this legislation will be for better protecting riparian habitats.

Development of Policy and Guidelines.—To help bring developers and municipalities into compliance with the water quality and habitat conservation intent of the Fisheries Act, urban land development guidelines were published in 1978 (DFO 1978). These guidelines recommended the first riparian zone protection setbacks and procedures required for development near streams. If landowners, developers, and municipal government complied with these guidelines, they would avoid violating the Fisheries Act and not be subject to prosecution.

A referral system was set up in the 1970s to carry out the mandate of federal and provincial legislation to protect fish habitat. Through the referral system, specific work proposed in or around a stream is referred to the appropriate agency, reviewed, and approval conditions (i.e., fisheries-sensitive zone setbacks, construction practices, mitigation measures, etc.) are set for the proponent. However, fish habitat in the LFV continued to be degraded and lost through the 1970s and 1980s. The referral system does not enforce compliance with approval conditions, so the system cannot guarantee that damage to fish habitat will be prevented. In addition, considering the number of unapproved works, and the fact that the referral system does not address the multitude of non-point source impacts associated with development throughout a watershed, habitat continues to be downgraded and lost.

In 1986, DFO released a national policy on habitat management. This policy document outlined how DFO would relate to the Fisheries Act so as to achieve a "net gain" in the productive capacity of Canada's fish habitat (DFO 1986). This national policy was an attempt to create certainty in the application of the Act and stem the continued loss of fish habitat in Canada. Key to this policy was a conservation principle of "no net loss" of the productive capacity of habitats.

The no net loss concept was incorporated into the 1992 revision of the Land Development Guidelines (Chilibeck et al. 1992), which was published jointly by both senior agencies (DFO and the B.C. Ministry of Environment, Lands and Parks). The revised guidelines clarify specifications for aquatic habitat protection in development situations.

Jurisdictional Issues.—Local government stream protection initiatives are complicated by the fact that municipal boundaries usually have no relationship to ecological or watershed planning needs. Consequently, in the LFV, stream protection at the local level must be negotiated with 27 different local governments. This is further complicated by the often competing interests of many First Nations and several other senior government agencies. There are no general provisions that require local governments to approach the protection of aquatic habitats consistently or on any schedule. Consequently, streams that flow from one municipality to another can have very different development requirements along each reach.

To overcome shortcomings in the Municipal Act, fishery agency staff often rely upon the B.C. Lands Act to obtain protective covenants along streams. Although covenants designed to protect riparian areas appear effective on paper, the will and resources to enforce the covenants is lacking (van Drimmelen 1996). As a result, variable success has been reported in terms of habitat protection. A recent survey of streams in Surrey and Maple Ridge has shown that the riparian zone in the protective covenant area has been encroached upon 75% and 63% of the time, with serious violations 15% and 29% of the time in each municipality, respectively (Inglis et al. 1996).

Recent Approaches.—To better implement the Land Development Guidelines and influence land developers and local government, in 1993, environmental agencies formed strategic partnerships. These partnerships resulted in a number of jointly managed and funded initiatives, including stream stewardship and greenways. These programs and products were designed to enable municipal planners, developers, and community groups to develop a proactive approach to protecting fish habitat (Lanarc Consultants 1994, 1996). Another priority has been to produce accurate and accessible technical information to enhance habitat management. For example, orthophotos* overlayed with maps identifying watercourses and the distribution of fish and critical habitats are now available in hard or digital copy (DFO 1996). Accompanied by written summaries, these maps provide municipalities in the LFV with information on the suitability of lands for development. Other partnership efforts include promotion of watershed councils and roundtables that involve the community, local government, and provincial and federal government agencies. This focus on awareness and educational programs has highlighted the concern for fish habitat in urban areas.

The need to continue these stewardship programs is supported by a recent assessment of techniques and processes available to protect urban streams in the LFV (Gardner 1996). This assessment found that considerable emphasis has been placed on regulatory approaches: legislation, regulations, bylaws, approvals, monitoring, and enforcement. The study also shows that economic instruments—such as taxes, subsidies, property rights and charges—have promise but have received relatively little attention. Given the strong role of the development community in the LFV, economic tools must be developed and tested. Most significantly, community stewardship initiatives show strong promise for the protection of aquatic and riparian resources, through activities such as surveillance and rehabilitation, as well as education, awareness-raising, and constituency building at the political level to promote environmental change.

It is expected that stewardship programs, as well as regulatory approaches and economic incentives, will continue to evolve in the LFV. However, history has shown that, unless means are found to effectively protect aquatic habitat on private land, such as through strong provincial legislation, the future of aquatic and riparian habitats will continue to be in jeopardy.

THE FUTURE OF FISH HABITAT IN THE LOWER FRASER VALLEY

Fractured Jurisdictions and Impacts of Growth.—Although we have made some progress in protecting urban fish habitat over the past 30 years, we have not made significant progress in a key area: coordination of jurisdictions. The current fragmented jurisdictional system has unfavorable consequences for streams. Environmentally sound land use practices within a watershed must be coordinated among municipalities if they are to be effective. Lack of coordination and strategic planning at the senior agency (federal and provincial) level has also contributed to poor management of resource issues. What gains have been made in the will to protect the environment and in technological advancements to mitigate impacts have been more than offset by human population growth and associated urban development.

Lack of a Conservation Ethic.—The conservation ethic has been advanced both locally and globally through a number of programs and movements. However, the right to develop private land as the owner sees fit remains jealously guarded. Existing land use law generally protects the rights and the investments of developers. The concerns for aquatic and other natural common property remain largely a secondary consideration. Today, with most of the LFV floodplain privately owned, fish habitat sustains increasing impacts as development continues.

Strategic Planning for Habitat Protection in the LFV.—Many improvements can and must be made to protect LFV streams if we are to make serious progress toward maintaining viable fish

---

* An orthophoto is a restructured image of an aerial photograph, which has ground feature locations that meet standard map accuracies.

habitat. In communities where land development is still occurring or about to occur, there is an opportunity to maintain healthy streams by implementing an ecosystem planning approach. We must consider the impacts of impervious areas, plan to concentrate development strategically instead of letting it sprawl, and protect natural watercourses (especially the headwaters). Without a watershed approach to providing the necessary level of stream protection, technical fixes such as sediment retention, stormwater detention, and riparian zone protection alone will not protect viable and productive streams in the long term. Where watersheds have been compromised by development, the remaining fish and fish habitat must be protected to maintain biological and genetic diversity, and quality of life values.

   Legislation and Political Will.—The opportunity to advance significant new laws to proactively protect fish and fish habitat from poor land and water use practices on private land does not appear to be high on the public agenda at this time. Government intervention to protect the environment is frowned upon by a strong pro-development lobby and would therefore be examined with grave reservation by many elected officials. Furthermore, the approach of relying on marketplace measures to provide environmental protection and the popularization of concepts of "sustainable development" and stewardship may be providing a false sense of progress. Strong political leadership is required to reverse habitat loss and effectively protect remaining habitats well into the 21st century. Politicians must find a better balance between private property rights and the public interest.

RECOMMENDED ACTION FOR SUSTAINABLE URBAN FISH HABITAT

The following recommendations reflect approaches to protecting streams and their fish habitat features in the face of LFV growth pressures. They represent proactive means of protecting fish habitat, involving partnerships as well as development of new habitat protection tools. They include measures that are already underway (undertaken by local governments, senior agencies, and community groups), as well as new actions that should be embraced to effectively maintain fish habitat in a rapidly urbanizing river valley.

1. Continue stream and wetland stewardship, educational, and partnership programs to create an appreciation of the vulnerability of fish habitat and what is needed to protect it.
2. Emphasize cooperation between various parties to improve habitat resources better protection and promote better conservation legislation.
3. Identify, and prioritize for protection, key watersheds that have not yet been compromised by urban development.
4. Implement a more thorough program of mapping aquatic habitats to encourage protection of these resources. Emphasis must be put on mapping key parts of watersheds that need to be protected to ensure the survival of viable habitat in the urban landscape.
5. Preserve adequate riparian areas along streams to protect fish habitat.
6. Introduce direct purchase and tax incentives and disincentives to encourage conservation of the private lands in the floodplain, where most fish habitat is located.
7. Implement a consistent approach to stream protection at the senior agency level.
8. Implement a consistent approach to stream protection at the municipal level.
9. Apply and enforce existing legislation to protect aquatic habitat more diligently and proactively.
10. Develop new legislation that embodies an aquatic ecosystem/watershed approach to protection of fish habitat.

Protection measures in the LFV have improved over the past 20 years. However, streams continue to be degraded at an alarming rate due to the many and complex impacts of population growth and urban development. Changes are required in growth and land use planning to ensure that healthy and productive streams can be sustained. New strategies must also be implemented by those who work to protect streams and fish habitat. If urban aquatic habitats and the associated fishery resource are important to society and our quality of life, a bold approach to protecting and restoring aquatic habitats is needed.

ACKNOWLEDGEMENTS

We thank S. Bourque, T. Brown, and T.G. Brown for reviewing this chapter. We are grateful for the assistance of T. Pinkerton, M. Mascarenhas, and Suzanne Richer with preparing digital figures and formatting this chapter.

REFERENCES

Baxter, David. 1992. Population Trends in the Metropolitan Vancouver Region, 1991 to 2021. Development Services, Greater Vancouver Regional District. Burnaby, British Columbia.

Birtwell, I., C.D. Levings, J.S. Macdonald, and I.H. Rogers. 1988. A review of fish habitat issues in the Fraser River system. Water Pollution Research Journal Canada, 23(1):1–30.

Butler, R.W., and R.W. Campbell. 1987. The birds of the Fraser River delta: populations, ecology, and international significance. Occasional Paper No. 65. Canadian Wildlife Service.

Cernetig, M. 1996. B.C. strains to cope with growth. Globe and Mail, July 22. Toronto, Ontario.

Chilibeck, B., G. Chislett, and G. Norris. 1992. Land development guidelines for the protection of aquatic habitat. Fisheries and Oceans Canada and B.C. Ministry of Environment, Lands and Parks, Vancouver, British Columbia.

DFO (Department of Fisheries and Oceans). 1978. Guidelines for land development and protection of the aquatic environment. Fisheries and Marine Service Technical Report No. 807, Vancouver, British Columbia.

DFO (Department of Fisheries and Oceans). 1986. The Department of Fisheries and Oceans policy for the management of fish habitat. Ottawa, Ontario.

Dorcey, A.H.J., and J.R. Griggs. 1991. Water in sustainable development: exploring our common future in the Fraser River Basin. Volume II. Westwater Research Centre. University of British Columbia. Vancouver, British Columbia.

Elliott, B., and N. Guppy. In press. On the edge: Demographic change in the lower Fraser basin. Pages 242–320 in Michael Healey, editor. Sustainability and human choices in the Lower Fraser Basin: Resolving the Dissonance. UBC Press, Vancouver, British Columbia.

Fishery Management Group. 1995a. Fraser River Action Plan. Fraser River Sockeye Salmon. Department of Fisheries and Oceans, Vancouver, British Columbia.

Fishery Management Group. 1995b. Fraser River Action Plan. Fraser River Pink. Department of Fisheries and Oceans, Vancouver, British Columbia.

Fishery Management Group. 1995c. Fraser River Action Plan. Fraser River Chinook. Department of Fisheries and Oceans, Vancouver, British Columbia.

Fishery Management Group. 1996a. Fraser River Action Plan. Fraser River Chum. Department of Fisheries and Oceans, Vancouver, British Columbia.

Fishery Management Group. 1996b. Fraser River Action Plan. Fraser River Coho. Department of Fisheries and Oceans, Vancouver, British Columbia.

Fraser River Action Plan. 1995. Lost Streams of the Lower Fraser River map poster. Department of Fisheries and Oceans. Vancouver, British Columbia.

Gardner, J. 1996. Urban stream stewardship: From bylaws to partnerships—an assessment of mechanisms for the protection of aquatic and riparian resources in the Lower Mainland. Prepared for the Fraser River Action Plan. Department of Fisheries and Oceans, Vancouver, British Columbia.

GVRD (Greater Vancouver Regional District). 1993. Managing Greater Vancouver's growth. Strategic Planning Department. Burnaby, British Columbia.

Harris, G. 1989. Vancouver's old streams. Vancouver Public Aquarium. Vancouver, British Columbia.

Inglis, S., P. Thomas, and E. Child. 1996. Protection of aquatic and riparian habitat on private land: Evaluating the effectiveness of covenants in the City of Surrey, 1995. City of Surrey and Department of Fisheries and Oceans (Fraser River Action Plan). Vancouver, British Columbia.

Kistritz, R.U. 1996. Habitat compensation, restoration and creation in the Fraser River estuary—Are we achieving a no-net loss of fish habitat? Canadian Manuscript Report of Fisheries and Aquatic Sciences No. 2349. Department of Fisheries and Oceans (Fraser River Action Plan). Vancouver, British Columbia.

Lanarc Consultants. 1994. Stream stewardship—a guide for planners and developers. Prepared for Department of Fisheries and Oceans (Fraser River Action Plan). Vancouver, British Columbia.

Lanarc Consultants. 1996. Community greenways: linking communities to country, and people to nature. 1996. Prepared for Department of Fisheries and Oceans (Fraser River Action Plan), Vancouver, British Columbia, and B.C. Ministry of Environment, Lands and Parks. Victoria, British Columbia.

Langer, O.E., R. U. Kistritz, and C.D. Levings. 1994. Evaluation of the no net loss compensation strategy used to conserve Fraser River estuary fish habitats. Submerged Lands Management Conference, Oct. 2–6, 1994. New Westminster, British Columbia.

Levings, C. et al. 1996. Assessment of restored fish habitat in the Fraser River. Prepared for Department of Fisheres and Oceans (Fraser River Action Plan). Vancouver, British Columbia.

Moore, K. December 1990. Urbanization in the Lower Fraser Valley. Technical Series No. 120. Canadian Wildlife Service/Pacific and Yukon Region. Delta, British Columbia.

Munro, H. 1996. Lower Mainland to need 500,000 new homes. The Vancouver Sun, September 9, 1996. Vancouver, British Columbia.

Northcote, T.G., and P.A. Larkin. 1989. The Fraser River: A major salmonine production system. Pages 172–204 in D.P. Dodge, editor. Proceedings of the international large river symposium. Canadian Special Publication of Fisheries Aquatic Sciences 106. DFO. Ottawa, Ontario.

Quadra Planning Consultants. 1995. Protection of aquatic and riparian habitat by local governments: An inventory of measures adopted in the Lower Fraser Valley, 1995. Prepared for Department of Fisheries and Oceans (Fraser River Action Plan). Vancouver, British Columbia.

Rood, K.M. and R. Hamilton, 1994. Hydrology and water use for salmon streams in the Fraser Delta Habitat Management Area, British Columbia. Canadian Manuscript Report of Fisheries and Aquatic Sciences No. 2238. Vancouver, British Columbia.

Ross, L.J. 1953. Richmond—Child of the Fraser. Hemlock Printers. Vancouver, British Columbia.

Roy, P. E. 1988. Vancouver. Pages 2239–2242 in The Canadian encyclopedia. Hurtig Publishers, Vancouver, B.C.

Schueler, T. R. 1996. Crafting better urban watershed protection plans. Watershed Protection Techniques 2:329–337.

van Drimmelen, B. 1996. Conservation convenants—Useless without teeth. Natural Resources Law Newsletter. Spring 1996:2–3.

Webb, C. 1996. Environmental stewardship in the Municipal Act. Department of Fisheries and Oceans (Fraser River Action Plan). Vancouver, British Columbia.

Section V

Artificial Production

23 Minimizing Ecological Impacts of Hatchery-Reared Juvenile Steelhead Trout on Wild Salmonids in a Yakima Basin Watershed

Geoffrey A. McMichael, Todd N. Pearsons, and Steven A. Leider

Abstract.—Adverse ecological effects on wild fish resulting from releases of hatchery-reared fish are increasingly being scrutinized and balanced against benefits afforded by hatchery programs. To improve understanding of the potential ecological effects of hatchery steelhead trout (anadromous form of *Oncorhynchus mykiss*) on wild trout (resident *Oncorhynchus* species) populations, we studied releases of 23,000 to 38,000 hatchery-reared steelhead trout smolts into a Yakima River, Washington, watershed from 1991 through 1994. In this chapter we synthesize results from many aspects of these studies as they relate to minimizing ecological risks to wild trout. We snorkeled in control and treatment streams to observe behavioral interactions between hatchery steelhead trout and wild salmonids. Movement of residual hatchery steelhead trout was examined using traps, and direct underwater observation and relative abundances of hatchery and wild fish were estimated by electrofishing. Instream enclosures were used to determine whether residual hatchery steelhead trout impacted growth of wild rainbow trout or spring chinook salmon (*O. tshawytscha*). Potential for adverse impacts resulting from ecological interactions among wild salmonids and hatchery steelhead trout was greatest when (1) hatchery fish did not emigrate quickly; (2) water temperatures were over 8°C; (3) hatchery fish were the same species as the wild salmonids; (4) hatchery fish were larger than the wild salmonids; (5) habitat and/or food were limiting; and (6) numbers of fish released was over about 30,000. Ecological interactions with wild salmonids could be reduced or minimized by releasing (1) only actively migrating smolts (no residuals); (2) hatchery fish that are smaller than wild fish; (3) the minimum number necessary to meet management objectives; (4) fish that are less likely to engage wild fish in agonistic encounters; (5) when water temperatures are relatively cold (less than 8°C); (6) in areas where wild salmonid populations are absent; and possibly (7) in areas where habitat diversity is complex. Management actions that encourage angler harvest of hatchery steelhead trout residuals while protecting coexisting wild species may also help minimize negative ecological impacts. Implementing these strategies may reduce the number of returning hatchery-origin adults of the target group but will help reduce risk to the sustainability of wild fish populations.

INTRODUCTION

Conservation of endemic wild fish populations in the Pacific Northwest has received increased emphasis in recent years (Nehlsen et al. 1991; NMFS 1995; Huntington et al. 1996; NRC 1996; Allendorf et al. 1997). Potential adverse impact of hatchery programs on wild salmonid populations has been one of many conservation issues (NRC 1996). Managers have acknowledged that competitive interactions with wild stocks as a result of hatchery stocking can have adverse consequences, and new management guidelines are beginning to emerge in efforts to reduce these effects (e.g., IHOT 1995). Due largely to the failure of many anadromous fish hatchery programs to meet expectations, and the continued decline of wild fish populations (Meffe 1992), the public is becoming more involved in decision-making processes regarding the use of hatchery fish (WDFW 1997). Most research on hatcheries has focused on ways to alter hatchery methods to increase post-release survival of hatchery fish (e.g., Jarvi and Uglem 1993; Wiley et al. 1993; Banks 1994). Few studies have directly examined the effects of hatchery fish on wild populations.

The Yakima River supports one of the premier wild resident rainbow trout *Oncorhynchus mykiss* fisheries in Washington State (Krause 1991; Probasco 1994). Concerns that proposed steelhead trout supplementation (Clune and Dauble 1991) might negatively impact this fishery prompted a 5-year study, beginning in 1991, of the potential ecological effects of releasing hatchery steelhead trout juveniles (judged to have reached the smolt stage) on coexisting wild salmonids in streams (McMichael et al. 1992; 1994; 1997; Pearsons et al. 1993; 1994; 1996). Results from this work helped us form hypotheses regarding the ecological impacts of hatchery steelhead trout releases on wild salmonids, particularly on resident rainbow trout. This chapter is a synthesis of our findings and recommendations; further details have been published (McMichael et al. 1992; 1994; 1997; Pearsons et al. 1993; 1994; 1996) or will be soon. Our specific objectives were to (1) determine whether releases of hatchery steelhead trout affected wild salmonids and (2) develop guidelines to reduce ecological impacts of hatchery steelhead trout releases on wild salmonids. We define ecological impacts as any or a combination of the following: behavioral dominance of hatchery fish over wild fish; displacement of wild fish by hatchery fish; reduced distribution of wild fish due to the presence of hatchery fish; decreased abundance of wild fish due to the presence of hatchery fish; and decreased growth of wild fish when hatchery fish are present. To meet our objectives we tested the general null hypothesis that releases of hatchery steelhead trout would have no ecological impacts on pre-existing wild salmonids. More detailed null hypotheses we tested were that hatchery steelhead trout will not (1) behaviorally dominate wild salmonids; (2) displace wild salmonids; (3) decrease abundance of wild salmonids; or (4) reduce the growth of wild salmonids.

METHODS

We used a control-treatment approach, in which streams not receiving hatchery steelhead trout were considered controls, while those that were stocked with steelhead trout were the treatment streams. We released 23,000–38,000 hatchery-reared smolt-sized steelhead trout in early May annually from 1991 through 1994 into Jungle Creek, a tributary of the North Fork of the Teanaway River, north of the Town of Cle Elum, Washington (Figure 23.1). Most of the hatchery steelhead trout smolts released into Jungle Creek (small treatment stream) moved downstream into the North Fork of the Teanaway River (large treatment stream). Jack Creek, a small tributary to the North Fork of the Teanaway River, served as a small control stream and was blocked off to prevent hatchery steelhead trout from moving up into it. The large control stream, the Middle Fork of the Teanaway River, was in an adjacent basin and was similar in size and other physical and biotic makeup to the large treatment stream (Figure 23.1). The West Fork of the Teanaway River was used as an additional control stream for wild salmonid abundance data only.

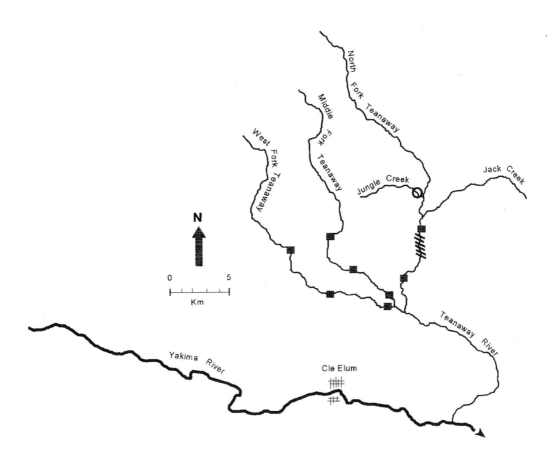

FIGURE 23.1   Map of the upper Yakima River basin showing study streams in the Teanaway River basin. The hatchery steelhead trout release location on Jungle Creek is marked with an open circle. Filled squares indicate 100 m-long population estimate sites. The cross-hatched area on the North Fork of the Teanaway River is where small enclosure growth experiments were conducted.

The hatchery steelhead trout were progeny of adult steelhead trout collected lower in the Yakima River basin. Table 23.1 shows the relevant data on adult steelhead trout used to produce fish for these experimental releases.

The various methods and sampling effort involved in testing for hatchery/wild interactions are summarized in Table 23.2. Snorkeling was used to (1) observe behavioral interactions between hatchery steelhead trout (recognized by excised adipose fins) and wild salmonids in index sites in two streams where hatchery steelhead trout were present (treatments) in comparison with two streams where no hatchery fish were released (controls) in 1991–1994 and (2) determine the distribution of residual hatchery steelhead trout within the North Fork of the Teanaway River in 1994. Direct underwater observation of fish behavior was performed by snorkeling in control and treatment streams as described by McMichael et al. (1992) and Pearsons et al. (1993). Each agonistic interaction was classified into one of the following five groups: threat, crowd, chase, nip, or butt. We defined

TABLE 23.1
Parentage (number, % hatchery and % wild) and loading density of smolts (kg of fish/l/min in the raceway immediately prior to release) for hatchery steelhead trout smolts released into Jungle Creek in 1991–1994. Estimated total number (survival) and percent of number released (in parentheses) to Prosser Dam (234 km downstream) during the 3 months following release are also shown.

| | | Parents | | | Progeny | Survival |
Brood Year	Release Year	Number	Percent Hatchery	Percent Wild	Loading Density	Number (percent)
1990	1991	106	0	100	0.46	648 (1.9)
1991	1992	24	63	37	0.19	575 (1.5)
1992	1993	26	100	0	0.16	5592 (24.9)
1993	1994	25	24	76	0.20	837 (2.6)

threats as overt signs of aggression, such as fin-flares and body arching (Taylor and Larkin 1986; Holtby et al. 1993). Crowding occurred when fish moved toward other fish laterally, causing a subordinate fish to move out of the way (Helfrich et al. 1982; Taylor and Larkin 1986; Holtby et al. 1993). Chases occurred when one fish pursued another fish for several body lengths without making physical contact (Keenleyside and Yamamoto 1962; Helfrich et al. 1982; Taylor and Larkin 1986). Nips were classified as physical contact in which one fish actively bit another fish (Stringer and Hoar 1955; Helfrich et al. 1982; Taylor and Larkin 1986). Physical contact made between two fish in which the mouth of the attacking fish was closed was classified as a butt. A contest may have included multiple interactions. For example, a hatchery steelhead trout and a wild rainbow trout could chase and nip each other several times during one contest. A fish was considered to be dominant if it displaced (defined below) its opponent in a contest.

Hatchery steelhead trout that failed to emigrate at the same time as wild steelhead trout were defined as residuals. Residuals may take up residence in streams or may emigrate in later years (Pearsons et al. 1993). To determine the extent of spatial overlap between residual hatchery steelhead trout and resident rainbow trout, cutthroat trout *O. clarki* and bull trout *Salvelinus confluentus* in the North Fork of the Teanaway River, we snorkeled pools and runs at 0.6-km intervals from the mouth of Jungle Creek upstream 13.4 km. Snorkeling took place on June 24, June 30, and July 12, 1994. A total of 22 sites were sampled. Each site, covering approximately 100 m of stream and separated by about 100 m, was sampled by two snorkelers. Snorkeling was repeated at the upper nine sites on the night of July 12, 1994, to better estimate bull trout presence. No bull trout were observed during daylight snorkeling in these areas, therefore data collected at night was used for the upper nine sites. The relative abundances of bull trout, cutthroat trout, rainbow trout, and hatchery steelhead trout were expressed as percentages of all salmonids observed at each site.

To determine whether juvenile hatchery-reared steelhead trout displaced wild fish, we examined displacement at three spatial scales using trapping and snorkeling methods described by McMichael et al. (1992) and Pearsons et al. (1993). We defined a displacement as one fish causing another fish to move at least two body lengths away from a preferred feeding or holding site (Brown 1975 as cited in Helfrich et al. 1982). Small-scale displacements were those that occurred within a channel unit of stream, such as a pool. Wild fish movement out of the release stream (determined by captures in a downstream migrant trap at the mouth of Jungle Creek) concurrent with large numbers of hatchery fish was considered a mid-scale displacement. Large-scale displacement was monitored

TABLE 23.2
Summary of methods used in examining the effects of hatchery steelhead trout releases on wild salmonids in the Teanaway River basin. Stream name, experimental function, year, and type of method are shown. Methods and levels of effort are shown for each stream and year. Snorkeling effort for behavioral observations is expressed in minutes. The number of 100-m long electrofishing sites in each stream is shown for each year. Trapping effort, to examine displacement, is expressed by the dates each trap was operated. NFT = North Fork of the Teanaway River, MFT = Middle Fork of the Teanaway River, WFT = West Fork of the Teanaway River.

Stream	Function	Year	Snorkeling minutes	Electrofishing sites	Trapping dates
Jungle Cr.	Small treatment	1991	1011	1	5/29–6/13
		1992	1847	1	5/9–6/2
		1993	722	1	4/30–7/13
		1994	972	1	4/29–7/11
		1995	—	1	5/6–7/25
NFT	Large treatment	1990	—	2	—
		1991	1931	2	4/22–5/31
		1992	1396	2	4/2–6/1
		1993	484	2	4/1–6/4
		1994	779	2	3/30–6/1
		1995	—	2	3/11–6/9
Jack Cr.	Small control	1992	739	—	5/5–6/2
		1993	488	—	4/30–7/13
		1994	823	—	4/29–7/11
		1995	—	—	5/10–7/25
MFT	Large control	1990	—	3	—
		1991	—	3	—
		1992	1558	3	4/3–5/27
		1993	554	3	3/31–6/4
		1994	756	3	3/29–5/31
		1995	—	3	3/11–6/9
WFT	Large control	1990	—	3	—
		1991	—	3	—
		1992	—	3	—
		1993	—	3	—
		1994	—	3	—
		1995	—	3	—

at a downstream migrant trap near the mouth of the North Fork of the Teanaway River, approximately 11 km downstream of the release site in Jungle Creek. Determination of small-scale displacements was more direct (because they were observed) than mid- and large-scale displacements, which had to be inferred from fish emigration information obtained by trapping.

To determine the influence of hatchery steelhead trout releases on rainbow trout abundance, population estimates were conducted by electrofishing in three study streams. Multiple-removal population estimates were conducted in 100-m long index sites in the North (N = 2), Middle (N = 3), and West (N = 3) forks of the Teanaway River each fall from 1990 (the year before the first release) to 1995 (the year after the last release) (Table 23.2).

TABLE 23.3
Estimated wild rainbow trout and residual hatchery steelhead trout number and biomass from electrofishing in Jungle Creek, 1991–1994 approximately 3 mo after release of steelhead trout. This index site, an example from all the index sites, was 100 m long and located immediately upstream of the confluence of Jungle Creek and the North Fork of the Teanaway River.

	Number		Biomass (g)	
Year	rainbow trout	steelhead trout	rainbow trout	steelhead trout
1991	5	9	47	530
1992	2	13	21	468
1993	40	11	477	453
1994	2	5	19	220

Growth experiments were performed in in-stream enclosures in 1993 and 1994 in the North Fork of the Teanaway River to determine the effects of residual hatchery steelhead trout on growth of wild rainbow trout and spring chinook salmon. Enclosures were wood-framed boxes, a cubic meter in size, and were divided in the middle and wrapped with 0.95-cm hardware cloth on the sides and bottom to allow free passage of invertebrates in and out of the enclosures (Cooper et al. 1990; McMichael et al. 1997). Enclosures were deployed with one wild rainbow trout (1993: n = 10, 1994: n = 20) or salmon (1993: n = 10) on one side (control) and one wild rainbow trout or salmon with a residual hatchery steelhead trout (treatment) on the other side for a 6-week period in mid-summer (McMichael et al. 1997). Specific growth rates of trout and salmon with and without residual hatchery steelhead trout were compared to determine impacts of residual hatchery steelhead trout on growth of wild rainbow trout and spring chinook salmon.

RESULTS

Residual steelhead trout were relatively abundant in our treatment streams in three of the four years we released hatchery steelhead trout (Figure 23.2). Although variable, abundance, and particularly biomass, of residual steelhead trout often significantly exceeded those of wild rainbow trout in the release stream during the summer rearing period (see, for example, Table 23.3; McMichael et al. 1994). Residual hatchery steelhead trout were also observed more than 12 km upstream from their release points, in an area containing populations of bull trout and cutthroat trout in addition to rainbow trout (Figure 23.3).

When smolt-sized (over 175 mm fork length-FL) hatchery steelhead trout failed to promptly migrate toward the sea and instead residualized in freshwater, they behaviorally dominated the typically smaller, coexisting wild salmonids (McMichael et al. 1992, 1994; Pearsons et al. 1993, 1996). Larger salmonids typically dominated smaller ones, regardless of hatchery or wild origin or species. The wild rainbow trout within our study area averaged 125 mm FL while hatchery steelhead trout we released averaged between 179 and 201 mm FL (McMichael et al. 1994). Despite this size disparity, rainbow trout and residual hatchery steelhead trout were often observed occupying the same or similar habitats. In 84% of the contests we observed between 1991 and 1994 (N = 538), larger fish were judged to be dominant. Hatchery steelhead trout also dominated over 75% of the contests between steelhead trout and wild rainbow trout (N = 147) (McMichael et al. 1994; Pearsons et al. 1996). Interactions observed in 1994, in streams where residual hatchery steelhead trout were present, more often involved physical contact (nips and butts; 25%) than those observed in control streams (14%) (Pearsons et al. 1996). In contrast, the most common type of behavior observed in contests in control streams was a threat (47%), while only 20% of the behaviors

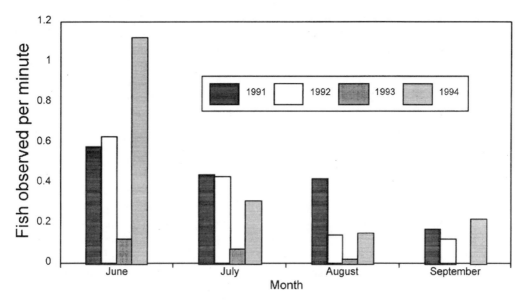

FIGURE 23.2   Frequency of residual hatchery steelhead trout observed during snorkeling activities in treatment streams during the summer and early fall of 1991, 1992, 1993, and 1994.

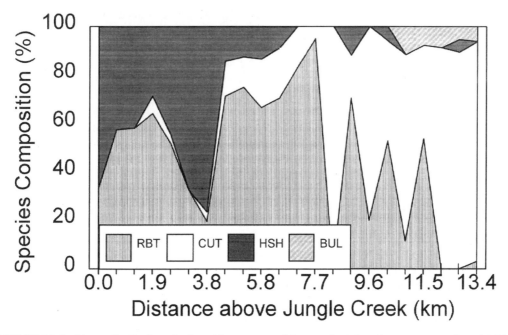

FIGURE 23.3   Linear distribution of salmonids upstream of the mouth of the release stream on June 24, 30, and July 12, 1994. RBT = rainbow trout, CUT = cutthroat trout, BUL = bull trout, and HSH = residual hatchery steelhead trout. An average of 27 salmonids were observed at each of 22 sites (0.6 km apart). Sites from 8.3 to 13.4 km were snorkeled at night, while all others were snorkeled during daylight.

displayed by fish in treatment streams were classified as threats (Pearsons et al. 1996). Threats are a more energy-efficient way to maintain established hierarchies than the more violent physical interactions commonly observed in treatment streams.

Wild trout were displaced from apparently preferred microhabitats by hatchery steelhead trout, but were generally not displaced over larger spatial scales (0.2 to 11.2 km) (Pearsons et al. 1996). Small-scale displacement of subordinate fish occurred in 47 to 59% of the contests observed in treatment streams in 1993 and 1994 (McMichael et al. 1994; Pearsons et al. 1996). Since wild fish were usually smaller than residual steelhead trout, this meant that residual steelhead trout were displacing wild salmonids more often than wild fish were displacing residual steelhead trout. In only one year (1994) of four, there was a correlation between wild trout and hatchery steelhead trout emigration from the release stream, indicating there may have been a mid-scale displacement that year. In all four years of snorkeling, we observed only two instances of pied-piper behavior (Hillman and Mullan 1989) in which a wild salmonid turned to follow a passing hatchery fish downstream.

Interaction rates were highest when stream temperatures were over 8°C. When water temperatures were below 8°C, we often had a difficult time locating wild salmonids, presumably because they were holding within the substrate (Hillman et al. 1992). In many locations throughout the upper Yakima basin we have seen far fewer fish when water temperatures were below 8°C than when temperatures were higher. Behavioral interactions between hatchery steelhead trout and wild salmonids were precluded at lower temperatures because wild salmonids were not in the water column. Furthermore, hatchery fish were not observed utilizing interstices at lower temperatures. It is possible that wild trout may have left the interstices during darkness, but we did not sample during darkness as lights were found to alter the behavior of hatchery steelhead trout (McMichael et al. 1992). In all four years, interaction rates (observed interactions/fish/minute) in the stream where the hatchery fish were released were higher after June 1 than before, though not significantly ($P = 0.233$) (McMichael et al. 1994; Pearsons et al. 1996). Water temperatures generally exceeded 8°C by June 1 of each year. June 1 is also the date after which we defined all remaining hatchery steelhead trout to be residuals.

In addition to higher interaction rates, the incidence of disease in both residual hatchery steelhead trout and wild fish was greater when water temperature was warmer. For example, in 1991, when water temperatures were very high in late May and June (often exceeding 20°C), *Saprolegnia* infections were commonly observed on hatchery fish and, though less common, were also observed on wild fish within the release stream (McMichael et al. 1992). The incidence of *Saprolegnia* on hatchery steelhead trout emigrating from Jungle Creek between May 29 and June 13, 1991, was 13.2% (N = 53). A sample of fish collected by electrofishing in Jungle Creek on June 25, 1991, showed 32.1% of the hatchery steelhead trout (N = 28) and 16.7% of the wild resident trout (N = 6) were infected. Infection sites were generally on the lateral body surface near the base of the dorsal fin which corresponds with the area where most of the violent agonistic attacks we observed were targeted (McMichael et al. 1992). It is likely that the combination of high densities of residual hatchery steelhead trout, elevated rates of agonistic interactions, and warm water temperatures combined to increase stress levels in many salmonids to the point that they were vulnerable to fungal infection by *Saprolegnia*.

The presence of hatchery steelhead trout reduced the abundance of rainbow trout where they were commingled. Wild rainbow trout in sites within the large treatment stream (the North Fork of the Teanaway River) were significantly less abundant than in sites in adjacent control streams ($P = 0.006$; McMichael et al. 1994; Pearsons et al. 1996) (Figure 23.4).

Hatchery steelhead trout residuals apparently had a greater impact on conspecifics than on other species, such as spring chinook salmon. For example, growth experiments revealed that rainbow trout paired with hatchery steelhead trout residuals had significantly lower specific growth rates than their unpaired counterparts (1993: $P = 0.019$; 1994: $P = 0.020$; McMichael et al. 1997). When spring chinook salmon were paired with residual hatchery steelhead trout, however, their growth was not significantly reduced ($P = 0.360$; McMichael et al. 1997).

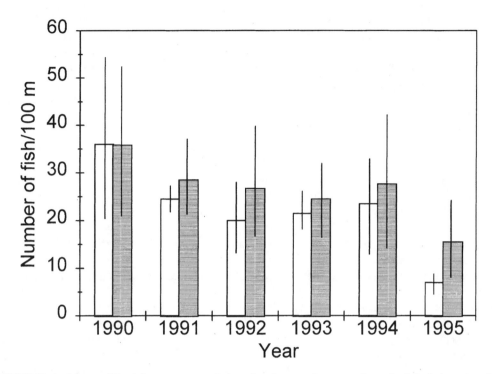

FIGURE 23.4   Mean wild rainbow trout population abundance estimates and standard deviations (number of fish per 100 m) for streams with (open bars) and without (cross-hatched bars) hatchery steelhead trout between 1990 and 1995. Index sites were 100 m long and were sampled by electrofishing. Hatchery steelhead trout smolts were released in 1991, 1992, 1993, and 1994.

DISCUSSION

Hatchery steelhead trout apparently adversely interact with wild salmonids under certain conditions. The ecological impacts from releases of hatchery steelhead trout smolts on wild salmonids were greatest when; (1) hatchery fish did not emigrate quickly; (2) water temperatures were above 8°C; (3) wild salmonids were the same species as the hatchery fish; (4) wild salmonids were smaller than hatchery fish; (5) habitat and/or food were limited (McMichael et al. 1997); and (6) numbers of fish released were at the high end of the range we investigated (McMichael 1994).

We fully appreciate that wild fish populations face a variety of risks besides those from interactions with hatchery fish (e.g., direct or incidental harvest, extreme environmental variation, habitat degradation and loss, genetic introgression, and hybridization). However, various measures could help reduce or minimize undesirable ecological effects of hatchery steelhead trout on wild fish. Those measures should reflect the information available about conditions under which interactions might be reduced. There are three general categories of actions culturists and managers can take to reduce ecological effects of hatchery steelhead trout on wild salmonids: (1) the hatchery product—*what* is released; (2) space—*where* the fish are released; and (3) timing—*when* the fish are released. Guidelines to reduce the impacts of released hatchery steelhead trout on wild salmonids are presented below.

THE HATCHERY PRODUCT—WHAT IS RELEASED

Hatchery fish slated for release should be ready to migrate, resulting in few residuals. Tools are becoming available with potential for reducing the number of hatchery steelhead trout that residualize after release. For example, Viola and Schuck (1995) successfully reduced the number of residual steelhead trout in the Tucannon River in southeastern Washington by periodically examining juveniles in an acclimation pond prior to release and closing the exit gate when most emigrants no longer exhibited smolt characteristics and were predominantly males; likely residuals were retained. Hatchery practices affecting rearing densities, size at release, and parentage may influence residualism in steelhead trout. In our studies, hatchery steelhead trout that were offspring of hatchery-origin parents and reared at low densities apparently residualized less and survived better than offspring of wild parents that were reared at higher densities (Table 23.1; McMichael 1994; Pearsons et al. 1996). We did not conduct paired tests of these factors and are therefore unable to discern the relative effects of parentage and rearing density on smolt survival.

Increased growth rates of fish reared in hatcheries have often been implicated in the production of residual anadromous salmonids (e.g., Rowe and Thorpe 1990; Mullan et al. 1992). In a study on the upper Salmon River, Idaho, Partridge (1986) found that hatchery steelhead trout released at a "large-size" (mean FL = 265 mm; number released = 40,322) constituted 92% of the residual steelhead trout catch by anglers, while the remainder were released at "normal-size" (mean FL = 202 mm; number released = 39,763).

It has been well established that a positive correlation exists between fish size and social dominance for salmonids (e.g., Abbott et al. 1985; Huntingford et al. 1990). Abbott et al. (1985) reported that a weight advantage of as little as 5% was sufficient to assure dominant status for larger steelhead trout juveniles. Behavioral data from our field studies indicated that in most contests (84%) socially dominant fish were larger than subordinates. To minimize the potential for hatchery fish to socially dominate wild fish, the size of hatchery steelhead trout released should be smaller than their wild counterparts. This smaller size may conflict with size criteria that maximize emigration, and should therefore be secondary to tactics that result in few or no residuals. In many, if not most, instances it may be difficult to meet competing objectives of smaller hatchery fish size, effective smoltification, and maximum long-term survival. For example, in our study area, where wild rainbow trout averaged 125 mm FL during the summer rearing period (McMichael et al. 1994), it would be difficult to produce steelhead trout juveniles in the hatchery that became smolts at a size smaller than 125 mm FL. Tipping et al. (1995) reported that hatchery-reared winter steelhead trout over 190 mm FL migrated at higher rates than smaller hatchery fish. It is conceivable that hatchery steelhead trout smolts much larger than their wild counterparts would occupy different habitats than smaller wild fish (Roper et al. 1994), thereby reducing interactions and their effects. There is a tradeoff in this approach, however, in that hatchery steelhead trout that are much larger than the wild fish may become predators of age-0 wild salmonids (Cannamela 1992; Martin et al. 1993). It is noteworthy that, even with the size disparity we observed between the hatchery fish and wild fish (a mean difference of about 50–70 mm), many hatchery and wild fish occupied the same habitats.

The number of hatchery fish released is a key influence on the level of ecological interactions experienced by wild fish. We suggest that only the minimum number of hatchery steelhead trout smolts necessary to meet management objectives should be released in any given situation. The likelihood of adverse ecological interactions resulting from hatchery steelhead trout releases is positively correlated with hatchery fish abundance. When factors affecting residualism rates are equal (e.g., parentage, size at release, precocity, rearing density), larger hatchery steelhead trout releases result in more residual steelhead trout in the freshwater rearing environment. In addition, hatchery managers are finding that fish reared at relatively low densities often produce the greatest percentage of returning adult hatchery fish (Banks 1994; Ewing and Ewing 1995).

Even when most hatchery smolts emigrate promptly, the combined effects of greater social interaction during even short spatiotemporal overlap increases the risk of adverse outcomes (McMichael et al. 1994). One such mechanism is the indirect effect of predation on wild fish through forced habitat shifts resulting from dominance of hatchery fish over wild fish (e.g., Abrahams and

Healey 1993; Walters and Juanes 1993). However, the greatest risk occurs when hatchery fish do not emigrate quickly and become residuals.

The number of residuals expected for any release group may vary considerably. We estimated that 26 to 39% of the fish we released (23,000–39,000) did not emigrate from the study area during the first month after release. Other researchers have reported nonmigrant portions of hatchery steelhead trout releases between 19.8% (Tipping et al. 1995) and 65.5% (Tipping and Byrne 1996). However, nonmigrant steelhead trout smolts do not always become residual steelhead trout; many are lost to predation or other sources of mortality and some may emigrate a year or more later as age-2 or -3 smolts (Pearsons et al. 1993). In our work, if we assume all nonmigrants became residuals, then an estimated 5,980–15,210 residual hatchery steelhead trout were in an area supporting an estimated 3,800 wild rainbow trout. Even applying a conservative residualism rate of 15% (derived from a nonmigrant rate of 30%, only half of which survived and remained as residuals) and our lowest release number, we would have expected nearly as many residual steelhead trout (3,450) as rainbow trout (3,800) in the 10-km reach of the North Fork of the Teanaway River below where the steelhead trout were released. Further, most steelhead trout releases involve larger numbers of smolts. For example, the average number of smolt-size steelhead trout released from State and tribal hatcheries in Washington watersheds in 1994 was 87,780 (range = 4,000 to 794,700) (WDFW 1995).

Some innovative fish culture strategies may help reduce undesirable behaviors expressed by many hatchery fish after release. These behaviors include agonistic tendencies with no perceptible benefit to either fish in the interaction (Ruzzante 1994). Maynard et al. (1995) reviewed the use of seminatural culture strategies for anadromous salmonids and concluded that some of these innovative techniques may improve behavior with respect to increasing survival of hatchery fish. These authors recommended such strategies be used in both enhancement and conservation hatcheries, even though the efficacy of the strategies remains to be tested at larger scales of production. While most refinements to standard hatchery protocols have focused on producing hatchery fish that survive better in the wild (e.g., Banks 1994; Olla et al. 1994; Jarvi and Uglem 1993; Wiley et al. 1993), little research has been conducted on means of producing hatchery fish that, following release, are ecologically compatible with wild fish. Cuenco et al. (1993) stated that, "There is a paucity of information on the potential competition between the supplemented salmon population and other fish populations inhabiting the target stream." Certainly, this is an area deserving further research.

SPACE—WHERE THE FISH ARE RELEASED

Viability of wild stocks would be impacted least if hatchery smolts were only released where recipient wild salmonid populations were absent although, even in areas where wild salmonid stocks are not present, potential effects on other species (e.g., non-salmonid fishes, aquatic invertebrates, and amphibians) should be recognized. A variety of management objectives may dictate when hatchery fish are to be released into a particular area. In areas containing depressed or critical steelhead trout stocks, managers may use supplementation to restore populations, while in areas that are considered to have healthy populations, they may use hatchery fish solely to augment harvest. Hatchery releases in areas containing depressed or critical stocks are not advisable simply because they may add ecological stress to precarious stocks. In addition, any impacts on depleted wild populations will have greater per capita effects than for healthy populations.

In situations where a wild steelhead trout stock is abundant and healthy (e.g., reaching escapement goals), hatchery releases should be considered only on a case-by-case basis. Some might argue that hatchery fish should be excluded from systems with abundant or healthy wild salmonid populations to avoid any ecological risk to wild stocks (Huntington et al. 1996). The merits of this argument will increasingly deserve attention, especially when managers are faced with decisions regarding new or proposed hatchery projects to augment fishery harvests where healthy wild stocks exist. For example, in areas that are known to have a density-dependent bottleneck after the smolt stage, it would not be advisable to release large numbers of hatchery steelhead trout smolts (Royal 1972, cited by Lichatowich and McIntyre 1987; Peterman 1984), because this would only exacerbate stresses on wild stocks, with negligible increase of returning adult hatchery fish. Therefore, we recommend that hatchery fish only be released in areas where there is minimal ecological risk to wild stocks.

Releases of hatchery steelhead trout smolts into areas that have complex habitat (e.g., a wide variety of depths, velocities, substrates, and cover) would be expected to have the least ecological impact on wild salmonids. Steelhead trout follow diverse and often intertwining life-history trajectories that require many types of stream habitat during their freshwater rearing phase. Therefore, areas with diverse and abundant habitat provide more opportunities for wild fish to segregate themselves from their hatchery counterparts. Partitioning the habitat enables fish to reduce the intensity of competition through differential resource use and/or visual isolation (Allee 1982; Hearn and Kynard 1986; Roper et al. 1994; Raddum and Fjellheim 1995). Conversely, an argument could be made that hatchery smolts should be released only in simple habitat (very little variation in habitat features). This approach might avoid wild salmonid "strongholds" and encourage hatchery fish to emigrate more rapidly. However, our field studies have shown that hatchery steelhead trout smolts and residuals dominated wild fish in all habitat types examined. Further study of this issue is necessary before establishing firm guidelines.

Martin et al. (1993) also recommended releasing hatchery steelhead smolts in locations that were easily accessible to anglers. Where hatchery fish residualize in streams, there may be ways of removing some of them without adversely impacting wild fish. Angling regulations could be adopted to encourage the harvest of hatchery steelhead residuals. A key consideration is that such activities should not inadvertently increase risks to wild fish. In our study, where hatchery steelhead were much larger than their wild salmonid counterparts, a minimum size limit of 18 cm (7 in) might have been an effective way to allow for the harvest of residual hatchery steelhead while protecting against harvest of the smaller wild fish. In areas where size ranges of wild and hatchery fish overlap, regulations could allow the retention of steelhead with clipped fins. In areas where minimum size limits might adversely impact larger wild salmonids, the fin clip regulation would be required. In any case, incidental mortality to wild fish is possible when undersized or unmarked fish are hooked and released (Ferguson and Tufts 1992). This factor should be seriously considered before angling is used to reduce the abundance of residual hatchery steelhead.

Timing—When the Fish Are Released

Release timing can have important effects on the extent to which adverse behavioral interactions occur between hatchery steelhead trout and wild salmonids. To minimize the chance for interactions, hatchery smolts should be released when they are naturally ready to migrate. To avoid release of nonmigrant individuals, access to streams should be precluded after a certain date, or when smolt characteristics decrease and sex ratios heavily favor males in volitional release facilities (e.g., >75% male). Dramatic reductions in the number of residual steelhead trout in streams may occur if nonmigrant hatchery steelhead trout are not forced from raceways and ponds into streams. The use of acclimation ponds or raceways with volitional exit capabilities can greatly enhance likelihood that released fish will tend to smolt rather than residualize (Viola and Schuck 1995). Most existing artificial propagation vessels may not be equipped for volitional releases. However, they can often be modified to include traps at the downstream ends for collection of actively emigrating smolts. These fish could then be released while other fish are retained until they move into the trap. This would help sort out nonmigrant from emigrant fish. Nonmigrant fish could be planted into lakes where ecological risks are not deemed unacceptable (Evenson and Ewing 1992), or the public could even be invited to fish in the acclimation pond itself (Viola and Schuck 1995).

Adverse interactions can be reduced by releasing hatchery steelhead trout smolts when water temperatures are relatively cold (less than 8°C) because we found most wild salmonids were on or in the substrate at low temperatures, while hatchery fish remained in the water column. Hillman et al. (1992) observed less than 20% of the known number of salmonids in Wenatchee River, Washington, sample sites when water temperatures were less than 9°C. Cooler water also reduces the possibility for transmission and outbreak of diseases (e.g., *Saprolegnia*). However, there is a possibility that releasing fish into cold water will adversely alter the smoltification process, and this should be examined.

Releasing hatchery fish at dusk, or shortly thereafter, can apparently reduce adverse interactions because over 90% of the hatchery steelhead trout emigration from treatment streams were captured in outmigrant traps during darkness (McMichael et al. 1992). Releasing fish in the dark should minimize competitive interactions with wild salmonids in the immediate release area because most hatchery fish should have moved downstream by morning, and should be less densely distributed than if released during the day.

SUMMARY

As with any general guidelines, decisions must be made regarding the relative importance of each factor in the final management decisions for a particular waterbody. Furthermore, implementation of guidelines should be coordinated with a monitoring program that is designed to determine whether the guidelines are effective in minimizing impacts. The guidelines presented in this chapter were developed for our specific location and species groups, and while some may be useful in other basins with the same or other species, others may not. Although these guidelines do not represent official Washington Department of Fish and Wildlife policy or address other risk factors associated with hatchery steelhead trout releases (e.g., genetic concerns), we believe they will be of value to those attempting to reconcile ecological risks associated with integrating hatchery and wild stock management programs. It is important to note that these guidelines pertain only to ecological interactions resulting directly from hatchery fish released in the freshwater rearing environment and do not address interactions in the ocean, in harvests, or at return.

Conservation of wild fish resources is a fundamental goal shared by fisheries management entities charged with stewardship responsibilities. However, in contrast to other factors (e.g., genetics—Busack and Currens 1995; Waples and Do 1995), the risk of adverse ecological interactions arising from the use of hatchery fish has received little attention. In situations where conservation of wild salmonid stocks is a high priority, explicit decisions must be made and steps taken to reduce the deleterious ecological effects of hatchery fish in the context of the many other threats to wild stocks. For the specific conditions of our study, the guidelines offered here should help reduce the impact of hatchery programs on wild salmonid populations, but by themselves, will not guarantee the viability of the wild salmonids. Squarely addressing various types of costs and benefits in the context of often contradictory management goals and objectives is necessary in order to make responsible management decisions. In many cases, such decisions will likely translate into reductions in the number of returning hatchery adults to the target area, or harvest opportunities, but will increase protection for wild stocks and the chances of their sustainability.

ACKNOWLEDGMENTS

Many individuals have contributed to the foundation of these ideas over the past 6 years. Discussions of hatchery/wild issues with WDFW staff in the Yakima Species Interactions Studies office, including Steven Martin, Eric Bartrand, Jim Olson, Marcia Fischer, Anthony Fritts, and others have helped shape some of these concepts. We sincerely appreciate the administrative support of Bill Hopley in facilitating the senior authors' attendance at the Sustainable Fisheries Conference. Funding for this work was provided to WDFW's Yakima Species Interactions Team by the Bonneville Power Administration.

REFERENCES

Abbott, J. C., R. L. Dunbrack, and C. D. Orr. 1985. The inheritance of size and experience in dominance relationships of juvenile steelhead trout (*Salmo gairdneri*). Behaviour 92:241–253.

Abrahams, M. V., and M. C. Healey. 1993. A comparison of the willingness of four species of Pacific salmon to risk exposure to a predator. Oikos 66:439–446.

Allee, B. A. 1982. The role of interspecific competition in the distribution of salmonids in streams. Pages 111–121 *in* E. L. Brannon and E. O. Salo, editors. Proceedings of the salmon and trout migratory behavior symposium. University of Washington Press, Seattle.

Allendorf, F. W., and nine coauthors. 1997. Prioritizing Pacific salmon stocks for conservation. Conservation Biology 11:140–152.

Banks, J. L. 1994. Raceway density and water flow as factors affecting spring chinook salmon (*Oncorhynchus tshawytscha*) during rearing and after release. Aquaculture 119:201–217.

Busack, C. A., and K. P. Currens. 1995. Genetic risks and hazards in hatchery operations: fundamental issues and concepts. Pages 71–80 *in* H. L. Schramm and R. G. Piper, editors. Uses and effects of cultured fishes in aquatic ecosystems, American Fisheries Society Symposium 15. Bethesda, Maryland.

Cannamela, D. A. 1992. Potential impacts of releases of hatchery steelhead trout "smolts" on wild and natural juvenile chinook and sockeye salmon. Idaho Fish and Game, Boise.

Clune, T., and D. Dauble. 1991. The Yakima/Klickitat fisheries project: a strategy for supplementation of anadromous salmonids. Fisheries 16:28–34.

Cooper, S. D., S. J. Walde, and B. L. Peckarsky. 1990. Prey exchange rates and the impact of predators on the prey populations in streams. Ecology 71:1503–1514.

Cuenco, M. L., T. W. H. Backman, and P. R. Mundy. 1993. The use of supplementation to aid in natural stock restoration. Pages 269–293 *in* J. G. Cloud and G. H. Thorgaard, editors. Genetic conservation of salmonid fishes. Plenum Press, New York.

Evenson, M. D., and R. D. Ewing. 1992. Migration characteristics and hatchery returns of winter steelhead volitionally released from Cole Rivers Hatchery, Oregon. North American Journal of Fisheries Management 12:736–743.

Ewing, R. D., and S. K. Ewing. 1995. Review of the effects of rearing density on survival to adulthood for Pacific salmon. The Progressive Fish-Culturist 57:1–25.

Ferguson, R. A., and B. L. Tufts. 1992. Physiological effects of brief air exposure in exhaustively exercised rainbow trout (*Oncorhynchus mykiss*): implications for "catch and release" fisheries. Canadian Journal of Fisheries and Aquatic Sciences 49:1157–1162.

Hearn, W. E., and B. E. Kynard. 1986. Habitat utilization and behavioral interaction of juvenile Atlantic salmon (*Salmo salar*) and rainbow trout (*S. gairdneri*) in tributaries of the White River of Vermont. Canadian Journal of Fisheries and Aquatic Sciences 43:1988–1998.

Helfrich, L. A., J. R. Wolfe, Jr., and P. T. Bromley. 1982. Agonistic behavior, social dominance, and food consumption of brook trout and rainbow trout in a laboratory stream. Proceedings of the Annual Conference of the Southeast Association of Fish and Wildlife Agencies 36:340–350.

Hillman, T. W., and J. W. Mullan. 1989. Effect of hatchery releases on the abundance and behavior of wild salmonids. Pages 265–285 *in* Don Chapman Consultants, Inc. Summer and winter ecology of juvenile chinook salmon and steelhead trout in the Wenatchee River, Washington. Final report to Chelan County Public Utility District, Washington.

Hillman, T. W., J. W. Mullan, and J. S. Griffith. 1992. Accuracy of underwater counts of juvenile chinook salmon, coho salmon, and steelhead. North American Journal of Fisheries Management 12:598–603.

Holtby, L. B., D. P. Swain, and G. M. Allan. 1993. Mirror-elicited agonistic behavior and body morphology as predictors of dominance status in juvenile coho salmon (*Oncorhynchus kisutch*). Canadian Journal of Fisheries and Aquatic Sciences 50:676–684.

Huntingford, F. A., N. B. Metcalfe, J. E. Thorpe, W. D. Graham, and C. E. Adams. 1990. Social dominance and body size in Atlantic salmon parr, *Salmo salar* L. Journal of Fish Biology 36:877–881.

Huntington, C., W. Nehlsen, and J. Bowers. 1996. A survey of healthy native stocks of anadromous salmonids in the Pacific Northwest and California. Fisheries 21:6–14.

IHOT (Integrated Hatchery Operations Team). 1995. Policies and procedures for Columbia basin anadromous salmonid hatcheries. Annual Report 1994. Bonneville Power Administration, Portland, Oregon.

Jarvi, T., and I. Uglem. 1993. Predator training improves the anti-predator behavior of hatchery reared Atlantic salmon (*Salmo salar*) smolt. Nordic Journal of Freshwater Research 68:63–71.

Keenleyside, M. H. A., and F. T. Yamamoto. 1962. Territorial behavior of juvenile Atlantic salmon (*Salmo salar*, L.). Behaviour 19:139–169.

Krause, T. 1991. The Yakima. Fly Fisherman 22:40–43, 76–78.

Lichatowich, J. A., and J. D. McIntyre. 1987. Use of hatcheries in the management of Pacific anadromous salmonids. Pages 131–136 *in* M. J. Dadswell and five coeditors. Common strategies of anadromous and catadromous fishes, American Fisheries Society Symposium 1. Bethesda, Maryland.

Martin, S. W., A. E. Viola, and M. L. Schuck. 1993. Investigations of the interactions among hatchery reared summer steelhead, rainbow trout, and wild spring chinook salmon in southeast Washington. Washington Department of Wildlife, Olympia.

Maynard, D. J., T. A. Flagg, and C. V. W. Mahnken. 1995. A review of seminatural culture strategies for enhancing the postrelease survival of anadromous salmonids. Pages 307–314 *in* H. L. Schramm and R. G. Piper, editors. Uses and effects of cultured fishes in aquatic ecosystems, American Fisheries Society Symposium 15. Bethesda, Maryland.

McMichael, G. A. 1994. Effects of parentage, rearing density, and size at release of hatchery-reared steelhead smolts on smolt quality and post-release performance in natural streams. Pages 194–213 *in* Pearsons, T. N., and five coauthors. 1994. Yakima River species interactions studies annual report 1993. Bonneville Power Administration, Portland, Oregon.

McMichael, G. A., and five coauthors. 1992. Yakima River species interactions studies, annual report 1991. Bonneville Power Administration, Portland, Oregon.

McMichael, G. A., T. N. Pearsons, and S. A. Leider. 1994. The effects of releases of hatchery-reared steelhead on wild salmonids in natural streams. Pages 143–165 *in* Pearsons, T. N., and five coauthors. 1994. Yakima River species interactions studies annual report 1993. Bonneville Power Administration, Portland, Oregon.

McMichael, G. A., C. S. Sharpe, and T. N. Pearsons. 1997. Effects of residual hatchery steelhead on growth of wild rainbow trout and chinook salmon. Transactions of the American Fisheries Society 126:230–239.

Meffe, G. K. 1992. Techno-arrogance and halfway technologies: salmon hatcheries on the Pacific coast of North America. Conservation Biology 6:350–354.

Mullan, J. W., A. Rockhold, and C. R. Chrisman. 1992. Life histories and precocity of chinook salmon in the mid-Columbia River. The Progressive Fish Culturist 54:25–28.

NMFS (National Marine Fisheries Service). 1995. Draft Snake River chinook recovery plan. National Marine Fisheries Service, Seattle.

NRC (National Research Council). 1996. Upstream: salmon and society in the Pacific Northwest. Report on the protection and management of Pacific Northwest anadromous salmonids, National Research Council of the National Academy of Sciences, National Academy Press, Washington, D.C.

Nehlsen, W., J. E. Williams, and J. A. Lichatowich. 1991. Pacific salmon at the crossroads: stocks at risk from California, Oregon, Idaho, and Washington. Fisheries 16:4–21.

Olla, B. L., M. W. Davis, and C. H. Ryer. 1994. Behavioral deficits in hatchery-reared fish: potential for effects on survival following release. Aquaculture and Fisheries Management 25:19–34.

Partridge, F. E. 1986. Effect of steelhead smolt size on residualism and adult return rates. Idaho Fish and Game, Boise.

Pearsons, T. N., and five coauthors. 1993. Yakima species interactions study, annual report 1992. Bonneville Power Administration, Portland, Oregon.

Pearsons, T. N., and five coauthors. 1994. Yakima River species interactions studies annual report 1993. Bonneville Power Administration, Portland, Oregon.

Pearsons, T. N., and five coauthors. 1996. Yakima River species interactions studies annual report 1994. Bonneville Power Administration, Portland, Oregon.

Peterman, R. M. 1984. Density-dependent growth in early ocean life of sockeye salmon (*Oncorhynchus nerka*). Canadian Journal of Fisheries and Aquatic Sciences 41:1825–1829.

Probasco, S. 1994. River journal, Vol. 2: Yakima River. Frank Amato Publications, Inc. Portland, Oregon.

Raddum, G. G., and A. Fjellheim. 1995. Artificial deposition of eggs of Atlantic salmon (*Salmo salar* L.) in a regulated Norwegian river: hatching, dispersal and growth of the fry. Regulated Rivers 10:169–180.

Roper, B. B., D. L. Scarnecchia, and T. J. La Marr. 1994. Summer distribution of and habitat use by chinook salmon and steelhead within a major basin of the South Umpqua River, Oregon. Transactions of the American Fisheries Society 123:298–308.

Rowe, D. K., and J. E. Thorpe. 1990. Differences in growth between maturing and non-maturing male Atlantic salmon, *Salmo salar* L., parr. Journal of Fish Biology 36:643–658.

Ruzzante, D. E. 1994. Domestication effects on aggressive and schooling behavior in fish. Aquaculture 120:1–24.

Stringer, G. E., and W. S. Hoar. 1955. Aggressive behavior of underyearling kamloops trout. Canadian Journal of Zoology 33:148–160.

Taylor, E. B., and P. A. Larkin. 1986. Current response and agonistic behavior in newly emerged fry of chinook salmon, *Oncorhynchus tshawytscha*, from stream- and ocean-type populations. Canadian Journal of Fisheries and Aquatic Sciences 43:565–573.

Tipping, J. M., R. V. Cooper, J. B. Byrne, and T. H. Johnson. 1995. Length and condition factor of migrating and non-migrating hatchery-reared winter steelhead smolts. The Progressive Fish Culturist 57:120–123.

Tipping, J. M., and J. B. Byrne. 1996. Reducing feed levels during the last month of rearing enhances hatchery-reared steelhead smolt emigration rates. The Progressive Fish Culturist 58:128–130.

Viola, A. E., and M. L. Schuck. 1995. A method to reduce the abundance of residual hatchery steelhead in rivers. North American Journal of Fisheries Management 15:488–493.

Walters, C. J., and F. Juanes. 1993. Recruitment limitation as a consequence of natural selection for use of restricted feeding habitats and predation risk taking by juvenile fishes. Canadian Journal of Fisheries and Aquatic Sciences 50:2058–2070.

Waples, R. S., and C. Do. 1995. Genetic risk associated with supplementation of Pacific salmonids: captive broodstock programs. Canadian Journal of Fisheries and Aquatic Sciences 51:310–329.

Wiley, R. W., R. A. Whaley, J. B. Satake, and M. Fowden. 1993. An evaluation of the potential for training trout in hatcheries to increase poststocking survival in streams. North American Journal of Fisheries Management 13:171–177.

WDFW (Washington Department of Fish and Wildlife). 1995. 1994–95 steelhead harvest summary. Washington Department of Fish and Wildlife, Olympia.

WDFW (Washington Department of Fish and Wildlife). 1997. Wild salmonid policy, draft environmental impact statement. Washington Department of Fish and Wildlife, Olympia.

24 Economic Feasibility of Salmon Enhancement Propagation Programs

Hans D. Radtke and Shannon W. Davis

Abstract.—The purpose of salmon propagation by artificial means is to replace natural production (i.e., mitigate) and/or augment levels of natural production (i.e., enhancement). Mitigation programs do not necessarily need to show economic feasibility, because their purpose is for replacing production. However, enhancement programs are primarily undertaken for economic development and their feasibility should be evaluated. In general, the feasibility of a program may be described in terms of (1) revenues received by the harvester and (2) total regional personal income generated in the local economy. While an evaluation of the first may be sufficient to decide whether a program is warranted, both measures must be evaluated to determine whether a more detailed cost-benefit analysis will provide needed information about a project's feasibility. In this chapter we illustrate, via a case study of the Terminal Fishery Project (TFP) on the Columbia River, an economic evaluation process that utilized both of the above measures. That is, comparisons were made (1) between the production costs of a Columbia River salmon enhancement program and the commercial fishing ex-vessel value (i.e., to the harvester) and (2) between the production costs and the total regional personal income as generated by recreational fishing and the commercial harvesting and processing of salmon. The analysis demonstrated the sensitivity of hatchery costs to a number of factors, including survival-to-fisheries rates and assumptions regarding responsibility for production costs. The analysis further illustrates the need to consider other factors when conducting an economic feasibility analysis, including international and Indian fishing treaties, hatchery mitigation program policies, harvest management regimes, environmental conditions, and regulatory resource protection programs. The TFP was used as an example in this chapter to determine the economic feasibility of salmon propagation for the purpose of enhancement. When all costs (hatchery, transportation, and marking) are included, and at recently experienced TFP survival rate of 2.24%, a coho enhancement program may return only $0.19 to the regional economy for every one dollar spent on hatchery production and acclimation operations. Since the hatchery costs, some transportation costs, and marking costs would have to be incurred anyway under hatchery mitigation agreements for the smolt supplied to the TFP, the return to the regional economy is $8.40 for every dollar spent on acclimation operations alone. If all other requirements of past mitigation agreements are satisfied, consideration could be given to implementing the TFP as a feasible project.

INTRODUCTION

In general, the artificial propagation of salmon is most often used to replace (mitigation) and/or augment natural production (enhancement). Because mitigation programs are designed to replace natural production lost due to human activities, it is not necessary to demonstrate their economic feasibility. However, enhancement programs are typically intended to support local or regional economic development. Therefore, it is important for such programs to have a net economic benefit to the target area. This chapter examines the feasibility of enhancement programs, using the lower Columbia River Terminal Fishery Project (TFP) as an example. The TFP is funded by the Bonneville Power Administration (BPA) and other participating agencies to support the local fishing industry in the lower Columbia River of Washington and Oregon.

Salmon hatcheries have been built on most of the major rivers of the west coast. Some of these hatcheries were built as mitigation for losses of salmon- and steelhead-producing streams due to dams. The basis of mitigation programs is usually specified in an agreement whereby dam operators are required to achieve salmon and steelhead production objectives to offset the negative effects of their activities on fish production. In this situation, the operation costs of the mitigation hatcheries, relative to measures of economic benefits, may be secondary to the attainment of production objectives.

Enhancement projects for economic development, on the other hand, may be evaluated in terms of traditional economic benefit analysis. However, few benefit–cost analyses have been completed on enhancement programs. In an evaluation of the Alaska Enhancement Program, Boyce et al. (1993) concluded that "the additional surplus generated by the pink and sockeye hatchery programs is estimated to be less than the costs of running these programs." In another study in Canada, Pearce (1994) estimated that the "lifetime cost of constructing and operating the enhancement facilities built under the Salmonid Enhancement Program exceeds the estimated benefits by $592 million, indicating a benefit–cost ratio of 0.6 for the program as a whole."

The need to evaluate hatchery programs is important in view of two emerging issues: (1) the potential negative effects of hatchery programs on wild stocks and (2) the strong worldwide growth of salmon aquaculture. Anderson (1984) and NRC (1996) concluded that a hatchery-based program and a mixed species ocean salmon fishery would have a negative effect on naturally spawning salmon stocks. Sylvia et al. (1996) reported that farmed salmon production costs are far lower than fishing fleet costs and will force drastic decrease of the market price for wild fish. Forster (1995) forecasted that farmed salmon in Chile can be produced for the U.S. market at $1.90 to $2.48 per pound (whole), and that ready-to-eat portioned salmon fillets will be available for $3.10 to $3.20 per pound. These two issues will affect future decisions about salmon production, enhancement, and management programs.

To make policy decisions about changes in existing production programs or evaluation of new projects, economic feasibility studies should be undertaken. Feasibility studies should identify and explain variables critical to the program's operation, including the following:

1. *Operational costs.*—This includes capital needs and operating costs (e.g., costs for equipment and hatchery components, labor, energy, fish food, etc.).
2. *Sources of revenue.*—Revenues from either subsidization or from the benefiting fisheries contributing through self-assessment.
3. *Type of project.*—Mitigation or enhancement; if mitigation, then hatchery rearing and trucking costs may not need to be included.
4. *Smolt-to-fisheries survival rates.*—Survival rates for various portions of the life cycle are basic indicators of factors that cause salmon mortality. In the case of enhancement programs, the objective is to produce salmon for harvest, and the smolt-to-fisheries survival rate is the most important.
5. *Ex-vessel and product sale price.*—The commercial culture of salmon has resulted in downward pressures on final product prices, but market niches remain for some products having very high prices, such as spring chinook *Oncorhynchus tshawytscha*.
6. *Fisheries interactions with the regional economy.*—Both recreational and commercial fishing need to be examined for inter-industry dollar flows.
7. *Access to fish.*—International and Indian treaties, recovery programs required by the federal Endangered Species Act (ESA), and other resource protection programs may preclude the harvest potential of a program.

In general, the feasibility of a program may be described in terms of (1) revenues received by the harvester and (2) total regional personal income generated in the local economy. While an evaluation of the first may be sufficient to decide whether a program is warranted, both measures must be evaluated to determine whether a more detailed cost–benefit analysis will provide the information needed to assess a project's feasibility.

In this chapter we illustrate, using a case study, an economic evaluation process that utilized both of the above measures. Comparisons were made (1) between the production costs of a Columbia River salmon enhancement program and the commercial fishing ex-vessel value (i.e., to the harvester) and (2) between the production costs and the total regional personal income as generated by recreational fishing and the commercial harvesting and processing of salmon. To illustrate the analysis, we also summarize salient historical economic and biological information about the Columbia River salmon fishery.

COLUMBIA RIVER SALMON FISHERY—THEN AND NOW

Salmon has always been an important part of life in the Pacific Northwest. To Native Americans on the west coast, salmon was their lifeblood—essential to their subsistence, their culture, and their religion. The return of the salmon was the time to congregate along the inland bays and rivers to trade, feast, and participate in games and religious ceremonies. In more recent times, salmon fishing was an important part of the economy of coastal communities. As early as 1828, various trading companies were purchasing and exporting salmon caught by the Indians in the Columbia River. Development of the canning process in the mid-1800s created a huge demand for salmon. In the 1860s, the process of canning salmon was perfected, permitting the fish to be transported over long distances, stored for extended periods, and kept palatable for consumers. By the 1880s as many as 55 canneries were operating on or near the Columbia River. In 1883, a total of 43 million pounds of spring chinook were harvested (Spranger and Anderson 1988). At this time, only the valuable chinook were canned; coho *O. kisutch*, sockeye, and chum *O. keta*, as well as steelhead *O. mykiss*, were not utilized by the canners.

The total salmon and steelhead harvested in the early 1890s ranged from 21 million to 33 million pounds (USCFF 1895). Chinook were generally about $1.00 per fish during that time, with other fish priced from $0.10 to $0.25 each (USCFF 1895). In the early 1890s the ex-vessel values were about $1 million. At today's prices, the ex-vessel value of these landings would be about $88 million, using price assumptions described by Radtke and Davis (1994).

As canning and transportation methods advanced, the major west coast salmon processing moved northward to Alaska. As for the Columbia, the declining abundance of salmon triggered an interest in artificial propagation as the means to solve the problem. According to Cone (1995) the words of Mr. W. A. Wilcox (as quoted in The Oregonian in 1896), an agent of the U.S. Fish Commission, reflect the contemporary thinking of the day regarding the Columbia River in general, and specifically the future of declining salmon stocks—

> The vast volume of fresh water coming down the Columbia will make it almost impossible ever to pollute it sufficiently to drive away the salmon, and it is hardly possible that civilization will ever crowd its banks to an extent that will endanger that [salmon industry], so I suppose it is safe to say that Columbia River salmon will always continue to be a choice dish in all parts of the world ... of course, the increased demand for fish and the growing scarcity of the same will call for more aid toward artificial propagation in order to keep up the supply.

The first artificial propagation operation in the Columbia Basin was established on the Clackamas River in 1877 (Hayden 1930; Mitchell 1949). Fish culture activity intensified between 1887 and 1894, and salmon populations increased in 1890. Although artificial propagation continued during the early 1910s, fish populations did not increase significantly until 1915 (Hayden 1930).

At that time, the Oregon Fish and Game Commission developed an improved hatchery system that included the use of feeding ponds to hold young fish until they were large enough to survive in the rivers. The new system was more expensive, but the later liberations led to an increase in the fish runs between 1917 and 1918. As a result, the federal government began to use the same system (Mitchell 1949). In 1938 the Mitchell Act provided funding for both state and federal hatcheries on the lower Columbia River, as a means to offset impacts to fish resulting from construction of Bonneville and Grand Coulee dams, and the effects of logging and pollution (USACE 1992).

Since the late 19th century, hatcheries have been viewed as a solution to the problem of declining salmon populations. However, even then scientists expressed caution regarding the effectiveness of hatcheries. For example, the Commissioner of Fisheries in 1937 stated that "… artificial hatching has definite limitations. At best it is only a supplement for natural spawning" (Bell 1937). Today's fish biologists more fully recognize the limitations of hatcheries, and, moreover, that hatcheries can impart potential problems to wild salmonid stocks, absent considerations of genetics, disease, and mixed stock fisheries (NRC 1995).

Today, the Columbia River Basin contains over 90 artificial production facilities producing about 166 million smolts annually (ODFW and WDFW 1996). Roughly three quarters of the basin's adult salmon and steelhead come from these hatcheries. These facilities are operated by several different entities with separate mitigation obligations. Hatchery managers are making changes to integrate hatchery rearing practices throughout the Columbia River Basin. These practices include controlling disease, maintaining genetic diversity, managing ecological interactions, standardizing hatchery operational procedures, and coordinating regional hatchery operations and programs (IHOT 1994). Fish managers are continuing to examine how hatcheries throughout the Columbia River Basin can best be managed to meet the different demands, including their role in augmenting natural production and supporting the recovery and conservation of severely depressed stocks (e.g., Nez Perce Tribe et al. 1996).

ENHANCEMENT PROGRAM PRODUCTION COSTS

An economic evaluation (funded by the BPA) of a proposal to expand the lower Columbia River Terminal Fishery Project (TFP) was completed in 1995–1996 by Radtke and Davis (1996) for Salmon For All, a trade association whose membership is largely comprised of owners of inriver gillnet fishing vessels. Parts of the final report of that study are used in this chapter to describe hatchery production costs and the economic contribution of harvested salmon.

The TFP receives most of its smolts from hatcheries operated by the Oregon Department of Fish and Wildlife (ODFW). Smolts are acclimated to salt water in net pens located in various places along Youngs Bay and the lower Columbia River using TFP facilities and operators. The purpose of the project is to replicate the biological changes in salmon as they transition from freshwater to saltwater, while reducing sources of mortality (e.g., avoiding predation) that occur during downstream passage of smolts released from upriver hatcheries. Harvesting of adults occurs off the Columbia River's main channel by a commercial fishing fleet that uses gillnets, so that mortality of upriver migrating salmon will be avoided. Sport angling in the Columbia River estuary also harvests some of the adults. For the economic analysis, the hatchery costs were taken from ODFW records, while acclimation costs were estimated using TFP facility records. Projected marking costs were also included because future salmon production in the Pacific Northwest may require all TFP fish to be fin-clipped and a portion coded wire-tagged (CWT).

The TFP operational plan calls for different acclimation periods and stocks, resulting in smolt releases throughout much of the year. The costs of acclimation and release were determined by species, release site, and by the length of acclimation time. A summary of TFP costs and relationship to expected commercial fishing vessel revenues is shown in Table 24.1. The acclimation and release costs on a species basis for all release sites, with or without marking, range from $0.03 per smolt for coho to $0.08 per smolt for both spring chinook (November to February rearing) and fall chinook (Table 24.1, line 2). Total costs per smolt vary from $0.18 for fall chinook to $0.62 for coho (Table 24.1, line 3).

Hatchery capture, rearing, and trucking costs are not included in the above TFP cost estimates, since they are borne by existing state and federal programs and may not need to be included in a feasibility statement. Mortality estimates during trucking and acclimation are also not included, which may reduce the total smolt-to-fisheries survival rates. The capital costs for state-operated hatcheries are about $50 per pound of smolts (R. Berry, ODFW, personal communication). At a 7% interest rate over a 30-year life period, the per smolt annualized capital cost ranges from $0.07 to $0.34, depending on the size of the smolts at release (Radtke and Davis 1996).

SURVIVAL TO FISHERIES

The projected economic performance of the TFP is sensitive to smolt-to-fisheries survival rates. Thus, it was important for the economic analysis to use rates representative of brood stock that would likely be used in the TFP operation. To evaluate this, a team of biologists (comprised of staff biologists from ODFW and the Washington Department of Fish and Wildlife (WDFW)), collectively called the Columbia River Management Team (CRMT), utilized the best available historical information to estimate a range of expected survival rates. The CRMT recommended using survival rates from brood stocks that showed a survival advantage (expressed as a percentage advantage in smolt survival-to-fisheries rates) based on past operations of the TFP. The recommendation included two fall chinook TFP stocks (Rogue River and Columbia River upper river brights) with a survival advantage of 1.77%, a single coho TFP stock (lower Columbia River early return) with a survival advantage of 2.33%, and a single spring chinook TFP stock (Willamette River stock) for which no survival advantage was determined.

Estimated catch rates for the Columbia River gillnet fishery range from 0.81 per 100 fall chinook smolts released to 3.09 per 100 coho smolts (Radtke and Davis 1996). The estimated survival to all west coast fisheries ranged from 2.81% for spring chinook (November to February acclimation and also for March acclimation) to 7.48% for coho (Table 24.1, line 4).

PRODUCTION COSTS PER HARVESTED FISH

With the projected survival rates, the hatchery costs calculated per fish harvested (anywhere on the west coast) range from $2.26 ($1.58 fixed and $0.68 variable) for Rogue River fall chinook stock to $16.73 for spring chinook (March release) (Table 24.1, line 5). The acclimation and release costs range from $0.97 for coho without marking, to $9.88 for fall chinook with marking (Table 24.1, line 6).

Estimated total hatchery, acclimation, and release costs per fish harvested anywhere on the west coast vary from $4.05 for fall chinook to $17.44 (Table 24.1, line 7) for spring chinook (March release). When the total costs are compared to harvest rates in the Columbia River gillnet fishery, the costs per harvested fish are $20.06 for coho, $12.33 to $22.37 for spring chinook, and $9.73 to $22.22 for fall chinook (Table 24.1, line 7). Inriver gillnetters, as a result of the acclimation and release project, could expect to receive ex-vessel value per smolt released from $0.20 for coho to $0.77 for spring chinook (Table 24.1, line 8).

The acclimation, release, and hatchery cost ratios will vary substantially, depending on survival rates. The expected survival rates used for this project were fairly high. If the lower range of survival rates were used (as noted in Table 24.1, line 4), hatchery cost ratios would increase substantially.

TABLE 24.1

Summary of costs associated with the lower Columbia River Terminal Fishery Project (TFP). Included are hatchery costs, acclimation and release costs, and expected ex-vessel value for in-river gillnet harvesters and other area harvesters. Costs are itemized with and without tagging/marking. Comparison costs between area fisheries are itemized for different survival rates.

| | | Spring Chinook | | Fall Chinook | |
		Nov–Feb Acclimation	March Acclimation	Upper River Brights	Rogue River
	Coho				
1. Hatchery costs per smolt in $					
Fixed costs/1	0.34	0.13	0.34	0.07	0.07
Variable costs	0.24	0.05	0.13	0.03	0.03
Total hatchery costs	0.57	0.19	0.46	0.10	0.10
2. Acclimation and release costs per smolt in $					
Without marking/2	0.03	0.08	0.03	0.06	0.06
With marking	0.05	0.08	0.03	0.08	0.08
3. Total costs per smolt in $					
Total variable costs with marking	0.29	0.13	0.16	0.11	0.11
Total costs with marking	0.62	0.27	0.49	0.18	0.18
4. Survival-to-fisheries rate in %					
In-river gillnet	3.09	2.19	2.19	1.85	0.81
Oregon-Washington	7.01	2.35	2.35	2.27	3.86
West Coast /3	7.48	2.81	2.81	3.42	4.44
5. Hatchery costs per west coast harvested fish in $					
Fixed cost	4.54	4.62	12.10	2.05	1.58
Variable cost	3.21	1.78	4.63	0.88	0.68
Total cost	7.75	6.40	16.73	2.93	2.26
6. Acclimation and release costs per in-river gillnet harvested fish in $					
Without marking	0.97	3.65	1.37	3.24	7.41
With marking	1.62	3.65	1.37	4.32	9.88
7. Total hatchery and acclimation costs per harvested fish without marking in $					
Total cost per west coast harvested fish	8.29	9.60	17.44	5.26	4.05
Total cost per in-river gillnet harvested fish	20.06	12.33	22.37	9.73	22.22
8. In-river gillnet ex-vessel revenues per smolt released in $					
Ex-vessel price	0.20	0.77	0.77	0.47	0.21
9. Costs as percent of in-river gillnet revenues and of west coast commercial fishing equivalent revenues					
Acclimation and release costs as percent of in-river gillnet revenues					
Without marking	15%	10%	4%	13%	29%
With marking	25%	10%	4%	17%	38%
Total cost to revenue ratios at assumed survival rates	310%	35%	64%	38%	86%
Total cost to revenue ratios at average historic Columbia River survival rates	720%	62%	114%	67%	136%
Total cost to revenue ratios at low historic Columbia survival rates	1033%	74%	136%	464%	184%

Notes: 1. This reflects capital costs (annualized over 30 years at 7%) of building a new facility. Hatcheries are owned by Oregon and have been depreciated. The maintenance costs are counted in the variable costs.

2. Marking costs are defined to include 100% fin clipping and CWT sampling.

3. West Coast includes Oregon, Washington, California, Alaska, and British Columbia.

4. Some numbers may not sum due to rounding.

Source: Radtke and Davis (1996).

PRODUCTION COSTS COMPARED TO EX-VESSEL VALUE

Enhancement programs such as hatcheries may be part of mitigation agreements resulting from a major change in salmon production capabilities. In such cases, the beneficiaries who may have a cultural or historical agreed-upon right to harvest fish may not be asked to bear the costs of producing the eggs and smolts. If a hatchery is operated for economic development, the fishing industry may be asked to pay at least a portion of the egg and smolt production costs. In the scenario where the TFP is expected to bear only the cost of acclimation at the assumed survival rates, and without marking costs, harvesters benefiting from the program would need to be assessed from 4% (i.e., $0.04 for every dollar received in ex-vessel value) for spring chinook (March release) to 29% ($0.29) for fall chinook (Table 24.1, line 9). Coho would require a 15% ($0.15) assessment to meet the program's costs. If hatchery costs as well as marking costs were to be borne by the gillnet fleet, coho production would require an assessment of 310% ($3.10), 35% ($0.35) for spring chinook, and 86% ($0.86) for fall chinook.

For the average historic Columbia River salmon survival rates, spring chinook (November to February acclimation) and fall chinook would generate one dollar of ex-vessel value for $0.62 and $0.67 of total hatchery and acclimation costs (i.e., assessment rates of 62 and 67%). At historic low survival rates, the assessment rates for coho would be as high as 1033% ($10.33 for every dollar received in ex-vessel value) (Table 24.1, line 9). Only spring chinook (November to February acclimation) would generate more ex-vessel value than costs ($0.74 for every ex-vessel dollar received). Figure 24.1 illustrates the sensitivity of survival rates to hatchery costs per harvested fish on breakeven points (i.e., hatchery costs compared to ex-vessel value) for TFP-produced coho salmon, at an assumed ex-vessel price of $0.90 per pound. In general, as survival rates decline, there is at first a gradual, and then a sharp, increase in breakeven total hatchery costs. For example, with a 9.9% smolt survival-to-fisheries rate, the breakeven hatchery cost would be around $6.50; at a survival rate of 2% breakeven costs would be over $30.00.

ENHANCEMENT PROGRAM ECONOMIC CONTRIBUTIONS

Salmon that are produced in the Columbia River system are harvested throughout the Pacific Northwest. The ODFW CRMT utilized recent harvest rates of the recommended TFP expansion stocks under average ocean conditions. The estimated smolt-to-fisheries survival rate for coho throughout the Pacific Northwest was 7.48% (7.48 adults survive to any and all fisheries for every 100 smolts released). Fall chinook were estimated to survive at 3.42 to 4.44% and spring chinook at 2.81% (Table 24.1, line 4). These rates were the best estimates using expected average conditions, and therefore will not occur every year. At present, abnormal ocean conditions are occurring that could negatively affect these projected survival rates. However, these estimated survival rates should be applicable to the TFP program over time.

The number of fish that would be available to the gillnet fishery is of primary importance to TFP investors. Based on CWT and other information, it was estimated that coho would return to the terminal areas to be harvested at 3.09%, spring chinook at 2.19%, and fall chinook at 0.81 to 1.85%, depending on the stock and brood used (Table 24.1, line 4). These returns and the amount of ex-vessel value generated by each harvested fish determine the total personal income that may be generated throughout the Pacific Northwest and to the lower Columbia River economy.

Potential economic contributions from the TFP in Northwest geographic regions are graphically displayed in Figure 24.2. These estimates were generated using the IMPLAN economic input/output (I/O) model developed by the U.S. Forest Service (Siverts et al. 1983). The model can be used to construct county or multi-county I/O models for any region in the U.S., and it was adjusted for the TFP so as to analyze the fishing industry and estimate the economic impacts of the project (Radtke and Davis 1996). Because spring chinook are expected to be harvested at a higher rate and to receive premium prices, the contribution in terms of total personal income to the Columbia River

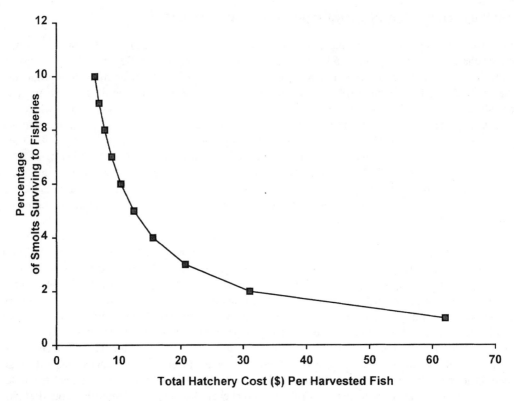

FIGURE 24.1 Hatchery costs per harvested fish as a function of the percentage of smolts surviving to fisheries. The line depicts the break-even points (hatchery costs compared to ex-vessel value) for the lower Columbia River TFP-produced coho salmon at an assumed ex-vessel revenue price of $0.90 per pound (source: Radtke and Davis 1996).

area economy per 100 smolts released is about $179. For fall chinook this may range from $80 to $119. Coho may generate up to $100 to the Columbia River area economy through the gillnet fishery. Estimated personal income for the other areas can be calculated from Figure 24.1 as the respective percentages of the total income for all areas. For the stocks of salmon evaluated, other areas that would receive significant income would be the recreational fishery off the Oregon and Washington coast (particularly for Columbia River coho and Rogue stock fall chinook), and the commercial fisheries in British Columbia and Alaska (for Upper River Bright fall chinook).

The amount of total personal income generated in an area from commercial and recreational fisheries can also be used as an indicator of the success of a program's objectives. Using the TFP as an example, the program may generate from $4.45 to $12.62 of total personal income to the west coast economies per dollar expended of operational costs (Table 24.2). For the lower Columbia gillnet fishery, if obligated to pay only for the acclimation and release costs, the TFP may return (as additional total personal income) from $4.63 for fall chinook to $44.67 for spring chinook (March release), for every one dollar spent on the program. However, if the commercial fishing gillnet industry was burdened with the total cost of coho smolt production, the region would spend one dollar for every $0.68 of total personal income that the people and industries receive from this program. The other species would return from $2.06 to $4.96 for every dollar invested. Thus, at a low historic survival rate of 2.24%, a coho enhancement program may only return $0.19 in total personal income to the area economy for every dollar spent on hatchery production and acclimation operations. Obviously, any enhancement program that uses coho as its base should investigate its feasibility carefully before investments are made. On economic grounds, such programs may only be "sustainable" if someone else is willing to pay to subsidize the program.

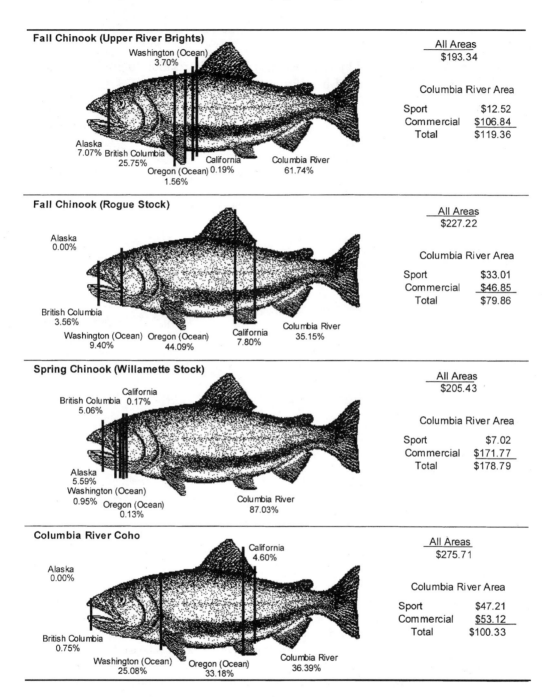

FIGURE 24.2 Total personal income generated per 100 smolts released for all west coast geographic areas and for the Columbia River area attributable to Columbia River acclimated and released salmon for commercial and sport fisheries. The percentages depicted for each fish represent the personal income that would be generated by a given stock produced by the Terminal Fishery Project for specific geographic areas (source: Radtke and Davis 1996).

TABLE 24.2
Economic impacts as measured by total personal income in relation to production costs for the Columbia River Terminal Fishery Project (TFP).

		Spring Chinook		Fall Chinook	
	Coho	Nov–Feb Acclimation	March Acclimation	Upper River Brights	Rogue River
1. Economic impacts (total personal income) per smolt acclimated and released and resulting adults harvested in $					
Impacts from Columbia River gillnets to Lower Columbia economies	0.42	1.34	1.34	0.83	0.37
Impacts to Oregon-Washington State economies	2.61	1.81	1.81	1.30	2.01
Impacts to West Coast economies	2.76	2.05	2.05	1.93	2.27
2. Ratios of economic impacts (total personal income) to hatchery and acclimation and release costs					
West Coast impacts to total hatchery and acclimation costs at assumed survival rates	4.45	7.59	4.18	10.72	12.62
In-river gillnet generated impacts to acclimation, marking, and release costs at assumed survival rates	8.40	16.75	44.67	10.38	4.63
In-river gillnet generated impacts to hatchery, acclimation, marking and release costs (total cost) at assumed survival rates	0.68	4.96	2.73	4.61	2.06
In-river gillnet generated impacts to hatchery acclimation and release costs (total cost) at low historic survival rates	0.19	2.35	1.29	0.38	0.97

Source: Radtke and Davis (1996).

SUMMARY

The TFP was used as an example in this chapter to determine the economic feasibility of salmon propagation for the purpose of enhancement. The TFP was designed to increase the smolt-to-fisheries survival rate by reducing the normal Columbia River downstream smolt passage mortality. By rearing and acclimating smolts to salt water conditions using off-channel net pens in the lower Columbia River estuary, an average smolt-to-fisheries survival rate advantage of 2.33% for coho has been found. When all costs (hatchery, transportation, and marking) are included, and at recently experienced TFP survival rate of 2.24%, a coho enhancement program may only return $0.19 to the regional economy for every one dollar spent on hatchery production and acclimation operations. Since the hatchery costs, some transportation costs, and marking costs would have to be incurred anyway under hatchery mitigation agreements for the smolt supplied to the TFP, the return to the regional economy is $8.40 for every dollar that is spent on acclimation operations alone. If all other requirements of past mitigation agreements are satisfied, consideration could be given to implementing the TFP as a feasible project.

This chapter has demonstrated that the economic contributions of enhancement programs can be calculated using past information about operational factors and knowledge of specific salmon stocks. However, there are other considerations that may affect future economic contributions, and thus the overall feasibility of a given enhancement program, such as those listed below.

1. International and Indian treaties may preclude access to program fish.
2. Mitigation agreements for hatchery production may be amended or abandoned. In March 1995, for example, the National Marine Fisheries Service announced a 197 million "cap" on smolt releases for the Columbia River system.
3. Harvest management policy and allocation agreements among user groups may affect the harvestability of program fish. The Pacific Fishery Management Council annually sets ocean harvest regulations that may intercept program fish.
4. Environmental variability can change smolt-to-fisheries survival rates. El Niño events and general downturns in favorable ocean conditions may increase marine mortalities of program fish.
5. Federal legislation, such as the ESA, and other resource protection regulations and recovery plans may eliminate "taking" of certain stocks mixed with program fish. Research about the feasibility and reliability of implementing selective fisheries has been inconclusive.

These considerations should be addressed when any feasibility analysis is performed for a specific enhancement program proposal.

REFERENCES

Anderson, J.L. 1984. Bioeconomic interaction between aquaculture and the common property fishery with application to northwest resources. Doctoral dissertation. University of California, Davis.

Bell, F.T. 1937. Report of the Commissioner of Fisheries, Bonneville Dam and Protection of the Columbia River Fisheries, 75th Cong., 1st Session, 1937, S. Exec. Doc. 87, SS 10104, p. 60.

Boyce, J., M. Hermann, D. Bischak, and J. Greenberg. 1993. The Alaska salmon enhancement program: a cost/benefit analysis. Marine Resource Economics 8:293–312.

Cone, J. 1995. A common fate, endangered salmon and the people of the Pacific Northwest. Henry Holt and Company, New York.

Forster, J. 1995. Cost trends in farmed salmon. The Alaska Department of Commerce and Economic Development, Juneau.

Hayden, M.V. 1930. History of the salmon industry of Oregon. University of Oregon, Eugene.

IHOT (Integrated Hatchery Operations Team). 1994. Policies and procedures for Columbia basin anadromous salmonid hatcheries. Bonneville Power Administration, Portland, Oregon.

Mitchell, H.C. 1949. The development of artificial propagation of salmon in the west. Oregon Fish Commission, Research Brief Volume 2, Number 1, Portland.

NRC (National Research Council). 1996. Upstream: salmon and society in the Pacific Northwest. Report of the Committee on Protection and Management of Pacific Northwest Anadromous Salmonids for the National Research Council of the National Academy of Sciences. National Academy Press, Washington, D.C.

Nez Perce Tribe, Confederated Tribes of the Umatilla Indian Reservation, Confederated Tribes of the Warm Springs Indian Reservation, and the Confederation Tribes and Bands of the Yakima Indian Nation. 1996. Anadromous fish restoration plan: Wy-Kan-Ush-Mi Wa-Kish-Wit: spirit of the salmon. Volumes I and II. Columbia River Inter-Tribal Fish Commission, Portland, Oregon.

ODFW and WDFW (Oregon Department of Fish and Wildlife and Washington Department of Fish and Wildlife). 1995. Status report, Columbia River fish runs and fisheries, 1938–94. Oregon Department of Fish and Wildlife and Washington Department of Fish and Wildlife, Portland, Oregon.

Pearce, P.H. 1994. Salmon enhancement: an assessment of the salmon stock development program on Canada's Pacific coast. Department of Fisheries and Oceans, Vancouver, British Columbia.

Radtke, H.D., and S.W. Davis. 1994. Some estimates of the asset value of the Columbia River gillnet fishery based on present value calculations and gillnetters' perceptions. Prepared for Salmon for All, Astoria, Oregon.

Radtke, H.D., and S.W. Davis. 1996. Lower Columbia River/Youngs Bay terminal fisheries expansion project. Prepared for Salmon for All, Astoria, Oregon.

Siverts, E., C. Palmer, and K. Walters. 1983. IMPLAN user's guide. U.S. Forest Service, Fort Collins, Colorado.

Spranger, M.S., and R.S. Anderson. 1988. Columbia River salmon. Washington Sea Grant Marine Advisory Services, WSG-AS-88-3, Seattle.

Sylvia, G., M.T. Morrissey, T. Graham, and S. Garcia. 1996. Changing trends in seafood markets: the case of farmed and wild salmon. Journal of Food Products Marketing 3(2):49–63.

USACE (U.S. Army Corps of Engineers). 1992. Hatchery vs. wild salmon. Salmon Passage Notes, Volume 1. U.S. Army Corps of Engineers, North Pacific Division, Portland, Oregon.

USCFF (United States Commission of Fish and Fisheries). 1895. Commissioner's Report, 1893. Pages 240–241, Washington, D.C.

25 The New Order in Global Salmon Markets and Aquaculture Development: Implications for Watershed-Based Management in the Pacific Northwest

Gilbert Sylvia, James L. Anderson, and Emily Hanson

Abstract.—Cultured product has become the dominant source of salmon in many regional and national seafood markets. During the last decade net-pen-farmed salmon supplies have grown at an average rate of 50,000 metric tons per year. By the year 2000, farmed product is expected to total over 1 million metric tons and comprise almost half of global salmon production. Private and public salmon ranching has also increased and is a major source of production in many Pacific Rim regions including Japan, Russia, the Columbia River, and Alaska's Prince William Sound. The rapid growth of salmon aquaculture, combined with production of wild Alaskan salmon, has resulted in global price decreases of 30 to 80%, major industry consolidation in both salmon aquaculture and commercial wild salmon fisheries, and increased emphasis on marketing and trade. Improvements in aquaculture technology are expected to continue through the next decade, resulting in increasing production and stable or decreasing prices for many species and product forms. Economic analysis suggests that, given these market realities and the constraints associated with rebuilding wild salmon resources, it will be difficult to sustain wild and cultured salmon industries in the Pacific Northwest. To increase success, regional governments and industry must recognize the new world order in global salmon markets, objectively assess industry strengths and weaknesses, and develop rights-based strategies—those which reward private and public entrepreneurs whose market-driven strategies are socially efficient and profitable, yet sustain salmon resources consistent with evolving ecosystem-based management systems. Market-driven and property rights-based strategies may include watershed-based user fees for terminal recreational and commercial fisheries, promotion of watershed "varieties" for the tourist trade and niche markets, marketing quality-assured frozen salmon for year-round distribution, providing ancillary sportfishing and ecotourism services, and eco-labeling and marketing of watershed-based salmon products and services.

INTRODUCTION

Today's global salmon markets are characterized by strong competition and rapidly growing supplies of cultured product. Between 1980 and 1995 annual harvests of farmed, ranched, and wild

1-56670-480-4/00/$0.00+$.50
© 2000 by CRC Press LLC

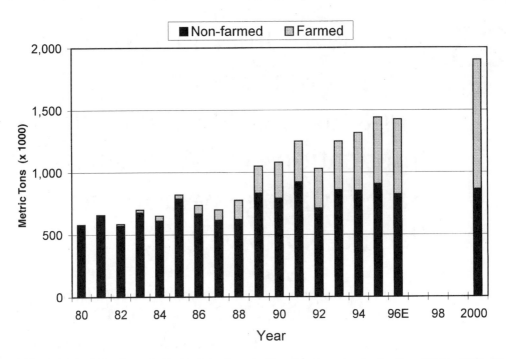

FIGURE 25.1   Farmed and non-farmed (wild and ranched) world salmon production from 1980 to 1995 with estimate for the year 2000. "E" indicates value was estimated (sources: FAO 1997; Salmon Market Information Service 1997).

salmon increased from less than 600,000 metric tons (mt) to over 1.4 million mt (Figure 25.1)*. Growth in salmon production is forecast to continue, reaching over 1.8 million mt by the year 2000. Farmed salmon supply has contributed substantially to world supplies, increasing from a negligible share in 1980 to almost 550,000 mt in 1995 (Figure 25.2). Atlantic salmon *Salmo salar*, the primary farmed species, increased from 3 to 35% of global production and by the year 2000 is expected to total over 1 million mt. Ranched salmon, particularly chum salmon *Oncorhynchus keta* and pink salmon *O. gorbuscha*, have also grown rapidly and today contribute approximately 40% of all non-farmed salmon harvests. This explosive growth in production, combined with improvements in aquaculture technologies and the development of efficient salmon marketing and distribution systems, has resulted in substantial price decreases of 30 to 80% for most cultured and wild salmon. The rapid increase in production and decrease in prices is expected to continue, compelling global and regional salmon industries to become efficient in production and marketing so they can successfully compete.

While world production has rapidly increased, production of various wild and publicly ranched (hatchery) salmon stocks in the Pacific Northwest south of Alaska have significantly decreased. This decrease, which is due to a complex set of natural and anthropogenic factors, has resulted in severe harvest reductions for many stocks (NRC 1996). However, while landings in the Pacific Northwest have decreased significantly, prices have also continued to decrease. For example, landings for troll-caught coho *O. kisutch* and chinook *O. tshawytscha* salmon in California, Oregon, and Washington were 60% lower in the period 1991–1995, as compared to 1981–1990 (PFMC 1997). Real prices, however, decreased by 36% during the same period, following the general trend of price decreases for coho, chinook, and Atlantic salmon in national and international markets.

---

* Salmon aquaculture, as discussed in this chapter, can refer to farming or ranching. Farmed salmon are raised in net-pens or tanks and spend their entire life in man-made structures. Ranched salmon are hatched and reared in private or public hatcheries, released into the wild as juveniles, and captured as adults when they return to spawn.

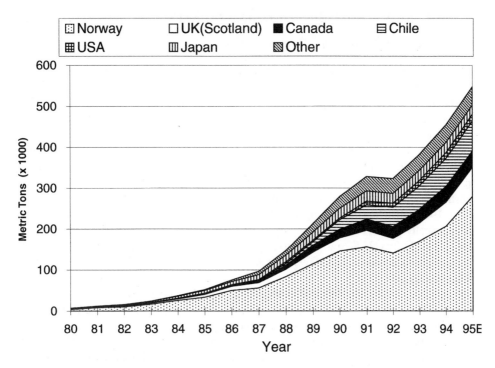

FIGURE 25.2 World farmed salmon production by major producing countries from 1980 to 1995 "E" indicates value was estimated (source: FAO 1997; Salmon Market Information Service 1997).

Over time some stocks of salmon in the Pacific Northwest may rebuild due to improvements in ocean and freshwater productivity, possibly providing for increases in harvests. However, due to efforts to sustain diversity in salmon resources and meet mandates of increasingly restrictive environmental laws, harvest rates on natural salmon and production of publicly and privately ranched salmon may not demonstrate a proportional increase relative to the historical averages of the last 25–50 years. In addition, evolution of ecosystem- and watershed-based management approaches may result in major institutional change and a shift in the legal and *de facto* property rights governing use of salmon resources.

For commercial salmon industries to survive and even grow in the Pacific Northwest, they will need to recognize the realities associated with (1) the new world order in global salmon markets; (2) the increasingly restrictive environmental policies designed to sustain the diversity and abundance of salmon resources; and (3) the evolution of ecosystem and watershed-based management systems. Based on these realities the industry must develop methods for producing, harvesting, and marketing wild and ranched salmon which is efficient and/or cost effective, sustains wild populations, and is compatible with emerging management systems. The challenge will be developing private and public policies which reward entrepreneurs whose strategies are consistent with market realities and opportunities, but also sustain the natural productivity and diversity inherent in salmon resources.

This chapter addresses these issues by describing global salmon markets and implications for developing sustainable salmon fisheries in the Pacific Northwest. Drawing largely from Anderson (1997) we describe the development of aquaculture and global salmon markets from 1970 to 1995, including the origin and growth of the farmed and ranched salmon industries and impacts on global marketing strategies. We then summarize economic evaluations of hatchery and salmon ranching industries in Alaska and the Pacific Northwest. We conclude by discussing the implications of the evolving order in marketing and management, and outline ideas, consistent with this new order, for sustaining salmon industries in the Pacific Northwest.

AQUACULTURE AND GLOBAL SALMON MARKETS

ORIGIN AND GROWTH OF NET-PEN REARED SALMON

The first attempts at salmon aquaculture occurred in Europe in the late 1700s (Folsom et al. 1992). The first hatchery propagation of Pacific salmon began in Canada in 1857 (Bardach et al. 1972). Salmon hatchery techniques were then adopted in the U.S. and introduced to Japan in 1877. However, it was not until the 1950s that hatchery-based enhancement programs were implemented on a significant scale. The Japanese Aquatic Resources Conservation Act, enacted in 1951, stimulated the growth of chum, pink, and cherry *O. masu* salmon ranching in Japan (Nasaka 1988).

In the early 1970s the USSR led the world in the artificial propagation of pink and chum salmon, followed closely by Japan (Bardach et al. 1972). With the enactment of the 200-mile limit and constraints on high-seas salmon fishing in the mid-1970s, Japan increased its aquaculture efforts. By 1980 Japan was harvesting 74,397 mt of hatchery ranched chum, which was over 45% of its domestically produced salmon supply.

Alaska was the first U.S. state to actively promote hatchery programs, creating the Fisheries Rehabilitation, Enhancement, and Development Division (FRED) in 1971. Three years later FRED authorized private non-profit hatcheries; the first harvests for these fisheries occurred in 1977 (Orth 1981). Private, for-profit salmon ranches were established in Oregon in 1980; Anadromous Inc. and Oregon Aqua-Foods were the largest operations (Anderson 1997). By the end of the 1970s many of the commercial coho and chinook salmon fisheries in Washington, Oregon, and California were dependent on salmon produced by public hatcheries. For example, in the Oregon Production Index* area, coho hatchery production accounted for less than 10% of total salmon harvest in 1960, but by 1979 hatcheries were responsible for 75% of total salmon harvest (ODFW 1982).

Private, farmed salmon industries also began to develop on an international scale in the early 1970s. Norway pioneered and led in production of pen-raised salmon, but significant production also occurred in Japan, Scotland, and Chile. In 1972 5 farms in Norway were producing a total of 46 mt; by 1980, 173 farms were producing a total of 4,300 mt (Heen et al. 1993). Japan was producing 1,855 mt of pen-raised salmon by 1980 (Japan Marine Products Importers Association, various years).

In North America the western coast led in development of pen-reared salmon. In 1969 the National Marine Fisheries Service conducted experiments with pen-reared salmon at the Manchester Field Station in Puget Sound, Washington. Ocean Systems, Inc. established coho and chinook cage systems in Puget Sound and harvested their first fish in 1971 (Sylvia 1989). By 1980 their farmed salmon production had reached an estimated 391 mt. The British Columbia net-pen salmon industry began in 1972 with surplus eggs from a government hatchery (Folsom et al. 1992), but remained essentially undeveloped with production totaling only 39 mt by 1979 (Heen et al. 1993).

Several salmon culture operations were also attempted on the eastern coast of North America in the 1970s (Bettencourt and Anderson 1990). Maine Salmon Farms began producing coho at a pen site in an estuary of the Kennebec River in 1970, but the company failed in the late 1970s. Fox Island Fisheries started production in 1973 in Vinalhaven, and was the first truly marine salmon farming operation on the U.S. east coast; it also went out of business in 1979. In New Brunswick, Canada, the first commercially viable salmon farming operation started in 1978 near Deer Island, and by 1979 had produced 6.3 mt of Atlantic salmon.

By the end of the 1970s only Norway had established a farmed salmon industry of any significance. At the same time, however, Japan was rapidly expanding production of ranched chum salmon. In 1980 global production of farmed and ranched salmon was approximately 13,321 mt (36% Atlantic salmon, 18% Pacific salmon, and 46% pen-raised trout).

---

* The Oregon Production Index area extends from south of Illwaco, WA to the Mexican border.

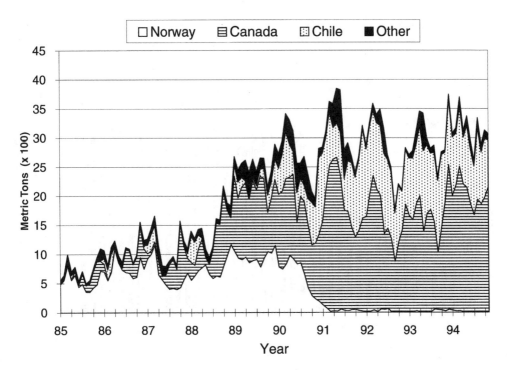

FIGURE 25.3   Source of U.S. imports of fresh/chilled salmon products from 1985 to 1995 (source: USDC, various issues).

Led by Norwegian producers the 1980s saw a dramatic expansion in production of farmed and ranched salmon. Scotland, Chile, Ireland, the Faroe Islands, New Zealand, and Australia joined Norway, Canada, and the U.S. in producing cultured salmon. The North American salmon industry, however, took little notice of these developments. Most participants in the traditional U.S. salmon fishery were preoccupied with public salmon management and development of non-profit ranching operations. As it turned out global production of farmed salmon would exceed all Alaskan salmon production by the end of the decade.

The 1980s also saw a reduction in foreign harvests of Alaskan salmon stocks. Japan and Europe were forced to rely increasingly on U.S. imports. The Alaskan salmon industry adjusted production and processing strategies to produce greater amounts of frozen product. For example, in the early 1970s over 70% of the Alaskan harvest was canned product; however, by the mid-1980s only 30% was canned (National Food Processors Association, various years). Throughout the 1980s most fresh and frozen salmon from the U.S. was exported, primarily to Japan (Japan Marine Importers Association, various years).

With most of their attention focused on exporting overseas the Alaskan industry neglected U.S. markets, leaving the door open for the farmed salmon industry. By the mid-1980s U.S. imports of fresh salmon accelerated rapidly, primarily consisting of farmed salmon from Norway (Figure 25.3). Initially the primary market destination was "white tablecloth" restaurants in the northeastern U.S., but other markets soon developed (Riely 1986). U.S. dominance in Europe and the eastern U.S. was undercut by exports of top quality, fresh, farmed salmon from Norway and other salmon farming nations. Owing to major marketing efforts by Norway, and the growing demand for salmon in North America, Japan, and Europe, farmed salmon prices began to climb significantly in the U.S. by 1989 (Figure 25.4).

With the establishment of farmed salmon aquaculture in Scotland, Ireland, Canada, Chile, and other regions, by the mid-1980s farmed salmon production exceeded the world's total commercial

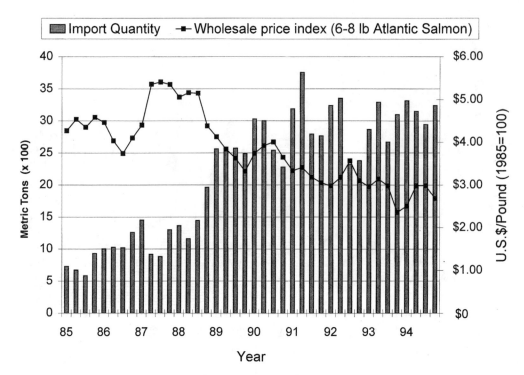

FIGURE 25.4 U.S. monthly imports of fresh Atlantic, chinook, and coho salmon from 1985 to 1995 and real U.S. wholesale prices for 6–8 lb Atlantic salmon (sources: USDC, various issues; UBP, various issues).

harvest of wild and ranched coho and chinook salmon. By 1990 global farmed salmon production exceeded the world's combined production of sockeye salmon *O. nerka.*

The developing U.S. farmed salmon industry faced a complicated and unclear regulatory environment and frequent public opposition. For example, in June 1987 Alaska imposed a temporary moratorium on private, for-profit, salmon and trout aquaculture, which became permanent in 1988. Alaska cited environmental concerns, spread of disease, pollution issues, and genetic degradation of native stocks as reasons for the ban. However, economic factors also played a role, specifically market competition and concern about control of the industry by multinational firms. Meanwhile, private non-profit aquaculture was growing rapidly in Alaska despite similar biological and genetic concerns. In 1979, for example, hatchery-reared pink salmon represented less than 3% of the harvest in Prince William Sound; by 1988 their contribution had jumped to 86% (Brady and Schultz 1988).

On both the east and west coasts of North America salmon aquaculture was opposed by environmentalists, local property owners, and fishermen. In Oregon salmon ranching was heavily protested, particularly by fishermen, even though private hatchery production contributed substantially to the commercial and recreational salmon harvests. Along with poor ocean conditions and low market prices, lack of public support and inadequate property rights (e.g., commercial and recreational salmon fishermen were not required to compensate private hatcheries) contributed to the demise of many U.S. salmon aquaculture projects in the early 1990s.

Due to a high unemployment rate and the decline of the herring fishery, one of the few areas in the U.S. that favored salmon aquaculture was the Eastport-Lubec region in Maine. Ocean Products, Inc. (OPI) began operation in 1982 with smolts produced by a Canadian hatchery, and continued operation in 1983 with 100,000 smolts acquired from the U.S. Fish and Wildlife Service. After development of their own hatcheries, OPI soon became the largest private salmon operation in the U.S. (Anderson and Bettencourt 1992).

The farmed salmon industry in the U.S. also faced new and more stringent regulations on effluent discharge, marine mammals, navigation, disease control, feed additives, and control of predatory birds. The escalating cost of regulatory compliance reduced salmon aquaculture profits and eroded U.S. competitiveness in global markets. By the end of the 1980s the U.S. accounted for only 2 to 3% of world farmed production, despite having developed much of the hatchery technology and health and nutritional requirements.

Farmed salmon production increased substantially in 1988 and 1989 and, in conjunction with large wild salmon harvests, contributed to rapid declines in salmon prices. The salmon industry had become a year-round and globally competitive industry. By 1988 farmed salmon held the dominant share of the fresh and frozen salmon market in Europe. In the U.S. imports of fresh farmed salmon more than doubled between 1988 and 1989. Even in Japan pen-raised salmon accounted for approximately 90% of fresh imports and 11% of frozen imports (Kusakabe 1992). In 1989 nearly 45% of Japan's total supply was from its hatchery-based chum fishery, and 6% from its domestic net-pen salmon industry. By the end of the decade, all farmed salmon accounted for over 20% of world production, and 40% of world trade.

Bankruptcies and industry restructuring resulted in consolidation. Between 1988 and 1990 the number of British Columbia salmon farms declined from 150 to 118 (Folsom et al. 1992). Industry reorganization also occurred in Ireland and Scotland although the largest impacts were in Norway. In the U.S., the 1989 price declines precipitated a petition from the Coalition for Fair Atlantic Salmon Trade (led by OPI), alleging that Norwegian producers had received subsidies and were dumping salmon in the U.S. market. In February 1991 the U.S. International Trade Commission ruled that the Norwegians were selling below fair market value, resulting in a countervailing duty of 2.3% and an antidumping duty ranging from 15.7 to 31.8%, depending upon the company. Norwegian salmon could no longer compete in the U.S. market, and by March of 1991, Norway's share of U.S. imports sank to less than 5%. This decline, however, was more than offset by imports of farmed salmon from Canada and Chile. An unsuccessful frozen-inventory program, establishment of U.S. countervailing duties, and implementation of trade barriers in Europe resulted in the bankruptcy of many Norwegian firms and Norway's main marketing organization (the Fish Farmer's Sales Organization).

Despite declining prices and reconsolidation there was tremendous growth in the farmed salmon industry. By the early 1990s world farmed salmon production was exceeding the entire U.S. wild and ranched salmon harvest, even though harvests were at record levels. The growth of the global farm industry, combined with record-breaking harvests of wild and ranched salmon, exerted strong downward pressure on prices throughout the 1990s.

IMPACT ON MARKETING

The downward pressure on prices over the last decade has resulted in increased emphasis on marketing and trade of all salmon. The industry has increased market recognition and consumer awareness through the use of brands, gill tags, and ancillary services. New products have been developed including portioned fillets, microwaveable entrees, salmon medallions, salmon ham, and marinated salmon. Chile has been successful in exporting fresh boneless fillets into the U.S. market. Recently, Chile, British Columbia, the International Salmon Farmers Association, and other producer groups have conducted generic marketing campaigns in the U.S. to stimulate demand for a wide range of salmon products. Smoked salmon suppliers have also worked to broaden the appeal of their products. Despite these efforts, however, the vast majority of farmed salmon is still sold fresh, whole, and head-on.

The emphasis on marketing by the farmed salmon sector has had a major influence on the wild salmon industry. A 1993 Alaskan strategic planning report stated "concerted action is urgently needed for the Alaskan salmon industry to regain its leadership position in the global marketplace."

(ADCED 1993). The six recommendations were to (1) improve market intelligence; (2) improve marketing efforts; (3) respond to consumer needs; (4) accelerate development of value-added technology; (5) stimulate value-added salmon production; and (6) improve quality.

In 1994 Alaskan salmon harvesters, funded by a 1% tax on harvest through the Alaska Seafood Marketing Institute, began an unprecedented marketing campaign in the U.S. The wild and ranched salmon sector also increased its efforts in product development including salmon nuggets, salmon burgers, and salmon surimi. However, due to the expected continued increase in supplies of aquacultured product, most analysts expect these marketing and product development efforts to potentially stabilize prices rather than lead to significant increases. If global prices remain 30 to 80% below 1970–1990 averages, salmon management systems in Alaska and the Pacific Northwest may not be socially or economically efficient. This may be particularly true for public and private non-profit salmon ranching operations whose expansion during the 1970s and 1980s was predicated on price assumptions that are no longer valid.

ECONOMIC EVALUATION OF PUBLIC AND PRIVATE RANCHING IN THE PACIFIC NORTHWEST

By the early 1990s private and public salmon ranching had become a significantly controversial issue in the Pacific Northwest and Alaska. Publicly ranched fish had been produced to augment wild production and mitigate damage caused by dams and loss of habitat. However, many biologists believed that hatchery-raised fish contributed to the demise of wild salmon populations through (1) direct and indirect competition with wild stocks for food and habitat and (2) non-selective, mixed-stock salmon fisheries (NRC 1996). Hilborn (1992) argued that large-scale salmon hatchery programs had failed to provide anticipated benefits and, rather than benefiting salmon, these programs posed a great threat to the long-term health of salmon populations.

Besides the possible impact on wild stocks the large decrease in world market prices had also raised concerns about the overall economic benefits of salmon ranching. Although price decreases in the farmed salmon industry have been somewhat offset by decreases in costs, it was unclear whether commercial salmon harvesting and ranching costs had also decreased. Environmental regulation to protect wild salmon, overcapitalization in the harvesting sector, and a decrease in the productivity of freshwater and ocean environments (particularly south of Alaska) may have significantly increased the costs of producing and harvesting ranched salmon. Given these conditions some private and public salmon ranching programs may not have been achieving economic, social, and biological objectives.

Before developing optimal strategies for salmon aquaculture in the Pacific Northwest, it is instructive to review some recent studies analyzing the economic effectiveness of salmon ranching programs. In his study of Pacific Canada's salmon enhancement program, Pearse (1994) noted that the program's objectives were to (1) double salmon harvests in 30 years from 70,000 to 140,000 mt; (2) generate economic and social benefits including jobs, income, recreation, and opportunities for native people and economically disadvantaged communities; and (3) recover program costs from harvesters. Management objectives shifted in the early 1980s from cost-effective production of sockeye and chum salmon to relatively more costly attempts at restoring depressed wild stocks of coho and chinook salmon. The cost–benefit analysis evaluated the harvesting and processing sectors, native fisheries, and recreational fisheries, but excluded effects on wild fish. Pearse showed that, after 17 years and $1.736 billion Canadian dollars, enhancement contributed only 13% to annual harvests. Costs exceeded benefits by $408 million for a cost–benefit ratio of 0.8, or approximately half of the original enhancement target of 1.5. Pearse concluded, however, that it may be possible to recover enhancement costs if new management strategies are implemented, including reconciling enhancement with conservation of wild salmon stocks, improving scientific information, assessing the management of individual enhancement projects, and directing beneficiaries to take responsibility for organizing and financing enhancement projects. Pearse proposed that control for the projects be transferred from the Department of Fisheries and Oceans to independent, non-profit organizations that would be managed and financed by fishing and regional communities.

In the second study Boyce et al. (1993) conducted a cost–benefit analysis of Alaska's private non-profit salmon enhancement program. Management goals of the program included providing long term employment and economic opportunity, protecting and supplementing wild stocks, and generating 100 million wild and hatchery fish in the Alaskan fishery. Six alternatives were analyzed within a regional cost–benefit analysis: (1) maintaining the status quo; (2) eliminating pink salmon enhancement; (3) eliminating sockeye salmon enhancement; and (4–6) both decreasing and increasing by 15% pink or sockeye enhancement. The authors then measured the net benefits accruing to fishermen in terms of producer's surplus, and the net benefit to the state given the costs of the six alternative enhancement programs. Although their findings are highly uncertain, results suggest that increasing salmon enhancement would not necessarily increase net benefits to the state. In fact, they found that a reduction in either pink or sockeye production could yield net benefits, primarily due to increasing output prices as harvests and market supply of these species decreased. The study, however, did not consider the costs or benefits of the six alternatives on wild fish, recreation, or subsistence. The authors also noted that maintaining production, although not economically efficient according to their analysis, would help maintain market share for Alaskan product by maintaining low prices and discouraging market entry by Russian or Chilean product.

In a third analysis Sylvia (1996) reviewed the cost effectiveness of the Youngs Bay net-pen acclimation program in the Columbia River near Astoria, Oregon. Management goals for the Youngs Bay project included the creation of terminal area gillnet fisheries, augmenting existing salmon production, protecting and supplementing wild stocks including compliance with the Snake River Salmon Recovery Plan (SRSRT 1994), and providing regional economic development. The regional cost effectiveness analysis compared the optimal design of the acclimation project (holding hatchery smolts of various species and stocks approximately 3 weeks in net-pens) with typical hatchery release practices. Analysis demonstrated a relative cost effectiveness advantage of 1.77 to 2.33 per returning adult, with cost savings of U.S. $8.12 to $14.72 per salmon harvested in the west coast salmon fishery, and $16.16 to $45.60 per salmon if harvested in the Columbia River gillnet fishery. Findings also showed that the project would yield significant cost savings to public agencies responsible for hatchery production and mitigation, ranging from $8 million to $14 million per year. Increases in cost effectiveness resulted from the higher return rates associated with optimal acclimation strategies, and relatively low costs of acclimation. No attempt was made to determine regional or national costs and benefits of acclimation and/or hatchery management practices in the Columbia River.

Although these three economic studies used different methodologies and did not compare the same costs or benefits, their findings raise a number of important issues. Their general conclusions suggest that, on balance, public and private non-profit salmon ranching programs are relatively costly and may not be generating net economic benefits for the Pacific Northwest. The results, however, also indicate that there can be wide differences in absolute and relative costs and benefits between projects depending on a myriad of factors, including the species or biological stocks, market demand, flesh and egg quality, harvest user group (sport, commercial, subsistence), hatchery location, wild stock interactions, ocean survival, hatchery return rates, regional infrastructure, hatchery management practices, and fishery management system. While these factors significantly complicate management and increase economic and social risks, they also provide a wide range of potential management alternatives for reducing costs and improving revenues and benefits. In many cases the success of private and public ranching operations will depend on the success of the beneficiary groups including commercial and recreational fishermen. Commercial salmon fishing industries that are highly overcapitalized may not be capable of compensating hatcheries for producing fish, no matter how reasonably cost-effective or efficient the hatchery. The success of ranching operations, therefore, will partially depend on the structure of the fishery management system and success in restructuring salmon fisheries so they are efficient, cost-effective, and competitive in global markets.

NEW WORLD ORDER—IMPLICATIONS FOR FUTURE SALMON INDUSTRIES IN THE PACIFIC NORTHWEST

MARKETING MANAGEMENT TRENDS

Expected increases in the global supply of salmon will continue to force prices downward and exert pressure on high-cost producers in both the wild and cultured salmon industries. In the farmed salmon industry low prices will lead to greater consolidation and integration. Small farms are expected to become contract growers for large facilities that integrate egg and smolt production, grow out, feed production, distribution, and marketing operations. Greater concentration, adoption of scale economies, and advanced husbandry techniques should decrease costs and increase profits for efficient salmon farmers.

Decreasing prices will also lead to greater marketing efforts by all salmon industries. Regional farmed salmon industries will likely source products from around the globe and supplement supplies with wild product. Domestic and international generic promotion of farmed fresh and frozen salmon products will increase as supply of homogenous products increase. Niche products, including high-quality, value-added specialty products for food service and retail, will expand. Production of pre-packaged boneless fresh and frozen fillets with stable and extended shelf life will increasingly compete with mid-priced beef, pork, and poultry products. Brand names, logos, and labels may be increasingly used to identify consumer-ready salmon products.

Forward contracts for value-added products based on predictable quantities of salmon may provide a reasonably stable revenue base for effectively designed private salmon aquaculture operations including salmon ranching. Other secondary salmon product markets can also be expected to expand, including eggs for smolt production, direct human consumption, or recreational bait, and carcasses for pet food and meal/compost products. Recreational services will benefit from more abundant and stable supplies of salmon including traditional sport fishing, ecotourism, fee-fishing, and ancillary services. Success for these types of services, however, will depend heavily on the commitment and cooperation of producers, users, and public management agencies.

Because of high start-up costs and long lag times between grow out and harvest, aquaculture salmon operators will increasingly search for methods to effectively minimize risk. Risk management will include forward marketing contracts, diversified production and marketing, and possible development of salmon futures markets.

MARKETING MANAGEMENT, ECONOMIC EFFICIENCY, AND WATERSHED-BASED PUBLIC MANAGEMENT

Evolving policies to protect weak stocks and establish sustainable and diverse salmon populations in the Pacific Northwest will continue to have major impacts on future salmon management. Current trends are toward ecosystem and watershed-based management approaches organized at the drainage basin level and involving all relevant local and regional groups (NRC 1996). These management systems, if implemented, are likely to limit or eliminate many types of commercial and recreational salmon harvesting practices. They will also significantly alter institutional structures and functions. Creating watershed councils comprised of government and non-government "investor" groups (e.g., timber, agriculture, environmental groups) may help in restoring salmon and habitats at local scales; but it will also create expectations for "returns on investment." These "returns" will take different forms depending on how watershed alliances are structured and operated. Whatever their forms, expectations of watershed-based groups will likely conflict with the needs of many (but not necessarily all) ocean and mixed stock salmon fisheries. Management and harvesting strategies will increasingly focus on river or tributary-specific recreational and commercial fisheries (see, for example, Booker 2000, Copes 2000).

These changes will have a major influence on salmon management and a significant impact on the salmon seafood industry. However, the evolution and design of quasi or formal property rights "assigned" to watershed-based groups will create new opportunities for creating and sustaining commercial and recreational salmon industries. In many cases watershed-based enhancement policies may include aquaculture as an important component in an overall salmon production strategy. Hatchery fish will continue to be used for generating fishing opportunities and augmenting natural production. The focus of hatcheries, however, will change from a "one size fits all" approach to one more compatible with watershed-based management (NRC 1992; Stickney 1994).

We believe there are four possible future salmon marketing management strategies and scenarios consistent with sustainability, the new order of global salmon markets, economic efficiency and cost effectiveness, Pacific Northwest demographic and economic trends, and development of watershed-based management systems.

Marketing and Sales of "Watershed" Salmon.—Watershed-based management groups will seek to recoup investment by controlling use and access of salmon resources through various fees, licenses, taxes, and other contractual arrangements. Control over "surplus" salmon will compel watershed groups to develop marketing management strategies consistent with optimizing their biological, economic, and social objectives. This would be similar to operations of private, non-profit regional salmon hatchery corporations in Alaska which charge user fees in the form of assessment taxes on commercial harvesters, and form long-term contracts with the state and private charter boat organizations for salmon targeted for sportfishing. Socially efficient managers would develop programs and contracts so that user fees offset the marginal costs of watershed restoration and protection.

Marketing Management Unique to Each Watershed.—Because biological, physical, and socio-economic circumstances vary from one watershed to the next, marketing management strategies will also vary. For example, salmon ranching may be used to supplement wild stocks in watersheds where cultured salmon are ecologically similar to wild salmon, or are selectively managed to avoid negative impacts on wild fish. Terminal fisheries may include commercial and/or recreational fishermen that pay user fees or purchase watershed licenses. Given the expected increase in tourism to the Pacific Northwest, targeting salmon to recreational anglers and providing ancillary services (e.g., cleaning, packaging, and smoking) may generate additional benefits to watershed management groups. Where commercial harvesting is determined to be the best management strategy, fishermen could be hired to manage terminal fisheries through cooperative arrangements. In other cases watershed managers may retain ownership and hire harvesters, a concept similar to the cost recovery programs operated by private, non-profit hatcheries in Alaska.

Branding and Marketing of Sustainable Watershed-Based Salmon Products.—Salmon resources that are managed and certified as sustainable would generate supplies of surplus salmon for commercial exploitation. In some cases individual stocks of salmon may possess unique quality characteristics that contrast favorably with typical farm product. Effective quality assurance and grading programs would help to maintain their distinctiveness. Watersheds could create name recognition (e.g., Copper River sockeye) and target products for niche markets, ranging from local tourists and restaurants, to upscale national and international food service and retail outlets. Effective marketing would include "green" promotion or "ecolabelling" which identify products consistent with sustainable practices and requirements. Other products not commonly marketed to tourists (such as salmon eggs) could be promoted with the support of local restaurants and retail outlets. Composts and fertilizers could be produced from salmon wastes for full utilization of a renewable resource.

Developing Inter-Watershed Management Cooperatives.—Many of the management and marketing issues faced by individual watersheds would be effectively managed through cooperative efforts of watershed groups. Cooperative strategies could include (a) cooperating in marketing and promotion of regional salmon varieties that minimize mutually destructive competition; (b) jointly managing and promoting quality assurance programs; (c) promoting quality frozen products for year-round distribution; (d) jointly promoting with other agricultural products (e.g., fruits and wines) that complement salmon varieties; (e) coordinating and supporting generic marketing programs that expand consumer demand and knowledge for mutually produced salmon products; (f) sharing technical resources; (g) negotiating contracts with fishing and aquaculture production groups; and (h) developing programs to manage risks by pooling resources and diversifying marketing.

CONCLUSION

Salmon industries in the Pacific Northwest face challenges on two fronts: contending with the implications of the new world order in global salmon markets and managing and sustaining native stocks of salmon. Watershed-based management councils will ultimately challenge the existing salmon management structure and contend for control over salmon resources. While this will threaten the existence of some ocean and mixed stock salmon fisheries, it will also result in formulation of more clearly defined salmon property rights—rights which ultimately provide the control entrepreneurs need to develop profitable industries. Such control helped propel the explosive growth of the global farmed salmon industry and shaped the current global salmon market. This control can also allow salmon industries in the Pacific Northwest to successfully compete within this new order.

As watershed councils develop over the next decade they should be encouraged by state and federal agencies to cooperate with salmon user groups and experiment with alternative management strategies that sustain salmon populations yet result in successful use of "surplus" salmon resources. These "experiments" will provide the experience necessary to ultimately forge successful property rights-based strategies consistent with market constraints and opportunities. This will require, however, policymakers who recognize the value of such experiments and use the results to formulate rational policy strategies—strategies which simultaneously sustain salmon resources yet exploit them for the long-run advantage of the citizens of Canada, the U.S., and the Pacific Northwest.

ACKNOWLEDGMENTS

We are grateful for the support of the University of Rhode Island Department of Resource Economics and Oregon State University Coastal Oregon Marine Experiment Station. The suggestions and comments of the session co-chair, Hans Radtke, and three referees are also gratefully acknowledged.

REFERENCES

ADCED (Alaska Department of Commerce and Economic Development). 1993. Scenario planning: developing a strategy for the future of the Alaska salmon industry. Juneau, Alaska.

Anderson, J.L. 1997. The growth of salmon aquaculture and the emerging new world order of the salmon industry. Pages 175–184 *in* K. L. Pikitch, D.D. Huppert, and M.P. Sissenwine, editors. Global Trends: Fisheries Management. American Fisheries Society, Bethesda, Maryland.

Anderson, J.L., and S.U. Bettencourt. 1992. Status, constraints, and opportunities for salmon culture in the United States: a review. Marine Fisheries Review 54:25–33.

Bardach, J.E., J.H. Ryther, and W.O. McLarney. 1972. Aquaculture—the farming and husbandry of freshwater and marine organisms. Wiley-Interscience, John Wiley & Sons, Inc., New York.

Bettencourt, S.U., and J.L. Anderson. 1990. Pen-reared salmonid industry in the Northeastern United States. Northeast Regional Aquaculture Center Report 100. University of Rhode Island, Kingston.

Booker, M. 2000. Integrating history into the restoration of coho salmon in the Siuslaw River, Oregon. Pages 625–636 *in* E. E. Knudsen, C. R. Steward, D. D. MacDonald, J. E. Williams, and D. W. Reiser, editors. Sustainable fisheries management: Pacific salmon. Lewis Publishers, Boca Raton, Florida.

Boyce, J., M. Herrman, D. Bischak, and J. Greenberg. 1993. The Alaska salmon enhancement program: a cost/benefit analysis. Marine Resource Economics 8:293–312.

Brady, J.A., and K.C. Schultz. 1988. Review of the Prince William Sound area commercial salmon fisheries, 1988. Alaska Department of Fish and Game, Division of Commercial Fisheries, Anchorage.

Copes, P. 2000. Aboriginal fishing rights and salmon management in British Columbia: matching historical justice with the public interest. Pages 75–91 *in* E. E. Knudsen, C. R. Steward, D. D. MacDonald, J. E. Williams, and D. W. Reiser, editors. Sustainable fisheries management: Pacific salmon. Lewis Publishers, Boca Raton, Florida.

Folsom, K., et al., 1992. World salmon culture. National Marine Fisheries Service, National Oceanic and Atmospheric Administration. United States Department of Commerce. Silver Spring, Maryland.

FAO (Food and Agriculture Organization of the United Nations). 1997. Yearbook of Fishery Statistics: Catches and Landings 1995. Rome.

Heen, K., and five coauthors. 1993. The distribution of salmon aquaculture. Pages 10–58 *in* K. Heen, R.L. Monahan, and F. Utter, editors. Salmon aquaculture. John Wiley & Sons, Inc., New York.

Hilborn, R. 1992. Hatcheries and the future of salmon in the Northwest. Fisheries 17(1):5–8.

Japan Marine Products Importers Association. Various years. Japanese imports of marine products (statistics). Tokyo, Japan.

Kusakabe, Y. 1992. A conjoint analysis of the Japanese salmon market. Ph.D. dissertation. University of Rhode Island, Kingston.

Nasaka, Y., 1988. Salmonid programs and public policy in Japan. Pages 25–31 *in* W.J. McNeil, editor. Salmon production, management, and allocation. Oregon State University Press, Corvallis.

National Food Processors Association. Various years. Canned salmon supply, stocks, shipment reports. Seattle, Washington.

NRC (National Research Council). 1992. Marine aquaculture: opportunities for growth. National Academy Press, Washington, D.C.

NRC (National Research Council). 1996. Upstream: salmon and society in the Pacific Northwest. National Academy Press, Washington, D.C.

ODFW (Oregon Department of Fish and Wildlife). 1982. Comprehensive plan for production and management of Oregon's anadromous salmon and trout—part II. Coho salmon plan. Fish Division, Anadromous Fish Section, Oregon Department of Fish and Wildlife, Portland.

Orth, F.L. 1981. Market structure of the Alaska seafood processing industry, Vol. II: finfish. University of Alaska Sea Grant Report 78–14, Fairbanks, Alaska.

PFMC (Pacific Fishery Management Council). 1994. Review of 1993 ocean salmon fisheries. Portland, Oregon.

PFMC (Pacific Fishery Management Council). 1997. Review of 1996 ocean salmon fisheries. Portland, Oregon.

Pearse, P.H. 1994. An assessment of the salmon stock development program on Canada's Pacific coast. Final Report of the Salmonid Enhancement Program Internal Audit and Evaluation Branch. Department of Fisheries and Oceans. Vancouver, British Columbia.

Riely, P.L. 1986. An economic analysis of the market for Atlantic salmon aquaculture. Doctoral dissertation. University of Rhode Island, Kingston.

Salmon Market Information Service. 1997. Database. University of Alaska, Fairbanks.

SRSRT (Snake River Salmon Recovery Team). 1994. Snake River Salmon Recovery Plan Recommendations. Final Report. National Marine Fisheries Service, Portland, Oregon.

Stickney, R.R. 1994. Use of hatchery fish in enhancement programs. Fisheries 19(6):6–13.

Sylvia, G. 1989. An economic policy for net-pen salmon aquaculture development: A dynamic multilevel approach. Doctoral dissertation. University of Rhode Island, Kingston.

Sylvia, G. 1996. A report on the economic significance of the Youngs Bay net-pen rearing project. Report prepared for the Oregon Economic Development Department.

UBP (Urner Barry Publications). Various issues. Seafood Price-Current.

USDC (United States Department of Commerce). Various issues. Bureau of Census, Monthly Trade Statistics. Washington, D.C.

26 Alaska Ocean Ranching Contributions to Sustainable Salmon Fisheries

William W. Smoker, Bruce A. Bachen, Gary Freitag,
Harold J. Geiger, and Timothy J. Linley

Abstract.—Alaska's salmon ocean ranching program began in the mid-1970s as an effort to develop sustainable fisheries after more than 25 years of frequent run failures. The failures were attributed to the effects of severe winters on embryo survival, particularly of pink and chum salmon *Oncorhynchus gorbuscha, O. keta.* Thirty-nine hatcheries, mostly run by private, non-profit corporations, now release over 1 billion fry and smolts, producing 25–50 million harvested salmon annually. A mark of the program's success is that some regional fisheries depend on hatchery-produced salmon in years when, or places where, wild stocks cannot be harvested. Protection of wild stocks is the first priority of fishery managers. Risks to wild stocks associated with the program, primarily of excessive harvest in mixed fisheries, of genetic introgression and lost fitness, and of disease dissemination, have been identified, addressed, and reduced (but not eliminated) throughout the history of the program. Tags and marks on hatchery-produced salmon have been used during annual harvests to discriminate wild salmon. These have included coded wire tags and recently mass-marking of hatchery salmon through otolith thermal marking; they have been used to provide managers of mixed stock fisheries with precise information about the relative abundances of hatchery and wild salmon during fishing seasons. Management of wild stocks in some instances has been made easier by diversion of fishing effort to hatchery stocks. Many releases of cultured salmon have been prohibited where interactions with significant or unique wild stocks were foreseen. Nearly all transfers of salmon into Alaska or between regions of Alaska have been prohibited. Strict disease-prevention inspections have been required before transplants within regions have been permitted. Risks associated with ecological interactions between Alaska hatchery salmon and wild salmon in the marine environment are potentially important but these interactions have not been detected.

INTRODUCTION

It is especially appropriate to review the Alaska salmon enhancement program because (1) recent abundance and harvest of salmon are at historic highs while market value is at historic low, calling into question the economic benefit of artificial enhancement (Hermann and Greenberg 1994; but see Wilen 1993 and Anderson 1997) and (2) questions have arisen about the biological effects of the Alaska program on the fitness and productivity of wild salmon (e.g., Thomas and Mathisen 1993; Gharrett and Smoker 1993). Indeed Hilborn (1992, p. 8) calls for the termination of the Alaska program because "there is no mitigative excuse for these facilities; they are the result of technology being sold by fisheries scientists to unwary fishers motivated by short-term greed."

It is evident that abundance and harvest of Pacific salmon in Alaska have risen to unprecedented highs since the mid-1970s. This has been an aggregate response to climate fluctuations (e.g., Hare

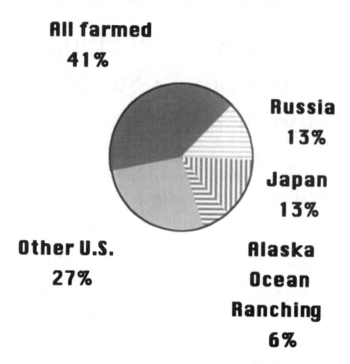

FIGURE 26.1 World production of salmon in 1995; 1.4 million metric tons. Combines Atlantic and Pacific salmon, farmed and "wild" (includes ocean-ranched production). Statistics from Salmon Market Information Service (1996) and McNair (1996).

and Francis 1995) and to effective harvest management (e.g., Royce 1988), and secondarily a response to salmon ocean ranching programs for fishery enhancement. It is also the case that, in the Pacific Northwest U.S. and in British Columbia, salmon hatcheries are widely assigned a share of fault in the loss of diversity and fitness of salmon stocks (reviewed by Campton 1995). We, however, argue contrary to many detractors (e.g., Hilborn 1992), that the Alaska program was motivated by a desire for long-term sustainability of fisheries, that the managers of the Alaska program have been conscious of the biological risks to diversity, fitness, and productivity identified in southern states and provinces, and that Alaska managers have acted to reduce those risks.

Alaska's salmon ocean ranching program is a system of hatcheries and other projects designed to enhance its salmon fisheries. Salmon ranching is "an aquaculture system in which juvenile fish are released to grow, unprotected, on natural foods in marine waters from which they are harvested at marketable size" (Thorpe 1980; Heard 1996). Unlike programs of other states and provinces, Alaska's was not primarily conceived to mitigate effects of habitat degradation on salmon stocks but was developed to augment fisheries otherwise dependent on unreliable wild salmon production. There are 39 hatcheries of various size in Alaska. They are located around the Gulf of Alaska but primarily in the southeastern archipelago, in Prince William Sound, and Cook Inlet (Figure 26.1 in McNair 1996). They produce Pacific salmon juveniles of all species, but mostly pink salmon *Oncorhynchus gorbuscha* and chum salmon *O. keta* fry that are released to sea early in their lives and are harvested immediately before maturity. Most of the hatcheries (31) are operated in the private sector, by regional or local non-profit corporations funded by revenues derived from the sale of fish or from a voluntary tax on harvests (Pinkerton 1994; Heard 1996). Members of the regional corporations are largely fishermen in the region; local corporations are formed in specific communities. Others are operated by public agencies and an Alaskan Native community (Metlakatla). A portion of salmon returning to the hatcheries at maturity are taken in common property fisheries.

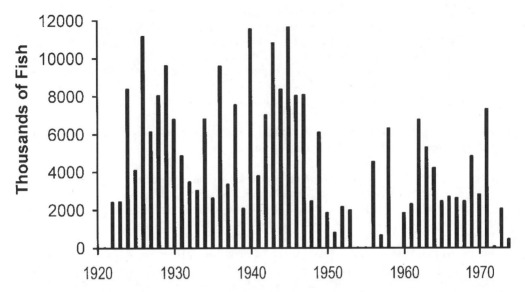

FIGURE 26.2   Total harvest of pink salmon in Prince William Sound before 1974 and the beginning of the ocean ranching program. No harvest was possible in 1954, 1955, 1959, 1972, 1974 (Alaska Department of Fish and Game statistics).

These fisheries are in principle managed so as to limit their harvest to protect the production of wild salmon stocks that are taken in the same fisheries. The other portion of hatchery salmon that escape the common property fishery are then harvested by the hatchery corporations and sold to support operations (reviewed in Pinkerton 1994).

Alaska's salmon ocean ranching program produces a significant but not predominant part of Alaska's annual harvest. The program annually produces 25–50 million adult fish, from releases of 1.3–1.5 billion juveniles (FRED 1990, 1991, 1992, 1993; McNair and Holland 1994; McNair 1995, 1996). In 1995 production exceeded 80,000 metric tons (statistics in McNair 1996); this was roughly 6% of the world salmon production. By way of comparison Alaska's salmon ocean ranching production was roughly half of the Japan salmon production, which was entirely supported by the other large salmon ocean ranching program in the North Pacific Ocean (Salmon Market Information Service 1996; Heard 1996; Figure 26.1).

In this chapter we (1) review the historical setting and motivation for development of the Alaska salmon ocean ranching program; (2) present, as case histories, three of the larger programs, showing how they have contributed to fisheries and how they illustrate the risks posed by hatchery production for sustained productivity and fitness of wild stocks; and (3) show in these same case histories how the risks are being managed and ameliorated.

PRINCE WILLIAM SOUND

HISTORICAL SETTING: RUN FAILURES ONE YEAR IN FIVE

The Alaska ocean ranching program has its roots in Prince William Sound (PWS), the region that is today the largest part of the program. Its developers were not "motivated by short-term greed" (Hilborn 1992) but by the need for a sustainable fishery: ocean ranching of salmon in PWS began in the mid-1970s after 25 years of low harvests (Figure 26.2). Wallace Noerenberg, formerly Alaska Department of Fish and Game Regional Biologist for Prince William Sound, architect of stock-based harvest management there in the 1960s, and later an Alaskan Commissioner of Fish and Game, described this history (Koernig and Noerenberg 1976):

wild stock were quite high for about 25 years (1920–45). Suddenly after 1945,…pinks…and chum… plunged to quite low levels…It led to total closure of fisheries for five different years in the last 25 years…In 1954–1955…1959, 1972 and 1974.…we were quite encouraged through the 1960s in the rising trend.…it is now clear that…adverse winter conditions in the streams which devastate eggs and alevins.…will continue to cancel much of this improved management.…So with this rationale behind us, and with a group of fishermen who have been a very stable group, many of them residents for 25, 30 or 40 years, there was a strong basis for self-help in fish enhancement.

Simpler (1976; then a commercial fisherman in PWS for 36 years) described the same rationale for artificial enhancement from a fisherman's point of view:

> in the past 36 years of fishing…we have never obtained a good return unless we have had an ideal winter, which we do not have very often…the fishermen and the newly formed Aquaculture Corporation must get the management and protection of the fisheries out of the political arena…only realistic answer…is to have a great deal more local input and local control.

From a modern vantage point many fishery scientists (e.g., Hare and Francis 1995) recognize that the post-1945 plunge was probably largely a consequence of a climate shift and that another shift, beginning the recent era of high salmon production, occurred in the mid-1970s just as the salmon enhancement program began in PWS. Even with the advantage of hindsight, however, modern analysts should recognize two things about the vision of the originators of the Alaska ocean ranching program: (1) they were not motivated by short-term greed but by a desire for a sustainable fishery; and (2) they gave first priority to protection of wild-spawning salmon ("The guideline is this: natural stocks must not be affected adversely by the hatchery operations"; FRED 1980). Modern analysts should also recognize that severe winters continue to be associated with reduced survival of wild salmon embryos in some years (Hoffmeister 1993).

Two Decades of Development

The program has grown in PWS until, after 1988, over 600 million fry have been released annually, primarily by Prince William Sound Aquaculture Corporation (PWSAC), a non-profit regional corporation (FRED 1989). They are mostly pink salmon. Since 1980 the contribution of the ocean ranching program to pink salmon harvest in PWS has grown considerably (Figure 26.3). In 1995 over 80% of the pink salmon harvest was attributed to ocean ranching production (McNair 1996). This trend since 1980 of increased contributions from hatchery stocks to the total harvest might be interpreted to illustrate one of the major concerns of many fishery scientists (e.g., Hilborn 1992)—that excessive harvest of wild stocks brought about by the simultaneous harvest of abundant hatchery stocks leads to displacement of wild stocks.

Considered in the context of a longer history, however, the evidence that wild production is being displaced in PWS is not convincing (Figure 26.3). Wild stocks of pink salmon in PWS have gone through frequent cycles of abundance and scarcity before the time the ocean ranching, probably largely caused by cycles of warm winters and freezing winters as described 20 years ago (Koernig and Noerenberg 1976; Simpler 1976) and described more rigorously by Hare and Francis (1995). The recent fluctuations of wild stocks are not obviously different. What seems to be different is that in recent years, when wild stocks have failed to produce a substantial harvestable surplus (1988, 1992), there have been pink salmon available for harvest that were produced in the PWS hatchery system.

This view of history vindicates Noerenberg's (Koernig and Noerenberg 1976) and Simpler's (1976) vision. They found that cold winters were associated with poor survival of pink salmon eggs and with run failures. They noted that over 25 years there had been five failures, years in which entire fisheries had been closed. In the most recent 15 years of PWS hatchery production there have also been years (at least two, 1988 and 1992) when wild stock returns were very low

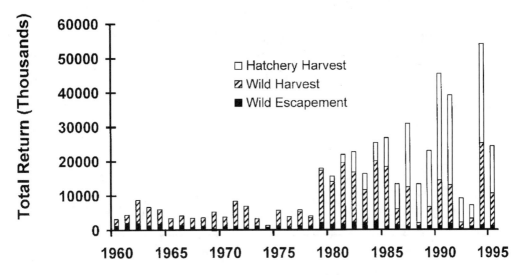

FIGURE 26.3 Pink salmon produced in Prince William Sound after escapement surveys began in 1960, including harvest of wild salmon, escapement of wild salmon to spawning streams, and, beginning in 1979, appreciable hatchery harvests (Alaska Department of Fish and Game statistics).

and there would have been little harvest without the availability of ocean ranching production. In those years the catches have instead been equal to some of the larger catches in history before the ocean ranching program was instituted (Figure 26.2).

RISKS

The program has not been free of risk, particularly of overharvest of wild stocks that transit a gauntlet of mixed stock fisheries on their return migration. These risks are a major concern of resource managers (Geiger et al. 1992) and it is worthwhile to examine them in more detail. Even without the presence of hatchery-produced stocks, fishery managers have a daunting problem to protect the reproductive capacity of a large number of wild stocks occurring simultaneously in a fishery. The ocean ranching program has increased the difficulty of the management problem.

Most pink salmon enter PWS through the southwest entrances and are harvested by the purse seine fleet in the southwest district (PWS is separated from the Gulf of Alaska by islands; its southeasternmost and southwesternmost entrances from the Gulf of Alaska are about 150 km apart). Mixtures of salmon stocks migrating to spawning grounds around PWS are harvested together in the southwest and other districts (Templin et al. 1996). The difficult task of fishery managers is to control the harvest in the districts where the mixture fisheries occur so that salmon returning to other districts are not excessively harvested. This is especially difficult, particularly in recent years because the fishery has moved farther out of the sound, into the southwest entrances near the Gulf of Alaska itself, in an effort to take pink salmon earlier in their lives, before they acquire nuptial color and a lesser value (Geiger et al. 1992).

In turn this has meant that the presence of large hatchery production has made difficult the managers' task of assessing the strength of wild stock production. During the weekly fishing seasons managers estimate wild stock strength by observing the catch-rate in the fishery—if the catch is high they infer that the stocks are abundant. If the manager cannot distinguish between wild and hatchery fish, he or she cannot know anything about the abundance of wild stocks and risks allowing too much fishing for wild stocks to bear.

Even though PWS wild stocks have in aggregate been maintained at or above historical levels, the manager's difficulty in distinguishing stocks in the mixed fishery of the southwest district has apparently reduced some wild stocks below desirable numbers recently. The history of escapement

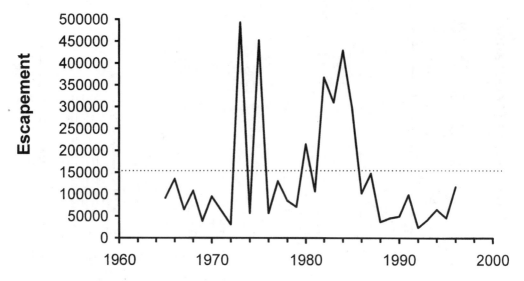

FIGURE 26.4 Escapement of pink salmon spawners to the Coghill District and the escapement goal (dotted line; data from Sharp et al. 1996, Table 12).

(the number of spawners escaping the fishery and on which the stock depends for reproduction) of one wild stock district, the Coghill district in the northwest corner of PWS, illustrates the concern (Figure 26.4). Pink salmon returning to the Coghill district must negotiate a series of mixed-stock fisheries, beginning with the interception fisheries in the entrance to the Sound and continuing with west side fisheries. Because they are exposed to a series of fisheries, they are at high risk of a cumulative harvest that is more than the Coghill stock can sustain, even in the absence of hatchery-produced salmon (Templin et al. 1996). Before the development of significant hatchery production in the early 1980s, escapements were often less than the escapement goal. In recent years the escapements have declined from historical highs in the mid-1980s and have been far below escapement goals (Figure 26.4). The risk of excessive harvest has probably been increased by the presence in the fisheries of large numbers of hatchery-produced salmon which can withstand a greater harvest rate than the Coghill stock. Whether or not hatchery production is the indirect cause of underescapement, it is clear that underescapement is caused by mixed-stock harvest along the migration pathway and that the presence of hatchery-produced fish does not make the management problem easier.

REDUCING THE RISK

PWS resource managers are using new tools for inseason harvest management to avoid excessive harvest of wild salmon in mixtures with abundant hatchery salmon. Since 1989 they have used coded micro wire tags and an associated mark, an excised adipose fin, to identify a portion (about 0.1%) of hatchery-produced fish (Peltz and Geiger 1990). In recent years, during the fishing season, harvest managers have used information from that tagging and marking to estimate relative wild stock and hatchery stock abundance (Geiger et al. 1992). The practice has helped managers but has not been entirely satisfactory. A shortcoming is that coded wire tags can only be applied to a small proportion of the hatchery fish so that estimates of relative abundance are imprecise. In contrast, thermal otolith marks can be applied to a large portion, even all, of the salmon embryos in a hatchery (Volk et al. 1990). Beginning in 1996 PWSAC and Valdez Fisheries Development Association (VFDA, a local non-profit aquaculture corporation) have marked all of their fry (over 600 million) with thermal otolith marks.

The effectiveness of this system of mass-marking of hatchery salmon has been demonstrated in southeast Alaska at the Douglas Island Pink and Chum, Inc. facility in Juneau, Alaska (a local non-profit aquaculture corporation). Information-rich marks can be economically applied to all salmon fry produced at a hatchery, even at hatcheries producing more than 100 million fry (Munk et al. 1993). Precise estimates of hatchery stock contribution to mixed fisheries can be obtained from reasonably attainable samples (Geiger 1994). The mark code can be read "inseason" i.e., quickly enough that fishery managers can know within several days after each weekly fishery opening how many hatchery fish were in the harvest, in time to decide whether to constrain the fishery the next week (Hagen et al. 1995).

In principle the risk of excessive harvest of wild salmon in mixed fisheries could also be avoided in PWS by either locating hatcheries in places where returning salmon will not commingle with wild salmon or by cultivating stocks that would return through the fishing districts only during times of the year when wild salmon are not present. Such opportunities in PWS are rare, however, because most appropriate watersheds are either on salmon migration corridors or because establishment of hatchery stocks when wild salmon were absent would require transplantation of stocks over long distances; such transplantations are proscribed by Alaska salmon genetics policy (ADFG 1985). Developing an early-returning pink salmon stock on the west side of PWS would entail, for instance, transplantation of early-returning salmon from the east side of PWS. A notable exception to these difficulties has been the development of an endemic early-returning stock of pink salmon in the northeast part of PWS, Valdez Arm, by VFDA. These salmon, nearly 7 million in 1995 (McNair 1996), return through the entrances to PWS about 2 weeks before substantial returns of wild salmon and are harvested in the confined waters of Valdez Arm in relative isolation from wild salmon.

SOUTHERN SOUTHEAST ALASKA

MANAGING HARVESTS OF MIXED HATCHERY AND WILD STOCKS

Another case history illustrates management of risks in a different setting. Southern Southeast Regional Aquaculture Association (SSRAA) in Ketchikan has relied on a central incubation facility which produces chum salmon fry released in remote locations. Adult salmon returning to the remote locations are harvested by particular fisheries. Nakat Inlet, in southeast Alaska at Dixon Entrance, is an example (Figure 26.5). This project illustrates two risks for wild stocks inherent in the ocean ranching program and demonstrates how the SSRAA program has managed the risk. The risks are the overharvest of wild stocks in a mixed stock fishery and the risk of straying and gene migration from an ocean ranched stock into wild populations. In this case the latter risk is increased because the program relies on release of fry originating and produced at a central incubation facility and released at a distant location, a practice which many would predict to engender less-faithful homing and possible outbreeding depression.

The project at Tree Point/Nakat Inlet would seem to be at high risk for excessive harvest of wild salmon because as many as half of the chum salmon taken in the weekly mixed-stock fisheries are produced from the project (Figure 26.5). However, SSRAA has marked a portion of the fry in each year with coded wire tags. These tags and the accompanying mark (an excised adipose fin) have been used during the fishing season to assess the relative abundance of hatchery and wild stocks in the fishery so that the fishery manager can control the harvest to protect wild stocks. There is no evidence of any decline of wild stock spawning escapement from surveys that have been made in the area over the years.

The project at Tree Point/Nakat Inlet would also seem to be at high risk for introgression of genes from hatchery into wild chum salmon because fry released there undergo embryonic development at

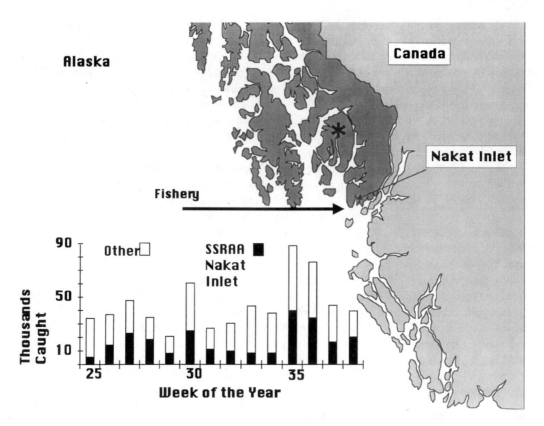

FIGURE 26.5 Chum salmon produced at SSRAA Neets Bay Hatchery on Revillagigedo Island (asterisk) are released at Nakat Inlet and harvested in a mixed-stock gillnet fishery nearby (heavy arrow.) Histogram is an example from 1995 of weekly counts of hatchery-produced chum salmon which were estimated in-season from identifying marks and tags in samples of salmon from the harvest.

a distant location and in an exotic water supply (i.e., Neets Bay Hatchery) so would be expected to lack appropriate embryonic experience of inlet freshwater and to stray into various streams within and outside the inlet. However, Nakat Inlet was in part chosen for the project because it has no substantial wild stock production so there is little or no risk to wild-stock fitness within the inlet. Further, any ocean ranched chums that return to the inlet, if they escape the common property fishery, are entirely harvested ("mopped up") by SSRAA, thereby minimizing the chances of these fish straying into chum salmon stocks outside the inlet. Anecdotal evidence supports the notion that these measures have been successful. Surveys of other streams and lakes in the region have detected very few tagged salmon from the Nakat Inlet releases. Some of those surveys, especially near Fish Creek at the head of Portland Canal, have been especially intense and have focused on recovery of wire tags.

NORTHERN SOUTHEAST ALASKA

FISHING OPPORTUNITY EASING HARVEST PRESSURE ON WILD STOCKS

The project at Hidden Falls (operated by Northern Southeast Regional Aquaculture Association-NSRAA) stands in contrast to Nakat Inlet. It was designed and located so as to avoid mixed-stock interactions with wild stocks and to provide fishing opportunities during a portion of the season when most wild stock fisheries were not open (Bachen and Linley 1995).

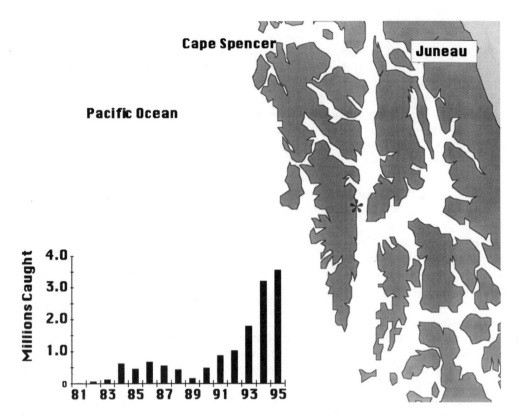

FIGURE 26.6 Chum salmon produced by NSRAA at the Hidden Falls Hatchery (asterisk) on Baranof Island. Substantial production is taken by the purse seine fleet from a single-stock fishery that is isolated temporally and spatially from other salmon stocks.

Historically the southeast Alaska seine fishery relied heavily on mixed-stock harvests of pink and chum salmon near Cape Spencer where many stocks of migrating salmon were concentrated in narrow passages as they moved eastward toward their natal streams throughout the northern part of southeast Alaska (Figure 26.6). Beginning in the 1970s harvest managers restricted those mixed-stock fisheries in an effort to manage fisheries more on the basis of discrete stocks—an important step toward conservation of diverse stocks of salmon in the region. This change restricted fishing opportunities for the seine fleet early in each summer's fishing season.

The Hidden Falls project was established on the east coast of Baranof Island in the late 1970s and became productive in the late 1980s. It is located on a very steep coastline that offers relatively little spawning habitat to salmon. The purpose of the project was to provide a fishing opportunity to the seine fleet at a place where wild stocks are relatively rare and at a time (early in the season) when relatively few stocks in the region are harvestable. Early-returning chum salmon were chosen for production; they are harvested in a terminal area before most pink salmon migrate through the area and before the chum salmon have fully matured and lost market value. The catch at Hidden Falls has risen substantially over the years and is a major resource for the industry (Figure 26.6).

An important feature of the Hidden Falls project is that it has eased the difficulty of wild-stock fishery management in the region. For example, over the 10 weeks of the 1995 season, 30 to 90% of the seine harvest effort in the entire region occurred at Hidden Falls. This simplified management of fisheries on both pink and chum wild stocks in the rest of the region because, with a substantial part of the fishing effort being directed at Hidden Falls, managers were able to open seasons in wild stock areas with some assurance that effort in those areas would not be excessive (A. McGregor, Alaska Department of Fish and Game, personal communication).

STRAYING AND GENETIC RISKS

Genetic risks, in contrast to risks of excessive harvest in mixed fisheries, are risks of depleted fitness and productivity associated with interbreeding of wild (locally adapted) salmon and straying hatchery salmon (possibly locally maladapted, either exotic or domesticated; Gharrett and Smoker 1993; Campton 1995). Even though direct evidence of such depleted fitness and productivity has been difficult to obtain (Campton 1995), risk-averse management requires that genetic risks be minimized. In Alaska these risks have been reduced by enforcement of the Alaska Genetic Policy (ADFG 1985; Holmes and Burkett 1996). It prohibits transport of salmon between regions of Alaska or from outside Alaska and permits transport within regions only with explicit consideration of genetic risks. The policy prohibits transport if there would be significant interaction with "significant or unique wild stocks" (ADFG 1985). The intended consequence is that any interbreeding between hatchery-produced salmon which might stray into wild spawning populations does not introduce extremely exotic genes into wild population and the risks of depleted fitness and productivity would be reduced. The policy has been enforced with enough rigor to prevent transfers of salmon from outside Alaska and between major zoogeographic regions of Alaska; some transfers within regions have not been permitted, some have been permitted.

The policy cannot prevent domestication effects, however. Theory predicts that hatchery culture, however careful, may cause genetically-based differences to arise between hatchery-bred populations and wild populations (reviewed, e.g., in Gharrett and Smoker 1993). Interbreeding between them will result in offspring genetically different from offspring of wild parents. Straying of mature fish from hatchery-bred populations into wild populations presents a risk of reduced fitness and productivity to wild populations.

Straying does occur but the degree of effects on the fitness and productivity of wild salmon in Alaska is unknown. Several marking and tagging studies provide direct evidence of straying:

- Even though chinook salmon hatcheries in southeast Alaska were purposefully sited more than 50 km from mouths of chinook spawning rivers, tagged hatchery salmon have been detected among some wild-spawning salmon in the region (Heard et al. 1995).
- Coded wire tags, implanted in over 1.2 million western PWS hatchery or wild pink salmon fry in 1990, were recovered in watersheds from post-spawning adults, suggesting that considerable portions of 1991 pink salmon spawning originated either in a) one of four hatcheries (hatchery strays); b) other streams as wild salmon (wild stock strays); or c) the stream of recovery (Sharp et al. 1993).
- Thermally marked otoliths from chum salmon originating in Gastineau Hatchery, Juneau, have been detected in chum salmon sampled in watersheds near the hatchery in 1995 and 1996 (L. Macaulay, Douglas Island Pink and Chum, Inc., personal communication). In earlier work near Juneau, however, Smoker and Thrower (1995) were unable to detect significant straying in coded wire tagged chum salmon; in the years of that study (1980–1985) chum salmon were much less abundant locally than in 1995 and 1996 (less than 100,000 vs. more than 1 million); straying may be more likely when chum salmon are very abundant.

In each known case of straying, with a few exceptions, the straying has occurred near the site where hatchery-produced juveniles were released or along the adult migration corridors leading back to the site. It is not known whether the straying hatchery salmon spawned successfully with spawning salmon, let alone whether serious loss of fitness and productivity has occurred, but the potential risk is a strong concern within Alaska's salmon enhancement program.

DISEASE DISSEMINATION

Risks of dissemination of infectious disease have been reduced by rigorous enforcement of Alaska's Disease Policies (State Pathology Review Committee 1988; Holmes and Burkett 1996), which restrict transfer of salmon and require inspection of facilities and examination of salmon for known diseases. Regulations require that the state pathologist approve any transfer of live salmon; permission is given only under strictly defined conditions designed to prevent dissemination of known pathogens. Regulations also require disinfection of all salmon eggs brought into any hatchery, frequent inspection of hatchery facilities, and reporting occurrences of critical diseases known only outside Alaska as well as certain endemic diseases.

ECOLOGICAL INTERACTIONS

That limitations of ocean carrying capacity might bring about interactions of hatchery-produced and wild salmon on the high seas if hatchery production in Alaska became large was recognized at the beginning of the Alaska salmon enhancement program in the 1970s (McNeil 1980). An increased risk of extinction posed by unlimited hatchery production has been predicted from models of density-dependent salmon population dynamics (Fagen and Smoker 1989). No evidence of high seas density effects on survival of Alaska stocks have been detected. Some observers have attributed a recent decades-long trend to smaller body size in several species to coincidental increases of abundance on the high seas due in part to increasing hatchery production since the mid-1970s (e.g., Bigler et al. 1996). However, trends toward smaller body size began as early as the 1950s, before recent increases of salmon abundance (e.g., Marshall and Quinn 1985; Ricker 1981), so explanation of trends of body size requires more than density dependence. Nonetheless the Alaska salmon enhancement program has been limited to production of fewer than 1.7 billion released fry (McNair 1997) considerably fewer than the projections of 2.5 billion released fry per year that were made in the 1970s (McNeil 1980). This restriction has been due in part to decisions by program managers to restrict hatchery production below planned capacity.

CONCLUSION

The Alaska ocean ranching program has successfully enhanced salmon harvests. It was developed because wild stocks could not provide an adequate harvest to sustain fisheries in particular years when severe weather reduced survival of pink salmon and chum salmon embryos. Even in the recent era of high wild salmon production, there have been some years in some regions in which a harvest was possible only because hatchery production was available. Risks, particularly of overharvest of wild stocks in mixed stock fisheries, have been avoided to some degree by siting facilities where harvests are not mixed (e.g., Hidden Falls), and by using tags to identify hatchery-produced fish in mixed harvests (e.g., Nakat Inlet). Where overharvest of wild fish has occurred (e.g., Prince William Sound) tagging and marking studies have helped identify the problem and large-scale marking programs have been initiated to contain the risk. Genetic risks associated with straying of hatchery salmon have been reduced even though not eliminated by requiring that hatchery broodstocks be derived from stocks within the same region. Risks of disease dissemination have been reduced by requiring disease sampling before any transport is permitted, by prohibiting transport of salmon between regions, and by requiring rigorous prophylaxis in fish culture facilities.

ACKNOWLEDGMENTS

We thank the University of Alaska for its support with funds appropriated by the state of Alaska and by the Alaska Science and Technology Foundation (WWS). We also acknowledge several anonymous reviewers for their helpful comments.

REFERENCES

ADFG (Alaska Department of Fish and Game). 1985. Genetic Policy. Alaska Department of Fish and Game, Juneau.

Andersen, J.L. 1997. Peer review of existing studies which address the impact of Alaskan hatcheries on the salmon industry. Pages 11–30 in Hatchery Policy Group Report to the Salmon Industry Response Cabinet. Alaska Department of Commerce and Economic Development, Division of Investments, Juneau.

Bachen, B.A., and T. Linley. 1995. Hidden Falls Hatchery chum salmon program. Pages 564–565 in H.L. Schramm, Jr. and R.G. Piper, editors. Uses and effects of cultured fishes in aquatic ecosystems. American Fisheries Society, Symposium 15, Bethesda, Maryland.

Bigler, B.S., D.W. Welch, and J.H. Helle. 1996. A review of size trends among North Pacific salmon (Oncorhynchus spp.). Canadian Journal of Fisheries and Aquatic Sciences 53:455–465.

Campton, D.E. 1995. Genetic effects of hatchery fish on wild populations of Pacific salmon and steelhead: What do we really know? Pages 337–353 in H.L. Schramm, Jr. and R.G. Piper, editors. Uses and effects of cultured fishes in aquatic ecosystems. American Fisheries Society, Symposium 15, Bethesda, Maryland.

Fagen, R.M., and W.W. Smoker. 1989. How large capacity hatcheries can alter interannual variability of salmon production. Fisheries Research 8:1–11

FRED (Fisheries Rehabilitation Enhancement and Development Division). 1980. FRED 1979 Annual Report to the Alaska State Legislature. Alaska Department of Fish and Game, Juneau.

FRED (Fisheries Rehabilitation Enhancement and Development Division). 1989. FRED 1988 Annual Report to the Alaska State Legislature. FRED Report 89. Alaska Department of Fish and Game, Juneau.

FRED (Fisheries Rehabilitation Enhancement and Development Division). 1990. FRED 1989 Annual Report to the Alaska State Legislature. FRED Report 101. Alaska Department of Fish and Game, Juneau.

FRED (Fisheries Rehabilitation Enhancement and Development Division). 1991. FRED 1990 Annual Report to the Alaska State Legislature. FRED Report 109. Alaska Department of Fish and Game, Juneau.

FRED (Fisheries Rehabilitation Enhancement and Development Division). 1992. FRED 1991 Annual Report to the Alaska State Legislature. FRED Report 117. Alaska Department of Fish and Game, Juneau.

FRED (Fisheries Rehabilitation Enhancement and Development Division). 1993. FRED 1992 Annual Report to the Alaska State Legislature. FRED Report 127. Alaska Department of Fish and Game, Juneau.

Geiger, H. 1994. A Bayesian approach for estimating hatchery contribution in a series of salmon fisheries. Alaska Fishery Research Bulletin 1:66–75.

Geiger, H., J. Brady, W. Donaldson, and S. Sharr. 1992. The importance of stock identification for management of the Prince William Sound pink salmon fishery. Regional Information Report No. 5J92-12. Alaska Department of Fish and Game, Division of Commercial Fisheries, Juneau.

Gharrett, A.J., and W.W. Smoker. 1993. A perspective on the adaptive importance of genetic infrastructure in salmon populations to ocean ranching in Alaska. Fishery Research 18:45–58.

Hagen, P., K. Munk, B. Van Alen, and B. White. 1995. Thermal Mark Technology for Inseason Fisheries Management: A Case Study. Alaska Fishery Research Bulletin 2:143–155

Hare, S.R., and R.C. Francis. 1995. Climate change and salmon production in the northeast Pacific Ocean. Pages 357–372 in R.J. Beamish, editor. Climate Change and Northern Fish Populations. Canadian Special Publication of Fisheries and Aquatic Sciences 121.

Heard, W., R. Burkett, F. Thrower, and S. McGee. 1995. A review of chinook salmon resources in southeast Alaska and development of an enhancement program designed for minimal hatchery-wild stock interaction. Pages 21–37 in H.L. Schramm, Jr. and R.G. Piper, editors. Uses and effects of cultured fishes in aquatic ecosystems. American Fisheries Society, Symposium 15, Bethesda, Maryland.

Heard, W.R. 1996. Ocean ranching: an assessment. Pages 833–869 in W. Pennell and B. A. Barton, editors. Principles of salmonid culture. Developments in Aquaculture and Fisheries Science 29, Elsevier, Amsterdam.

Hermann, M., and J. Greenberg. 1994. A revenue analysis of the Alaska pink salmon fishery. North American Journal of Fisheries Management 14:537–549.

Hilborn, R. 1992. Hatcheries and the future of salmon in the northwest. Fisheries 17:5–18.

Hoffmeister, K. 1993. Southeast Alaska winter air temperature cycle and its relationship to pink salmon harvest. Pages 111–122 in Proceedings of the 16th Northeast Pacific Pink and Chum Salmon Workshop. Alaska Sea Grant College Program Report No. 94-02. Referenced by permission.

Holmes, R.A., and R.D. Burkett. 1996. Salmon stewardship: Alaska's perspective. Management Essay. Fisheries 21:36–38.

Koernig, A., and W. Noerenberg. 1976. First year activities of the Prince William Sound Aquaculture Corporation. Pages 163–170 in D.H. Rosenberg, editor. Proceedings of the conference on salmon aquaculture and the Alaskan fishing community. Sea Grant Report 76-2. Alaska Sea Grant College Program, Fairbanks.

Marshall, R.P., and T.J. Quinn II. 1985. Estimation of average weight and biomas of pink, chum, sockeye and coho salmon in southeast Alaska commercial harvests. Alaska Fishery Research Bulletin 88-07.

McNair, M., and J.S. Holland, editors. 1994. Alaska Fisheries Enhancement Program 1993 Annual Report. Alaska Department of Fish and Game, Commercial Fisheries Management and Development Division, Juneau.

McNair, M. 1995. Alaska Fisheries Enhancement Program 1994 Annual Report. Regional Information Report 5J95-06. Alaska Department of Fish and Game, Commercial Fisheries Management and Development Division, Juneau.

McNair, M. 1996. Alaska Fisheries Enhancement Program 1995 Annual Report. Regional Information Report 5J96-08. Alaska Department of Fish and Game, Commercial Fisheries Management and Development Division, Juneau. See also http://www.state.ak.us/local/akpages/FISH.GAME/cfmd/geninfo/finfish/salmon/salmhome.htm#enhancement

McNair, M. 1997. Alaska Fisheries Enhancement Program 1996 Annual Report. Regional Information Report 5J97-09. Alaska Department of Fish and Game, Commercial Fisheries Management and Development Division, Juneau.

McNeil, W.J. 1980. Salmon ranching in Alaska. Pages 13–28 in J.E. Thorpe, editor. Salmon Ranching. Academic Press, London.

Munk, K.M., W.W. Smoker, D.R. Beard, and R.W. Mattson. 1993. A hatchery water heating system for use in 100% thermal marking of incubating salmon. Progressive Fish-Culturist 55:284–288.

Peltz, L., and H. Geiger. 1990. Pilot studies in tagging Prince William Sound hatchery pink salmon with coded-wire tags. Alaska Department of Fish and Game, Fishery Research Bulletin No. 90-02. Juneau.

Pinkerton, E. 1994. Economic and management benefits from the coordination of capture and culture fisheries: The case of Prince William Sound pink salmon. North American Journal of Fisheries Management 14:262–277.

Ricker, W.E. 1981. Changes in the average size and average age of Pacific salmon. Canadian Journal of Fisheries and Aquatic Sciences 38:1636–1656.

Royce, W.F. 1988. Centennial Lecture IV: The historical development of fishery science and management. Marine Fisheries Review 50:30–39.

Salmon Market Information Service. 1996. Tables "Estimated World Production of Farmed Salmon" and "Estimated World Production of Wild Salmon." Salmon Market Information Service. Institute of Social and Economic Research. University of Alaska, Anchorage.

Sharp, D.S., S. Morstad, and J. Johnson. 1996. Prince William Sound management area salmon report to the Alaska Board of Fisheries. Regional Information Report 2A96-41. Alaska Department of Fish and Game, Commercial Fisheries Management and Development Division, Anchorage.

Sharp, D. S., S. Sharr, and C. Peckham. 1993. Homing and straying patterns of coded wire tagged pink salmon in Prince William Sound. Pages 77–82 in Proceedings of the 16th Northeast Pacific Pink and Chum Salmon Workshop. Alaska Sea Grant College Program Report No. 94-02. Referenced by permission.

Simpler, C. 1976. The historic role of fisheries management from the fishermen's point of view. Pages 1–6 in D.H. Rosenberg, editor. Proceedings of the conference on salmon aquaculture and the Alaskan fishing community. Sea Grant Report 76-2. Alaska Sea Grant College Program, Fairbanks.

Smoker, W.W., and F.P. Thrower. 1995. Homing propensity in transplanted and native chum salmon. Pages 575–576 in H.L. Schramm, Jr. and R.G. Piper, editors. Uses and effects of cultured fishes in aquatic ecosystems. American Fisheries Society, Symposium 15, Bethesda, Maryland.

State Pathology Review Committee. 1988. Regulation changes, policies and guidelines for Alaska fish and shellfish health and disease control. FRED Special Report. Alaska Department of Fish and Game. Juneau, Alaska.

Templin, W., J. S. Collie, and T.J. Quinn II. 1996. Run construction of the wild pink salmon fishery in Prince William Sound 1990–1991. Pages 499–508 in S.D. Rice, R.B. Spies, D.A. Wolfe, and B.A. Wright, editors. Proceedings of the Exxon Valdez Oil Spill Symposium. American Fisheries Society, Symposium 18, Bethesda, Maryland.

Thomas, G.L. and O. A. Mathisen. 1993. Biological interactions of natural and enhanced stocks of salmon in Alaska. Fisheries Research 18:1–18.

Thorpe, J.E. 1980. The development of salmon culture towards ranching. Pages 1–12 *in* J.E. Thorpe, editor. Salmon Ranching. Academic Press, London.

Volk, E.C., S.L. Schroder, and K.L. Fresh. 1990. Inducement of unique otolith banding patterns as a practical means to mass-mark juvenile Pacific salmon. Pages 203–215 *in* N.C. Parker, A.E. Giorgi, R.C. Heidinger, D. B. Jester, Jr., E. D. Prince, and G. A. Winans, editors. 1990. Fish-Marking Techniques. American Fisheries Society, Symposium 7, Bethesda, Maryland.

Wilen, J.E. 1993. Technical review of a benefit-cost analysis of the Alaska salmon enhancement program. Alaska Department of Fish and Game, Commercial Fisheries Management and Development Division, Contract IHP 93-044, Juneau.

Section VI

Modeling Approaches

27 The Proportional Migration Selective Fishery Model

Peter W. Lawson and Richard M. Comstock

Abstract.—Ocean fishing seasons for Pacific salmon *Oncorhynchus* spp. are set to limit fishing impacts on certain naturally reproducing stocks. Fishing-related mortalities are estimated using harvest models based on historic salmon catch distributions in fisheries. Current models assume that, in a fishery, all stocks are caught at the same rate. Each stock is assumed to be a single, uniformly mixed population. New fisheries have been proposed in which most hatchery fish would receive an external mark and only marked fish would be retained. These "selective" fisheries would allow fishing access to hatchery runs while reducing mortalities on naturally reproducing stocks. In selective fisheries, stocks would not be equally vulnerable to the fishing gear, and would suffer several forms of gear-related mortality. A new family of harvest models is needed to permit managers to assess differential impacts of selective fisheries on marked and unmarked salmon. We describe such a model and use it to evaluate effects of a proposed selective fishery using 126 stocks of coho salmon *O. kisutch* in 183 fisheries from California to Alaska. Our results suggest that selective fisheries will have lower total exploitation rates on unmarked stocks compared with marked stocks. Modeled stocks that were caught in a mixture of traditional mixed-stock and selective fisheries showed less benefit than those caught primarily in selective fisheries. In the fisheries we modeled, catches were lower because many wild fish that would have been landed in a traditional fishery were released. Mass marking and selective fishing may provide fishing opportunity and access to hatchery fish while reducing mortalities of wild fish. However, non-catch mortalities of unmarked fish in selective fisheries may be substantial, and must be accurately accounted for if conservation goals are to be achieved.

INTRODUCTION

Traditional ocean fisheries for Pacific salmon *Oncorhynchus* spp., generally harvest mixtures of wild and hatchery fish. In the ocean off California, Oregon, and Washington, where fisheries are managed by the Pacific Fishery Management Council (PFMC), landing quotas and time/area restrictions are used to achieve exploitation rates that are calculated to allow sufficient spawner escapements for certain naturally reproducing salmon stocks (PFMC 1995). In an effort to reduce fishing impacts on wild stocks of chinook salmon *O. tschawytscha* and coho salmon *O. kisutch*, additional restrictions have been placed on many fisheries. While these strategies have reduced exploitation rates on wild stocks, they have also limited access to hatchery stocks.

One major obstacle to the conservation and sustainability of wild Pacific salmon runs is their vulnerability to incidental harvest in fisheries that are directed at hatchery runs. Fishery managers are seeking techniques for allowing harvest of hatchery fish while simultaneously achieving better wild escapements. Since 1998, most adult hatchery-produced coho salmon from Grays Harbor, Washington, the Columbia River, and the Oregon coast have been marked with an adipose fin clip. The intention is to allow fishers to identify and selectively retain marked (hatchery) coho while releasing unmarked (wild) fish (PSMFC 1992; PSC 1995). In such "selective" fisheries, the unmarked stocks suffer mortalities from encounters with fishing gear. These "non-catch" mortalities are difficult to quantify because they are, by their nature, unobservable in normal fishery sampling programs. Current fisheries models were not designed to provide estimates of non-catch mortality. Fishery managers need a new set of modeling tools to enable them to negotiate fishing seasons that provide access to marked hatchery stocks in selective fisheries while minimizing mortality of unmarked salmon.

Harvest models are used extensively by the PFMC for setting salmon harvest regulations in the waters off Washington, Oregon, and California. In March and April of each year the PFMC meets with representatives of the fishing industry, conservation groups, and state, tribal and federal agencies to negotiate fishing seasons, which typically start in May and continue through September. These negotiations rely on models designed to relate time and area patterns of fishing effort and catch to exploitation rates on a large number of salmon runs grouped into "stocks" for management purposes. The models that were used through the 1990s were designed for "mixed-stock" fisheries with the assumption that all stocks in a fishing area are equally vulnerable to fishing gear.

The model currently used by the PFMC for setting coho salmon fishing seasons is the Fishery Resource Assessment Model (FRAM). This model treats each stock as a single population of fish, vulnerable to all fisheries in proportion to catch rates in a base period. A base scenario is established using stock-specific landings data from coded-wire tag (CWT) recoveries, estimated pre-season stock abundances, and escapements from the "base years" of 1979–1981. Stock sizes from the base years are scaled to projected abundances in the year to be modeled. Mortalities in each fishery are adjusted in linear proportion to the effort expected. Changes in mortality are used to adjust the stock size in the next time step. One limitation of this "single pool" approach is that fish escaping a fishery in Canada, for example, are available in the next time step to a fishery in California (i.e., there is no discrimination among fisheries from one time period to the next). A second limitation is that mortalities, including incidental mortalities, are estimated as proportions of the total stock size, and are linear functions of effort. This approach can substantially underestimate non-catch mortality in selective fisheries, especially when local harvest rates on marked (fin-clipped) fish are relatively high (> 0.5) (Lawson and Sampson 1996).

Lawson and Sampson (1996) described a method to estimate gear-related mortality in a single fishery with two stocks, one marked and one unmarked. In this chapter we (1) describe a model designed to apply the Lawson and Sampson (1996) mortality algorithm to West Coast coho salmon fisheries including 126 stocks (marked and unmarked, hatchery and wild) in 183 fisheries from Alaska to California in three time periods; (2) apply the model in a test case; and (3) discuss potential utility of the model in harvest management programs.

MODEL DESCRIPTION

The current family of fishery models uses a single estimate of total stock abundance. In contrast, the algorithms from the Lawson and Sampson (1996) model include modified catch equations (Parzen 1962), which require, as input, the number of marked and unmarked fish available to each fishery. This is a more stringent data requirement, and leads to an underlying model of stock distribution and abundance that is fundamentally different from the harvest models currently used to manage West Coast salmon fisheries. Estimating fish available to a fishery is a difficult problem that we have not solved to our satisfaction. The basic approach used here, however, is to divide catch in a fishery, available from a coastwide cohort run reconstruction (B. Tweit, Washington Department of Fish and Wildlife, personal communication), by an assumed fishery-specific harvest rate. Because fishery-specific harvest rates are difficult to estimate, there is likelihood of relatively large errors in the numbers of fish estimated to be available. Lawson and Comstock (1995) found low sensitivity of estimated catch and mortality to this factor when harvest rates in selective fisheries were low.

In the course of fishery modeling, the change in fishery mortality under a changed management scenario (e.g., selective vs. mixed-stock) is used to estimate changes in stock abundances in the next time period. Maturing salmon, in their migrations through the ocean, are generally understood to travel at moderate rates of speed, and to converge on their natal streams. Unconstrained, the migration algorithm employed here is similar to the single pool model used in FRAM in that it could allow changes in mortalities to affect distant fisheries in subsequent time steps. Ideally, a fisheries model would incorporate the specific migration patterns of each stock in the model, but for the majority of salmon these patterns are not known. An intermediate solution, employed here, was to constrain the migration of salmon using a movement map, partitioning fisheries into broad geographical areas, and linking receiving areas with source areas for each time step. Because of our uncertainty in estimating local abundances, we have more confidence in our ability to predict proportional changes in mortality rather than numerical changes. In the model described here, abundances in the subsequent time period are adjusted in proportion to the change in mortality in the current time period. Consequently, we refer to this as the "proportional migration" or "PM" model.

"New" Modeling Procedure

Prior to fishery modeling, a stock reconstruction is performed for each stock. Inputs are catch and harvest rate from a base period for each fishing area and time period. Base catch (*BaseC*) by stock, fishery, and time was constructed from the coastwide CWT database for the years 1986–1991. The 1991 data were used for the simulation presented here. Outputs are total mortalities in each fishery, total abundance available to each fishery, and an effort coefficient. Scaling factors are applied to the base period abundances to estimate the starting abundances available to each fishery in the modeled year. Fisheries are then simulated in a series of time steps. In each time step, catch and total mortality are based on (1) abundances from the stock reconstruction; (2) fishery input parameters specifying effort or catch levels; and (3) whether the fishery is selective or non-selective. Mortalities in selective fisheries usually differ from mortalities in the input fisheries, and they differ for marked and unmarked stocks. These differences in mortalities are transferred to the abundances available to fisheries in the subsequent time step using a migration model. The process is repeated through several time steps. Resulting simulated catch and mortality for each stock can be compared with catch and mortality in the input (non-selective) fisheries to evaluate the effects of the proposed selective fishery regime.

Stock Reconstruction

The following is a general description of the model. Subscripts are used to indicate the level of resolution of each set of calculations. The non-catch fraction of total mortality (*S*) for each fishery and time period is estimated using the algorithm from Lawson and Sampson (1996):

$$S_{jt} = 1 - (1-a) / [(1-a) + a \cdot b], \tag{27.1}$$

where a = dropoff probability, b = dropoff mortality rate, j = fishery, and t = time period. Dropoff probability is the probability that a fish encounters a hook but escapes or is removed by a predator before being brought to the boat. Dropoff mortality rate is the proportion of dropoffs that die as a result of encountering fishing gear. The relationship of these factors to total mortalities is treated in detail in Lawson and Sampson (1996).

Base period abundances (*BaseA*) available to each fishery are calculated from base period catch (*BaseC*), a fishery-specific harvest rate (*H*), and the non-catch mortality fraction (*S*).

$$BaseA_{ijt} = BaseC_{ijt} / [H_{jt} \cdot (1 - S_{jt})], \tag{27.2}$$

where i = stock.

Base period mortality (*BaseM*) is base catch plus non-catch mortality,

$$BaseM_{ijt} = BaseC_{ijt} \big/ \left(1 - S_{jt}\right). \tag{27.3}$$

Catch, mortality, and abundance are scaled to the modeled year with stock-specific scaling factors (R):

$$C_{ijt} = BaseC_{ijt} \cdot R_i, \tag{27.4a}$$

$$M_{ijt} = BaseM_{ijt} \cdot R_i, \tag{27.4b}$$

$$A_{ijt} = BaseA_{ijt} \cdot R_i. \tag{27.4c}$$

For this analysis R was set at 1 for all stocks. In practice, R would be estimated for each stock from the ratio of abundance in the base period to abundance for the scenario being modeled. Each stock is then separated into as many as three stocks: wild unmarked, hatchery marked, and hatchery unmarked. These are the stocks, i, that are carried through the model. Abundance is then recombined into categories of marked (N_M) and unmarked (N_U) fish.

The original fishing rate (E) for each fishery is found iteratively by solving a modified catch equation for *BaseM* until a value of E is found which satisfies the equality

$$BaseM_{jt} = BaseA_{jt} \cdot \left\langle 1 - \exp\left\{1 - \left[(1-a) + a \cdot b\right] \cdot E_{jt}\right\}\right\rangle . \tag{27.5}$$

Note that E is the time-integrated rate coefficient for an encounter by a fish with the gear, and is a function of total fishing effort in the time period and catchability. The rate was computed as the appropriate value required to produce the observed mortality given the starting abundance. We assumed that catchability was constant in each fishery.

The original number of fish escaping from the fishery (O) is abundance (A) minus mortalities (M).

$$O_{ijt} = A_{ijt} - M_{ijt}. \tag{27.6}$$

This number will be used for estimating the proportional change in stock abundance at the end of the time step.

FISHERY MODELING—FIRST TIME STEP

New mortality rates for marked and unmarked fish are calculated based on effort scalars (F) or catch ceilings (quotas), using algorithms from Lawson and Sampson (1996). In selective fisheries the mortality rates for marked and unmarked fish will differ. For marked fish, the probability of mortality (P_M) is:

$$P_{Mjt} = 1 - \exp\left\langle -E_{jt} \cdot F_{jt} \cdot \left\{(1-a) \cdot \left[g_M + (1 - g_M) \cdot d\right] + a \cdot b\right\}\right\rangle, \tag{27.7}$$

where g_M = mark recognition rate for marked fish, and d = release mortality rate. The mark recognition rate may differ for marked and unmarked stocks. Catch ceilings are modeled by iteratively adjusting F until the specified catch results.

For unmarked fish the probability of mortality (P_U) is:

$$P_{Ujt} = 1 - \exp\left\langle -E_{jt} \cdot F_{jt} \cdot \left\{ (1-a) \cdot \left[g_U \cdot d + (1-g_U) \right] + a \cdot b \right\} \right\rangle,$$ (27.8)

where g_U = mark recognition rate for unmarked fish.

New fishery-specific mortalities and catch are calculated from the new mortality rates. Total marked fish mortality (D_M) is:

$$D_{Mjt} = P_{Mjt} \cdot N_{Mjt}.$$ (27.9)

Likewise, total unmarked fish mortality (D_M) is:

$$D_{Ujt} = P_{Ujt} \cdot N_{Ujt}.$$ (27.10)

Catch is always less than total mortalities. The fraction of marked fish mortalities that are landed (L_M) is:

$$L_{Mjt} = (1-a) \cdot g / \left[(1-a) \cdot g + (1-a) \cdot (1-g) \cdot d + a \cdot b \right].$$ (27.11)

The fraction of unmarked fish mortalities that are landed (L_U) is:

$$L_{Ujt} = (1-a) \cdot (1-g) / \left[(1-a) \cdot g \cdot d + (1-a) \cdot (1-g) + a \cdot b \right].$$ (27.12)

Therefore, catch of marked fish (C_{Mjt}) is:

$$C_{Mjt} = D_{Mjt} \cdot L_{Mjt},$$ (27.13)

and catch of unmarked fish (C_{Ujt}) is:

$$C_{Ujt} = D_{Ujt} \cdot L_{Ujt}.$$ (27.14)

The new number of fish escaping from the fishery is calculated using the modeled abundances (A_{ijt}) along with the appropriate mortality probability (P_{Mjt} or P_{Ujt}) to compute the modeled number of survivors (O'_{ijt}). For marked stocks:

$$O'_{ijt} = A_{ijt} \cdot \left(1 - P_{Mjt} \right).$$ (27.15)

For unmarked stocks:

$$O'_{ijt} = A_{ijt} \cdot \left(1 - P_{Ujt} \right).$$ (27.16)

The modeled abundance will include any abundance changes carried forward from previous time steps. Modeled mortality (D) differs from original mortality (M), as it results from applying any user-requested changes (e.g., effort increase, ceiling change, etc.) to the fisheries, and may be different for marked and unmarked fish.

PROPORTIONAL MIGRATION

The abundance calculation for a fishery in time step $t+1$ uses the O_{ijt} and O'_{ijt} values. For each stock s_i in fishery j in time ($t+1$), the original abundace ($O_{i,t}$) is the sum of O_{ijt} over all contributing fisheries in time (t). Contributing fisheries are defined in the movement map. Likewise, the new number of fish escaping from contributing fisheries ($O'_{i,t}$) is the sum of O'_{ijt} over all contributing fisheries in time (t). If stock i did not occur in fishery j during time step t, then $O_{ijt} = 0$ and $O'_{ijt} = 0$. The abundance used for modeling ($A'_{ij(t+1)}$) is calculated as the proportional change in fish contributing to fisheries in the next time step:

$$A'_{ij(t+1)} = A_{ij(t+1)} \cdot O'_{i,t}/O_{i,t}. \qquad (27.17)$$

The total fishery abundance ($A'_{j(t+1)}$) is the sum of the individual stock abundances ($A'_{ij(t+1)}$).

FISHERY — SUBSEQUENT TIME STEPS

The original number of fish escaping from the fishery, $O_{ijt,}$ is original abundance minus original mortality. It does not include abundance changes from previous fisheries. New mortality rates for marked and unmarked fish are calculated based on effort scalars or catch ceilings (quotas), and using the new abundance, A'_{jt}. New fishery-specific catch and non-catch mortalities are calculated from the new mortality rates. The new number of fish escaping from the fishery is calculated using the modeled mortality rates and the new abundance. For marked stocks:

$$O'_{ijt} = A'_{ijt} \cdot \left(1 - P_{Mjt}\right). \qquad (27.18)$$

For unmarked stocks:

$$O'_{ijt} = A'_{ijt} \cdot \left(1 - P_{Ujt}\right). \qquad (27.19)$$

INPUTS AND MODEL SPECIFICATION

STOCKS, FISHERIES, AND TIME STEPS

For modeling purposes, stocks were considered to be aggregates of CWT releases with similar ocean catch distributions. These groups were initially defined in the coastwide coho cohort run reconstruction project. Stock groups, for reporting, are combinations of stocks that originate from the same area and may need to be managed as a unit. For example, the "Hood Canal" stock consists of hatchery and wild CWT groups from the Skokomish and Quilcene rivers, Port Gamble net pens, and some CWT experiments on wild runs in Hood Canal tributaries.

Fisheries, like stocks, are based on fishery definitions from the coastwide coho cohort run reconstruction project. Fisheries generally coincided with common management units, such as "Southwest Vancouver Island Troll," "Newport Sport," or "Area 10 Net." Three types of fisheries were specified: sport, troll, and net. Each type was associated with a unique set of selective fishery parameters. For each, a fishery-specific harvest rate (H) was specified. Three time steps were used for the modeling presented here: January–May, June–August, and September–December.

TABLE 27.1
Parameters used for estimating mortalities in selective fisheries.

Type of fishery	Release mortality rate (d)	Mark recognition rate (g_M)	Unmark recognition rate (g_U)	Dropoff probability (a)	Dropoff mortality rate (b)
Sport	0.11	1.00	1.00	0.30	0.10
Commercial troll	0.25	0.94	0.94	0.30	0.20
Commercial net	0.70	0.94	0.94	0.40	0.50

SELECTIVE FISHERY SIMULATIONS

Specific selective fishery scenarios were defined in terms of fishery definitions, stock definitions, and selective fishery parameters. Each fishery was flagged as either selective or non-selective, and with either an effort or a ceiling control. For effort-controlled fisheries, an effort scalar (F) was specified. For ceiling fisheries, a catch ceiling was specified, which then acted as a quota.

Stocks were defined at the finest level of the coastwide coho cohort run reconstruction data set. For each stock three parameters, an abundance scalar, the percent of wild fish, and percent of unmarked hatchery fish were specified.

Selective fisheries parameters are the encounter rate and mortality rate parameters required by the Lawson and Sampson (1996) algorithms to estimate fishery-related mortalities. These parameters, with values used for three fishery types, are listed in Table 27.1. A mass marking mortality rate of 0.05 was used for all stocks.

MOVEMENT MAP

The migration algorithm transfers proportional changes in mortalities from fisheries in time (t) to fisheries in time (t+1) using a movement map which defines "receiving areas" and "source areas" for each time step. The movement map constrains movements of fish to biologically reasonable geographic areas. Changes in stock size for each source area algorithm are applied to each receiving area. For example, the Washington coast, the Alaska/Canada outside fisheries, and the Oregon/California coast are source areas for the Washington coast receiving area. The Oregon/California coast is also a source area for the Columbia River, but not for the Georgia Strait.

AN EXAMPLE APPLICATION: THE OREGON AND WASHINGTON PROPOSAL

THE MANAGEMENT PROPOSAL

To gain insights into some of the potential implications of selective fishing to harvest management, we used the PM model to simulate the selective fisheries and mass marking program proposed by the Washington Department of Fish and Wildlife (WDFW) and the Oregon Department of Fish and Wildlife (ODFW) (P. Pattillo, Washington Department of Fish and Wildlife, personal communication). This is not intended to be a complete evaluation of the ODFW/WDFW proposal. Rather, we wish to test the PM model's ability to evaluate future annual and long-term management plans that incorporate mass marking and selective fishing.

The ODFW/WDFW proposal assumed that most hatchery coho salmon from Grays Harbor, south through Oregon, would be mass marked, as would portions of several Strait of Juan de Fuca and Puget Sound stocks, including the large hatchery runs from South Puget Sound. Selective fisheries were proposed in Puget Sound and Washington coastal sport fisheries, in the Buoy 10 sport fishery at the mouth of the Columbia River, and in all sport and troll fisheries along the Oregon and California coasts. In Washington and the Columbia River, fishing effort was to be held

TABLE 27.2
Catch and non-catch mortalities in five coho salmon
fisheries under traditional and selective fishing regimes.
Results are from combined sport and commercial fisheries
unless specified. Selective fishing is patterned after the
ODFW/WDFW proposal (in thousands of fish).

Fishery	Traditional		Selective	
	Catch	Non-catch	Catch	Non-catch
Canada	699	75	702	74
Puget Sound Area 5&6 sport	94	4	52	9
North of Cape Falcon	291	12	256	17
Columbia River terminal net	265	88	267	89
South of Cape Falcon	413	27	329	45

the same as in the base year with the exception of the Puget Sound recreational fishery in Areas 5 and 6 where effort would be doubled. In Oregon, selective fisheries were designed to achieve an exploitation rate on unmarked stocks of less than 15%. Selective fisheries with an effort reduction to 65% of base year levels achieved this goal.

Comparison of Selective and Traditional Fisheries Based on 1991 Catch and Abundance

We compared catch and non-catch mortalities and total exploitation rates on several stock aggregates in several broad fishery areas in traditional (mixed-stock) and selective fisheries (as proposed by ODFW/WDFW) using 1991 fisheries data as a basis (Tables 27.2, 27.3, and 27.4). Non-catch mortalities in traditional fisheries were estimated by applying the dropoff probability and dropoff mortality rates.

Catch and non-catch mortalities in five major fishery areas are compared in Table 27.2. Within the selective fishery scenario, areas without selective fisheries (Canada and Columbia River terminal net) had slight increases in catch. This resulted from reduced mortalities in selective fisheries, which increased the numbers of fish available in subsequent fisheries. In contrast, Puget Sound Area 5&6 Sport had only about half the total catch (despite a doubling of effort), but twice the non-catch mortality. This fishery had a high proportion of unmarked fish, so most of the fish brought to the boat were released. The North of Cape Falcon fishery was modeled with selective recreational fisheries and traditional commercial troll fisheries. With no change in effort, catch was reduced by about 12%, and non-catch mortalities increased by 50%. All South of Cape Falcon fisheries were selective and effort was reduced to 0.65 of the base period. Catch was reduced 25%, with non-catch mortalities increased by about 60%. In general, terminal fisheries and fisheries without selective fishing benefited slightly, while areas with selective fishing experienced reduced overall catch and mortality, but increased non-catch mortality under the selective fishing regime.

We then compared catch and non-catch mortalities in five major stock groups under the traditional and selective fishing scenarios (Table 27.3). In every case, catch was reduced and non-catch mortality was either unchanged or increased under the selective fishing regime. The greatest decrease in catch was in the Oregon/California stock group, which contributed primarily to the South of Cape Falcon fisheries, where effort was reduced and selective regulations were in effect. Smaller changes were seen for the other four stocks.

The total exploitation rates on marked and unmarked stocks were also compared under the modeled traditional and selective fishery scenarios (Table 27.4). In traditional fisheries exploitation rates differed between marked and unmarked fish for two reasons. For most of these stock groups there are separate CWT groups representing hatchery and wild stocks. Different CWT groups will

TABLE 27.3
Catch and non-catch mortalities for five coho salmon stock groups under traditional and selective fishing regimes. Selective fishing is patterned after the ODFW/WDFW proposal (in thousands of fish).

Stock	Traditional		Selective	
	Catch	Non-catch	Catch	Non-catch
Skagit River	35	5	30	5
South Puget Sound	615	118	580	126
Grays Harbor	174	41	166	43
Columbia River	909	128	865	137
Oregon/California	94	6	46	26

TABLE 27.4
Total exploitation rates for marked and unmarked components of five stock groups under traditional and selective fishing regimes. Selective fishing is patterned after the ODFW/WDFW proposal.

Stock	Traditional		Selective	
	Marked	Unmarked	Marked	Unmarked
Skagit River	0.78	0.59	0.78	0.50
South Puget Sound	0.84	0.91	0.84	0.85
Grays Harbor	0.64	0.68	0.64	0.65
Columbia River	0.68	0.64	0.68	0.39
Oregon/California	0.47	0.47	0.47	0.14

have different ocean catch distributions and different exploitation rates. The Skagit River stock has a large difference between marked (0.78) and unmarked (0.59) exploitation rates. This may reflect harvest management strategies (other than selective fishing) designed to reduce impacts on wild fish. Natural stocks in the Oregon/California stock group are represented by hatchery CWTs, so we show no difference in exploitation rate in the traditional fishery. Comparing between traditional and selective fisheries, the exploitation rates on marked stocks were unchanged, while unmarked stock exploitation rates decreased with selective fishing. For stocks north of the Columbia River this decrease was slight (i.e., only 3 to 9 percentage points reduction). Columbia River unmarked stocks saw exploitation rates reduced from 0.64 to 0.38, and the Oregon/California stock showed a threefold reduction from 0.47 to 0.14. Recall that the effort reduction of 0.65 in South of Cape Falcon fisheries was chosen to achieve a less than 0.15 exploitation rate on these stocks.

DISCUSSION

Total catch was reduced in all selective fisheries examined under the selective fishing scenario. Catch reductions resulted from releasing unmarked fish that would otherwise have been retained. In fisheries with large proportions of wild (and hence unmarked) fish, such as the Puget Sound Area 5&6 Sport, catches were reduced even with a doubling of effort. Reduced catch and mortality of unmarked stocks may result in larger numbers of unmarked fish available to late-season and terminal mixed-stock fisheries, and escapements. This has implications for the Tribes, which largely harvest in terminal areas.

We have produced a fishery model that is operational at the scale needed for evaluating proposals and negotiating fishing seasons. Model results reported here are considered to be "reasonable," as they reflect patterns of change that fit our understanding of selective fisheries. While we have confidence in the patterns of change reported by the model, the magnitude of changes is more difficult to interpret. Results presented here are for 1991 only, without confidence limits. Our plans for further development of the model include using each year from 1986 to 1991 as a base year. We would then scale modeled year abundances to each base year before simulating fisheries. The range of outcomes would serve as empirical confidence limits, facilitating uncertainty analysis.

Behavior of the underlying fishery model (Lawson and Sampson 1996) and its sensitivity to input parameters are well documented. The more difficult problem is associated with modeling the distribution and migration behavior of a large number of stocks through a comparably large number of fisheries in several time steps. Once several base years have been established, the task will be to explore the sensitivity of the model to a variety of parameters including fishery-specific harvest rates, effort scalars, and movement map constraints. Effort should then be directed at improving estimates of high sensitivity factors. Results will need to be compared with other fishery models such as FRAM. However, there is currently no good basis for comparing predictive capabilities of two models because selective fisheries are yet to be implemented.

Usefulness of the model, at this stage, is to explore the kinds of changes expected with selective fisheries. Results reported here suggest that selective fisheries are most effective at reducing mortalities on unmarked stocks when those stocks are not also subjected to large mixed-stock fisheries. Catch per effort in selective fisheries is predicted to be lower than in comparable mixed-stock fisheries. Depending on the local stock composition, catch per effort may be considerably lower. Mixed-stock fisheries, including terminal net fisheries, that occur after selective fisheries, saw higher numbers of fish, although the effect modeled here was slight. If selective fisheries were designed to exploit marked fish at a higher rate, numbers available to terminal fisheries could decrease. These results are useful to fishery managers seeking to understand how to design selective fisheries to increase access to hatchery fish while reducing mortalities on natural stocks. Various parties negotiating fishery allocations can use this kind of modeling to predict the possible changes they would have to respond to under selective fishing regulations.

Selective fisheries, by their nature, will differ from traditional salmon fisheries on the West Coast. Benefits from the mass-marking program include improved ability to recognize hatchery juveniles in estuaries and the ocean, and adult salmon in fisheries and spawning grounds. Selective fishing for marked salmon can provide additional angling opportunity for recreational fishers, while maintaining or reducing unmarked fish mortality rates. The trade-off would be in fewer fish retained and more fish released. Emphasis on catch and release fishing is likely to increase, as will concerns about hook and release mortality rates. The need to examine each fish caught, and to release some, will change the fishing operation for both recreational and commercial fishers. Landings, and catch per unit effort, will decrease in selective fisheries. Marked and unmarked stocks will experience different fishery mortality rates in selective fisheries. The greatest benefits may occur for unmarked, wild stocks which, because of the potential to reduce exploitation rates (Table 27.4), could experience improved escapements. The general effects of selective fisheries toward improving Pacific Salmon sustainability are demonstrated by the model results presented here.

ACKNOWLEDGMENTS

C. Cook-Tabor and S. Caromile compiled data and worked with us throughout the model development and manuscript preparation. R. Moore participated in analysis and manuscript preparation. P. Pattillo worked with us in simulating the ODFW/WDFW proposal. T. Wainwright and R. Kope reviewed early drafts of our manuscript. We thank R. Lincoln and two anonymous reviewers for many helpful comments.

REFERENCES

Lawson, P. W., and R. M. Comstock. 1995. Potential effects of selective fishing on stock composition estimates from the mixed-stock model: Application of a higher dimension selective model. Information Report 95-2. Oregon Department of Fish and Wildlife, Portland.

Lawson, P. W., and D. B. Sampson. 1996. Gear-related mortality in selective fisheries for ocean salmon. North American Journal of Fisheries Management 16:512–520.

Parzen, E. 1962. Stochastic processes. Holden-Day, San Francisco.

PFMC (Pacific Fishery Management Council). 1995. Review of 1994 ocean salmon fisheries. Portland, Oregon.

PSC (Pacific Salmon Commission). 1995. Selective fishery evaluation. Report of the Ad-Hoc Selective Fishery Evaluation Committee. Pacific Salmon Commission. Vancouver, British Columbia.

PSMFC (Pacific States Marine Fisheries Commission). 1992. Mass marking anadromous salmonids: Techniques, options, and compatibility with the coded wire tag system. Pacific States Marine Fisheries Commission, Portland, Oregon.

28 A Simulation Model to Assess Management and Allocation Alternatives in Multi-Stock Pacific Salmon Fisheries

Norma Jean Sands and Jeffrey L. Hartman

Abstract.—A fisheries simulation model was developed to evaluate different management regimes on multiple salmon stocks harvested by multiple fisheries. The model also assesses the economic effects on individual and collective fisheries. The model is tested using the sockeye salmon *Oncorhynchus nerka* and pink salmon *O. gorbuscha* fisheries of northern British Columbia and southern Southeast Alaska. Four stock groups (U.S. and Canadian pink and sockeye stocks) and 12 fisheries (five Alaskan and seven Canadian intercepting and terminal commercial fisheries) are included. Fishing effort, in terms of harvest rate for these simulations, is the exogenous variable and stock size and net economic benefit over time are among the output variables. Criteria were developed to reflect different management schemes; given the criteria, the model simulates the effort needed in each fishery to implement the management policy. Three management schemes were assessed for this study: maximizing sustainable yield of the stocks, balancing interceptions by the two countries, and maintaining a fixed harvest rate per fishery. The simulation model suggests that stock production and economic benefits to the fisheries may be reduced significantly when the two countries allocate according to a system that equalizes fishery interceptions rather than maximizes the size of the aggregate harvest.

INTRODUCTION

Successful management of Pacific salmon *Oncorhynchus* spp. stocks involves considerations of both sustainable yield of the stocks over time and economic and social consequences to the users. If these latter aspects are not considered in making allocations, the results may be detrimental to both the users and the stocks. Development of the fishery model presented here was inspired by a desire to better understand some of the conservation and economic consequences of salmon allocation proposals having been discussed within Pacific Salmon Treaty talks. The model is applied to the sockeye *O. nerka* and pink *O. gorbuscha* salmon fisheries of the North Pacific area near the border between southern Southeast Alaska and northern British Columbia.

Management of Pacific salmon stocks is complicated by the migratory patterns of salmon stocks along the entire U.S. and Canadian west coast, by the often large annual fluctuations in salmon abundance, and by the desire of both the U.S. and Canada to maintain traditional fisheries and harvest opportunities. In 1985 the U.S. and Canada signed the Pacific Salmon Treaty to jointly manage the salmon stocks. That treaty, which formally recognized that conservation, rational management, and optimum production must be taken into account in joint management, states that harvest sharing should "provide for each Party to receive benefits equivalent to the production of

1-56670-480-4/00/$0.00+$.50
© 2000 by CRC Press LLC

salmon originating in its waters" and that this is to be done while taking into account the desirability to reduce interceptions, the desirability of avoiding undue disruptions to fisheries, and the annual variations that occur in salmon abundance. This has proved to be difficult. Various proposals have been made to implement this principle including balancing interceptions using numbers of fish landed or their monetary worth, allocation based on negotiated percents of the return or total allowable harvest, and allocations based on negotiated or sustainable harvest rates. Interceptions, as used by the treaty, are catches by one country of stocks that spawn in the other country's waters. Harvest rate, as used in this study, is the proportion of fish available to a fishery that is taken by that fishery, which is a function of number of boats, number of days open, and gear efficiency.

Interest in fishery analyses and models that incorporated socioeconomic factors expanded in the 1970s. In the U.S., the need for appropriate modeling techniques was formally addressed in the Magnuson Fisheries Conservation and Management Act of 1976, which stated that fisheries were to be managed according to optimum yield based on maximum sustainable yield (MSY), as modified by relevant social, economic, or ecological considerations. Textbooks have since stressed the need to consider socioeconomic factors in determining fishery harvest policies (Clark 1985, Hilborn and Walters 1992).

Many and varied fishery simulation models have been developed to incorporate ecological and socioeconomic factors into the analysis and assessment of fishery management regimes. McGlade (1989) referred to these as integrated fisheries management models. One purpose of such simulation models is to determine the degree to which conflicting goals jeopardize management success. Both Pikitch (1987) and McGlade (1989) developed simulation models that included stock recruitment; fishing mortality based on gear characteristics; market parameters of price, cost, and demand; and factors describing the influence of fishermen's behavior and sociology on system dynamics. Both models were used to assess implications of management alternatives; Pikitch's model was developed for the multiple-species flatfish fishery off Oregon, and McGlade's model was developed for a single-stock fishery—the haddock fishery off Nova Scotia. The Alaska Department of Fish and Game (ADF&G) contracted the Institute of Social and Economic Research with the University of Alaska to develop a model for determining economic effects of proposed management changes for sockeye salmon on the Kenai River (ISER 1996). This model incorporated both commercial and sport fisheries and evaluated weekly management options to determine how openings of varying duration would allocate net social benefits among the users.

Another type of management assessment model is based on decision analysis. This type of assessment attempts to score or rank various alternatives as to their effect on various biological and socioeconomic aspects of the fishery. Both Healey (1984) and Sprout and Kadowaki (1987) described the decision-making process that was used in developing the management measures for the commercial, sport, and native Indian sockeye fisheries in the Skeena River. Not only were the welfare of the stocks and the fishing groups considered, but also that of other regional groups tied to fishing, such as processors and lodge operators. Ault and Fox (1989) developed a simulated decision model that addressed multiple objectives in the management of a large-scale, multi-user tropical marine fishery, incorporating stock recruitment, growth function, maturity fraction, fecundity, value of catch, costs, and decision factors on when a user would leave the fishery. The model was determined to be useful in creating integrated management structures to test various policy strategies.

A third type of analysis examines how management alternatives are developed in comanagement situations, specifically in managing shared resources under stochastic conditions, such as is the case for Pacific salmon. This is based on game theory (Munro 1990, Miller 1996, McKelvey 1997). Game theory models attempt to gain insights into how decisions are made and opportunities are gained or lost, why stalemates and fish wars develop, and how cooperative agreements, in theory, can be crafted.

The model presented here is an integrated fisheries management model that simulates the movement of multiple salmon stocks through multiple fisheries of two comanagement entities, the

U.S. and Canada, and assesses the impacts of specific management proposals. Stock production is modeled based on density dependence, which is common for salmon stocks (Collie and Walters 1987; Eggers and Rogers 1987; Walters and Staley 1987). As with all the models described above, this simulation model is meant to be a tool that can be used by managers and policy makers to help understand consequences of allocation actions. Simulation models can further be used to develop and evaluate new allocation proposals that may rectify any adverse consequences of earlier tested proposals.

Our model was developed to assess specific management proposals in terms of their effects on salmon stock production and economic viability of the fisheries of both countries. While three management scenarios are assessed and presented here, it is important to note that the model does not provide "the solution" for the fisheries and stocks modeled, as there is no single solution that will satisfy all the needs and demands on the resource. Instead, the model provides a tool for managers and policy makers to assess the consequences of, and to better evaluate, proposed solutions. Simulation model runs presented in this work show that some proposed allocation schemes might have undesirable consequences for the long-term production of stocks and reduced economic benefits from the fisheries.

METHODS

MODEL STRUCTURE

The model was designed, and written in Microsoft Excel, to simulate multiple fisheries exploiting multiple salmon stocks over a multi-year period. The model simulates future year salmon abundance based on parent escapement levels; determines the catch per fishery based on fishing effort (either as boat-days or harvest rate); and determines the net economic benefits to each fishery based on landed fish prices and average operating costs. Each salmon stock is defined by its own production function. Each stock is assumed to migrate through several spatial levels of fisheries: first through the outer fisheries of the country where it does not spawn; then the inner fisheries of that country; then the outer fisheries of the country where it does spawn; then the inner and terminal fisheries of its spawning country. Terminal fisheries in each country, by definition, do not intercept stocks that spawn in the other country. The model is designed with an annual time step; however, within the year, catches in outer fisheries do influence the number of fish available to the inner fisheries in the model calculations.

Production of salmon is commonly considered to be density-dependent on the number of spawners (Burgner 1991; Heard 1991; Hilborn 1976, 1985). We used the Ricker function (Ricker 1975), which assumes density dependence, to generate stock production based on parent escapement:

$$R_{j+\Delta} = \alpha S_y e^{-\beta s_y} \tag{28.1}$$

where
$R_{y+\Delta}$ = returning adult run size in year $y + \Delta$,
Δ = life cycle in years,
α = fecundity parameter,
S_y = number of spawners in year y, and
β = density dependent parameter.

A separate function is defined for each stock in the model. Since not all stocks migrate through all fisheries in equal proportions, the model includes a migration parameter for each stock in each fishery which indicates the proportion of the stock leaving the prior spatial level and entering the given fishery. Fishery spatial levels, or strata, include outer fisheries, which are encountered first, inner fisheries, and terminal fisheries. The distribution of a stock through the fisheries and the abundance of the stock minus previous catches (catches in previous spatial levels) were used to determine the availability of the stock in each fishery. Availability, A, was expressed as

$$A_{ij} = \mu_i \cdot m_{ij} \cdot \left[R_i - (\textit{any catch in previous strata}) \right] \tag{28.2}$$

if stock i and fishery j are from different countries and as

$$A_{ij} = m_{ij} \cdot \left[R_i - (\textit{any catch in previous strata including other country}) \right] \tag{28.3}$$

if stock i and fishery j are from the same country,

where
μ_i = proportion of stock i that migrates into the other country's fishing waters, and
m_{ij} = migration parameter indicating the proportion of stock i approaching the spatial level of
      fishery j that goes to fishery j.

Since the model operates on a yearly time scale, availability was determined first for the outer, then for the inner, fisheries of the country the stock does not spawn in, and then, in order, the outer, inner, and terminal fisheries of the country it does spawn in. The portion of a stock that migrates into the waters of the non-natal country mixes freely with the stocks of that country and is harvested by each fishery at the same harvest rate as the commingled local stocks. There is no ability within a given fishery to target on a specific stock within the mix of stocks available.

Fishing effort may be expressed in the model as the number of boats and days fished or as the harvest rate per fishery. If boats and days are used as the input, the model calculates harvest rate (H) as

$$H_j = b_j \cdot d_j \cdot q_j \tag{28.4}$$

where
b_j = number of boats in fishery j,
d_j = number of days fishery is open in fishery j, and
q_j = catchability parameter for fishery j.

Fishing effort is the exogenous variable in the model. When fishing effort is provided for each fishery, annual catch per stock per fishery can be determined as

$$C_{ij} = H_j \cdot A_{ij} \tag{28.5}$$

where
C_{ij} = catch per i^{th} stock and j^{th} fishery,
H_j = harvest rate per j^{th} fishery, and
A_{ij} = available abundance of ith stock in waters of jth fishery.

Interceptions are catches by fisheries of one country of the stocks that spawn in the other country's waters. To compare total interceptions by each country, a "sockeye equivalent" parameter is used to convert pink salmon catches to sockeye equivalents for each fishery. The sockeye equivalent parameter (θ) is calculated as

$$\theta = \sum_{k} \rho_k \cdot \frac{w_{pink,\,k} \cdot P_{pink,\,k}}{w_{sockeye,\,k} \cdot P_{sockeye,\,k}} \tag{28.6}$$

where
ρ_k = proportion of pink salmon caught by gear k,
p = price per pound for indicated species caught by gear k, and
w = average weight per fish for indicated species caught by gear k.

The number of spawners per stock is determined as the number of fish remaining from the total run size after all catches are removed:

$$S_i = R_i - \sum_{j} C_{ij}, \tag{28.7}$$

where the returning adult run size (R_i) is the number of adult fish returning to the coastal waters where the fisheries take place. No natural mortality is assumed between when the fish enter the fishing grounds and when they spawn.

In this analysis, the net benefits of the fishery were estimated as economic profits to the harvesters of the salmon and did not include benefits to the consumers or other components of society. The price per landed fish was fixed per species per gear type as if controlled by international market factors, not by local abundance of stocks or catch. Market prices were estimated at the point of landing based on the assumption of competitive buying and selling of salmon in the marketplace (Edwards 1990).

The net benefit for each fishery was determined by subtracting fishing costs from gross earnings at landing (Anderson 1986). Net profit (N) for a fishery was expressed as

$$N_j = p_{sk} w_{sk} C_{ij} (1 - \kappa_k) - v_k \tag{28.8}$$

where
p = price per pound for species s caught by gear k,
w = average weight per fish for species s caught by gear k,
κ = proportion of gross revenue that represents cost of fishing for gear type k, and
v = fixed cost for operation of gear k.

Catch and net benefit were expressed as annual values per fishery and/or summed over all fisheries, over gear types, and all fisheries of each country. Net benefits were cumulated over many annual time steps of the simulation. The dollar values, all expressed as U.S. dollars, were discounted to be expressed in dollars of the first year of the simulation. When the model was run for n years, cumulated net profits were determined as

$$N_{\cdot}^{n} = \sum_{y=1}^{n} N_{\cdot y} (1 - r)^{y} \tag{28.9}$$

where
$N_{\cdot y}$ = the net benefits for year y summed over the pertinent fisheries (e.g., all of one gear type, all from one country), and
r = the discount rate.

MODEL APPLICATION

The northern salmon fisheries located in the border area between Alaska and Canada were used as a first test of our fisheries model. The scope of our prototype simulation model run includes the sockeye and pink salmon stocks from the area. Four salmon stock groups were defined: southern Southeast Alaska pink salmon, southern Southeast Alaska sockeye salmon, northern British Columbia pink salmon, and northern British Columbia sockeye salmon. The commercial fisheries of the area were segregated into 12 fisheries by gear type and amount of interceptions: five in southern Southeast Alaska and seven in northern British Columbia, including gillnet, purse seine net, and troll fisheries. A minimum of 12 fishery groups was needed to realistically simulate the catch and interception levels experienced in the area in recent years (1990–1993).

The model simulates the migratory movement of the four salmon stock groups through a gauntlet of fisheries in the waters of both the U.S. and Canada. Each country also has limited non-interception fisheries; these are terminal area fisheries that have limited access to the total return of the country's spawning stocks. The unit of time is one year; the model does not consider seasonal shifts in species and stock compositions.

Harvest rates per fishery are the exogenous variables in these model simulations. Stock size, catch, interceptions, and net revenues are among the output variables. The model was designed so that allocation or harvest conditions can be set (e.g., specific harvest sharing or production requirements) and the model will perform an iterative search for a solution of harvest rates to meet those conditions.

PARAMETER ESTIMATION

Ricker parameters for the Alaska pink and sockeye and the Canadian pink salmon production equations were estimated from 19 years of catch and escapement data (1978–1996) and from 27 years of data (1970–1996) for Canadian sockeye production (NBTC 1997). Pink salmon have a 2-year life cycle. Sockeye salmon stocks in this northern area return to spawn mostly (around 60 to 70%) as age 5 (Foerster 1968; McDonald and Hume 1984; McGregor et al. 1984), with the remainder returning as age 4 or 6. For this exercise, we assumed the sockeye all returned as 5-year-olds. Escapement data for the U.S. stocks are provided as index counts (NBTC 1997); the pink index counts were multiplied by 2.5 and the sockeye counts were multiplied by 3.9 to estimate total escapement (Ben Van Alen, ADF&G, personal communication). The Canadian escapements were used as provided (NBTC 1997).

The fisheries in the northern boundary area were grouped into 12 commercial fisheries: Alaska district 4 seine; Alaska district 6 and 8 gillnet; Alaska district 1, 2, and 3 seine; Alaska district 1 gillnet; Alaska terminal area seine; Canadian area 1 and 3 net; Canadian area 1 and 3 troll; Canadian area 2 west net; Canadian area 2 west troll; Canadian area 4 and 5 net; Canadian area 4 and 5 troll; and Canadian terminal area net. Most of the Canadian net catch is from seine nets, so, although the total catch represents all net catch, the fishery was given the characteristics of a seine fishery in the model (e.g., prices for catch were based on seine landings).

The proportion of a stock that migrates through the waters of the non-natal country was determined by assuming (a) that the harvest rate is the same on natal and non-natal stocks in each fishery and (b) that the entire run of natal stocks minus interceptions by the non-natal country is available to the natal-country outer fisheries. The harvest rate for each of the outer fisheries was determined by dividing the fishery's catch of natal stocks by the stocks' total run size minus any interceptions by the other country. Applying those harvest rates to the catch of the non-natal stocks (taken from interception estimates, JIC 1993) for each outer fishery, the number of salmon present from the non-natal stocks in each outer fishery could be determined.

Migration parameters describing the distribution of the four stock groups through the 12 fisheries were determined using iterative runs of the fishery component of the model, until catch, interception, and escapement levels for each stock were similar to those observed in 1990–1993, given that the harvest rate is the same on all stocks in a given fishery and that the stocks move in a linear fashion from the outer fisheries to the inner fisheries to the terminal fisheries. Stocks that are subject to interception pass through the other country's fisheries first and then through the spawning country's fisheries. Interception data, expressed as numbers of fish, were taken from the Pacific Salmon Commission report on salmon interceptions (JIC 1993).

Prices used in the model simulations for sockeye and pink salmon in Alaska net fisheries are based upon Commercial Fishery Entry Commission (CFEC) data for pricing in two southern statistical areas of Southeast Alaska. An average price was generated for the years 1990 to 1996 and mean price computed after adjusting for inflation. Prices for landed sockeye salmon in the net fisheries of Southeast Alaska were compared with data in Knapp (1994) and found to have very small variance between British Columbia and Southeast Alaska. For Canadian troll fisheries a market price was generated from personal communications with Canadian fishery management staff. A midpoint price was computed from data for the 1995 troll season based upon low and high reported prices for fresh and frozen troll-caught sockeye and pink salmon. This midpoint price was compared with Southeast Alaskan troll prices from the CFEC data set and found to be approximately equal for both sockeye and pink salmon, after adjustment for exchange rates and weight conversions from kilograms to pounds. Because there were relatively small differences in salmon prices between these regions of Alaska and British Columbia, final prices for the simulations were based upon the CFEC data. Terminally caught pink salmon were set at half the price of ocean caught fish; for sockeye salmon the price was the same for terminal and ocean-caught fish.

The cost of fishing was taken to be a fixed percentage of the gross income and the fixed cost parameter was set to zero; these cost percentages were taken from a Canadian study on the British Columbia salmon fleet (PPEB 1992). The rate used for discounting and estimating a net present value over 30 years was 3.5%.

Once the first 2 years of escapement for pink salmon and the first 5 years for sockeye salmon are given, the future runs of salmon are self-generating based on escapement levels and the associated Ricker function. Escapements from 1992–1996 were used to start the simulation runs.

FISHING REGIME SCENARIOS

To evaluate a fishing regime, the regime was quantified (e.g., for a fixed escapement policy, catch must be limited such that the resulting escapement equals the escapement goal), a testable criteria was identified (e.g., the difference between actual escapement and the escapement goal was minimized), and the model was run to solve for the harvest rates in each fishery that will satisfy the given criteria. The EXCEL solver function was used using the tangent estimates, forward derivatives, and Newton search. The harvest rates were constrained to be between 0 and 1.

The three fishing regime scenarios and the criteria used to evaluate their impacts are identified below.

1. Interception balancing in the numbers of fish (i.e., forcing the difference in interceptions by the two countries to be zero). The test criterion is

$$\left(\dot{C}_{A,CS} + \theta \cdot \dot{C}_{A,CP}\right) - \left(\dot{C}_{C,AS} + \theta \cdot \dot{C}_{C,AP}\right) = 0 \qquad (28.10)$$

where
$\dot{C}$ = catch summed over party and stock group,
A = Alaskan fisheries,
C = Canadian fisheries,
AS = Alaskan sockeye stock group,

TABLE 28.1
Estimates of the Ricker parameters (Δ, α, β) for the four salmon stock groups and the proportion (μ) of each stock group that migrates through, and is susceptible to, fisheries in the other country before entering waters of its natal country. Also given are each stock group's calculated maximum sustainable yield (MSY) production (total return), escapement goal, and total allowable catch (TAC).

| Stock group | Stock parameters | | | | Millions of salmon at MSY levels | | |
	Δ	α	β	μ	Total return	Escapement goal	TAC
Alaska pink	2	8.99	−6.90 E–8	20%	46.335	11.0042	35.293
Alaska sockeye	5	6.31	−2.43 E–6	6%	0.899	0.283	0.616
Canadian pink	2	7.30	−3.10 E–7	42%	8.250	2.318	5.932
Canadian sockeye	5	6.47	−6.29 E–7	55%	3.565	1.100	2.465

AP = Alaskan pink stock group,
CS = Canadian sockeye stock group,
CP = Canadian pink stock group, and
θ = sockeye equivalent.

2. Maximizing total salmon production (i.e., minimizing the absolute difference between the actual escapements and the MSY escapement goals summed over all stock groups). Minimization was used rather than solving for zero, because with mixed-stock fisheries, it is generally impossible to attain MSY escapement goals for all stocks simultaneously. The test criterion was to minimize

$$ABS\left(S_{AS} - G_{AS}\right) + ABS\left(S_{AP} - G_{AP}\right) + ABS\left(S_{CS} - G_{CS}\right) + ABS\left(S_{CP} - G_{CP}\right) \qquad (28.11)$$

where
ABS = absolute function,
S = spawning escapement, and
G = spawning escapement goal.

3. Using fixed harvest rates per fishery equaling those used in the Northern Boundary area fisheries in the period 1990–1993. A test criterion was not needed; the harvest rates per fishery were held fixed over time.

RESULTS

As evidenced by the estimated stock parameters (Table 28.1), the four stock groups were found to have distinct characteristics in terms of production potential and migration into non-natal waters. Approximately 50% of the Canadian stocks included in this study migrate into Alaska waters, while few Alaskan sockeye are found in Canadian waters and about 20% of the Alaskan pink salmon are estimated to cross the border into Canadian waters (Table 28.1). The distribution of fish through the fisheries of each country varies from stock to stock (Table 28.2). To approximate the catch and interception data from the Alaskan fisheries, a percentage of the salmon passing through both the second and third strata was estimated to bypass the fisheries (given in Table 28.2 as "escape fisheries"). The proportions within each migration strata for each stock group sum to one (Table 28.2).

TABLE 28.2
Estimates of model parameters (m_{ij}) for stock distribution through the Alaskan (AK) district (D) and Canadian (CA) area (A) fisheries based on catch, interceptions, and escapements during 1990–1993.

Fishery	Migration strata	Stock group distribution within each migration stratum			
		AK pink	AK sockeye	CA pink	CA sockeye
AK D 4 seine	1	0.90	0.50	0.98	0.90
AK D 6,8 gillnet	1	0.10	0.50	0.02	0.10
AK D 1,2,3 seine	2	0.78	0.72	0.40	0.18
AK D 1 gillnet	2	0.08	0.12	0.10	0.11
Escape fisheries	2	0.14	0.16	0.50	0.71
AK inside seine	3	0.40	0.80	0.00	0.00
Escape fisheries	3	0.60	0.20	1.00	1.00
CA A 1,3 seine	1	0.60	0.80	0.30	0.30
CA A 1,3 troll	1	0.40	0.20	0.31	0.05
CA A 2W seine	1	0.00	0.00	0.20	0.05
CA A 2W troll	1	0.00	0.00	0.19	0.60
CA A 4,5 seine	2	0.01	0.10	0.04	0.66
CA A 4,5 troll	2	0.01	0.01	0.06	0.01
CA inside seine	2	0.00	0.00	0.88	0.31
Escape fisheries	2	0.98	0.89	0.02	0.02

TABLE 28.3
Economic parameter estimates used in the model. Parameters include price per pound per species, average weight per fish per species, and cost expressed as a proportion of gross benefits. The cost parameter is taken from PPEB 1992.

Gear type	Price/lb. (US$) ($p$)		Average lb./fish (w)		Cost parameter (κ)
	Sockeye	Pink	Sockeye	Pink	
Purse seine	$1.25	$0.18	5.8	3.0	0.630
Gillnet	$1.25	$0.19	6.4	3.7	0.711
Troll	$1.34	$0.24	5.2	3.0	0.604

Estimates of price per pound varied both with species and gear type (Table 28.3). Since the model uses numbers of fish, the average pound per fish was also determined (Table 28.3). The cost of fishing was represented as a proportion of the gross benefits of the fleet and is taken directly from the literature (PPEB 1992, Table 28.3). The sockeye equivalent factor for pink salmon was calculated as 0.076 based on the price and weight parameters in Table 28.3.

Thirty years of simulation runs were found to be sufficient for stock production, catch, and economic benefits to stabilize for each of the three management regimes. The steady state was reached most quickly under the MSY management scenario, after 2 years for pink salmon (with a 2-year cycle) and after 5 years for sockeye stocks (with a 5-year cycle). The interception-balancing regime took the longest to reach a steady state; Alaska pink salmon reached steady production only after 20 years.

TABLE 28.4

Annual catches, escapements, and total run sizes for each of the four stock groups under three fishing regimes after 30 years.

Fishing regime	Salmon stock group			
	Alaska pink	Canadian pink	Alaska sockeye	Canadian sockeye
Interception balancing				
catch	13,403,387	5,931,410	614,356	2,460,667
escapement	25,559,947	2,318,321	299,865	1,040,951
total run	38,963,334	8,249,731	914,221	3,501,618
MSY production				
catch	35,293,131	5,931,409	612,845	2,463,156
escapement	11,041,406	2,318,103	258,497	1,060,283
total run	46,334,537	8,249,512	871,342	3,523,439
Fixed harvest rate				
catch	26,328,771	5,815,787	539,446	2,298,343
escapement	19,403,934	1,928,252	417,542	1,224,657
total run	45,732,705	7,744,039	956,988	3,523,000

Under mixed stock fisheries, not all individual stocks can be expected to reach individual MSY production levels (TAC in Table 28.1); in these simulation runs, the MSY production was reached for the pink salmon stocks, but was not achieved for the sockeye salmon stocks (Table 28.4).

The annual catch, escapement, and total run by the end of the 30-year simulation has the most variation between management regimes for the Alaska pink salmon and least for Canadian pink and sockeye stocks (Table 28.4). The MSY production regime gives the highest total catch of salmon in the region, with Alaska pink and Canadian pink and sockeye being caught nearly at MSY and Alaska sockeye salmon catch being slightly under MSY (Tables 28.1 and 28.4). The fixed harvest-rate regime gives the next highest total catch for the region with Canadian stocks being caught near MSY levels and the Alaskan stocks being caught at lower levels. The interception-balancing regime gives the lowest total catch; this is due mostly to the low catches of Alaska pink salmon, as the other three stocks are caught at levels approaching MSY.

Because the fixed harvest-rate regime approximates the actual 1990–1993 fishing management, that regime was used as the basis of comparison for the other two fishery management regimes. All three simulations started with the fixed-rate harvest rates (i.e., those approximating 1990–1993 harvest rates); the fixed-rate management scenario used the same harvest rates throughout the 30-year simulation run (Table 28.5). The other two scenarios started with the fixed-rate harvest rates and the harvest rate per fishery were changed each year to satisfy the management scenario criteria. The harvest rates at the end of the 30-year simulations represent the steady state solution (Table 28.5). Under the MSY management regime, the harvest rates for three of the five Alaska fisheries increased from year one to year 30, one decreased slightly, and the terminal fishery (AK inside seine) harvest rate remained at the same low rate of 3% (Table 28.5). For the Canadian fisheries under MSY management, all but the terminal fishery harvest rate remained nearly constant; the terminal fishery (CA inside seine) harvest rate experienced a drop from 86% in the first year to 65% in year 30 (Table 28.5). Under the interception-balancing regime, all the mixed stock fisheries were closed (e.g., a harvest rate of zero) and the non-intercepting fisheries experienced increased harvest rates over the fixed harvest-rate management scenario (Table 28.5).

The annual year 30 catch by each country under the three fishing regimes is given in Table 28.6. It is compared with the idealized, but unattainable, situation of MSY catch for each stock and no interceptions (i.e., the individual MSY total allowable catch for each stock, as if each country were catching all the surplus production of its spawning stock). The idealized situation illustrates the

TABLE 28.5
Harvest rates for each of the 12 fisheries in year 30 under three fishing regimes. Harvest rate is the percentage of fish available to the fishery that is taken by the fishery. A "—" indicates the fishery would be closed.

| | Fishing regime | | |
Fishery	Interception balancing	MSY production	Fixed harvest rate
AK D 4 seine	—	55%	36%
AK D 6,8 gillnet	—	35%	39%
AK D 1,2,3 seine	—	54%	32%
AK D 1 gillnet	—	40%	35%
AK inside seine	100%[a]	3%	3%
CA A 1,3 seine	—	27%	26%
CA A 1,3 troll	—	42%	43%
CA A 2W seine	60%	33%	35%
CA A 2W troll	98%	43%	43%
CA A 4,5 seine	—	75%	73%
CA A 4,5 troll	—	50%	50%
CA inside seine	77%	65%	86%

[a] Although this fishery takes 100% of the fish available to it, only a small portion of the total run of stocks is available to this terminal fishery and, under the criteria of this fishing management scenario, escapement goals are met.

upper limit of sustainable catch for each species. For pink salmon, the maximum catch is realized under the MSY management regime; however, in this scenario Canada catches more of the total pink salmon production than under the idealized situation of no interceptions (Table 28.6). For sockeye salmon, the idealized maximum catch of 3.08 million fish is approached, but not fully attained, by both the interception-balancing and MSY-management regimes (Table 28.6); however, under the MSY scenario, the U.S. is catching a greater share of the total surplus production than under the idealized situation.

Cumulative net economic benefits accrued by each country are not very different for the MSY production and the fixed harvest-rate regimes, about $95/96 million for Canadian and $209 million for Alaskan fisheries (Table 28.7). The total cumulative benefit for both countries' fisheries for either of these two scenarios is $305 million, compared to $203 million under the interception-balancing regime. Interception-balancing results in greatly reduced revenue for the Alaskan fisheries and increased revenue for Canadian fisheries over the other two regimes (Table 28.7). The net cumulative earnings in the last 5 years is less than half of that during the first 5 years (Table 28.7) due to the decline in the value of real dollars over time.

DISCUSSION

The simulation run proved to be a useful tool for assessing three specific fishing policies for the sockeye and pink salmon fisheries of the border area between Alaska and Canada. These salmon stocks have been healthy for many years and fishing regimes are based on allocation rather than conservation of depressed stocks. Parameter estimations were based on data from years of healthy stock production (1990–1996). The fixed harvest-rate regime used in the model approximated the fishing regime already in place during these years. The purpose of the model exercise was to compare other fishing policies with the early 1990s fishing regime. This model could also be used

TABLE 28.6
Annual total catch in millions of fish by both countries under the three fishing regimes in the 30th year of the simulation and, for comparison purposes, under the idealized and unattainable situation of MSY catch for each stock with no interceptions.

Catch by	Fishing regimes			Idealized situation (from Table 28.1)
	Interception balancing	MSY production	Fixed harvest rate	
		Pink salmon		
Alaska	13.403	34.499	24.803	35.293
Canada	5.931	6.725	7.341	5.932
Total	19.334	41.224	32.144	41.224
		Sockeye salmon		
Alaska	0.614	1.745	1.376	0.616
Canada	2.461	1.312	1.609	2.465
Total	3.075	3.057	2.985	3.081

TABLE 28.7
Cumulative net benefits to the fisheries of each country under three scenarios of allocation at 5-year intervals. Net benefits given in millions of U.S. dollars.

Years	Interception balancing		MSY production		Fixed harvest rates	
	Alaska	Canada	Alaska	Canada	Alaska	Canada
5	$13.4	$ 37.6	$ 48.6	$24.7	$ 49.4	$23.7
10	$24.8	$ 68.3	$ 93.1	$44.6	$ 93.5	$43.6
15	$34.5	$ 93.8	$130.3	$61.2	$130.6	$60.1
20	$42.6	$115.0	$161.4	$75.0	$161.7	$74.0
25	$49.4	$132.8	$187.4	$86.6	$187.7	$85.5
30	$55.1	$147.6	$209.2	$96.3	$209.5	$95.2
Combined						
30	$202.7		$305.5		$304.7	

to evaluate different management regimes for rebuilding depressed stocks, but that was not done with the current simulations.

The interception-balancing fishing regime provided the lowest total net benefits (Table 28.7), due largely to the reduced catch of both pink and sockeye salmon in the Alaska fisheries (Table 28.6). The model solution found for balancing interceptions was to eliminate all interception fisheries and continue fishing only in the non-intercepting fisheries (Table 28.5). Harvest rates in the non-intercepting fisheries were allowed to increase as much as possible without letting escapements go below goal levels defined by the production function for maximum yield. Canada had three non-intercepting fisheries (Table 28.5) and could catch most of their spawning stocks within these areas. Alaska had only one non-intercepting fishery in this simulation and that terminal fishery did not have the ability to access all stocks spawning in Alaska. Southeast Alaska pink salmon spawn in thousands of small streams along the entire coast area and are not easily accessible by terminal fisheries, resulting in escapements above MSY levels. Overescapement resulted in forgoing the revenue from fish that are surplus to escapement needs.

The MSY fishing regime gave the highest combined net benefits (Table 28.7), although the benefits were not much better than under the fixed harvest-rate regime. The MSY production fishing regime allowed for catching pink salmon close to MSY (41 million fish, Table 28.6), while sockeye salmon were underharvested (3.057 million fish compared to 3.081) compared to individual MSY levels. At this point, model users might consider testing the influence on net benefits by harvesting the sockeye at MSY levels and letting the pink salmon be harvested secondarily to achieving sockeye MSY harvest. Part of the power of the simulation model as a management tool is to discover relationships and test new proposals. The simulation model will then become a learning tool for the user (Hilborn and Walters 1987; Getz et al. 1987).

Parameter estimation was conducted to provide a fairly realistic stock and fishery situation; it was not done to predict stock productivity or actual net benefits to the fisheries. The goal of the modeling exercise was to compare relative benefits among the fishing policies tested. Some preliminary sensitivity studies on stock parameters showed no difference in the relative results with minor changes in productivity parameters. The range of parameter values tested maintained the relative stock sizes and productivity. Some preliminary tests of sensitivity to price data show that changes in relative price of the two species can change the balance of net benefits. Under the model runs presented here, Alaska derives most of its net benefits from pink salmon, while Canada derives most of its net benefits from sockeye salmon. If pink salmon prices were to double and sockeye prices were to fall by 25%, the Alaskan fisheries would realize an increase in net benefits, while Canadian fisheries would remain about the same or lose, depending on the scenario. It is highly unlikely that such price changes would hold for a 30-year run. The prices we used in our 30-year simulation represent average prices over time.

Our model design has not included annual fluctuations in recruitment as we were looking for steady-state solutions. For some uses of the model as a management tool, it would be worth incorporating random variability in stock recruitment to see how much annual variability there would be in catches and net benefits. It is unlikely that random variability would affect the outcome of modeling the three fishing regimes in terms of relative catch and benefits between the two countries, but it would certainly add variability to the annual catches and net benefits. One goal of a fishing regime policy might be to minimize annual fluctuations in net benefits while maintaining a sustainable yield; in this case annual variability in returns would be important to add to the model structure.

For this first phase of model development, the time step of a year allowed us to examine management regimes only at the annual level of regulation. For more finely tuned management regimes, which would include testing time-wise closures, a finer time step would be needed. However, this would involve more assumptions on migration patterns of the stocks with time through the fisheries. Some interception data are available on a finer time-area scale (JIC 1993). If random variability in stock production were added, it could be done on a stock or cohort basis, or annually, to reflect marine survival. In the first case, an entire cohort, even if returning in multiple years, could be affected by freshwater survival or predation upon entering salt water. In the latter case, marine survival would affect all cohorts present in a year and a more realistic age structure for sockeye salmon than used in the present simulations would be recommended (i.e., returning adults could be 4- to 6-year-olds, not just 5-year olds as presently modeled). Recruitment variability could be added either at the cohort level (affecting all age groups of one cohort as when predation affects an emerging cohort) or at the annual level (affecting all cohorts present in a given year as in an oceanic weather effect). Different stock-recruitment functions could also be tested for the different stocks. This would not affect the relative results of catch and net benefits of our tested fishing regimes, but could change the amount of uncaught surplus production in regimes that restrict catching the full total allowable catch of a stock in order to protect another stock.

CONCLUSIONS

The following conclusions are based on results of the simulation runs for the northern U.S. and Canadian Pacific sockeye and pink salmon fishery management schemes.

1. The fixed harvest-rate regime retained all the identified fisheries with allocation (harvest-rate distribution) similar to that during the 1990–1993 period. The fixed harvest-rate regime was effective in maintaining relatively high stock productivity and net economic benefits.

2. Model simulations suggest that the only way to obtain equal interceptions of salmon by the two parties would be to close all mixed-stock fisheries and concentrate the fleet in non-interception fisheries. This is due to the highly migratory nature of the stocks, the different relative proportion of each stock in each fishery, and the inability of a mixed-stock fishery to successfully target on one stock and avoid catching other stocks. Alaska's only non-interception fishery in this simulation is a terminal area fishery and does not fully access all stocks spawning in domestic rivers in the study area. This lack of access results in reduced catches and, therefore, reduced net economic benefits. Since Canada has an outer non-interception fishery (Area 2), its fisheries could better access the stocks spawning in its domestic rivers.

3. Eliminating interceptions in Alaska fisheries led to decreased catches relative to the fixed harvest rate regime, escapement levels greater than MSY goals and, therefore, decreased economic benefits and stock productivity (assuming Ricker density dependent productivity). Catch reductions were relatively larger than productivity reductions. Catch restrictions in the mixed-stock fisheries resulted in not allowing full utilization of the MSY surplus production, particularly on pink salmon. Allocation schemes that allowed for interceptions by both parties allowed greater stock productivity in the model runs.

4. Managing for MSY increased only slightly the overall stock production and the combined net benefits from all fisheries over the fixed harvest-rate regime. While Canada's net benefits were similar under MSY and fixed harvest-rate management regimes, the benefits were less for both these scenarios than those realized under the interception-balancing regime (due to greater numbers of the less valuable pink and fewer of the more valuable sockeye salmon in their catch).

5. Regarding the economic paramteters, the fixed harvest-rate regime provided the highest net benefits to Alaska, the interception-balancing regime provided the highest net benefits to Canada, and the MSY management regime provided the highest total net benefits to the two countries combined.

Although the model is restrictive in scope of stocks and fisheries from a coastwide perspective, it demonstrates the nature of the tradeoffs faced in mixed-stock, multi-fishery systems. With increased complexity of a model, through adding more stocks, fisheries, and time-area strata, and including stochasticity, the number of parameters to estimate often increases at a greater rate than the available data. With increased model complexity, the number of possible fishing scenarios that may be assessed increases with the consequence of focusing on individual events rather than overall consequences. We believe the broad type of simulation model presented here provides a useful tool to evaluate the consequences of fishing management strategies and could help developers of fishery allocation plans understand the difficulties of balancing resource conservation, economic, and political concerns and the consequences of proposed policies in terms of these concerns.

ACKNOWLEDGMENTS

We would like to acknowledge the assistance of Elaine Denniford at the Commercial Fisheries Entry Commission, who contributed data reports and data interpretation of price files, and Ben Van Alen for technical and editorial review of the manuscript. Milo Adkison, Stan Carlson, Dana Schmidt, and an anonymous reviewer provided helpful comments on the manuscript. This chapter is contribution PP-153 of the Alaska Department of Fish and Game, Commercial Fisheries Management and Development Division, Juneau.

REFERENCES

Anderson, L. G. 1986. The Economics of Fisheries Management. The Johns Hopkins University Press, Baltimore.

Ault, J. S., and W. W. Fox, Jr. 1989. FINMAN: Simulated decision analysis with multiple objectives. Pages 166–179 *in* E. F. Edwards and B. A. Megrey, editors. The mathematical analysis of fish stock dynamics. American Fisheries Society Symposium 6.

Burgner, R. L. 1991. Life history of sockeye salmon (*Oncorhynchus nerka*). Pages 1–118 *in* C. Croot and L. Margolis, editors. Pacific Salmon Life Histories. UBC Press, Vancouver, British Columbia.

Clark, C. W. 1985. Bioeconomic modelling and fisheries management. John Wiley and Sons, New York.

Collie, J. S., and C. J. Walters. 1987. Alternative recruitment models of Adams River sockeye salmon, *Oncorhynchus nerka*. Canadian Journal of Fisheries and Aquatic Sciences 44:1551–1561.

Edwards, S. F. 1990. An economics guide to allocation of fish stocks between commercial and recreational fisheries. National Oceanic and Atmospheric Administration Technical Report NMFS 94. Woods Hole, Massachusetts.

Eggers, D. M., and D. E. Rogers. 1987. The cycle of runs of sockeye salmon (*Oncorhynchus nerka*) to the Kvichak River, Bristol Bay, Alaska: cyclic dominance or depensatory fishing? Pages 343–366 *in* H. D. Smith, L. Margolis, and C. C. Wood, editors. Sockeye salmon (*Oncorhynchus nerka*) population biology and future management. Canadian Special Publication in Fisheries and Aquatic Sciences 96.

Foerster, R. E. 1968. The sockeye salmon, *Oncorhynchus nerka*. Bulletin of the Fisheries Research Board of Canada 162.

Getz, W. M., R. C. Francis, and G. L. Swartzman. 1987. On managing variable marine fisheries. Canadian Journal of Fisheries and Aquatic Sciences 44:1370–1375.

Healey, M.C. 1984. Multiattribute analysis and the concept of optimum yield. Canadian Journal of Fisheries and Aquatic Sciences 41:1393–1406.

Heard, W. R. 1991. Life history of pink salmon (*Oncorhynchus gorbuscha*). Pages 119–230 *in* C. Croot and L. Margolis, editors. Pacific Salmon Life Histories. UBC Press, Vancouver, British Columbia.

Hilborn, R. 1976. Optimal exploitation of multiple stocks by a common fishery: a new methodology. Journal of the Fisheries Research Board of Canada 33:1–5.

Hilborn, R. 1985. Apparent stock recruitment relationships in mixed stock fisheries. Canadian Journal of Fisheries and Aquatic Sciences 42:718–723.

Hilborn, R., and C. J. Walters. 1987. A general model for simulation of stock and fleet dynamics in spatially heterogeneous fisheries. Canadian Journal of Fisheries and Aquatic Sciences 44:1366–1369.

Hilborn, R., and C. J. Walters. 1992. Quantitative fisheries stock assessment: choice, dynamics and uncertainly. Chapman & Hall, New York.

ISER (Institute of Social and Economic Research). 1996. Economic effects of management changes for Kenai River late-run sockeye. Prepared for: Alaska Department of Fish and Game. University of Alaska, Anchorage.

JIC (Joint Interceptions Committee). 1993. Third report on the parties' estimates of salmon interceptions—1980–1991. Pacific Salmon Commission Report: JIC (93)-1.

Knapp, G. 1994. A comparison of salmon prices in Alaska and Canada. Contract report prepared for: Division of Economic Development, Alaska Department of Commerce and Economic Development, Juneau.

McDonald, J., and J. H. Hume. 1984. Babine Lake sockeye salmon (*Oncorhynchus nerka*) enhancement program: testing some major assumptions. Canadian Journal of Fisheries and Aquatic Sciences 41:70–92.

McGlade, J. M. 1989. Integrated fisheries management models: understanding the limits to marine resource exploitation. Pages 139–165 *in* E. F. Edwards and B. A. Megrey, editors. The mathematical analysis of fish stock dynamics. American Fisheries Society Symposium 6.

McGregor, A. J., S. A. McPherson, and J. E. Clark. 1984. Abundance, age, sex, and size of sockeye salmon (*Oncorhynchus nerka* Walbaum) catches and escapements in southeastern Alaska in 1983. Alaska Department of Fish and Game, Division of Commercial Fisheries, Technical Data Report 132, Juneau.

McKelvey, R. 1997. Game-theoretic insights into the international management of fisheries. Natural Resource Modeling 10:129–171.

Miller, K. A. 1996. Salmon stock variability and the political economy of the Pacific Salmon Treaty. Contemporary Economic Policy 14:112–129.

Munro, G. R. 1990. The optimal management of transboundary fisheries: game theoretic considerations. Natural Resource Modeling 4:403–426.

NBTC (Northern Boundary Technical Committee). 1997. U.S./Canada northern boundary area 1996 salmon fisheries management report and 1997 preliminary expectations. Pacific Salmon Commission Report TCNB(97)-1. Vancouver, British Columbia.

Pikitch, E. K. 1987. Use of a mixed-species yield-per-recruit model to explore the consequences of various management policies for the Oregon flatfish fishery. Canadian Journal of Fisheries and Aquatic Sciences 44(Supplement 2):349–359.

PPEB (The Program Planning and Economic Branch). 1992. The British Columbia salmon fleet 1986–1990. The Department of Fisheries and Oceans Pacific Region Report: July 1992. Vancouver, British Columbia.

Ricker, W. E. 1975. Computation and interpretation of biological statistics of fish populations. Bulletin of the Fisheries Research Board of Canada No. 191. Ottawa, Ontario.

Sprout, P. E., and R. K. Kadowaki. 1987. Managing the Skeena River sockeye salmon (*Oncorhynchus nerka*) fishery—the process and the problems. Pages 385–395 *in* H. D. Smith, L. Margolis, and C.C. Wood, editors. Sockeye salmon (*Oncorhynchus nerka*) population biology and future management. Canadian Special Publication in Fisheries and Aquatic Sciences 96.

Walters, C. J., and M. J. Staley. 1987. Evidence against the existence of cyclic dominance in Fraser River sockeye salmon (*Oncorhynchus nerka*). Pages 375–386 *in* H. D. Smith, L. Margolis, and C. C. Wood, editors. Sockeye salmon (*Oncorhynchus nerka*) population biology and future management. Canadian Special Publication in Fisheries and Aquatic Sciences 96.

29 Defining Equivalent Exploitation Rate Reduction Policies for Endangered Salmon Stocks

James G. Norris

Abstract.—I used the Pacific Salmon Commission Chinook Model to help define equivalent exploitation rate reduction policies for endangered Snake River fall chinook salmon *Oncorhynchus tshawytscha*. This stock is harvested in gauntlet fashion by a number of mixed-stock fisheries from Alaska to California. The overall exploitation rate on Snake River fall chinook can be reduced by various means, each having different economic consequences for the individual fisheries. I consider eight general types of policies. Four reduce harvests in single geographic regions: Alaska, British Columbia, Washington, and Oregon ocean fisheries, and the Columbia River. Two policies reduce harvests in all regions in equal or scaled amounts, and two policies reduce harvests only in U.S. waters by equal or scaled amounts. Scaled policies reduce each fishery's legal harvests in proportion to that fishery's estimated share of the total adult equivalent fishing mortalities of Snake River fall chinook during the period 1979–1993. Policies are deemed equivalent when the overall adult equivalent exploitation rate on Snake River fall chinook is reduced by the same percentage. Under equilibrium conditions equivalent policies are shown to be independent of assumptions about Snake River fall chinook productivity, thus eliminating a major source of uncertainty in recovery planning. The methodology described in this chapter can be incorporated into recovery plans for other depressed fish stocks.

INTRODUCTION

Many Pacific salmon *Oncorhynchus* spp. stocks are depressed and at risk of extinction (Nehlsen et al. 1991). Reducing directed and incidental fishing mortalities to a biologically acceptable level is usually the first management action taken to protect such stocks. Since most salmon stocks are harvested by several fisheries, many different combinations of individual fishery harvest reductions can achieve the same biological goal—an acceptable total exploitation rate on the threatened stock. Allocating harvest reductions among competing fisheries is primarily a political decision, because each combination of reductions has different social and economic consequences for individual fleets. Fishery science can help remove some uncertainties from the political process by objectively defining combinations of individual fishery harvest reductions that achieve the same biological objective. Clearly defining available options for competing stakeholders improves their understanding of how their individual sacrifices contribute to the overall conservation effort and thereby increases the likelihood of reaching agreement on a sustainable fishery management plan. I use the case of Snake River fall chinook salmon *O. tshawytscha* to illustrate how mathematical models can be used to define equivalent harvest rate reduction policies once the target exploitation rate reduction has been determined by other fisheries population analyses.

The U.S. National Marine Fisheries Service (NMFS) listed Snake River spring/summer and fall chinook salmon "threatened" under the Endangered Species Act (ESA) in April, 1992 (57 FR

FIGURE 29.1 Map showing the major components of the URB stock and mainstem dams on the Columbia and Snake rivers.

14653). A proposed rule to reclassify these species as "endangered" was published in December, 1994 (59 FR 66784) but was withdrawn in January, 1998 (63 FR 1807). These stocks spawn upstream from four mainstem dams on the Columbia River and another four dams on the Snake River (Figure 29.1). In addition to the upstream and downstream hazards created by these dams, other threats to recovery include habitat destruction due to logging, grazing, mining, and water diversions, commercial and recreational harvests, freshwater and marine predation, competition with hatchery fish, loss of genetic diversity, natural factors such as droughts and El Niño events, and dysfunctional management institutions (NRC 1996; NMFS 1995; SRSRT 1993).

Snake River fall chinook are part of a larger group of fall chinook runs that spawn above the confluence of the Columbia and Snake Rivers, commonly referred to as the upriver bright (URB) stock. Most production for the URB stock comes from Priest Rapids hatchery and from naturally spawning fish in the Hanford Reach located between Priest Rapids dam and the confluence with the Snake River. These two components of the URB stock are considered healthy (CTC 1994). Natural production in the Snake River is now limited primarily to the region between Lower Granite and Hells Canyon dams. Some Snake River hatchery production comes from Lyon's Ferry Hatchery (LYF) located between Lower Monumental and Little Goose dams (Figure 29.1).

A variety of recovery actions have been proposed for Snake River fall chinook salmon, all of which include some form of exploitation rate reduction (e.g., NRC 1996; NMFS 1995; SRSRT 1993). Exploitation rate reduction options for Snake River fall chinook are complicated because this stock is harvested in mixed-stock ocean troll fisheries from Alaska to California and in mixed-stock sport and net fisheries in the Columbia River. Thus, reducing Snake River fall chinook harvest will likely require harvest reductions for stronger stocks as well. Since harvests occur in gauntlet fashion, any reductions by ocean fisheries make more fish available for harvest by subsequent river fisheries. Alaska and British Columbia residents argue that the Snake River stock declines are due mostly to habitat degradation resulting from land-use policies adopted in Washington, Oregon, and

Idaho, and they therefore resent being forced to reduce catches of healthy stocks to help a stock threatened by factors outside their control.

There is no single salmon management institution to deal with all aspects of these complicated harvesting issues. Ocean fisheries are managed primarily by the Pacific Salmon Commission (PSC) and the Pacific Fishery Management Council (PFMC), while Columbia River fisheries are managed by state and tribal agencies which must follow the guidelines set forth in the court-decreed Columbia River Fish Management Plan (Rutter 1997). Historically, ocean fisheries have been managed under fixed catch policies (also termed catch ceilings), while river fisheries have been managed under fixed escapement or fixed harvest rate policies. These policies have become more complicated in recent years. All salmon harvesting plans within U.S. waters must be approved by NMFS as part of the Section 7 consultation requirements of the ESA.

There are two fundamental questions with regard to harvesting options for Snake River fall chinook salmon: (1) What overall exploitation rate, if any, should be permitted? and (2) How should the allowable harvest be allocated among the different fleets? These questions are germane for a single year or over an extended time horizon. Powers (1996) suggests that recovery plans should take an adaptive management approach in which a recovery trajectory is established and annual exploitation rates are determined each year by rules based on the stock's current status with respect to the trajectory. For stocks threatened by a single fleet's overfishing, as many marine species are, the Powers (1996) plan may be adequate. But for stocks threatened by multiple fleets, the Powers (1996) plan is incomplete because it does not address the allocation issue. The political and economic pressures associated with allocating harvest reductions among several fleets can make it difficult to make any reductions at all.

Recovery plans for species listed under the ESA can be far more complex because threats other than overfishing, such as habitat destruction, must be addressed. Lichatowich et al. (1995) proposed an ecosystem approach to restoring Pacific salmon *Oncorhynchus* spp. stocks. They argue that "Sustainable restoration cannot be achieved through programs that focus entirely on numbers of fish." For these stocks, other recovery measures, such as improving downstream survival of smolts and upstream survival of adults, will determine the amount of exploitation rate reduction necessary. For example, Figure 29.2 (from Norris 1995) illustrates various combinations of three general control variables—overall exploitation rate reduction (expressed as a percentage change from status quo rates: age 3 = .215, age 4 = .422, age 5 = .368; from Schaller and Cooney 1992), upstream survival rate, and downstream survival rate—required to achieve an escapement goal of 3,000 spawners by year 2017 (this objective was one option being considered by NMFS staff during preliminary recovery planning). This type of analysis does not answer specific questions regarding alternative methods of improving downstream survival, reducing exploitation rates, or improving upstream survival. Other models and analysis methods must be used to decide among those alternatives.

This chapter focuses on the second question—the allocation of harvests among different fisheries. I assumed that the desired overall exploitation rate reduction is given, and that management seeks a combination of harvest rates or catches in individual fisheries that results in the desired overall rate reduction. While the allocation of catch among different fisheries is inherently a political decision, the decision process can be improved by narrowing the choices to those that will meet the desired criteria. Current modeling efforts to analyze recovery strategies utilize downstream passage (e.g., CRiSP, Anderson et al. 1996; FLUSH, Wilson 1994) and life-cycle models (e.g., SLCM, Lee and Hyman 1992; ELCM, Schaller and Cooney 1992; and PSC Chinook Model, CTC 1993) to predict the effects of management options that usually include interventions at all life-history stages. This approach often makes it difficult for competing stakeholders to see how their sacrifices contribute to overall recovery.

This chapter presents a method for defining equivalent exploitation rate reduction policies. I use the term "policy" to mean any one of an infinite number of ways of combining harvest reductions among existing fleets. I deem policies equivalent when they result in the same percentage reduction

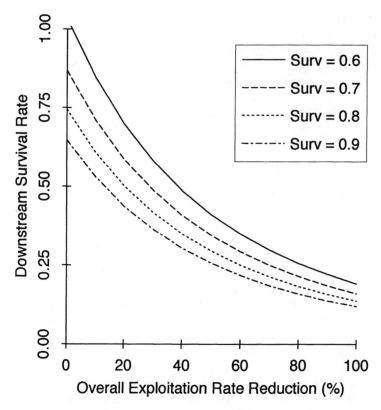

FIGURE 29.2 Combinations of downstream survival rate, exploitation rate reduction, and upstream survival rate (labeled "Surv" in the legend) that gives 3,000 spawners in year 2017.

in the overall exploitation rate over some base period rate. Although I demonstrate the technique by using the PSC Chinook Model to determine equivalent policies for Snake River fall chinook, the same general technique can be used for other stocks, fisheries, and models.

METHODS

PSC Chinook Model

Rebuilding depressed chinook salmon stocks was a major goal of the 1985 Pacific Salmon Treaty between Canada and the U.S. During negotiations leading up to the treaty, several single-stock computer models were developed to analyze how proposed management actions would affect the rebuilding process. The single-stock models evolved into a multiple-stock model that could simulate the effects of the primary PSC management tool—fixed catch ceilings imposed on mixed stock ocean fisheries. The long-term management strategy was to hold ocean catches at low enough levels to underharvest strong stocks, thereby increasing their relative contribution to the ocean fisheries and decreasing the exploitation rate on the weaker stocks.

Over the past decade, more stocks were added to the model, along with other features designed to more accurately simulate incidental mortalities, size limit changes, changes in enhancement activities, and the effects of time/area closures. The model configuration used in this analysis included 30 stocks and 25 fisheries (Table 29.1) and simulated catches and escapements through the year 2017.

The LYF stock was used as a surrogate for the Snake River fall chinook stock. Since 1984, large numbers of releases from the LYF stock have been coded wire-tagged (CWT) and the analysis

TABLE 29.1
Fisheries and stocks included in the Pacific Salmon Commission Chinook Model.

Fisheries	Stocks
* Alaska Troll	Alaska South SE
* Northern B.C. Troll	Northern/Central B.C.
* Central B.C. Troll	Fraser River Early '
* West Coast Vancouver Island Troll	Fraser River Late
* Washington/Oregon Troll	West Coast Vancouver Island Hatchery
* Strait of Georgia Troll	West Coast Vancouver Island Natural
* Alaska Net	Upper Strait of Georgia
* Northern B.C. Net	Lower Strait of Georgia Natural
* Central B.C. Net	Lower Strait of Georgia Hatchery
West Coast Vancouver Island Net	Nooksack River Fall
Juan de Fuca Net	Puget Sound Fingerling
North Puget Sound Net	Puget Sound Natural Fingerling
South Puget Sound Net	Puget Sound Yearling
Washington Coast Net	Nooksack River Spring
Columbia River Net	Skagit River Wild
Johnstone Strait Net	Stillaguamish River Wild
Fraser River Net	Snohomish River Wild
* Alaska Sport	Washington Coastal Hatchery
* North/Central B.C. Sport	Columbia River Upriver Brights
* West Coast Vancouver Island Sport	Spring Creek Hatchery
* Washington Ocean Sport	Lower Bonneville Hatchery
North Puget Sound Sport	Fall Cowlitz River Hatchery
South Puget Sound Sport	Lewis River Wild
* Strait of Georgia Sport	Willamette River
Columbia River Sport	Spring Cowlitz Hatchery
	Columbia River Summers
	Oregon Coastal
	Washington Coastal Wild
	Snake River Wild Fall
	Mid Columbia River Brights

Note: Fisheries marked with an "*" were modeled as catch ceiling fisheries (i.e., catches controlled by fixed catch quotas) for all analyses reported in this chapter. All others were modeled as harvest rate fisheries (i.e., catches controlled by fixed harvest rates).

of subsequent recoveries is considered to provide the best available information on the Snake River wild fall chinook stock. The PSC Chinook Technical Committee uses selected fingerling release groups from the LYF stock as the indicator stock for Snake River fall chinook. I used these same fingerling release groups for all analyses reported in this chapter.

The original PSC Chinook Model was written for the PC platform and the computation speed was relatively slow. During 1994–1995 the Columbia River Salmon Passage (CRiSP) Project at the University of Washington School of Fisheries created a C++ version of the model to run on Sun Workstations (Norris 1995). The underlying computing engine remained the same, but the computation speed was greatly improved. The increased speed made possible the numerous model runs required for this analysis. A detailed description of the model theory can be found in CTC (1993) and Norris (1997). The following outlines the salient features of the model:

The PSC Chinook Model has seven important general assumptions.

1. Age specific natural ocean mortality rates are the same for all stocks in all years.
2. Stocks not represented in the model are constant in size.
3. The only changes in harvest rates are those that result from management actions being modeled.
4. Stock distribution and fishing patterns (e.g., fishery specific time/area closures that might affect harvest rates on stocks with different migration timing through the fishing areas) are identical from year to year.
5. Indicator stocks adequately represent natural stocks.
6. Stock productivities and optimum escapements do not change over the rebuilding period.
7. All age four and older fish captured in net fisheries are mature.

Life cycle computations in the PSC Chinook Model are performed on an annual basis. The sequence of computations reverses the procedures employed in the cohort analysis used to generate the stock-specific input data. The annual computation sequence is similar to that used by Hankin and Healey (1986) and is illustrated in Figure 29.3. Model runs are divided into two time periods: (1) a calibration period and (2) a management simulation period. The calibration period runs from 1979 through the last year for which parameters can be estimated (usually one year behind the current year). The simulation period runs from the current year to any future year (year 2017 for analyses in this chapter).

Natural ocean survival rates for ages 1 through 5 are assumed fixed (at 0.5, 0.6, 0.7, 0.8, and 0.9, respectively) for all stocks. Ocean and terminal fishing mortalities include legal catches and incidental mortalities, such as "shakers" (sublegal-sized fish caught and released during chinook fisheries) and "CNRs" (legal and sublegal-sized fish caught and released during "chinook non-retention" fisheries directed at other salmon species).

Production parameters for both hatchery and natural stocks are estimated from historical data. For each stock, the relationship between spawners and progeny the following year is perhaps the most critical component of the model. It is through this relationship that time dynamics are incorporated into the analysis of alternative stock rebuilding strategies. In general, hatchery production is modeled as a simple linear relationship between spawners and age-1 fish the following year, while natural production is modeled by a truncated Ricker curve. For both types of stocks, progeny predicted by the spawner/recruit relationship (SRR) are adjusted to make allowances for recruitment variability by incorporating "Environmental Variability" (EV) scalars, as follows:

$$R_{s,y+1} = SRR(S_{s,y})\, EV_{s,y}$$

where $S_{s,y}$ is the number of spawners of stock s in year y, $EV_{s,y}$ is the environmental variability associated with stock s in brood year y, and $R_{s,y+1}$ is the recruitment of stock s age one fish in year $y+1$. The $EV_{s,y}$ are assumed independent, with no correlations between stocks within a year or between years for a single stock. Note that the EV scalars account for variability in both freshwater and marine survival during the first year of life, as well as any biases in the assumed SRR.

The model is calibrated by finding a suite of stock- and year-specific EV scalars resulting in model outputs that most closely match observed terminal run sizes, escapements, or catches for individual stocks during the calibration period. The user specifies the EV scalars for the simulation period, often taken to be the mean, or some other statistic, from the calibration period values. For this chapter, I assumed the EV scalars were independent and log-normally distributed and estimated the median of each log-normal distribution from the calibration period values. The model results are known to be sensitive to the selection of the EV scalars for the simulation period. Typically, the PSC Chinook Technical Committee updates the input data and recalibrates the model once each year. All analyses in this chapter were conducted using the November 1994 calibration of the model.

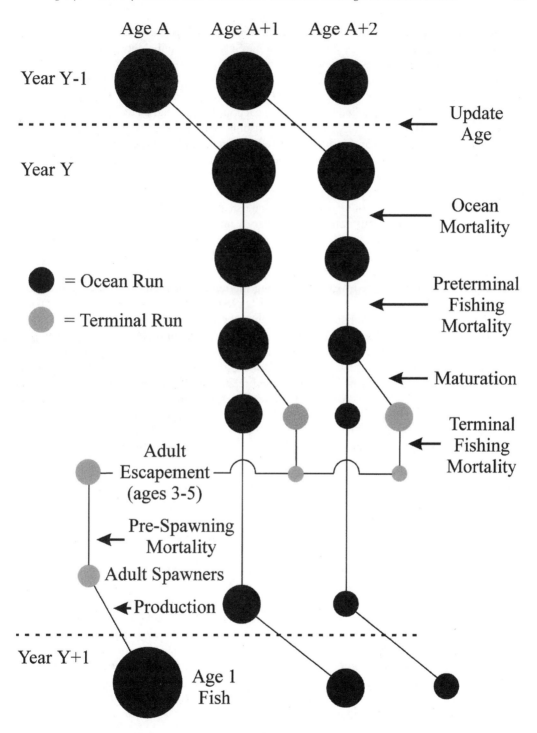

FIGURE 29.3   Computation flow in the PSC Chinook Model.

HARVEST STRATEGIES

I defined a status quo harvest policy to simulate the long-term effects of current harvesting practices and eight general exploitation rate reduction policies. The status quo policy was defined to be the average catch, or catch ceilings (for fisheries under catch ceiling management) and average harvest rates (for fisheries under harvest rate management) over the 4-year period 1990–1993 (Table 29.1). I simulated the long-term effects of this policy on the Snake River fall chinook stock through the year 2017.

Four exploitation rate reduction policies reduce harvests in single geographic regions—Alaska, British Columbia, Washington/Oregon coast, and Columbia River. Analyzing the impacts of regional catch reductions provides insight into the relative impacts each region has on the overall exploitation rate and into the dynamics of the gauntlet style of fishing. Two general policies reduce catches in all fisheries by equal or scaled amounts, and two reduce catches in U.S. fisheries by equal or scaled amounts. The scaled policies reduce each fishery's legal catch in proportion to that fishery's estimated contribution to the total adult equivalent (AEQ) fishing mortalities on the LYF stock during the period 1979–1993, as estimated by the PSC Chinook Model. "Adult equivalence" refers to the proportion of one adult fish represented by one fish at an earlier stage of development. Thus, AEQ mortalities more accurately reflect each fishery's impact on the adult spawning stock. The basic idea of the scaled policies is to reduce fisheries during the recovery period in proportion to their historical contribution to harvesting impact, as measured by that fishery's historical AEQ total mortalities on the Snake River fall chinook stock.

Table 29.2 lists the estimated AEQ total fishing mortalities (including incidental mortalities) during the period 1979–1993 for each fishery that harvests the LYF stock using the November, 1994 calibration of the PSC Chinook Model. Eight fisheries—Alaska troll, northern B.C. troll, central B.C. troll, west coast Vancouver Island (WCVI) troll, Washington/Oregon troll, Washington ocean sport, Columbia River net, and Columbia River sport—accounted for 96.1% of the total AEQ mortalities. Table 29.3 lists the AEQ mortalities for these eight fisheries and the scaling factors used in the analysis. Scale factors were determined relative to the fishery with the largest AEQ mortalities. For example, for the "All Scaled" policy, the WCVI troll fishery had the largest AEQ mortalities. Thus, the scale factor for the Alaska troll fishery is the Alaska AEQ mortalities divided by the WCVI troll AEQ mortalities (2,589/11,310 = 0.23). For the "U.S. Scaled" policies, the largest U.S. fishery AEQ mortalities was the Columbia River net fishery, so all other U.S. fisheries were scaled to that fishery.

ANALYSIS

For each of the eight exploitation rate reduction policies, I adjusted the appropriate catch ceilings and/or exploitation rates in the PSC Chinook Model such that simulated legal catches in year 2017 were reduced by 20, 40, 60, 80, and 100% from the legal catches in year 2017 under the status quo harvesting policy. Policies involving both catch ceiling and exploitation rate reductions (e.g., U.S. Scaled policies) required using an iterative approach and several model runs to determine the exploitation rates necessary to reduce the legal catches by the desired amounts.

In addition to simulated legal catches, each model run produced AEQ total fishing mortalities (including incidental mortalities), adult escapements for the Snake River fall chinook indicator stock (LYF), and the total AEQ exploitation rate as computed by the "cohort method" of the PSC Chinook Model. This method computes the ratio of total AEQ fishing mortalities (legal plus all incidentals) divided by the sum of the age 2 through age 5 AEQ cohort sizes at the start of the ocean fisheries. Thus, these exploitation rates are based on synthetic, rather than true, cohorts. I summarized the results by plotting the AEQ exploitation rates in year 2017 against percent reductions in the legal catches.

As noted earlier, results from the PSC Chinook Model are known to be sensitive to the selection of EV scalar values during the simulation period. Figure 29.4 illustrates the simulated LYF escapement trajectories from seven values of the EV scalars under the status quo policy. To examine the effects of different assumptions about the productivity of the LYF stock, I did the analysis using two EV scalar values, EV = 1 and EV = 3. The total analysis required over 100 model runs.

TABLE 29.2
Total adult equivalent (AEQ) mortalities of the LYF stock during the period 1979–1993 by fishery as modeled by the PSC Chinook Model using the 10/94-calibrated parameters. Eight fisheries (Alaska Troll, Northern B.C. Troll, Central B.C. Troll, West Coast Vancouver Island Troll, Columbia River Net, and Columbia River Sport) account for 96.1% of the total AEQ mortalities.

Region	Fishery	AEQ Mortalities	Percentage
Alaska	Troll	2,589	7.9
	Net	10	0.1
	Sport	130	0.4
	Total	2,729	8.4
British Columbia	Northern Troll	2,419	7.4
	Central Troll	796	2.4
	WCVI Troll	11,310	34.7
	Northern Net	126	0.4
	Central Net	129	0.4
	Johnstone Strait Net	82	0.3
	WCVI Sport	173	0.5
	Georgia Strait Sport	30	0.1
	Total	15,065	46.2
WA/OR Coast	WA/OR Troll	4,374	13.4
	WA Ocean Sport	1,282	3.9
	Total	5,656	17.4
Puget Sound	Juan de Fuca Net	187	0.6
	Northern Net	23	0.1
	Northern Sport	300	0.9
	Southern Sport	11	0.1
	Total	521	1.6
Columbia River	Net	7,931	24.3
	Sport	682	2.1
	Total	8,613	26.4
Total All Regions		32,584	100.0

RESULTS

Under status quo harvesting conditions, the model predicts continued declines in the LYF stock for EV values less than about 3.5 (Figure 29.4). For LYF EV = 3, legal catches within the four fishing regions are shown in Figure 29.5. Note that, under the status quo policy, the Columbia River fisheries actually have higher catches than during the 1990–1993 period, whereas the ocean fisheries have the average 1990–1993 catches. This is because the river fisheries are modeled under a fixed harvest rate policy. The status quo policy predicts increased abundance of the other URB stock components resulting in higher harvests in the river. Since the LYF stock is a minor component of the total coastwide system, the total legal catches for all fisheries are virtually identical whether EV = 1 or EV = 3 for the LYF stock.

Results from the four single-region exploitation rate reduction policies using two LYF EV values (1.0 and 3.0) are presented in Figure 29.6. In each graph, the y-axis represents the total AEQ exploitation rate on the LYF stock as computed by the "cohort method" of the PSC Chinook Model.

TABLE 29.3
Scaling factors used in scaled multiple-region exploitation rate reduction policies. The base fisheries for the "All Scaled" and "U.S. Scaled" policies are the WCVI Troll and Columbia River Net fisheries, respectively.

| | | Scale factors | |
Fishery	AEQ mortalities	"All scaled"	"U.S. scaled"
Alaska Troll	2,589	0.23	0.33
Northern B.C. Troll	2,419	0.21	N/A
Central B.C. Troll	796	0.07	N/A
WCVI Troll	11,310	1.00	N/A
WAOR Troll	4,374	0.39	0.55
WA Ocean Sport	1,282	0.11	0.16
Columbia River Net	7,931	0.70	1.00
Columbia River Sport	682	0.06	0.09

Note: Scaling factors are equal to the AEQ mortalities divided by those for the appropriate base fisheries. For example, the scale factor for the AK Troll fishery in the "U.S. Scaled" policy is 2,589/7,931 (= 0.33). Scaled exploitation rate reduction policies reduce individual fishery catches in proportion to their scaling factor. For example, in the "U.S. Scaled" policies the Alaska Troll fishery is reduced one third as much as the Columbia River Net fishery.

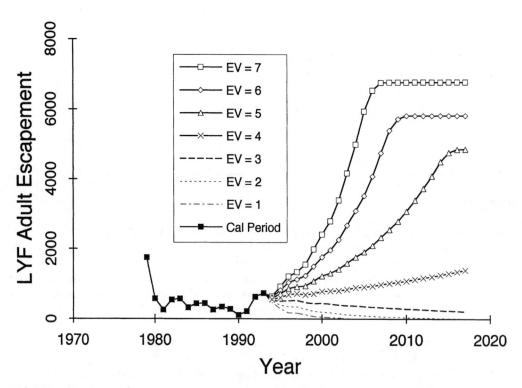

FIGURE 29.4   Simulated escapements from the PSC Chinook Model for the LYF stock under the status quo harvesting strategy assuming seven different values for the LYF Environmental Variability (EV) scalars. The model calibration period is 1979–1993.

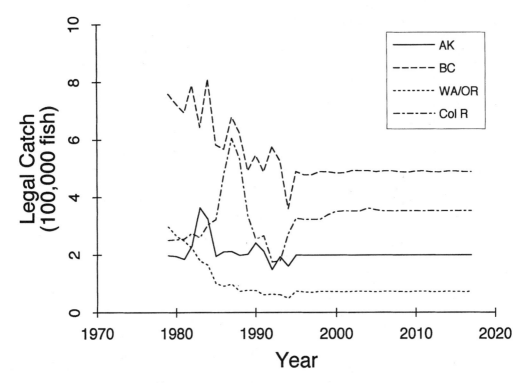

FIGURE 29.5 Simulated catches from the PSC Chinook Model for the four major fishing regions under the status quo harvesting strategy.

The negatively sloping lines in Figure 29.6 define the amount of catch reduction by a fishing policy (expressed as a percentage of legal catch under status quo management) required to achieve a given overall exploitation rate on the LYF stock. The intersection of a horizontal line drawn at the desired exploitation rate with each fishing policy line determines equivalent fishing policies (i.e., policies that result in the same overall exploitation rate). The specific catch reductions required by each fishing strategy can be read off the x-axis.

Note that if one ignores the scale of the y-axes in Figure 29.6, the slopes of the lines in the left and right graphs appear to be nearly identical. Thus, if the y-axis is rescaled to a maximum value of 1.0 (representing the overall exploitation rate under the status quo policy), other values on the y-axis represent proportional exploitation rates, relative to the status quo rate. Under this representation, it is apparently unnecessary to assume a given level of future brood year survival rates. That is, when exploitation rates are expressed as a percentage of the status quo policy, equivalent policy definitions are not affected by the choice of the LYF EV scalar.

Relative legal catches for eight managed fisheries (expressed as a percentage of status quo catches) required to reduce the LYF exploitation rate by 10% are given in Table 29.4 and illustrated in Figure 29.7. The results for single-region reduction policies indicate that reducing the B.C. troll fisheries by 29% is equivalent to reducing the Washington/Oregon troll and ocean sport fisheries by 84% or the Columbia River net and sport fisheries by 19%. Completely eliminating the Alaska troll fishery alone will result in only a 5% reduction in overall exploitation rate (Figure 29.7). Note that when only the Alaska, British Columbia, or Washington/Oregon troll fisheries are restricted, the Columbia River net and sport catches increase by 8% and 3%, respectively. This is due to more fish "passing through" the ocean fisheries and being made available for the in-river fisheries.

For the All Equal and All Scaled policies, reducing all fisheries by 8% is equivalent to reducing the WCVI troll fishery by 13%, and other fisheries by lesser amounts (in proportion to their historic catch of the LYF stock). Similarly, reducing all U.S. fisheries by 13% is equivalent to reducing the

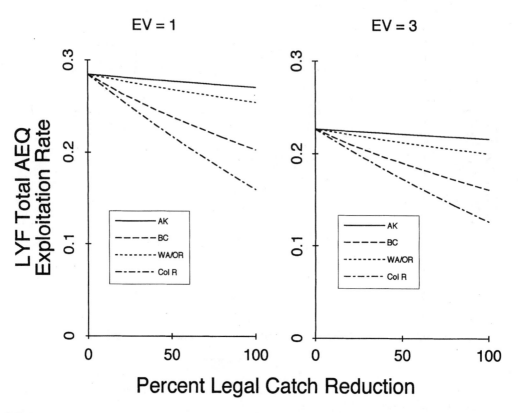

FIGURE 29.6   Simulated total AEQ exploitation rates from the PSC Chinook Model for the LYF stock for single region policies using two values of LYF EV scalars (EV = 1 and EV = 3).

TABLE 29.4
Relative legal catches for eight managed fisheries (expressed as a percentage of base case catches) required to lower the Lyon's Ferry adult equivalent exploitation rate (LYF AEQ HR) by 10% under eight types of harvest reduction policies. Also shown are the percentage changes in LYF adult escapements.

Fishery	Single-region reductions				All equal	All scaled	U.S. equal	U.S. scaled
	AK	B.C.	WA/OR	ColR				
Alaska Troll	0	100	100	100	92	97	87	95
Northern B.C. Troll	100	71	100	100	92	97	100	100
Central B.C. Troll	100	71	100	100	92	99	100	100
WCVI Troll	100	71	100	100	92	87	100	100
WA/OR Troll	100	100	16	100	92	95	87	91
WA Ocean Sport	100	100	16	100	92	99	87	97
Columbia River Net	108	108	108	81	92	92	87	83
Columbia River Sport	103	103	103	81	92	99	87	99
LYF Adult Esc	136	191	186	208	198	199	204	204
LYF AEQ HR	95	90	90	90	90	90	90	90

Note: The reference fisheries for the All Scaled and U.S. Scaled policies are the WCVI Troll and Columbia River Net fisheries, respectively.

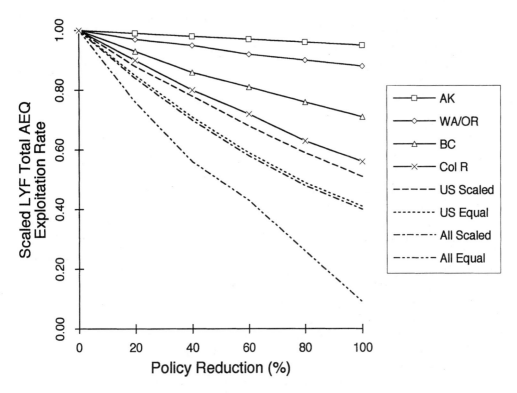

FIGURE 29.7   Simulated total AEQ exploitation rates from the PSC Chinook Model for the LYF stock for all exploitation rate reduction policies.

Columbia River net fishery by 17% and other fisheries by lesser amounts. Note also that equal exploitation rate reductions do not lead to equal changes in LYF adult escapements. This is due to differences in the age-specific exploitation rates between fisheries.

DISCUSSION

The results of this analysis do not identify an optimal Snake River fall chinook exploitation rate for any given year, or years, because other techniques must be used to determine target harvest rates and escapement goals for each stock. Instead, these results simply illustrate which combinations of legal catch reductions among competing fisheries produce the same level of overall exploitation rate reduction from the status quo policy under long-term equilibrium conditions.

To illustrate how the results may help promote agreement during the decision-making process, consider the question of selecting a "best" policy. From a biological perspective, it does not matter which policy is used, as long as it reduces the total exploitation rate by an amount believed to be needed to recover and sustain the Snake River fall chinook stock. Thus, once a harvest reduction is identified, all policies are equally "good." Selecting an appropriate total harvest reduction is an "amount of harvest" problem. The more difficult management decision is selecting the "method of harvest," because it involves allocating total harvest among competing fleets. From an Alaskan's perspective, any policy that does not reduce Alaskan fisheries is "best"; and from a Columbia River gillnetter's perspective, any policy that does not reduce Columbia River fisheries is "best." Each stakeholder wants to minimize their share of the total stock rehabilitation cost, and, as Easley and Prochaska (1987) have noted, fishery managers attempt to "minimize the heat on the agency from irate fishermen."

My premise is that the political heat is minimized and the decision-making process is more likely to achieve agreement if the biological "amount of harvest" and the political "method of harvest" decisions are clearly separated and analyzed in an objective manner. Once an exploitation rate reduction goal is established, negotiators can focus on finding a "fair" allocation of the reductions. As I characterized the Snake River fall chinook problem, this amounts to selecting one of the eight policies I defined. The four single-region policies would undoubtedly be rejected because they would be perceived as being "unfair" to the region shouldering all of the harvest reduction. More likely, the debate would focus on whether it is fair to reduce Canadian fisheries and/or whether it is more or less fair to reduce fisheries by equal or scaled amounts. The scaled policies I defined were based on one objective measure, namely each fishery's contribution to the Snake River fall chinook catch during the recent past. This concept of fairness would likely be challenged and replaced by new policies.

The model and computation methods used in the analysis is also likely to be challenged. For example, the specific results reported here are dependent on the PSC Chinook Model and its assumptions. One important concern is the suitability of the LYF hatchery stock to represent the Snake River wild fall chinook stock. If the hatchery and wild stocks have different fishery contribution rates due to different ocean distributions or maturation rates, the results reported here would not be appropriate for the wild stock. This is a common problem for depressed salmon stocks. The lack of CWT recoveries from wild stocks forces researchers and managers to use closely related hatchery stocks as surrogates.

My decision to compute AEQ exploitation rates using the "cohort method" of the PSC Chinook Model was made primarily for convenience—it is a standard output of the model. Other methods are often used and may give quite different rates. For example, some methods do not express exploitation rates in adult equivalents, and therefore appear larger (because each age 2, 3, and 4 fish suffering fishing mortality represents less than one adult equivalent fish). Also, rates computed for river, or terminal, fisheries may be expressed as a fraction of the fish present in the river at the time of fishing, whereas the "cohort method" expresses the exploitation rate as a fraction of the total adult equivalent cohort size at the start of the ocean fishing season. The specific computation method is not important to this chapter—any method can be specified. What I am illustrating is how to define equivalent exploitation rate reduction strategies, once a particular computation method has been selected.

I have demonstrated that the PSC Chinook Model can be used to generate a single graph depicting equivalent exploitation rate reduction policies for a single species. Although the graph produced for this demonstration involved long-term exploitation rate reductions under equilibrium conditions, the same technique could be used for any given year and for any given species. Thus, graphs could be computed at the start of each season for each weak stock to help negotiators evaluate different options.

In broader context, the techniques illustrated in this chapter can be incorporated into other more general recovery plan components, such as those suggested by Powers (1996) and Lichatowich et al. (1995). Powers (1996) proposes four recovery plan components: (1) a threshold measure (or measures) of the overfished state and periodic monitoring of the fishery resource relative to that measure; (2) a recovery period; (3) a recovery trajectory for the interim stock status relative to the overfished state; and (4) transition from a recovery strategy to an "optimal yield" strategy. The key component is the recovery trajectory and, more importantly, how to deal with the inevitable deviations from that trajectory. Powers' suggestion to determine annual harvest rates based on the stock's current status with respect to the recovery trajectory should be expanded to include a formal policy on how harvests will be allocated among competing fleets. The techniques described in this chapter can be used to help define such a policy.

To generalize the Powers' approach even further, I suggest replacing the term "overfished" with "depressed" to indicate that stocks can be threatened by factors other than overfishing. As noted in the introduction, this may not be necessary for marine species whose primary threat is overfishing.

However, for salmon stocks (and more generally, any stocks listed under the ESA), overfishing is only one of many threats. At some point it becomes necessary to decide how much effort to place on each type of threat. The type of analyses illustrated in Figure 29.2 can be helpful in making these difficult public policy decisions. Clearly, threats other than overfishing greatly complicate the decision rules necessary to keep a recovering stock on the trajectory. However, without such rules, the political and economic pressures to maintain status quo make recovery less likely.

ACKNOWLEDGMENTS

I would like to thank the PSC Chinook Technical Committee for providing the code for the PSC Chinook Model and in particular Jim Scott, Gary Morishima, and Jim Berkson for their suggestions and comments regarding analysis and interpretation of results. I also thank an anonymous reviewer for helpful comments and Peter Dygert for an especially thorough review. This project was funded by the Bonneville Power Administration (Contract Number: DE-BI79-89BP02347; Project Number: 890-108).

REFERENCES

Anderson, J., J. Hayes, and R. Zabel. 1996. Columbia River salmon passage model: theory, calibration and validation. University of Washington, Seattle.

CTC (Chinook Technical Committee). 1993. Users guide for the Pacific Salmon Commission Chinook Technical Committee chinook model. Pacific Salmon Commission, Vancouver, British Columbia.

CTC (Chinook Technical Committee). 1994. 1993 annual report. Pacific Salmon Commission, Report TCCHI-NOOK (94)-1. Pacific Salmon Commission, Vancouver, British Columbia.

Easley, J. E., and F. J. Prochaska. 1987. Allocating harvests between competing users in fishery management decisions: appropriate economic measures for valuation. Marine Fisheries Review 49(3):29–33.

Hankin, D.G., and M.C. Healey. 1986. Dependence of exploitation rates for maximum yield and stock collapse on age and sex structure of chinook salmon (*Oncorhynchus tshawytscha*) stocks. Canadian Journal of Fishery and Aquatic Sciences 43:1756–1759.

Lee, D., and J. B. Hyman. 1992. The stochastic life-cycle model (SLCM): simulating the population dynamics of anadromous salmonids. U.S. Department of Agriculture, Forest Service, Intermountain Research Station, Research Paper INT-459.

Lichatowich, J., L. Mobrand, L. Lestelle, and T. Vogel. 1995. An approach to the diagnosis and treatment of depleted Pacific salmon populations in Pacific Northwest watersheds. Fisheries 20(1):10–18.

Nehlsen, W., J. E. Williams, and J. A. Lichatowich. 1991. Pacific salmon at the crossroads: stocks at risk from California, Oregon, Idaho, and Washington. Fisheries 16(2):4–21.

NMFS (National Marine Fisheries Service). 1995. Proposed recovery plan for Snake River salmon. National Marine Fisheries Service, Northwest Region, Seattle.

NRC (National Research Council). 1996. Upstream: salmon and society in the Pacific Northwest. National Academy Press, Washington, D.C.

Norris, J.G. 1995. A simple spreadsheet model for evaluating recovery strategies for Snake River fall chinook salmon. Report of Columbia Basin Research to Bonneville Power Administration, Portland, Oregon.

Norris, J.G. 1997. Documentation manual for the CRiSP Harvest version of the Pacific Salmon Commission Chinook Model. Columbia Basin Research, Seattle.

Powers, J. E. 1996. Benchmark requirements for recovering fish stocks. North American Journal of Fisheries Management 16:495–504.

Rutter, L. 1997. Salmon fisheries in the Pacific Northwest: How are harvest management decisions made? Pages 355–374 in D.J. Stouder, P.A. Bisson, and R.J. Naiman, editors. Pacific Salmon and Their Ecosystems: Status and Future Options. Chapman & Hall, New York.

Schaller, H., and T. Cooney. 1992. Draft Snake River fall chinook life-cycle simulation model for recovery and rebuilding plan evaluation. Oregon Department of Fish and Wildlife, Portland, and Washington Department of Fisheries, Olympia.

SRSRT (Snake River Salmon Recovery Team). 1993. Draft Snake River salmon recovery plan recommenda-
    tions. Report of the Snake River Salmon Recovery Team to National Marine Fisheries Service, Northwest
    Region, Seattle.
Wilson, P. 1994. Spring FLUSH (Fish Leaving Under Several Hypotheses), version 4.5. Draft documentation.
    Columbia Basin Fish and Wildlife Authority, Portland, Oregon.

30 Decadal Climate Cycles and Declining Columbia River Salmon

James J. Anderson

Abstract.—This chapter explores the interactive effects of anthropogenic trends and climate cycles on salmon declines in the Columbia and Snake river basins. A basic population model—including anthropogenic and environmental factors—is discussed, and literature relating decadal-scale climate patterns and the response of the North Pacific ecosystem is reviewed. From this background a ratchet-like decline in Columbia and Snake river salmon production has resulted from the interactions of human activities and climatic regime shifts. These interactions are illustrated using hundred-year patterns in spring chinook salmon *Oncorhynchus tshawytscha* catch, the Columbia River hydroelectric generating capacity, and a climate index characterizing the shifts between a cool/wet regime favorable to salmon and a warm/dry regime unfavorable to salmon. A half-century correlation of the climate index and chinook catch suggests that a favorable climate regime counteracted detrimental impacts of hydrosystem development between 1945 and 1977, while an unfavorable climate regime negated beneficial effects of salmon mitigation efforts after 1977. This hypothesis is elaborated by a comparison of changes in the climate index relative to changes in Snake River salmon survival indicators. Proposed Snake River salmon restoration plans are considered in terms of this counteractive effects hypothesis. The recent declines of salmon stocks have led a number of groups to propose plans that discontinue the present recovery actions, especially transportation of juvenile salmon around the dams. This chapter hypothesizes that salmon recovery efforts, in part, have been limited by recent poor climate/ocean conditions. If this hypothesis is true, then eliminating the transportation program could be detrimental to fish. If the hypothesis is false, then eliminating transportation may be a viable recovery measure. In either case resolving the issue of counteracting processes is essential prior to making major changes to hydrosystem operations. In a larger perspective the influence of climate cycles on the Columbia River illustrates that in achieving sustainability we do not achieve stability. The better we understand that stocks will fluctuate by factors outside our control, the better chance we have to avoid the ratchet-to-extinction—and this is the first goal of sustainability.

INTRODUCTION

The catch of Columbia River chinook salmon *Oncorhynchus tshawytscha* has declined over this century (Figure 30.1) and recently has reached record low levels, with many stocks extinct and others on the brink of extinction (Stouder et al. 1997). The decline, from over 10 million to under 2 million adult fish, has been oscillatory and is the result of both natural and anthropogenic factors (Lichatowich and Mobrand 1995). Three major anthropogenic factors are generally mentioned: (1) loss by fishing; (2) loss of spawning and rearing habitat from agriculture, grazing, logging, mining, and dams; and (3) loss during river migration related to passing through the dams and reservoirs of the hydrosystem (Ebel et al. 1989; Wissmar et al. 1994). In addition, the decline has been

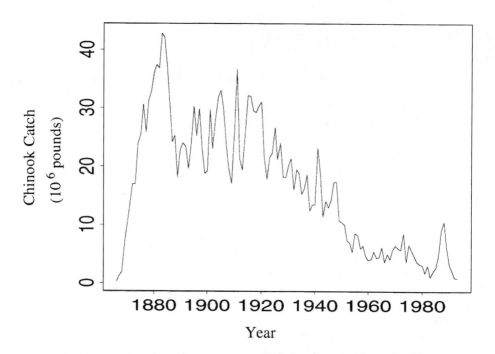

FIGURE 30.1   Total chinook catch in the Columbia River system (1880–1990).

affected—sometimes enhanced and sometimes diminished—by the natural cycles of the climate and ocean. In this chapter I consider the interactions of these factors. I first discuss the theoretical basis for the interaction of anthropogenic and natural processes, especially those involving climate; then review information on climate regime shifts and the response of the North Pacific ecosystem; and finally focus on the Columbia River system and how the climate has confounded our understanding of the decline of the salmon stocks in the system.

THE RELATION BETWEEN ANTHROPOGENIC AND NATURAL EFFECTS

To establish a conceptual framework for the problem consider that a salmon stock decline is the result of cumulative and cyclic factors. Assume that the cumulative degradation of a stock is the result of anthropogenic processes, while the cyclic processes are typically the result of natural environmental processes driven ultimately by a climate/ocean interaction. In terms of a stock recruitment equation, natural and anthropogenic factors are added to traditional density dependent and independent mortality terms giving:

$$R_t = a\, S_t\, e^{-\alpha(t)-\beta(t)-bS} \tag{30.1}$$

where R_t is the recruitment of fish into the stock from brood year t, S_t is the stock size in year t, $\alpha(t)$ is a mortality rate due to anthropogenic factors, $\beta(t)$ is a rate of decline from natural environmental time-varying factors and a and b are the Ricker density independent and density dependent mortality factors, respectively, which are taken to be time independent for this example.

In the traditional fisheries management paradigm $\alpha(t)$ and $\beta(t)$ are ignored and the parameters a and b are assumed constant with variations due to random processes. This simplified two parameter model (based on a and b) has outlived its usefulness for stock management, especially when a need exists to separate the impacts of human actions from climatic variations. In its place a number of approaches have been suggested to include additional factors. Adkison et al. (1996) outlined three

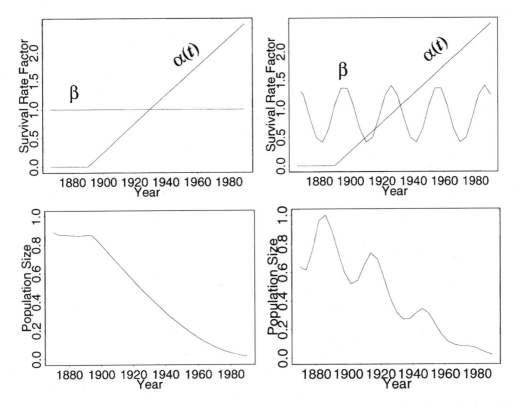

FIGURE 30.2 Example of salmon decline under slide and ratchet processes. A population extinction slide is driven by increasing cumulative anthropogenic impacts, $\alpha(t)$, and a constant climatic effect, β. The population ratchet-to-extinction is driven by increasing cumulative anthropogenic impacts, $\alpha(t)$, and a climatic cycle $\beta(t)$.

approaches to evaluate climate effects on stocks. The first approach, designated as the "effect of climate variables method" explicitly includes climate variables in the spawner recruitment curve. In terms of Eq. (30.1), this is formulated as $\beta = c_i E_t$, where i designates stock-specific variables and t designates time-specific variables; c is a stock-specific coefficient; and E is a time-specific climate variable, such as temperature. The second method, designated the "one-time shift method," involves using one set of the stock recruitment variables before a climate regime shift and a second set of variables after the shift. The equation then has a set of (a,b) for each regime and $\beta = 0$. The third method, called the "common influence method," involves defining a common climate factor for a number of stocks and then fitting Eq. (30.1) in a multiple regression for stocks sharing the common effect through $\beta(t)$, which varies over time.

The model must be expanded to account for the anthropogenic effects. In the Columbia River the cumulative impacts of dams, overfishing, irrigation and intensive land use are significant factors contributing to the decline of stocks. To illustrate how these factors, in principle, alter the stock recruitment system assume in Eq. (30.1) that $\alpha(t)$ is a linearly increasing function with time so the population exhibits a gradual "slide" toward extinction (Figure 30.2). This description is inadequate, however, to assess the significance of cumulative anthropogenic processes when they occur along with climatic cycles. When cyclic natural and cumulative anthropogenic variations combine, the stock decline will assume a "ratchet"-like character where the stock increases and decreases with the periodicity of the natural cycles. Over a number of cycles, however, the population declines (Lawson 1993; Lichatowich and Mobrand 1995). This pattern is illustrated in Eq. (30.1) with a linear increase in $\alpha(t)$ and a sine wave in $\beta(t)$. The resulting population has a ratchet pattern as it declines to extinction (Figure 30.2). If the population is only observed within the period of one environmental

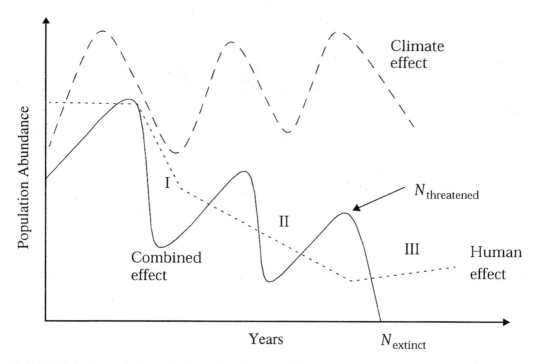

FIGURE 30.3 A population extinction ratchet (Lawson 1993) results from a natural cycle combining with cumulative anthropogenic use. Prior to exploitation the population varies in a natural cycle. Human use alters the population in three stages: (I) a period of expanding use; (II) a period of regulated use; and (III) a period of threatened use, where actions may be taken to recover the population. The population enters a threatened status when the most recent peak falls below $N_{threatened}$. At population level $N_{extinct}$ it is unlikely that the stock can be sustained.

cycle, identifying the contributions of the natural and the anthropogenic changes to the decline are difficult to assess.

To explore this system in the context of a fishery, consider that the cumulative human impact results from direct exploitation of the population and by indirect impacts, such as degradation of the habitat. In either case, the anthropogenic-based trend typically evolves in three stages (Figure 30.3). The first stage (I) is expansion in which the population is directly or indirectly affected without regard to the consequences. In the second stage (II), activity is controlled through regulation to achieve some purpose such as to maximize yield or to distribute the economic benefits of the resource. If the regulation does not sustain the population, the third stage (III) follows in which the population's existence is threatened and mitigation actions are taken to recover the population.

For populations declining in a ratchet pattern, the threatened stage is entered after the anthropogenic factors, acting over a long term, finally depress the population to a critically low level at which the natural fluctuation can drive it toward extinction. Under this scenario the natural place to define the beginning of the threatened stage is when the population is either at a high point or appears to have stabilized its abundance. From this threatened abundance, $N_{threatened}$, the decline in the natural factors drive the population toward $N_{extinct}$, where genetic and ecological factors (Shaffer 1981; Soulé 1987; Thompson 1991) affect its sustainability. Thus $N_{threatened}$ is an early warning mark that, because of the uncontrolled nature of the natural cycle, is inextricably linked to $N_{extinct}$. Defining $N_{threatened}$ is not an easy matter, however, because it depends on the magnitude of $\beta(t)$, the continued rate of decline of anthropogenic factors, the effectiveness of mitigation actions to reverse $\alpha(t)$, and the periodicity in the natural cycle. This example illustrates that extinction should be considered in the context of the cyclical variations and not just when a population is near extinction. This point was also emphasized by Lawson (1993).

THE NORTH PACIFIC ECOSYSTEM AND DECADAL CLIMATE PATTERNS

Considerable evidence indicates cycles in environmental conditions occur at many temporal scales and interact with many points in the salmon's life cycle. Before considering these interactions, it is important to first consider evidence that oceanic and climatic fluctuations play major roles in the natural variations of the North Pacific ecosystem of which salmon are an integral member.

FISH AND CLIMATE PATTERNS

Multidecadal fluctuations in fish stock abundance have been observed for centuries (Rothschild 1995), but an appreciation of the importance of climate–fish fluctuations is relatively recent. The longest record of fish population fluctuations was obtained from a 2,000-year sedimentary record off California. A surrogate record for the abundance of Pacific sardines and northern anchovy, inferred from scales in sediment cores, exhibited strong fluctuations over two millennia. A spectral analysis of the records revealed a peak at periods of about 60 years (Smith 1978; Baumgartner et al. 1992; Sharp 1992). The contribution of climate to these types of fluctuations was inferred from a similar pattern of climate indicators and sardine *Sardinops* catches around the Pacific basin (Kawasaki 1984). One of the earliest papers on fishery oceanography documented the impact of the 1972–1973 El Niño on the crash of the Peruvian anchovy fishery (Valdivia 1978). The general significance of climate on fisheries variability has also been the focus of other treatises (e.g., Smith 1978; Gantz 1992; Beamish 1995).

OCEAN/CLIMATE MECHANISMS

Studies detailing the North Pacific ecosystem and its decadal-scale ocean variability have mostly emerged since 1990 and were motivated by a striking shift in oceanic and biological conditions that occurred between 1976 and 1977. The oceanographic literature reported the shift as a change in the intensity of the wintertime Aleutian Low pressure regime between 1976 and 1977, which was associated with a major and persistent change in oceanographic conditions (Figure 30.4). The regime shift occurred as an abrupt change in the large-scale boreal winter patterns over the North Pacific. The change was marked by a southward shift and intensification of the Aleutian Low and prevailing westerlies over the mid-central and eastern Pacific (Graham 1994; Miller et al. 1994). In particular, the sea surface temperatures warmed.

The mechanisms driving these decadal interactions of the atmosphere and currents are not clearly understood. Studies have suggested they involve ocean–atmosphere couplings (Graham 1994), an 11-year variation in solar radiation (van Loon and Labizke 1994; Kerr 1995), and even the 18-year lunar cycle (Currie and Fairbridge 1986; Parker et al. 1995). Lagerloef (1995), following an idea proposed by Reed (1984), demonstrated that short-period random forcing of the winds, decoupled from any long-term cycles, can also produce a bidecadal periodicity in winds and ocean temperature anomalies. Theoretical papers have also discussed the importance of changes in the ocean heat flux, the thermocline depth (Graham 1994; Miller et al. 1994) and northward advection of warmer water along North America (Emery and Hamilton 1985; Bakun 1996). The warm coastal water of the warm/dry regime has been associated with a strong Alaska Gyre circulation which may be induced by weakening of upwelling on the west coast of North America (Hollowed and Wooster 1992). In contrast, the temperature change can also be considered as a cooling of the coastal water in the cool regime resulting from changes in nearshore winds related to the intensity of the Aleutian Low pressure system (Bakun 1996).

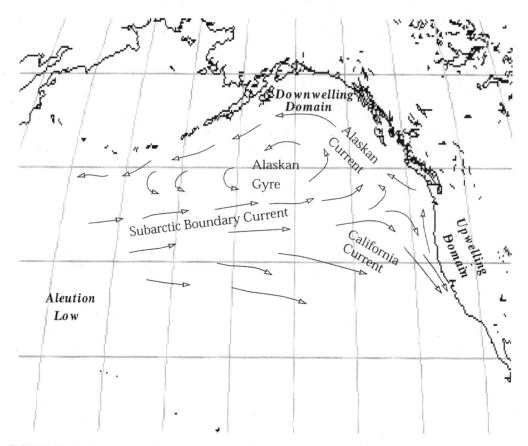

FIGURE 30.4 Currents and production Domains. Years with an intense winter Aleutian Low shift the Subarctic Current northward, strengthen the Alaskan Current and increase the Downwelling Domain production. Years with a weak Aleutian Low shift the Subarctic Current southward and strengthen the California Current and the Upwelling Domain production. Shifts occur on decadal scales.

Although the warm and cool regimes appear to persist over about two decades, other periods have also been noted. Using a spectral analysis of 21 climate records over about 100 years, Ware (1995) identified four dominant time scales: a 2- to 3-year (quasi-biennial oscillation), 5- to 7-year (El Niño–Southern Oscillation), 20- to 25-year bidecadal oscillation, and a poorly resolved, very low frequency oscillation with a 50- to 75-year period. Of his indicators, the bidecadal scale oscillation is most identified with the 1977 regime shift. A 100-year simulation of Northeast Pacific winter drift patterns and a 500-year record of tree ring growth data from Oregon exhibited cycles with a dominant period of 18 and 17.1 years, respectively (Ingraham et al. 1988). The tree ring data, similar to Ware's analysis, had significant spectral peaks at 4 and 50 years (Richard A. Hinrichsen, University of Washington School of Fisheries, personal communication). Mantua et al. (1997) noted a decadal periodicity with major climate shifts in 1925, 1947, and 1977.

PLANKTON RESPONSE

Climate regime shifts also have had significant impacts on Northeast Pacific plankton. The 1977 regime shift allowed warm sea surface water to move further northward and increased phytoplankton production in the Alaskan Downwelling Domain (Figure 30.4). At the same time the upwelling intensity in the West Coast Upwelling Domain and the associated phytoplankton production both decreased (Venrick et al. 1987, Ware and Thompson 1991). Current patterns also affected the advection of

Subarctic Current and Alaskan Gyre zooplankton into the Domains (Figure 30.4). Abundant zooplankton levels in the Downwelling Domain have occurred with intense Aleutian Lows (warm regime) and abundant zooplankton in the west coast Upwelling Domain have occurred with weak Aleutian Lows (cool regime) (Wickett 1967; Brodeur and Ware 1992; Roemmich and McGowan 1995).

Several models have been proposed for the connection between climate and phytoplankton production, which in part drives changes in fish production. Venrick et al. (1987) postulated increased phytoplankton production is driven by increased vertical mixing during intense Aleutian Low. Roemmich and McGowan (1995) also hypothesized that, in the California current system, decreased zooplankton abundance after the 1977 regime shift was in part the result of suppression of nutrient supply by enhanced stratification. Gargett (1997) expanded on these hypotheses connecting stability, decadal fluctuations, phytoplankton growth, and salmon. Because increased stability limits the surface layer nutrient flux, which limits plankton growth but restricts phytoplankton mixing out of the photic zone, vertical mixing has counteracting effects and so plankton growth should be optimized at some optimum "window" of surface water stability. The window of optimum stability would change with latitude in Gargett's hypothesis to generate out-of-phase variations between northern and southern salmon stocks. Wong et al. (1995) proposed an alternative explanation for the phytoplankton growth relationship with the Aleutian Low intensity. The hypothesis is based on the recent findings that iron can become a limiting nutrient for phytoplankton production in the open ocean. The iron source is terrestrial, so North Pacific phytoplankton production could be correlated with atmospheric transport of iron from Asia. In this scenario, larger iron transport during an intense Aleutian Low would increase phytoplankton production in the Alaskan Gyre. Although the contributions of physical (Gargett 1997) vs. chemical (Wong et al.1995) processes in driving the phytoplankton cycles has not been rigorously evaluated, the preponderance of evidence favors dominantly physical processes.

SALMON RESPONSE

Climate regime shifts are also associated with variations in salmon stocks. Trends in total salmon catches in the warm regime, post 1977, improved primarily from increased production of salmon from Alaska (Ware and McFarland 1989; Francis and Hare 1994; Mantua et al. 1997). The pattern was particularly strong with Bristol Bay sockeye *O. nerka*, which jumped from catches on the order of 1 million fish in the early seventies, to a record catch of 44 million sockeye in 1995 (van Amerongen 1995). The general trend was the same for Alaskan pink salmon *O. gorbuscha*, coho salmon *O. kisutch*, and chinook salmon *O. tshawytscha*, with upward trends beginning in 1976–1977 and continuing into the 1980s (Beamish 1993). These Alaskan catches remained high in the early 1990s. The pattern (comparing the 1950s to the 1980s) is also evident over a wide range of the Gulf of Alaska for sockeye, pink, chum *O. keta*, coho, and chinook salmon and steelhead trout *O. mykiss* (Brodeur and Ware 1995).

A trend in Alaskan chinook size also correlates with the regime shift, with decreasing sizes from 1951 to 1975 followed by an abrupt shift with increasing average chinook weight in 1977–1978. This pattern first appeared in the northern stocks, with some southern stocks showing the increasing trend beginning in 1983, while other southern chinook stocks did not exhibit the increase (Beamish 1993). In contrast, declining size and increasing age at maturity were observed for chum salmon stocks from western North America between 1972 and 1992 (Helle and Hoffman 1995), and in the North East Pacific, halibut have exhibited a decline in weight since 1976 (Clark et al., 1999).

LATITUDINAL PATTERNS

Evidence indicates that regime shifts have opposite effects on high- and mid-latitude biological populations. A century-long record of sardine catch exhibited a striking coherence among several stocks in the mid-latitudes of the Pacific Ocean. Isolated stocks of *Sardinops* spp., from Chile, Asia, and California had high catches about 1940 and 1990 with low catches in the 1970s. The catch statistics closely followed the sea surface temperature anomaly for the Pacific Basin (Kawasaki 1984; Sharp 1992). After 1977 the higher latitudes of the North Pacific exhibited increasing salmon catch trends from Alaska, Russia, Canada, and Japan (Beamish and Bouillon 1993). West Coast salmon populations, however, exhibited evidence of decline after the 1977 regime shift (Richards and Olsen 1993; Francis and Hare 1994). Furthermore, a 60-year data record between 1925 and 1985 showed that Gulf of Alaska pink and sockeye salmon catches were in phase, but they varied inversely with catches of Washington/Oregon/California coho (Francis 1993). The pattern was clear; Alaskan and West Coast salmon stocks oscillated out of phase. In the dry/warm regime Alaskan catches increased and the West Coast catches decreased and during the wet/cool regime the Alaskan catches decreased and the West Coast catches increased. This inverse Alaska and West Coast pattern was also evident in marine survival of hatchery-reared salmon. Coronado-Hernandez (1995) estimated marine survival using data on 8,596 coho, 11,051 chinook, and 1,389 steelhead tag groups from Alaska to California. In general, survivals declined over most of the geographical range and were particularly poor in the late 1970s and in the late 1980s. Alaska coho exhibited the opposite trend with marine survival increasing in the late 1970s and 1980s.

Ecosystem Mechanisms

The impact of regime shifts on salmonid marine survival and production is complex. In general, the Alaska stocks appear to be favored by the warm regime because of increased phytoplankton production, which increases the forage base for salmon. Factors that increase the fish forage base in the Downwelling Domain during the warm regime appear to decrease it in the West Coast Upwelling Domain. During the cool regime the same factors appear to have the opposite effect. Gargett (1997) hypothesized the inverse relationship between salmon productivity in the Upwelling and Downwelling Domains results from opposite response of plankton productivity in the two Domains. Increased stability in the Upwelling Domain decreases phytoplankton productivity by decreasing nutrient flux into the surface layer. Increased stability in the Downwelling Domain increases phytoplankton productivity by confining the plants to the photic zone. To connect the dynamics to decadal scale processes Gargett (1997) hypothesized a mechanism relating water column stability to climate. In the North, water column surface stability is dominated by salinity and so a strong winter Aleutian Low increases Alaska rainfall and runoff which increases the salinity gradient and surface layer stability. In the South a strong winter Aleutian Low shifts winds to a more southerly direction which may increase temperature and inhibit upwelling.

Predators are also a factor, and environmental changes between warm and cool water regimes may affect the distribution and abundance of predators. In particular, the movement of warm water mackerel northward during a warm regime may decrease survival of West Coast salmon smolts during their ocean entry. Mackerel may move north with a stronger West Coast Davidson current associated with a strong winter Aleutian Low.

Finally, the climate regime shifts also affect the freshwater habitat of salmon. The interaction can be either positive or negative depending on the patterns of rainfall, snowpack, temperatures, and runoff. For example, floods during the fall spawning can disturb redds reducing egg survival, while floods in late spring may increase turbidity and smolt migration speed which together decrease exposure to predators.

Regime Shift Indicator

Relationships between climate and salmon are suggested in a variety of records. In these characterizations, a number of climate indicators have been used including the Kodiak winter air temperature (Francis and Hare 1994), the Central North Pacific winter atmospheric pressure index (Cayan and Peterson 1989), and the sea surface temperature anomaly-based Pacific Decadal Oscillation (PDO) (Mantua et al. 1997) to name a few. Many other environmental changes, both physical and biological, are also correlated with decadal scale climate changes. Ebbesmeyer et al. (1991) found changes in 40 environmental indicators correlated with the 1976 regime shift.

For comparison with Columbia River salmon, the Pacific Northwest Index (PNI) developed by Ebbesmeyer and Strickland (1995) is useful because it is a century-long composite index that characterizes Pacific Northwest climate patterns in both coastal waters and freshwater habitats. The PNI uses three measures: air temperature at Olga in the San Juan Islands, averaged annually from daily data; precipitation at Cedar Lake in the Cascade Mountains, averaged annually from daily data; and snowpack depth at Paradise on Mount Rainier on March 15 of each year. For each measure, annual averages are normalized by subtracting annual values from the average of all years and dividing by the standard deviation about the average for all years. The three measures are averaged, giving a relative indicator of climate variations. Positive PNI values indicate warmer and dryer years than the average, and negative values indicate cooler and wetter years than the average. Over a 90-year record the running 5-year PNI average switches between warm/dry and cool/wet regimes on about a 20-year period. The cycle has a distinctive double-peak pattern with a strong regime shift followed by two weaker regime shifts and then another strong regime shift. The strong regime shifts in the PDO in 1925, 1947, and 1977 (Mantua et al. 1997) are also evident in the PNI. In support of an impending climate shift, Ingraham et al. (1998), evaluating a 500-year tree ring growth record and 100-year Gulf of Alaska winter surface drift simulation, suggest that the Northeastern Pacific climate recently shifted, or is due to shift, to the cool regime. It is not clear, though, whether a shift would be long lasting. If the double-peak pattern were to hold, the shift to cooler, wetter Pacific Northwest weather may in fact be followed by a more intense period of warm, dry weather. On the other hand, based on the tree ring data, the recent warm regime has been one of the longest in history which would support the contention that the Pacific Northwest is due for a major shift into a cool, wet regime (J. Ingram, National Oceanic and Atmospheric Administration, personal communication).

PNI and Columbia River Salmon

The evidence discussed above strongly suggests that salmon catch and climate cycles are correlated. Although catch is not a direct measure of abundance or survival, it does reflect abundance and is the only population information that spans the century. The PNI relationship with the Columbia River spring chinook catch (WDF and ODFW 1992) is especially compelling because of a strong coherence (Figure 30.5). The evidence suggests that warm/dry Pacific Northwest climate regimes produce poor conditions for some Columbia River salmon and good conditions for Alaska salmon. The opposite effect, favoring Columbia River salmon, appears to occur in cool/wet climate regimes.

Climate and the Hydrosystem

With these patterns established we can evaluate the combined impact of climate and the development of the Columbia River hydrosystem on salmon populations. Over the last hundred years, three major climate shifts have occurred (1925, 1947, 1977), which in turn have significantly altered salmon survival. The hydrosystem was developed over a 50-year interval embedded within these cycling climate shifts. As hypothesized and illustrated in Figure 30.3, climatic variations can mask anthropogenic impacts creating a complex history of salmon decline that is misunderstood if the climatic cycle is ignored. In particular, the patterns suggest that the climate cycle may have masked both the negative impact of the Columbia/Snake River hydrosystem during its development and the positive impact of Columbia River mitigation measures implemented post-development.

To illustrate this idea of counteracting processes, consider the decadal scale cycle as represented by the PNI, the anthropogenic use of the river system as represented by the hydroelectric generating capacity, and the fish population as represented by the total Columbia River chinook catch (Figure 30.6). To understand these trends consider that, in the initial years of the Columbia River fishery, prior to 1925, the total harvest was relatively stable and the cool/wet climate pattern was

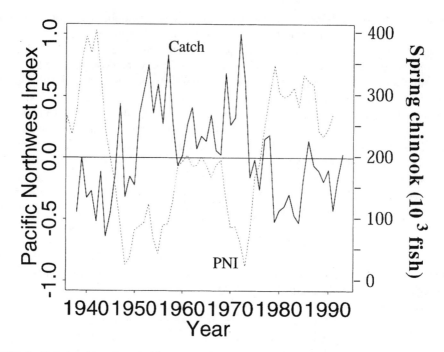

FIGURE 30.5 Relationship of catch of Columbia River spring chinook vs. the PNI. Both indicators are expressed as 5-year running averages. Periods when the PNI are > 0 tend to be warmer and drier; when the PNI < 0, cooler and wetter.

favorable to fish. An approximate balance between chinook fish production and harvest likely existed. In 1925, when the climate changed to the unfavorable warm/dry pattern, the productivity declined, and in response salmon catch began its long decline. About 1947 the climate shifted to the cool/wet pattern favorable to fish but the rapid development of the hydrosystem commenced, undoubtedly detrimental to fish. The climate shifted back to the unfavorable regime in 1977, which was one year after the completion of the last dam on the Snake River.

After completion of the dams, attention was given to reducing fish mortality caused by the hydrosystem. Some of the mitigation measures implemented included replacing lost habitat with hatcheries and installing fish ladders to pass adult salmon upriver. To improve survival of juveniles migrating through hydrosystem reservoirs, additional water is released from storage reservoirs to increase flow and predators are removed in a bounty harvest program. To lessen the impacts of both dams and reservoirs, a significant fraction of the juvenile fish are collected at dams (starting 1975 in the Snake and 1979 in the lower Columbia) and barged to below Bonneville Dam, the lowest dam on the river. A series of changes have also been implemented to reduce turbine mortality, including operating turbines near peak efficiency to reduce cavitation and removing the debris accumulated in the trashracks at turbine entrances (Williams and Matthews 1995). These changes increased survival of fish migrating from the Snake River, but the effect has likely in part been masked by the shift to an unfavorable climate regime in 1977.

INTERACTION OF PROCESSES

The scenario that the impacts of climatic and anthropogenic factors have been countervailing for a number of decades is based on a significant number of patterns observed over the entire North Pacific. Specific to the Columbia River, the counteracting-factors hypothesis is also supported by patterns of modeled smolt survival and redd count-based survival estimates which are correlated with the PNI pattern.

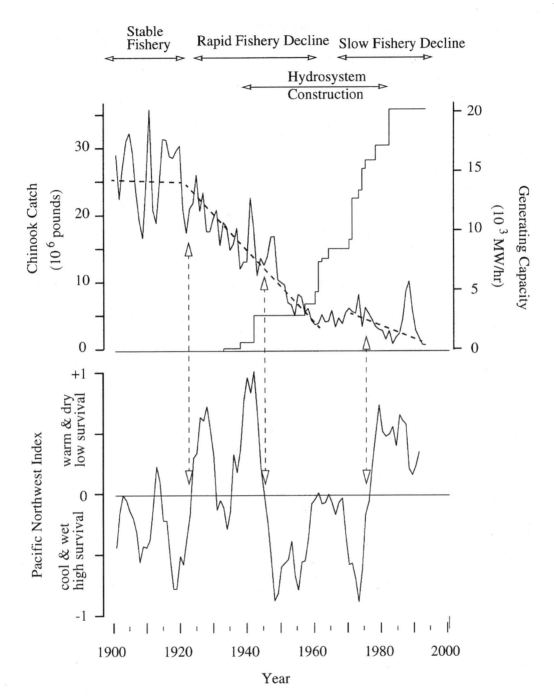

FIGURE 30.6   Relationship of the PNI climate index, Columbia River spring chinook catch, and hydroelectric capacity. Vertical dashed lines represent major climate regime shifts of 1925, 1947, and 1977.

SALMON SURVIVAL AND CLIMATE PATTERNS

The annual smolt hydrosystem passage survival rate over the past three decades was estimated using the CRiSP mainstem passage model (Anderson et al. 1996). The model accounts for smolt migration mortality from exposure to predators, dissolved gas supersaturation, and dam passage. The model characterizes hydrosystem operations, fish transportation schedules, flow, temperature, and dissolved gas for each day from 1966 through 1995. The model's reservoir mortality algorithm was calibrated with data from predator consumption studies conducted in the 1980s (Rieman et al. 1991); dam passage mortality was calibrated with studies conducted over three decades (compiled in Anderson et al. 1996); and mortality from gas bubble disease was calibrated to laboratory studies (Dawley et al. 1976). Fish migration rate was calibrated with fish travel time data collected from PIT tag studies conducted between 1989 and 1996 (Zabel and Anderson 1997). Model predictions were compared to over 50 spring chinook smolt survival estimates derived from mark recapture studies conducted between 1966 and 1996 over different reaches of the river and for different stages of construction of the hydrosystem. The predicted vs. observed survivals have a one-to-one relationship with $r^2 = 0.7$.

To characterize the hydrosystem impact on juvenile salmon over the past three decades, the CRiSP model was used to simulate annual survivals between a Snake River tributary and the Columbia River estuary. The survival, including barge- and river-migrating fish, was high in the late 1960s, dropped to very low values in the 1970s, and then increased (Figure 30.7). Between 1966 and 1969 fish encountered only Ice Harbor dam on the Snake River and three Columbia River dams. The model estimated survival was about 40%. In 1969 Lower Monumental Dam was put in service above Ice Harbor Dam. Because the dam was initially operated without turbines the entire river flow was spilled, creating gas supersaturation levels well above 130%. High spill and dissolved gas levels also occurred when Little Goose and Lower Granite dams were put in service with only half their turbines in 1970 and 1976, respectively. During these years hydro operations were not optimized for fish survival (Raymond 1979; Williams and Matthews 1995). To complicate matters, 1973 had extremely low flows resulting in longer fish travel times and increased exposure to predators. As a result of these construction and operation conditions, the system survival of juvenile fish dropped to under 10%. Following this period, modeled system survival increased for two reasons. First, the supersaturation problems were reduced at Snake River dams when a full complement of turbines were installed. This allowed water to pass through turbines instead of being spilled. Second, survival of smolts to the estuary was increased to 60 to 70% by increased collection and transportation of smolts around dams. In terms of river passage alone, the model indicates that the survival in the 1990s is equal to the survival in the 1960s (Figure 30.7).

These estimated high survivals depend on the high fish survival in barge transportation. This estimate is based on experiments that compared the percent survival-to-adults of fish that were transported to the percent survival-to-adults of fish that migrated through the river (Mundy et al. 1994). The ratio of the two survivals, when combined with in-river survival, provides an estimate of transportation survival. Spring chinook transportation survival was estimated to be 89% in this analysis. That is, 89% of the fish transported through the hydrosystem are estimated to survive juvenile passage.

To explore the counteracting factors hypothesis, a measure of smolt-to-adult survival for Snake River spring chinook has been used (Raymond 1988; Schmitten et al. 1995). The smolt outmigration population size is derived from counts of adult spawners estimated from redd counts on the spawning grounds. The returning adults are estimated from adult escapement counts. The resulting smolt-to-adult survival includes smolt migration survival, survival of adults in the ocean, survival in upriver migration and their spawning success. This survival was high in 1965. It declined and remained low between 1970 and the early 1990s, which are the most recent stock recruitment data (Figure 30.7).

Finally, Figure 30.7 shows a normalized form of the PNI where high values represent the cool/wet regime and low values represent the warm/dry regime. Although it is impossible to identify what parts of the life history are most affected by conditions characterized by the PNI, studies of climatic effects on the North Pacific ecosystem (Pearcy 1992) suggest a survival correlation with the PNI may be the result of conditions affecting smolts during ocean entry.

Over the past three decades, climate and river conditions had nearly opposite trends relative to smolt survival (Figure 30.7). In the early 1970s the PNI climatic index indicated a cool/wet Pacific

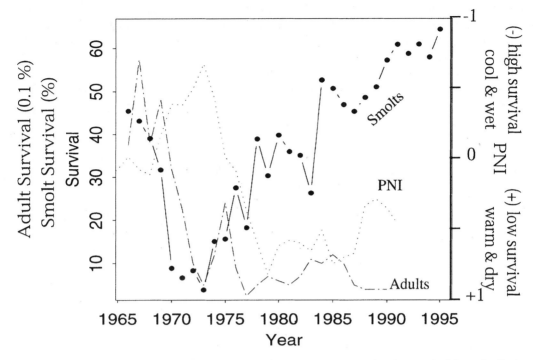

FIGURE 30.7 Trends of PNI climate indicator, CRiSP-estimated smolt migration survival from the Snake River Basin to the Columbia River Estuary and a measure of the smolt-to-adult (×10) survival (%) of Snake River fish.

Northwest climate, favorable to spring chinook, while the smolt river passage survival was at its all time low. After 1975 the climate indicators declined, indicating a shift to a warm/dry, unfavorable climate, while the river survival steadily increased. The opposite trends in the PNI and smolt survival suggest that climate and hydrosystem impacts on salmon have counteracted each other over the past several decades. When smolt river survival was lowest, the climate was most favorable to ocean survival, and when the river survival was highest, the climate was least favorable to salmon survival. The counteraction between hydrosystem and climate survivals should be reflected in the smolt-to-adult survival. This is in fact the case, with the smolt-to-adult survival trend following the smolt passage survival through 1975 and then remaining low with simultaneous warm/dry PNI values but good in-river smolt survivals.

Quantifying Climate and Anthropogenic Processes

Although evidence strongly suggests that decadal-scale climate processes and anthropogenic changes interact and complicate the interpretation of causes for fish stock variations, it is another matter to separate the effects of each process using historical data or new experiments. To evaluate the impact of climatic and anthropogenic variables the climate variables can, in principle, be defined by the three methods identified by Adkison et al. (1996). The impact of anthropogenic actions, though, needs to be quantified by an "anthropogenic index." For example, the hydrosystem impact might be quantified by the number of dams fish pass, an estimate of passage mortality, or, specific to the Columbia River system, the percent of fish transported. We may express the anthropogenic impact of dams on mortality by setting $\alpha(t) = c_i D_{it}$ where D is the number of dams a fish passes, and this may change as dams are constructed. Anthropogenic and climate impacts might also be separated through a "differential influence" method incorporating stocks with a common climate influence but different anthropogenic influences. For the Columbia River, this method requires information on stocks passing through different numbers of dams but experiencing a common climate effect. The two methods are also distinguished in the grouping of stocks used in the analyses. The anthropogenic index method groups stocks with common climate and anthropogenic influences and seeks to resolve the two factors by the temporal changes in the indices. For the Columbia River this means stocks from a single subbasin are used in an analysis. In the differential influence method, stocks from several subbasins are combined. This requires spawner recruitment data from, for example, the Snake River and the lower Columbia stocks.

Several problems make separating climatic and anthropogenic factors difficult with these techniques. An anthropogenic effects method may demonstrate a relationship between stock declines and increases in dam indices but a correlation over time does not, of itself, imply the dams are responsible for stock decline. As such, if actions were taken to remove dams, it is not clear that the stocks decline would reverse. A biological basis is required to connect indices to processes that alter stock productivity. In the differential method the assumption that stocks experience similar climate influences is required, but if the assumption is violated any differential climate impacts can bias estimates of the anthropogenic impact. These types of problems become significant when identifying the potential effect of mitigating the anthropogenic factors through actions. A cautious approach is to apply both methods and evaluate the results by their statistical properties and biological rationales.

CONCLUSION

This chapter illustrates the importance of resolving the interaction of anthropogenic and natural factors when exploring the historical causes of a population decline and in planning stock recovery actions. This analysis shows that the ratchet-like extinction process proposed by Lawson (1993) and Lichatowich and Mobrand (1995) is relevant to Columbia River salmon. It is clear that Columbia River salmon declines are the result of a long history of development of the river system which has been modified by three major climate regime shifts. A significant drop in Columbia River salmon harvest about 1925 (Figure 30.1) marked the beginning of a long stock decline well before the construction of the hydrosystem. This event was likely the result of overharvest being accentuated by the 1925 climate/ocean shift from favorable to unfavorable conditions. The shift of the climate/ocean regime to fish-favorable conditions in 1947 counteracted the detrimental effects of hydrosystem construction between 1932 and 1977. A shift back to fish-unfavorable conditions counteracted the mitigation efforts over the last 20 years.

This interaction of climate and dam operations has more than historical significance: by ignoring climate cycles salmon managers may misinterpret the effectiveness of the recent stock recovery efforts, and particularly they underestimate the benefits of smolt transportation. Several proposed recovery plans for the endangered Snake River salmon (Schmitten et al. 1995; CRITFC 1995; Williams et al. 1996) directly or indirectly assume that the past 20 years of salmon mitigation efforts have been ineffective, since salmon runs have continued to decline. These plans, to differing degrees, propose to de-emphasize fish transportation and advocate inriver passage using higher flows from storage reservoirs, spilling water at dams, and drawing down the reservoirs behind the dams to improve water velocity and fish habitat. The premise is that these actions will mimic the natural river conditions that occurred in the past when fish runs were larger. Whatever the effect of these proposed actions, it is unlikely that they will approach the benefit of a climatic shift back to a cool/wet regime favoring Columbia River salmon. Plans that eliminate beneficial programs, such as fish transportation, will only worsen the present situation.

In a larger perspective the consequences of ignoring the complex and long-term interactions of climatic cycles and anthropogenic actions and focusing instead on simple short-term explanations are significant. Particularly when dealing with endangered species, a narrow focus may lead to misguided beliefs on the success and failures of restoration efforts. It is only through a long-term ecological perspective, embracing natural and anthropogenic interactions, that society can assess the actual impact of human activities and realistically identify options and limitations faced when correcting environmental damage accumulated over decades.

Beyond the issue of rebuilding damaged populations, an understanding of climate cycles is essential for long term sustainable management of fisheries resources. Since climate cycles have such a large impact on stock productivity, an improved understanding of the cycles and their coupling to fisheries should eventually improve our ability to sustain fisheries over periods of scarcity and abundance. For today though, an awareness of the cycles teaches us that achieving sustainability does not mean achieving stability. Stock size will fluctuate by factors outside our control, and through this understanding we have a better chance of avoiding the ratchet-to-extinction.

ACKNOWLEDGMENTS

This work was supported by Bonneville Power Administration under contract DE-BI79-89BP02347. I wish to thank Nathan Mantua and Richard Zabel for their helpful comments.

REFERENCES

Adkison, M.D., R.M. Peterman, M.F. Lapointe, D.M. Gillis, and J. Korman. 1996. Alternative models of climate effects on sockeye salmon, *Oncorhynchus nerka,* productivity in Bristol Bay, Alaska, and the Fraser River, British Columbia. Fisheries Oceanography 5:137–152.

Anderson, J.J., J.A. Hayes, and R.C. Zabel. 1996. Columbia River Salmon Passage Model. CRiSP1.5: Theory, Calibration and Validation. University of Washington, Seattle.

Bakun, A. 1996. Patterns in the Ocean—Ocean Processes and Marine Populations Dynamics. California Sea Grant College System, NOAA.

Baumgartner, T. R., A. Soutar, and V. Ferreira-Bartrina. 1992. Reconstruction of the history of Pacific sardine and northern anchovy populations over the past two millennia from sediments of the Santa Barbara Basin, California. CalCOFI Report 33:24–40.

Beamish, R. J. 1993. Climate and exceptional fish production off the west coast of North America. Canadian Journal of Fisheries and Aquatic Sciences 50:2270–2291.

Beamish, R. J. 1995. Climate Change and Northern Fish Populations. Canadian Special Publication of Canadian Fisheries and Aquatic Sciences 121.

Beamish, R. J., and D. R. Bouillon. 1993. Pacific salmon production trends in relation to climate. Canadian Journal of Fisheries and Aquatic Sciences 50:1002–1016.

Brodeur, R. D., and D. M. Ware. 1992. Long-term variability in zooplankton biomass in the subarctic Pacific Ocean. Fisheries Oceanography 1(1):32–38.

Brodeur, R. D., and D. M. Ware. 1995. Interdecadal variability in distribution and catch rates of epipelagic nekton in the Northeast Pacific Ocean. Pages 329–356 *in* R. J. Beamish, editor. Climate change and northern fish populations. Canadian Special Publication of Canadian Fisheries and Aquatic Sciences 121.

Cayan, D.R., and D.H. Peterson. 1989. The influence of North Pacific atmospheric circulation on the streamflow in the west. Pages 375–398 *in* D. H. Peterson, editor. Aspects of Climate Variability in the Pacific and Western Americas. American Geophysical Union Geophysical Monograph 55.

Clark, W. 1996. Long-term changes in halibut size at age. Pages 55–62 *in* IPHC Report of Assessment and Research Activities, International Pacific Halibut Commission.

Clark, W. G., S. R. Hare, A. M. Parma, P. J. Sullivan, and R. J. Trumble. 1999. Decadal changes in growth and recruitment of Pacific halibut (*Hippoglossus stenolepis*). Canadian Journal of Fisheries and Aquatic Sciences 56:242–252.

Coronado-Hernandez, M. C. 1995. Spatial and temporal factors affecting survival of hatchery-reared chinook, coho and steelhead in the Pacific Northwest. Doctoral Dissertation. University of Washington, Seattle.

CRITFC (Columbia River Intertribal Fisheries Commission). 1995. Wy-Kan-Ush-Mi Wa-Kish-Wit: Spirit of the Salmon. The Columbia River Anadromous Fish Restoration Plan of the Nez Perce, Umatilla, Warm Springs, and Yakama Tribes. Volume I. Portland, Oregon.

Currie, R. G., and R. W. Fairbridge. 1986. Periodic 18.6-year and cyclic 11-year signals in northeastern United States precipitation data. Journal of Climatology 8:255–281.

Dawley, E. M., B. Monk, M. Schiewe, F. Ossiander, and W. Ebel. 1976. Salmonid bioassay of supersaturated dissolved air in water. EPA Ecological Research Series, Report #EOA-600/3-76-056.

Ebbesmeyer, C.C., D. R. Cayan, D. R. McLain, F. H. Nichols, D. H. Peterson, and K.T. Redmond. 1991. 1976 step in the Pacific climate: forty environmental changes between 1968–1975 and 1977–1984. Pages 120–141 in J.L. Betancourt and V.L. Sharp, editors. Proceedings Seventh Annual Pacific Climate (PACLIM) Workshop, April 1990. California Department of Water Resources. Interagency Ecological Studies Program Technical Report 26.

Ebbesmeyer, C.C., and R.M. Strickland. 1995. Oyster condition and climate: evidence from Willapa Bay. Publication WSG-MR 95-02, Washington Sea Grant Program, University of Washington, Seattle.

Ebel, W. J., C. D. Becker, J. W. Mullan, and H. L. Raymond. 1989. The Columbia River—toward a holistic understanding. Pages 205–219 in D.P. Dodge, editor. Proceedings of the International Large River Symposium. Canadian Special Publication of Fisheries and Aquatic Sciences 106.

Emery, W. J., and K. H. Hamilton. 1985. Atmospheric forcing of interannual variability in the North Pacific Ocean; connections with el Niño. Journal of Geophysical Research 90:857–868.

Francis, R.C. 1993. Climate change and salmon production in the North Pacific Ocean. Pages 33–43 in K.T. Redmond and V.J. Tharp, editors. Proceedings of the Ninth Annual Pacific Climate (PACLIM) Workshop, April 21–24, 1992. California Department of Water Resources. Interagency Ecological Study Program Technical Report 34.

Francis, R.C., and S. R. Hare. 1994. Decadal-scale regime shifts in the large marine ecosystems of the Northeast Pacific: a case for historical science. Fisheries Oceanography 3(4):279–291.

Gargett, A.E. 1997. The optimal stability "window": a mechanism underlying decadal fluctuations in North Pacific salmon stocks? Fisheries Oceanography 6:109–117.

Gantz, M. H. 1992. Climate variability, climate change, and fisheries. Cambridge University Press, Cambridge, United Kingdom.

Graham, N. E. 1994. Decadal-scale climate variability in the tropical and North Pacific during the 1970s and 1980s: Observations and model results. Climate Dynamics 10:135–162.

Helle, J.H., and M. S. Hoffman. 1995. Size decline and older age at maturity of two chum salmon (Oncorhynchus keta) stocks in the western North America, 1972–92. Pages 245–260 in R. J. Beamish, editor. Canadian Special Publication of Canadian Fisheries and Aquatic Sciences 121.

Hollowed, A. B., and W. S. Wooster. 1992. Variability of winter ocean conditions and strong year classes of Northeast Pacific groundfish. ICES Marine Science Symposia 195:433–444.

Ingraham, W. J., Jr., C. C. Ebbesmeyer, and R. A. Hinrichsen. 1998. Imminent climate and circulation shift in Northeast Pacific Ocean could have major impact on marine resources. Eos 79(6):197.

Kawasaki, T. 1984. Why do some fishes have wide fluctuations in their numbers? A biological basis of fluctuation from the viewpoint of evolutionary ecology. Pages 1065–1080 in G. D. Sharp and J. Csirke, editors. Proceedings of the Expert Consultation to Examine Changes in Abundance and Species Composition of Neritic Fish Resources. FAO Fisheries Report 291(3).

Kerr, R. A. 1995. A fickle sun could be altering earth's climate after all. Science 269:633.

Lagerloef, G.E. 1995. Interdecadal variations in the Alaska Gyre. Journal of Physical Oceanography 25:2242–2258.

Lawson, P. W. 1993. Cycles in ocean productivity, trends in habitat quality, and the restoration of salmon runs in Oregon. Fisheries 18(8):6–10.

Lichatowich, J. E., and L. E. Mobrand. 1995. Analysis of chinook salmon in the Columbia River from an ecosystem perspective. Bonneville Power Administration, U.S. Department of Energy Publication DOE/BP-251-5-2, Portland, Oregon.

Mantua, N.J., S.R. Hare, Y. Zhang, J.M. Wallace and R.C. Francis. 1997. A Pacific Interdecadal climate oscillation with impacts on salmon production. Bulletin of the American Meteorological Society 78:1069–1079.

Miller, A. J., D.R. Cayan, T. P. Barnett, N. E. Graham, and J. M. Oberhuber. 1994. The 1976–77 climate shift of the Pacific Ocean. Oceanography 7:21–26.

Mundy, P. R., and nine coauthors. 1994. Transportation of juvenile salmonids from hydroelectric projects in the Columbia River Basin; an independent peer review. Final Report. U.S. Fish and Wildlife Service, Portland, Oregon.

Parker, K. S., T. C. Royer, and R. B. Deriso. 1995. High-latitude climate forcing and tidal mixing by the 18.6-year lunar nodal cycle and low-frequency recruiting trends in Pacific halibut (Hippoglossus stenolepis). Pages 447–459 *in* R. J. Beamish, editor. Climate change and northern fish populations. Canadian Special Publication of Canadian Fisheries and Aquatic Sciences 121.

Pearcy, W. 1992. Ocean ecology of the North Pacific salmonids. University of Washington Press, Seattle.

Raymond, H. L. 1979. Effects of dams and impoundments on migrations of juvenile chinook salmon and steelhead from the Snake River, 1966 to 1975. Transactions American Fisheries Society 108(6): 505–529.

Raymond, H. L. 1988. Effects of hydroelectric development and fisheries enhancement on spring and summer chinook salmon and steelhead in the Columbia River basin. North American Journal of Fisheries Management 8:1–24.

Reed, R.K. 1984. Flow of the Alaskan Stream and its variations. Deep-Sea Research 31, 369–386.

Richards, K. J., and D. Olsen. 1993. Inter-basins comparison study: Columbia River salmon production compared to other West Coast production areas. Phase I. Study sponsored by the U.S. Army Corps of Engineers, Kennewick, Washington.

Rieman. B.E., R.C. Beamesderfer, S. Vigg, and T.P. Poe. 1991. Estimated loss of juvenile salmonids to predation by northern squawfish, walleyes and small mouth bass in John Day Reservoir, Columbia River. Transactions American Fisheries Society 120:448–458.

Roemmich, D., and J. McGowan. 1995. Climate warming and the decline of zooplankton in the California current. Science 267:1324–1326.

Rothschild, B. J. 1995. Fishstock fluctuations as indicators of multidecadal fluctuations in the biological productivity of the ocean. Pages 107–117 *in* R.J. Beamish, editor. Climate change and northern fish populations. Canadian Special Publication of Canadian Fisheries and Aquatic Sciences 121.

Schmitten, R., W. Stelle, Jr., and R. P. Jones. 1995. Proposed recovery plan for Snake River salmon. U.S. Department of Commerce National Oceanographic and Atmospheric Administration, National Marine Fisheries Service. March 1995.

Shaffer, M. L. 1981. Minimum population sizes for species conservation. Biosicence 31:131–134.

Soulé, M. E., editor. 1987. Viable populations for conservation. Cambridge University Press, Cambridge, United Kingdom.

Smith, P. E. 1978. Biological effects of ocean variability: time and space scales of biological response. Rapports Pour-von Reunion de conseil internacionale Exploracion du Mer 173:117–127.

Sharp, G. D. 1992. Fishery catch records, El Niño/southern Oscillation, and longer-term climate change as inferred from fish remains in marine sediments. Pages 379–417 *in* H. E. Diaz and V. Markgraf, editors. El Niño: historical and paleoclimatic aspects of the southern oscillation. Cambridge University Press, Cambridge, United Kingdom.

Stouder, D. J., P. A. Bisson, and R. J. Naiman, editors. 1997. Pacific Salmon and their ecosystems status and future options. Chapman & Hall. New York.

Thompson, G. G. 1991. Determining minimum viable populations under the Endangered Species Act. NOAA Technical Memorandum NMFS F/NWC-198.

Valdivia, J. 1978. Anchoveta and El Niño. Rapports Pour-von Reunion de conseil internacionale Exploracion du Mer 173:196–202.

van Amerongen, J. 1995. Bay harvest sets new record. Alaska Fisherman's Journal 18(9):6.

van Loon, H., and K. Labizke. 1994. The 10–12 year atmospheric oscillation. Meterologie Zeitschrift 3:259.

Venrick, E. L., J. A. McGowan, D. R. Cayan, and T. L. Hayward. 1987. Climate and chlorophyll a: Long-term trends in the central North Pacific Ocean. Science 238:70–72.

Ware, D. M., and G. A. McFarlane. 1989. Fisheries production domains in the Northeast Pacific Ocean. Pages 359–379 *in* R. J. Beamish and G.A. McFarlane, editors. Effects of ocean variability on recruitment and an evaluation of parameters used in stock assessment models. Canadian Special Publication of Canadian Fisheries and Aquatic Sciences 108.

Ware, D. M., and R. E. Thompson. 1991. Link between long-term variability in upwelling and fish production in the northeast Pacific ocean. Canadian Journal of Fisheries Aquatic Sciences 48(12):2296–2306.

Ware, D. M. 1995. A century and a half of change in the climate of the NE Pacific. Fisheries Oceanography 4(4):267–277.

WDF and ODFW (Washington Department of Fisheries and Oregon Department of Fish and Wildlife). 1992. Status report: Columbia River fish runs and fisheries, 1938–91.Washington Department of Fisheries and Oregon Department of Fish and Wildlife, Portland, Oregon.

Wickett, W. P. 1967. Ekman transport and zooplankton concentration in the North Pacific Ocean. Journal of the Fisheries Research Board of Canada 24(3):581–594.

Williams, J. G., and G. M. Matthews. 1995. A review of flow and survival relationships for spring and summer chinook salmon, *Oncorhynchus tshawytscha*, from the Snake River Basin. Fishery Bulletin 93:732–740.

Williams, R. N., and 11 coauthors. 1996. Return to the river: restoration of salmonid fishes in the Columbia River ecosystem. Development of an alternative conceptual foundation and review and synthesis of science underlying the Fish and Wildlife Program of the Northwest Power Planning Council. Northwest Power Planning Council 96-6, Portland, Oregon.

Wissmar, R. C., J. E. Smith, B. A. McIntosh, H. W. Li, G. H. Reeves, and J. R. Sedell. 1994. A history of use and disturbance in riverine basins of eastern Oregon and Washington (Early 1800s–1900s). Northwest Science 68:1–35.

Wong, C. S., F. A. Whitney, K. Iseki, J.S. Page, and J. Zeng. 1995. Analysis of trends in primary productivity and chlorophyll-a over two decades at Ocean Station P (50°N, 145°W) in the Subarctic Northwest Pacific Ocean. Pages 107–117 *in* R. J. Beamish, editor. Climate change and northern fish populations. Canadian Special Publication of Canadian Fisheries and Aquatic Sciences 121.

Zabel, R., and J.J. Anderson. 1997. A model of the travel time of migrating juvenile salmon, with an application to Snake River spring chinook. North American Journal of Fisheries Management 17:93–100.

31

The Effect of Environmentally Driven Recruitment Variation on Sustainable Yield from Salmon Populations

Steven P. Cramer

Abstract.—Use of the deterministic maximum sustainable yield (MSY) concept has led to unrealistic expectations for harvest and escapement of salmon populations. Fits of stock-recruitment data to a Ricker curve for Pacific salmon typically show that most variation in recruitment is environmentally driven. Frequency distributions for the number of recruits produced per spawner from several coho salmon *Oncorhynchus kisutch* and chinook salmon *O. tshawytscha* populations show that recruitment varies among years by more than an order of magnitude, and that this variation is strongly skewed with a preponderance of below-average recruitment rates. The same is true for smolt-to-adult survival of hatchery coho and chinook populations, which indicates that a large share of the variability in survival is generated while salmon are in the ocean. Stochastic simulations, based on Ricker parameter values estimated from naturally produced coho in the Nehalem River, Oregon, were used to illustrate how environmental variation affects the sustainability of catch and escapement. This chapter does not develop new stock-recruitment theory but instead illustrates often-neglected aspects of its application to harvest management. The skewed, or log-normal, nature of environmental variation causes catch and escapement to be less than the deterministically predicted equilibrium in about 60% of all years. Options for adjusting harvest rate annually to reflect variations in survival for a specific stock are few, because recruitment varies asynchronously among populations harvested predominantly in mixed stock fisheries. By reducing the harvest rates on Oregon coho to about two thirds of the theoretical MSY level, total catch over the long term would remain near the maximum, but spawning escapement would double. Increased escapement would reduce genetic risks and sustain greater marine nutrient supply to rearing areas for juvenile salmonids.

INTRODUCTION

Recruitment data for anadromous salmonids show large inter-annual variation that is distinct from the variation caused by density-dependent mechanisms. This variation appears to be driven by environmental factors in both fresh and saltwater, and it often substantially exceeds the density-dependent variation in recruitment. For example, Pearcy (1992) summarizes a large body of evidence that variation in marine survival of anadromous salmonids often exceeds the variation in freshwater survival. Yet, management of harvest and the extent of supplementation from hatcheries for most anadromous salmonid populations on the West Coast has been based primarily on the dome-shaped recruitment models of Ricker (1954) or Beverton and Holt (1957). The Ricker and Beverton-Holt models were formulated to account for variation in recruitment controlled by density-dependent factors, particularly in freshwater spawning habitats and, in their standard forms, these models do not incorporate parameters for environmental variation.

Visual inspection of the Ricker or Beverton-Holt curves fitted to examples of actual stock-recruitment data for Pacific salmon shows that a substantial portion of the variation in recruitment is typically not accounted for by these curves (Figure 31.1). The repeated visual imprint left in our minds by plotting these curves can easily lead us to an expectation that fish populations will behave just like the theoretical curve, rather than like the less intuitive scatter of data points. The history of overharvest that is prevalent throughout major fisheries of the world is a testimony to how quickly fisheries managers forget that the number of recruits produced per spawner (recruitment rate) is not simply a function of parental stock size, but is also heavily influenced by environmental variation. Too often, harvest rates are set based on an expectation that future recruitment rates will mimic above-average recruitment rates that have occurred in the past. The usual practice of predicting future recruitment as a point estimate has also fostered false expectations of consistently high yield among resource agencies and stakeholders, followed by conflict between these entities when the realized yield each year frequently falls short of expectations.

Recent declines in exploited fish populations, such as salmon along the West Coast (Nehlsen et al. 1991), have fueled discussions for developing new strategies and goals for harvest management (e.g., Malvestuto and Hudgins 1996). Many workers have concluded that adherence to a goal of achieving maximum sustained yield (MSY) in the face of large scale changes in climate and ocean productivity along the West Coast have led to widespread overexploitation and decline of salmon and steelhead populations. Many of the mathematical approaches being offered to account for environmental variation in recruitment (i.e., Bayesian statistics; McAllister et al. 1994; Walters and Punt 1994) are unfamiliar to fisheries managers and are not being rapidly integrated into management strategies. Although there is now a widespread awareness that environmental variation has a substantial influence on recruitment, few biologists have a clear understanding of how this variation influences the sustainability of various harvest levels and spawner escapements.

In this chapter, I explore examples of actual time-series recruitment data from coho and chinook salmon populations on the West Coast to show the magnitude and pattern of environmentally caused variation in recruitment rate. The Ricker function was then used to account for density-dependent variability, with a factor for variable environmental effects on survival added to simulate the variation in catch and escapement that will be realized at various harvest rates. This chapter does not develop new stock-recruitment theory, but illustrates often-neglected aspects of its application to harvest management. The simulation outcomes can be used to compare the frequency and range of different levels of catch and escapement that should be expected across a set of years, given a fixed harvest rate policy.

METHODS

To establish the level of variation common to the recruitment rate of anadromous salmonids, indices of recruitment rate were calculated for several coho, spring chinook, and fall chinook populations on the West Coast of the U.S. Age-3 recruits produced per spawner was used as an index of recruitment rate for coho salmon. Age-3 coho recruits were defined as the number of age-3 fish included in ocean and freshwater catch plus spawner escapement. Cohort analysis (PSC 1988) was used with recoveries of coded wire tagged (CWT) fish to back-calculate an estimate for survival of hatchery smolts until they reached age 2 in the ocean. This survival estimate was used as an index of ocean survival for Cowlitz River spring chinook, fall chinook, and coho. Because natural mortality rate after age 2 was assumed to be fixed, all variation in natural survival was expressed in the estimate of survival from smolt to age 2. Inasmuch as the assumption of constant, age-specific mortality was accurate, the estimate of survival to age 2 was independent of the estimates of harvest rate. Furthermore, Pearcy (1992) demonstrated that year-class success of coho was largely determined within their first month in the ocean.

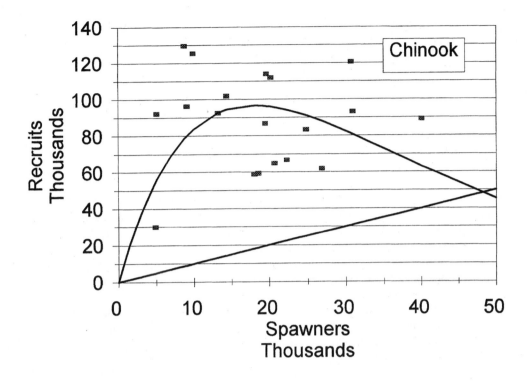

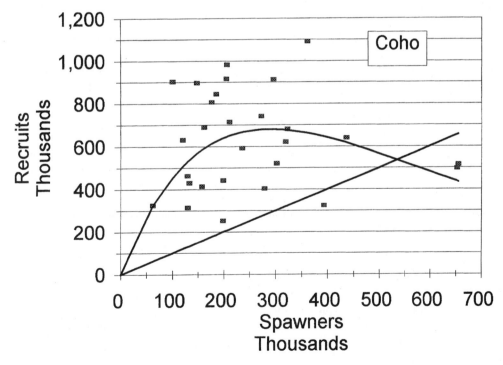

FIGURE 31.1 Examples of Ricker stock-recruitment curve fitted to actual data on spring chinook from the Rogue River, Oregon (top graph from Cramer et al. 1985), and to naturally-produced coho in Oregon coastal streams (bottom graph adapted from ODFW 1982).

The number of age-3 recruits per coho spawner was estimated as follows:

$$R_{i+3} = \{[S_{i+3}/(1-Rharv_{i+3})]/(1-Oharv_{i+3})\}$$ (31.1)

where
R_{i+3} = recruits at age 3 before annual harvesting begins;
S = spawners;
i = year;
$Rharv$ = river harvest rate (proportion of run harvested); and
$Oharv$ = ocean harvest rate (proportion of ocean cohort harvested).

For coho in three coastal basins of Oregon, I used the peak count of spawners per mile as an index of spawners in the above equation. Spawners per mile have been counted by staff of the Oregon Department of Fish and Wildlife (ODFW) since 1949 (Cooney and Jacobs 1994), and I summed the peak counts in standard stream sections across the tributaries surveyed within a coastal basin. Only tributaries reported to receive little or no contribution from hatchery coho spawners by ODFW (Cooney and Jacobs 1994) were included; therefore indices were limited to five streams in the Nehalem Basin, two streams of the Umpqua Basin, and four streams in the Coquille Basin.

For the process of estimating past recruitment rates, sport and commercial harvest in freshwater also had to be accounted for. I assumed freshwater sport harvest was 7% for the Nehalem and Umpqua rivers, and 3.5% for the Coquille River. These rates were the approximate average for 5 years during the 1980s, as determined from ODFW estimates of spawner escapement (Cooney and Jacobs 1994) and the angler punch card estimates of coho catch (data available from Fisheries Information Systems, ODFW, Portland). Commercial harvest, conducted only in tidewater portions of the rivers, was closed after 1956 in the Nehalem and Coquille rivers, and after 1947 in the Umpqua River. I assumed the harvest rate by in-river commercial fisheries during 1950–1956 was 40% in the Nehalem River (based on Wallis 1961b), and 50% in the Coquille River (based on Wallis 1961a). These harvest rates are slightly lower than the mark-recapture estimates of Wallis (1961a and 1961b) to account for a 50% reduction in commercial effort on the Nehalem River after 1950, and a 40% reduction in commercial effort on the Coquille River after 1950 (Mullen 1981).

Ocean harvest rates were assumed to be equal to those for the Oregon Production Index (OPI), as explained below. However, for the Nehalem and Umpqua rivers, in the 8 years for which data were available, I substituted the harvest rates estimated by cohort analysis of CWT recoveries from coho released in the North Fork of the Umpqua River. Coho returning to the North Umpqua River are counted as they pass Winchester Dam, and these counts are the most accurate escapement data of any stock on the Oregon coast. Thus, ocean harvest rate can be estimated through cohort analysis most accurately for the Umpqua stock. The ocean harvest distribution of Umpqua coho is similar to all other Oregon coastal streams north of there, including the Nehalem River (Lewis 1994). Coho from the Coquille River, south of the Umpqua, tend to have a more southerly ocean distribution (Lewis 1994), so I used the OPI harvest rate in all years for the Coquille River stock.

I revised the estimates of OPI harvest rate reported by PFMC (1995). The OPI represents the Pacific coastal water bounded on the north by Leadbetter Point, WA, and to the south by Monterey Bay, CA. The abundance of adult coho in the OPI is estimated as the sum of sport and commercial landings at all ports ($OCATCH_{OPI}$) plus the sum of all spawners (excluding jacks) escaping to public hatcheries, private hatcheries, and natural spawning areas ($ESCAPE_{OPI}$). The OPI harvest rate ($OHARV_{OPI}$) is the proportion of the total adults that are landed. That is:

$$OHARV_{OPI} = OCATCH_{OPI}/(ESCAPE_{OPI} + OCATCH_{OPI})$$ (31.2)

The portion of the $ESCAPE_{OPI}$ term made up by Oregon's coastal natural (OCN) stocks each year since 1950 has been estimated by ODFW using peak counts of spawning coho on standardized surveys of 48 stream segments from 17 different coastal river basins (Cooney and Jacobs 1994). Extrapolations of spawning fish survey counts have been used for estimating the spawning escapement of Oregon coastal natural (OCN) stock since 1981. Cooney and Jacobs (1994) found from expanded spawning surveys that many coastal streams counted as producing coho had none, so I reduced the number of stream miles used to calculate OCN escapement by the 9% of miles found not to have coho. Cooney and Jacobs (1994) also found that the index streams had an average spawner density that was 4.8 times higher than in randomly selected sections of coho streams along the Oregon coast in 1991 and four times that of randomly selected sections in 1992. Spawner escapements were low in 1991 and 1992, so the difference between spawner densities in standard and random survey areas may have been greater than in years of high escapement when coho are likely to be more evenly distributed. Accordingly, I assumed that true natural escapement across all years was one third of the extrapolated ODFW estimates when the OPI harvest rates were calculated. This assumption had little influence on the estimate of OPI harvest rate, because OCN escapement was generally less than 20% of the OPI escapement.

A complete data set for both catch and spawning escapement of coho in the Oregon Production Index (OPI) area, from which to estimate ocean harvest rate, extends back to 1970. I extended harvest rate estimates for the OPI area back to 1950, based on harvest rates estimated from mark–recapture studies, and on ancillary data for harvesting effort. Early mark–recapture studies indicated that the ocean harvest rate of Alsea Hatchery coho was 47% in 1947 (Wallis 1963) and of Coos River Hatchery coho was 39% in 1952 and 54% in 1953 (Wallis 1961). These estimates are probably biased high, because there is no indication in the reports that fish spawning below the hatchery rack were accounted for. Based on these data, I assumed ocean harvest rates were 45% between 1950 and 1960 (Table 31.1), a period of stable harvesting effort (Johnson 1984).

Data on harvesting effort were used to interpolate harvest rates in the ocean between 1960 and 1970. Indices of harvest effort by the ocean troll fleet and recreational anglers approximately doubled during the 1960s (Johnson 1984). Because harvesting effort increased at a roughly linear rate between 1960 and 1976, I assumed that harvest rates increased linearly also. The estimated OPI harvest rate in 1976 was 89%, so the average increase in ocean harvest rate was 2.75% per year between 1960 and 1976, and I applied this increase to the years 1961–1970 when harvest rate was not estimated (Table 31.1).

Estimated survival from smolt-to-age-2 in the ocean was used as an index of ocean survival for Cowlitz River Hatchery spring chinook, fall chinook, and coho. I used cohort analysis of CWT recoveries from 76 groups of spring chinook, 19 groups of fall chinook and 160 groups of coho released from Cowlitz Salmon Hatchery to estimate mean smolt-to-age-2 survival for most of the 1967–1991 broods.

The cohort analysis used here is simply an expanded accounting system of the number of fish from each CWT-marked release group that were caught in ocean fisheries, caught in river fisheries, escaped to spawn in the river, or escaped to spawn in a hatchery. Additionally, estimates of the number of fish that died between each of these events are incorporated into the accounting. The procedure begins with the oldest age group and works back through time to reconstruct the population at successively younger ages. Calculations are based on the relationship:

$$\text{Recruit}(i) = \text{Remain}(i+1) + \text{Spawn}(i) + \text{Stray}(i) + \text{Rcatch}(i) + \text{Ocatch}(i) \qquad (31.3)$$

where
$\text{Recruit}(i)$ = number of fish alive at age i before harvest begins;
$\text{Remain}(i)$ = number of fish remaining alive in the ocean after fish maturing at that age left to spawn;
$\text{Spawn}(i)$ = number of fish at age i returning to the hatchery;
$\text{Stray}(i)$ = number of fish at age i that spawn in the river or other hatcheries;
$\text{Rcatch}(i)$ = number of fish at age i caught in river fisheries; and
$\text{Ocatch}(i)$ = number of fish at age i caught in ocean fisheries.

TABLE 31.1

Age-3 coho ocean harvest rates used in estimates of coho recruits per spawner for Oregon coastal stocks. Harvest rates were derived from the OPI index as described in the text. Values for Umpqua were estimated by cohort analysis of CWT recoveries from North Umpqua stock, and those values were used for Nehalem and Umpqua river coho.

Year of catch	% Harvested		Year of catch	% Harvested	
	Umpqua	OPI		Umpqua	OPI
1950	—	45.0%	1972	—	83.3%
1951	—	45.0%	1973	—	81.6%
1952	—	45.0%	1974	—	82.6%
1953	—	45.0%	1975	—	80.1%
1954	—	45.0%	1976	—	89.2%
1955	—	45.0%	1977	—	87.9%
1956	—	45.0%	1978	—	81.4%
1957	—	45.0%	1979	—	76.6%
1958	—	45.0%	1980	—	70.9%
1959	—	45.0%	1981	—	76.4%
1960	—	45.0%	1982	—	58.1%
1961	—	47.8%	1983	81.0%	70.5%
1962	—	50.5%	1984	26.9%	27.9%
1963	—	53.3%	1985	36.8%	34.3%
1964	—	56.0%	1986	32.8%	28.7%
1965	—	58.8%	1987	77.0%	60.1%
1966	—	61.5%	1988	74.5%	54.8%
1967	—	64.2%	1989	61.7%	53.8%
1968	—	67.0%	1990	71.6%	67.0%
1969	—	69.7%	1991	42.0%	42.3%
1970	—	72.5%	1992	—	50.0%
1971	—	81.5%	1993	—	39.0%

I began the analysis at age 6 for spring chinook, by assuming that no fish would remain alive after age 6, and at age 3 for coho. At each life stage in the accounting, factors were incorporated for unaccounted fates of fish, such as natural mortality and non-landed fishing mortality. Natural mortality during winters in the ocean for both chinook and coho was assumed to be 0.4 between age 2 and 3, 0.3 between ages 3 and 4, 0.2 between ages 4 and 5, and 0.1 between ages 5 and 6, following the assumptions adopted by Pacific Salmon Commission (PSC 1996). Pre-spawning mortality after adults had entered freshwater was assumed to be 20% for spring chinook, 10% for fall chinook, and negligible for coho. Data on the number of CWT fish spawning naturally were derived from records maintained by Washington Department of Fish and Wildlife (WDFW; G. Norman, WDFW, personal communication) for the Cowlitz River, and by Oregon Department of Fish and Wildlife (ODFW; C. Corarino, ODFW, personal communication) for coastal coho stocks in Oregon. Non-landed hooking mortality in the ocean was incorporated into Ocatch, and was derived for Cowlitz spring and fall chinook from age-specific estimates by PSC (1996) for Cowlitz fall chinook, and was assumed to be 5% of the landed catch for coho (Stohr and Fraidenburg 1986).

Recruitment data were used to view patterns of interannual variability in recruits per spawner, to establish and compare frequency distributions of recruits per spawner between stocks, and to estimate

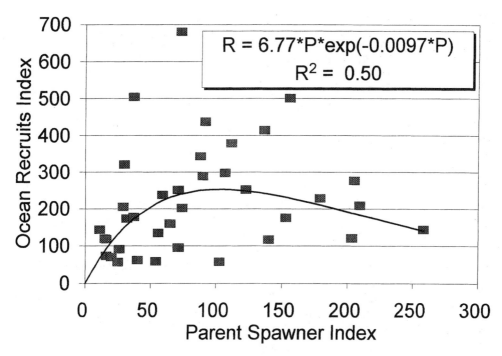

FIGURE 31.2 Fit of spawner and recruitment data to Ricker curve for Nehalem River coho, 1950–1990 broods.

stock recruitment functions. Parameters of the Ricker stock-recruitment curve were estimated for Nehalem coho by regressing ln(R/P) on P, and the residual error variance from the regression was used in simulations as the measure of environmental variation.

The effects of varying recruitment rates on sustainable yield were evaluated with a simple stochastic simulation of escapement and harvest from a hypothetical coho population in Oregon. The model assumed that recruitment was density dependent, following a Ricker curve, and that environmental variation was lognormal. That is

$$R = \alpha P e^{-\beta P} e^{\varepsilon} \tag{31.4}$$

where R = number of recruits, P = number of parents, and ε is a normally distributed variate with mean of 0.0 and a standard deviation of σ_{ε}. The value of α was set at 6.6, as estimated by ODFW (1982) for Oregon coastal coho, and which is similar to the value estimated here for Nehalem coho. The value of β was set arbitrarily at 1/10,000, which would represent a spawning escapement of 10,000 fish to produce maximum recruitment. As recommended by Hilborn (1985), stochastic environmental variation was estimated by the residual error variance from the regression of ln(R/P) on P, which in this case was $\sigma_{\varepsilon} = 0.632$ for Nehalem coho, 1950–1990 broods (see Figure 31.2). Thus, recruitment in each year was multiplied by an environmental scalar e^{ε} for which ε varied randomly within a normal distribution of mean 0.0 and $\sigma_{\varepsilon} = 0.632$. This simulated environmental variation reflects the magnitude of variation actually observed for Nehalem coho.

Simulations were repeated to generate 50 trials of 100 future broods for each of three fixed harvest rates. Those harvest rates at age 3 in the ocean were 70%, 50%, and 30%, while harvest rate in freshwater remained fixed at 15%. Harvest rates in freshwater were not varied in the simulations, because regulatory change has typically been focused on the much higher harvest rates in the ocean. Each trial of 100 years was initiated by recruitment of 7,000 coho for the first 3 years.

RESULTS

All indices of recruitment rate for salmon varied widely among years. Among OCN coho stocks, the rate of recruitment peaked in different years on the different streams, and varied more than 50-fold (Figure 31.3). Recruitment rate for coho appeared to be most stable on the Umpqua River and most variable on the Nehalem River. Further, all frequency distributions of observed recruits per spawner were skewed (protracted) to the right (Figure 31.4), and were approximated by a lognormal distribution. Thus, there were more years of below-average recruitment rate than there were years above average.

Smolt-to-age-2 survival estimated by cohort analyses for fish from Cowlitz Salmon Hatchery also showed wide variation that followed a lognormal distribution (Figure 31.5). Brood year means varied 10-fold for spring chinook, 20-fold for fall chinook, and 30-fold for coho during the 1967–1991 broods (Figure 31.5). A Kolmogorov-Smirnov test to determine whether recruits per spawner or smolt-to-age-2 survival for each of the above data sets was adequately represented by a lognormal distribution, indicated no lack-of-fit ($P > 0.5$).

Of the coho data I examined, only on the Nehalem River was there an indication that variation in recruits per spawner was related to spawner abundance, and might be described by a Ricker stock-recruitment curve (Figure 31.2). A least-squares fit to the Ricker curve accounted for only 50% of the variation in recruits per spawner. This is the regression from which I used the estimate of $\sigma_\varepsilon = 0.632$ for the simulation modeling. The $\alpha = 6.77$ value from this curve is similar to the estimate of $\alpha = 6.6$ for OPI coho (ODFW 1982).

The coho runs generated by simulation ranged widely for both catch and spawner escapement at each of the three harvest rates simulated. At all harvest rates, spawner escapement and catch each dropped below 500 in some years and exceeded 20,000 in others (Figures 31.6 and 31.7). Thus, there was nearly complete overlap in the range of observed catches and escapements at each of the three harvest rates, but the frequency that each catch or escapement level was observed differed between harvest rates.

Median values of catch and escapement across the 50 trials for each run year varied within a stable range throughout the 100-year simulation period that was characteristic for each harvest rate tested. Median catches were always less at the 30% ocean harvest rate than at either the 50% or 70% harvest rates (Figure 31.8), but median catches from the 50% and 70% ocean harvest rates varied within similar ranges and often crossed one another (Figure 31.8). In contrast, the drop in ocean harvest rate from 70% to 50% resulted in substantially greater median spawner escapements, and spawner escapements were greatest when ocean harvest rate was limited to 30% (Figure 31.8).

The median and the average differed substantially from each other in all simulation outcomes. Because of the skewed nature of the frequency distributions for both catch and spawner escapement, the median was several thousand fish less than the arithmetic average for each set of 5,000 runs (50 trials × 100 broods) (Table 31.2). The medians of simulated catches and escapements were about 20% lower than the average, but the directly calculated equilibriums from the Ricker function were close to the average calculated from stochastic simulations (Table 31.2). The directly calculated equilibrium is an average, and given a large enough sample size, the average of the stochastic simulations should be the same as the deterministic equilibrium. Stochastically simulated catch and escapement for *individual* years fell below the deterministic equilibrium values in about 60% of years at each of the three harvest rates simulated (Figures 31.6 and 31.7).

DISCUSSION

The cited examples of actual data indicate that the lognormal distribution of survivals is probably typical of coho and chinook salmon populations. Peterman (1981) showed that variation in marine survival for coho, chum, and sockeye salmon was consistent with a lognormal distribution, and Hilborn and Walters (1992) recommended that a lognormal distribution be the starting assumption

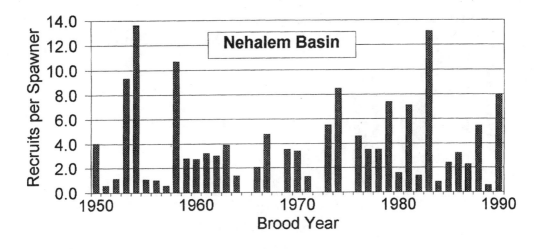

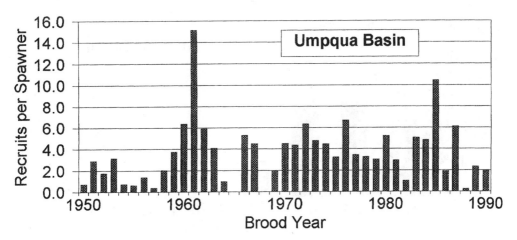

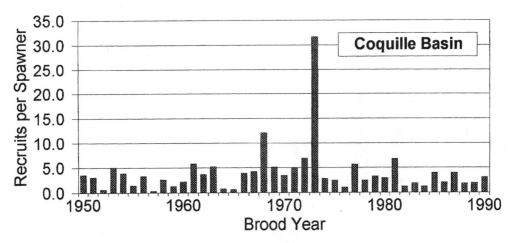

FIGURE 31.3 Annual estimates of age-3 recruits produced per coho spawner in the Nehalem River, Umpqua River and Coquille River, 1950–1990 broods. Estimates are based on peak spawner counts in standard sections of the following creeks within each river basin. Nehalem River: North Cronin, West Humbug, Hamilton, Oak Ranch, and North Wolf creeks. Umpqua River: Buck and Schofield creeks. Coquille River: North Coquille River, Cherry, Steel, and Salmon creeks.

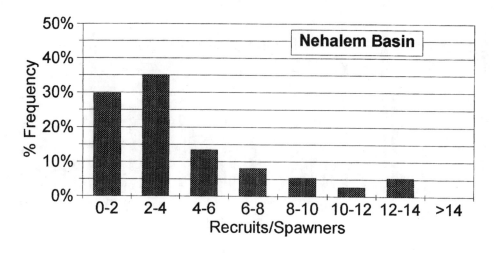

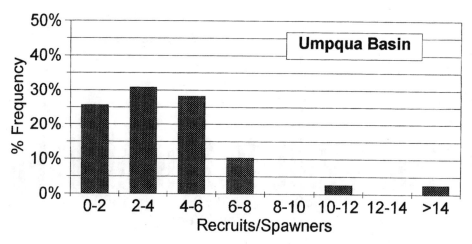

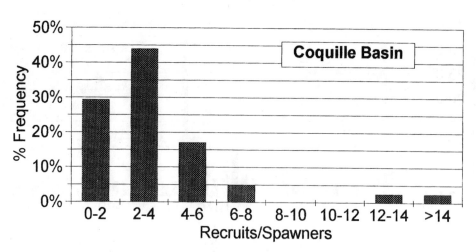

FIGURE 31.4 Frequency histogram of recruits produced per spawner from the 1950–1990 broods of coho in index streams of the Nehalem, Umpqua, and Coquille rivers.

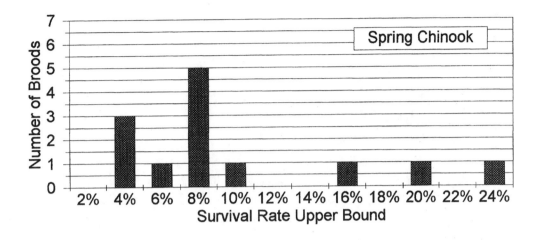

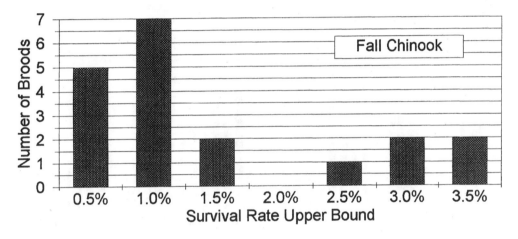

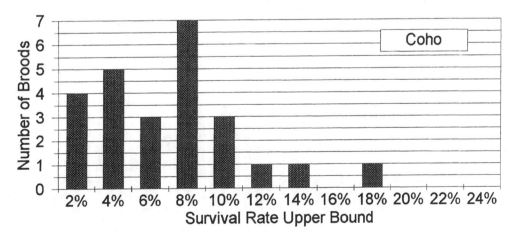

FIGURE 31.5 Frequency histogram of brood-year means for smolt-to-age-2 survival rates of spring chinook, fall chinook, and coho from Cowlitz Salmon Hatchery, as estimated by cohort analysis of CWT groups from the 1967–1991 broods (from Cramer 1996).

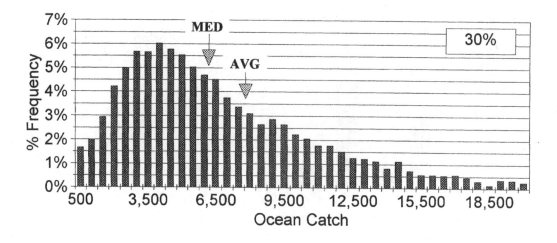

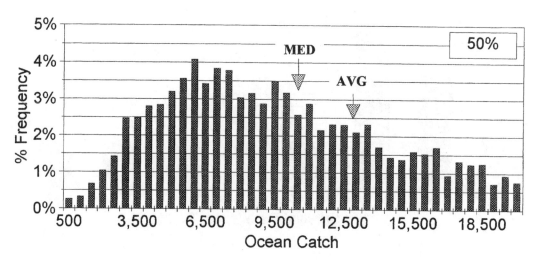

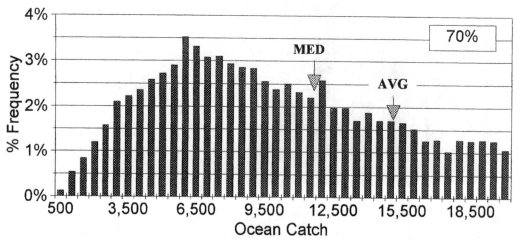

FIGURE 31.6 Frequency histograms of annual ocean catches simulated in 50 trials of 100 years each for a coho population with ocean harvest rate (Oharv) set at 30% (top graph), 50% (middle graph), and 70% (bottom graph). Median (MED) and average (AVG) values are indicated.

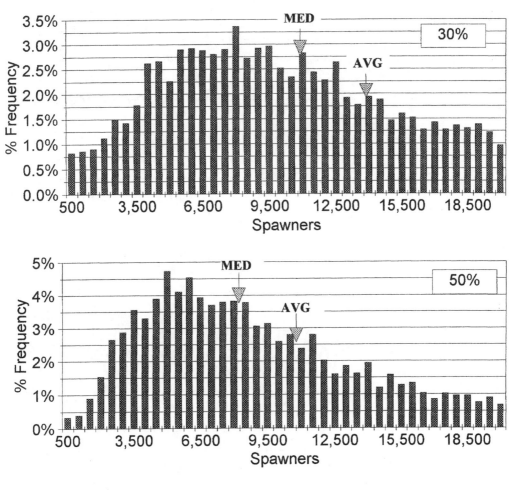

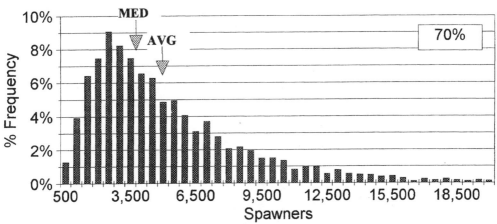

FIGURE 31.7 Frequency histogram of annual spawner abundance simulated in 50 trials of 100 years each for a coho population with ocean harvest rate (Oharv) set at 30% (top graph), 50% (middle graph), and 70% (bottom graph).

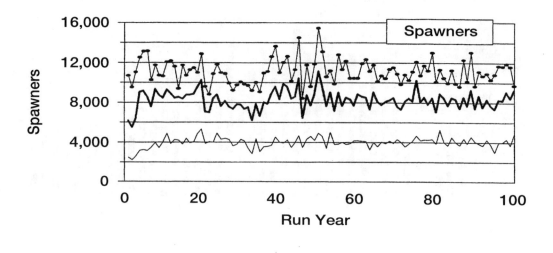

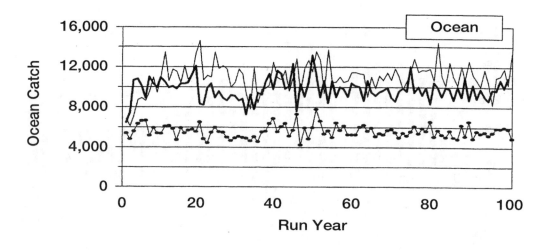

FIGURE 31.8 Median spawner escapement (top graph) and ocean catch (bottom graph) of coho each year across 50 simulation trials with ocean harvest rate (Oharv) set at 30%, 50%, and 70%.

for stock-recruitment analyses. Given this form of variation, the majority of years produce lower catches and escapements at a given harvest rate than the simple point estimates calculated from the Ricker curve. Because most fishery managers are unaware of this phenomenon, they have tended to overestimate harvestable surpluses in a majority of years.

TABLE 31.2
Values for three statistics used to express long-term expectations for catch and spawner escapement of the simulated coho population. Catch includes ocean and freshwater. Median and average were calculated from the same simulations.

| Harvest rates | | Catch | | | Spawners | | |
| | | Deterministic equilibrium | Stochastic | | Deterministic equilibrium | Stochastic | |
Ocean	Total		Median	Average		Median	Average
30%	40.5%	9,276	7,443	9,366	13,700	11,022	13,827
50%	57.5%	13,970	11,207	14,056	10,300	8,307	10,437
70%	74.5%	15,204	11,786	15,399	5,200	4,039	5,297

The stochastic simulations also provide a vivid illustration of a point that Hilborn (1985) noted analytically, i.e., the arithmetic average recruitment at any stock size is greater than the mode. The average is also greater than the median. This is because the average is elevated by the few very high values in the right tail of the distribution. It is important that fishery managers and harvesters recognize from these examples that realized catches and escapements will be less than the value calculated by plugging parent spawners into a Ricker equation in about 60% of all years. Further, catch and escapement must be expected to vary between years by an order of magnitude, even when harvest rates are held constant. Both alarmingly low and surprisingly high recruitments should be expected periodically, even in the absence of a change in human influences. Because of this variation, it is doubtful that the concept of "equilibrium yield" conveys a meaningful picture of what can realistically be achieved. Rather, fishery managers should seek to convey the notion of a frequency distribution of outcomes when they forecast catch and escapement. This notion should reduce harvester conflicts arising from false expectations, and should remind fishery managers of the inherent variability of the system when they try to interpret trends in catch or escapement between years.

More of the variation in survival among years for the example data sets presented was attributable to environmental factors rather than to density dependence. This was most clearly demonstrated by the variation in smolt-to-adult survival rate, a life phase that takes place almost entirely in the ocean. Smolt-to-adult survival rates for salmon from Cowlitz Salmon Hatchery varied 10-fold for spring chinook, 20-fold for fall chinook, and 30-fold for coho during the 1967–1991 broods. There are no dams or unusual obstacles to migration between the hatchery and the ocean. At most, there is evidence of only very weak density-dependent survival in the ocean for chinook and coho salmon (Emlen et al. 1990), so most of the smolt-to-adult variation must have been driven by environmental factors. The variation in recruitment assignable to density dependent factors according to the Ricker formula can be calculated from the expression, $e^{-\beta P}$), from Eq. (31.4). Seventy percent of all spawner escapements for the simulations, with ocean harvest set at 50%, fell between 4,000 and 12,000 fish. Substituting these values into the above expression gives 0.67 and 0.30. These scalars represent the proportionate reduction from the maximum recruitment rate per spawner, α, that would occur at these spawner levels. In comparison, the variation assignable to environmental factors for two thirds of all observations (plus or minus one standard deviation), can be calculated from the expression, e^{ε}, of Eq. (31.4). The calculated value of this environmental scalar ranged from 0.549 to 1.822. Thus, the ratio of high-to-low value for the density dependent scalar was 2.23, while the ratio of high-to-low value for the environmental scalar was 3.32.

The wide variation in recruitment rate of coho and chinook salmon, particularly in the life stage between smolting and recruitment to ocean fisheries, points to fluctuations in the ocean environment as a major cause of the lognormal variation in recruitment. There is substantial evidence to indicate that variations in ocean conditions are the largest single factor influencing production of Oregon coho (see Pearcy 1992 for a review). Variation in survival of coho between years in the ocean is substantially greater than in freshwater (Pearcy 1992), and this variation, at least for hatchery fish, has been highly correlated to ocean upwelling and sea surface temperatures (Nickelson 1986; Emlen et al. 1990). Ware and Thomson (1991) presented data suggesting a link of upwelling to the magnitude of primary and secondary production off southern California. A mid-1970s shift in the California current has caused generally less desirable upwelling and sea surface temperatures for coho since 1976 (Pearcy 1992; Emlen et al. 1990). Extremely low ocean survivals of coho in 1983 and 1984 were directly attributable to the El Niño event in the ocean (Pearcy 1992).

The stochastic simulations I presented assume variation was random and, therefore, present a more optimistic picture of outcomes than for the actual situation in which variation is also periodic. During a sequence of low ocean survival years, failure to reduce high harvest rates could quickly lead to depletion of the spawning stock. Because of high interannual variability, regime shifts in survival rate are difficult to detect until 5–10 years after the shift started. Therefore, it would be prudent for harvest managers to adopt a more risk-averse strategy, rather than attempting to achieve the MSY harvest rate. Most ocean salmon fisheries on the West Coast of the U.S. have been managed for relatively constant harvest rates near the MSY level. The harvest rates for OPI coho are a case in point, where harvest rates during 1965 to 1992 generally remained between 55% and 80%, except for 3 years following the 1983 El Niño event. The Central Valley Index for harvest rate of chinook salmon off California has remained between 60% and 75% in nearly all years since 1972 (PFMC 1995). The overall harvest rates on Willamette Basin spring chinook have remained near 70% for the past two decades (Cramer et al. 1996). These consistently high harvest rates may reflect a consistent opportunity for commercial fishermen to fish, but the simulations presented here indicate that substantially lower harvest rates would allow greater spawning escapement while foregoing little catch over the long term. This conclusion applies to naturally-produced fish, but not to hatchery fish for which smolt production is generally not limited by spawning escapement.

It is clear from the past track record of salmon fisheries management that harvest has been targeted to achieve the theoretical MSY for a fixed stock-recruitment curve, rather than some lesser harvest rate that would accommodate the effects of environmental variation. The Oregon Department of Fish and Wildlife (ODFW) did an admirable job of thoroughly documenting their plan for managing coho salmon (ODFW 1982), so that plan can be used as an example here. The Oregon Coho Salmon Plan (1982) states: "Fisheries will be managed to obtain optimum yield from the resource." The Plan recommended that ocean harvest rate be maintained at 69% to achieve the objectives of the Plan, and that harvest rate was calculated from a Ricker stock-recruitment function with an α parameter (recruits per spawner) of the same value (6.6) used in simulations presented here. The Plan stated that harvest in addition to the 69% in the ocean would be allowed in freshwater. The simulations I completed assumed a freshwater harvest rate of 15%. The Oregon Coho Salmon Plan called for harvest management that corresponded almost exactly with the scenario I have simulated for a 70% ocean harvest rate. Natural production of coho in Oregon was depleted following the regime shift to years of lower ocean survival in 1977 (Pearcy 1992), and the decline in spawner escapement led the National Marine Fisheries Service (NMFS) to propose Oregon coastal coho for listing as "threatened" under the Endangered Species Act. Only an aggressive plan of action by ODFW, that included closure of all ocean fisheries off Oregon to direct take of coho, temporarily averted the federal listing.

The observed values of recruits per spawner and smolt-to-adult survival demonstrate that occasional years of outstanding recruitment could support harvest rates of 90% or greater for a given population of coho or chinook. Unfortunately, those years occur at different times for different populations. Peaks in recruits per spawner occurred in different years for coho in the Nehalem, Umpqua, and Coquille rivers (see Figure 31.2). Cramer et al. (1996) found that variation in survival for the fall and spring releases of spring chinook smolts from Willamette Basin hatcheries was similar for broods during the 1980s, but differed during the mid-1970s. Survival was better than normal for fall smolt releases during the mid-1970s, but below normal for yearling releases. These

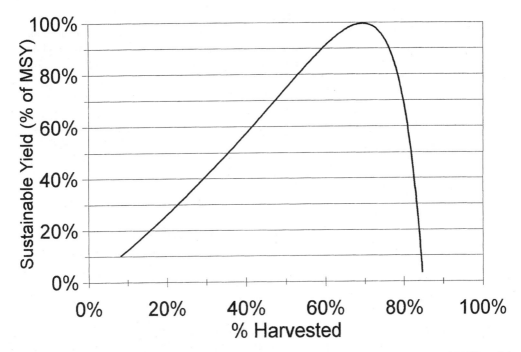

FIGURE 31.9   Relationship between average sustainable yield, expressed as a percentage of MSY, and the harvest rate, based on the Ricker stock-recruitment function employed in the coho simulations of this chapter.

differences indicate that key factors affecting survival in the ocean can change sharply between the fall in one year and the spring of the next year. Because ocean fisheries harvest multiple stocks at the same time, the ocean fishery cannot be designed to harvest localized surpluses. Given the dilemma of density independent variation in recruitment, coupled with the lack of synchrony in recruitment between populations, it will be difficult to manipulate ocean harvest to take advantage of high recruitment in one population without overharvesting other populations. Solving of this dilemma may necessitate creative strategies to direct harvest toward more terminal areas.

The loss of future yield (catch) that results from overharvesting stocks in years when their survival is low is accentuated by a sharp drop in sustainable yield, predicted by the Ricker function, as harvest rate increases beyond the MSY level. The graph of average sustainable yield as a function of harvest rate (Figure 31.9), shows the slope of the ascending limb is flatter than the descending limb. Thus, a harvest rate that slightly exceeds the MSY level causes a greater reduction in future yield than a harvest rate that falls slightly under the MSY level. For example, an absolute increase of 15% in harvest rate above the 69% MSY harvest rate for the coho population simulated here would drive the sustainable yield to zero and the stock to extinction, while an absolute decrease of 15% in the harvest rate below the 69% MSY harvest rate would only reduce sustainable yield to 85% of MSY (Figure 31.9). In a mixed stock fishery where some stocks are experiencing above average survival years while others are experiencing below average survival years, the consequences of overharvesting the lagging stocks are likely to exceed the gains from harvesting the growing stocks (assuming carrying capacity and inherent productivity are similar among stocks).

The fixed, but lower, harvest rate policy simulated in this chapter may be the best approach to managing mixed-stock fisheries in which variation in recruitment rate is poorly correlated between stocks. Walters and Parma (1996) demonstrated with stochastic simulations that fixed harvest rate policies generally achieve a long-term yield that is within 10% of the yield from a variable harvest rate adjusted annually to take the full harvestable surplus. The simulations I presented indicate that reducing the harvest rates to about two thirds of the theoretical MSY level would still sustain total catch near the maximum over the long term, and would double spawning escapement for stocks with productivity factors (α) similar to the one modeled here. Thus, there is little to lose and much to gain by reducing harvest rate to about two thirds of the MSY rate. The greater spawner escapements at reduced harvest rates would provide a buffer against overharvest when ocean survival shifts to a lower regime and would contribute to the maintenance of high genetic diversity and stream ecosystem function. Further, presently depleted stocks could increase while other stocks remain strong, increasing the total production. Increased escapements may also increase total salmon production by raising organic-based carrying capacity (via more carcasses; Bilby et al. 1996) and by filling vacant habitat. This increase would likely exceed that predicted by a Ricker-type model based on recent escapements during years of high harvest rates. Thus, the benefits of increased escapement following a reduction in harvest rate are likely to be even greater than modeled in this chapter.

CONCLUSIONS AND MANAGEMENT RECOMMENDATIONS

- The number of recruits produced per spawner among Pacific salmon populations varies more than an order of magnitude between years, and more of this variation results from environmental factors than from density-dependant mechanisms.
- Environmentally induced variation in recruitment is log-normally distributed with a preponderance of below-average recruitment rates. Much of this variation is induced in the ocean.
- The skewed, or log-normal, nature of environmental variation causes catch and escapement to be less than the deterministically predicted equilibrium in about 60% of all years.
- Recruitment often varies asynchronously between populations that are harvested predominantly in mixed stock fisheries. Therefore, harvesting regulations generally cannot be manipulated to take advantage of recruitment surpluses without overharvesting some stocks.
- Harvest rates should be targeted for two thirds or less of the theoretical MSY level for salmon stocks with Ricker $\alpha \approx 6.8$. This strategy will sustain total catch over the long term near the maximum, but spawning escapement will double compared to that at MSY. Increased spawning escapement is highly desirable for supplying marine-derived nutrients to juvenile rearing areas and for maintaining genetic diversity.

ACKNOWLEDGMENTS

I am grateful to J. G. Norris for his careful critical review and discovery of analytic errors in two drafts of the manuscript. He also offered valuable advice for correcting the errors. I thank T. A. Williams who completed much of the computer work for this project. Useful suggestions for improving the manuscript were offered by T. E. Nickelson and J. M. Emlen.

REFERENCES

Beverton, R.J.H., and S.J. Holt. 1957. On the dynamics of exploited fish populations. Fishery Investment Series 2. Vol. 19. United Kingdom Ministry of Agriculture and Fisheries, London.

Bilby, R.E., B.R. Fransen, and P.A. Bisson. 1996. Incorporation of nitrogen and carbon from spawning coho salmon into the trophic system of small streams: evidence from stable isotopes. Canadian Journal of Fisheries and Aquatic Sciences 53:164–173.

Cooney, C.X., and S.E. Jacobs. 1994. Oregon coastal salmon spawning surveys, 1992. Oregon Department of Fish and Wildlife, Fish Division, Information Reports, Number 94-2, Portland.

Cramer, S.P. 1996. Contribution to catch and spawning escapement of salmon produced at Cowlitz Salmon Hatchery and steelhead produced at Cowlitz Trout Hatchery. Report prepared by S.P. Cramer & Associates, Inc. for Harza Consulting Engineers and Scientists, Bellevue, Washington.

Cramer, S. P., T. Satterwaite, R. Boyce, and B. McPherson. 1985. Impacts of Lost Creek Dam on the biology of anadromous salmonids in the Rogue River. Phase 1 Completion Report, Oregon Department of Fish and Wildlife, Volume 1, Submitted to U.S. Army Corps of Engineers. Portland, Oregon.

Cramer, S.P., C.F. Willis, D.P. Cramer, M. Smith, T. Downey, and R. Montagne. 1996. Status of Willamette River Spring Chinook Salmon in regards to the Federal Endangered Species Act, Part 2. Submitted to National Marine Fisheries Service. Portland, Oregon.

Emlen, J.M., R.R. Reisenbichler, A.M. McGie, and T.E. Nickelson. 1990. Density-dependence at sea for coho salmon (*Oncorhynchus kisutch*). Canadian Journal of Fisheries and Aquatic Sciences 47:1765–1772.

Hilborn, R.B. 1985. Simplified calculation of optimum spawning stock size from Ricker's stock recruitment curve. Canadian Journal of Fisheries and Aquatic Sciences 42:1833–1834.

Hilborn, R., and C.J. Walters. 1992. Quantitative fisheries stock assessment: choice, dynamics, and uncertainty. Routledge, Chapman & Hall, New York.

Johnson, S.L. 1984. The effects of the 1983 El Niño on Oregon's coho and chinook salmon. Oregon Department of Fish and Wildlife, Information Reports (Fish) 82-5, Corvallis.

Lewis, M.A. 1994. Stock assessment of anadromous salmonids. Oregon Department of Fish and Wildlife, Fish Division Progress Reports, Project Number AFC-134, Portland.

Malvestuto, S.P., and M.D. Hudgins. 1996. Optimum yield for recreational fisheries management. Fisheries 21(6):6–17.

McAllister, M.K., E.K. Pikitch, A.E. Punt, and R. Hilborn. 1994. A Bayesian approach to stock assessment and harvest decisions using the sampling/importance resampling algorithm. Canadian Journal of Fisheries and Aquatic Sciences 51:2673–2687.

Mullen, R.E. 1981. Oregon's commercial harvest of coho salmon, *Oncorhynchus kisutch* (Walbaum), 1892–1960. Oregon Department of Fish and Wildlife, Information Reports (Fish) 81-3, Corvallis.

Nehlsen, W., J. E. Williams, and J. A. Lichatowich. 1991. Pacific salmon at the crossroads: Stocks at risk from California, Oregon, Idaho, and Washington. Fisheries 16(2):4–21.

Nickelson, T.E. 1986. Influences of upwelling, ocean temperature, and smolt abundance on marine survival of coho salmon (*Oncorhynchus kisutch*) in the Oregon Production Area. Canadian Journal of Fisheries and Aquatic Science 43:527–535.

ODFW (Oregon Department of Fish and Wildlife). 1982. Comprehensive plan for production and management of Oregon's anadromous salmon and trout. Part II. Coho salmon plan. Oregon Department of Fish and Wildlife, Portland.

Peterman, R.M. 1981. Form of random variation in salmon smolt-to-adult relations and its influence on production estimates. Canadian Journal of Fisheries and Aquatic Sciences 38:1113–1119.

PFMC (Pacific Fishery Management Council). 1995. Review of 1994 ocean salmon fisheries. Portland, Oregon.

PSC (Pacific Salmon Commission). 1988. Joint chinook technical committee 1987 annual report. Report TCCHINOOK (88)-2, Vancouver, British Columbia.

PSC (Pacific Salmon Commission). 1996. Joint chinook technical committee 1994 annual report. Report TCCHINOOK (96)-1, Vancouver, British Columbia.

Pearcy, W.G. 1992. Ocean ecology of north Pacific salmonids. University of Washington Press, Seattle.

Ricker, W.E. 1954. Stock and recruitment. Journal of Fisheries Research Board of Canada 11:559–623.

Stohr, A. H., and M. E. Fraidenburg. 1986. A delphi assessment of chinook and coho salmon hooking mortality. Technical Report 94, Washington Department of Fisheries, Olympia.

Wallis, J. 1961a. An evaluation of the Coos River Hatchery. Oregon Fish Commission Research Laboratory, Clackamas.

Wallis, J. 1961b. An evaluation of the Nehalem River Salmon Hatchery. Oregon Fish Commission Research Laboratory, Clackamas.

Wallis, J. 1963. An evaluation of the Alsea River Salmon Hatchery. Oregon Fish Commission Research Laboratory, Clackamas.

Walters, C., and A.M. Parma. 1996. Fixed exploitation rate strategies for coping with effects of climate change. Canadian Journal of Fisheries and Aquatic Sciences 53:148–158.

Walters, C., and A. Punt. 1994. Placing odds on sustainable catch using virtual population analysis and survey data. Canadian Journal of Fisheries and Aquatic Sciences 51:946–958.

Ware, D.M., and R.E. Thomson. 1991. Link between long-term variability in upwelling and fish production in the northeast Pacific Ocean. Canadian Journal of Fisheries and Aquatic Sciences 48: 2296–2306.

32 Using Photosynthetic Rates to Estimate the Juvenile Sockeye Salmon Rearing Capacity of British Columbia Lakes

Ken S. Shortreed, Jeremy M. B. Hume, and John G. Stockner

Abstract.—We describe refinements to a simple sockeye salmon *Oncorhynchus nerka* rearing capacity model, the photosynthetic rate (PR) model, which was first described in an earlier paper. The model is based on a correlation between photosynthetic rate expressed as metric tons of carbon per year and sockeye salmon smolt biomass. Estimates of optimum escapements and spring fry recruitment required to produce maximum smolt numbers and biomass were taken from the Alaskan euphotic volume (EV) model. We define rearing capacity as the point at which the maximum number and biomass of smolts are produced and optimum escapement as the number of spawners that results in maximum smolt production. We compare model predictions to direct estimates of optimum escapements (developed from fry models—the relationship between numbers of spawners and numbers of fall fry) from two British Columbia (B.C.) lakes and discuss assumptions and limitations of the model. Although we currently have direct estimates of optimum escapement (e.g., fry models) for only two lakes that make up 16% of the total B.C. sockeye salmon nursery lake area, PR data are currently available for 57% of B.C.'s nursery lake area. We provide estimates of optimum escapements and smolt production from those lakes where suitable data are available. By making assumptions about productivity of lakes where PR is unknown, we also provide estimates of optimum sockeye salmon escapement to all major regions of B.C. Although more research and data are needed, the PR model is a promising tool to help managers make decisions regarding sockeye salmon escapement and enhancement.

INTRODUCTION

British Columbia (B.C.) and Alaska have highly valued stocks of Pacific salmon *Oncorhynchus* spp. Of the five salmon species which occur in western North America, sockeye salmon *O. nerka* are the most economically valuable, with an annual catch worth several hundred million dollars (Burgner 1991). Sockeye salmon are planktivorous throughout their 4–5 year life cycle and reside in lakes for 1–2 years before they migrate to the ocean. Consequently, lakes are an important nursery area for this species.

For much of the 20th century there have been efforts in many parts of B.C. to increase adult sockeye salmon numbers. Without appropriate methods to estimate rearing capacity, it was assumed that sockeye salmon numbers could be increased beyond ambient numbers to some unspecified level. In some lakes this assumption was validated by historic escapement data (returning spawners that are not caught in the fishery), while in others no data were available to indicate the lake could support additional sockeye salmon. Enhancement efforts were undertaken without a clear understanding of the rearing capacity of a particular sockeye salmon stock's nursery lake. Sockeye salmon enhancement focused on increasing fry recruitment by increasing escapements (Roos 1989); by increasing egg-to-fry survival using spawning channels (McDonald 1969); by direct additions of sockeye salmon fry to lakes (Diewert and Henderson 1992); and by increasing freshwater growth and survival through lake fertilization (Hyatt and Stockner 1985). Enhancement efforts such as spawning channels have been highly successful in some locations and have had little or no effect on stock size in others (Hilborn 1992). Efforts to increase escapement and subsequent smolt production through harvest management have been very successful on some Fraser system stocks (Roos 1989).

Predicting the production capacity for fish in a particular body of water has long been an objective of freshwater research in North America (see Leach et al. 1987 for a review). It has relevance to management of recreational and commercial fisheries (sustainable yield) and to enhancement (amount that recruitment to a lake can be increased). There have been numerous attempts to develop empirical relationships between lake productivity and fish yield. Since a direct measure of productivity (i.e., photosynthetic rate) was not usually available, investigators used a number of other limnological variables as surrogates for photosynthetic rate (PR). These included mean depth and total dissolved solids (Ryder 1965); summer average chlorophyll concentration (Oglesby 1977; Jones and Hoyer 1982); lake area (Youngs and Heimbuch 1982); and total phosphorus concentration (Stockner 1987; Downing et al. 1990).

Fee et al. (1985) and Downing et al. (1990) reported that PR measurements were positively correlated to fish yield. Downing et al. (1990) also found that PR was more closely correlated to fish yield than some other variables commonly used as indices of lake productivity (chlorophyll, total phosphorus). While these surrogates may be correlated to PR, using abiotic or biomass variables instead of PR in empirical relationships with fish yield will introduce additional uncertainty. Furthermore, an improved understanding of energy flow between lake trophic levels is more likely when rate measurements at each trophic level are used.

Development of empirical models in earlier studies was hampered by difficulties in reliably measuring fish yield. Model development was also confounded by resident fish populations which were often multi-species, with varying life histories (i.e., different degrees of planktivory and piscivory). To a large extent, selection of lakes on which estimates of fish yield were available was done qualitatively, with the main criterion being that the lakes must have had "moderately intensive to intensive fishing effort on a spectrum of species for a number of years" (Ryder et al. 1974). If a lake met these criteria for fish yield data, two assumptions had to be made: first, that the quality of the data (e.g., creel census, commercial landings) was good enough for model development, and second, that catches actually did represent maximum sustainable yield. In most cases, these assumptions could not be tested.

When developing rearing capacity (fish yield) models, sockeye salmon nursery lakes in B.C. and Alaska offer a number of advantages over most other lakes. Adult sockeye salmon spawn in the fall, fry enter the lake the following spring, reside in the lake for 1–2 years (in B.C. lakes, residence for most sockeye salmon is 1 year), and the following spring migrate to the ocean as smolts. Prior to or during migration, numbers and biomass of juvenile sockeye salmon can be accurately estimated from midwater acoustic and trawl surveys (fry) or from counts at fences (smolts). On a given lake, if juvenile sockeye salmon numbers and biomass resulting from a wide interannual range of adult spawner escapements are available, maximum juvenile production (e.g., rearing capacity) can be determined by measuring the point where juvenile numbers and biomass are greatest. If spawner numbers increase past this point where escapements are optimum, juvenile sockeye salmon production will not increase and may decrease (Koenings and Burkett 1987; Hume et al. 1996). While sockeye salmon nursery lakes support a variety of other fish species, their biomass is often small relative to maximum juvenile sockeye salmon biomass (Hume et al. 1996). Consequently, maximum biomass of juvenile sockeye salmon is a reliable indicator of rearing capacity or annual fish yield. Sockeye salmon fry are almost exclusively limnetic planktivores, so they are more strongly coupled to limnetic zooplankton production than fish species with a wider dietary range that may include terrestrial and benthic organisms. In addition, determination or prediction of maximum biomass in a sockeye salmon nursery lake also permits calculation of spawner numbers required to produce that biomass (i.e., optimum escapements). Given the very high economic value of the stocks and the suitability of available data, fish yield models are perhaps an even more important management tool in sockeye salmon nursery lakes than they are in other North American lakes.

While it is generally assumed that most of the over 90 sockeye salmon nursery lakes in B.C. are recruitment limited (i.e., greater spawning escapements would produce additional smolts), some recent escapements to several major B.C. sockeye salmon lakes have been at or above optimum levels (Hyatt and Steer 1987; Hume et al. 1996). Consequently, a reliable rearing capacity model for sockeye salmon nursery lakes is of even greater importance now than it was in the past. Better predictions of optimum escapements would enable managers to determine whether current escapements are above or below optimum. If escapement of a particular stock is below optimum, the amount of feasible or desirable enhancement to bring lake production up to full rearing capacity could be quantified and the economic value of a successful enhancement program could be estimated. Conversely, the cost in lost production from not enhancing a stock could be assessed. Furthermore, when escapements exceed that required to maximize smolt production, the economic cost of the foregone catch can also be determined. In short, a reliable rearing capacity model would be a powerful tool for fisheries managers concerned with maximizing and sustaining B.C. sockeye salmon.

The euphotic volume (EV) model (Koenings and Burkett 1987; Koenings and Kyle 1997) was developed using data from a number of Alaskan lakes and was the first rearing capacity model developed specifically for sockeye salmon. It provided predictions of optimum escapement, optimum spring fry recruitment, and maximum smolt output. In Hume et al. (1996) we modified the EV model so that it could be used in B.C. lakes. Our model was based on photosynthetic rate and was called the PR model. Objectives of this chapter are to describe a revised version of the PR model, to explain its derivation from the EV model and from the original PR model, to test model predictions in B.C. lakes where appropriate sockeye salmon data are available, to discuss assumptions of the model, and to present model predictions for all B.C. lakes for which suitable data are available.

BRITISH COLUMBIA SOCKEYE SALMON NURSERY LAKES

Sockeye salmon nursery lakes occur in all regions of B.C. with the exception of the Peace River drainage basin in northeastern B.C. (east of the Cassiar mountains). Extensive dam construction in the Columbia River drainage basin has blocked access to anadromous fish in lakes in the central and eastern portion of southern B.C. (parts of the Okanagan and Kootenays), although remnant sockeye salmon stocks still reach some lakes. B.C.'s sockeye salmon nursery lakes occur in most of the province's varied climatic and geologic regions, with corresponding large variations in latitude and elevation. Some lakes occur virtually at sea level while the elevation of others exceeds 1,200 m. B.C. nursery lakes occur over a north-south range of >1,000 km. This results in wide differences in thermal regimes, water clarity, water residence times, nutrient loading, and trophic status. Surface area of the nursery lakes range over more than two orders of magnitude, from <2 km^2 to >400 km^2. Mean depths range from <6 m to >150 m and water residence times range from several days to >20 years. Trophic status ranges from ultra-oligotrophic (e.g., Chilko Lake; Stockner and Shortreed 1994) to meso-eutrophic (e.g., Fraser Lake; Shortreed et al. 1996). Some nursery lakes have very high water clarity while others are turbid from either glacial or organic inputs. Distance from the ocean is also highly variable, resulting in freshwater migration distances of <1 km for some coastal systems to >1,000 km for some interior stocks. The very wide range of lake types that sockeye salmon inhabit confirm the "elastic" nature of their habitat requirements.

The total surface area of B.C.'s approximately 90 sockeye salmon nursery lakes is about 3,800 km^2. The size of individual sockeye salmon stocks is highly variable. In some years adult returns (catch and escapement) to major producers such as Quesnel or Shuswap Lakes exceed 10 million and spawning escapements exceed 1 million. Returns of some smaller stocks (or of major producers in non-dominant brood years) are as low as a few hundred to a few thousand fish. The nursery lakes contain over 590 spawning streams, of which over half have less than 1,000 spawners. About 4% of the streams have peak spawning populations in excess of 100,000 (Williams and Brown 1994). The Fraser River system has nine lakes where escapements have exceeded 100,000 in recent years (Chilko, Fraser, Harrison, Lillooet, Quesnel, Shuswap, Stuart, Takla, and Trembleur). Of northern B.C. stocks, only Babine Lake on the Skeena River and Meziadin Lake on the Nass River have had recent escapements exceeding 100,000. Of coastal B.C. lakes, five (Great Central, Long, Nimpkish, Owikeno, and Sproat) have had recent escapements that exceeded 100,000.

METHODS

MEASUREMENT OF PRODUCTIVITY VARIABLES

We used data on sockeye salmon smolts and fall sockeye salmon fry to support and validate our model. Smolt data were from Alaskan lakes (Koenings and Burkett 1987) and from two B.C. lakes (Babine and Chilko). Smolt numbers and size have been determined using fence counts and mark-recapture estimates since 1949 at Chilko Lake (Roos 1989) and since 1961 at Babine Lake (MacDonald et al. 1987). Since the 1970s we have estimated sockeye fry numbers for three B.C. lakes (Fraser, Quesnel, and Shuswap) using hydroacoustic and trawl techniques as described in Hume et al. (1996). All sampling was done at night when the fish were dispersed and within the working range of the midwater trawl and hydroacoustic system (McDonald and Hume 1984; Burczynski and Johnson 1986). Hydroacoustic and trawl data presented in this chapter were collected in the fall (October and early November).

The PR data used in this chapter were collected from 33 lakes during 1977–1995. Data were collected from spring (April to May) to fall (October to November) and sampling frequency varied from once weekly to once monthly. PR data were collected using *in situ* incubations of light and dark bottles inoculated with ^{14}C. PR data collected prior to 1994 have been reported and methods described elsewhere (Stockner and Shortreed 1979; Stockner et al. 1980; Shortreed and Stockner 1981; Hume et al. 1996; Shortreed et al. 1996). In 1994 and 1995 we sampled a number of lakes in the Skeena River drainage basin. Methods used for determination of PR were identical to those reported in earlier studies, with the following exceptions. After the 1.5–2.0-h *in situ* incubations were completed, samples were filtered onto 25-mm diameter Micro Filtration Systems GF75 glass fiber filters, which are equivalent to Whatman GF/F filters. Filters were placed in scintillation vials containing 0.5 mL of 0.5 N HCl. Lids were not put on the vials for 6–9 h. In the laboratory, 10 mL of scintillation cocktail (Fisher Scientific's Scintiverse II) were added to the vials, which were then counted in a Packard scintillation counter. Methods used to calculate volumetric, integrated, and daily (mg C·m^{-2}·d^{-1}) rates were described in the previously mentioned papers. Prior to 1980, we used scintillation cocktails which were not alkalized. Consequently, PR data we reported for those years overestimated actual PR by a factor of 1.49 (Kobayashi 1978). We divided PR data collected prior to 1980 by this factor to ensure compatibility with more recent data.

Seasonal average daily PR (PR$_{\bar{x}}$) in mg C·m^{-2}·d^{-1} for each lake was computed by integrating daily PR (measured at least monthly from May to October) and dividing by the length of the growing season, which we defined as May 1 to October 31. Total seasonal PR in metric tons C/lake (PR$_{total}$) was calculated by multiplying PR$_{\bar{x}}$ by the length of the growing season and by lake area. Where multiple years of data were available for a lake, we averaged all years to obtain a single PR estimate. Alaskan PR data used in development and verification of our model were taken from Figure 9 in Koenings and Burkett (1987). In Koenings and Burkett (1987) PR data were presented as $PR_{\bar{x}}$. To convert to PR_{total}, we multiplied $PR_{\bar{x}}$ by the length of the growing season, which we assumed extended from May 1 to October 31, and by lake area.

MODEL DEVELOPMENT

The PR model presented in this chapter has evolved from the Alaskan EV model (Koenings and Burkett 1987; Koenings and Kyle 1997) and from an earlier version of the PR model (Hume et al. 1996). The EV model is a rearing capacity model developed specifically for sockeye salmon. It uses seasonal average euphotic zone depth ($EZD_{\bar{x}}$) as an analog for seasonal average PR ($PR_{\bar{x}}$) in mg $C \cdot m^{-2} \cdot d^{-1}$ and correlates it to maximum juvenile sockeye salmon numbers and biomass and optimum escapements. This was possible because, in the Alaskan study, lakes with widely varying water clarities caused by glacial or organic stain resulted in a positive linear correlation between $EZD_{\bar{x}}$ and $PR_{\bar{x}}$. This positive correlation does not occur in most B.C. sockeye salmon nursery lakes or in most North American lakes, where lake productivity is generally negatively correlated to $EZD_{\bar{x}}$ and positively correlated with nutrient loading (Wetzel 1975; Hume et al. 1996). Consequently, in its original form, the EV model was not suitable for use in B.C. sockeye salmon lakes.

The EV model utilized EV units (1 EV unit was defined as $10^6 m^3$ of euphotic volume) which were calculated using the equation:

$$EV \text{ units} = EZD_{\bar{x}}(Area)/10^6 \qquad (32.1)$$

where
$EZD_{\bar{x}}$ = seasonal average euphotic zone depth (m)
Area = lake surface area (m^2)

In Hume et al. (1996) we modified the calculation of EV units to use $PR_{\bar{x}}$ instead of $EZD_{\bar{x}}$, which made the EV model useful in a wider range of lake types (including B.C. lakes). We did this by using the correlation between $EZD_{\bar{x}}$ and $PR_{\bar{x}}$ in Alaskan lakes ($EZD_{\bar{x}} = 0.0583PR_{\bar{x}}+3.25$; $r^2 = 0.81$) to substitute $PR_{\bar{x}}$ for $EZD_{\bar{x}}$ in the calculation of EV units. Because the revised model calculated EV units using $PR_{\bar{x}}$ instead of $EZD_{\bar{x}}$, we renamed them PR units, and also renamed the model the PR model (Hume et al. 1996). PR units are equivalent to EV units. PR units are calculated by the equation

$$PR \text{ units} = (0.0583PR_{\bar{x}}+3.25)Area/10^6 \qquad (32.2)$$

where
$PR_{\bar{x}}$ = seasonal average daily PR (mg $C \cdot m^{-2} \cdot d^{-1}$)
Area = lake surface area (m^2)

In the original EV model, recommended escapements were 800–900 adult spawners/EV unit, which were observed to produced an average of 23,000 smolts with a mean weight of 2.0 g (Koenings and Burkett 1987; Koenings and Kyle 1997). However, adult sockeye salmon production was greater at spawner densities approximately one half that necessary to produce 2.0-g smolts (Koenings and Burkett 1987; Koenings et al. 1993). At these lower spawner densities, juvenile sockeye salmon freshwater survival and growth were both greater, resulting in the production of similar numbers of larger (4–5 g) smolts (Koenings and Burkett 1987; Koenings et al. 1993). Since we wanted the PR model to be a method for estimating escapements that would produce maximum smolt numbers and biomass and, subsequently, optimum adult production, our PR model used the following:

Optimum escapement = 425 adult spawners/PR unit
Maximum smolt numbers = 23,000/PR unit
Maximum smolt biomass = 103.5 kg/PR unit

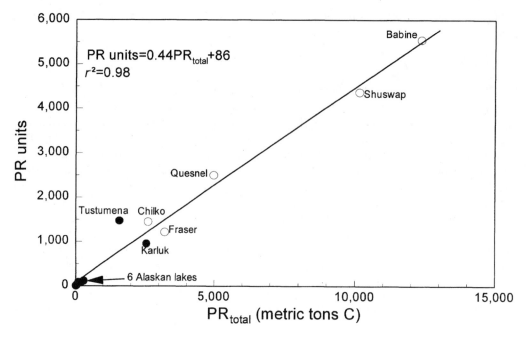

FIGURE 32.1 Relationship between photosynthetic rate (PR) units and total seasonal carbon production (PR_{total}) in Alaskan (solid circles) and B.C. (open circles) lakes.

where

Optimum escapement = Number of spawners needed to maximize smolt production.

Maximum smolt numbers = Maximum number of smolts a lake can produce, based on observed maximum production in Alaskan lakes (Koenings and Burkett 1987).

Maximum smolt biomass = Maximum number of smolts × a mean smolt weight of 4.5 g, also from observed maximum production in Alaskan lakes (Koenings and Burkett 1987).

The primary reasons for developing the PR model were to simplify the use of production models and to enhance their utility for understanding energy flow between lake trophic levels. Consequently, we revised the model to use PR_{total} (units are metric tons C/lake) rather than PR units. PR units from both Alaskan and B.C. lakes were significantly correlated to PR_{total} (PR units = $0.44PR_{total}+86$; $r^2 = 0.98$; Figure 32.1). In this relationship the intercept was not significantly different than zero and forcing the intercept through zero resulted in the equation:

$$PR \text{ units} = 0.44PR_{total} \qquad\qquad (32.3)$$

With this relationship we modified the PR model to use PR_{total} instead of PR units. Resulting predictions were

Optimum escapement = $187 \times PR_{total}$

Maximum smolt numbers = $10,120 \times PR_{total}$

Maximum smolt biomass (kg) = $45.5 \times PR_{total}$

where

PR_{total} = Total seasonal (May to October) carbon production (metric tons)

With these simple equations the revised PR model generates predictions of optimum escapement to and smolt production from any lake where suitable PR data are available.

RESULTS AND DISCUSSION

TESTING THE PR MODEL

The two variables ($PR_{\bar{x}}$ and lake surface area) used in computing PR_{total} strongly affect PR model predictions. Surface areas of sockeye salmon nursery lakes in B.C. vary over two orders of magnitude, while $PR_{\bar{x}}$ in lakes for which we have data varies approximately one order of magnitude. If $PR_{\bar{x}}$ was similar in all lakes then surface area alone would cause interlake variation in model predictions. Conversely, if all lakes were similar in size, then $PR_{\bar{x}}$ alone would be sufficient to explain the variation. Since variation in surface area is substantially greater than that in $PR_{\bar{x}}$, we expected it to explain a substantial proportion of the variation in model predictions and, hence, in a lake's measured rearing capacity. Youngs and Heimbuch (1982) reported that surface area alone explained 94% of the variability in fish yield for a large suite of lakes located in many different parts of the world. However, the surface area of lakes in their data set varied over four orders of magnitude. As the range in area decreases in the set of lakes being analyzed, the proportion of variation explained by area alone would also be expected to decrease.

To test the relative contribution of surface area and productivity in determining rearing capacity, we selected data from Alaskan and B.C. lakes where both $PR_{\bar{x}}$ and maximum juvenile sockeye salmon biomass were known. Data from Alaskan lakes came from Koenings and Burkett (1987). Hume et al. (1996) determined maximum fall fry biomass in two B.C. lakes. In these lakes smolt biomass was unavailable, so we assumed that smolt biomass was equal to fall fry biomass (i.e., overwintering mortality was balanced by overwintering growth). We also used data in this analysis from three additional B.C. lakes (Babine, Chilko, and Fraser) that have recently had record high numbers of adult spawners which produced record high juvenile numbers and biomass. While these three lakes have not had sufficient years of high escapements to conclusively demonstrate that maximum juvenile production has been reached, we decided that maximum recorded juvenile biomass was sufficiently high to merit being used in a test of the PR model. For these Alaskan and B.C. sockeye salmon nursery lakes, maximum observed juvenile sockeye salmon biomass was significantly correlated ($P < 0.05$, $r^2 = 0.65$) with lake area (Figure 32.2a). However, when PR_{total} was used, the relationship improved substantially ($r^2 = 0.91$) (Figure 32.2b). Removing data points from the three British Columbia lakes where maximum juvenile production has not been conclusively demonstrated did not affect the significance of the relationships, but did reduce the slopes. In addition, observed and predicted (from the PR model) maximum juvenile sockeye salmon biomass were highly correlated (Figure 32.2b, $r = 0.95$, $n = 11$), although the predicted biomass slope was significantly lower (Figure 32.2b).

Further confirmation of the importance of PR in determining rearing capacity is the strong log relationship ($r^2 = 0.87$) between maximum fry or smolt biomass and total seasonal PR, when the variables are normalized to area (Figure 32.3a). Using published data from lakes covering a wide range of trophic levels, Downing et al. (1990) found that total fish production and PR were highly correlated. To compare productive capacity of North American sockeye salmon nursery lakes with data presented by Downing et al. (1990), we reproduced one of their figures and included data from Alaska and B.C. sockeye salmon nursery lakes (Figure 32.3b). Despite wide differences in techniques used in data collection, both PR and fish production in sockeye salmon nursery lakes fall near the middle of the range found in many other lakes around the world. While the slope of the relationship in Alaskan and B.C. lakes appears to differ from that for the other lakes, an analysis of covariance indicated those differences were not significant ($F = 1.82$, $P > 0.05$). However, intercepts were different ($F = 8.54$, $P < 0.05$). We do not have sufficient information to suggest reasons for the difference in intercepts.

Since PR data are positively correlated to observed maximum smolt or fry biomass, it should be possible to develop a rearing capacity model solely from B.C. data. However, for several reasons, sufficient or suitable data are not yet available to do this. First, we have data for only two B.C.

a.

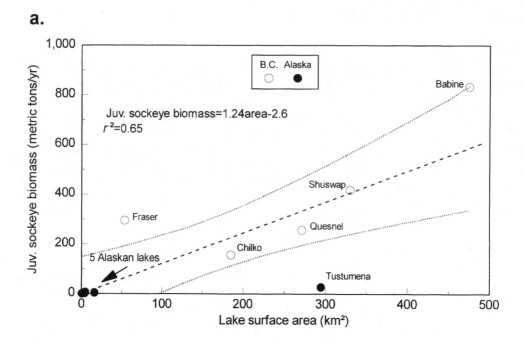

b.

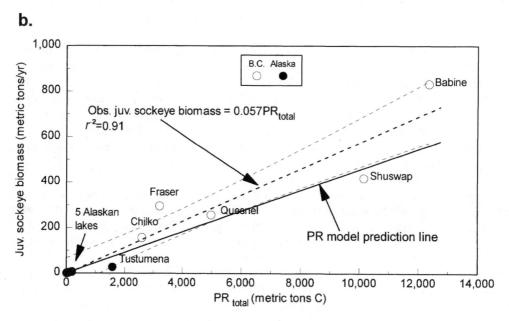

FIGURE 32.2 Observed maximum juvenile sockeye biomass correlated with (a) lake area and with (b) total seasonal carbon production (PR_{total}). Data are smolt biomass except for Fraser, Quesnel, and Shuswap lakes, which are fall fry biomass. Data for Alaskan lakes are from Koenings and Burkett (1987). The regression equation and 95% confidence intervals between biomass and area or PR are shown (dashed and dotted lines). On (b) the solid line is the predicted maximum smolt biomass from the PR model.

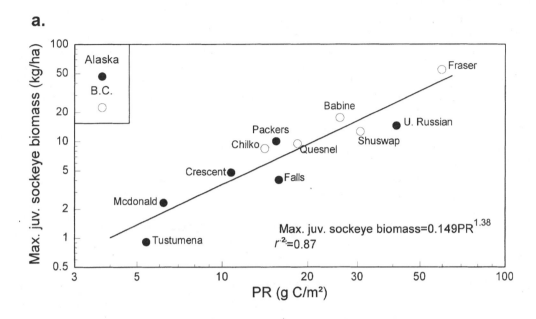

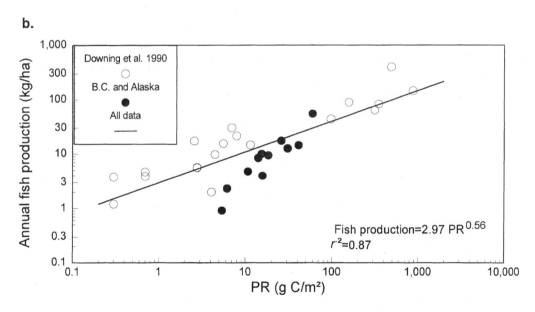

FIGURE 32.3   Variation in maximum juvenile sockeye production (kg/ha) with total seasonal PR. To standardize units with those of Downing et al. (1990), PR data are presented as g $C \cdot m^{-2} \cdot year^{-1}$. Figure 32.3(a) presents data from B.C. and Alaskan lakes (Koenings and Burkett 1987) and 3(b) the same data in addition to data from lakes from a wide range of geographic areas (Downing et al. 1990).

sockeye salmon nursery lakes (Quesnel and Shuswap) which conclusively show that maximum juvenile sockeye salmon biomass has been reached (Hume et al. 1996, Figure 32.4). Second, even on these lakes, smolt biomass was not available, so we assumed that fall fry biomass was equivalent to smolt biomass. Third, on two other B.C. lakes where smolt counts and biomass are available (Babine and Chilko), trends in the data suggest, but do not conclusively show, that maximum smolt biomass has been reached. While we believe current relationships between observed maximum juvenile biomass and PR in B.C. lakes are useful in validating the PR model, current data are too limited to allow independent development of a B.C. rearing capacity model.

The PR model also provides predictions of optimum escapements, and we wished to test these predictions in B.C. lakes. If sufficient data are available, relationships between effective female spawners (successfully spawned female sockeye salmon as determined by examination of carcasses) and fall fry numbers provide a way of estimating optimum escapements (Hume et al. 1996). Currently, we have such data for two B.C. lakes (Quesnel and Shuswap). These relationships were originally published in Hume et al. (1996), but for this chapter we added more recent data and also calculated total fry biomass (Figure 32.4). PR model predictions of optimum escapements (solid vertical lines on the figure) agree quite well with observed optimum escapements, since no increase in fry production is seen at escapements in excess of PR model predictions.

PR MODEL ASSUMPTIONS

In the EV model, spring fry recruitment necessary to produce maximum smolt biomass and/or numbers was determined by a series of fry stocking experiments (Koenings and Burkett 1987). Spawner numbers necessary to produce this fry recruitment (e.g., optimum escapements) were determined from multiple years of data from 12 Alaskan nursery lakes (Koenings and Kyle 1997). Since sufficient data on all freshwater life-history stages of juvenile sockeye (egg deposition, spring fry recruitment, fall fry, smolt numbers) are not available for most B.C. lakes, we chose to adopt those published for Alaskan lakes and used in the EV model (Koenings and Burkett 1987; Koenings and Kyle 1997). This assumes that the components (sex ratio, fecundity, spawning success, egg-to-fry survival) of the spawner-to-fry relationship are similar (and constant) in B.C. and Alaska lakes. Undoubtedly, this assumption is not always met, since the components can vary between years for a given population and also between populations (Burgner 1991; Bradford 1995). Nevertheless, in the two B.C. lakes (Quesnel and Shuswap) where maximum fry production has been observed, PR model predictions of optimum escapement correspond closely to escapements which produce maximum fry numbers and biomass (Figure 32.4). Also, PR model equations predict smolt production of 54 smolts/adult spawner or 108 smolts/female spawner. In Chilko Lake, which has the only natural (non-enhanced) sockeye stock in B.C. with long-term data for both spawners and smolts, an average of 91 smolts/female spawner are produced at high spawner densities. Over a wide range of spawner densities, an average of 120 smolts/female spawner are produced (Hume et al. 1996).

Another assumption of the PR model is that lake rearing capacity and not spawning ground capacity controls smolt production. However, some B.C. lakes have less spawning ground capacity than lake rearing capacity. In these cases the PR model will overestimate optimum escapements and may be more useful for estimating the amount of enhancement (e.g., spawning channel construction, fry outplants) required to maximize production.

The PR model assumes that planktivores other than age-0 juvenile sockeye salmon are present in low numbers. While this is true of most nursery lakes, some contain large populations of limnetic planktivores such as kokanee (lake rearing *O. nerka*), stickleback *Gasterosteus aculeatus*, longfin smelt *Spirinchus thaleichtys*, and age-1 juvenile sockeye salmon. While other planktivores sometimes occur at higher densities than juvenile sockeye salmon (Simpson et al. 1981; Henderson et al.

a.

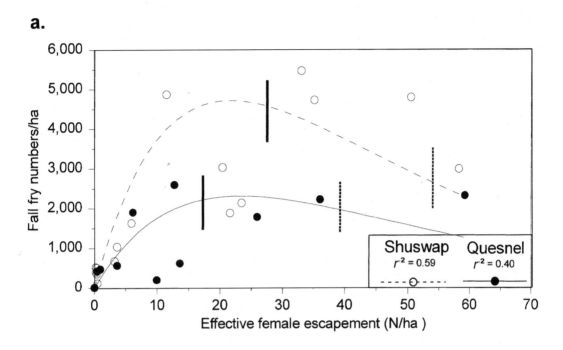

b.

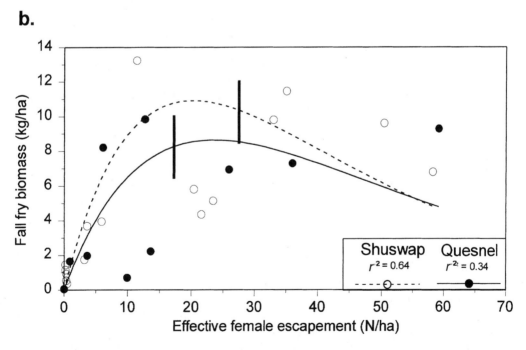

FIGURE 32.4 Density (a) and biomass (b) of fall fry in Quesnel and Shuswap lakes produced by effective female spawners (Hume et al. 1996). Ricker stock-recruit functions are fitted to the data. Optimum escapement predictions from the PR model are shown by the solid vertical lines. EV model optimum escapements are indicated by the dashed vertical lines on (a).

1991), not all planktivores are competitors with juvenile sockeye salmon (Diewert and Henderson 1992). However, when competitive planktivores are numerous, estimates of optimum escapements and maximum smolt production must be reduced proportionately to the relative biomass of age-0 sockeye salmon and their competitors.

Both the PR and EV models assume that smolt weights will average 4.5 g at maximum productivity (Koenings and Burkett 1987; Koenings et al. 1993). However, some B.C. lakes do not produce smolts this large even at low escapements (Hyatt and Stockner 1985). We suggest that the models' assumption of a constant relationship between phytoplankton productivity (PR) and juvenile sockeye salmon production is not always valid. Efficiency of energy transfer from phytoplankton to zooplankton to juvenile sockeye can vary (Stockner 1987; Stockner and Shortreed 1989). However, methods to routinely quantify variation in this efficiency are not available. B.C. sockeye salmon nursery lakes for which we have data cover a relatively narrow trophic range (almost all are oligotrophic). Within this group of lakes, those with a theoretically lower energy transfer efficiency (i.e., longer food chain) are the most oligotrophic (Stockner 1987). In oligotrophic lakes, higher efficiencies occur in lakes near the upper range of oligotrophy; so varying efficiencies theoretically should affect the positive relationship between PR and sockeye production only by changing its slope. Further work is needed to incorporate varying energy transfer efficiencies into PR model predictions.

Given that zooplankton provide the linkage between phytoplankton and juvenile sockeye, some correlation should exist between zooplankton biomass and both PR and juvenile sockeye salmon production. Under some circumstances, these correlations can be detected. In lakes where planktivore densities are low, zooplankton biomass is strongly correlated to lake productivity (Hanson and Peters 1984; Shortreed and Stockner 1986; Shortreed et al. 1996). In lakes where sockeye salmon densities exhibit substantial annual variability, a negative correlation between plankton biomass and juvenile sockeye salmon numbers occurs (Hume et al. 1996). When planktivores are numerous, they can strongly influence zooplankton biomass and community composition (Kyle et al. 1988; Koenings and Kyle 1997; Hume et al. 1996). Because the zooplankton community can be suppressed by sockeye salmon grazing, we found that zooplankton biomass is most closely correlated to PR in lakes and years where grazing pressure is minimal (i.e., low planktivore densities) (Figure 32.5). If available, zooplankton biomass from years of low grazing pressure may be an indicator of underutilized rearing habitat which could support further development of rearing capacity models.

Zooplankton productivity and community structure strongly affect a lake's ability to rear juvenile sockeye. In the presence of continuous high grazing pressure, a lake may develop a predator-resistant, less productive zooplankton community (Kyle et al. 1988). This community will support fewer juvenile sockeye, even though PR remains the same. Adult escapements (800–900/EV unit) suggested in the EV model can produce this situation, with small (2.0 g) smolts, lowered freshwater and marine survival, reduced adult sockeye production, and a predator-resistant zooplankton community (Koenings and Burkett 1987; Koenings and Kyle 1997; Koenings et al. 1993). Data from B.C. lakes where high escapements have been observed support these findings, since EV model predictions of optimum escapement are much higher than those needed to maximize juvenile sockeye salmon production (Figure 32.4). Most recorded escapements to B.C. lakes have been much smaller than recommended by either rearing capacity model, and development of predator-resistant zooplankton communities has not been observed. Further, in many lakes, including those where excessive escapements have occurred (e.g., Chilko, Quesnel, Shuswap), sockeye returns tend to be cyclical, with both high and low adult returns in any 4-year cycle. Consequently, consecutive over-escapements have not occurred, allowing the zooplankton community to recover from any year of severe grazing pressure.

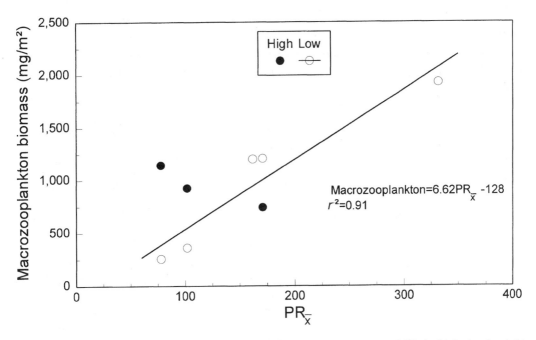

FIGURE 32.5 Differences in the relationship between macrozooplankton and PR in high density (>14 EFS/ha) and low density (<2 EFS/ha) brood years, where EFS = effective female spawners. Data are seasonal averages from Fraser River lakes. (Hume et al. 1996 and Shortreed et al. 1996). Regression line is for low density years only.

PREDICTIONS FOR B.C. LAKES

Of the 90 B.C. sockeye salmon nursery lakes, we have data to directly estimate maximum smolt production and optimum escapements (i.e., fry models) only from Quesnel and Shuswap lakes. These lakes make up 16% of the total nursery lake surface area in B.C. and 29% of the total surface area in the Fraser River system. We have PR data for a much larger number of B.C. lakes (Tables 32.1 and 32.2). The Fraser River drainage basin is one of the world's most important sockeye salmon producers and contains 66% of B.C.'s nursery lake area (Table 32.2). Currently, PR data are available for 43% of the Fraser's nursery area. Of sockeye salmon nursery lakes in other regions of B.C., we have PR data for the majority of nursery lakes near the mainland coast, on Vancouver Island, and in the Skeena River drainage basin. No PR data are available for lakes on the Queen Charlotte Islands or for lakes north of the Nass River drainage basin (Table 32.2).

To make a first estimate of total productive capacity of B.C. sockeye salmon lakes, we assumed that lakes where no PR data were available had PR equal to the average of other lakes within the region, or of adjacent regions if no data were available within the region (Table 32.2). While this assumption may not be valid, we believed it was the best option until better data are available. This resulted in estimates of optimum escapement of 12.3 million to the Fraser River drainage basin and 16.9 million to all B.C. sockeye salmon lakes. Maximum smolt numbers (assuming 4.5-g smolts) from optimum escapements were estimated to be 667 million from the Fraser River drainage basin and 914 million from all B.C. sockeye salmon lakes (Table 32.2). If a smolt-to-adult survival of 10 to 15% is assumed (Hume et al. 1996), maximum adult returns to the Fraser system would range from 67 to 100 million. This is similar to the estimated historical return of 100 million sockeye salmon during dominant cycle years suggested by Ricker (1987). If the same survivals are assumed for all B.C. sockeye salmon, maximum adult returns would range from 91 to 137 million. Current dominant-year returns of adult sockeye salmon to the Fraser River system are 15 to 25 million (PSC 1996).

TABLE 32.1

British Columbia sockeye salmon nursery lakes where PR data are available and resulting PR model predictions (optimum escapement = $187 \times PR_{total}$; maximum smolt number = $10,120 \times PR_{total}$; maximum smolt biomass = $45.5 \times PR_{total}$)

				PR model predictions		
Region	Lake	Surface area (km²)	PR_{total} (metric tons C)	Optimum escapement (thousands)	Max. smolt number (millions)	Max. smolt biomass (metric tons)
Central coast	Bonilla	2.3	44	8.2	0.45	2.0
	Curtis[a]	3.0	62	12	0.63	2.8
	Devon	1.8	22	4.1	0.22	1.0
	Kitlope	12	99	19	1.0	4.5
	Long	21	240	45	2.4	11
	Lowe	3.7	52	9.7	0.5	2.4
	Owikeno	91	904	170	9.2	41
	Simpson	8.7	100	19	1.0	4.6
Fraser River	Chilko	185	2,614	490	26	119
	Francois	250	7,247	1,400	73	330
	Fraser	54	3,215	600	33	146
	Quesnel	270	4,985	930	50	227
	Shuswap	296	10,157	1,900	103	462
Nass River	Fred Wright	3.9	84	16	0.85	3.8
	Meziadin	36	991	190	10	45
Skeena River	Alastair	6.9	244	46	2.5	11
	Babine	470	12,355	2,300	125	562
	Bear	19	491	92	5.0	22
	Johanson	1.4	17	3.2	0.17	0.8
	Kitsumkalum	18	109	20	1.10	5.0
	Kitwanga	7.8	372	70	3.8	16.9
	Lakelse	13	173	32	1.75	7.9
	Morice	96	1,488	280	15	68
	Morrison	13	255	48	2.6	12
	Sustut	2.5	40	7.5	0.40	1.8
	Swan	18	369	69	3.7	17
Vancouver Isl.	Great Central[a]	51	889	170	9.0	40
	Henderson[a]	15	504	94	5.1	23
	Hobiton[a]	3.6	51	9.5	5.2	2.3
	Kennedy	64	627	120	6.4	28
	Nimpkish	37	412	77	4.2	19
	Sproat	41	325	61	3.3	15
	Woss	13	209	39	2.1	10

[a] PR estimates for these lakes were collected during whole-lake fertilization experiments. Under normal conditions all estimates would be lower.

CONCLUSIONS

The EV model is a useful predictor of rearing capacity in some Alaskan lakes. By using photosynthetic rate instead of euphotic zone depth, our PR model makes EV model predictions usable in a wider range of lakes and conditions. Although there are few B.C. lakes where predictions can be tested, in Quesnel and Shuswap lakes, where maximum sockeye production has been observed,

TABLE 32.2
Variation in total sockeye salmon nursery area within major regions or drainage basins of British Columbia, the proportion of each region where photosynthetic rate (PR) data are available, and PR model predictions for the regions. For the predictions, we assumed that PR for lakes where no data were available was similar to PR in other lakes in the region. Where no regional data were available, PR data from adjacent regions were used.

Region	Approximate total lake area (km²)	Percent of total B.C. nursery area	Percent of area with PR data	PR model predictions		
				Optimum escapement (millions)	Maximum smolt biomass (metric tons)	Maximum smolt number (millions)
Fraser River	2,501	66.3	43	12.3	3060	667
Skeena River	666	17.6	100	2.9	724	158
Vancouver Island	257	6.8	91	0.6	153	33
Central coast	172	4.6	83	0.34	86	18
Nass River	75	2.0	53	0.38	94	21
Queen Charlottes	56	1.5	0	0.11	28	6.1
Taku River	28	0.8	0	0.12	30	6.6
Stikine River	19	0.5	0	0.08	21	4.5
Total	3,774	100		16.9	4,196	914

PR model predictions correspond well to observed optimum escapements (Figure 32.4). In Chilko Lake, PR model predictions of 108 smolts/female spawner were similar to observed smolt production of 91 smolts/female spawner at maximum sockeye production.

The PR model allows predictions of optimum escapement and maximum smolt production to be made after 2–3 years of data collection. Estimates can be made with only 1 year of data (for a number of lakes we have only 1 year of PR data), but annual variability can be substantial and multiple years of data ensure greater confidence in annual estimates. Predictions can be made when the only available data on sockeye nursery lakes are seasonal carbon production and lake area, but model predictions can be readily modified and/or verified as more data are available about spawning area, rearing environment, and the life-history stages of a sockeye stock.

PR model predictions of optimum escapement are for the number of spawners required to fully utilize lake rearing capacity. These predictions must be modified if key assumptions are not met. For example, if spawning ground capacity is less than lake rearing capacity, the PR model will overestimate optimum escapements. If planktivores other than age-0 sockeye salmon are present and compete with sockeye salmon fry, model predictions must be adjusted for the proportion of the rearing capacity which they utilize. If a lake has a predation-resistant plankton community because of extreme oligotrophy (Stockner and Shortreed 1989) or because of high grazing pressure (Kyle et al. 1988), its rearing capacity may be lower than PR model predictions.

In lakes which are below rearing capacity (the majority), the amount of additional sockeye production that is biologically feasible can be estimated using the PR model. Increased production could be accomplished by increasing escapements or by some form of enhancement (e.g., spawning channels or fry stocking). In a lake which is currently producing large numbers of adult sockeye salmon, the PR model enables escapement targets to be set based on the lake's productive capacity. If escapements exceed the amount required to maximize smolt production, the economic cost of the foregone catch can be determined. The PR model would be useful in maintaining productive stocks at optimum levels and in determining the amount that unproductive stocks can be increased, making it a valuable tool in sustaining and enhancing B.C. and other sockeye salmon stocks.

ACKNOWLEDGMENTS

We thank the large number of dedicated people who participated in this project during the many years of data collection and analysis. They are too numerous to mention, but prominent among them are H. Enzenhofer, E. MacIsaac, S. MacLellan, K. Masuda, K. Morton, and B. Nidle. Dana Schmidt provided data from Alaskan lakes which we used in some of our analyses. Reviewers Mazumder Asit, Gary Kyle, and Chris Luecke provided many helpful suggestions.

REFERENCES

Bradford, M. J. 1995. Comparative review of pacific salmon survival rates. Canadian Journal of Fisheries and Aquatic Sciences 52:1327–1338.

Burczynski, J. J., and R. L. Johnson. 1986. Application of dual-beam acoustic survey techniques to limnetic populations of juvenile sockeye salmon (*Oncorhynchus nerka*). Canadian Journal of Fisheries and Aquatic Sciences 43:1776–1788.

Burgner, R. L. 1991. Life history of sockeye salmon (*Oncorhynchus nerka*). Pages 1–118 *in* C. Groot and L. Margolis, editors. Pacific salmon life histories. UBC Press, Vancouver, British Columbia.

Diewert, R. E., and M. A. Henderson. 1992. The effect of competition and predation on production of juvenile sockeye salmon (*Oncorhynchus nerka*) in Pitt Lake. Canadian Technical Report of Fisheries and Aquatic Sciences 1853.

Downing, J. A., C. Plante, and S. Lalonde. 1990. Fish production correlated with primary productivity, not the morphoedaphic index. Canadian Journal of Fisheries and Aquatic Sciences 47:1929–1936.

Fee, E. J., M. P. Stainton, and H. J. Kling. 1985. Primary production and related limnological data for some lakes of the Yellowknife, NWT area. Canadian Technical Report of Fisheries and Aquatic Sciences 1409.

Hanson, J. M., and R. H. Peters. 1984. Empirical prediction of crustacean zooplankton biomass and profundal macrobenthos biomass in lakes. Canadian Journal of Fisheries and Aquatic Sciences 41:439–445.

Henderson, M. A., and five coauthors. 1991. The carrying capacity of Pitt Lake for juvenile sockeye salmon (*Oncorhynchus nerka*). Canadian Technical Report of Fisheries and Aquatic Sciences 1797.

Hilborn, R. 1992. Institutional learning and spawning channels for sockeye salmon (*Oncorhynchus nerka*). Canadian Journal of Fisheries and Aquatic Sciences 49:1126–1136.

Hume, J. M. B., K. S. Shortreed, and K. F. Morton. 1996. Juvenile sockeye rearing capacity of three lakes in the Fraser River system. Canadian Journal of Fisheries and Aquatic Sciences 53:719–733.

Hyatt, K. D., and G. J. Steer. 1987. Barkley Sound sockeye salmon (*Oncorhynchus nerka*): Evidence for over a century of successful stock development, fisheries management, research, and enhancement effort. Pages 435–457 *in* H. D. Smith, L. Margolis, and C. C. Wood, editors. Sockeye salmon (*Oncorhynchus nerka*) population biology and future management. Canadian Special Publication of Fisheries and Aquatic Sciences 96.

Hyatt, K. D., and J. G. Stockner. 1985. Response of sockeye salmon (*Oncorhynchus nerka*) to fertilization of British Columbia coastal lakes. Canadian Journal of Fisheries and Aquatic Sciences 42:320–331.

Jones, J. R., and M. V. Hoyer. 1982. Sportfish harvest predicted by summer chlorophyll-*a* concentration in midwestern lakes and reservoirs. Transactions of the American Fisheries Society 111:176–179.

Kobayashi, Y. 1978. Counting of ^{14}C-bicarbonate solutions. Page 31 *in* Y. Kobayashi and W. G. Harris, editors. Liquid scintillation counting applications notes. New England Nuclear, Boston, Massachusetts.

Koenings, J. P., and R. D. Burkett. 1987. Population characteristics of sockeye salmon (*Oncorhynchus nerka*) smolts relative to temperature regimes, euphotic volume, fry density and forage base within Alaskan lakes. Pages 216–234 *in* H. D. Smith, L. Margolis, and C. C. Wood, editors. Sockeye salmon (*Oncorhynchus nerka*) population biology and future management. Canadian Special Publication of Fisheries and Aquatic Sciences 96.

Koenings, J. P., H. J. Geiger, and J. J. Hasbrouck. 1993. Smolt-to-adult survival patterns of sockeye salmon (*Oncorhynchus nerka*): effects of smolt length and geographic latitude when entering the sea. Canadian Journal of Fisheries and Aquatic Sciences 50:600–611.

Koenings, J. P., and G. B. Kyle. 1997. Consequences to juvenile sockeye salmon and the zooplankton community resulting from intense predation. Alaska Fishery Research Bulletin 4:120–135.

Kyle, G. B., J. P. Koenings, and B. M. Barrett. 1988. Density dependent trophic level responses to an introduced run of sockeye salmon (*Oncorhynchus nerka*) at Frazer Lake, Kodiak, Alaska. Canadian Journal of Fisheries and Aquatic Sciences 45:856–867.

Leach, J. H., and five coauthors. 1987. A review of methods for prediction of potential fish production with application to the Great Lakes and Lake Winnipeg. Canadian Journal of Fisheries and Aquatic Sciences 44 (Supplementary 2):471–485.

MacDonald, P. D. M.,H. D. Smith, and L. Jantz. 1987. The utility of Babine smolt enumerations in management of Babine and other Skeena River sockeye salmon (*Oncorhynchus nerka*) stocks. Pages 280–295 *in* H. D. Smith, L. Margolis, and C. C. Wood, editors. Sockeye salmon (*Oncorhynchus nerka*) population biology and future management. Canadian Special Publication of Fisheries and Aquatic Sciences 96.

McDonald, J. G. 1969. Distribution, growth, and survival of sockeye fry (*Oncorhynchus nerka*) produced in natural and artificial stream environments. Canadian Journal of Fisheries and Aquatic Sciences 26:229–267.

McDonald, J., and J. M. B. Hume. 1984. Babine Lake sockeye salmon (*Oncorhynchus nerka*) enhancement program: testing some major assumptions. Canadian Journal of Fisheries and Aquatic Sciences 41:70–92.

Oglesby, R. T. 1977. Relationship of fish yield to lake phytoplankton standing crop, production, and morphometric factors. Journal of the Fisheries Research Board of Canada 34:2271–2279.

PSC (Pacific Salmon Commission). 1996. Report of the Fraser River Panel to the Pacific Salmon Commission on the 1993 Fraser River sockeye and pink salmon fishing season. Pacific Salmon Commission, Vancouver, British Columbia.

Ricker, W. E. 1987. Effects of the fishery and of obstacles to migration on the abundance of Fraser River sockeye salmon (*Oncorhynchus nerka*). Canadian Technical Report of Fisheries and Aquatic Sciences 1522.

Roos, J. F. 1989. Restoring Fraser River salmon. Pacific Salmon Commission, Vancouver.

Ryder, R. A. 1965. A method for estimating the potential fish production of north-temperate lakes. Transactions of the American Fisheries Society 94:214–218.

Ryder, R. A., S. R. Kerr, K. H. Loftus, and H. A. Regier. 1974. The morphoedaphic index, a fish yield estimator—review and evaluation. Journal of the Fisheries Research Board of Canada 31:663–668.

Shortreed, K. S., J. M. B. Hume, and K. F. Morton. 1996. Trophic status and rearing capacity of Francois and Fraser lakes. Canadian Technical Report of Fisheries and Aquatic Sciences 2151.

Shortreed, K. S., and J. G. Stockner. 1981. Limnological results from the 1979 British Columbia Lake Enrichment Program. Canadian Technical Report of Fisheries and Aquatic Sciences 995.

Shortreed, K. S., and J. G. Stockner. 1986. Trophic status of 19 subarctic lakes in the Yukon Territory. Canadian Journal of Fisheries and Aquatic Sciences 43:797–805.

Simpson, K., L. Hop Wo, and I. Miki. 1981. Fish surveys of 15 sockeye salmon (*Oncorhynchus nerka*) nursery lakes in British Columbia. Canadian Technical Report of Fisheries and Aquatic Sciences 1022.

Stockner, J. G. 1987. Lake fertilization: the enrichment cycle and lake sockeye salmon. Pages 198–215 *in* H. D. Smith, L. Margolis, and C. C. Wood, editors. Sockeye salmon (*Oncorhynchus nerka*) population biology and future management. Canadian Special Publication of Fisheries and Aquatic Sciences 96.

Stockner, J. G., and K. S. Shortreed. 1979. Limnological studies of 13 sockeye salmon lakes in British Columbia, Canada. Fisheries and Marine Service Technical Report 865.

Stockner, J. G., and K. S. Shortreed. 1989. Algal picoplankton production and contribution to food-webs in oligotrophic British Columbia lakes. Hydrobiologia 173:151–166.

Stockner, J. G., and K. S. Shortreed. 1994. Autotrophic picoplankton community dynamics in a pre-alpine lake in British Columbia, Canada. Hydrobiologia 274:133–142.

Stockner, J. G., K. S. Shortreed, and K. Stephens. 1980. The British Columbia lake fertilization program: limnological results from the first 2 years of nutrient enrichment. Fisheries and Marine Service Technical Report 924.

Wetzel, R. G. 1975. Limnology. W.B. Saunders Company, Philadelphia.

Williams, I. V., and T. J. Brown. 1994. Geographic distribution of salmon spawning streams of British Columbia with an index of spawner abundance. Canadian Technical Report of Fisheries and Aquatic Sciences 1967.

Youngs, W. D., and D. G. Heimbuch. 1982. Another consideration of the morphoedaphic index. Transactions of the American Fisheries Society 111:151–153.

Section VII

Habitat Protection and Restoration

33 Protecting and Restoring the Habitats of Anadromous Salmonids in the Lake Washington Watershed, an Urbanizing Ecosystem

Kurt L. Fresh and Gino Lucchetti

Abstract.—Numbers of naturally spawning sockeye salmon *Oncorhynchus nerka*, coho salmon *O. kisutch*, chinook salmon *O. tshawytscha*, and steelhead *O. mykiss* have recently declined to historic lows in the Lake Washington Watershed (LWW), a heavily populated watershed located in the Puget Sound lowlands of Washington State. A major reason for these declines is degradation of spawning, rearing, and migratory habitats by urbanization, the conversion of the landscape to residential, commercial, and industrial uses. Urbanization has altered the physical nature of stream habitats in the LWW primarily by modifying hydrologic disturbance regimes (floods have become more frequent and of greater magnitude) and riparian zones (forest cover has been eliminated). The resultant loss of large woody debris (LWD) in stream channels has been especially significant. Stream habitats have become simpler (e.g., loss of pool habitats, straighter and wider channels) and less suited for anadromous salmonids. Increases in nutrient loadings (especially phosphorus) are also a major concern because they can change the limnology and fish resources of Lake Washington and Lake Sammamish, two large lakes that are important to all anadromous salmonids that use the watershed. To increase the chances of rebuilding LWW salmon and steelhead populations, we believe that an ecosystem-based approach to protection and restoration of freshwater habitats is needed. Of primary importance is the need to direct protection and restoration efforts at large scales (e.g., watersheds) because most processes that create and maintain aquatic habitats occur at such scales. While the current basin-oriented approach is logical for stream habitat management, there is a need for agencies to more effectively coordinate management actions at a regional scale. This would help direct resources to areas where they will be most effective and help ensure that water quality in Lake Washington and Lake Sammamish is protected. We recommend that three categories of basins be established in the LWW based upon the status of their anadromous salmonid populations, the condition of their aquatic habitats, ecological integrity, and how heavily developed they are. The first category consists of those basins with the most diverse and abundant populations of salmon and steelhead, the highest quality habitat, the greatest ecological integrity, and least amount of development. A major focus of management efforts in these basins should be prevention of further ecosystem degradation. The second category consists of those basins that are lightly to moderately developed. Management priorities in these basins would be to upgrade the condition of these basins by preventing any further degradation from occurring and restoring damaged ecological processes. The third category consists of those basins that are the most heavily developed. These basins would be managed primarily to reduce their nutrient levels to help protect the water quality of Lake Washington and Lake Sammamish.

INTRODUCTION

The abundance of many naturally spawning populations of Pacific salmon and steelhead *Onco-rhynchus* spp. in the Northwestern U.S. and California has declined to critically low levels (Nehlsen et al. 1991; Slaney et al. 1996; Stouder et al. 1997). The most common factor associated with declines of anadromous salmonids has been impacts on spawning, rearing, and migratory habitats by human activities (Nehlsen et al. 1991; Bisson et al. 1992; Gregory and Bisson 1997). As a result, fish management agencies and local governments have substantially increased their efforts to protect and restore anadromous salmonid habitats. Many scientists and resource managers argue that an ecosystem-based approach to these protection and restoration efforts is needed to improve chances of successfully rebuilding anadromous salmonid populations (e.g., Moyle and Yoshiyama 1994; Lichatowich et al. 1995; Reeves et al. 1995; Bottom 1997).

Ecosystem approaches to salmon habitat management have primarily been applied in landscapes where the dominant land-use practice is timber harvest. This emphasis is understandable given the large land base and large numbers of anadromous populations at risk to effects of timber harvest (Gregory and Bisson 1997). While ecosystem management is a promising approach, it is still largely experimental (Lichatowich et al. 1995; Noss and Scott 1997). Thus, it needs to be widely applied in a variety of land use situations in order to better understand its strengths, weaknesses, and chances of success. Each land-use practice presents different challenges since they differ in their impacts and the portions of the landscape that are altered.

One land use practice that we believe deserves more attention is urbanization. Urbanizing ecosystems present a formidable challenge to salmon conservation because they are so dramatically altered and fragmented. As a result of the inevitable increase in the human population, the amount of developed land will expand. Although urban lands make up less than 2% of the land base in the Pacific Northwest (Pease 1993), they exert a disproportionate influence on salmon production because they typically encompass prime spawning, migratory, and rearing habitats. Anadromous populations found in the low elevation streams and lakes that lie within these areas represent an important genetic resource; their unique adaptations and life history characteristics should be preserved (Wood 1995; Holling and Meffe 1996).

This chapter examines protection and restoration of aquatic habitats in the Lake Washington Watershed (LWW), an urbanizing ecosystem located in the central Puget Sound basin of western Washington (Figure 33.1). Our primary purpose is to describe an ecosystem-based approach to the management of anadromous salmonid habitats in this watershed. Specific objectives of the chapter are to characterize the life history and status of anadromous salmonid populations of this system; describe the major impacts of urbanization on freshwater habitats; and describe an ecosystem-based approach to habitat management in the LWW.

DESCRIPTION OF THE LAKE WASHINGTON WATERSHED

The Lake Washington Watershed encompasses about 1,600 km² (Figure 33.1). Elevations in the LWW vary from sea level to nearly 1,585 m. Annual rainfall averages about 75 cm in the western portions of the watershed and over 250 cm in the headwaters of the Cedar River and Issaquah Creek. Precipitation is heaviest from late fall to early spring when freshets and floods commonly occur; snowfall occurs in the higher elevations of the Cedar River and Issaquah Creek basins*. The LWW includes about 740 km of streams, 13 primary drainage basins, and numerous minor ones. Flows in the largest river, the Cedar, are regulated by Masonry Dam located at RK (River Kilometer) 53. Water stored behind the dam is used by the City of Seattle for municipal purposes and is diverted by the impassable Landsburg Dam at RK 35. Two major lakes, Sammamish and Washington, with a

---

* In this chapter, watershed refers to the catchment or area that drains into a common estuary. It is composed of basins which, in turn, are composed of subbasins.

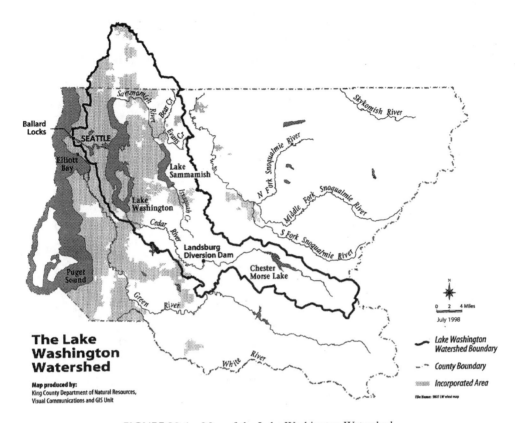

FIGURE 33.1   Map of the Lake Washington Watershed.

combined surface area of about 109 km², and many smaller lakes and wetlands occur within the watershed. All major basins discharge into either Lake Washington or Lake Sammamish.

Residential and commercial land uses dominate in most basins (Table 33.1). Land cover ranges from nearly 90% developed in the Thornton Creek basin to 93% forested in the Issaquah Creek basin; in 1992, 86 and 14% of the LWW was classified as forested and developed, respectively (King County, Department of Natural Resources, unpublished data). Most forests are either Puget Sound lowland, second growth forests consisting of hemlock *Tsuga heterophylla*, Douglas fir *Pseudotsuga menziesii*, and western red cedar *Thuja plicata*, or deciduous forests composed of alder *Alnus rubra* and maple *Acer* spp.

The human population of the watershed in 1992 was an estimated 1.02 million people living in more than 20 incorporated cities, including Seattle, which is the largest city in the state, and two counties, King and Snohomish. Population growth has been dramatic in the last several decades and this trend is expected to continue. For example, the population of the Swamp Creek basin, located at the north end of Lake Washington, increased 50% between 1986 and 1990 (State of Washington, unpublished data) while the city of Kirkland is anticipating a 27% increase in population between 1990 and 2010 (Cleveland 1995). Changes in land use and cover will occur as the population expands, such as in the lower Cedar River basin where the coverage of developed lands are predicted to increase from 15 to over 50% during the next 20 years (King County 1993).

ANADROMOUS SALMONIDS OF THE LAKE WASHINGTON WATERSHED

Five species of anadromous salmonids currently utilize aquatic habitats within the LWW: sockeye salmon *Oncorhynchus nerka*, coho salmon *O. kisutch*, chinook salmon *O. tshawytscha*, steelhead *O. mykiss*, and cutthroat trout *O. clarki*. We have not considered sea-run cutthroat trout in this

TABLE 33.1

Characteristics of major drainage basins in the Lake Washington Watershed.

Basin	Basin Size (km²)	Stream len (km)	Wetland Area (km²)	% Land Cover Developed	% Land Cover Forest/Open	% Stream km lost	Mean Flow m³/s	%EIA
Cedar River								
Upper	316	>26	—	1	99	—	24.7	—
Lower	171	209	6.35	15	85	—	17.8	6
Sammamish River	68	119	1.02	40	60	—	11.3	—
Issaquah/Tibbets	157	344	3.45	7	93	—	3.7	4
Bear Creek	132	128	8.65	24	76	10	0.8	7
North Creek	77	40	—	50	50	17	1.1	27
Swamp Creek	63	34	0.50	75	25	24	0.9	24
May Creek	36	61	1.72	17	83	12	0.7	9
Coal Creek	17	26	0.03	30	70	43	0.7	13
Kelsey Creek	42	23	—	67	23	—	0.6	30
Little Bear	41	19	0.16	—	—	12	—	10
Juanita Creek	17	18	0.13	76	20	54	0.3	31
Lyon Creek	5	5	0.02	50	50	61	0.2	18
McAleer Creek	8	5	0.72	60	40	56	0.4	25
Thornton Creek	28	10	0.05	90	10	30	0.3	34

Source: Geographic data is primarily from analysis of 1992 LANDSAT photographs but also includes information from basin plans and conditions reports. Wetland information is unpublished data from King County and flow data is from the most recent USGS reports that had data for that basin. Data on percent stream km lost is from Lucchetti and Fuerstenberg (1993). EIA is Effective Impervious Area.

chapter because little is known about their abundance, life history, or population structure within the LWW.

LIFE HISTORY AND STATUS OF ANADROMOUS SPECIES

Anadromous salmonids in the LWW are of both hatchery and wild origin (i.e., naturally produced). There is an extensive history of stocking hatchery-produced salmon and steelhead in the LWW; many of these hatchery fish have originated from outside the watershed. The most important introduction of non-native salmon occurred between 1935 and 1945 when sockeye salmon eggs and fry from a hatchery on the Skagit River were liberated into the watershed (Shaklee et al. 1996). As a result of these introductions, a large sockeye salmon run was established in the Cedar River that represents one of the best examples of a self perpetuating salmon run originating from transplants (Wood 1995). Currently, coho and chinook salmon smolts are released from two hatcheries in the watershed while coho fry are planted annually in all basins, with the exception of the Cedar River. A sockeye hatchery program that uses only local broodstock has been in operation on the Cedar River since 1991 and small numbers of hatchery-produced steelhead are occasionally released in the LWW.

Each LWW anadromous species possesses characteristic morphological, physiological, behavioral, and ecological attributes that separates it from the other species. Major differences between species are summarized in Table 33.2. Spatial and temporal patterns of habitat use are especially important because these relate directly to how vulnerable each species is to effects of urbanization. For example, juvenile sockeye are less vulnerable to alterations of stream habitats because they do not rear in streams but instead rear solely in lakes Washington and Sammamish (Table 33.2). Juvenile coho and steelhead are the most vulnerable to degradation of stream habitats because they spend >1.5 years rearing in streams.

TABLE 33.2
Important life history characteristics of anadromous species in the Lake Washington Watershed.

| Attributes | Anadromous Species | | | |
	Sockeye	Coho	Steelhead	Chinook
Mean escapement (1987–1996)	166,500	3,450	598	901
Lowest recorded escapement	26,000	200	70	245
Year observed	1995	1994	1994	1993
Escapement goal	350,000	15,000	1,600	1,500
Spawning				
Duration	Sept–Feb	Oct–Jan	March–June	Sept–Nov
Peak	Mid-Oct–Mid-Nov	Nov	May	Oct
Primary habitat	Mainstem	Small Tributary	Mainstem/Tribs	Mainstem
Emergence				
Duration	Jan–May	March–April	May–Aug	Feb–May
Peak	March–April	March	July–Aug	Feb–March
Freshwater residence				
Habitat	Lakes	Small Tributaries	Streams	Streams
Duration (years)	0.5–2.3	1.5	2.0	<1.0
Dominant type (years)	1.3	1.5	2.0	<1.0
Timing of smolt outmigration				
Duration	Late Apr–Early June	Late Apr–Early May	April–May	Late May–Early July
Peak	May	Early May	Early May	June
Marine residence				
Range (years)	1–3	0.5–1.5	2–3	2–5
Most common (years)	2	1.5	2–3	4

Source: Escapement data is from Washington Department of Fish and Wildlife (WDFW) spawner survey databases. Other data is from the following sources: sockeye (D. Seiler, WDFW; Hendry 1995; Fresh, WDFW), coho (B. Tweit, WDFW; Fresh, WDFW), steelhead (R. Leland, WDFW; S. Foley, WDFW), and chinook (D. Seiler, WDFW; S. Foley, WDFW).

To evaluate the status of anadromous salmonids in the LWW, estimates of the escapement of naturally spawning fish were used (WDF et al. 1993; WDFW unpublished data). While there are shortcomings to utilizing escapement as a measure of population status (Bisson et al. 1992), it represents the longest and most consistently collected type of data available for all species in the LWW. In general, most survey efforts have been concentrated in basins that have historically had the most fish (Cedar River, Big Bear Creek, and Issaquah Creek). The Cedar River has the largest escapements of anadromous salmonids in the LWW, accounting for about 83% of the sockeye salmon, 15% of the coho, 70% of the chinook, and 50% of the steelhead (based upon mean annual escapements from 1987–1996). Because hatchery fish are released throughout the watershed, some naturally reproducing fish are of hatchery origin. This is especially true for coho salmon because of the large numbers of fry planted annually throughout the watershed.

Escapements of all four species have declined in recent years, with the lowest recorded escapements of all species occurring between 1993 and 1995 (Table 33.2). For example, the escapement of sockeye salmon averaged 213,000 between 1967 and 1976 and 167,000 between 1987 and 1996. In recent years, all species have rarely met escapement goals that were established by fish managers to represent full seeding of the watershed with naturally reproducing fish (Table 33.2). However, it is unclear how relevant these goals are under current conditions since they were established over 25 years ago for some species.

Many factors have contributed to declines in abundance of natural spawners in the LWW including overharvest, marine mammal predation, and introductions of exotic species (Fresh 1994). As we discuss in the following section, freshwater habitats have been severely degraded throughout the LWW. Because salmon require high quality habitats for spawning, rearing, and migration, degradation of these habitats has undoubtedly played a role in declines in abundance of all species in this system.

LIFE HISTORY DIVERSITY

Considerable life history diversity exists within species in the LWW (e.g., Hendry 1995; Hendry and Quinn 1997). This diversity is the variable use of spawning, rearing, and migratory habitats through space and time. It represents a strategy to fully utilize a range of habitats and thereby reduce the risk of extinction or reduced production due to environmental fluctuations such as floods (NRC 1996). Using phenotypic and genotypic characteristics, each species can be divided into stocks which are groups of individuals that are more or less reproductively isolated from other groups and differ in their spatial and temporal use of habitats. Stocks in turn are composed of populations and subpopulations. Because populations and subpopulations are difficult to distinguish using currently available methods, stocks are typically used as the primary salmon management unit. A recent inventory of salmon and steelhead in Washington (WDF et al. 1993) identified three sockeye stocks, two coho stocks, three chinook stocks, and one steelhead stock in the LWW; these are the primary units of salmon conservation in the LWW.

IMPACTS OF URBANIZATION

Urbanization is the conversion of a landscape from a rural, an agricultural, or a forested condition to one dominated by residential, commercial, and industrial land use. This process is complex and includes changes in vegetation, changes in soils, increases in the amount of impervious surface, dredging and filling of wetlands and stream channels, streambank rip-rapping, pollution, and so on. The amount of impervious surface area (defined as the amount of roads, parking lots, rooftops, and other impermeable surfaces in a drainage area) is widely employed as a measure of development intensity (Booth 1991; Schueler 1994; May 1996); as a landscape becomes more developed, the percent of imperviousness increases. The major impacts of urbanization are described in the following sections.

ELIMINATION OF STREAM HABITAT

The amount of stream habitat available for anadromous salmonids in the LWW streams has declined due to filling of stream channels, channelization, and the placement of blockages to fish migrations. Lucchetti and Fuerstenberg (1993) calculated that losses of formerly accessible habitat in individual basins in the LWW ranged from 10% to 61% (Table 33.1).

IMPACTS ON PHYSICAL HABITAT QUALITY

The physical characteristics of stream habitats are a function of interactions between channel geomorphology, hydrologic pattern, the channel, and the surrounding riparian forest (Naiman et al. 1992). Urbanization degrades the physical characteristics of stream habitats primarily by altering hydrologic regimes and riparian zones (Booth 1991; May 1996).

Changes to Hydrologic Regimes.—The hydrologic regime of a drainage basin refers to how water is collected, moved, and stored. The frequency and magnitude of floods in streams are especially important since floods are the primary source of disturbance in streams and thus play a key role in how they are structured and function (Reice et al. 1990; Naiman et al. 1992; Reeves et al. 1995). In ecologically healthy systems, the physical and biotic changes caused by natural disturbances are not usually sustained and recovery is rapid to predisturbance levels (Resh et al.

1988; Yount and Niemi 1990). If the magnitude of change is sufficiently large, however, permanent impacts can occur (Reice et al. 1990).

Urbanization increases the magnitude and frequency of floods and creates new peak runoff events (Booth 1991; King County 1993; Schueler 1994; May 1996). Moscrip and Montgomery (1997) found that basins in the Puget Sound lowlands (including several basins from the LWW) that had experienced a significant increase in the amount of urbanized area also exhibited an increased frequency of floods. The 10-year recurrence interval discharges in these basins under pre-urbanized conditions corresponded to 1–4 year recurrence events following urbanization; no discernible shift was observed in basins where there had been little urbanization. Alterations in basin hydrology are caused by changes in soils, decreases in the amount of forest cover, increases in imperviousness, elimination of riparian and headwater wetlands, and changes in landscape context. Hydrologic impacts occur even at low levels of development (<2% imperviousness) and generally increase in severity as more of the landscape is converted to urban uses (Booth 1991; Schueler 1994; May 1996).

In the Cedar River basin, which supports the largest naturally spawning populations of salmon and steelhead, other hydrologic changes have occurred that are indirectly related to urbanization. These hydrologic changes are the result of using the Cedar River as a municipal water source for the ever increasing human population of the region. The City of Seattle withdraws about one-third of the mean annual flow of the Cedar River at RK 35 for municipal uses. Consequently, there is now less water available for use by anadromous fish, especially during the outmigration of juvenile salmon and steelhead and in the early fall when chinook and sockeye salmon are spawning. One important effect of water withdrawal (in combination with channelization of the river) has been a reduction in the wetted width of the mainstem Cedar River by about 56% (King County 1993).

Changes to Riparian Zones.—Urbanization alters riparian zones, the area of living and dead vegetative material adjacent to a stream. Riparian zones provide hydraulic diversity, add structural complexity, provide a refuge from predators and extreme environmental events, buffer the energy of runoff events and erosive forces, moderate temperatures, and provide a source of nutrients (Gregory et al. 1991; Naiman et al. 1992; Bisson et al. 1992). Riparian zones are especially important as the source of large woody debris (LWD) in streams which directly influences several habitat attributes important to anadromous species (Bisson et al. 1987; Naiman et al. 1992; Gregory and Bisson 1997). In particular, LWD helps control the amount of pool habitat. Pools provide a refuge from predators and high-flow events for juvenile salmon, especially coho, that rear for extended periods in streams. LWD also directly influences the quality of spawning and incubating gravels.

As basins become more urbanized, riparian forests become increasingly fragmented as they are either cut down to the stream-bank or broken by stream crossings (May 1996). In addition, encroaching development can reduce width of riparian zones. Even when riparian forests are not removed to the channel bank, their species composition can be dramatically altered when native, coniferous trees are replaced by exotic species, shrubs, and young deciduous species. Urbanization also causes a decline in the frequency, volume, and quantity of LWD due to washout, removal, and reduced recruitment from forested areas. May (1996), for example, found that the frequency of LWD in Puget Sound lowland streams averaged 210 pieces/km in lightly urbanized basins (<5% imperviousness) and was <100 pieces/km in basins with more imperviousness.

Overview of Impacts on Physical Habitat Quality.—The combined effects of alterations in the natural disturbance regime, changes in riparian areas, loss of LWD, and in some cases channelization, are profound changes in stream channels. Channels widen, deepen, and incise; streambanks become more unstable; erosion and scour increase; the amount of pool habitat is reduced; the proportion of fines in gravels increases; the quantity of off-channel areas declines; and suspended sediment loads increase (Bisson et al. 1987, 1992; Booth 1991; Booth and Reinelt 1993; May 1996). Significant degradation in physical habitat quality occurs even at low levels of development (<5% imperviousness) and becomes more severe as basins become increasingly urbanized (Booth 1991; Booth and Reinelt 1993; May 1996).

Stream channel degradation ultimately results in simplification of the habitat. Complex and diverse aquatic habitat is critical to the persistence of anadromous species. Persistence of salmon and steelhead depends upon their ability to withstand extreme environmental events such as winter floods and landslides (NRC 1996). This is largely a function of the life history diversity (e.g., variability in spatial and temporal use of particular habitats) that they exhibit. In essence, some portion of the population is able to survive the events because of the existence of spatial, temporal, or behavioral refuges. The life history diversity that is expressed is a function of the genotype interacting with its environment (Healey and Prince 1995). Thus, complex and diverse freshwater habitats help create the array of genotype-phenotype combinations that have allowed salmon to persist.

WATER QUALITY CHANGES

Although water quality degradation results from both point and non-point sources, most attention in urban landscapes has focused on non-point sources, or those derived from a broad area. Non-point sources in urban landscapes include failing septic systems, roadways and parking lots, changes in riparian zones, stream channels (i.e., erosion), and application of pesticides and herbicides on the surrounding landscape. In rural portions of urbanizing systems, agricultural and forest practices can also degrade water quality.

The primary approach used by regulatory agencies to evaluate the extent of water quality degradation in streams of the LWW has been to measure various indicators during baseflow conditions (chronic) and flood events (acute) and then compare results to water quality standards, criteria, and historical data. The frequency and degree to which standards or criteria are exceeded is a measure of the extent of degradation. Some of the commonly used indicators of water quality degradation are nutrient levels, animal-generated bacteria (fecal coliforms), water temperatures, dissolved oxygen, heavy metals in both sediments and the water column, and suspended sediments (e.g., Behnke et al. 1981; Scott et al. 1986; Karr 1991).

Water quality monitoring has occurred in the LWW since 1971 and has identified a variety of localized (i.e., site-specific) problems that are largely a function of conditions adjacent to and immediately upstream of the sampling site; degradation often becomes less downstream of the site due to dilution. One example of a localized water quality problem is an increase in stream temperature due to local riparian forest clearing.

Basin-scale water quality problems are usually indicated by sampling near the mouth of the basin. In general, basin-scale water quality degradation is primarily an issue in the most developed basins. Based upon measurements of water quality in Puget Sound lowland streams (including some LWW systems), May (1996) found that violations of chemical water quality standards during baseflow and storm events occurred infrequently in basins with low to moderate levels of development but consistently occurred in heavily urbanized systems.

Of the basin-scale water quality issues identified by monitoring, much attention has focused on nutrients because of their linkage to water quality in Lake Washington and Lake Sammamish (Edmondson 1991). For example, a 57% increase in bioavailable phosphorus loadings from the Issaquah Creek basin is projected to occur in the future under likely growth scenarios; this represents 70% of the total increase expected in Lake Sammamish (King County 1994). Nutrient loadings are a function of a number of factors including nutrient levels (i.e., concentration), basin size, flow, and amount of development (Richey 1982; King County 1993, 1995).

EFFECTS ON STREAM BIOTA

The cumulative effects of urbanization causes changes in stream biotic communities (Schueler 1994). Recent work by May (1996) in Puget Sound lowland streams suggests that biological changes begin at low levels of development (<2% imperviousness) and become more severe as basins become more urban. Impacts have been documented on diversity, richness, and composition of fish and invertebrate communities (Behnke et al. 1981; Scott et al. 1986; Weaver and Garman 1994); reproductive success rates (Limburg and Schmidt 1990); size and growth rates (Scott et al. 1986); food web relationships (Weaver and Garman 1994); and overall biotic integrity (Steedman 1988; Kleindl 1995; Wang et al. 1997). In a Trenton, New Jersey stream, Garie and McIntosh (1986) found that the number of benthic macroinvertebrate taxa declined from 13 in a relatively undeveloped upstream area to 4 in a heavily developed downstream area. Similarly, Jones and Clark (1987) found that diversity of aquatic insects in urban streams in northern Virginia decreased sharply at human population densities over 4 persons/acre.

The specific responses of anadromous species to urbanization have been examined in several instances. Scott et al. (1986) evaluated the fish communities of Kelsey and Bear creeks, two streams in the LWW. They found that total fish biomass in the two systems was comparable but that there were relatively fewer coho juveniles and more cutthroat trout in the more urbanized Kelsey Creek than in the more rural Bear Creek. They suggested this was likely due to simplification of stream habitat in Kelsey Creek, especially a reduction in the number of pools (Richey 1982). Lucchetti and Fuerstenberg (1993) found that salmonid assemblages in streams in the LWW tended to become less diverse as the extent of urbanization increased; in particular, the proportion of juvenile coho salmon in the assemblage declined and the proportion of juvenile cutthroat trout increased as basin imperviousness increased. Using spawner survey data, Moscrip and Montgomery (1997) found systematic declines in salmon abundance in two urbanizing basins and no evidence for a similar decrease in a control basin. They suggested these declines were related to either increased flood frequency or habitat degradation in the urbanizing basins.

CHANGES IN LAKE WASHINGTON AND LAKE SAMMAMISH

Lakes Washington and Sammamish play a key role in anadromous salmonid production in the LWW since all adult and juvenile salmon and steelhead must migrate through one or both of these lakes on their way to and from ocean feeding grounds. Further, the lakes provide rearing habitat for juveniles of all species, particularly sockeye which typically rear for about 15 months in the lakes. We focus here on impacts to the littoral and open water zones.

Littoral Zone Changes.—Littoral zones generally change more dramatically and rapidly than limnetic areas because they are influenced by factors originating from both the lake as a whole and from the surrounding landscape (Crowder et al. 1996). Considerable changes have occurred in the littoral areas of both lakes. The 1916 completion of the Chittendon (Ballard) Locks, to create a direct water route between Lake Washington and Puget Sound, and diversion of the Cedar River into Lake Washington lowered the lake's level about 3.0 m. This exposed 5.4 km² of previously shallow water habitat, reducing the lake's surface area 7.0%, decreasing the shoreline length by 16.9 km (a 12.8% reduction), and eliminating much of the lake's wetlands (Chrzastowski 1983). Lake levels are regulated by the Locks and kept within about a 1 m range. Historically, lake elevations varied up to 2.1 m during flood events (Chrzastowski 1983). The highest lake levels now occur in summer and seasonal lows occur in winter which is the opposite pattern exhibited by unregulated lakes in the region. Because distribution of aquatic macrophytes in lakes can be limited by the occurrence of extremely low water levels (Cooke et al. 1993), the stable lake levels have probably promoted the expansion of aquatic macrophytes, which are now dominated by the non-native Eurasian watermilfoil *Myriophyllum spicatum*.

Dredging, filling, bulkheading, removal of shoreline vegetation, and the construction of piers, docks, and floats have also occurred as the watershed has been developed. For example, an estimated 82.0% of the Lake Washington shoreline has been bulkheaded and approximately 4.0% of the lake's surface area within 33 m of shore has been covered with residential piers (R. Malcom and E. Warner, Muckleshoot Indian Tribe, unpublished data). As a result of these changes and the lowering of water levels, little natural shoreline remains in either lake, which has likely reduced the amount of woody debris in littoral areas (Christensen et al. 1996).

One probable impact of littoral zone changes has been a decrease in sockeye beach spawning areas due to elimination of their spawning beds by aquatic macrophytes. Shoreline development may also have contributed to elimination of beach spawning habitat if it altered substrate composition and water circulation patterns of spawning areas.

Based upon results of studies in other lakes (Bryan and Scarnecchia 1992; Beauchamp et al. 1994; Christensen et al. 1996; Weaver et al. 1997), it is likely that the types of littoral zone changes observed in Lake Washington and Lake Sammamish have altered the composition, diversity and abundance of fish communities. The amount and spatial patterning of macrophytes can directly affect littoral zone fish abundance and assemblage structure (Bryan and Scarnecchia 1992; Weaver et al. 1997). For example, moderate amounts of macrophytes usually increase abundance of fish (Bryan and Scarnecchia 1992) while high densities of macrophytes can cause localized mortalities of fish due to D.O. depletion (Frodge et al. 1995). In addition, the type of complex habitat provided by piers, docks, and bulkheads can increase fish abundance (Beauchamp et al. 1994). It is difficult, however, to predict the net effect of changes in littoral zones. For example, while shoreline development and macrophytes may result in more habitat for juvenile fish (such as salmonids) in the two lakes, it may also enhance habitat for their predators, such as the non-native smallmouth bass *Micropterus dolomieu* (Bryan and Scarnecchia 1992).

Changes in Water Quality.—The water quality of both lakes has undergone dramatic changes over the last 50 years, primarily due to fluctuations in phosphorus loadings. In Lake Washington, sewage that was discharged into the lake from 1945 to the mid-1960s dramatically increased phosphorus loadings (Edmondson 1991). As a result, blue-green algae dominated the phytoplankton community and suppressed production of some species of zooplankton (e.g., *Daphnia*). By 1966, the discharge of sewage had been eliminated and the lake's limnology reverted to a less eutrophic state. Fluctuations in phosphorus loadings due to wastewater releases and subsequent diversion also occurred in Lake Sammamish (King County 1994). The main difference in response of the phytoplankton community in the two lakes was that recovery was slower in Lake Sammamish (Welch et al. 1980).

Although nutrient increases from sewage discharges are no longer a problem in either lake, non-point sources are. Because it is smaller, shallower, and more rapidly developing than Lake Washington, nutrients are more of a concern in Lake Sammamish (King County 1994). Nutrients in Lake Sammamish cause anoxic conditions during late summer in the hypolimnion that has the potential to either directly kill fish and zooplankton or constrain them to a smaller region of the lake. Alkalinity levels in Lake Washington are also of concern because they have increased from an annual mean of 28.6 mg of calcium carbonate/l in 1963 to >40.0 by 1990; implications of this increase are presently unknown but under investigation (A. Litt, University of Washington, personal communication).

Impacts of water quality changes on fish populations in LWW are poorly understood. Studies in other lake systems have demonstrated that strong linkages occur between nutrient levels and phytoplankton, zooplankton, and fish (e.g., Stockner and MacIssac 1996). Changes in nutrients can directly kill fish by creating anoxic conditions or can alter assemblage structure and abundance of fish by changing the composition of available zooplankton. Thus, while it is likely the fish community responded to nutrient fluctuations, we can only speculate on what the changes may have been.

AN ECOSYSTEM APPROACH TO THE PROTECTION AND RESTORATION OF THE HABITATS OF ANADROMOUS SALMONIDS IN THE LWW

ECOLOGICALLY RELEVANT MANAGEMENT UNITS

A key element of ecosystem management is definition of management units. To have the greatest potential of providing effective and lasting protection of aquatic biota, management units must be defined at large spatial scales (Franklin 1993; Slocombe 1993; Hansen et al. 1993; Grumbine 1994;

Crow and Gustafson 1997). A major implication of such a "big picture" approach is that boundaries of management units will cross different ownerships and political boundaries, such as cities and counties. Managing at large spatial scales thus requires coordination and cooperation across these boundaries to resolve such issues as property rights, conflicting mandates, growth management, and intergovernmental trust.

Basins are a logical scale at which to establish stream habitat management units (Gregory et al. 1991; Bisson et al. 1992; Naiman et al. 1992) and have been adopted by local and state governments within the LWW as the fundamental unit of land and water management; plans have been developed in a number of LWW basins to guide land and water management (e.g., King County 1991, 1994, 1996). All plans are based upon a report that describes current basin conditions (e.g., hydrology and habitat) and forecasts likely future conditions. Plans subdivide each basin into sub-basins based upon hydrologic and geologic factors, identify the major problems that exist, why they exist, and propose and prioritize a set of actions to address problems.

While basins represent a logical and ecologically sound unit for stream management in the LWW, there is also a need to have an overarching management unit that encompasses the entire watershed. One reason watershed-scale management is needed is that basin plans do not target management of limnetic and littoral areas of Lake Washington and Lake Sammamish. While some basin plans recognize the need to manage water quality of stream basins to protect lake water quality, the stream basin planning areas typically stop at the stream mouth. Successful conservation of salmon in the watershed will require managing the lakes' littoral and open water habitats which can only be addressed at a watershed scale.

A second reason for a watershed-scale management unit is to provide a way to prioritize, coordinate, and organize management actions in the watershed as a whole (i.e., between basins). The funds that that can be committed to restoration efforts will be considerably less than what will be needed. For example, cost estimates of 82 restoration projects in the Cedar River basin exceed $50 million (Lucchetti, unpublished data). The Waterways 2000 program of King County has about $8 million budgeted for property acquisition in the watershed. A watershed perspective would help provide a way of allocating these scarce resources.

One encouraging opportunity for regional coordination and cooperation in the LWW is the recently formed Watershed Forums. Two forums were established by King County in the LWW: the Lake Washington/Cedar River Watershed Forum and the Lake Sammamish Watershed Forum. The Watershed Forums were designed to enhance cooperative, local management of fish habitat, water quality, and flooding issues. The implementing legislation (Motion RWQC 95-05 of King County) directs each forum to identify important issues, coordinate among forum members to address the issues, and develop and implement action programs.

CONSIDERATION OF CONTEXT AND SCALE

To have viable and self-sustaining populations of anadromous species in the LWW, it will be necessary to manage the landscape as a whole (Jensen et al. 1996; Christensen 1997). Scientists have learned that ecosystems are complex, integrated systems where landscape patterns (e.g., size of habitat units, connectivity, and composition of adjacent habitats) help determine ecosystem structure and functions (e.g., O'Neill et al. 1989; Noss 1990; King 1993; Jensen et al. 1996). For example, anadromous salmonid production depends upon complexity of aquatic habitats, which is determined by location of the habitats, their relative size, composition and quantity of habitats in an area, and their connectivity in space and time (Frissell and Bayless 1996).

One particularly important landscape element that affects function is scale. Because landscapes are heterogeneous, their structure, function, and dynamic character are especially dependent upon spatial and temporal scale (Turner 1989; Frissell and Bayless 1996). Scales are not isolated from each other but are organized hierarchically so that what occurs at one scale is influenced by what occurs at larger scales (Noss 1990; King 1993). There is not a single correct spatial or temporal scale at which to direct actions. Rather, the success of an action directed at one scale will depend to a considerable extent upon the constraints or enablers placed on it by other scales (Rabeni and Sowa 1996). In streams, the influence of processes at higher spatial scales such as regional land use patterns determines presence or absence of biota while processes operating at smaller scales (stream reach) define distributional patterns within a stream (Roth et al. 1996; Rabeni and Sowa 1996).

EMPHASIZE PROTECTION AND ECOLOGICAL RESTORATION

Protection, or the avoidance of further habitat degradation, must be the cornerstone of management efforts in the LWW. It is more likely to succeed, is of lower risk, and provides long-term and lasting results because it typically involves policy and administrative actions directed at larger spatial scales (Bisson et al. 1992; Frissell and Nawa 1992; Naiman et al. 1992; Kauffman et al. 1997). Because major sources of habitat degradation are the result of impacts to large-scale ecological processes by land use practices, directing management actions at these large scales is an essential part of salmon conservation.

Many management agencies in the LWW emphasize protection of existing resources as a matter of policy using a variety of actions that includes, depending on the context, avoidance, habitat creation, preservation, and compensatory mitigation. As evidenced by the habitat loss and degradation that has occurred despite the policy emphasis on protection, improvements are needed in how habitat is protected. For example, a more even application and enforcement of existing laws and statutes, such as size of riparian buffer strips, is clearly needed. In addition, approaches to compensatory mitigation techniques need to be rethought since in many cases they have not worked (e.g., Booth 1991; Booth and Jackson 1994). While new tools and approaches to mitigation can help (Ferguson 1991; Schueler 1994), there are also limits to what can be successfully mitigated, suggesting avoiding impacts is the only sure way of resource protection in urban environments (May 1996).

In addition to protection, restoration will be needed to improve habitat quality for anadromous salmonids. The focus of restoration has shifted recently from an emphasis on small-scale, site-specific actions to ecological restoration, or the repair of ecological processes (Ebersole et al. 1997; Bisson et al. 1997; Roper et al. 1997; Kauffman et al. 1997). This is because site-specific restoration actions have generally not been successful as they do not address the degraded condition of the ecosystem processes that originally led to the need for the restoration actions (Frissell and Nawa 1992; House 1996; Roper et al. 1997). There is an increasing emphasis on ecological restoration as the conceptual framework needed to implement ecological restoration (e.g., Hobbs and Norton 1996; Ebersole et al. 1997; Roper et al. 1997). Ecological restoration of aquatic habitats basically calls for reestablishing pre-development processes, functions, and related biological, chemical, and physical linkages between land and water. The specific restoration actions implemented will depend upon characteristics of each basin such as the condition of the aquatic habitat, condition of processes, habitat fragmentation, and condition of the surrounding landscape (Turner 1989; Sheldon 1988; House 1996; Frissell and Bayless 1996). Thus, while LWD placement may be appropriate in a system that is not heavily developed or as a temporary fix, it will not likely work as a permanent solution in a more heavily developed basin with degraded and fragmented habitat. It may still, however, be useful for dissipating energy and reducing bedload transport.

GOALS AND OBJECTIVES

Setting goals is a critical part of an ecosystem approach since it is goals and objectives that provide the vision and direction for what needs to be accomplished (Slocombe 1993; Grumbine 1994; Christensen 1997). They must be stated in terms of desired future conditions of the ecosystem's structure, functions, patterns, and processes rather than focusing exclusively on deliverables such as numbers of returning salmon or on generic issues such as improving habitat or water quality.

Goals and objectives will have to be tailored to conditions in individual basins and subbasins. In addition, they will have to be realistic, attainable, and amenable to direct measurement and monitoring. Having realistic goals is especially critical in urbanizing ecosystems where fully restoring habitats or salmon abundance to historic conditions is unlikely.

AGENCY CHANGE

Agencies are the heart and soul of natural resource management since they must convert ecological, cultural, and social knowledge into resource management decisions and actions. To manage salmon and steelhead habitats from an ecosystem perspective, changes in the internal structure and functioning of these agencies will be required (Lichatowich 1997). Agencies will have to (1) more effectively use the best available information; (2) more rigorously and consistently evaluate their performance; (3) more effectively cooperate and coordinate; (4) change institutional values; (5) use a more participatory decision-making process; (6) more effectively involve citizens in management actions; and (7) become more flexible (e.g., Grumbine 1994; Steedman and Haider 1993; Holling and Meffe 1996; Lichatowich 1997). Of these, increased coordination and cooperation among agencies will be especially important in implementing ecosystem management because of the large number of agencies involved in habitat management.

There is a number of excellent examples of agencies working cooperatively and coordinating across jurisdictional boundaries in the LWW. For example, basin plans have been developed and implemented with the involvement of many organizations. The Cedar River Basin Plan includes three cities, one county, one Native American Indian Tribe, one federal agency, and three state agencies.

Arguably the greatest challenge to interagency coordination and cooperation is at the watershed scale where >20 cities, two counties, five state agencies, three federal agencies, and one tribal government, can be potentially involved in management decisions. Although such regional cooperation is a huge hurdle, there are examples of successful coordination and cooperation at this scale. The formation of METRO in the 1950s to clean up the water quality of Lake Washington by rerouting sewage discharges (Edmondson 1991), provides one of the best examples anywhere of a successful, large-scale, regional, restoration program.

ADAPTIVE MANAGEMENT

Uncertainty and risk go hand in glove with natural resource management decision-making (Ludwig et al. 1993). The process of ecosystem management is especially fraught with uncertainty and risk because we cannot predict with certainty the outcome of our decisions (Frissell and Bayless 1996; Christensen 1997). As a result, an adaptive approach to ecosystem management is needed (Walters 1986; Walters and Holling 1990). In adaptive management, one learns by doing; future decisions and actions are adjusted in response to feedback obtained by monitoring previous actions. Monitoring is essential since it provides the continuous feedback loop that managers require to identify and weigh their options. Without feedback obtained by monitoring, learning and flexibility are compromised (Walters 1986). Unfortunately, monitoring is often one of the first victims of budget cuts because of the general political and public perception that it is not accomplishing anything productive.

Monitoring must be directed at determining how well goals and objectives are being met. It should employ a diverse array of tools and approaches to assess ecological integrity, the condition of critical ecological processes such as basin hydrology (e.g., frequency and magnitude of floods), riparian forest condition, the status (both abundance and life history diversity) of salmon populations, and the biogeographic context or spatial patterning of habitat (Noss 1990; Karr 1991; Steedman and Haider 1993). Monitoring must be sensitive to scale and context. The temporal aspects of scale are especially important since detecting the outcomes of specific actions will in many cases depend upon the time scale over which monitoring is conducted. Monitoring should be scientifically based and directed at answering specific questions (i.e., hypothesis testing). Comparative approaches that examine multiple sites at the same scale, such as comparing between basins, may prove useful (Frissell and Bayless 1996). Finally, monitoring must be consistent both within and across basins. The same type of information (e.g., similar measures) on physical processes, habitat quality, water quality changes, and salmon usage is needed in all basins to provide the watershed-scale context.

We suggest that monitoring in the LWW be guided by comprehensive plans that are developed at the basin and watershed scale. Plans should not exist in place of monitoring and research but as a guide to these efforts; they should be regularly updated and developed and implemented as a collaborative effort between scientists and managers. Both disciplines are essential to the process since managers know what types of data are needed while scientists possess the background to design and implement such programs.

ESTABLISHING SALMON CONSERVATION AREAS

Given the inevitable increase in the human population and the impacts expected from this increase, we must ask whether people and salmon can coexist in the LWW. We suggest that the prospects of sustainable salmon populations in this urban environment would be considerably enhanced if salmon preserves or reserves were established. The concept of forming reserves or spatial refuges to conserve genes, populations, and landscapes is not new, having its roots in national parks and reserves, and it enjoys widespread support (e.g., Moyle and Yoshiyama 1994; Reeves et al. 1995; Frissell and Bayless 1996). Most authors, however, regard reserves as short-term solutions to be applied in conjunction with other, longer-term management actions since natural disturbance will eventually dramatically alter the reserve (e.g., Reeves et al. 1995).

Our model of reserves envisions areas within the urban landscape where development would be managed, not halted, with salmon and steelhead conservation as the priority. It is not a reserve, national park, or sanctuary concept in the traditional sense. It is consistent with the intent of the state Growth Management Act (1990) that encourages the designation and establishment of areas critical to fish and habitat conservation. It is also consistent with local land use policies in King County, especially Significant Resource Areas (SRA) and Sensitive Areas. A major difference between critical/sensitive areas and our concept of salmon conservation areas is that we would establish conservation units at the basin scale because it is a logical and ecologically relevant scale for managing aquatic habitat in the LWW. However, we do not propose to abandon the establishment of critical and sensitive areas but rather to fit these areas into the larger-scale salmon conservation units (basins).

Our method of establishing conservation basins in the LWW was to classify them into three categories using information on such attributes as ecological integrity, condition of headwaters, wetland conditions, landscape context of habitat, status of salmon populations, patterns of life history diversity, condition of riparian zones, and amount of impervious area (Table 33.3). Category I represents the salmon conservation basins and includes the Cedar River, Big Bear Creek, and Issaquah Creek. They are the least developed basins and in general have the highest overall ecological integrity, the highest quality habitat, lowest amounts of impervious surface, and healthiest and most diverse naturally spawning anadromous populations. Aquatic habitat protection would be the management priority in these basins (Schueler 1994; May 1996) and employ a variety of approaches such as acquisition of property and development rights; lower zoning; wide buffer strips; retention of the maximum amount of forest cover; better incentives for land owners; use of a wide variety of techniques to reduce impervious surfaces; Best Management Practices (BMPs); and onsite infiltration of stormwater. Restoration efforts would target improving the condition of ecological processes by reconnecting rivers to floodplains, improving condition of surrounding forests, and increasing habitat complexity.

TABLE 33.3
Proposed classification of major basins within the Lake Washington Watershed.
Category I basins represent salmon conservation basins. Basins were classified
based upon an evaluation of all available information and not solely on the
basis of a single parameter.

	Category		
	I	**II**	**III**
Major basins included	Cedar River	May Creek	Thornton Creek
	Issaquah Creek	Little Bear Creek	McAleer Creek
	Big Bear Creek		Juanita Creek
			Kelsey Creek
			Swamp Creek
			North Creek
			Coal Creek
			Lyon Creek
Amount of development	Low	Moderate	Heavy
Imperviousness	< 10%	10–15%	> 15%
Overall condition of anadromous populations			
Abundance	Healthy	Moderate	Low
Diversity	High	High to Moderate	Low
Overall ecological integrity	High	Moderate	Low
Condition of headwaters	Mostly intact	Somewhat intact	Largely lost
Overall condition of wetlands	High	Moderate	Degraded, lost

Source: Data comes from a variety of sources including basin plans and conditions reports, theses, and dissertations (e.g., Kleindl 1995; May 1996), and unpublished files such as salmon spawner surveys.

Category II basins are those that are moderately developed and still maintain more than remnant populations of naturally reproducing salmon and steelhead (Table 33.3). To a large degree, we regard these basins as *potential* salmon conservation areas where management priorities would focus on preventing further degradation and then restoring and rehabilitating degraded ecological processes (May 1996).

Category III basins would consist of the most urbanized basins. It seems doubtful that we have the technological ability or economic resources to maintain self-sustaining populations of salmon and steelhead in these systems. Further, by trying to implement costly restoration programs in these basins, we run the risk of spreading our limited resources too thinly and thus reduce chances of success in other, less degraded systems. Management priorities in these basins should focus on maintaining healthy populations of cutthroat trout which appear to be sustainable in urban settings (Lucchetti and Fuerstenberg 1993). There should also be an emphasis on improving quality of stream waters in these heavily degraded basins (May 1996) to help maintain high quality water in Lake Washington and Lake Sammamish.

While the emphasis of the salmon conservation area plan proposed here is the individual basin, we must not forget about Lake Washington and Lake Sammamish and their role in the life history of salmon and steelhead in the LWW. It is absolutely essential that we maintain the critical functions of these lakes in order to have sustainable populations of anadromous species in the LWW. Clearly, regional approaches will be needed to protect and restore these functions. While managing nutrient loadings is an obvious priority, conservation of the littoral habitats used by beach spawning sockeye which have all but disappeared in the LWW is also needed. We also recommend further research be conducted on impacts of urbanization on the lakes.

While establishing conservation basins is reasonable from a scientific perspective, the concept will undoubtedly face significant political and social obstacles. There will be economic and social costs associated with any conservation approach that many citizens and politicians will not find acceptable. Regardless, it is clear that traditional approaches to management of anadromous salmonid habitats have not been successful and that to have salmon in urban environments we must be willing to do things differently, to take chances, to be innovative, and to be realistic. Business as usual will only result in more of the same.

ACKNOWLEDGMENTS

The insights, comments, and suggestions of R. Fuerstenberg, R. Little, and R. Wissmar are gratefully acknowledged. Critical comments on drafts of this manuscript by D. Booth, E. Warner, R. Francis, and two anonymous reviewers are greatly appreciated. King County's graphics staff helped prepare the figure.

REFERENCES

Beauchamp, D.A., E.R. Byron, and W. Wurtsbaugh. 1994. Summer habitat use by littoral-zone fishes in Lake Tahoe and the effects of shoreline structures. North American Journal of Fisheries Management 14:385–394.

Benke, A.C., G. Willeke, F.K. Parrish, and D.L. Stites. 1981. Effects of urbanization on stream ecosystems. Georgia Institute of Technology ERC 07-81, University of Georgia, Atlanta.

Bisson, P.A., T.P. Quinn, G.H. Reeves, and S.V. Gregory. 1992. Best Management Practices, cumulative effects, and long-term trends in fish abundance in Pacific Northwest River systems. Pages 189–232 in R.J. Naiman, editor. Watershed management. Balancing sustainability and environmental change. Springer-Verlag, New York.

Bisson, P.A., and eight coauthors. 1987. Large woody debris in forested streams in the Pacific Northwest: past, present and future. Pages 143–190 in E.O. Salo and T.W. Cundy, editors. Streamside management: forestry and fishery interactions. Contribution 57, Institute of Forest Resources, University of Washington, Seattle.

Bisson, P.A., G.H. Reeves, R.E. Bilby, and R.J. Naiman. 1997. Watershed management and Pacific salmon: desired future conditions. Pages 447–474 in D.J. Stouder and P.A. Bisson. Pacific salmon and their ecosystems: status and future options. Chapman & Hall, New York.

Booth, D. 1991. Urbanization and the natural drainage system—impacts, solutions, prognoses. The Northwest Environmental Journal 7:93–118.

Booth, D., and C.R. Jackson. 1994. Urbanization of aquatic systems — degradation thresholds and the limits of mitigation. Pages 425–434 in R. Marston and V.R. Hasfurther, editors. Effects of human induced changes on hydrologic systems. Proceedings of the Annual Summer Symposium of the American Water Resources Association. American Water Resources Association, Bethesda, Maryland.

Booth, D., and L.E. Reinelt. 1993. Consequences of urbanization on aquatic systems- measured effects, degradation thresholds, and corrective actions. Pages 545–550 in Watershed '93. A national conference on watershed management. Superintendent of Documents, U.S.G.P.O., Alexandria, Virginia.

Bottom, D. 1997. To till the waters: a history of ideas in fisheries conservation. Pages 569–598. in D.J. Stouder, P. Bisson, and R. Naiman, editors. Pacific salmon and their ecosystems: status and future options. Chapman & Hall, New York.

Bryan, M.D., and D.L. Scarnecchia. 1992. Species richness, composition, and abundance of fish larvae and juveniles inhabiting natural and developed shorelines of a glacial Iowa lake. Environmental Biology of Fishes 35:329–341.

Christensen, D.L., B.R. Herwig, D.E. Schindler, and S.R. Carpenter. 1996. Impacts of lakeshore residential development on coarse woody debris in north temperate lakes. Ecological Applications 6:1143–1149.

Christensen, N.L. 1997. Implementing ecosystem management: Where do we go from here. Pages 325–341 in M.S. Boyce and A. Haney, editors. Ecosystem management. Applications for sustainable forest and wildlife resources. Yale University Press, New Haven, Connecticut.

Chrzastowski, M. 1983. Historical changes to Lake Washington and route of the Lake Washington ship Canal, King County, Washington. U.S. Geological Survey, Water Resources Investigations Report WRI 81-1182.

Cleveland, S. 1995. Watershed management study for the city of Kirkland, Washington. Master's thesis. University of Washington, Seattle.

Cooke, G.D., E.B. Welch, S.A. Peterson, and P.R. Newroth. 1993. Restoration and Management of Lakes and Reservoirs. Lewis Publishers, Boca Raton, Florida.

Crow, T.R., and E.J. Gustafson. 1997. Concepts and methods of ecosystem management: Lesson from landscape ecology. Pages 54–67 *in* M.S. Boyce and A. Haney, editors. Ecosystem management. Applications for sustainable forest and wildlife resources. Yale University Press, New Haven, Connecticut.

Crowder, A.A., J.P. Smol, R. Dalrymple, R. Gilbert, A. Mathews, and J. Price. 1996. Rates of natural and anthropogenic change in shoreline habitats in the Kingston Basin, Lake Ontario. Canadian Journal of Fisheries and Aquatic Sciences 53 (Suppl. 1):121–135.

Ebersole, J.L., W.J. Liss, and C.A. Frissell. 1997. Restoration of stream habitats in the Western United States: Restoration as reexpression of habitat capacity. Environmental Management 21:1–14.

Edmondson, W.T. 1991. The uses of ecology. Lake Washington and beyond. University of Washington Press, Seattle.

Ferguson, B.K. 1991. Taking advantage of stormwater control basins in urban landscapes. Journal of Soil and Water Conservation 46:100–103.

Franklin, J.F. 1993. Preserving biodiversity: Species, ecosystems, or landscapes? Ecological Applications 3:202–205.

Fresh, K.L. 1994. Lake Washington fish: a historical perspective. Lake and Reservoir Management 9:148–151.

Frissell, C.A., and D. Bayless. 1996. Ecosystem management and the conservation of aquatic biodiversity and ecological integrity. Water Resources Bulletin 32:229–240.

Frissell, C.A., and R.K. Nawa. 1992. Incidence and causes of physical failure of artificial habitat structures in streams of western Oregon and Washington. North American Journal of Fisheries Management. 12:182–197.

Frodge, J.D., D.A. Marino, G.B. Pauley, and G.L. Thomas. 1995. Mortality of largemouth bass (*Micropterus salmoides*) and steelhead trout (*Oncorhynchus mykiss*) in densely vegetated littoral areas tested using *in situ* bioassay. Lake and Reservoir Management 11:343–358.

Garie, H.L., and A. McIntosh. 1986. Distribution of benthic macroinvertebrates in a stream exposed to urban runoff. Water Resources Bulletin 22:447–455.

Gregory, S.V., and P. Bisson. 1997. Degradation and loss of anadromous salmonid habitat in the Pacific Northwest. Pages 277–314 *in* D.J. Stouder, P. Bisson, and R. Naiman, editors. Pacific salmon and their ecosystems: status and future options. Chapman & Hall, New York.

Gregory, S.V., F.J. Swanson, W.A. McKee, and K.W. Cummins. 1991. An ecosystem perspective of riparian zones. BioScience 41:540–551.

Grumbine, R.E. 1994. What is ecosystem management? Conservation Biology 8:27–38.

Hansen, A.J., S.L. Garman, B. Marks, and D. Urban. 1993. An approach for managing vertebrate diversity across multiple-use landscapes. Ecological Applications 3:481–496.

Healey, M.C., and A. Prince. 1995. Scales of variation in life history tactics of Pacific salmon and the conservation of phenotype and genotype. Pages 176–184 *in* J.L. Nielsen, editor. Evolution and the aquatic ecosystem: defining unique units in population conservation. American Fisheries Society Symposium 17. Betheseda, Maryland.

Hendry, A.P. 1995. Sockeye salmon (*Oncorhynchus nerka*) in Lake Washington: an investigation of ancestral origins, population differentiation and local adaptation. Master's thesis. University of Washington, Seattle.

Hendry, A.P., and T.P. Quinn. 1997. Variation in adult life history and morphology among Lake Washington sockeye salmon (*Oncorhynchus nerka*) populations in relation to habitat features and ancestral affinities. Canadian Journal of Fisheries and Aquatic Sciences 54:75–84.

Hobbs, R.J., and D.A. Norton. 1996. Towards a conceptual framework for restoration ecology. Restoration Ecology 4:93–110.

Holling, C.S., and G.K. Meffe. 1996. Command and control and the pathology of natural resource management. Conservation Biology 10:328–337.

House, R. 1996. An evaluation of stream restoration structures in a coastal Oregon stream, 1981–1993. North American Journal of Fisheries Management 16:272–281.

Jensen, M.E., P. Bourgeron, R. Everett, and I. Goodman. 1996. Ecosystem management: a landscape ecology perspective. Water Resources Bulletin 32:203–216.

Jones, R., and C. Clark. 1987. Impact of watershed urbanization on stream insect communities. Water Resources Bulletin 23:1047–1056.

Karr, J.R. 1991. Biological integrity: a long-neglected aspect of water resource management. Ecological Applications 1:66–84.

Kauffman, J.B., R.L. Beschta, N. Otting, and D. Lytjen. 1997. An ecological perspective of riparian and stream restoration in the Western United States. Fisheries 22(5):12–24.

King, A.W. 1993. Considerations of scale and hierarchy. Pages 19–45 in S. Woodley, J. Kay, and G. Francis, editors. Ecological integrity and the management of ecosystems. St. Lucie Press, Delray Beach, Florida.

King County. 1991. Issaquah Creek basin current/future conditions and source identification report. King County Department of Public Works, Surface Water Management Division, Seattle.

King County. 1993. Cedar River Current and future conditions report. King County Department of Public Works, Surface Water Management Division, Seattle.

King County. 1994. Issaquah Creek watershed management committee proposed basin and non-point action plan. King County Department of Public Works, Surface Water Management Division, Seattle.

King County. 1995. May Creek current and future conditions report. King County Department of Public Works, Surface Water Management Division, Seattle.

King County. 1996. Cedar River basin and non-point pollution action plan. King County Department of Public Works, Surface Water Management Division, Seattle.

Kleindl, W.J. 1995. A benthic index of biotic integrity Puget Sound lowland streams, Washington, U.S.A. Master's thesis. University of Washington, Seattle.

Lichatowich, J.L. 1997. Evaluating salmon management institutions: The importance of performance measures, temporal scales, and production cycles. Pages 69–87 in D.J. Stouder, P. Bisson, and R. Naiman, editors. Pacific salmon and their ecosystems: status and future options. Chapman & Hall, New York.

Lichatowich, J., L. Mobrand, L. Lestelle, and T. Vogel. 1995. An approach to the diagnosis and treatment of depleted Pacific salmon populations in Pacific Northwest Watersheds. Fisheries 20(1):10–18.

Limburg, K., and R. Schmidt. 1990. Patterns of fish spawning in Hudson river tributaries — response to an urban gradient. Ecology 7:1231–1245.

Lucchetti, G. and R. Fuerstenberg. 1993. Management of coho salmon habitat in urbanizing landscapes of King County, Washington, U.S.A. Pages 308–317 in L. Berg and P. Delaney, editors. Proceedings of a workshop on coho salmon. Canadian Department of Fisheries and Oceans, Vancouver, British Columbia.

Ludwig, D., R. Hilborn, and C.J. Walters. 1993. Uncertainty, resource exploitation, and conservation: lessons from history. Science 260:17, 36.

May, C.W. 1996. Assessment of cumulative effects of urbanization on small streams in the Puget Sound Lowland Ecoregion. Doctoral dissertation. University of Washington, Seattle.

Moscrip, A.L., and D.R. Montgomery. 1997. Urbanization, flood frequency, and salmon abundance in Puget Lowland streams. Journal of the American Water Resources Association 33:1289–1297.

Moyle, P.B., and R.M. Yoshiyama. 1994. Protection of aquatic biodiversity in California: a five-tiered approach. Fisheries 19(2):6–18.

Naiman, R.J., and eight coauthors. 1992. Fundamental elements of ecologically healthy watersheds in the Pacific Northwest Coastal Ecoregion. Pages 127–188 in R.J. Naiman, editor. Watershed management. Balancing sustainability and environmental change. Springer-Verlag, New York.

Nehlsen, W., J.E. Williams, and J.A. Lichatowich. 1991. Pacific salmon at the crossroads: stocks at risk from California, Oregon, Idaho, and Washington. Fisheries 16(2):4–21.

Noss, R.F. 1990. Indicators for monitoring biodiversity: a hierarchical approach. Conservation Biology 4:355–364.

Noss, R.F., and J.M. Scott. 1997. Ecosystem protection and restoration: The core of ecosystem management. Pages 239–264 in M.S. Boyce and A. Haney, editors. Ecosystem management. Applications for sustainable forest and wildlife resources. Yale University Press, New Haven, Connecticut.

NRC (National Research Council). 1996. Upstream. Salmon and society in the Pacific Northwest. Committee on Protection and Management of Pacific Northwest Anadromous Salmonids. National Academy Press, Washington, D.C.

O'Neill, R.V., A.R. Johnson, and A. King. 1989. A hierarchical framework for the analysis of scale. Landscape Ecology 3:193–205.

Pease, J.R. 1993. Land use and ownership. Pages 31–39 *in* P.L. Jackson and A.J. Kimerling, editors. Atlas of the Pacific Northwest. Oregon State University Press, Corvallis.

Rabeni, C.F., and S.P. Sowa. 1996. Integrating biological realism into habitat restoration and conservation strategies for small streams. Canadian Journal of Fisheries and Aquatic Sciences 53 (Suppl. 1): 252–259.

Reeves, G.H., L.E. Benda, K.M. Burnett, P.A. Bisson, and J.R. Sedell. 1995. A disturbance-based ecosystem approach to maintaining and restoring freshwater habitats of evolutionarily significant units of anadromous salmonids in the Pacific Northwest. Pages 334–349 *in* J.L. Nielsen, editor. Evolution and the aquatic ecosystem: defining unique units in population conservation. American Fisheries Society Symposium 17. Bethesda, Maryland.

Reice, S.R., R.C. Wissmar, and R.J. Naiman. 1990. Disturbance regimes, resilience, and recovery of animal communities and habitats in lotic ecosystems. Environmental Management 14:647–659.

Resh, V.H., and nine co-authors. 1988. The role of disturbance in stream ecology. The Journal of the North American Benthological Society 7:433–455.

Richey, J.S. 1982. Effects of urbanization on a lowland stream in western Washington. Doctoral dissertation. University of Washington, Seattle

Roper, B.B., J.J. Dose, and J.E. Williams. 1997. Stream restoration: Is fisheries biology enough? Fisheries 22(5):6–11.

Roth, N.E., J.D. Allan, and D.L. Erickson. 1996. Landscape influences on stream biotic integrity assessed at multiple spatial scales. Landscape Ecology 11:141–156.

Schueler, T. 1994. The importance of imperviousness. Watershed Protection Techniques 1:100–111.

Scott, J.B., C.R. Steward, and Q.J. Steward. 1986. Effects of urban development on fish population dynamics in Kelsey Creek, Washington. Transactions of the American Fisheries Society 115:555–567.

Shaklee, J.B., J. Ames, and L. LaVoy. 1996. Genetic diversity units and major ancestral lineages for sockeye salmon in Washington. State of Washington. Department of Fish and Wildlife Technical Report. Number RAD 95-02/96.

Sheldon, A. 1988. Conservation of stream fishes: patterns of diversity, rarity, and risk. Conservation Biology 2:149–156.

Slaney, T.L., K.D. Hyatt, T.G. Northcote, and R.J. Fielden. 1996. Status of anadromous salmon and trout in British Columbia and Yukon. Fisheries 21:20–35.

Slocombe, D.S. 1993. Implementing ecosystem-based management. BioScience 43:612–622.

Steedman, R.J. 1988. Modification and assessment of an index of biotic integrity to quantify stream quality in Southern Ontario. Canadian Journal of Fisheries and Aquatic Sciences 45:492–501.

Steedman, R., and W. Haider. 1993. Applying notions of ecological integrity. Pages 47–60 *in* S. Woodley, J. Kay, and G. Francis, editors. Ecological integrity and the management of ecosystems. St. Lucie Press, Delray Beach, Florida.

Stockner, J.G., and E.A. MacIssac. 1996. British Columbia lake enrichment programme: Two decades of habitat enhancement for sockeye salmon. Regulated Rivers Research and Management 12:547–561.

Stouder, D.J., P. Bisson, and R. Naiman, editors. 1997. Pacific salmon and their ecosystems: status and future options. Chapman & Hall, New York.

Turner, M.G. 1989. Landscape ecology: The effect of pattern on process. Annual Review of Ecology and Systematics 20:171–197.

Walters, C.J. 1986. Adaptive management of natural resources. McGraw-Hill, New York.

Walters, C.J., and C.S. Holling. 1990. Large-scale management experiments and learning by doing. Ecology 71:2060–2068.

Wang, L., J. Lyons, P. Kanehl, and R. Gatti. 1997. Influences of watershed land use on habitat quality and biotic integrity in Wisconsin streams. Fisheries 22(6):6–12.

WDF et al. 1993. 1992 Washington state salmon and steelhead stock inventory. Washington Department of Fisheries, Washington Department of Wildlife, Western Washington Treaty Indian Tribes, Olympia.

Weaver, L.A., and G.C. Garman. 1994. Urbanization of a watershed and historical changes in a stream fish assemblage. Transactions of the American Fisheries Society 123:162–172.

Weaver, M.J., J.J. Magnuson, and M.K. Clayton. 1997. Distribution of littoral fishes in structurally complex macrophytes. Canadian Journal of Fisheries and Aquatic Sciences 54:2277–2289.

Welch, E.B., C.A. Rock, R.C. Howe, and M.A. Perkins. 1980. Lake Sammamish response to wastewater diversion and increasing urban runoff. Water Research 14:821–828

Wood, C.C. 1995. Life history variation and population structure in sockeye salmon. Pages 195–216 *in* J.L. Nielsen, editor. Evolution and the aquatic ecosystem: defining unique units in population conservation. American Fisheries Society Symposium 17. Bethesda, Maryland.

Yount, J.D., and G.J. Niemi. 1990. Recovery of lotic communities and ecosystems from disturbance—a narrative review of case studies. Environmental Management 14:547–569.

34 Rehabilitating Stream Channels Using Large Woody Debris with Considerations for Salmonid Life History and Fluvial Geomorphic Processes

Larry G. Dominguez and C. Jeff Cederholm

Abstract.—Pacific salmon *Oncorhynchus* spp. exist in fluvial systems that are physically and biologically dynamic. Salmonid life history characteristics and associated habitat requirements vary widely by species. Some species use the freshwater environment almost solely for incubation, while others use it for both incubation and extended rearing. Salmon species have evolved into several life history patterns that maximize their potential for survival and minimize their spatial and temporal overlap. To rehabilitate salmon habitat and thereby strengthen wild runs requires a knowledge of fish life histories and the aquatic system's potential range of conditions. Using large woody debris to rehabilitate stream channels is a popular management activity in the Pacific Northwest. Prior knowledge of factors such as spawning distribution and timing, incubation environment quality, seasonal rearing habitat needs (i.e., summer/winter), limiting factors in freshwater production, and the relative habitat quality and availability is imperative for successful projects. We review woody debris ecology in streams and provide planning information for woody debris placement projects. In addressing limiting aspects of properly functioning aquatic and riparian ecosystems, instream and riparian habitats can be created that provide the interim structural framework for streams until riparian and upland forests recover from past disturbances. The discussion is based on a decision flow diagram that guides the need assessment process and suggests appropriate rehabilitation technique. Visits to several stream rehabilitation projects, combined with information from the literature and our own experience, led us to a number of conclusions supporting stream restoration for future sustainability of Pacific salmonids.

INTRODUCTION

Recent status reviews indicate that many Pacific Northwest salmon stocks are either threatened or have already gone extinct. The causes for decline fall into four general impact categories: (a) weak stock overharvesting; (b) hatcheries; (c) hydropower facilities; and (d) habitat loss. Nehlsen et al. (1991) concluded that there is a need for a paradigm shift that "... advances habitat restoration and ecosystem function rather than hatchery production, ... for many of these stocks to survive and prosper into the next century." The need for taking an ecosystem perspective when planning anadromous salmon rehabilitation* is now widely recognized. Attempts to restore habitats and

---

* Rehabilitate (to restore to a former state) is the preferred term as compared to restore, reclaim, or enhance. There is a potential wide range of historical changes and conditions that the stream channel has manifested. Rehabilitation assumes that the means would be provided to the channel to steer it toward that range of conditions.

ecosystem function are already occurring throughout the Pacific Northwest (e.g., House et al. 1988; Cowan et al. 1995; P. Slaney, British Columbia Ministry of Environment, personal communication); with substantial efforts focused on instream large woody debris (LWD) placement.

There are several definitions for LWD. For this discussion, LWD is described as root wads and tree stems (>10 cm dia. and >1m long) that provide overhead cover and flow modifications that contribute to effective creation and maintenance of fish spawning and rearing habitat. Instream LWD abundance has been greatly diminished in the past due to streambank logging and stream clearance practices. These activities have created the potential for streams to reach minimum LWD abundances within 100 years after these original impacts, about the middle of the next century. A main challenge for watershed and fish rehabilitation work is to provide an interim period of artificially constructed instream LWD habitat, while planning for the self-rehabilitation of river basins through forest and riverine ecosystems recovery. Although interim rehabilitation work is vital to sustaining many weakened fish stocks, it is misleading to believe that it is the complete answer to recovering salmon habitats to pre-intensive land management productivity.

Several salmon habitat and watershed rehabilitation manuals exist (Anonymous 1980; House et al. 1988; Adams and White 1990; Spence et al. 1996; Slaney and Zaldokas 1997), along with extensive related literature. Many of these articles caution the user to be aware of the stream dynamics and establish the importance of recognizing unique site-specific characteristics during rehabilitation. Theoretically, each piece of wood that is introduced into stream channels will respond differently due to the variables associated with that particular system. Every attempt at artificially placing LWD into a stream should be approached with caution and with a goal of learning. A "failed" project (one that does not accomplish what was "engineered" on paper) can be a valuable addition to our knowledge of stream habitat rehabilitation. Also, a "failed" project may even be an unrecognized success, positively contributing to a different part of the stream ecosystem, that the researcher or rehabilitation worker has overlooked (e.g., stabilizing a stream feature or trapping gravel elsewhere).

Despite significant research and improved understanding of LWD and its associated processes (Harmon et al. 1986; Maser and Sedell 1994), we are in the early stages of understanding the role of LWD from an ecosystem perspective. As a template for understanding the basic physical and biological processes at work in watersheds, we recommend studying the river continuum concept put forth by Vannote et al. (1980) and modified by Triska et al. (1982). Many abuses of our landscapes have occurred because of incomplete understanding of stream ecology, and have left scientists with few opportunities to examine LWD processes in unmanaged (unharvested) forest environments.

Stream rehabilitation is a very complex subject, especially as research continues to generate information about what factors contribute to healthy watersheds. There is a range of opinion about the overall benefit of LWD-addition projects due to the mixed results when fish productivity has been evaluated. Some professionals would opt to do nothing in streams and eliminate riparian disturbances while the riparian forests recover. This is partly due to the realization that stream systems are highly dynamic, undergoing continual disturbance-recovery cycles. These cycles are significant in creating habitat features. Bisson et al. (1997) and Benda et al. (1998) provide exceptional discussions on disturbance ecology and the potential for a wide range of conditions within stream channels and watersheds. We recommend that stream workers become familiar with that information as they proceed with large-scale rehabilitation work.

THE PHILOSOPHY OF LARGE WOODY DEBRIS REHABILITATION

Many Pacific Northwest watersheds are in a state of early recovery from habitat degradation caused by past land use practices, and it may take centuries for LWD-associated processes to recover. Rehabilitation projects should be viewed as interim measures until riparian forests recover from past management activities. Habitat rehabilitation, restoration, or enhancement must never be viewed as a substitute for habitat protection (Reeves et al. 1991; Koski 1992).

Stream channel rehabilitation begins with upslope management. Timber harvest can influence snow accumulation and rain-on-snow storm events (Harr 1981, 1986; Coffin and Harr 1992). These influences on the hydrologic processes can affect the frequency and magnitude of channel-disturbing events. With many of our lands under intensive management for natural resource extraction, the riparian forest remains the last line of defense for preserving stream health. Van Cleef (1885) emphasized the necessity of protecting the structure of streambanks and riparian areas over 100 years ago. Today, the message is as fresh as when it was written. We should begin the process of habitat rehabilitation by protecting those watersheds and riparian ecosystems that still remain in relatively good condition. Communication between fish managers and land managers is also an essential element of successful habitat rehabilitation. Instream rehabilitation efforts need complementary upslope rehabilitation for long-term benefits.

THE ECOLOGICAL ROLE OF LARGE WOODY DEBRIS IN STREAMS

PHYSICAL FUNCTION

LWD provides important physical functions in a wide variety of channel configurations. In small, first- and second-order channels, abundant LWD can be single pieces or multiple pieces bound together in continuous matrices. In small (<5 m bankfull width (BFW*)) to intermediate size (5-20 m BFW) streams, LWD contributes to channel stabilization, energy dissipation, and sediment storage (Bisson et al. 1987; Harmon et al. 1986; Heede 1986), causes local bed and bank scour (Robison and Beschta 1990b), and creates a stepped-channel profile (Heede 1972). These factors contribute to the development of downstream fish-bearing habitat and should be recognized as part of rehabilitation programs.

LWD also provides important biological functions such as predator and high velocity escape cover for adult and juvenile salmonids, storage sites of biological and chemical activity (Naiman and Sedell 1979), and a food base for many aquatic macroinvertebrates (Cummins et al. 1982). Spawned out salmon carcasses trapped and held in place by LWD later contribute to riparian animal communities and aquatic productivity (Cederholm and Peterson 1985; Cederholm et al. 1989). Single or multiple pieces of conifer trees may hold log jams together in large streams (Abbe and Montgomery 1996) or play similar roles in small streams (Beschta and Platts 1986). Accumulations of LWD are also buried out of sight below the gravel surface, providing important structure and nutrient supply to the hyporheic zone. Scour from floods or channel reconfigurations often expose a legacy of this old, buried LWD which can provide function for many years.

Intermediate stream orders (3rd–5th, generally 10–20 m BFW) have less continuous accumulations of LWD, but have drift jams wood accumulations mobilized during high flows) that can block the flow and occasionally redirect the stream course, resulting in evacuated floodplain channel features that can become important overwintering habitat for juvenile salmonids (Peterson and Reid 1984). It is common to see full-sized trees laying within and across these streams and huge chunks of wood protruding out of the gravel. Large gravel wedges are held back by logs, and plunge pools and dam pools are commonly associated with this feature. Typically, many species of resident and anadromous salmonids inhabit these streams, as well as a diversity of non-salmonids. These streams have some of the highest salmonid productivity.

The highest order channels (>5th order) are characterized by infrequent, but occasionally massive accumulations of LWD. Gradients tend to average less than 0.5%. The LWD tends to be of large diameter and often complete trees with rootwads and branches accumulate smaller logs and other wood debris. Much of this debris has floated down from upstream during floods or has entered the channel from bank undercutting of relatively young stands of trees growing on gravel

---

* Bankfull width is defined as the distance between two points on the right and left bank of a stream channel where perennial vegetation exists (synonymous with Mean Annual Flood). The transect formed by the two points is perpendicular to the stream channel.

bars. LWD contributes to major habitat-forming processes such as bar stability, side-channel development, island formation, and channel feature formation (Gregory et al. 1993; Nakamura and Swanson 1993). Extensive LWD accumulations in the lower floodplain of coastal rivers can direct the flow into meander loops and result in formation of riverine ponds and swamp features called "wall-base channels"*. These habitats are used heavily by immigrant juvenile coho and cutthroat (Cederholm and Scarlett 1982; Peterson and Reid 1984), as well as other aquatic species (dace, sculpins, crayfish, salamanders) seeking to escape the main river discharge and turbidity. Large rivers function both as fish migration corridors to access upper basin spawning and rearing habitats and as mainstem spawning and rearing habitats for a variety of fishes.

LONG-TERM EFFECTS ON LWD BY STREAM CLEARANCE AND LOGGING

The practice of stream clearance and channelization has contributed to LWD deficiency in many streams. These practices occurred earlier in the century to make rivers more navigable (Sedell and Luchessa 1982); to make streams more accessible to migrating salmon (Kramer 1953); and to reduce the threat of flooding (Cederholm 1972; Brookes 1985). Clearcut logging to the stream edge, often with minimal buffer strips, was common practice throughout Oregon and Washington until the early 1980s and in British Columbia until the late 1980s. Reductions of LWD material have been noted soon after streamside clearcut logging (Bilby and Ward 1991). Lammel (1972) reported that a 50% reduction in volume of coarse debris (LWD) occurred immediately after logging in stream channels that were unprotected by buffer strips in western Oregon. Toews and Moore (1982) noted that significant but less dramatic reductions occurred following log yarding in Carnation Creek, British Columbia. Bryant (1980) found a reduction in the number of log jams about 20 to 25 years after logging in Maybeso Creek, Southeastern Alaska. Cross-stream yarding and widely accepted practices of salvage logging within the stream further depleted LWD.

Ralph et al. (1994) analyzed stream channel morphology and LWD abundance in 101 western Washington streams, and concluded that intensive streamside logging leads to simplification and homogenization of stream habitat. Lestelle (1978) and Bilby (1984) observed large changes in channel structure in western Washington after removal of LWD. Sullivan et al. (1987) reported that pool area was significantly reduced after stream clearance, declining from 70% to 20%. Generally, there is an initial release of large quantities of sediment immediately following woody debris removal (Beschta 1979). Thus, within a stream reach, these adjustments may result in a smoothing of channel gradient and filling of the deepest pools.

Logging along streams causes a long-term decline in the abundance of large conifers (Grette 1985; Murphy and Koski 1989) available for recruitment as LWD. A century or more may be required for the forest to regrow to the point where it again produces this type of LWD (Grette 1985). Second growth LWD is often made up of greater proportions of small diameter deciduous and conifer trees than the old-growth forest (Grette 1985; Ralph et al. 1994). These materials have poorer long-term structural value as LWD because they degrade more easily than the large, old-growth conifers.

BIOLOGICAL FUNCTION: SALMONID USE OF NATURAL AND ARTIFICIALLY PLACED LWD

The benefits of LWD for juvenile rearing salmonids change seasonally. Rearing coho, steelhead, and cutthroat juveniles are known to evacuate their summer stream habitats and seek refuge in deep pools with woody debris as winter temperatures decline and stream discharges increase (Bustard and Narver 1975; Tschaplinski and Hartman 1983; Murphy et al. 1984). Grette (1985) found that

---

* This is an off-channel habitat that is located outside of the ordinary high water mark on the second or third terrace of the river. They are much less subject to damage during major flood events (Peterson and Reid 1984).

the rearing densities of juvenile coho in Olympic Peninsula streams were closely associated with LWD in winter. Murphy et al. (1986) found no correlation between summer coho densities and debris in southeast Alaska streams, but a close association during the winter. On the Keogh River, B.C., coho densities were not closely correlated with woody debris in the summer, except in shallow pools or "flats" (P. Slaney, British Columbia Ministry of the Environment, personal communication).

In Oregon, large expanses of gravel and fines were deposited upstream of full spanning logs placed in the Nestucca River (House et al. 1991). Many of these sites were readily used by chinook and steelhead spawners the first winter after rehabilitation. In Washington, "spawning pads" made of logs and planks have been used to sort and retain gravels in Willapa Bay and Grays Harbor tributaries. These sites were subsequently used by spawning chum and coho salmon (L. Cowan, Washington Department of Fish and Wildlife, personal communication). Full crossing logs installed in Porter Creek, an experimental stream along the Washington coast, trapped gravels that were previously transported through the system. The retained gravels were used by spawning coho and steelhead the first winter after enhancement (Cederholm et al. 1997).

Pools formed by log structures in Oregon streams resulted in a threefold increase in juvenile coho numbers in summer (House and Boehne 1986). Nickelson et al. (1992a,b) found that adding bundles of small trees to dammed pools and artificially constructed "alcoves,"* resulted in significant increases in coho density during winter. However, recent flood events have breached most of these protected areas indicating that this strategy is susceptible to major storm events. Conclusions of Quinn and Peterson (1994) at Big Beef Creek, in Puget Sound, indicate a positive correlation between end of summer LWD volume and coho overwinter survival. Research in Oregon on instream rehabilitation shows that addition of conifer logs to debris-poor streams can increase salmonid smolt production several fold (Solazzi and Johnson 1994), and in Washington careful placement and anchoring of LWD in Porter Creek had increased coho smolt yield two- to eight-fold under various treatments (Cederholm et al. 1997).

In Washington's Olympic Peninsula, experimental addition of small debris bundles to Clearwater River pools was effective in attracting substantial numbers of summer rearing juvenile coho and trout; many of these same fish later immigrated into wintering ponds as temperatures dropped and discharge increased (Peters et al. 1992). Debris addition and depth creation in off-channel overwintering ponds or "wall base channels" along the Clearwater River produced significant increases in juvenile coho winter survival (Cederholm et al. 1988; Cederholm and Scarlett 1991).

In the last few decades, hundreds, perhaps thousands, of low-cost rehabilitation and enhancement projects have been implemented throughout the Pacific Northwest. However, the most effective monitoring of these approaches has occurred at the experimental level. Therefore, large-scale rehabilitation projects that include monitoring can be costly (Table 34.1).

SITE SELECTION, STREAM ASSESSMENTS, AND DESIGN OF REHABILITATION PROJECTS

The goal of fish habitat rehabilitation is to produce watershed-level improvements in fish productivity, not just to redistribute fish within the watershed by attracting them to project areas. This can be accomplished by gearing rehabilitation efforts to a minimum of a reach level up to the watershed scale. Hartman and Miles (1995) strongly emphasize the need to integrate both physical and biological considerations when doing stream channel enhancement work. With few exceptions, they determined that factors affecting biological success were physical in nature. They listed several factors that contributed to successful or outstanding projects:

---

* This is an off-channel habitat that is located within the ordinary high water mark and highly subject to damage during major flood events (T. Nickelson, Oregon Department of Fish and Wildlife (OFDW), personal communication).

TABLE 34.1
Types, techniques, and associated costs of LWD rehabilitation in several Pacific Northwest streams.

Name/Location	Year	Scope[a]	Rehabilitation prescription	Equipment	Dist. (km)	Cost (US$) per km	Comments
Upper Keogh River, B.C. (Ward and Slaney 1993)	1977	Experimental Management	Various boulder and LWD attachments	N/A	1.0	26,400[b]	Limited LWD, emphasis on boulder clusters
Fish Ck., OR (Heller 1993)	1988	Experimental	Attached LWD	N/A	14.0	44,000	N/A[c]
Paradise Pond (Cederholm et al. 1988)	1988	Hab. Enhancement Experimental	Flooded wetland with LWD placement	Explosives, hand tools, import LWD	0.25	49,280[d]	Limited planning, no engineering
Swamp Creek Beaded Channel (Cederholm and Scarlett 1991)	1991	Hab. Enhancement Experimental	Series of interconnected pools with LWD placement	Explosives	0.2	20,250[d]	Limited planning, no engineering, readily available LWD
Porter Ck., WA (Cederholm et al. 1997)	1997	Experimental	Free-fall, cabled LWD throughout a stream reach	Tree-felling equipment, work crews	0.5	12,900[d]	Limited planning, "loggers choice" reach
Porter Ck., WA (Cederholm et al. 1997)	1997	Experimental	Attached LWD using cable and epoxy resins throughout a stream reach	Heavy equipment, extensive log anchoring	0.5	164,500[d]	Includes hydraulic and topographic analysis
Shop Ck., B.C. (B.C. Watershed Restoration Program)	1994	Experimental Management	Log "V" weirs, boulder clusters, groundwater side-channels and pond	Hand crews and tracked excavators	1	40,000	Additional 0.5 km rehab. outside of creek; 20% in-kind money excluded
Hoh River, WA, Lewis Ranch (Washington Dept. of Natural Resources)	1994	Management	LWD placement in groundwater-fed spawning/rearing channel	Hand crews and tracked excavators	0.5	277,500	Hydraulic and construction engineering
Skagit R., WA, Park Slough and Extension (Cowan et al. 1995)	1994	Experimental Management	LWD placement in groundwater-fed spawning/rearing channel	Heavy equipment tracked excavators	1.1	170,200	Extensive engineering and planning
Coquitlam R., Oxbow project (B.C. SEP-DFO)	1994	Hab. Enhancement	Side-channel/pond development	N/A	1.5	59,200	1 km, included a relic side channel
East and Lobster Cks., OR (T. Nickelson, ODFW)	unpub.	Management Hab. Enhancement	LWD and "alcoves"	N/A	3.4	21,500	Siltation problems during recent flooding

a Experimental had extensive planning and monitoring; Management are broad applications with some monitoring of habitat/fish productivity; Habitat enhancement are projects that have minimum evaluation.
b Pre-1990 British Columbia projects converted at 1990 rate. All amounts are from costs at time of project.
c Information not available.
d Does not include monitoring costs.

- appropriate physical locations
- small stream size and gentle gradient
- small post-construction floods
- regulated water supply
- successful emulation of natural structures
- experienced design and construction team
- proper installation
- the ability of the structure to collect large volumes of debris and withstand freshets
- provision of adequate cover
- managed by people who have an ongoing interest in the project and
- regular monitoring and maintenance

The following discussion outlines a systematic approach to using LWD for watershed rehabilitation. This discussion is also outlined in a decision matrix format (Figure 34.1-A,B) to graphically present procedures for identifying and implementing restoration projects using LWD.

SITE SELECTION

Large Wood Function in Upslope Landscape Rehabilitation (I)*

Just as streams exhibit stepped profiles within which sediment-storage elements are provided, so does LWD function as storage elements on hillslopes (Wilford 1982; Harmon et al. 1986). Most watershed rehabilitation efforts should begin in upslope areas where unnatural activities have been the cause of their instability. Carrying out instream work, without first addressing unstable slopes, has very short-term benefits and rarely contributes to the restoration of watershed processes. Hillslope LWD faces similar crises as instream LWD. Future sources are being eliminated due to management activities resulting in soil structure and nutrient cycling alterations. Although the focus of this chapter is on inchannel LWD, upslope areas are the first places to look when addressing the needs of a watershed. LWD can be placed below roads, or in overland runoff areas, to catch and store sediments.

Riparian Zone Rehabilitation and Protection (II)

Determining how to recover riparian ecosystems by management intervention is in an experimental phase. Hardwood-dominated riparian areas are generally susceptible to poor conifer regeneration due to low light, poor seedbed conditions, and a general lack of a conifer seed source. Underplanting and thinning of deciduous trees to re-establish native conifers along streams that have been logged in the past is being tested in many areas of the Pacific Northwest (Emmingham and Hibbs 1997; T. Nickelson, ODFW, personal communication). Silvicultural thinning of overly dense conifer stands to increase diameter growth has been successful in upland forests (G. Hoyer, Washington Department of Natural Resources, personal communication), and could be applied to riparian corridors to increase size and abundance of potential LWD trees. Regulated girdling of deciduous trees, in combination with conifer planting and instream LWD placement, is being tried in some previously harvested Oregon coastal watersheds (T. Nickelson and M. Solazzi, ODFW, personal communication). The increased susceptibility to windthrow expedites input of deciduous LWD into the streams which provide short-term LWD functions. Instream LWD provides interim habitat while the riparian areas restore conifer dominance.

Riparian degradation associated with livestock impacts can be remediated by livestock exclusion. In remote, forested rangelands where fencing of hundreds of miles of riparian corridors is not feasible, range managers are experimenting with placement of local, downed woody material

---

* Bold Roman numerals in parentheses are included in Figure 34.1-A,B to correspond to this discussion.

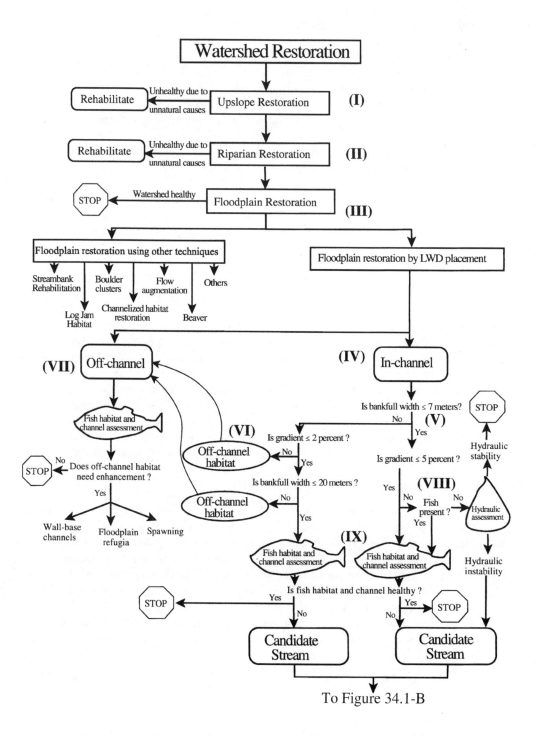

FIGURE 34.1-A　Flow chart for determining candidate streams for rehabilitation.

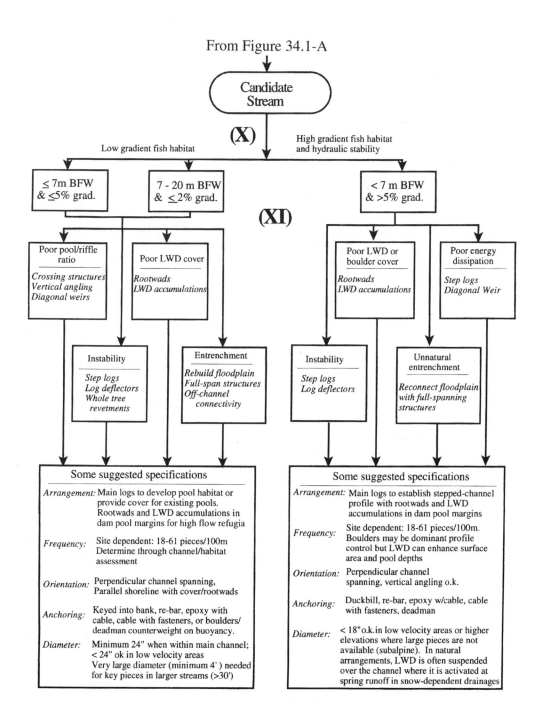

FIGURE 34.1-B Flow chart for determining candidate streams for rehabilitation (cont.) with emphasis on potential applications.

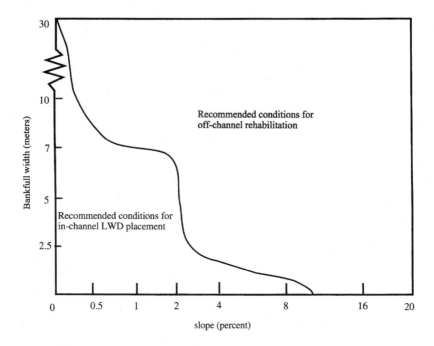

FIGURE 34.2   General physical conditions recommended for LWD fish habitat rehabilitation work.

in heavily impacted areas to discourage multiple watering/crossing sites in streams, thus restoring bank vegetation (W. Keller, Natural Resources Conservation Service, personal communication).

FLOODPLAIN REHABILITATION (III)

At the floodplain decision node in Figure 34.1-A (**III**), one option is to conduct various non-LWD rehabilitation techniques. Slaney and Zaldokas (1997) compiled a variety of applications for rehabilitation which should be considered. In addition to traditional LWD projects, there are several other ways to approach offsetting habitat impacts in large, unstable stream channels, including streambank rehabilitation, placement of boulder clusters, flow augmentation, encouragement of beaver activity, and large-scale logjam habitat formation.

If a stream is deficient in LWD, and the desired rehabilitation technique is instream wood placement, then carry the previously-selected stream's characteristics through the flow chart to further confirm its candidacy (**IV**). In recent years, the refinement of channel and landform classification systems are providing information that could be applied to understanding a stream channel's potential to accommodate rehabilitation techniques (Frissell et al. 1986; Rot 1995; Rosgen 1996; Montgomery and Buffington 1997). Beechie and Sibley (1997) describe various relationships between LWD, channel characteristics, and habitat features in 28 streams ranging in bankfull width from 4.4 to 22 meters. The results of their study indicate that minimum pool-forming LWD diameters generally increased with increasing bankfull width. Although it is important to understand the need to have larger-diameter pieces in larger streams for habitat formation, further investigation is needed to describe necessary LWD piece characteristics per channel type. Figure 34.2 demonstrates the general physical conditions recommended for LWD-based fish habitat rehabilitation work. There may be exceptions to these criteria when dealing with larger, naturally regulated, lake-headed systems since they tend to be more hydraulically stable.

The recommended width and gradient criteria are the next decision node in Figure 34.1-A (**V**). A wide stream with greater than 2% gradient leads to the off-channel decision tree (**VI**) eventually leading to an assessment of those habitats (**VII**). Narrow streams with higher gradients will lead to a hydraulic assessment (**VIII**). The remaining streams not characterized by the previous features lead to habitat and channel assessments (**IX**) to determine their potential for rehabilitation.

The most significant part of the planning phase are the assessments conducted to determine candidate streams for restoration (**X**). Specific wood placement strategies can be considered after stream assessments (**XI**).

CANDIDATE STREAM ASSESSMENTS

LARGE WOODY DEBRIS ASSESSMENT

Bilby and Ward (1991) found that the number of LWD pieces varied significantly with channel width, with larger streams having fewer pieces than smaller streams. In an extensive review of the literature on LWD abundance, Peterson et al. (1992) found that the number of LWD pieces in streams flowing through unmanaged forests varies between 180 and 610 pieces per km in small streams (< 5 m wide), and 30 to 460 pieces per km in large streams (greater than 23 m wide). In research within 28 medium-sized (5–8 m wide) western Olympic Peninsula old growth streams, Grette (1985) documented 400 to 800 pieces of LWD per km of stream, of which 100 to 510 pieces were from the old growth forest that was logged up to 62 years prior to his study. Cederholm et al. (1997) found as few as 100 pieces of conifer LWD per km still remaining in a stream that was logged and cleaned over the previous 50 years. Similar-sized Shop Creek, in British Columbia, contained approximately 40 pieces per km prior to LWD rehabilitation (P. Slaney, British Columbia Ministry of the Environment, personal communication).

LWD categories such as "key pieces" and "functional pieces" are now used in assessments to emphasize LWD function more than quantity (WDNR 1993). These categories of wood are described as pieces that are independently stable in the stream, and perform habitat-forming functions or retain other pieces of LWD. In a typical debris jam or accumulation, there may be a few "key" pieces that are responsible for the initial formation of the LWD complex. It is the quantity of these "key" pieces (usually larger diameter and/or length than the average) that can give indications of a stream's stability or potential stability. Washington State's watershed analysis process (WDNR 1993) recognizes key LWD pieces as contributors to channel stability during floods, suggesting that the lack of such pieces is a cause for concern.

Because simple computation of LWD volume per stream length or area may not adequately describe spatial complexities, Robison and Beschta (1990a,b) recommend that LWD pieces be differentiated by zones of influence. Ungrouped LWD may not benefit fish habitat, per se, in the form of pools or sediment storage, but they may have a greater overall effect on channel morphology because resistance to flow is spread over a longer distance. Peterson et al. (1992) and the Washington Department of Natural Resources' revision to Watershed Analysis (WDNR 1993), suggest that LWD surveys should not only be limited to fish-bearing waters, and that the type and amount of information collected during LWD surveys should be expanded.

Many survey techniques are available in habitat assessment manuals, which include protocol for LWD assessment (Johnston and Slaney 1996; Anonymous 1996; Overton et al. 1997). One of the most comprehensive, management-oriented, LWD survey methods available is the LWD module included in the Timber-Fish-Wildlife (TFW) Ambient Monitoring Manual (Schuett-Hames et al. 1994). This procedure, which describes LWD abundance (pieces and volume), distribution, type, size, and potential contribution to channel stability and habitat formation (key piece), is an important element in the LWD rehabilitation project planning and design. Some of these procedures have been utilized in the procedure adapted for use in British Columbia as the Fish Habitat Assessment

Procedure (FHAP) (Johnston and Slaney 1996) where LWD frequency has been scaled to channel width (BFW) and rated as poor (<1piece/BFW), fair (1–2 pieces/BFW), and good (>2 pieces/BFW). Unfortunately, an assessment only gives a characterization of the current situation. We caution the assessors when determining the appropriate key piece size for a given stream reach. The LWD diameter distribution characterizes the wood that is present and may not reflect the potential diameter distribution of LWD if that forest was unmanaged. Since there can be such a wide range of LWD abundance even in unmanaged watersheds, other factors such as channel geomorphology, hydraulic stability, and riparian vegetation potential need to be considered when determining wood loading needs.

Given the wide range of LWD abundance within a stream network, it is difficult to recommend a certain number of pieces necessary to rehabilitate a reach. This will depend on what information is gathered in the assessments and the goals that are set. For example, the final LWD diameter distribution may exhibit a different diameter distribution needed if the project goal was to accumulate gravel and restore floodplain connection as compared to providing cover habitat in pools or restoring pool-riffle sequences.

CHANNEL ASSESSMENT

Analysis and evaluation of the physical components of a LWD project must be completed by a salmon biologist and an engineer with experience in salmon biology, hydrology, natural channel flow, construction methodologies, and construction supervision. This blend of expertise provides both a biological and engineering perspective necessary for successful project completion. Reliance on one perspective or allowing one perspective to dominate will usually result in projects not providing the anticipated results.

The Channel Assessment Procedure (CAP) can be used to classify stream channels (British Columbia Forest Practices Code 1996a,b). These protocols are designed to identify past occurrences of channel changes in a consistent, repeatable manner across various landscapes. With information acquired from CAP and the FHAP, candidate watersheds can be identified for rehabilitation needs and possibilities. Information on channel stability, available habitat, limiting factors, etc., can then be synthesized to determine what applications of LWD rehabilitation are appropriate.

HYDRAULIC ASSESSMENT

Streamflows in the potential project reach must be evaluated to determine the range of conditions the proposed LWD structures will be exposed to during each season. Characteristic flows to be identified include the 7-day average low flow, average annual flow, and average peak flow. These flows define the anticipated summer conditions, the size of the stream, and the typical high flow conditions, respectively. Determining these flow levels for a LWD habitat rehabilitation project will require streamflow records for the stream or a regional hydrologic model.

Large woody debris installed in the channel will create localized changes in the streamflow pattern. If these changes in streamflow characteristics are severe, the water surface profile may change significantly and cause bank erosion and/or new channel formation during high flows. The hydraulic assessment is mainly conducted to (1) make planners aware of the potential to create "flashy" streams—streams having more frequent flood events and more rapid stage increases than historically observed, despite little change in precipitation patterns and (2) estimate the extent of fastening needed to withstand buoyancy and shear factors during high flow events. The British Columbia Watershed Restoration Program (P. Slaney, British Columbia Ministry of the Environment, personal communication) recommends that designs be targeted to withstand a 50-year flood event. Our concern is that altered watersheds are currently experiencing more frequent floods at this magnitude. We recommend that designs for 50-year events be applied to relatively stable streams. In altered watersheds, designs should accommodate between 75- to 100-year flood events.

FISH HABITAT ASSESSMENT

Developing effective LWD habitat rehabilitation designs that maintain stream channel characteristics and create fish habitat features requires knowledge of the target fish species and their habitat utilization periods within the project area. Each of the target species may utilize the project area in a different manner. Fish species of primary concern vary regionally; for example coho and cutthroat are ubiquitous along the coastal areas; while chinook salmon, steelhead, and resident trout may have greater distribution in interior regions. Knowing the species of fish you are dealing with is important, as is knowing the age class of a species, like steelhead and cutthroat which have multiple ages, each with a different critical habitat need. The seasonal movement behavior of juvenile fish will also differ regionally. For example, immigration into wall-base channel tributaries occurs in the fall and spring along the coast, while spring immigrations to tributaries are common in interior rivers.

There are also several habitat assessment protocols with various applications in the Northwest. We recommend the Fish Habitat Assessment Guidebook (Johnston and Slaney 1996). This document provides a systematic approach to evaluating fish habitat, where key questions are asked to pinpoint habitat limitations and quantitative data are utilized to rate habitat features as either poor, fair, or good.

Reeves et al. (1991) put forth the idea that fish production is limited by discrete factors, or "bottlenecks." When planning habitat rehabilitation projects, care must be taken to identify aspects of habitat that limit production, and attention should be focused on improving those elements. The timing of life history events is also an important consideration. Using the concept of the bottleneck effect as a guide to choice of life history stage to enhance, may achieve the greatest success in improving ultimate smolt yield. We believe that recent management considerations for multiple-species welfare and ecosystem function may limit the efforts taken to identify species-specific production "bottlenecks." However, as management moves in this direction, these "bottleneck" concerns will generally be addressed by the process of identifying the properly functioning condition of a stream channel.

DESIGN

Once the candidate stream has been selected, one should consult Figure 34.1B (**X**) to determine whether the stream will require low-gradient fish habitat rehabilitation or high-gradient fish habitat and hydraulic rehabilitation. Section (**XI**) displays potential treatment types (Figure 34.1-B).

Work to reestablish LWD within stream channels may date back to the 1940s, but the development of appropriate design and installation techniques to benefit salmonids is continuing. Some past efforts at instream enhancement of LWD structures have been criticized as ineffective (Beschta et al. 1991). Frissell and Nawa (1992) found some widespread damage to LWD structures in 15 Oregon and Washington streams with recent watershed disturbance, high sediment loads, and unstable channels. Cross-stream (channel-spanning) structures were the most vulnerable. In an evaluation of enhancement projects, prepared for the British Columbia Ministry of Environment, Lands and Parks, Hartman and Miles (1995) found only a 50 to 55% success rate.* Some projects in Oregon have been successful at rehabilitating inchannel salmonid habitat. House et al. (1991) and Crispin et al. (1993) report a good success with over 1,000 coastal Oregon instream habitat structures evaluated 1–8 years after placement. In Washington, only seven of over 250 LWD pieces placed in Porter Creek showed instability (Cederholm et al. 1997), and this was after several significant floods during a 6-year period. However, it was noted that, when engineered structures are used, some long-term maintenance may be required. Retention of wood pieces is only one way to rate a project's success. Habitat structure and fish populations have been monitored over a long

---

* Many of these projects were conducted as research or pilot projects and success is generally defined as a wood piece meeting its placement objectives (i.e., trapping gravel, scouring a pool, piece orientation retained, etc.).

term in a large-scale rehabilitation program in Fish Creek, Oregon (G.H. Reeves, USDA Forest Service, personal communication), where initial results show an increase in pool habitat. However, despite an initial increase in juvenile coho, smolt numbers have not increased. Many of the 500+ boulder/large wood structures placed between 1986 and 1988 were displaced during a 100-year flood in 1996.

Many types of instream structures have been used for stream rehabilitation in the Pacific Northwest (Crispin 1988; Koski 1992; Cederholm et al. 1997). Through site visits and review of stream characteristics within these reports, we determined that most LWD structures had been installed in streams of fourth- to fifth-order with normal peak flows of 6–60 m^3/s and channel gradients of 1–3%. This may have been because streams were readily accessible by roads and had strong likelihood of containing fish populations. Instream projects should not be limited to these stream types, although rehabilitation objectives may need to be modified for other stream types. For example, work in large rivers may best be carried out along the stream periphery, such as in more stable "wall-base channel" habitats located on the second or third terrace above the active river floodplain (Peterson and Reid 1984).

MONITORING AND EVALUATING PROJECTS

Documenting fish presence in areas of restored habitat is a common approach to evaluating habitat rehabilitation projects. Increased fish presence in previously little-used areas may indicate that favorable habitat conditions were created as a result of certain woody debris arrangements. However, apparent increases in fish must be interpreted cautiously when evaluating productivity. Since many small-scale rehabilitation projects occur on a site-by-site basis, it is very difficult to determine their actual contribution to overall stream productivity. The matter becomes even more complicated if multiple species are involved and/or if there are artificial influences (hatcheries, stocking, artificial flow conditions, variable harvest rates). The best way to evaluate the effects of a rehabilitation project on anadromous salmonid fish abundance is to quantify basinwide smolt output. Often winter habitat limits fish production and, therefore, it can be important to gain knowledge on the smolt yield benefits (with a viable control) before concluding what improvements have been achieved. Because of the proven seasonal mobility of juvenile salmonids during the winter, it can be misleading to evaluate enhancement projects using summer standing crop estimates. For evaluating resident fish responses in interior regions, enumerations over extended reaches are typically required.

PROJECT COSTS

Over the past century, stream rehabilitation has occurred at all levels of project scales from strict volunteer work within a habitat unit to multi-decade, million-dollar/year watershed rehabilitation programs. Since funding is often limited and temporary, rehabilitative work should be concentrated on a subbasin or stream reach level. Habitat evaluations are more likely to show benefits from concentrated rehabilitative efforts as compared to scattered, basin-wide projects where existing conditions may mask rehabilitation efforts. Table 34.1 lists examples of rehabilitation project costs of various stream reaches. The costs per km of rehabilitated stream reaches can vary. Generally, higher costs are associated with rehabilitation projects that have undergone extensive planning and engineering and/or were conducted for experimental purposes.

CONCLUSIONS

Providing habitats and processes that form habitats can help watersheds maintain or increase their production potential. Synchronizing salmonid life history patterns with the naturally occurring instream processes during project design phase will assist the restoration worker in site selection and interpreting project success.

We have visited several fish habitat rehabilitation sites throughout western Washington, British Columbia, and Oregon. A general conclusion is that the majority of the work has taken place in streams that are too large and/or unstable. This may be the main reason instream rehabilitation projects have often not reached their expected benefits. Our professional judgment and the habitat rehabilitation and improvement-related literature has provided some consistent recommendations that will improve likelihood of long-term success.

- Habitat protection is the first priority in maintaining healthy aquatic systems; LWD-related fish habitat rehabilitation is an interim strategy appropriate until maturing forests provide the functions to maintain viable habitats.
- Rehabilitation efforts should focus on underlying processes, not on superficial improvements; active use of existing natural templates or "analogues" will assist in focusing on processes.
- Priority upslope and riparian rehabilitation work can occur simultaneously with instream work. However, in cases of extremely degraded upslope and riparian conditions, it is advisable to rehabilitate these areas before initiating instream work.
- In addressing instream rehabilitation, understanding the habitat requirements and the limiting factors of existing fish communities contributes to the development of a successful project.
- It is important to understand the existing state of channel disturbance or recovery, and design projects that consider these conditions; avoid unstable geomorphic settings.
- Do not necessarily limit LWD placement strategies to existing manuals and designs, but attempt to simulate natural abundance and arrangements.
- Biologists and engineers should complement each other by considering the physical and biological components of LWD rehabilitation.
- For key-piece LWD, large diameter (>0.6 m minimum) conifer pieces are preferred in streams with <7 m BFW width. Diameters should increase with stream size up to 1.2 m minimum in streams up to 20 m BFW.
- We recommend **off-channel** (wall-base channels) LWD work in fish habitat rehabilitation when the site has either of the following characteristics: (a) the channel is greater than 7 m BFW and has a gradient greater than 2%; or (b) the channel has a BFW greater than 20 m.*
- We recommend LWD work targeting fish habitat rehabilitation when the site has either (a) a channel less than 7 m BFW and gradient less than 5%; (b) a channel 7 to 20 m BFW and gradient less than 2%.*
- We recommend LWD work targeting hydraulic conditions when the site is less than 7 m BFW and the gradient is greater than 5%.
- Monitoring, as a form of learning, is essential for the development of any watershed program concerned with maintaining or improving the existing level of aquatic productivity.

ACKNOWLEDGMENTS

Thanks to Pat Slaney, E. Sue Kuehl, Nick Gayeski, and one anonymous reviewer for their helpful comments on the manuscript. Thanks to U.S. Forest Service Hood Canal and Mt. Baker Ranger Districts, Oregon Department of Fish and Wildlife, Washington Department of Fish and Wildlife, and British Columbia Ministry of the Environment personnel for field tours.

---

* Note: these recommendations are for Pacific Northwest coastal and West Cascade streams. Richmond and Fausch (1995) characterize LWD functions in some medium-sized (4–10 m BFW) Northern Colorado Rocky Mountain streams.

REFERENCES

Abbe, T.B., and D.R. Montgomery. 1996. Large woody debris jams, channel hydraulics and habitat formation in large rivers. Regulated Rivers 12:201–221.

Adams, M.A., and I.W. White. 1990. Fish habitat enhancement: a manual for freshwater, estuarine, and marine habitat. Department of Fisheries and Oceans Canada, Report 4474.

Anonymous. 1980. Stream enhancement guide. Salmonid Enhancement Program, Government of Canada, Department of Fisheries and Oceans, Province of British Columbia, Ministry of Environment, Kerr Wood Leidal Associates Ltd., and D.B. Lister & Associates Ltd., Vancouver, British Columbia.

Anonymous. 1996. Stream Inventory Handbook, Level I and II. USDA Forest Service, Pacific Northwest Region, Corvallis, Oregon.

Beechie, T.J., and T.H. Sibley. 1997. Relationships between channel characteristics, woody debris, and fish habitat in northwest Washington streams. Transactions of the American Fisheries Society 126:217–229.

Benda, L.E., D.J. Miller, T. Dunne, G.H. Reeves, and J.K. Agee. 1998. Dynamic landscape systems. Pages 261–288 in R.J. Naiman and R.E. Bilby, editors. River ecology and management. Springer-Verlag, New York.

Beschta, R.L. 1979. Debris removal and its effects on sedimentation in an Oregon Coast Range stream. Northwest Science 53:71–77.

Beschta, R.L., and W.S. Platts. 1986. Morphological features of small streams: significance and function. Water Resources Bulletin 22:369–379.

Beschta, R.L., W.S. Platts, and B. Kauffman. 1991. Field review of fish habitat improvement projects in the Grande Ronde and John Day River Basins of Eastern Oregon. U.S. Department of Energy, Bonneville Power Administration, Division of Fish and Wildlife, DOE/BP-21493-1, Portland, Oregon.

Bilby, R.E. 1984. Removal of woody debris may affect stream channel stability. Journal of Forestry 82:609-613.

Bilby, R.E., and J.W. Ward. 1991. Characteristics and function of large woody debris in streams draining old-growth, clear-cut, and second-growth forests in southwestern Washington. Canadian Journal of Fisheries and Aquatic Sciences 48:2499-2508.

Bisson, P.A., and eight coauthors. 1987. Large woody debris in forested streams in the Pacific Northwest: past, present, and future. Pages 143–190 in E.O. Salo and T.W. Cundy, editors. Streamside Management: forestry and fisheries interactions. College of Forest Resources, University of Washington, Seattle.

Bisson, P.A., G.H. Reeves, R.E. Bilby, and R.J. Naiman. 1997. Watershed management and pacific salmon: desired future conditions. Pages 447–474 in D.J. Stouder, P.A. Bisson, and R.J. Naiman, editors. Pacific salmon and their ecosystems: status and future options. Chapman & Hall, New York.

British Columbia Forest Practices Code. 1996a. Channel assessment procedure guidebook. Province of British Columbia, Victoria.

British Columbia Forest Practices Code. 1996b. Channel assessment procedure field guidebook. Province of British Columbia, Victoria.

Brookes, A. 1985. River channelization: traditional engineering methods, physical consequences, and alternative practices. Progress in Physical Geography 9:44–73.

Bryant, M.D. 1980. Evolution of large organic debris after timber harvest: Maybeso Creek, 1949–1978. USDA, Forest Service, Pacific Northwest Forest and Range Experimental Station, General Technical Report PNW-101, Portland.

Bustard, D.R, and D. W. Narver. 1975. Aspects of the winter ecology of juvenile coho salmon (*Oncorhynchus kisutch*) and steelhead trout (*Salmo gairdneri*). Journal of the Fisheries Research Board of Canada 32:667–680.

Cederholm, C.J. 1972. The physical and biological effects of stream channelization at Big Beef Creek, Kitsap County, WA. Master's thesis. University of Washington, Seattle.

Cederholm, C.J., and N.P. Peterson. 1985. The retention of coho salmon (*Oncorhynchus kisutch*) carcasses by organic debris in small streams. Canadian Journal of Fisheries and Aquatic Sciences 42:1222–1225.

Cederholm, C.J., and W.J. Scarlett. 1982. Seasonal immigrations of juvenile salmonids into four small tributaries of the Clearwater River, Washington, 1977–1981. Pages 98–110 in E.L. Brannon and E.O. Salo, editors. Proceedings of the salmon and trout migratory behavior symposium. University of Washington, Seattle.

Cederholm, C.J., and W.J. Scarlett. 1991. The beaded channel: a low-cost technique for enhancing winter habitat of coho salmon. Pages 104–108 *in* J. Colt and R.J. White, editors. Fisheries Bioengineering symposium. American Fisheries Society Symposium 10, Bethesda, Maryland.

Cederholm, C.J., W.J. Scarlett, and N.P. Peterson. 1988. Low-cost enhancement technique for winter habitat of juvenile coho salmon. North American Journal of Fisheries Management 8:438–441.

Cederholm, C.J., D.B. Houston, D.L. Cole, and W.J. Scarlett. 1989. Fate of coho salmon (*Oncorhynchus kisutch*) carcasses in spawning streams. Canadian Journal of Fisheries and Aquatic Sciences 46:1347–1355.

Cederholm, C.J., and six coauthors. 1997. Response of juvenile coho salmon and steelhead to placement of large woody debris in a coastal Washington stream. North American Journal of Fisheries Management 17:947–963.

Coffin, B.A., and R.D. Harr. 1992. Effects of forest cover on volume of water delivery to soil during rain-on-snow. Washington Department of Natural Resources, Timber/Fish/Wildlife Report SH1-92-001, Olympia.

Cowan L., D. King, and C. Detrick. 1995. Wild salmonid habitat enhancement and restoration project. Salmonid Screening, Habitat Enhancement, and Rehabilitation Division (SSHEAR). Washington Department of Fish and Wildlife, Olympia.

Crispin, V. 1988. Main channel structures. Pages 1–28 *in* R. House, J. Anderson, P. Boehne, and J. Suther, editors. A training in stream rehabilitation emphasizing project design, construction, and evaluation. Oregon Chapter of the American Fisheries Society, Corvallis.

Crispin, V., R. House, and D. Roberts. 1993. Changes of instream habitat, large woody debris, and salmon habitat after the restructuring of a coastal Oregon stream. North American Journal of Fisheries Management 13:96–102.

Cummins, K.W., and seven coauthors. 1982. Organic matter budgets for stream ecosystems: problems in their evaluation. Pages 299–253 *in* G.W. Minshall and J.R. Barnes, editors. Stream ecology: application and testing of general ecological theory. Plenum Press, New York.

Emmingham, B., and D. Hibbs. 1997. Riparian area silviculture in Western Oregon: research results and perspectives, 1996. Coastal Oregon Productivity Enhancement (COPE) Program Report 10(Numbers 1 and 2):24–27. Oregon State University Forestry Sciences Laboratory, Corvallis.

Frissell, C.A., W.J. Liss, C.E. Warren, and M.D. Hurley. 1986. A hierarchical framework for stream habitat classification: viewing streams in a watershed context. Environmental Management 10:199–214.

Frissell, C.A., and R.K. Nawa. 1992. Incidence and causes of physical failure of artificial habitat structures in streams of western Oregon and Washington. North American Journal of Fisheries Management 12:182–197.

Gregory, K.J., R.J. Davis, and S. Tooth. 1993. Spatial distribution of coarse woody debris in the Lymington Basin, Hampshire, U.K. Geomorphology 6:207–224.

Grette, G.B. 1985. The role of large organic debris in juvenile salmonid rearing habitat in small streams. Master's thesis. University of Washington, Seattle.

Harmon, M.E., and twelve coauthors. 1986. Ecology of coarse woody debris in temperate ecosystems. Advances in Ecological Research 15:133–302.

Harr, R.D. 1981. Some characteristics and consequences of snowmelt during rainfall in Western Oregon. Journal of Hydrology 53:277–304.

Harr, R.D. 1986. Effects of clearcutting on rain-on-snow runoff in western Oregon: A new look at old studies. Water Resources Research 22:1095–2000.

Hartman, G. F., and M. Miles. 1995. Evaluation of fish habitat improvement projects in BC and recommendations on the development of guidelines for future work. British Columbia Ministry of the Environment, Lands and Parks, Fisheries Branch, Vancouver.

Heede, B.H. 1972. Influences of a forest on the hydraulic geometry of two mountain streams. Water Resources Bulletin 8:523–529.

Heede, B.H. 1986. Designing for dynamic equilibrium in streams. Water Resources Bulletin 22:351–357.

House, R.A. and P.L. Boehne. 1986. Effects of instream structures on salmonid habitat and populations in Tobe Creek, Oregon. North American Journal of Fisheries Management 6:38–46.

House, R. A., J. Anderson, P. Boehne, and J. Suther, editors. 1988. Stream rehabilitation manual emphasizing project design-construction-evaluation. Organized by Oregon Chapter, American Fisheries Society, Corvallis.

House, R.A., V.A. Crispin, and J.M. Suther. 1991. Habitat and changes after rehabilitation of two coastal streams in Oregon. Pages 150–159 *in* J. Colt and R.J. White, editors. Fisheries Bioengineering Symposium, American Fisheries Society Symposium 10, Bethesda, Maryland.

Johnston, N.T., and P.A. Slaney. 1996. Fish habitat assessment procedure. Provincial Ministry of Environment, Lands and Parks and Ministry of Forests Watershed Restoration Program Technical Circular 8, Vancouver, British Columbia.

Koski, K.V. 1992. Restoring stream habitats affected by logging activities. Pages 343–403 *in* G.W. Thayer, editor. Restoring the nation's marine environment. Maryland Sea Grant, College Park.

Kramer, R. 1953. Completion report by stream clearance unit on Ozette and Big Rivers. Washington Department of Fisheries, Stream Improvement Division, Olympia.

Lammel, R.F. 1972. Natural debris and logging residue within the stream environments. Master's thesis. Oregon State University, Corvallis.

Lestelle, L.C. 1978. The effects of debris removal on cutthroat trout. Master's thesis. University of Washington, Seattle.

Maser, C., and J.R. Sedell. 1994. From the forest to the sea: the ecology of wood in streams, rivers, estuaries, and oceans. St. Lucie Press, Delray Beach, Florida.

Montgomery, D.R., and J.M. Buffington. 1997. Channel-reach morphology in mountain drainage basins. Geological Society of America Bulletin 109:596–611.

Murphy, M.L., and K.V. Koski. 1989. Input and depletion of woody debris in Alaska streams and implications for streamside management. North American Journal of Fisheries Management 9:427–436.

Murphy, M.L., K.V. Koski, J. Heifetz, S.W. Johnson, D. Kirchhofer, and J.F. Thedinga. 1984. Role of large organic debris as winter habitat for juvenile salmonids in Alaska streams. Pages 251–260 *in* Proceedings, 64th Annual Conference of the Western Association of Fish and Wildlife Agencies.

Murphy, M.L., J. Heifetz, S.W. Johnson, K.V. Koski, and J.F. Thedinga. 1986. Effects of clear-cut logging with and without buffer strips on juvenile salmonids in Alaskan streams. Canadian Journal of Fisheries and Aquatic Sciences 43:1521–1533.

Naiman, R.J., and J.R. Sedell. 1979. Benthic organic matter as a function of stream order in Oregon. Archiv für Hydrobiologie 87:404–422.

Nakamura, F., and F.J. Swanson. 1993. Effects of coarse woody debris on morphology and sediment storage of a mountain stream system in western Oregon. Earth Surface Processes and Landforms 18:43–61.

Nehlsen, W., J.A. Lichatowich, and J.E. Williams. 1991. Pacific salmon at the crossroads: stocks at risk from California, Oregon, Idaho, and Washington. Fisheries 16(2):4–21.

Nickelson, T.E., J.D. Rodgers, S.L. Johnson, and M.F. Solazzi. 1992a. Seasonal changes in habitat use by juvenile coho salmon (*Oncorhynchus kisutch*) in Oregon coastal streams. Canadian Journal of Fisheries and Aquatic Sciences 49:783–789.

Nickelson, T.E., M.F. Solazzi, S.L. Johnson, and J.D. Rodgers. 1992b. Effectiveness of selected stream improvement techniques to create suitable summer and winter rearing habitat for juvenile coho salmon (*Oncorhynchus kisutch*) in Oregon coastal streams. Canadian Journal of Fisheries and Aquatic Sciences 49:790–794.

Overton, C.K., S.P. Wollrab, B.C. Roberts, and M.A. Radko. 1997. Northern/Intermountain Regions fish and fish habitat standard inventory procedures handbook. USDA Forest Service Intermountain Research Station, General Technical Report INT-GTR-346, Ogden, Utah.

Peters, R.J., E.E. Knudsen, C.J. Cederholm, W.J. Scarlett, and G.B. Pauley. 1992. Preliminary results of woody debris use by summer-rearing juvenile coho salmon (*Oncorhynchus kisutch*) in the Clearwater River, Washington. Pages 323–339 *in* L. Berg and P.W. Delaney, editors. Proceedings of the coho workshop, Department of Fisheries and Oceans, Vancouver, British Columbia.

Peterson, N.P., and L. M. Reid. 1984. Wall-base channels: their evolution, distribution, and use by juvenile coho salmon in the Clearwater River, Washington. Pages 215–225 *in* J.M. Walton and D.B. Houston, editors. Olympic wild fish conference, Peninsula College, Port Angeles, Washington.

Peterson, N.P., A. Hendry, and T. Quinn. 1992. Assessment of cumulative effects on salmonid habitat: some suggested parameters and threshold values. University of Washington, Center for Streamside Studies, Seattle.

Quinn, T.P., and N.P. Peterson. 1994. The effects of forest practice on fish populations. Washington Department of Natural Resources, Timber-Fish-Wildlife Report F4-94-00, Olympia.

Ralph, S.C., G.C. Poole, L.L. Conquest, and R.J. Naiman. 1994. Stream channel morphology and woody debris in logged and unlogged basins of western Washington. Canadian Journal of Fisheries and Aquatic Sciences 51:37–51.

Reeves, G.H., J.D. Hall, T.D. Roehlofs, T.L. Hickman, and C.O. Baker. 1991. Rehabilitating and modifying stream habitats. Pages 519–557 in W.R. Meehan, editor. Influences of forest and rangeland management on salmonid fishes and their habitats. American Fisheries Society, Special Publication 19, Bethesda, Maryland.

Richmond, A.D., and K.D. Fausch. 1995. Characteristics and function of large woody debris in subalpine Rocky Mountain streams in northern Colorado. Canadian Journal of Fisheries and Aquatic Sciences 52:1789–1802.

Robison, G.E., and R.L. Beschta. 1990a. Characteristics of coarse woody debris for several coastal streams of southeast Alaska, U.S.A. Canadian Journal of Fisheries and Aquatic Sciences 47:1684–1693.

Robison, G.E., and R.L. Beschta. 1990b. Coarse woody debris and channel morphology interactions for undisturbed streams in southeast Alaska, U.S.A. Earth Surf. Proc. Landforms 15:149–156.

Rosgen, D. 1996. Applied River Morphology. Wildland Hydrology, Pagosa Springs, Colorado.

Rot, B.W. 1995. The interaction of valley constraint, riparian landform, and riparian plant community size and age upon channel configuration of small streams of the western Cascade Mountains, Washington. Master's thesis. University of Washington, Seattle.

Schuett-Hames, D., A. Pleuss, L. Bullchild, and S. Hall. 1994. Ambient monitoring program manual. Northwest Indian Fisheries Commission, Timber-Fish-Wildlife Report TFW-AM9-94-001, Lacey, Washington.

Sedell, J. R., and K. J. Luchessa. 1982. Using the historical record as an aid to salmonid habitat enhancement. Pages 210–223 in N.B. Armantrout, editor. Acquisition and Utilization of Aquatic Habitat Inventory Information. American Fisheries Society Symposium, Western Division, Portland, Oregon.

Slaney, P.A., and D. Zaldokas, editors. 1997. Fish rehabilitation procedures. Watershed Restoration Technical Circular 9. Ministry of Environment, Watershed Restoration Program, Vancouver, British Columbia.

Solazzi, M.F., and S.L. Johnson. 1994. Development and evaluation of techniques to rehabilitate Oregon's wild salmon coho. Oregon Department of Fish and Wildlife, Fish Research Project F-125-R, Annual Progress Report, Portland.

Spence, B.C., G.A Lomnicky, R.M. Hughes, and R.P. Novitzki. 1996. An ecosystem approach to salmonid conservation. ManTech Environmental Research Services Corp., TR-4501-96-6057, Corvallis, Oregon. (Available from the National Marine Fisheries Service, Portland, Oregon).

Sullivan, K., T.E. Lisle, C.A. Dolloff, G.E. Grant, and L.M. Reid. 1987. Stream channels: the link between forests and fishes. Pages 39–97 in E.O. Salo and T.W. Cundy, editors. Streamside management: forestry and fisheries interactions. University of Washington, Institute of Forest Resources, Contribution 57, Seattle.

Toews, D.A.A., and M.K. Moore. 1982. The effects of streamside logging on large organic debris in Carnation Creek. Ministry of Forests, Publication R 28-81071, Victoria, British Columbia.

Triska, F.J., J.R. Sedell, and S.V. Gregory. 1982. Coniferous forest streams. International Biological Program Synthesis Series 14:292–332.

Tschaplinski, P.J., and G.F. Hartman. 1983. Winter distribution of juvenile coho salmon (*Oncorhynchus kisutch*) before and after logging in Carnation Creek, British Columbia, and some implications for overwinter survival. Canadian Journal of Fisheries and Aquatic Sciences 40:452–461.

Van Cleef, J.S. 1885. How to restore our trout streams. Pages 51–55 in Report of the Fourteenth Annual Meeting of the American Fisheries Society.

Vannote, R.L., G.W. Minshall, K.W. Cummins, J.R. Sedell, and C.E. Cushing. 1980. The river continuum concept. Canadian Journal of Fisheries and Aquatic Sciences 37:130–137.

WDNR (Washington Department of Natural Resources). 1993. Forest practice board manual: standard methodology for conducting watershed analysis. Washington Department of Natural Resources, Forest Practices Division, Version 2.1, Olympia.

Wilford, D.J. 1982. The sediment-storage function of large organic debris at the base of unstable slopes. Pages 115–129 in W.R. Meehan, T.R. Merrell, Jr., and T.A. Henley, editors. Fish and wildlife relationships in old-growth forests. American Institute of Fishery Research Biologists, Northwest Section, Juneau, Alaska.

35 Development of Options for the Reintroduction and Restoration of Chinook Salmon into Panther Creek, Idaho

Dudley W. Reiser, Michael P. Ramey, and Paul DeVries

Abstract.—The Panther Creek drainage (east central Idaho) historically supported substantial runs of spring and summer chinook salmon *Oncorhynchus tshawytscha* and steelhead trout *O. mykiss*. Since the early 1900s, mining activities within the drainage (and associated releases of heavy metals) have eliminated chinook runs from the basin. In 1995, a biological restoration and compensation program (BRCP) was formulated to identify measures that would lead to development of a self-sustaining run of chinook salmon in Panther Creek. Because of the Endangered Species Act (ESA) endangered status of spring chinook salmon stocks in Idaho, the restoration program was developed with consideration for the source of brood stock, required numbers of adult pairs, rearing practices, and monitoring and evaluation. The development of the BRCP was completed in three steps: selection of restoration/enhancement options; time series analysis to provide a restoration/recovery curve of chinook populations in Panther Creek based on the selected measures; and cost estimation (assigning costs to the BRCP elements). Four general categories of restoration options were identified and evaluated including (1) restoration of populations by hatchery supplementation techniques (preferably using wild donor stocks); (2) creation of additional habitats (within or outside of the Panther Creek drainage) to promote increased survival of fish and increase carrying capacities; (3) restoration and/or enhancement of existing habitats (within or outside of the Panther Creek drainage) to increase fish survival; and (4) reduction in fish losses such as those occurring in irrigation diversions. Specific engineering concepts were identified and evaluated for each category. In addition to engineering options, various biological strategies for reintroducing chinook salmon into the system were evaluated. Although the effectiveness of the program has yet to be determined in Panther Creek, we believe the BRCP as outlined provides the best opportunity for successful reintroduction and restoration of chinook salmon into the system. The general steps completed and considerations addressed in developing this BRCP included those likely to be necessary in other drainages where restoration of depleted or extirpated stocks of salmon and/or steelhead is desired. In those cases, this BRCP process may provide guidance in developing successful and economically feasible reintroduction and restoration programs.

INTRODUCTION

The Panther Creek drainage lies within the Salmon National Forest in east central Idaho (Figure 35.1). As one of the larger tributaries in the upper Salmon River system (drainage area = 530 mi^2), Panther Creek historically supported substantial runs of spring and summer chinook salmon *Oncorhynchus tshawytscha* and steelhead trout *O. mykiss*. According to Smith and Seaberg (1991), the Panther Creek drainage was the fourth most important producer of chinook salmon of

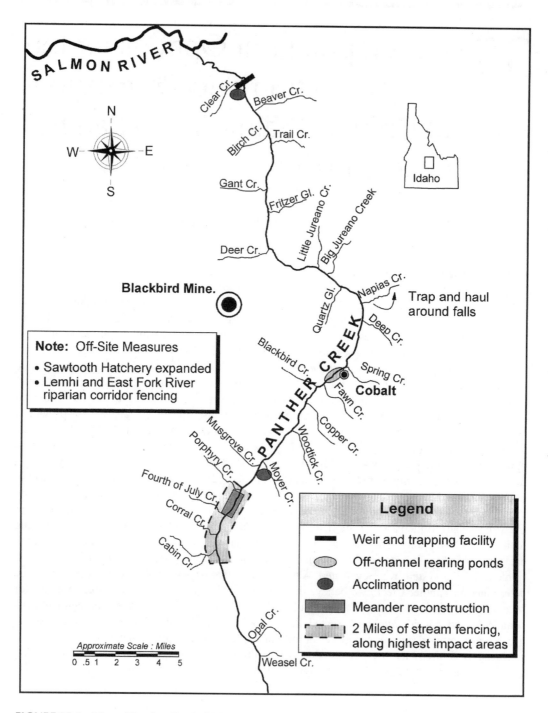

FIGURE 35.1 Map of Panther Creek, Idaho with measures proposed for restoration to baseline and compensation of interim losses.

the Salmon River tributaries, surpassed only by the Middle Fork Salmon River, Lemhi River, and North Fork Salmon River. Major Panther Creek tributaries of historical importance to anadromous fish production likely included Musgrove, Moyer, Deep, Deer, Beaver, and Clear creeks (Figure 35.1) (Reiser 1986). Historic runs of chinook salmon were estimated at approximately 2,000 spawners (IDFG 1992). However, since the early 1900s, mining at the Blackbird Mine site has contributed to the decline and ultimately elimination of runs of chinook salmon in the Panther Creek system (Platts et al. 1979; R2 1995).

In 1992, the State of Idaho, followed by the National Oceanic and Atmospheric Administration (NOAA) and the U.S. Forest Service (USFS), filed a natural resource damages claim as provided for under the Comprehensive Environmental Response, Compensation and Liability Act (CERCLA) against the mine owner. A series of studies were subsequently completed which (1) determined the extent and amount of injury to the resources of the area (RCG/Hagler Bailly 1994) and (2) developed plans for remediating and restoring water quality conditions to levels capable of supporting all salmonid life stages in Panther Creek (RMC 1995). In 1994, NOAA commissioned a third study to develop a program specifically targeted on the restoration of chinook salmon into the Panther Creek drainage. The study addressed two primary objectives of the agencies: (1) restoration of spring/summer chinook salmon to a baseline condition in the Panther Creek watershed (baseline defined as the condition of the resource absent the release of hazardous substances); and (2) compensation for the chinook salmon injured by Blackbird Mine releases of hazardous substances, as estimated from the time (1957) the runs of salmon were considered to have been essentially eliminated from the system until chinook salmon numbers are expected to return to baseline. The study resulted in the development of a Biological Restoration and Compensation Program (BRCP) which identified, evaluated, and selected various engineering and management options for meeting the stated objectives (R2 1995).

This chapter describes various components of the Panther Creek BRCP as they relate to the restoration of Snake River chinook salmon, a species listed as endangered under the federal ESA. We illustrate the multitude of biological considerations that must be addressed when dealing with restoration/reintroduction options of an endangered or threatened species in the context of mandated recovery plans and critical habitat designations. Several methods and techniques for assessing restoration options are also illustrated.

ASSUMPTIONS

The development of the BRCP was based on certain technical considerations and assumptions for establishing restoration goals. These assumptions and a brief rationale for each are provided below:

- *Water quality non-limiting.* It was assumed that water quality conditions in Panther Creek would be capable (by the year 2005) of sustaining salmonids throughout all life phases. This assumption was contingent on the implementation of the remediation program designed to eliminate water quality problems related to heavy metals and acid mine drainage (RMC 1995).
- *Chinook salmon as target resource.* The Panther Creek watershed historically supported runs of chinook salmon, steelhead, and resident species, such as rainbow trout, whitefish *(Prosopium williamsoni)*, and bull trout *(Salvelinus confluentus)*. All species were affected by releases of hazardous substances from the Blackbird Mine. Snake River spring/summer chinook salmon were selected as the biological target for developing the restoration (physical as well as biological) program, since they are at record low numbers within Idaho and have been listed as endangered under the ESA. We concluded that recovery of this species within Panther Creek could not be expected in a reasonable time without implementation of additional restoration measures and a reintroduction strategy.
- *Target baseline condition = 200 adult chinook salmon.* The restoration program developed as part of this analysis assumed a target baseline condition of 200 adult chinook salmon returning to the Panther Creek drainage. The number of annual chinook smolts that would be produced under present habitat capacity was calculated to be 250,000. These numbers were derived as part of the injury quantification studies performed by RCG/Hagler Bailly (1994) and represent the approximate size of the adult chinook salmon run which would exist in Panther Creek, absent the release of hazardous materials.

The baseline target is the number of adults produced by 250,000 smolts, taking into account the effects of downstream dams, ocean harvest, disease, predation, and habitat loss (both within Panther Creek and along the entire migratory pathway).

- *Replacement of interim losses.* In the context of the BRCP, interim losses constitute the total number of adult chinook salmon that would have been lost (as calculated on an annual basis) since 1957 until the time baseline conditions of 200 adult spring/summer chinook salmon are restored. To estimate these losses, we developed and used a chinook salmon population model which calculates smolt production relative to various restoration options.
- *Off-site replacement.* The evaluation of components in the BRCP included measures both within and outside the Panther Creek drainage.

FEDERAL AND STATE STATUTES AND POLICIES

The BRCP was developed with consideration for a number of federal and state laws and policies, and the objectives of the Shoshone Bannock Indian Tribes. These specifically included the ESA; the Snake River Recovery Plan (Bevan et al. 1994) as commissioned by the National Marine Fisheries Service (NMFS); the Idaho Anadromous Fishery Management Plan (1992); the sub-basin plan for the Salmon River Sub-basin (containing Panther Creek) (IDFG et al. 1990), representing 1 of 31 separate plans specified by the Northwest Power Planning Council's Columbia River Basin Fish and Wildlife Program; and the USFS Salmon National Forest Long Range Management Plan, which lists a number of objectives relative to the Panther Creek system (Smith and Seaberg 1991).

Specific considerations related to restoration, reintroduction, and recovery protocols of the NMFS and IDFG, as listed and discussed in Lichatowich and Watson (1993), were also addressed (where applicable) in developing the BRCP (Table 35.1). These included elements related to broodstock source, impacts on donor populations, number and type of adults mated, fertilization protocol, rearing practices, release procedures, precautions against straying, and monitoring and evaluation.

BIOLOGICAL RESTORATION AND ENHANCEMENT OPTIONS

Numerous restoration and enhancement options were identified and evaluated. These included options directed specifically toward restoration of a baseline condition of 200 adult salmon returning to the Panther Creek drainage, as well as measures focused on compensation for cumulative interim losses. For the latter, the options encompassed locations both within and outside of the Panther Creek drainage, inasmuch as compensation for interim losses could not be achieved realistically if measures were limited only to the Panther Creek drainage.

Measures were considered that would:

- meet logistical and biological requirements for reintroducing chinook salmon into an unoccupied drainage;
- assist the initial population to reach self-sustenance, either naturally or artificially;
- increase stream carrying capacity over existing conditions; and
- improve egg-to-smolt survivals in the drainage.

Enhancement and restoration options were not considered for Blackbird Creek due to the extremely degraded condition of physical habitats in that system. Likewise, no measures were considered for Big Deer Creek due to a steep falls located about 0.6 mi above its confluence with Panther Creek; the falls represent a barrier to anadromous fish passage, although good habitat for resident species occurs in Big Deer Creek.

TABLE 35.1
**Institutional considerations related to chinook salmon restoration
(c.f. Lichatowich and Watson 1993).**

Specific element	National Marine Fisheries Service	Idaho Department of Fish and Game
Program development	Statutory requirements for elements of recovery program, Sect. 7 and 10 consultation, permitting requirements; listing and delisting criteria; requirement to change program when "appreciable" hatchery/wild differences develop, or adverse ecological impacts of hatchery fish on wild occur.	No description of discrete development/planning program; program developed under Drainage Management Plans consistent with long-range goals, policies and principles; Idaho Fish and Game Commission has approval authority; no required risk analysis mentioned.
Broodstock source	Exclusively from ESU; intra-ESU distinctions; extensive subsampling rules.	From targeted natural population or from adjacent, environmentally similar drainage.
Impacts on donor population	Methods to maintain genetic variability of supplemented population; conditions for subsampling vs. total capture of ESU for broodstock.	Extensive effort to prevent broodstock mining, provide for effective size of donor populations in broodstock collections at weirs.
Number and type of adults mated	Minimum size re: drift; methods to maintain N_0 by reducing family size variability re: mating protocols.	Type, but not minimum number, discussed; 1:1 male/female mating, non-selective broodstock sampling re: all characteristics except disease.
Fertilization protocol	Gamete mixing to minimize family size variation (one female with overlapping pairs of males).	1:1, male/female; detailed handling of gametes (e.g., sperm pooling) not discussed.
Rearing practices	Trade-offs between numerical increase, and genetic/ecological divergence of hatchery from wild; recommendation and description of naturalistic rearing procedures.	Charged to produce hatchery fish behaviorally/ecologically compatible with wild/natural; developing some elements of naturalistic rearing.
Release procedures	Mimic spatiotemporal movements with hatchery releases; mimic distribution of measurable wild attributes; provide prolonged acclimation (consider release of pre-smolts).	Time and place of release manipulated to reduce straying and residualism (few details); release size determined primarily by survival data (wild size distribution not mimicked).
Precautions against straying	Mark all hatchery fish; provide prolonged acclimation or release pre-smolts.	All hatchery fish marked, can be excluded at weirs (few details); no releases of hatchery fish in wild streams; time/place manipulation to reduce straying; ban on more than 50% hatchery fish on spawning ground in some drainages (few enforcement details).
Monitoring and evaluation (M&E)	Monitor genetic variability, phenotypes, life history traits of hatchery and wild fish for development of "appreciable" differences; monitor production of progeny from naturally spawning fish; monitor egg/smolt, smolt/adult survival of hatchery fish, by family if possible.	Much more M&E implied than described; no explicit genetic monitoring discussed; no "unevaluated" supplementation permitted; intent to monitor natural production, survival, straying, natural spawner composition (few details).
Compliance monitoring and reporting	NMFS oversight of recovery program; regulations of directed and incidental take.	

TABLE 35.2

Applicable concepts for restoration and enhancement of chinook salmon populations in Panther Creek. Concepts include those for restoration of baseline and compensation of interim losses.

Concept	Measure	BRCP[a] goal	Inclusion in BRCP
Restore chinook populations	Hatcheries	Restoration	Yes
	Weir and adult trapping facility	Restoration	Yes
	Acclimation ponds	Restoration	Yes
Create additional habitat	Channel meander reconstruction	Compensate for interim losses	Yes
	Instream structures	Compensate for interim losses	No
	Off-channel habitat construction	Compensate for interim losses	Yes
	Barrier removal	Compensate for interim losses	No
Restore/enhance existing habitat	Riparian corridor fencing	Compensate for interim losses	Yes
	Spawning gravel cleaning	Compensate for interim losses	No
	Bank stabilization	Compensate for interim losses	No
	Surfacing of Panther Creek Road	Compensate for interim losses	No
	Land purchase (change use to restore habitat)	Compensate for interim losses	Yes
	Instream flow improvement opportunities	Compensate for interim losses	No
	Improvement of irrigation efficiency	Compensate for interim losses	No
	Storage reservoir construction	Compensate for interim losses	No
	Off-site habitat enhancement projects	Compensate for interim losses	Yes
Reduce fish losses	Irrigation diversion screening	Compensate for interim losses	No

[a] Biological restoration and compensation program.

Four general categories of restoration options were identified which represented both direct and indirect measures for restoring chinook salmon populations in Panther Creek. The categories included: (1) restoration of populations using hatchery supplementation techniques (with a focus on wild donor stocks); (2) creation of additional habitats (within or outside of the Panther Creek drainage) to promote increased survival of fish and increase carrying capacities; (3) restoration and/or enhancement of existing habitats (within or outside of the Panther Creek drainage) to increase fish survival; and (4) reduction in fish losses, such as those occurring in irrigation diversions. Specific engineering concepts were identified and evaluated for each of these four categories. These were evaluated based on the following criteria:

- technical feasibility/likelihood of success;
- relationship of expected costs of the measure to the expected benefits relative to the return to baseline conditions and replacement of interim losses;
- reliability of the measure/application of proven technology;
- potential for additional injury to the resource or complications resulting from the proposed measure; and
- consistency with relevant federal, state, and tribal management policies and plans, in particular, consistency with ESA mandated recovery plans for the Snake River spring/summer and fall chinook salmon.

This review resulted in the selection and integration of various options into a comprehensive program for achieving the stated objectives (Table 35.2). In general, the measures not included in the BRCP were judged as having limited or uncertain biological benefits relative to construction/implementation costs (e.g., instream structures, barrier removal, spawning gravel cleaning, bank stabilization, and road resurfacing).

CHINOOK SALMON REINTRODUCTION STRATEGIES

There are no chinook salmon runs in the Panther Creek system, not even vestigial runs. The native stocks of summer and spring chinook were likely extirpated in the early 1960s, and continued water quality problems have prevented colonization and the establishment of natural runs. In our analysis, we assumed that water quality conditions would be improved to levels suitable for all life history stages of salmonids by the year 2005. This was premised on the timely implementation of measures proposed by RMC (1995) that are focused on pollution abatement from the Blackbird Mine and restoration of water quality characteristics conducive to salmonid production. Thus, the reintroduction of chinook salmon would not be constrained by differential life stage sensitivities to heavy metals toxicity.

DONOR STOCK SELECTION

An important and necessary step in the development of a reintroduction strategy is the identification and selection of a suitable donor stock from which to build the Panther Creek chinook runs. Broodstock selection and supplementation criteria have been developed by the BPA for transplanting salmon within the Columbia River Basin. The Regional Assessment of Supplementation Project (RASP) Summary Report Series (BPA 1992) summarized salmon supplementation theory within the Columbia Basin and emphasized that donor stocks should exhibit similarity to recipient stocks with respect to genetics, life history, and ecology. Miller (1990) stressed the need to understand the ecology of the area to be supplemented, the factors limiting fisheries production, and the unique qualities of the donor stock. Bjornn (1978) also recommended that consideration be given to stocks from closely adjoining streams and to those with similar run timings to the population to be supplemented. The RASP developed a template/patient analysis designed to match donor stocks for supplementing endangered stocks using criteria related to each life history stage of the salmon in terms of habitat, timing, survival, and demographics. Since no chinook stocks currently exist within Panther Creek, characteristics of recipient stocks cannot be matched. Furthermore, information on the life history and ecology of Panther Creek chinook salmon is sparse. However, length of migration, basin geomorphology, and run timing provide useful criteria for matching potential chinook salmon donor stocks for introduction to the Panther Creek drainage.

The first criterion is the *length of the migration* that outmigrating smolts and returning adults will encounter. For Columbia River chinook salmon stocks, only fish in Idaho waters migrate distances similar to the Panther Creek migration distance (to the Pacific Ocean). Therefore, the search for donor stock was limited to areas within Idaho, particularly from the Salmon River basin. Stocks with migration distances greater than that of the historic Panther Creek stock would be expected to have an advantage over other stocks with shorter migrations.

The second criterion for selecting an appropriate donor stock is *similarity of drainage basins*. The genetic attributes of native chinook stocks that once existed in Panther Creek were influenced by ecological and environmental survival pressures characteristic of the Panther Creek basin. Chinook salmon transplanted or introduced to the Panther Creek drainage will have adapted to slightly different pressures. Donor stocks from drainages with ecological and geomorphological characteristics similar to Panther Creek should be the best to adapt and survive there. Schreck et al. (1986) concluded that stocks of Columbia River chinook salmon from geographically proximal areas tend to be genetically similar. Smith et al. (1985) recommended that broodstock for supplementation be from a closely-adjoining stream. Results of genetic monitoring of Columbia River chinook salmon populations in the Snake River Basin indicated that genetic similarity between stocks was related to geographical proximity (Waples et al. 1991). These results suggest that strong preference should be given to fish from drainages as close to Panther Creek as possible, preferably from drainages within the Salmon River basin.

In emphasizing the need to select donor stock from close geographic proximity, it is assumed that fish from similar areas will experience similar selective pressures. Due to the natural selective pressures they encounter throughout their entire life cycle, wild stocks should be better adapted to a wild environment and better able to produce a self-sustaining population than hatchery stocks. Within the Salmon River Basin, the Middle Fork chinook salmon stocks contain the highest degree of wild genetics (M. Reingold, 1994, Idaho Department of Fish and Game (retired), personal communication); these stocks have ostensibly been unsupplemented. Chamberlin Creek, Middle Fork Salmon River, and North Fork Salmon River are the chinook salmon streams geographically closest to Panther Creek.

The third criterion for selection of donor stock is *run-timing*. Waples et al. (1991) found that stocks of summer-run chinook salmon from within the South Fork Salmon River (Secesh, Johnson Creek, and McCall Hatchery) were genetically closer to summer-run chinook in the Imnaha River in Oregon than to fall-run chinook salmon from the Snake River which were geographically closer. These results suggest that run timing can be as strong an indicator of genetic similarity as geographic proximity. Historically, the Panther Creek Basin contained both spring- and summer-run chinook salmon stocks. Spring-run chinook in the Salmon River subbasin generally cross Bonneville Dam on the Columbia River between March 1 and May 31; the summer run crosses between June 1 and July 31 (IDFG et al. 1990). The early run in Panther Creek reportedly spawned in the headwaters, between Moyer Creek and Porphyry Creek, with the peak activity occurring around mid-August. The later run reportedly spawned below Napias Creek around the first of September (Welsh et al. 1965). Information is not available as to which run was more abundant and it is therefore difficult to select one run over the other. However, either spring- or summer-run chinook salmon from nearby basins are likely to do well in Panther Creek. Characteristics of Salmon River Basin chinook salmon stocks critical to donor stock selection are summarized in Table 35.3.

An important supplementation consideration is genetic outbreeding depression (i.e., loss of fitness of the native stock resulting from introduction of genes of inferior fitness; Waples et al. 1991), between native and transplanted stocks. This does not appear to be critical in Panther Creek because chinook salmon stocks are virtually extinct there. However, it does factor into how donor stocks are utilized in the restoration process, since multiple donor stocks may be needed (because of low numbers) to meet recommended founding population sizes (see below). Specifically, crossing of geographically distant and therefore genetically less similar stocks should be minimized to avoid outbreeding depression.

Another important supplementation factor to consider to help ensure a self-sustaining population of chinook salmon within Panther Creek is the optimal number of fish necessary to initiate supplementation. For example, inbreeding becomes a risk when there is a limited number of available parents for donor stock. The effect of inbreeding is to limit the genetic diversity of offspring and therefore the fitness of the stock. With a small population size, recessive genes or rare genes may occur in frequencies greater than if the breeding population is large. If such genes produce unfavorable traits which reduce the fitness of the individual, the fitness of successive generations will be further reduced (Steward and Bjorn 1990). Inbreeding depression may follow from genetic inbreeding, resulting in reduced fitness and loss of stock vigor. Increasing the number of parents in the donor stock will help to reduce risks of genetic inbreeding depression in progeny of the donor stock.

Minimum stock sizes have been estimated for maintaining genetic diversity. The proper number depends on the environment and reproductive biology. Steward and Bjornn (1990) reviewed a number of studies in which recommended founding population sizes ranged from 50 to 200 fish. Because these numbers are representative of fish that spawn naturally, fewer fish could likely be utilized in an artificial breeding program where eggs from each female are mixed with sperm from many males until runs rebuild. Maintenance of genetic quality of the broodstock can also be assisted by taking eggs throughout the duration of a run rather than at one time, and by spawning multiple age classes and sizes (Smith et al. 1985). As noted above, for Panther Creek, it may be necessary to utilize donors from a number of geographically proximal (and therefore assumed genetically similar) stocks to meet a minimum founding population size. Although this invokes concerns relative to outbreeding depression (Waples et al. 1991), careful selection of the donor stocks may in this case be beneficial in reducing the risk of inbreeding and loss of genetic diversity. Straying can also add to genetic diversity, although Snake River spring and summer chinook salmon appear to stray at relatively low rates compared to fall chinook as indicated by tagging studies and carcass counts (Dauble and Mueller 1993).

In our analysis, we assumed that 50 adult fish (consisting of a 50:50 mix of male:females; i.e., 25 males and 25 females) from appropriate donor stocks would be used to initiate the reintroduction of chinook salmon to Panther Creek, and that the removal of these fish from other drainages would not impact donor populations. The latter assumption requires further evaluation but is not directly relevant to Panther Creek restoration.

Our review of available information (Table 35.3) suggested that the following drainages (in decreasing order of preference) would provide suitable donor stocks of chinook salmon for reintroduction into Panther Creek:

- Middle Fork Salmon River
- North Fork Salmon River
- Chamberlin Creek/Bargamin Creek
- Other local tributaries on the mainstem Salmon River (within the river reach extending from French Creek to Middle Fork of the Salmon River)
- Lemhi River, Pahsimeroi River, or East Fork Salmon River
- Upper Salmon River and Yankee Fork
- South Fork Salmon River

REINTRODUCTION—RESTORATION STRATEGIES

The reintroduction-restoration of chinook salmon to Panther Creek can theoretically be accomplished by natural recolonization or through supplementation techniques. However, because of the extremely depressed spring and summer chinook stocks throughout Idaho, and the predicted continuance of these low levels over the next 5 to 10 years, we concluded that natural colonization (via straying from other drainages) would not provide sufficient seed stock to build the runs back to baseline conditions within a reasonable time (projections based on a computer life cycle model indicated over 150 years would be required).

Supplementation methods are broadly comprised of the trapping of returning anadromous salmonid adults, artificial spawning, incubation and rearing at hatcheries, and subsequent release of pre-smolts or smolts to the stream. Miller (1990) reviewed the literature on salmon and steelhead supplementation within the Columbia River drainage. Of 316 projects aimed at rebuilding self-sustaining anadromous fish runs with hatchery fish, only 25 were judged to have been successful at supplementing natural existing runs. Of the 25, some of the most significant results were obtained for projects aimed at reestablishing runs, or introducing runs to formerly unused areas (i.e., cases similar to Panther Creek).

Supplementation efforts have only recently occurred within the Salmon River basin (Miller 1990). A number of satellite fish rearing and trapping stations have been developed specifically for supplementation/augmentation. Most attention has focused on the South Fork Salmon River drainage where adults are trapped at a weir and taken to the McCall hatchery for spawning and subsequent rearing of offspring. Fry have been outplanted in both the South Fork and in Johnson Creek, a tributary to the East Fork of the South Fork of the Salmon River. Semi-natural rearing habitats have been constructed in the Yankee Fork drainage in the upper Salmon River basin using off-site dredge mine ponds (Reiser et al. 1994). These have been seeded with hatchery fry, which then rear in the ponds until they become smolts (Richards et al. 1992). Bevan et al. (1994) recommended that, where appropriate, artificial propagation be implemented to aid recovery of ESA-listed species. Specifically, captive broodstock, gene banks, supplementation, reintroduction, and research should be considered to help preserve, maintain, and recover discrete populations. However, existing supplementation programs have not been successful in increasing natural production (Chapman et al. 1991). For hatcheries to contribute to recovery of the species, improvements are necessary

TABLE 35.3
Description and status of current stocks of chinook salmon in the Salmon River Basin.

Drainage [Approximate distance from Panther Creek in km]	Run Timing	Basin Geomorphology	Stock Origin	Comments and Stock Status
Panther Creek [0]	Spring-run in headwaters, summer-run in mainstem	Volcanic rocks Lateral and terminal moraines, with boulders and swampy areas throughout Gradients vary[d]	Extinct	Estimated 50 to 60% of fry overwinter and smolt the following spring
Middle Fork Salmon River [26]	Spring-run in tributaries, summer-run in mainstem	Glacial valley with alpine lakes, hanging valleys, glacial till and moraines Granitic and volcanic soils Gradient increases downstream[d]	Wild[c]	
North Fork Salmon River [36]	Spring-run[a]	Narrow mountain valley, with little glaciation Quartzites, slates, and volcanic rocks Steep gradient with little meandering[d]	Small hatchery supplementation and limited natural production[a]	
Salmon River (French Creek to Middle Fork) [63]	Spring-run[a]	Deep canyon, high gradient, with some lower gradient tributaries Granitic soils[d]	Wild (Chamberlin Ck)[a] Wild (Bargamin Ck)[b]	
Lemhi River [71]		Meandering, low gradient, spring-fed, and flows through broad valley[a] Tributaries with steep gradients, productive alluvial and glacial valley[d]	Natural	Planting of native and Rapid River broodstock from 1920s to 1980s. Currently no supplementation; Hayden Creek hatchery currently closed—has water quality, disease, and homing problems[a]
Salmon River (North Fork to East Fork) [120]		Low gradient, braided channels and alluvial plains[a]	Wild/natural[c]	

Location	Run type	Habitat	Stock status	Notes
South Fork Salmon River (includes McCall Hatchery) [130]	Summer-run[a]	Extensive meadows in headwaters, steep rocky canyons, located within the Idaho batholith, and soils of decomposed granite[a]	Natural and hatchery (summers)	McCall hatchery chinook outplanted to East Fork, South Fork, Johnson Creek, and Pahsimeroi Hatchery
Pahsimeroi River [144]	Spring-run and summer-run[a]	Meandering, alluvial fan in upper valley with springs in lower valley[a]-Similar to Pahsimeroi valley[d]	Natural and hatchery (summers and springs)[a]	Summer-run South Fork Salmon River stocks supplemented
Little Salmon River (includes Rapid River Hatchery) [198]	Spring-run and summer-run[a]	Steep gradient, narrow valley	Hatchery (springs) from Snake River above Hells Canyon-Wild (summers)	
East Fork Salmon River [207]	Spring-run in upper reaches, summer-run in lower reaches[a]	Lower drainage bisects highly erosive volcanic soils Mostly volcanic with some sedimentary rocks Steep to moderate gradient[d]	Natural and hatchery springs Natural and hatchery summers[a]	Spring chinook captured and raised in Sawtooth Hatchery, no surplus for other supplementation activities[a]
Salmon River (East Fork to Yankee Fork) [230]	Spring-run and summer-run[a]	Deep canyons and narrow alluvial bars[a]	Wild/natural[c]-Wild summers[a]	
Lower Salmon River (mouth to French Creek) [240]	Spring-run[a]	Deep, rocky canyon, with high gradient Unstable granitic soils[d]	Strays from Rapid River Hatchery	
Yankee Fork [250]	Spring-run[a]	Steep to moderate gradient with erosive sandy and clay-loam soils[d]	Rapid River and Sawtooth stock	
Upper Salmon River (Yankee Fork to Sawtooth Hatchery) [264]	Summer-run[a]	Intermontane valley, mostly granitic and igneous bedrock with sedimentary outcrops[d]	Wild summers[a]	
Sawtooth Hatchery [272]	Spring-run[a]		Natural and hatchery[a]	
Salmon River headwaters [280]	Spring-run[a]	Meandering meadow stream in a subalpine valley[a]	Natural and hatchery[a]	

[a] Idaho Department of Fish and Game (1992).
[b] Idaho Department of Fish and Game (1991).
[c] Mel Reingold, Idaho Department of Fish and Game (retired), personal communication.
[d] Idaho Department of Fish and Game (1990).

to address the reduced survival of hatchery smolts to adults, problems with genetic fitness and diversity, and some of the other potentially deleterious effects of artificial propagation.

We considered four reintroduction strategies based on supplementation: (1) adult outplantings—natural spawning; (2) hatchery production—egg plantings; (3) hatchery production—fry outplants; and (4) hatchery production—smolt outplants. We selected the latter for the BRCP. The first three options, adult outplantings, egg plantings, and fry outplantings, have all been successfully used as stock supplementation measures. However, based on prior attempts to utilize those methods (Kiefer and Lockhart 1993; C. Petrosky, Idaho Department of Fish and Game, personal communication) they were considered less efficient and reliable than the outplanting of smolts, which was the supplementation method we selected for Panther Creek.

BIOLOGICAL RESTORATION AND COMPENSATION PROGRAM

Based on our review of restoration options and reintroduction strategies, we developed a restoration program deemed most suitable for restoring chinook salmon into Panther Creek. The BRCP includes those components to restore chinook salmon in the Panther Creek watershed to baseline population levels (200 adults per year) and to compensate for interim losses of fish.

BASELINE RESTORATION PROGRAM

The program to restore baseline conditions centers on the use of hatchery supplementation to culture appropriate wild donor stock fish for introduction into the Panther Creek drainage. The program would use an appropriate donor stock, preferably within either the Middle Fork Salmon River drainage, North Fork Salmon River drainage, or Chamberlin Creek/Bargamin Creek. Although we considered the construction of an onsite hatchery, we concluded that the use of an existing facility would be as reliable and yet less expensive. Consequently, the program specified development of additional capacity (spawning, incubation, and rearing) at an existing hatchery (Sawtooth Hatchery, located on the upper Salmon River) to accommodate donor stock adults and individuals taken from adult returns to Panther Creek once a run has been initiated. The program included the construction of a weir and adult trapping facility on lower Panther Creek near the mouth of Clear Creek and two acclimation ponds, one at the Clear Creek location and the other adjacent to Moyer Creek (Figure 35.1).

The use of a hatchery as the initial reintroduction strategy was based on seeding the Panther Creek drainage with the maximum number of smolts that can be produced from a limited number of wild donor stock adults; we assumed 50 adult fish. Thus, for the first 5 years, all donor stock adults would be transported to and spawned at the hatchery. Eggs would be incubated and resulting fry and parr reared to pre-smolt size. The pre-smolts would then be transported via truck to the Panther Creek system and placed in the two acclimation ponds for grow-out and smolting; the ponds would be opened in the spring and smolts allowed to volitionally emigrate into Panther Creek. By the sixth year, adult salmon will be returning to Panther Creek, so reliance on the donor sources would be reduced or eliminated. This will depend largely on the overall survival rates of the donor fish back to Panther Creek. If sufficient numbers of adult salmon (50 pairs of adults) do not return to Panther Creek during the sixth year, the deficit would be made up from one or more of the identified donor sources, pending availability and a determination of "no-impact" to the donor stocks. Assuming sufficient returns to Panther Creek (e.g., 100 pairs), all adult fish would be trapped at the weir, with half of the adults transported to the hatchery for spawning and fry/parr rearing, and the remaining half allowed to migrate upstream to spawn naturally. The fish transported to the hatchery would be managed like donor stocks, with adult fish spawned and resulting fry and parr reared to pre-smolt size, then to smolt size in Panther Creek acclimation ponds. The process of trapping and transporting 50% of the adults to the Sawtooth Hatchery would continue until the number of returning adults reaches or exceeds the baseline condition. At that time, reliance on and use of the hatchery would cease, because a self-sustaining population of chinook salmon would be established.

To evaluate potential benefits of the measures included in the BRCP for restoration of chinook salmon into Panther Creek, we developed a spreadsheet-based computer life cycle model. The model estimates annual adult returns and smolt production for the Panther Creek drainage, based on assigned input parameters which collectively represent the prescribed restoration options contained in the BRCP, as well as various life stage survival rates. These included survival estimates of egg to smolts (hatchery production of 55%, based on studies reported by Howell et al. 1985; Scully and Petrosky 1991; Bowles and Leitzinger 1991), smolt to adult returns (SAR) for hatchery production (0.15%, based on studies of Bowles and Leitzinger 1991; Miller 1990), and SARs for natural production (0.55%, based on studies of Cramer and Neeley 1993; Scully et al. 1989). The model used a fecundity estimate of 4,800 eggs per female, computed as the average of estimates reported by a number or investigators including Petrosky and Holubetz (1988), Scully et al. (1989), and Kiefer and Lockhart (1993). The model tracks the adult returns to baseline, and the cumulative losses from 1957 (date from which interim losses are calculated, measured in adult-years) for estimating interim losses. In the model, the age structure of returning adults was estimated to include 20% 2-year ocean fish, 60% 3-year ocean fish, and 20% 4-year ocean fish. These estimates approximate but are slightly lower than those reported by Bjornn et al. (1964) and IDFG (1990) for both 2-year ocean (31%) and 3-year ocean (67%) fish for spring chinook salmon in the Middle Fork drainage. As noted above, differential survival rates for both the egg-to-smolt and smolt-to-adult return were considered for natural spawners compared to hatchery operations. We did not include differential rates to address inter-annual variability in smolt and ocean survival. A more detailed description of the model and input variables (and the justification for selecting specific model parameters), and results from specific model runs are provided in R2 (1995).

Assuming that the hatchery operations begin in 2005, implementation of the above program would result in an estimated return to baseline condition by the year 2021. At that time, use of the hatchery would cease (Figure 35.2). The model projects that the run of salmon would continue to build for an additional 5 years, reflecting the continuing effects of the initial five years of hatchery supplementation. Subsequently, the runs would trend downward to, and stabilize, near 200 adult fish.

PROGRAM FOR COMPENSATION OF INTERIM LOSSES

The compensation for interim losses was evaluated with the same population model used for evaluating baseline restoration. The measure of loss used for this analysis was a chinook salmon adult-year, defined as one returning chinook to the Salmon River drainage in one year. For example, assuming a baseline of 200 adult salmon, for the period 1975 to 1995 when no fish were present, the interim loss would be 20 years × 200 adults = 4,000 adult-years. A model run calculated the total numbers of Panther Creek adult salmon lost from the date when injuries to chinook salmon were first quantified (1957) to the time when baseline conditions are restored. In this calculation, we assumed different baseline conditions between 1957 and 1968, to reflect fewer dams in the Columbia and Snake rivers. The programs for compensating for interim losses were based on developing measures, both within and outside of the Panther Creek drainage, that would fulfill (based on life cycle model projections) the deficit by the year 2055; total deficit estimated at 47,333 adult-years (R2 1995). Measures outside the basin were included in the BRCP, since the model projected that in-basin measures would not be sufficient to completely compensate for interim losses within the specified period.

The measures within the Panther Creek basin included channel meander reconstruction, riparian corridor fencing, and construction of off-channel rearing ponds (Figure 35.1). These measures represented those judged most biologically beneficial and cost-effective for compensating interim losses directly within the Panther Creek drainage. We considered additional within-basin measures

Panther Creek
Biological Restoration

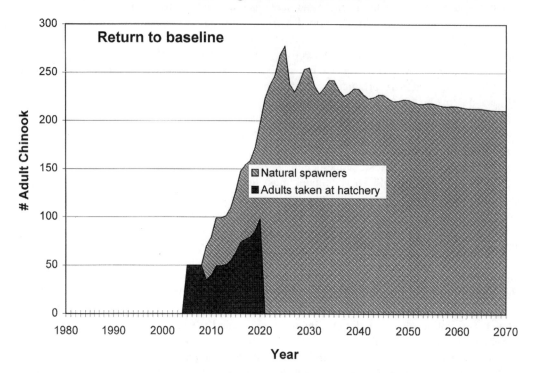

FIGURE 35.2 Number of returning adults to the Panther Creek, Idaho drainage projected from proposed plan for return to baseline. Baseline of 200 adults achieved in year 2021.

including spawning gravel cleaning, erosion control and bank stabilization, and barrier removal. However, the benefits achievable from implementing such measures were relatively low (based on habitat quality and quantity gains and resulting adult returns) compared to others obtainable through implementation of out-of-basin measures.

In-basin measures were not sufficient to completely compensate for interim losses. We therefore selected two basins outside of Panther Creek and within the upper Salmon River drainage in which to implement enhancement measures designed to compensate for further interim losses. These included the Lemhi River drainage and the East Fork Salmon River drainage. These basins were selected based on the type and extent of land-use impacts occurring in each, the degree to which such impacts may be limiting existing salmon production, and how readily the impacts could be reduced or eliminated via engineering measures. We focused all out-of-basin measures on eliminating/reducing the effects of livestock grazing. Riparian corridor fencing on private lands was identified as the measure that would provide the greatest benefits relative to costs. Benefits would include: (1) reduction in sediment loading (via increased bank stability) leading to improved quality of spawning gravel and higher egg-to-fry survivals; (2) provision of streamside cover in the form of overhanging vegetation; (3) reduced water temperatures due to increased shading; and (4) increased food availability both from increased production due to cleaner substrates and from terrestrial insects recruited from riparian vegetation. We subsequently estimated (based on IDFG smolt production estimates for both basins) potential increases in smolt production and adult returns that could be achieved through riparian corridor fencing within selected reaches of each system.

COSTS OF RESTORATION AND COMPENSATION PROGRAM

The total costs for restoring a self-sustaining run of chinook salmon into Panther Creek was estimated to be approximately $2,916,000 (net present value). This included total capital and operation and maintenance costs of $2,046,464, engineering costs of $196,769, and contingency costs of $673,000. These costs were developed assuming that the facilities would be constructed in the year 2005 and operated until the year 2025. The salmon life cycle model suggested that baseline levels will be achieved in 2021 but operation until the year 2025 was assumed to ensure return to baseline. The greatest capital cost is associated with the construction of a weir and adult trap on Panther Creek. Operations and maintenance costs are primarily associated with hatchery production activities. A more detailed description and discussion of costs associated with the BRCP is found in R2 (1995).

The total estimated cost for enhancement projects designed to compensate for interim losses was about $16,227,000. This included total capital and operation and maintenance costs of $11,390,000, engineering costs of $1,092,292, and contingency costs estimated at $3,745,000. The highest capital and operations and maintenance costs are associated with the off-site construction of 78 mi of riparian corridor fencing in the Lemhi and East Fork Salmon River basins. These costs were developed assuming that facilities within Panther Creek would be constructed in the year 2005 and facilities outside of Panther Creek would be constructed in the year 1996 and operated until the year 2055 when the interim loss is fully compensated. Biological monitoring and Trustee oversight were estimated as $1,260,520 and $3,041,356, respectively, and the combined total cost of the Restoration and Compensation Program is $22,445,307.

DISCUSSION

In May of 1995 a settlement was reached in the lawsuit between NOAA, the USFS, and the State of Idaho vs. the mining companies regarding damages to the natural resources affected by the Blackbird Mine. The settlement represented an important next step toward implementation of the BRCP and ultimate restoration of chinook salmon runs into the Panther Creek drainage.

It will be many years before the success of the Panther Creek program can be evaluated, with much being dependent on the overall success of the Snake River chinook salmon recovery program. That program is much broader in scope and attempts to address the myriad factors — both within Idaho and downstream (e.g., adult upstream passage and downstream passage of smolts through dams, water quality and quantity, etc.) — that presently impact Snake River stocks of chinook salmon. As long as those factors continue to impact existing Salmon River stocks, the effectiveness of the BRCP measures in restoring runs into Panther Creek will be more difficult to judge and evaluate.

It should also be noted that the proposed BRCP assumes that 50 adult chinook salmon can be obtained from suitable donor streams. Given the current depressed status of chinook salmon stocks in the Salmon River basin, availability of 50 adult donors coincident with implementation of the BRCP does not seem likely. Despite best efforts to select Panther Creek donor stocks based on selection criteria, the biggest obstacle may simply be availability and abundance of any chinook stocks, wild or hatchery, within the State of Idaho. At the extreme end in which there would be no donor stock available (i.e., natural recovery only), the life cycle model predicted it would take over 150 years to achieve the baseline condition of 200 returning adult chinook salmon. This contrasts with the predicted return to baseline of 16 years (from the time of program implementation) based on the use of 50 adults from donor streams. Thus, the timelines predicted for restoring "baseline" runs of salmon to the system will vary (between 16 to 150 years) depending upon the sources and actual numbers of adult salmon that can be obtained for the program. However, we believe that the program described herein provides the best opportunity for the successful restoration of chinook salmon into Panther Creek, and we ascribe to it a high degree of probable success, especially if the major downstream impacts in the lower Snake and Columbia rivers are reduced or eliminated.

This case study demonstrates a process that can be used to identify, evaluate, and select appropriate restoration options for salmon and steelhead stocks in any watershed where such runs have been depleted or extirpated. The general steps completed and considerations addressed include many of those likely necessary in other drainages where restoration of depleted or extirpated stocks of salmon and/or steelhead stocks is desired. In those cases, the BRCP process presented herein may provide guidance in developing successful and economically feasible reintroduction and restoration programs.

ACKNOWLEDGMENTS

The authors wish to acknowledge the time and commitment of a number of individuals who contributed substantially to this work. Nick Iadanza (NMFS) provided both solid technical advice as well as project management oversight. James Templeton, Stuart Beck, Glen Anderson, Randall Jeric, and Karee Oliver were involved in the development and evaluation of various restoration options. Jody Breckenridge and Loyce Panos were responsible for report and manuscript production. The authors also wish to acknowlege the support and input provided by the technical staffs from the Idaho Department of Fish and Game, U.S. Forest Service, and the U.S. Fish and Wildlife Service. Thoughtful and constructive reviews of an earlier manuscript were provided by Peter Hassemer, David Hanson, and one anonymous reviewer, to whom we extend our appreciation. Funding for this project was provided by the National Oceanic and Atmospheric Administration (NOAA) under a contract with Industrial Economics Inc. pursuant to Contract No. 50-DGNC-1-00007.

REFERENCES

Bevan, D., and six coauthors. 1994. Snake River Salmon Recovery Team: Final recommendations to the National Marine Fisheries Service.

Bjornn, T.C., D. Ortmann, D. Corley, and W. Platts. 1964. The gathering and compilation of relevant data for Idaho salmon and steelhead runs. Job No. 3, Salmon and Steelhead Investigations—1963. Idaho Department of Fish and Game, Boise.

Bjornn, T.C. 1978. Survival, production, and yield of trout and chinook salmon in the Lemhi River, Idaho. University of Idaho, College of Forestry, Wildlife and Range Sciences Bulletin 27, Moscow, Idaho.

Bowles, E., and E. Leitzinger. 1991. Salmon supplementation studies in Idaho rivers (Idaho supplementation studies), experimental design. Bonneville Power Administration, Division of Fish and Wildlife, Project No. 89-098. Portland, Oregon.

BPA (Bonneville Power Administration). 1992. Supplementation in the Columbia basin. Summary Report Series, Final Report. Division of Fish and Wildlife, Project No. 85-62, Portland, Oregon.

Chapman, D., and 10 coauthors. 1991. Status of Snake River chinook salmon. Report prepared for the Pacific Northwest Utilities Conference Committee. Don Chapman Consultants, Inc., Boise, Idaho.

Cramer, S.P., and D. Neeley. 1993. Evaluation of delisting criteria and rebuilding schedules for Snake River spring/summer chinook, fall chinook, and sockeye salmon: recovery issues for threatened and endangered Snake River salmon. Technical Report 10. Bonneville Power Administration, Division of Fish and Wildlife, Project No. 93-013, Portland, Oregon.

Dauble, D.D., and R.P. Mueller. 1993. Factors affecting the survival of upstream migrant adult salmonids in the Columbia River basin: recovery issues for threatened and endangered Snake River salmon. Technical Report 9. Bonneville Power Administration, Division of Fish and Wildlife, Project No. 93-026, Portland, Oregon.

Howell, P., K. Jones, D. Scarnecchia, L. LaVoy, W. Kendra, and D. Ortmann. 1985. Stock assessment of Columbia River anadromous salmonids. Volume I: chinook, coho, chum and sockeye salmon stock summaries. Final Report. Bonneville Power Administration, Division of Fish and Wildlife, Project No. 83-335, Portland, Oregon.

IDFG (Idaho Department of Fish and Game). 1990. Salmon River Subbasin, Salmon and Steelhead Production Plan. Columbia Basin System Planning. Northwest Power Planning Council, Portland, Oregon.

IDFG (Idaho Department of Fish and Game). 1991. Fisheries management plan 1991–1995. Idaho Department of Fish and Game, Boise.

IDFG (Idaho Department of Fish and Game). 1992. Anadromous fish management plan 1992–1996. Idaho Department of Fish and Game, Boise.

Kiefer, R.B., and J.N. Lockhart. 1993. Idaho habitat and natural production monitoring: part II. Annual Report 1992. Bonneville Power Administration, Division of Fish and Wildlife, Project No. 91-73, Portland, Oregon.

Lichatowich, J., and B. Watson. 1993. Use of artificial propagation and supplementation for rebuilding salmon stocks listed under the Endangered Species Act: recovery issues for threatened and endangered Snake River salmon. Technical Report 5. Bonneville Power Administration, Division of Fish and Wildlife, Project No. 93-013, Portland, Oregon.

Miller, W.H. 1990. Analysis of salmon and steelhead supplementation. Bonneville Power Administration, Division of Fish and Wildlife, Project No. 88-100, Portland, Oregon.

Petrosky, C.E., and T.B. Holubetz. 1988. Idaho habitat evaluation. Annual Report 1987. Bonneville Power Administration, Division of Fish and Wildlife, Project No. 83-7, Portland, Oregon.

Platts, W. S., S. B. Martin, and Edward R. J. Primbs. 1979. Water quality in an Idaho stream degraded by acid mine waters. U.S. Forest Service, General Technical Report INT-67, Intermountain Forest and Range Experimental Station, Ogden, Utah.

R2 (R2 Resource Consultants, Inc. and Industrial Economics, Inc.). 1995. Panther Creek biological restoration and compensation program. Report prepared for the National Oceanic and Atmospheric Administration (NOAA), Department of Commerce, Washington, D.C.

RCG/Hagler Bailly. 1994. Blackbird mine injury studies, draft documentation of aquatic habitat sections. Boulder, Colorado.

Reiser, D.W. 1986. Panther Creek, Idaho, habitat rehabilitation. Final Report. Bonneville Power Administration, Division of Fish and Wildlife, Project No. 84-29, Portland, Oregon.

Reiser, D.W., M.P. Ramey, P. Cernera, and C. Richards. 1994. Conversion of remnant dredge mine ponds into chinook salmon rearing habitat: from feasibility to construction. Pages 208–225, *in* I.G. Cowx, editor. Rehabilitation of freshwater fisheries. Fishing News Books, Oxford, United Kingdom.

Richards, C., P. Cernera, M. Ramey, and D. W. Reiser. 1992. Development of off-channel habitats for use by juvenile chinook salmon. North American Journal of Fisheries Management. 12:721–727.

RMC (Rocky Mountain Consultants, Inc. and Industrial Economics, Inc.). 1995. Physical restoration analysis, Blackbird Mine, Lemhi County, Idaho. Final Draft Report. Prepared for National Oceanic and Atmospheric Administration (NOAA), Department of Commerce, Washington, D.C.

Schreck, C.B., H.W. Li, R.C. Hjort, and C.S. Sharpe. 1986. Stock identification of Columbia River chinook salmon and steelhead trout. Final Report. Bonneville Power Administration, Division of Fish and Wildlife, Project No. 83-451, Portland, Oregon.

Scully, R.J., E.J. Leitzinger, and C.E. Petrosky. 1989. Idaho habitat evaluation for off-site mitigation record. part I, subproject I, Annual Report 1988. Bonneville Power Administration, Division of Fish and Wildlife, Project No. 83-7, Portland, Oregon.

Scully, R.J., and C.E. Petrosky. 1991. Idaho habitat/natural production monitoring, part I, general monitoring subproject. Annual Report 1989. Bonneville Power Administration, Division of Fish and Wildlife, Project No. 83-7, Portland, Oregon.

Smith, B.H., and G.A. Seaberg. 1991. Panther Creek fisheries habitat improvement project. Cobalt Ranger District, Salmon National Forest. Lemhi County, Idaho.

Smith, E.M., B.A. Miller, J.D. Rodgers, and M.A. Buckman. 1985. Outplanting anadromous salmonids—a literature survey. Bonneville Power Administration, U.S. Department of Energy, Project 85-68, Portland, Oregon.

Steward, C.R., and T.C. Bjornn. 1990. Supplementation of salmon and steelhead stocks with hatchery fish: a synthesis of published literature. Bonneville Power Administration, Division of Fish and Wildlife, Project No. 88-100, Portland, Oregon.

Waples, R.S., D.J. Teel, and P.B. Aebersold. 1991. A genetic monitoring and evaluation program for supplemented populations of salmon and steelhead in the Snake River basin. Annual Report of Research. Bonneville Power Administration, Project No. 89-096, Portland, Oregon.

Welsh, T.L., S.V. Gebhards, H.E. Metsker, and R.V. Corning. 1965. Inventory of Idaho streams containing anadromous fish including recommendations for improving production of salmon and steelhead. Part I: Snake, Salmon, Weiser, Payette, and Boise River Drainages. Idaho Department of Fish and Game, Boise, Idaho.

36 Effectiveness of Current Anadromous Fish Habitat Protection Procedures for the Tongass National Forest, Alaska

Calvin H. Casipit, Jeff Kershner, Tamra Faris, Steve Kessler, Steve Paustian, Lana Shea, Max Copenhagen, Mason Bryant, and Richard Aho

Abstract.—This chapter is a condensed version of a report that was prepared for the U.S. Congress, which is available from the primary author. We studied the effectiveness of current procedures for protecting fish habitat in the Tongass National Forest. The resulting information was used to determine whether any additional measures are needed to protect anadromous fish populations. The multidisciplinary study team consisted of fishery biologists and hydrologists from the Forest Service management and research branches, a fishery biologist from the National Marine Fisheries Service, and a habitat biologist from the Alaska Department of Fish and Game. The results of this study indicate that, while current procedures have clearly improved the treatment of anadromous fish streams and provided improved protection for stream habitat compared to previous procedures, the existing procedures are not entirely effective in protecting fish habitat or in precluding increased risk to some anadromous fish stocks over the long term. It was also concluded that additional protection for fish habitat is needed to reduce risk to fish habitat quality in the Tongass National Forest. An ecosystem approach is recommended for evaluating and protecting watershed processes and functions at the landscape scale as a precursor to timber sales and other management activities that could significantly influence fish habitat. A broadened perspective for conservation of fresh-water salmon habitat is also needed to provide for sustainable salmon fisheries.

INTRODUCTION

In many coastal Pacific areas, naturally reproducing stocks of Pacific salmon *Oncorhynchus* spp., steelhead *O. mykiss*, and sea-run cutthroat trout *O. clarki* are at risk of extinction. Of the more than 400 stocks from California, Idaho, Oregon, and Washington recently evaluated, 214 were considered to be at "moderate" or "high" risk of extinction or of "special concern," 106 were extinct, and only about 120 were considered secure (Nehlsen et al. 1991). The results of a similar study in British Columbia revealed that 702 stocks were at moderate or high risk of extinction, 230 were of special concern, and 142 were extinct (Slaney et al. 1996). While a statewide assessment has not been conducted in Alaska, recent information suggests that some salmon stocks in this region may be declining (Halupka et al. 1994). In southeastern Alaska, sufficient information was available on 907 stocks of salmon and steelhead to support escapement trend analysis (Baker et al. 1996). Of these, 39 stocks showed declining trends.

Historically, Alaska salmon harvests have fluctuated over a 20- to 30-year cycle (Marshall 1992). During the mid- to late 1980s, favorable oceanic habitat conditions have contributed to excellent salmon survival and growth (Beamish and Bouillon 1993). For example, survival from smolts to returning adults has averaged 5% over that period. In recent years the survival of coho salmon has been much higher. In one location, Deer Lake on South Baranof Island, smolt to adult survival has been as high as 24% and averaged 15% from 1990 to 1997 (S. Riefensthul, Northern Southeast Regional Aquaculture Association, personal communication). Although salmon populations have been increasing, concerns regarding populations of steelhead and cutthroat trout have resulted in increased harvest restrictions on those species to prevent further population declines and recover declining populations (ADF&G 1994).

The U.S. Forest Service (USFS) has an important role to play in managing watersheds and fish habitat in coastal Alaska. The lands administered by the USFS contain more than 80% of the freshwater anadromous fish spawning and rearing habitat in southeast Alaska. The Nehlsen et al. (1991) report, coupled with the Endangered Species Act (ESA) listing of the Snake River sockeye salmon *O. nerka* and fall chinook salmon *O. tshawytscha* as endangered and the Snake River spring/summer chinook salmon as threatened, compelled the USFS to develop a proposed strategy for managing Pacific anadromous fish and their habitat (PACFISH). The goal of the strategy was to provide habitat conditions that contribute to the recovery and sustained natural reproduction of Pacific anadromous fish stocks within National Forests. The draft Environmental Analysis for PACFISH concluded that applying interim direction to Alaska was unnecessary at that time because, in general, the potential for salmon stock extinction in the near future was small in Alaska compared to the lower 48 states (USDA and USDI 1994).

The FY 1994 Appropriations Act for Interior and Related Agencies included specific direction that prohibited implementing the PACFISH strategy in the Tongass National Forest. The Act also directed the USFS to "proceed with studies and review procedures related to the PACFISH strategy, to assess the effectiveness of current procedures to provide protection for salmon and steelhead habitat, and to determine if any additional protection is needed." The Alaska Region and the Pacific Northwest Research Station of the USFS were asked to complete the requirements of the 1994 Appropriations Act. Stated more simply, the Appropriations Act implied two questions:

Are current procedures for protecting salmon and steelhead habitat effective?

Is additional protection needed?

This chapter summarizes the results of a study that was conducted by an interdisciplinary study team to answer these two questions. The team consisted of fishery biologists and hydrologists from the USFS management and research branches, a fishery biologist from the National Marine Fisheries Service, and a habitat biologist from the Alaska Department of Fish and Game. More detailed information on this study is presented in a report to the U.S. Congress (USDA 1995).

METHODS

The Fish Habitat Analysis Team (the study team) was chartered by the Alaska Region and Pacific Northwest Research Station to answer two Appropriations Act questions, as identified above. While it was anticipated that ongoing USFS monitoring and evaluation would produce answers to these questions over time, the current practices have not been in effect or been monitored long enough to provide them in the near-term. For this reason, the study team was asked to utilize existing information to answer the two questions and to support the development of conclusions and recommendations.

Direction for land management in the Tongass National Forest results from many sources of law, regulation, and policy. The Tongass Land Management Plan, developed in 1979, was significantly amended in 1986 and again in 1991 to respond to the requirements of the Tongass Timber Reform Act. Among the requirements is a 100-ft minimum, no-commercial-timber-harvest buffer along salmon streams (class I) and resident fish-bearing streams that flow directly into anadromous fish streams (class II). For the purposes of addressing these questions, the study team interpreted "current procedures" as those habitat protection measures that have been in use since the passage of the Tongass Timber Reform Act. Therefore, only the procedures and resultant practices that have been implemented on the ground between 1991 and 1994 were evaluated. We understood that practices and procedures that were used in 1994 might be different from those used in 1991; however, we had to have a sufficiently long period of implementation of procedures to evaluate as "current."

Providing definitive answers in the time allotted required us to create a process that integrated the professional knowledge and experience of the study team members in three concurrent studies, including:

- A literature review of existing knowledge;
- An expert review of selected watersheds where management activities were initiated after passage of the Tongass Timber Reform Act; and
- The reports of area and regional Watershed Analysis Teams.

These elements of our evaluation of current procedures for protecting anadromous fish habitat in the Tongass National Forest are described below.

LITERATURE REVIEW

The existing knowledge on anadromous fish habitats and related topics was acquired by conducting a review of the available literature. This review drew upon published and unpublished papers on fish habitat and management effects from Alaska, the coastal Pacific Northwest, and British Columbia. Unpublished reports from the Tongass National Forest's District and Supervisor Offices were also collected and compiled during this review. This information acquisition effort resulted in the compilation of 1,542 citations into a computerized literature reference database. Documents not readily available were copied, then filed at the library in the Forestry Sciences Laboratory in Juneau.

EXPERT FIELD REVIEW

We used experts in a structured analysis process to assess the effects on fish habitat resulting from current timber management procedures. This expert analysis also assessed the potential effects of additional timber management activities that are possible under current procedures. The study was designed following a modification of the Delphi technique (Linstone and Turoff 1975; Martino 1983).

The group of experts assembled consisted of six individuals with expertise in fish biology, hydrology, or watershed processes. We designed the study to allow individual experts to observe the procedures that were being used to protect fish habitat and to provide an assessment of their effectiveness based on their professional experience. The goal was to gather a range of anonymous opinions, not to reach consensus on any of the findings. The rationale for this structure was that valid reasons can exist for differences in professional opinions and ratings, and we did not wish to lose any valuable insights.

Maps and other factual information were provided to the experts. Map information included contour lines, stream classes and channel types, forested land considered tentatively suitable for timber harvest, wetlands, unstable slopes, very unstable slopes, specified roads, and harvest units (both pre- and post-Reform Act). Other information usually available included unit and road-layout cards, environmental impact statements, and personal interaction with local managers. An overflight of each watershed was completed before the field inspection was conducted on the basin.

A total of seven watersheds were evaluated in the field. These watersheds were classified into two categories, including:

- watersheds harvested following general procedures implemented since the November 1990 passage of the Tongass Timber Reform Act (i.e., requiring a minimum 100-foot, unharvested buffer on both sides of class I streams and class II streams that flow directly into class I streams); and
- watersheds studied with watershed analysis described later in this chapter.

The watersheds were identified and evaluated so as to avoid conclusions about management effects on fish habitat associated with pre-Reform Act timber harvest practices. These watersheds are distributed across the major timber-producing areas of the forest and represented the range of conditions found in southeast Alaska for many watershed attributes, such as topography and fish production. The experts also visited undisturbed areas before assessing the watersheds to calibrate their thoughts to the natural condition.

A form was designed by consensus for use during the field review. The forms requested the experts to rate risk to fish habitat (1 to 5, from low to high) for several different habitat variables (i.e., upslope processes, hydrology, and habitat), based on their observations. The ratings provide a basis for estimating the probability of observing detectable adverse change in fish habitat quality. Experts were asked to rate only the categories for which they felt qualified. The experts also commented on probable causes of risk to fish habitat, and made recommendations for monitoring, rehabilitation, and stream buffer design. Both current risk and anticipated future risk (i.e., 30 and 100 years hence) resulting from the cumulative effects of multiple timber harvests with similar habitat protection measures were evaluated.

WATERSHED ANALYSES

Watershed analysis is a systematic procedure for developing an understanding of physical and biological processes that occur within a watershed. This type of analysis is conducted by measuring key land and water attributes and describing the geomorphic and fluvial processes.

An Area Watershed Analysis Team (i.e., Area Team) was formed for each Tongass National Forest Administrative Area to conduct the watershed analyses. An overview team (i.e., Regional Team) with interagency representation was also formed to facilitate inter-team coordination and technology transfer. The Regional Team adapted the procedures outlined in A Federal Agency Guide for Pilot Watershed Analysis (Anonymous 1994) to meet our needs for responding to the congressional study request. This analysis covered fish-related resources but did not include wildlife or social considerations.

Several criteria were established to govern the selection of watersheds for conducting this analysis, including that they must

- be covered by a post-Reform Act Record of Decision or a pre-Reform Act Record of Decision modified to meet the requirements of the Act;
- have a substantial amount of class I stream habitat; and
- be more than 10 square miles in area.

Forty-eight watersheds were considered for this analysis. Application of the selection criteria resulted in the identification of three watersheds that would be appropriate for conducting a comprehensive watershed analysis, including the Game Creek, Kadake Creek, and Old Franks Creek watersheds.

The Game Creek (Hoonah Ranger District, Chatham Area) watershed is about 48 square miles, with the lower 12 square miles in private ownership. Approximately 1,545 acres (5%) of the watershed had been harvested at the time that the analysis was conducted, with timber sales having occurred on 29 harvest units on National Forest lands. Five of these units were harvested before the Reform Act and 24 were harvested in 1993 after passage of the Reform Act. Minimal harvest has been done on the private lands in the watershed.

The Kadake Creek (Petersburg Ranger District, Stikine Area) watershed has a drainage area of about 50 square miles. Roughly 4,700 acres (15%) of the watershed had been harvested when this study was initiated. Twelve post-Reform Act units have recently been authorized for timber harvest under the North and East Kuiu Final Environmental Impact Statement and Record of Decision. A portion of these units has been harvested and the remainder may be available as independent timber sales in the future. About 50 harvest units and their associated roads have been developed under pre-Reform Act Records of Decision.

The Old Franks Creek (Craig Ranger District, Ketchikan Area) watershed is about 25 square miles in area, with the lower 9 square miles (below the lakes) in mixed ownership. About 540 acres (6%) of National Forest lands in the upper watershed had been harvested when this analysis was undertaken. Eight post-Reform Act units were harvested in 1992 and 1993, in accordance with the Record of Decision for the 1989–1994 long-term sale. Thirteen more units in the upper watershed are proposed for harvest in Alternative 5 of the Polk Inlet Environmental Impact Statement.

The watershed analyses were intended to provide a "snapshot" of conditions after harvest of relatively small portions of the watersheds that included a range of completed timber harvest intensities and geoclimatic conditions typical of the Tongass National Forest. However, they did not assess the potential for cumulative effects. Without baseline data, cause and effect relations are difficult to establish. For this reason, the expert field review was used to help address possible long-term changes in these watersheds. The riparian habitat conservation strategy (A Federal Agency Guide for Pilot Watershed Analysis; Anonymous 1994) identifies riparian management areas, provides guidance for planning activities within these areas, describes restoration and monitoring needs, and identifies information needed for project analysis. The focus of these analyses was primarily to delineate Riparian Management Areas and describe their functions and sensitivities.

The analyses used an existing database stored in a geographic information system that was adjusted by field review to validate and update existing inventories. The data layers used in these analyses included digital elevation models, geology, soils, harvest units, streams, roads, and vegetation type.

Fish habitat data on each watershed were compared to the interim Tongass-wide Fish Habitat Objectives. These interim objectives were viewed primarily as benchmarks to use in assessing fish habitat conditions in a given stream segment. Three sets of habitat objectives were defined and used for the pilot watershed analyses — pieces of large woody debris within a 1,000 m^2 channel area, percentage of pool area, and channel width-to-depth ratio. A specific set of habitat objectives was developed for some channels and groups of similar channel types, called stream process groups.

Several factors were considered in comparing the fish habitat objectives with habitat measurements. Some of the benchmark data sets from pristine watersheds are small and were not always representative. Pool area estimates have a relatively high sampling error. Width-to-depth indices were given the most weight, and percentages of pool area were given the least weight, in assessing fish habitat condition. A qualitative rating of habitat condition (ranging from poor to excellent) was assigned to each habitat objective by comparing measured values from each watershed (compiled by channel type or process group) to 25th, 50th, and 75th percentile values for large woody debris frequency, percentage of pool area, and width-to-depth ratios measured in pristine watersheds.

RESULTS

LITERATURE REVIEW

No studies were found in the literature which provided information that could be used to directly assess the effectiveness of Best Management Practices (BMPs) or buffers as applied on the Tongass National Forest (USDA 1992, 1993). However, the results of studies conducted in landscapes similar to Southeast Alaska showed declines in salmonid habitat capability after timber harvest of more than 25% of the watershed (Reeves et al. 1993). The results of these studies provide important information for evaluating the potential effects of timber harvesting activities on anadromous fish habitat in southeast Alaska for several reasons.

First, anadromous fish have existed for centuries in Southeast Alaska with episodic natural disturbances such as floods, large-scale windthrow, and landslides. These disturbances are generally short-term, infrequent, and primarily limited to small areas within watersheds. Although disturbances may affect anadromous fish in short reaches, high-quality refuges near disturbed areas ensure that source populations can redistribute fish to them as habitat recovers (Sedell et al. 1990, Reeves et al. 1993). Populations subjected to disturbance are generally resilient and recover in time. However, when human-caused disturbances are added to natural disturbances, both the frequency and magnitude of the effects on anadromous fish can be affected (Reeves et al. 1993, Reid 1993).

Managed forests typically have harvest units and roads throughout a watershed. Harvest units are generally managed to grow and harvest trees every 100 years on suitable soils. Timber stands may be thinned twice before final harvest. When combined with natural disturbances, the effects from human-caused disturbances may be greater and may also affect an entire watershed. Frequent disturbances may limit habitat recovery, reduce the availability of key refuges in streams, and limit the ability of local populations to recover (Sedell et al. 1990). This problem is particularly critical to small populations or stocks where minor changes may pose a risk of extinction (Rieman and McIntyre 1993). Small populations and stocks likely occur in many small island streams in southeast Alaska, which makes this area particularly vulnerable to forest management activities.

Expert Field Review

The results of the expert analysis indicate that the procedures currently used in Tongass National Forest have increased the level of protection afforded to anadromous fish habitats (Table 36.1). Of the seven watersheds visited by the expert team, two were rated as having low risk of a detectable adverse change in fish habitat and watershed condition, five were rated as having moderately low risk, and one was rated as having moderate risk. Experts' ratings for future risk showed greater consistency than those for current risk. All of the watersheds examined were rated as having either moderate or moderately high risk to fish habitat in 30 or 100 years if timber harvest and road building continue in the basin. In contrast, all watersheds were rated as having low or moderately low risk to fish habitat in 30 years with no additional timber harvest and roads. All of the watersheds were rated as having low risk to fish habitat in 100 years with no further timber harvest and roads.

Five of the six experts were available for closeout interviews at the conclusion of the field review. At the interview, each was asked the question, "Are current management procedures on the Tongass National Forest effective in protecting fish habitat?" They all indicated that additional protection measures are necessary to fully protect fish habitat. The unanimity of this opinion was also evident in their statements and ratings for each of the individual watersheds. Concerns expressed by one or more experts are listed below to explain the reasons for their evaluations of effectiveness. Although the good practices observed by the experts were discussed, this listing is not comprehensive because the experts were not specifically asked to list good practices; rather, they were asked to list their concerns.

Class I and II Stream Buffers.—Stream class I and II buffers were sometimes observed to be too narrow for optimum stream protection. The experts stated that many buffers on class I and II streams need to exceed the minimum 100-ft requirement, particularly on large flood-plain channel types where the streams meander over time and often cover distances exceeding 100 ft from the main channel. Buffer design must match the site topography and remain windfirm, or buffer function will be reduced through changes in shading for microclimate control, nutrient input, and long-term recruitment of large woody debris. Significant windthrow was observed in many areas harvested during the last 1 to 3 years. In contrast to the observations of less than adequate buffers, many other buffers were found to be sufficient to protect the interactions between the streams and adjacent

---

TABLE 36.1

Experts' summary ratings of the increased risk to fish habitat over natural background risk. This table shows ratings for average risk to fish habitat given by the experts for each of the watersheds. It represents the probability of a detectable adverse change in fish habitat quality.

Watershed	Now	In 30 Years With	In 30 Years Without	In 100 Years With	In 100 Years Without
Old Franks	2	3	2	3	1
Upper Thorne	1	3	1	3	1
Kadake Creek	1	3	1	3	1
Frosty Creek	2	3	2	3	1
Whale/Buckhorn	2	4	2	4	1
Game Creek	2	3	2	4	1
Seagull Creek	3	4	2	4	1

A scale of 1 to 5 was used, with the following risk ratings: 1 = low (< 20%); 2 = moderately low (21 to 40%); 3 = moderate (41 to 60%); 4 = moderately high (61 to 80%); 5 = high (>81%). Now refers to current risk and risk in the near future. The with and without refer to the experts' evaluation of the potential future risk with additional roading or logging and without additional roading and logging under current post-Reform Act management practices, given the identification of approximate areas considered tentatively suitable for commercial timber harvest.

---

terrestrial systems. This observation was particularly true at those sites where field personnel were given the most flexibility by managers in designing buffers.

Class III Stream Buffers.—The experts believed that more no-harvest buffers were needed on class III streams and some unclassified channels to prevent adverse effects on fish habitat. Impacts are largely associated with the reduction in the supply of large woody debris important for the entrapment of sediment, slowing movement of sediment through stream systems, and as a source of large wood and nutrient input to downstream fish habitat. One expert stated: "Because class III's are open conduits to downstream fish habitat, sediment will flush down to fish habitat during big events and affect fish habitat. It may take a long time to get there, but it will get there." The experts were unanimous in the need for more buffers on class III streams, but no consistent recommendations were made on the design of the buffers (such as exactly how wide the buffers need to be or the specific circumstances where they are needed).

Activities on Unstable Slopes.—Harvesting timber on unstable slopes with soils rated high or very high for mass movement will generally accelerate mass wasting and transport sediment to fish streams, primarily via the small intermittent channels. Risk to fish habitat is from aggradation in streambeds and channel widening. Surface erosion processes seemed well protected through harvest techniques and best management practices, but shallow mass wasting is the dominant process likely to deliver significant amounts of sediment to downstream channels. Loss of root strength after clearcut harvest and increased runoff from reduced evapotranspiration are the factors primarily contributing to anticipated loss of slope stability and increases in mass wasting. Concern was most often expressed about activities on the steep slopes observed on the Chatham Area of the Tongass National Forest.

Road Location, Design, and Management.—Mid-slope roads cause changes in channel morphology and substrate through chronic and acute (such as landslides) introduction of road material from cuts and fills to stream systems. Location and design of roads through wetlands has resulted in roads intercepting and re-routing wetland runoff. This re-routing causes concentration of water in road ditches and may affect wetland water tables and increase peak flows. Failure to implement some of the best management practices associated with road drainage design and maintenance was identified as an important factor that increased risk to fish habitats.

Problems with road culverts that adversely affect the productivity and accessibility of fish habitat were also identified. These included undersized culverts installed in ways that did not assure fish passage, use of round culverts which often impede fish passage, inadequate numbers of road drainage structures, and the accumulation of materials above culverts that can lead to the failure of crossing structures and road beds during large runoff events. Many times, the experts mentioned inadequate maintenance of road drainage systems and inadequate closeout of temporary roads as potential mechanisms for road failure and delivery of sediment to fish habitat.

Variability in Implementing Guidance.—Several experts expressed concern over the apparent variability in interpreting guidance by different staff implementing timber-sale and road building projects. Included in this concern were problems associated with recognizing and classifying stream channels for buffer protection and determining where to start buffer measurements. Most districts had a process to incorporate new information, such as data on previously uninventoried streams, and placed them on harvest-unit cards; however, this practice was not followed by every district. Inaccurate inventory information and untimely updating of the corporate database with information gathered during harvest layout were identified as problems.

WATERSHED ANALYSES

The watershed analyses provided an assessment of existing watershed conditions in three watersheds in the Tongass National Forest. These analyses included evaluations of instream habitat conditions, headwater conditions, and riparian management area conditions.

Instream Condition Assessment

Fish habitat objectives have been established to identify the conditions needed to support anadromous fish in the Tongass National Forest. Specifically, habitat standards have been established for several indicators of fish habitat quality, including large woody debris and pool area. The results of the watershed analyses indicated that habitat conditions were generally rated as average or above average for the three watersheds. With the possible exception of Kadake Creek, the stream survey data suggest that current fish habitat is in good condition in the three watersheds and that the habitat has not been degraded by current management practices.

Game Creek watershed has very large amounts of woody debris and favorable width-to-depth values for all channel types surveyed (Tables 36.2 and 36.3). Three Game Creek channel types have average or above average pool area while three channel types have below average pool area. However, stream survey data indicate a relatively high pool frequency for most stream segments in Game Creek.

Pool habitat on Kadake Creek was consistently above average, but large woody debris was generally below average (Table 36.2). Width-to-depth ratios for flood plain channel types on Kadake Creek (Table 36.3) were well below average. We speculate that high width-to-depth values and low large woody debris accumulation in the main stem and major valley tributaries of Kadake Creek may be attributed to recent large floods. We cannot determine whether past or current management activities may have exacerbated the effects of these floods because of limited pre-harvest data.

Pool habitat on Old Franks Creek was above average on all stream segments (Table 36.2). Large woody debris on Old Franks Creek was slightly below average for three channel types or process-group segments, but abundant in one channel type (Table 36.2). Width-to-depth ratios for Old Franks Creek were below average habitat, indicating potential sensitivity to sediment loading (Table 36.3).

Windthrow, a natural process on the Tongass National Forest due to frequent gales, is a major cause of forest disturbance in both Game Creek and Kadake Creek watersheds. Some of the high woody debris loading in streams was due to buffer windthrow. The amount of large woody debris

TABLE 36.2
Comparison between watershed-analysis habitat measurements and the interim Tongass fish habitat standards for pools and large woody debris.

Fish Habitat Objective	Process Group/ Channel Type[a]	Interim Tongass Habitat Standard (percentiles)[b]			Measured Habitat Values[c]		
		25th	50th	75th	Game Creek	Kadake Creek	Old Franks
Large Woody	FP3	10	32	54	96 (1.3)	**30** (1.4)	
Debris,	FP4	8	24	34	46 (1.5)	**10** (3.3)	**12** (1.3)
pieces/1000m²	FP5	4	5	6	21 (2.8)	**2** (7.7)	18 (2.1)
	LC and MC	6	15	22	20 (0.4)	**11** (2.4)	**8** (0.5)
	MM1	27	45	82	96 (0.4)	110 (2.0)	
	MM2	33	35	44	42 (1.7)	**26** (4.2)	
Pool Area %	FP3	20	53	76	**45** (1.3)	75 (1.4)	
	FP4	35	47	59	50 (1.5)	69 (3.3)	71 (1.3)
	FP5	47	51	60	**37** (2.8)	69 (7.7)	58 (2.1)
	LC	8	20	27	39 (0.4)	64 (1.2)	62 (0.3)
	MC	11	22	39		29 (1.1)	22 (0.2)
	MM1	28	40	52	**18** (0.4)	56 (2.0)	
	MM2	2	22	39	33 (1.7)	39 (4.2)	

[a] Category, and numeric codes define distinct channel types within each group. Channel types listed include: FP3, small flood plain; FP4, medium flood plain; FP5, large flood plain; LC, large contained; MC, medium contained; MM1, moderate-gradient, mixed-control; and MM2, medium moderate-gradient unconstrained.

[b] Fish habitat objectives are expressed as a range of values based on the 25th, 50th (median), and 75th percentiles for a given channel type or process group. Key for interpreting interim Tongass pool and large woody debris fish habitat standards: <25th, poor habitat; >25th and <50th, below average habitat; >50th and <75th, above average habitat; >75th, excellent habitat.

[c] Values are watershed averages for all surveyed stream process groups or channel-type stream segments. Numbers in parentheses represent surveyed stream length in miles. Blanks represent channel types not sampled. Habitat values in bold represent below average or poor habitat condition based on interim standards.

in some segments of the Game Creek flood plain is well beyond the 75th percentile that was calculated using data from all stream reaches surveyed. The edges of some harvest units on Old Franks Creek have also had some windthrow.

Criteria have been established for upstream passage of juvenile salmon and steelhead in the Tongass National Forest. Conditions suitable for upstream passage of juvenile salmon, steelhead, and trout were generally assumed to also permit upstream passage of adult fish. No major obstructions to fish passage resulting from management activities were noted in any of the three watersheds. However, minor obstructions were identified in the Kadake Creek watershed where two culverts on pre-Reform Act road segments did not meet the passage criteria.

Water quality was indirectly assessed in two watersheds by collecting and analyzing the insect communities (macroinvertebrates) in the stream gravel. Most streams with reasonable water quality and substrate conditions support a diverse population of aquatic insects and other macroinvertebrates. Macroinvertebrates are generally accepted as useful indicators of water quality. From our macroinvertebrate analyses, no indication was found that water quality in these watersheds was impaired. Trends were unknown because no prior measurements were available.

Headwater Condition Assessment

Landslides are a common and natural source of sediment in southeast Alaska watersheds. Relatively few large landslides were observed in either Game Creek or Old Franks Creek watersheds. In Game

TABLE 36.3
Comparison between watershed-analysis habitat values and the interim Tongass fish habitat standards for channel width-to-depth ratio.

Process Group/ Channel Type[a]	Interim Tongass Habitat Standard (percentiles)[b]			Measured Habitat Values[c]		
	25th	50th	75th	Game Creek	Kadake Creek	Old Franks
FP3	8	13	18	11 (1.3)	**48** (1.4)	
FP4	16	25	35	20 (1.5)	**48** (3.3)	**31** (1.3)
FP5	30	45	70	32 (2.8)	**55** (7.7)	**49** (2.1)
MM1	9	12	18	**7** (0.4)	**13** (2.0)	
MM2	17	24	33	**15** (1.7)	**28** (4.2)	

[a] Category, and numeric codes define distinct channel types within each group. Channel types listed include: FP3, small flood plain; FP4, medium flood plain; FP5, large flood plain; MM1, moderate-gradient, mixed-control; and MM2, medium moderate-gradient unconstrained.

[b] Fish habitat objectives are expressed as a range of values based on the 25th, 50th (median), and 75th percentiles for a given channel type or process group (see appendix C.1). Key for interpreting interim Tongass channel width to depth ratio fish habitat standards: >75th, poor habitat; >50th and <75th or <25th, below average habitat; >25th and <50th, above average habitat.

[c] Values are medians for all channel-type stream segments surveyed. Blanks represent channel types not sampled. Numbers in parentheses represent surveyed stream length in miles. Habitat values in bold represent below average to poor habitat condition based on interim standards which may indicate some channel aggradation associated with sediment loading from recent landslides in the watershed and less than optimal fish habitat conditions.

Creek, although small landslides (i.e., smaller than the 0.5-acre minimum size established in the landslide protocol) have occurred in headwater basin harvest units, none of them appear to have reached or entered streams. An increase in sediment delivery is expected over the next few decades in Game Creek, as the strength of dead tree roots declines and the potential for small landslides increases.

Five landslides were observed in the upper Old Franks Creek watershed. However, only two of these have occurred in the last 25 years. One landslide associated with a harvest unit appears to be related to a 1993 flood. This slide did not reach any stream, but fine sediment is entering streams by way of the road drainage system. Rills and other signs of active erosion and sediment delivery are common on unstable soils of road cut-slopes. Landslides may be a future concern based on the frequency of observed unstable soils. Field analysis has not been sufficient either to verify or change soil mass-movement hazard ratings or to assess how serious this concern is.

Landslides appear to be a major source of sediment in the Kadake Creek watershed. Fifty-seven landslides were inventoried during this study. Four of these originated in harvest units and one consisted of spoil material stripped from a rock pit. The remaining 52 landslides occurred outside of managed areas. Kadake Creek appears to have a high risk of accelerated sediment deposition in class I and class II channels, based on about 40 (70%) of the landslides reaching streams, on the presence of unstable soils in portions of the upper watershed, and on apparent aggrading conditions in the main stem. Landslides associated with a large rain-on-snow flood in 1988 appear to be a major source of sediment.

Erosion associated with road cut and fill slopes appears to be less of a concern than the existing or potential erosion from landslides. Poor cross-drain function on 7 to 13% of the road culverts surveyed in Game Creek and Kadake Creek watersheds could result in increased sediment delivery to streams in the future.

Riparian Management Areas

A riparian management area was delineated in each watershed to identify the riparian, wetland, and sediment-source areas directly or indirectly affecting salmonid habitat. These areas are fundamental components of a specific riparian habitat conservation strategy for each watershed. Current management procedures were compared with fish habitat-protection measures that could have come from applying this strategy.

Most timber harvest in the Game Creek watershed has occurred since passage of the Reform Act. About 27% of the area harvested is on riparian or potentially unstable soils in the riparian management area. Currently, fish habitat does not appear to be impaired although continued management activity in riparian management areas could increase the risk of cumulative watershed effects.

On Game Creek, some prescriptions that would have been derived from the riparian habitat conservation strategy could have been different from the ones implemented. For example, a silvicultural prescription in the riparian management area could have recommended that the buffer strip be feathered to help maintain the core riparian areas (i.e., where stand conditions allow). In addition, timber harvest and road construction could have been minimized on potentially unstable soils.

In the Kadake Creek watershed, about 6% of the post-Reform Act harvest-unit area is within the riparian management area. The results of the watershed analysis conducted in this system indicated that applying a riparian habitat conservation strategy for Kadake Creek could have caused some adjustments on post-Reform Act units (i.e., if the strategy had been available during the original planning). For example, one unit would probably have received more consideration for preventing windthrow in the buffer along a moderate-gradient, mixed-control class II stream perpendicular to the prevailing wind. The buffer could have been feathered and tapered on the abrupt windward face. Alternatively, the unit could have remained unharvested.

The Old Franks Creek watershed has had a small amount of post-Reform Act harvest. About 6% of the upper Old Franks Creek watershed has been harvested, but 52% of the area harvested is within the riparian management area on soils that were shown to be potentially unstable. Most post-Reform Act units lie partially or totally within the riparian management area, which indicates that the sale would probably have had a different unit layout and timber harvest prescriptions would have provided more fish habitat protection, had a riparian habitat conservation strategy been implemented. In addition, the total area in sale units would probably have decreased. Most of the change would likely have been in one particular harvest unit due to mass-movement and sediment transport concerns. In addition, application of the strategy would likely have modified the road layout and prescriptions on potentially unstable soils and changed leave-tree requirements along class III streams.

DISCUSSION AND RECOMMENDATIONS

EFFECTIVENESS OF EXISTING PROCEDURES

The procedures currently being used for protecting fish habitat in Tongass National Forest have clearly improved the treatment of anadromous fish streams and provided improved protection for valuable stream habitat, as compared to the procedures that had been used previously. However, the effectiveness of current procedures depends on how well existing guidelines have been interpreted and practiced on the ground. Analysis of our results indicates there are deficiencies in current practices and procedures for protecting fish habitat that could have long-term adverse effects on some Tongass National Forest salmonid populations. Specifically, the existing guidelines have not been implemented consistently throughout the Forest. In addition, the Tongass Timber Reform Act and current procedures do not address fish habitat and watershed processes over long time frames or over large landscape scales. Our review and comparison of current procedures to various proposed procedures also pointed to a lack of measurable fish habitat objectives against which to measure the effectiveness of existing guidelines.

We recognized that efforts are being made to clarify the stream protection requirements of the Tongass Timber Reform Act to better and more consistently implement buffers on class I and II streams (i.e., as evidenced by the Chatham Area 1993 policy on Reform Act buffers). In some areas, flood-plain habitat, streams, riparian areas, and important wetlands have not been well protected because they were misclassified or simply did not show up on inventories used for sale planning and layout. Large woody debris and wetland functions were sometimes not fully recognized in buffer designs during timber-sale layout. Many buffers were designed too narrow to withstand high wind. Some streams, classified as class III and not given a buffer, should have been classified as class II and buffered, based on habitat characteristics for resident fish.

The Tongass Timber Reform Act does not require minimum buffers on class III streams but does require the application of best management practices to them. Class III streams comprise about half of a typical watershed's channel network. These streams act as conduits for sediment, bedload, and woody debris to downstream fish habitat. We observed that class III streams were sometimes not given buffers or sufficient protective measures to maintain all of their important functions.

The landslide hazard and mitigation measures that were applied to timber harvest and road activities on steep unstable slopes were not always adequate. Slopes with the highest mass-movement rating — based on the Tongass landslide rating system — are considered unsuitable for timber production in the proposed Forest Plan Revision but may be available for timber harvest under the current Plan. The moderately high mass soil movement risk areas are considered suitable, but these slopes are required to have further site-specific inventory and prescriptions before timber harvest and road activities are planned. Inadequate field checking of the soils database to verify the hazard associated with the mass-movement classification was apparent from the field review, although our team recognizes that decisions to operate on high-risk soils may have been accepted by USFS regulators in some locations. While the current procedures for operating on MMI3 areas are generally adequate for minimizing short-term soil displacement and effects of surface erosion, the long-term effects (5 to 25 years) are not well addressed by these procedures.

The most serious long-term effects of the current procedures on high mass movement-hazard soils will occur following the loss of root strength in harvested areas. When these steep slopes are clear-cut, the roots of the harvested trees slowly decompose and no longer hold the soil in place. Peak loss of root strength (i.e., before the strength associated with regrowth of trees returns) usually occurs about 10 to 15 years after timber harvest.

During periods of intense rainfall, landslides can deliver large amounts of sediment and debris to streams and footslope areas. Negative effects to fish habitat are not always immediate, but accelerated rates of mass wasting are thought to cause long-term, chronic effects (Swanston 1971; Swanston et al. 1987; Hicks et al. 1991; Swanston and Erhardt 1991). Other soil productivity concerns accompany these mass-wasting events, but they are not directly related to fish habitat. Given current procedures, we expect that future timber harvesting and road building will continue to access increasing acreage of high mass movement areas. Therefore, it is likely that complex and unpredictable negative effects on downstream fish habitat will occur if widespread harvest activities on these soils continue over the long term.

Problems in design, construction, maintenance, mitigation, and closure of roads were observed throughout the harvested portion of the Tongass National Forest, especially on steep, unstable slopes. Stream crossings were sometimes designed for less than the critical flow and ditch relief culverts were sometimes not sufficient to maintain the hydrology of steep slopes, hollows, and wetlands. In at least one instance, a road was constructed on highly erosive soils through fall storms, which undoubtedly contributed to the sedimentation of streams. Culvert crossings of roads on steep mountain-slope channels was another concern identified during the study. These culverts have a tendency to fail and plug with bedload, becoming persistent maintenance problems.

Based on the expert field review and our own observations, the road construction problems that required more timely and consistent mitigation included erosion control at stream crossings, grass seeding on cut-and-fill slopes, and fish passage at class I and II culvert crossings. The failure to act on decisions to close road segments was also identified as a significant concern in some locations.

Road maintenance was identified as a concern identified by some of the field-review experts and our study team. Funds for maintaining the many miles of open roads on the Tongass National Forest seem to be inadequate. Low-use roads typically are not stabilized or "put to bed"—such as by removing culverts, constructing waterbars, outsloping road surfaces, and seeding—after timber harvest.

The study team and the expert reviewers expressed general concerns with the planning process. Specific problems included not evaluating potential cumulative watershed effects thoroughly (although we recognize that the Stikine Area has had a process for analyzing cumulative effects thresholds for a few years), lack of a holistic approach in describing the important watershed functions and processes and how they should be protected, lack of a long-term view over a large watershed landscape, lack of contingency planning for large, stochastic events (such as floods and windstorms), and minimal concern for aquatic species other than salmon. All of these examples of planning deficiencies could be due to the pressure to produce timber sale offerings.

NEED FOR ADDITIONAL PROTECTION

Based on the results of this study, it is apparent that additional measures are needed to effectively protect anadromous fish habitat in the Tongass National Forest. Provision of additional protection for fish habitat requires two parallel efforts:

- Implementation of an ecosystem approach for evaluating and protecting watershed processes and functions at the landscape scale; and
- Full implementation of existing best management practices in planning and implementing activities that could affect aquatic ecosystems.

The study team developed a number of recommendations for addressing research, institutional, and information needs associated with fish habitat protection in the Tongass National Forest. While it is recognized that risks to fish habitat can never be eliminated, we believe that additional risks associated with timber management activities will be minimized and the goal of preserving the biological productivity of fish streams will be met if the following recommendations are implemented in their entirety.

1. Given our concerns about the inconsistent application of riparian management guidelines in class III and smaller streams, the inadequate identification and consideration of high mass-movement soils, and a need for better planning of timber management activities that may influence fish habitat, we believe watershed analysis, using the concepts presented in A Federal Agency Guide for Pilot Watershed Analysis (1994), should be implemented as a precursor to timber sales and other management activities that could significantly influence fish habitat. The cornerstone of this approach is an ecosystem analysis applied at the watershed scale. Although some procedures in watershed analysis are currently implemented as part of timber-sale planning, we believe that the watershed analysis process provides important new information about fish habitat needs and the important factors that influence habitat. Watershed analysis places each stream in the context of a continuum where small stream processes provide input into successively larger streams throughout the river system. Maintaining this connectivity is important for protecting healthy watersheds and fish habitat (Hicks et al. 1991).

2. Riparian management areas should be defined and the area within them managed to fully protect fish habitat in the long term. Some specific recommendations include the following.

   • Riparian zones adjacent to unconfined alluvial flood plain channels, alluvial fan channels, and glacial outwash channels (Paustian 1992) should not be subject to timber harvest unless they are fully evaluated. The entire floodplain should be considered as the interim Riparian management area. In these channel types, riparian zones may extend beyond the minimum width specified under current procedures because the stream is often dissected into a main low-flow channel with several side channels. These side channels are important fish habitat. Ecosystem-scale analysis should consider the whole riparian area, as defined by riparian soils and vegetation. Site-specific harvest is only allowable when these riparian areas are fully evaluated and the benefit of timber harvest is consistent with maintaining riparian health.

   • The distance equivalent to the maximum height a tree growing in that area could attain should be used to determine the riparian management area width (assuming it is greater than 100 feet) for confined alluvial channel types of class I and II streams. The riparian management areas should never be less than 100 feet wide and any timber harvest in these areas should fully protect riparian values.

   • Minimum 100-foot buffers on each side of class III streams should be established until individual needs are identified during watershed analysis. Class III streams have important water quality values. These streams are also sources for woody debris recruitment and litter, and they deliver nutrient and sediment inputs into larger streams (Hicks et al. 1991; Gregory et al. 1991; FEMAT 1993). These streams are typically high-gradient streams, which are often associated with steep, unstable terrain. Timber management opportunities within the buffers should be evaluated case by case, considering mass-movement hazards, as well as debris, litter, nutrient, and sediment input.

   • A new category, class IV streams, should be established to define the intermittent or ephemeral colluvial channels and small, perennial spring-fed rill channels that are not dominant sediment-transport streams. These streams are typically very small, high-gradient streams draining mountain slopes and should be managed primarily to protect water quality. While they rarely need buffer strips, special provisions for felling, yarding, and determining where to place landings and roads are often required. These types of streams are currently either unclassified and/or misclassified as class III streams

   • Consistent forest-wide definitions, inventory standards, and interpretations of mass-movement-hazard areas should be developed. In addition, a full inventory and analysis of high- and very high-hazard soils should be conducted. As part of an ecosystem approach to watershed planning, mass-movement hazard should be incorporated into the design of all riparian management areas. Given the results of the comparison between post-Reform Act buffers and watershed analysis recommendations, we believe that the watershed analysis more accurately blends riparian area concerns with high-hazard geomorphic conditions to identify appropriate riparian management areas.

   • Cumulative-effects procedures should be included as part of watershed analysis to assess the effects from past management. This analysis should identify where current operating thresholds exist that could influence fish habitat and aquatic resources.

   • A set of objectives for fish habitat management should be developed and adopted. Such objectives should be measurable, reflect the diversity of fish habitat needs, and serve as key monitoring indicators. Therefore, it is recommended that the current set of interim objectives be improved and expanded to include other measures. It is further recommended that future inventories be conducted on a watershed scale and that these objectives be included as a starting point; regional training will be necessary to ensure consistency in collection methods and interpretation.

3. We recommend some specific measures to reduce risks to fish habitats associated with road construction, maintenance, and use.
    *   Stream crossings should be designed and maintained to ensure the upstream and downstream movement of all life stages of anadromous fish. Similar passage criteria are desirable for streams which support resident fish species.
    *   Best management practices for mitigating effects of erosion and sedimentation from roads should be implemented consistently throughout the National Forest.
    *   Building roads on steep slopes and through flood plains and fens should be avoided. When roads are built in these areas, stringent implementation of best management practices should be used to determine the appropriate construction timing, engineering standards, season and condition of use, and maintenance schedule.
    *   A program for annually checking and maintaining culverts should be implemented for the entire Tongass National Forest. Where culverts have a high risk of failure during large flow events because of potential debris inputs, contingency designs for road overtopping should be implemented to prevent damage to fish habitats.
    *   All roads not essential for forest transportation and management needs should be identified and closed. Timely closure of unneeded roads and immediate road drainage and erosion mitigation measures should be vigorously pursued.
    *   Open-bottom stream-crossing structures or bridges should be used more frequently on low-to-moderate gradient streams where fish passage is required.
4. It is recommended that focused research and monitoring programs be conducted to increase our understanding of the long-term effects of forest management on fish habitat. Studies should be implemented on
    *   evaluating the effectiveness of riparian area prescriptions and management practices for protecting fish habitat;
    *   the basic life-history requirements of anadromous and resident fish in streams;
    *   the effects of windthrow on fish habitat in both the short and long terms and how negative effects can be minimized;
    *   movement of large woody debris through stream systems;
    *   sediment routing, to determine the risk resulting from mass failures on fish habitat;
    *   key measures of habitat and ecosystems to be used as management indicators for aquatic ecosystems;
    *   hydrological and biological effects of roads and timber harvest on fens and other wetlands;
    *   cumulative effects of management activities on watershed processes and fish habitat;
    *   population viability assessment for anadromous fish stocks in Southeast Alaska; and
    *   adaptive management areas for "learning to manage by managing to learn" (Rapport 1992).

CONCLUSION

The results of this study show that existing procedures are insufficient for fully protecting anadromous fish habitats in the Tongass National Forest. As such, additional measures are needed to assure long-term protection for fish habitats. The results of this study have been used to revise the Tongass Land Management Plan in a manner that promotes ecosystem-based management and should support sustainable salmon fisheries. The recommendations developed by the study team are intended to provide a basis for conserving freshwater habitat using a broad perspective of whole watersheds over expanded time periods. If these recommendations are implemented, it is anticipated that the higher level of protection afforded anadromous fish habitats will support sustainable salmon fisheries in the Tongass National Forest.

ACKNOWLEDGMENTS

We appreciate the careful reviews of P. DeVries, S. Elliott, and one anonymous reviewer.

REFERENCES

ADF&G (Alaska Department of Fish and Game). 1994. Report to the Alaska Board of Fisheries on imple-
    menting special regulations for harvesting steelhead and cutthroat trout in Southeast Alaska. January
    1994 Board Meeting, Alaska Department of Fish and Game, Sportfish Division, Ketchikan.
Anonymous. 1994. A Federal Agency Guide for Pilot Watershed Analysis. Version 1.2. Portland, Oregon.
Baker, T.T., and eight coauthors. 1996. Status of Pacific salmon and steelhead escapements in southeastern
    Alaska. Fisheries 21(10):6–18.
Beamish, R. J., and D. R. Bouillon. 1993. Pacific salmon production trends in relation to climate. Canadian
    Journal of Fisheries and Aquatic Sciences 50:1002–1016.
FEMAT (Forest Ecosystem Management Assessment Team). 1993. Forest ecosystem management: an eco-
    logical, social, and economic assessment. U.S. Department of Agriculture; U.S. Department of the Interior
    [and others], Portland, Oregon.
Gregory, S. V., F. J. Swanson, W. A. McKee, and K. W. Cummins. 1991. An ecosystem perspective of riparian
    zones: focus on links between land and water. BioScience 41:540–551.
Halupka, K. C., J. K. Troyer, M. F. Wilson, and F. H. Everest. 1994. Identification of unique and sensitive
    salmonid stocks of southeast Alaska. Unpublished. Pacific Northwest Research Station, Forestry Sciences
    Laboratory, Juneau, Alaska.
Hicks, B. J., J. D. Hall, P. A. Bisson and J. R. Sedell. 1991. Responses of salmonids to habitat changes. Pages
    483–518 in W. R. Meehan, editor. Influences of forest and rangeland management on salmonid fishes
    and their habitats. American Fisheries Society Special Publication 19. Bethesda, Maryland.
Linstone, H. A., and M. Turoff, editors. 1975. The Delphi method: Techniques and applications. Addison-
    Wesley, Reading, Massachusetts.
Marshall, R.P. 1992. Forecasting catches of Pacific salmon in commercial fisheries of southeast Alaska.
    Doctoral dissertation. University of Alaska, Fairbanks.
Martino, J. P. 1983. Technological forecasting for decision-making. North-Holland, New York.
Nehlsen, W., J. E. Williams, and J. A. Lichatowich. 1991. Pacific salmon at the crossroads: Stocks at risk from
    California, Oregon, Idaho, and Washington. Fisheries 16(2):4–21.
Paustian, S. J., editor. 1992. A channel type users guide for the Tongass National Forest, southeast Alaska.
    R10TP26. U.S. Department of Agriculture, Forest Service, Alaska Region, Juneau.
Rapport, D. J. 1992. Evaluating ecosystem health. Journal of Aquatic Ecosystem Health 1:15–24.
Reeves, G. H., F. H. Everest, and J. R. Sedell. 1993. Diversity of juvenile anadromous salmonid assemblages
    in coastal Oregon basins with different levels of timber harvest. Transactions of the American Fisheries
    Society 122:309–317.
Reid, L. M. 1993. Research and cumulative watershed effects. U.S. Department of Agriculture, Forest Service,
    Pacific Southwest Research Station. General Technical Report PSW-GTR-141. Albany, California.
Rieman, B. E., and J. D. McIntyre. 1993. Demographic and habitat requirements for conservation of bull trout
    (Salvelinus confluentus). U.S. Department of Agriculture, Forest Service. Intermountain Research Station.
    General Technical Report INT-GTR-302. Ogden, Utah 38 pp.
Sedell, J. R., G. H. Reeves, R. R. Hauer, J. A. Stanford, and C.P. Hawkins. 1990. Role of refugia in recovery
    from disturbances: Modern fragmented and disconnected river systems. Environmental Management
    14:711–724.
Slaney, T.L., K.D. Hyatt, T.G. Northcote, and R.J. Fielden. 1996. Status of anadromous salmon and steelhead
    trout in British Columbia and Yukon. Fisheries 21(10):20–35.
Swanston, D. N. 1971. Judging impact and damage of timber harvesting to forests soils in mountainous regions
    of western North America. Western Forestry Conference Proceedings 62:14–19.
Swanston, D. N., and R. Erhardt. 1991. Short-term influence of natural landslide-dams on the structure of
    low-gradient channel. Pages 34–38 in T. Brock, editor. Proceedings of Watershed '91: a conference on
    the stewardship of soil, air and water resources. U.S. Department of Agriculture, Forest Service.
    R10MB217. Juneau, Alaska.

Swanston, D. N., G. W. Lienkaemper, R. C. Mersereau, and A. B. Levno. 1987. Effects of timber harvesting on progressive hillslope deformation in Southwest Oregon. Pages 141–142 *in* R. L. Beschta, T. Blinn, G. E. Grant, F. J. Swanson, and G. G. Ice, editors. Erosion and sedimentation in the Pacific Rim. International Association of Hydrological Sciences.

USDA (U.S. Department of Agriculture, Forest Service). 1992. Best management practices. Implementation monitoring report 1992. U.S. Department of Agriculture, Forest Service, Juneau, Alaska.

USDA (U.S. Department of Agriculture, Forest Service). 1993. Best management practices. Implementation monitoring report 1993. U.S. Department of Agriculture, Forest Service, Juneau, Alaska.

USDA (U.S. Department of Agriculture, Forest Service). 1995. Report to Congress: Anadromous Fish Habitat Assessment. Report R10-MB279. USDA Forest Service, Pacific Northwest Research Station and the Alaska Region, Juneau.

USDA and USDI (U.S. Department of Agriculture, Forest Service and U.S. Department of Interior, Bureau of Land Management). 1994. Draft environmental assessment for the implementation of interim strategies for managing anadromous fish-producing watersheds in eastern Oregon, Washington, Idaho, and portions of California. U.S. Forest Service and U.S. Bureau of Land Management, Washington, D.C.

37 Using Watershed Analysis to Plan and Evaluate Habitat Restoration

Neil B. Armantrout

Abstract.—Watershed analysis is a tool to evaluate the condition of the resources and processes of a watershed. This chapter describes the use of watershed analysis to plan and evaluate aquatic habitat restoration projects in Wolf Creek, a low-gradient tributary of the Siuslaw River, western Oregon. Its runs of coho salmon *Oncorynchus kisutch*, chinook salmon *O. tschawytscha,* and steelhead *O. mykiss* have declined over 90% from historic levels. The watershed analysis for Wolf Creek was prepared by the Eugene District of the Bureau of Land Management using inventory information, historic data, and remote sensing. Wolf Creek is a naturally sediment-rich and rock-poor basin. Landslide and road impacts vary across the basin. Historic salmonid habitat depended on the presence of large woody material. Loss of instream woody structure from timber harvest and settlement beginning in the 1800s contributed to the loss of salmonid spawning and rearing habitat and increased water temperatures. Over the past 28 years the Eugene District of the Bureau of Land Management has completed extensive habitat restoration that includes placing in-channel structures in more than 19 km of third- to sixth-order streams, closing roads, removing or replacing culverts, placing rock and gravel into the channel, and increasing the percentage of conifers in riparian areas. Additional work has been performed on private lands through cooperation of the landowners and the Oregon Department of Fish and Wildlife. Monitoring includes spawning counts, snorkeling, smolt trapping, vegetation plot analysis, and photographic documentation of channel changes. Evaluation of restoration projects, particularly following flooding in 1996, indicated the aquatic system was responding to the restoration efforts as anticipated and shows positive responses. Although initial results are promising, additional monitoring is needed to determine the full impact of restoration efforts at the stream reach and watershed level.

INTRODUCTION

Aquatic systems are nested in the landscape. Water enters the system through precipitation, flowing through watersheds as surface and groundwater. The moving water interacts with the landscape to create the variety of flowing and standing water habitats. Because the aquatic system is inter-related through a watershed, actions in one part of the watershed may influence the nature of aquatic habitats elsewhere in the basin. Human activities that modify the physical and biological features of the landscape may also modify the aquatic system. Watershed analysis is a method for evaluating the landscape, processes, and human influences throughout the basin to determine basic conditions and recent modifications. The results of watershed analysis help to identify locations of habitat problems and opportunities for habitat improvement. The watershed analysis process also provides the framework for evaluating the impact of restoration activities within the watershed. Since the aquatic system is nested within the landscape, having a better understanding of the landscape features and processes facilitates development of restoration strategies that produce the desired responses. The purpose of this chapter is to show how watershed analysis was used to (1) summarize information on hydrologic patterns, delivery of materials to the stream, water quality, and habitat suitability; (2) guide habitat restoration; and (3) evaluate the responses of restoration activities compared to anticipated changes.

BACKGROUND

Wolf Creek is a sixth-order (Strahler 1957) tributary of the Siuslaw River in central Lane County, Oregon. It originates on the eastern margins of the Coast Range Mountains, flowing 41 km westward to join the Siuslaw River. The basin is separated in the east from Coyote Creek, a tributary of the Willamette River, by a low ridge less than 300 m above sea level. The difference in elevation between headwaters of the two basins is less than 22 m. To the north, a higher ridge separates Wolf Creek from the Long Tom River, also part of the Willamette River system. The southern ridge contains the highest elevations (>600 m) in the basin, and separates Wolf Creek from the Siuslaw River.

Native Americans have used the basin for at least 10,000 years (Mike Southard, Bureau of Land Management (BLM), personal communication; USDI 1995a) for hunting, gathering, and fishing for salmon and lamprey. Europeans first began settling on the coast of the Siuslaw River basin in the 1860s but settlement in Wolf Creek did not occur until the end of the 19th and beginning of the 20th centuries. However, many homesteads were later abandoned because of the isolation and poor farming conditions.

Today ownership in the basin is primarily private forest lands and intermingled forest lands managed by the BLM. The private timber lands are managed for forest products, while the federal lands are managed under several prescriptions as designated in the Eugene Record of Decision (USDI 1995b) and includes both reserve forests and general forest product lands. A small percentage of land concentrated in the upper reaches of the basin are privately owned and managed for residential or agricultural purposes.

Historically, the basin had runs of chinook salmon *Oncorhynchus tshawytscha*, coho salmon *O. kisutch*, steelhead *O. mykiss*, resident and searun cutthroat trout *O. clarki*, and Pacific lamprey *Lampetra tridentatus*. As a result of the loss of aquatic habitat, together with events occurring outside the basin, the populations of the these fishes have declined. Other native fishes in the basin include brook lamprey *Lampetra richardsoni*, sculpins *Cottus* sp., dace *Rhinichthys* sp., shiners *Richardsonius balteatus*, suckers *Catastomus macrocheilus* and pikeminnow *Ptychocheilus oregonensis*.

Timber harvesting was initiated during the early period of settlement. Most of the early harvest was associated with the settlements and was concentrated in the upper half of the basin. Since World War II, extensive harvest has occurred in the lower half of the basin, and now includes second and third growth areas throughout the basin. Initially, timber was moved by floating down streams and rivers. As technology improved after World War II, an extensive network of roads was extended throughout the basin.

At the time of European settlement the Siuslaw River was described as full of large woody material, creating extensive wetlands and channels. Although we do not have a description specific to Wolf Creek, we can assume conditions were the same. Much of the large woody material in larger streams was removed to facilitate the floating of logs and to reduce flooding of roads and agricultural lands. The removal of larger conifers from riparian areas left scant large woody material to replace what was lost from stream channels.

As a result of the loss of woody material that historically created channel structure, aquatic habitat declined. Together with poor ocean conditions, high harvest rates, and other factors, runs of anadromous fish declined, with coho in particular falling to levels less than 5% of historic highs.

METHODS

WATERSHED ANALYSIS

The BLM Wolf Creek Watershed Analysis was prepared by a team of specialists in the BLM Eugene District Office using existing inventory data, historical information, expert knowledge, and remote sensing. Guidelines for doing the watershed analysis were provided by the Regional Ecosystem Office (1994). A geographic information system (GIS) was used for data analysis and display. For analytic purposes, the resources in the basin were divided into a series of resource modules including climate, geology and soils, geomorphology, hydrology, riparian vegetation, wildlife, upslope vegetation, forestry, fisheries, and human use. While the resource modules are each discussed separately, it is the interaction that determines the nature and condition of aquatic resources.

ARC/INFO was used for integrating most data layers. The one exception was aquatic habitat, which is not geographically referenced so could not be used in a GIS layer. Storage and analysis of aquatic and riparian habitat data for individual streams and stream reaches, and the basin as a whole, was completed with CARP, a database developed in the Eugene District.

The analysis included past and current conditions, landscape interactions, and management history and options. Processes, such as precipitation, sediment production, and water quality, were summarized across the basin. More specific resource discussion, such as forest stand types, is limited to public lands. Summaries of information for each resource model are found in the watershed analysis documentation (USDI 1995b; see also Weyerhaeuser Corporation 1996).

Geology and Soils.—Geological descriptions were based on information from the Oregon Department of Geology. The information was available in publications and on maps. Soil information was obtained from a soil survey conducted by the State of Oregon and the U.S. Soil Conservation Service (now National Resource Conservation Service) which was available as maps and a GIS layer.

Climate.—There are no weather-recording stations in the Wolf Creek basin. Information on rainfall in a comparable part of the Coast Range, 25 km to the north near Triangle Lake, is available from a non-government recording station. Rainfall patterns in the Wolf Creek basin are extrapolated from existing data. Temperature information from the Wolf Creek basin includes instantaneous readings from throughout the basin for over 20 years, as well as readings from the Triangle Lake station.

Hydrology.—There are no stream flow gauging stations in the Wolf Creek basin. Hydrologic information was calculated using discharge data from stations on the Siuslaw River, in conjunction with information on drainage area.

Geomorphology.—Land forms were described using topographic maps in a GIS layer, supplemented by profiles described using a digital terrain model (DTM). Erosion from roads and hillslopes was estimated by District soil and hydrology specialists. Landslide potential was estimated from models developed for the existing geologic type by Dietrich et al. (1993). Gradient, slope stability, and erosion rates were used to estimate the production rate of rock and sediment to the stream channel.

Stream Channel.—Inventories of fish-bearing streams on public lands in the basin were conducted in 1971–1973, and again beginning in 1985. Habitat inventories included Wolf Creek and its larger tributaries. Analysis of this information is the basis for describing current stream conditions. Original conditions are not known directly, but were extrapolated from early descriptions within the Siuslaw Basin, and from comparison to the few remaining unaltered reaches of stream. General habitat in larger streams was traced for half a century using a series of aerial photographs.

Vegetation.—Historic vegetation information was extracted from notes of early land and forest surveys by state and federal personnel reaching back into the 1800s. Beginning in the 1940s, aerial photographs became available from which forest units can be described. Earlier information is quite general, and defines the forest in terms of dominant tree species. The forest units are generally described by dominant species and general age classes, or seral stages. Recent inventories using remote sensing and ground-based data have provided greater detail. Records of current vegetation conditions are maintained in the Eugene District Forests Operations Inventory file.

Riparian Vegetation.—None of the early vegetation inventories separated riparian and upslope communities. Current information on riparian communities is based on data gathered during stream inventories, vegetation transects, interpretation of remote sensing data, and interpretation of aerial photographs. Wolf Creek was used by the Eugene District to test the use of remote sensing for determining riparian communities and conditions. While early work was encouraging (see Armantrout 1997; Armantrout, in press), attempts to automate the determination of vegetation communities did not provide consistent results due, in part, to high heterogeneity and fragmentation of the riparian community.

Fish Distribution.—General distribution of spawning salmonids and lamprey in the basin was determined using extensive spawning ground surveys and fish sampling (unpublished data in files of Eugene District BLM; Armantrout 1990). Spawning index areas have been established where suitable habitat exists on public lands. Annual counts of the indexes for each of the anadromous species were used to determine between-year trends and to assist in evaluating habitat restoration projects.

A digital terrain model (DTM), using gradient, valley confinement, and drainage area, was used to predict the location of potential habitat (Armantrout, in press). The results were similar to the distribution patterns in the GIS fisheries layer, developed from expert knowledge of District personnel and information from Oregon Department of Fish and Wildlife (ODFW) personnel, confirming the basic distribution patterns.

Sampling of fish has been done using electrofishing and seining since 1971, and snorkeling since 1995. Sites have been sampled to evaluate changes in populations as a result of habitat improvement projects, the impact of forest management practices, and to determine distribution of fish in headwaters. Production of salmonid smolts from the basin is now being monitored using a rotary fish trap. Sampling has provided information on the presence, absence, and abundance of non-salmonids, but data are insufficient to characterize distribution patterns or population trends in the basin.

Wildlife.—No detailed inventories of wildlife were available. The best documented species are northern spotted owls *Strix occidentalis* and marbled murrelets *Brachyramphus marmoratus*; both are listed as threatened species. Random observations of species of special interest, such as beaver *Castor canadensis*, elk *Cervus elaphus*, amphibians, and raptors have been recorded, but the information is insufficient to determine population distribution and status.

Human Use.—Information on human use was obtained from a variety of federal, state, county, and private records, summarized in both the Eugene District Resource Management Plan EIS and Record of Decision (USDI 1995a) and the Wolf Creek Watershed Analysis (USDI 1995b).

Aquatic Habitat Restoration Projects

In 1969 BLM, in cooperation with the State of Oregon, initiated a program to improve aquatic habitat. A number of pools were blasted in bedrock, about 15 gabions were built, and one spawning channel was constructed. A second set of aquatic habitat improvement projects was initiated in 1985, involving placement of small log and boulder structures, large log and boulder structures, and two gabions. With the exception of the larger boulder structures, all work was done by hand. An evaluation of this project work was provided by Armantrout (1991).

After 1984, project work by the Eugene District in the Wolf Creek Basin was based on a comprehensive summary of aquatic resources in an unpublished draft Wolf Creek Aquatic Habitat Management Plan (HMP). The information in the draft HMP was subsequently incorporated into the Wolf Creek Watershed Analysis, which describes the basin in a much broader scope. The information in the watershed analysis was used in the ongoing monitoring and review of existing projects, the development of new projects, and the establishment of criteria for evaluating ongoing restoration projects.

Guidelines.—The Aquatic Conservation Strategy (ACS from the President's Forest Plan, FEMAT 1993, USDA and USDI 1994), included in the Eugene District Resource Management Plan and Environmental Statement (USDI 1995b), provided basic guidelines for management and restoration activities in the Wolf Creek basin. In addition to the ACS, the Forest Plan land-use designations include retention of Riparian Reserves and Late Successional Reserves (LSR), and descriptions of Standards and Guidelines for management actions.

Differential stream flows, sediment delivery, and the location of potential rock sources were the primary factors considered in designing project activities for individual streams and reaches. Projects were planned for those locations on public lands most lacking in specific habitat elements. In addition to potential projects on lands administered by BLM, opportunities were also documented on private lands in cooperation with the private landowner and the ODFW. The project design took into account activities upstream and downstream of the project. Larger main stem projects differed from activities in the smaller headwater and tributary streams.

The Eugene District used the following criteria for developing restoration efforts, based on previous work in the Eugene District (Armantrout 1991), projects in systems with similar conditions (House et al. 1990), and opportunities summarized in the watershed analysis:

- Reduce road impacts on hydrology, sediment production, and blockages to fish migration.
- Using silvicultural practices, increase the abundance of riparian conifers to provide future large wood sources.
- Target restoration in the larger streams. Larger streams provide a greater amount and diversity of habitats, and are used by a greater number of aquatic species. In addition, project work in the larger streams has the potential to reach into smaller tributary streams through changes in hydrology and materials transport.
- Integrate a series of actions along extended reaches, as individual structures have a lower likelihood of survival. The structures at the upstream and downstream ends of a series are important water and material transport controls. A series of structures creates a synergistic effect, with interaction among structures increasing the complexity of habitats.
- Place emphasis on controlling the movement of debris, water, and sediment and reduce emphasis on creating specific types of habitat.
- As much as possible, re-connect the stream to the riparian area.
- Incorporate both fixed structures and structures which can be moved by the stream. Natural woody debris is dominated by long boles attached to root wads which can anchor the material. Most of the debris entering the stream is shorter logging debris that does not remain in the channel but is carried out during high flows. Well-anchored larger logs and boulders are able to intercept the shorter debris and help retain it in the channel.
- Use larger, more substantial structures as controls; anchor structures well and provide protection where structures connect to the banks. The key control structures need to be large enough to function during periods of peak flow.
- Increase overall channel complexity and diversity for a variety of aquatic species. While the emphasis in project selection is on anadromous salmonids, the projects also take into account other resident vertebrate and invertebrate species.

Project Design and Implementation.—Five restoration activities met most or all of the project design criteria and, hence, were utilized in the Wolf Creek basin.

1. Logs and boulders were placed in the stream to increase channel structure and complexity, and to control flow, sediment, and debris. Structures included full-spanning weirs, artificial cascades, jetties, log and boulder clusters, single boulder placements, and placement of logs on or projecting into the channel. Key logs and boulders were anchored using cable and epoxy.
2. Cobble, rubble, boulders, and spawning gravel were added to the channel at pre-determined locations using heavy equipment. Gravel was placed in the channel to allow the current to distribute it along the channel.
3. Roads were closed using subsoiling and planting the road bed, placing a barrier, and allowing the road to re-vegetate naturally. Other roads were upgraded through pull-back of material, improving drainage, and hardening of surfaces.
4. Inappropriate culverts were replaced with larger culverts or culverts with baffles to facilitate passage of aquatic animals.
5. After preparation of planting sites in the riparian area, bare-root trees were planted and then protected with tubing to reduce animal browsing. Planting was done both in conjunction with channel restoration, and as separate projects in other suitable locations.

Riparian Restoration.—From the start, it was recognized that instream projects are of short-term benefit. The historic natural habitat was created by large trees that fell into the stream channel. Because of logging and stream cleaning, little natural wood was present in the channel or growing in the riparian zone as replacement wood. Restoration of riparian vegetation to increase the number of conifers for future stream structure was a part of all stream restoration projects. Openings were created in the red alder *Alnus rubra*–dominated riparian areas and conifer seedlings were planted. Evaluation of the riparian restoration projects was based on tree survival and growth with different tree species and site treatments, such as degree of canopy opening and control of competing vegetation.

Channel Restoration.—Habitat restoration projects initiated in 1969 included blasting holes in bedrock and construction of gabions to create pools and spawning areas. Sites were primarily bedrock-dominated reaches on public lands.

Since 1984 restoration projects have been designed based on a more comprehensive analysis of the basin. Increasing emphasis has been given to working with natural processes, such as increasing groundwater storage and retention of woody material. Projects included placement of log and boulder structures in streams, with designs utilizing the results of monitoring and evaluation of previous work in the Eugene District (Armantrout 1991), Salem District BLM (House and Boehne 1987) and Coos Bay District BLM (Anderson 1981). In smaller streams, structure was placed on by hand or rubber-tired tractors using small boulders, logs, and gabions, although more recent projects have not used gabions due to their short life span. Evaluation of the projects was based on a range of criteria, including changes in structure and stream channel characteristics in response to storm events, changes in channel complexity and stability, and changes in fish communities associated with the project areas.

Roads and Culverts.—Culverts that were barriers to fish migrations were replaced or removed. In addition, some culverts that were not barriers to fish, but inadequate for stream flows, were replaced with larger culverts. Passage for small fish, amphibians, and invertebrates is now used as part of the design criteria. Some roads were closed using gates, tank traps, or subsoiling and planting. Because of the mixture of private and public land, and shared road ownership, opportunities for road closures are limited. Evaluation of culverts was based on modification of the channel with the new culverts and the presence of fish upstream of the culvert. Road evaluation was primarily based on a reduction in sediment and road ditch stream flows.

MONITORING AND EVALUATION

Monitoring and evaluation are being conducted on the physical and biological features of both old and new projects, although intensity varies with the availability of personnel and funding.

Monitoring the success of in-channel structures was done by a variety of methods. Some photographs, but no written records, are available for projects prior to 1983. From 1983 to 1994 both written and photographic records were made of each project site both before and after project work was implemented. From 1994 to the present, both a written and photographic record were made, using both digital and film cameras. In the latter period, a GPS recorder was used to provide a geographic location for entry of information into a GIS database. Additional photographs were taken in subsequent years. Physical habitat inventory and photographs were used to document the changes in the stream channel over time. Most of the evaluation was subjective, comparing the survivability and functionality of structures and physical changes in the channel.

Monitoring of vegetation was done with standard vegetation plot analysis. Additional data was gathered during stream inventories, using vegetation transects, and by interpreting remote sensing data and aerial photographs. Changes in fish populations resulting from restoration projects have been monitored through spawner counts, snorkeling, and operation of a smolt trap near the mouth of Wolf Creek.

RESULTS AND DISCUSSION

WATERSHED ANALYSIS

Summaries of information from the Wolf Creek watershed analysis (USDI 1995a) findings are discussed in the following modules.

Geology.—Wolf Creek originates on the eastern margins of the Coast Range Mountains, formed by the eastward tilting of the Coast Range Block. Wolf Creek is in the northern end of the Tyee Basin formation. The dominant rock are Siuslaw members of the Flournay formation, thick-bedded sandstones, with thinly bedded siltstone. There are some localized basaltic intrusions and feeder dikes.

The sedimentary sandstones and siltstones weather into fine particles. They produce few larger rocks, such as rubble and boulders. The gravels, rocks, and boulders that reach the stream channel often break down over short periods into sand and silt. Harder gravels, rocks, and boulders are produced by the basaltic intrusions, but only in a very limited range in the basin. As a result, the dominant size of materials delivered to the stream channel is <64 mm. Harder basaltic boulders are found primarily in the lower end of the Wolf Creek basin and between the middle and upper part of the basin.

Overall, soils are more stable and less likely to erode in the upper reaches of the basin than in the lower reaches. The upper reach soils are also more xeric than soils near the mouth.

Geomorphology.—Because the basin tilts due to the continued upthrusting, the highest elevations are in the middle and lower parts of the basin. The upper stream channels in the basin arise near the ridge tops, but have gradients of less than 10%. From its origins to its mouth, Wolf Creek has an average elevation drop of little more than 13 m/km, or less than 1%. Except for tributary headwaters in the middle and lower reaches of Wolf Creek, the channel is moderately confined to unconfined (defined as the valley floor width being 12 or more times the average channel width).

Delivery of rock and gravel to the channel is limited, particularly in the upper reaches of the basin, where moderate slopes and stable soils limit erosion. Highest rates of erosion, and hence delivery of sediments to the channel, occur in the lowest reaches of the basin. Structure, primarily large woody material, retain gravels and rock in the channel; in the absence of structure, materials flush from the system and the stream channel frequently incises to bedrock.

TABLE 37.1
Current and 1956 vegetation classes for all lands in the Wolf Creek watershed.

Vegetation Classes	Hectares (total)		Hectares (% of total)		Patches (number)		Patches (% of total)		Average Patch size (hectares)	
	1956	Current	1956	Current	1956	Current	1956	Current	1956	Current
Unclassified		14		0.1		1		0.1		14.2
Nonforest	199	153	1.3	1.0	15	21	6.6	2.6	13.3	7.3
Hardwoods	147	950	1.0	6.2	10	56	4.4	6.8	14.8	17
Mixed conifer/hardwoods	131	743	0.9	4.8	2	99	0.9	12.1	65.6	7.5
Clearcut	590	3706	3.9	24.2	21	159	9.2	19.4	28.8	23.3
Sapling-pole	4068	1688	26.5	11.0	33	97	14.4	11.8	123.3	17.4
Pole-young	1779	4166	11.4	27.2	57	119	24.9	14.5	30.7	35
Mature over young	304	1064	2.0	6.9	12	85	5.2	10.4	25.3	12.5
Mature	3590	141	23.4	0.9	58	30	25.3	3.7	61.9	4.7
Old over young		749		4.9		71		8.7		10.6
Old Forest	4542	1960	29.6	12.8	21	81	9.2	9.9	216.3	24.2
Total	15350	15334	100	100	229	819	100	100		

Climate.—Wolf Creek basin has a maritime climate, with relatively cool summers and warm winters. The average winter minimum is 6°C and the average summer maximum is 24°C. Seventy-five percent of the precipitation falls as rain in October-April. There is some winter snowfall, but it does not accumulate. The rainfall differs considerably from east to west; in the eastern headwaters, rainfall is about 100 cm/year, while near the mouth it is 200 cm/year.

Greatest rainfall occurs with periodic winter cyclonic storms which may drop 3 cm or more of rain in less than 24 h. These storms are often accompanied by winds exceeding 100 km/h. These storms provide most of the energy for channel-modifying flows and contribute woody material and sediments from adjoining riparian and upslope areas.

Stream flow is highly influenced by precipitation events. Groundwater storage capacity is limited by the underlying geology and shallow soils. As a result, runoff increases rapidly during storm events and is considerably reduced during extended dry periods. Secondary incision in valley deposits is postulated to further reduce effective groundwater storage by lowering the surface of the groundwater table. The generally low summer flows and slow velocities are also thought to contribute to the higher water temperatures found in the basin. Summer water temperatures over 21°C are common in the main stem Wolf Creek and the lower reaches of larger tributaries.

Vegetation.—The dominant vegetation is the western hemlock *Tsuga heterophylla*–cedar *Thuja plicata* climax community. This community became dominant about 5,000–7,000 years ago as conditions became warmer and wetter after the last ice age. Both natural fires and those set by Native Americans and early settlers have influenced forest age and structure (Teensma 1987). The forest was typically a patchy mosaic of trees of different ages and composition, with about 40% being old growth (over 200 years of age). More recently, accelerated harvesting and shorter growing periods has reduced the number of older trees, leaving the basin dominated by younger age classes (Table 37.1).

Before World War II, most logging was concentrated in the upper half of the basin, where older trees were more abundant and where the most settlement took place. Following World War II, logging in the lower half of the basin increased. Most of the older timber throughout the basin has now been harvested. Although conifer species continue to dominate, hardwoods, particularly big leaf maple *Acer macrophyllum,* have become more common.

Riparian Vegetation.—Historically, riparian communities were dominated by conifer, particularly hemlock and Douglas fir *Pseudotsuga menziesii*, with some cedar and grand fir *Abies grandis*. Earlier logging along stream bottoms harvested trees over 1 m diameter at breast height, leaving

TABLE 37.2
Dominant Riparian species in Wolf Creek Watershed, by percent of habitats where present.

Overstory Species (%)		Midstory Species (%)		Ground Species (%)	
Red alder	44	Vine Maple	44	Sword fern	21
Douglas fir	25	Salmonberry	34	Other fern	17
Big leaf maple	15	Willow	6	Oxalis	10
Western hemlock	6	Hazel	1	Grass	8
Western red cedar	2	Ninebark	1	Forbs	6
Total	92		86		62

Salmonberry—*Rubus spectabilis*
Hazel—*Corylus cornuta*
Ninebark—*Physocarpus capitatus*
Sword fern — *Polystichum minitum*
Oxalis—*Oxalis oregana*
Ferns, Grass, Forbs—various genera

smaller trees. Hardwoods replaced many of the conifers, leaving communities of scattered older, larger conifer, red alder, and big leaf maple.

Until forest practice rules were changed in the 1970s, logging, particularly after World War II, harvested all conifers to the stream edge. Forest practice rules provided for streamside vegetation buffers to protect fish-bearing streams. Most of the buffers were less than 50 ft wide. Informal monitoring of these riparian areas showed variable tree mortality, reaching as much as 90%, of the retained trees in the decade following harvest. In most instances, the trees were replaced by brush, particularly vine maple *Acer circinatum*.

During the past 15 years willow *Salix* sp. has expanded rapidly along the stream channels, especially the larger fifth- and sixth-order channels. Where it was once scarce, willow is rapidly becoming the dominant brush species along Wolf Creek and larger tributaries. One hypothesis is that the rapid increase in beaver populations led to an increase in willow shoots downstream from beaver activity, and the lack of major flood events between November 1982 and February 1996 allowed more shoots to become rooted without removal by flooding.

Presently, hardwoods dominate riparian areas (USDI 1995a, Table 37.2), with younger conifers being the next most common overstory group. Most of the young conifers are associated with timber harvest units. While some patches of old growth and reaches with larger trees border Wolf Creek (more commonly in headwaters), there is a very limited supply of larger woody material. Some larger trees have been observed falling into Wolf Creek, and following major storm events there are accumulations of woody debris, but little of this material is retained. Lack of retention appears related to the lack of channel structure, (e.g., large trees or boulders) needed to trap woody debris.

Wildlife.—Spotted owls and marbled murrelets are present in the basin. Deer *Odocoilesu hemionus* are common, and elk present, as well as predators and a range of small mammals (Brown 1985). At least seven species of amphibians are present but numbers are unknown. Two of the three frog species observed during aquatic inventories appear to have declined since 1983. Beavers have increased in numbers during the past two decades. The number of beaver dams, an important refuge habitat for salmonids during both winter and summer, have increased.

Hydrology.—Stream flow patterns are closely tied to precipitation events. Water yield, as measured by the cubic meters per second per square km of basin area, is considerably less than in most of the Siuslaw River basin. The low water yield is likely a result of the low elevation of the hills in the upper basin, reduced precipitation in the upper reaches, and limited groundwater storage

TABLE 37.3
Wolf Creek stream habitat types, by percentage.

Stream	Pool	Glide	Riffle	Rapids	Cascade	Falls	Run	Other
Bill Lewis	86	4.1	9.9	0	0	0	0	0
Eames	49.5	13	33	4.2	0.1	0	0	0
Gall	69.8	3.1	27.6	0.03	0.45	0.02	0	0
Grenshaw	54	10.8	34	0.02	0	0.03	0	1.15
Pittenger	41	32.3	26.6	0.2	0	0	0	0
Saleratus	20.7	12.5	54.5	6.4	0.63	0	4.4	0.87
Salmon	18.9	0	41	10.5	0	0	0	29.6
Swamp	78.3	0	21.7	0	0	0	0	0
Wolf Swamp	66.8	7.5	21.2	0	0	0	0.19	4.31
Swine Log	87.2	6.5	6.2	0	0.11	0	0	0
Van Curen	36.0	6.4	56.0	0.97	0	0	0.61	0.02
Wolf 1985	70.8	15	12.7	1.1	0.37	0	0	0.03
Wolf 1986	61.5	2.6	13	8	0	0.2	14.8	0
Wolf 1993	17.7	31.7	48.8	0	0	0	0	1.8

throughout the basin. Peak flows occur during winter storms. Summer flows can be very low, with some small and medium sized tributaries losing surface flow during extended dry periods.

Limited water quality testing has been done in the basin. The most complete information relates to water temperature. Temperatures in Wolf Creek and the lower reaches of some tributaries exceed 21°C, and often reach 27°C (temperatures lethal to salmonids), during summer low flow periods. Low flows, extensive bedrock exposure, reduced groundwater interchange, and reduced shading are thought to be the principal causes of the high temperatures.

Sediments are naturally abundant in the basin because of the nature of the parent substrate materials. Highest levels of sediment production occur during winter storms, and are greater in the lower basin than the upper reaches. Human activities associated with timber management and associated roads also increase sediment production, particularly in the steeper, less stable, lower portions of the basin.

Stream Channel.—The stream channel is dominated by glides, shallow pools, and riffles (Table 37.3). Rapids and cascades are uncommon, being most prevalent near basaltic intrusions and regions of recent mass soil movements. Bedrock, silt, and sand are the dominant substrates (USDI 1995a), while gravel, cobble, rubble, and boulders are sparse in many reaches. The absence of gravels and larger size rock is most pronounced in the upper reaches. Production of spawning gravels is more abundant in the lower basin, but lack of instream structure limits its retention.

The dominant structural material in the basin has historically been large wood. The second most common structural feature was beaver dams. Large boulders provide localized structure but are uncommon. Both large wood and beaver populations are below historic levels. Even at reduced levels, they create many of the pools, including the deepest and largest pools (Table 37.4). Lack of structure contributes to channel incision and faster currents, and limits the channel's ability to retain woody debris or sediments. Pools, habitat diversity, and the connection of the stream to riparian areas have all been reduced.

Fish.—Because of the low gradient, the Siuslaw Basin, including Wolf Creek, is considered best suited for production of coho and chinook, and for resident and searun cutthroat trout. The abundance of coho recorded for the Siuslaw Basin probably included Wolf Creek. Steelhead, which favor swifter, rockier habitats, were probably never abundant in the basin, but have shown reduced numbers in recent years. Fish runs in Wolf Creek mirror changes elsewhere in the Siuslaw River and other coastal streams (Jacobs and Cooney 1997, ODFW 1997). Information for other species is inadequate to determine present condition or trends.

TABLE 37.4
Causes of pool creation in Wolf Creek (%).

Stream	Large wood	Curve	Beaver dam	Boulder	Bedrock	Falls	Project	Debris Torrent	Other
Bill Lewis	38	31	15	6	1	0	0	0	9
Eames	22	16	20	10	28	4	0	0	0
Gall	33	33	7	4	12	0	0	<1	11
Grenshaw	39	12	12	1	13	1	11	1	10
Pittenger	47	34	9	0	4	0	2	0	4
Saleratus	12	43	6	0	6	0	6	0	27
Swamp	6	26	33	0	20	0	6	6	3
Wolf Swamp	34	25	12	<1	8	0	1	9	11
Swine Log	38	16	36	0	6	1	0	1	2
Van Curen	50	16	0	16	7	0	0	0	11
Wolf 1985	5	19	5	17	35	0	0	0	19
Wolf 1986	4	9	1	2	69	1	0	0	14
Wolf 1993	41	25	5	1	25	0	0	0	3
Basin Summary	30	21	11	4	19	<1	3	2	10

TABLE 37.5
Summary of BLM spawning ground counts for Wolf Creek basin (fish per km).

Year	Chinook	Coho	Steelhead
1983	~	4.5	~
1984	~	0.9	~
1985	6.9	3.9	0.4
1986	1.1	1.9	3.0
1987	1.7	1.9	2.9
1988	9.6	1.7	4.0
1989	1.9	1.4	2.3
1990	2.1	2.8	0.7
1991	4.2	6.8	1.5
1992	7.0	2.3	2.8
1993	1.3	8.5	1.6
1994	2.2	0.7	4.6
1995	~	4.8	1.1*
1996	7.7	5.6	0.9*
1997	2.7	0.4	5.0

*incomplete counts

Anadromous fish escapements show no distinctive patterns as a result of habitat restoration efforts (Table 37.5). Non-habitat issues operating along the coast have had a greater influence on escapements. Nevertheless, habitat use has changed, with an increased percentage of spawners using project sites. The smolt trap operated in lower Wolf Creek since 1988 has shown a general upward trend in smolt production of all anadromous salmonids which suggests that freshwater survival rates are improving. Populations of resident fish, including cutthroat trout, dace, sculpins, and suckers, have shown notable increases at project locations.

Land Use.—The basin has been used for at least 10,000 years by native people. There is little indication that the natives established permanent residence in the basin. Instead, they apparently used the basin seasonally for harvest of salmon and lamprey, and probably for vegetative foods. The basin was part of the lands of the Confederated Tribes of Coos, Curry, Umpqua, and Siuslaw. According to one Siletz tribal elder, sections of Wolf Creek were divided and fished for Pacific lamprey by specific families or clans.

The first permanent colonization occurred in 1866, in the headwaters of Panther Creek, a tributary of the upper basin. Most of the settlement remained in the upper part of the basin, with some scattered homesteads in the middle of the basin and near the mouth. Most of the homesteads were abandoned by 1960, and reverted to timber management. Currently, some private farms and residences are found in the basin, all in the upper portion, and most along Panther Creek. The basin has an extensive road network, and is managed for timber.

Synthesis of Aquatic Habitat Results.—Watershed analysis identified a number of factors which directly impact the availability and quality of aquatic habitat in the Wolf Creek basin. Aquatic habitat is the result of interaction among land form, climate, and vegetation. Human activities also influence these factors. In the Wolf Creek basin, the available water is less than in similar Coast Range streams. Large wood falling into the stream historically provided stream structure that retained sediments, including spawning gravels, and created a diversity of habitats. In the absence of large woody material, the stream channel became simplified, groundwater capacity was reduced, temperatures increased, and overall habitat quality declined. Human activities continue in the basin. Roads particularly contribute to changes in hydrology through interception of groundwater and sediment delivery to the stream, and create fish passage problems where roads and streams intersect.

Watershed analysis was used to describe these patterns across the basin. It was also used to identify locations where restoration would improve aquatic and riparian habitat, and helped define the types of restoration activities that were most appropriate. The factors that influence aquatic habitat are discussed in much more detail in the watershed analysis (USDI 1995a).

Wolf Creek originates on the eastern margin of a tilting fault block, so that the highest points in the basin are in the lower reaches. Because Wolf Creek lacks a typical headwater area, stream gradients are quite low, being steeper in the tributaries near the mouth than in the headwaters.

The basin is mostly underlain by sandstone and siltstones, producing mostly silts and sand to the stream channel. Gravels and larger particles are mostly produced from basaltic intrusions and fracture lines. Soils are deeper and less erodible in the upper basin, resulting in limited delivery of materials to the stream channel in the upper reaches of the watershed. The highest levels of erosion and sediment delivery are in the lower reaches of the basin. Landslides are almost non-existent in the upper basin, reducing the potential delivery of materials to the channel. The incidence of landslides increases to the west, and is greatest in the lowest part of the basin. Much more material is delivered to the stream system at the mouth.

Precipitation is twice as high at the mouth as at the headwaters. Flows are directly dependent on precipitation since there is little snow melt and only limited groundwater storage capacity. Stream flows fluctuate very closely with storm precipitation patterns. Water production per square km is well below that of similar adjoining basins, resulting in flows that are lower than might be predicted for the drainage area. The highest levels of water production are in the lower reaches, not the headwaters.

The limited delivery of sediments and other materials to the stream channel in the upper reaches of the basin reduces the potential source for channel and valley floor accumulations. Even though delivery is slow, the low gradients and relatively low flows do not generate enough energy to continually flush materials through the basin, especially when channel structure is present. As a result, habitat structures created in the restoration project were generally small and simple. Gravel was placed in the channel to help create spawning beds since natural delivery of gravel was very slow.

In contrast, the lower reaches of the basin have much more dramatic and frequent flows. The amount of material delivered to the channel is high, but is quickly flushed in the absence of structure because of the high energy flows. Channel structures placed in the lower reaches were therefore much larger and better anchored to accommodate the peak flows. In both the upper and lower reaches, the structure designs anticipated the delivery of silt and sand as primary bedload material, so structures were designed to promote deposition in both the stream channel and riparian areas.

Water temperatures are naturally higher than in adjoining basins. A reduction in streamside shading, low gradients, lowering of groundwater levels due to downcutting of the channel, and extensive exposure of bedrock also contribute to higher water temperatures. Water temperatures are limiting to fish in the Wolf Creek main stem and some parts of tributaries during summer low flow periods.

Channel structure was primarily dependent on large wood, and, to a lesser extent, on beaver dams. Boulders provided only limited structure. Loss of structure led to channel incision and loss of salmonid habitat. Spawning gravels are adequate for current fish runs, but well below those needed to maintain historic populations. Deeper pools, refuge pools, and nursery areas are also very limited.

Secondary channel confinement due to downcutting has increased water velocities, decreasing the likelihood of sediments or woody material accumulating in the stream channel. Placement of the larger structures in the stream channel have created velocity discontinuities that promoted deposition of materials. While they were not designed to create specific habitat types, it was anticipated that permitting the channel to retain sediments and smaller woody materials would lead to increased channel complexity, increased groundwater interchange, and lower temperatures.

Potential sources of large wood for stream structure replacement are lacking. Riparian communities are currently fragmented, highly heterogeneous, and dominated by red alder and big leaf maple. While the abundance of large trees will increase over time on public lands, the process will take over a century and there may still be inadequate amounts of large woody material from non-Federal lands under current State forest practice rules.

HABITAT RESTORATION

Gabions built between 1969 and 1974 have all broken through in the middle, but the ends of many remain in place and provide some channel structure. Blasted pools are still present but none provide very useful habitat because of the lack of cover for fish. The spawning channel is no longer used, although it did provide spawning for several years.

Road restoration projects have reduced the amount of sediment reaching the streams. All closed roads have increased vegetation, reduced surface runoff, and reduced ditch runoff. All 21 culvert removals and replacements are functioning as expected, with anadromous fish documented above several of the culverts that were previously barriers. No information has been collected to document increased use by amphibians or other aquatic organisms although expert opinion suggests habitat connectivity for these species has been improved.

Channel structuring, involving hand-placed logs and boulders, has had mixed results (see Armantrout 1991). Retention of structures for more than two or three years was low, but the structures did create good to excellent habitat while in place. Because of their low cost they are cost effective but require frequent replacement over time. Primary benefits were to increase spawning habitat, increase the depth and quality of pools, and increase the amount of juvenile rearing habitat. Two gabions built during the same period are still functional after 10 years.

Larger log and boulder structures, placed with mechanized equipment, were built starting in 1994. They have had a retention rate approaching 90% even after three flood events. Particularly effective have been larger structures designed to raise streambed levels and improve groundwater storage. No hydrologic measurements have been made, but increases in riparian wetlands and changes in vegetation to a more water-tolerant community suggests the groundwater table has been raised. Pool habitat has increased, particularly for pools deeper than 2 m. Additional channel changes are expected as structures mature. Over 19 km of stream habitat has been restored in third- through sixth-order stream channels on both public and private lands. Woody material and sediments have deposited above structures, effectively extending the influence of structures further upstream. This benefit was anticipated but the amount of additional channel structuring was uncertain.

Riparian vegetation conversion has been carried out in association with channel structure projects. Survival and growth of trees has been greatest where canopy opening was greatest, trees were protected from animal grazing, and competing vegetation was vigorously controlled. Survival rates of planted trees have varied from less than 5% to over 90%, with highly variable growth rates.

Monitoring and Evaluation

The existing conditions, as described in the watershed analysis (USDI 1995a), are used as the baseline for interpreting results. For some actions, the evaluation is more qualitative than quantitative due to limited resources.

In-Channel Structures.—Smaller, hand-built, wood structures had the shortest life span, most lasting less than 3 years. Boulder weirs functioned the longest; some have been in place for 28 years. Some of the log and boulder structures are still providing habitat after 12 years. Gabions, on average, survived 12 years. Blasted pools are still present after 28 years but usually provide little habitat because of lack of cover. Over 90% of more recent, larger structures have maintained integrity even after three major floods. Based on previous experience, we expect structures to undergo a maturing process for 5 to 7 years.

Channel changes generally occurred as predicted. Structures accumulated smaller woody debris, larger woody material, and sediments. Retention time for the accumulated woody material varied by gradient and structure type. The deposition of sediment extended into riparian areas as planned. Channel changes include deepening of water, increased channel habitat complexity, increased off-channel rearing habitat, increased cover, increased gravel accumulations, and increased channel diversity.

Discharges.—In the absence of a gauging station, no information is available on the impact of projects on flows or sediment yields. An increase in riparian wetlands suggests groundwater levels have been raised. Temperature gauges in Wolf Creek in 1997 recorded lower temperatures compared to previous years; temperatures did not exceed lethal levels in middle and upper Wolf Creek. Higher than normal stream flows in 1997 probably contributed to the lower temperatures, as well as habitat structures.

Roads and Culverts.—Data are not available to assess the impact of road improvements on sediment yields or hydrology. Closed roads developed heavy vegetation and showed no indication of erosion. Coho salmon were documented above replaced culverts indicating fish are able to reach previously inaccessible habitat.

Vegetation.—Trees in riparian conversion plots showed variable survival rates. Where the canopy was opened, competing vegetation controlled, and tubing used to reduce ungulate predation, tree survival after 2 years generally exceeded 90% and growth exceeded 30 cm/year the first year. Western hemlock typically had the best survival and growth. Western red cedar was more likely to be browsed by ungulates, retarding growth. Douglas fir growth and survival was more dependent on site preparation and location than the other two species. The two treatments that most influenced survival and growth were the degree of canopy opening and the control of competing vegetation.

Fish.—Both the number of spawning salmonids and rearing juvenile salmonids increased in project areas. For spawning adult salmonids, this may represent a shifting in habitat selection, rather than an increase in numbers of adults, both because of a lag time before projects would be expected to increase fish numbers and the general coastal decline in spawning adults. Juvenile salmonids, particularly coho, favored project structures. Sculpins, dace, and lamprey also increased following placement of structures. The smolt trap documented a general increase in the number of emigrating smolts of all salmonid species, although numbers of steelhead and cutthroat trout smolt were small compared to coho and chinook.

Monitoring and evaluation considers how well we are achieving our goal of improving aquatic habitat. Since project work has already been carried out and additional project work is planned, we are using a continual feed-back to improve the design of projects. Experience in other streams has been very useful in designing and evaluating projects (Armantrout 1991). Although only limited quantitative information is collected on the physical aspects of the projects, the subjective information has proven useful for developing new proposals from previous experience.

SUMMARY

Watershed analysis is not a single process, but an ongoing program for integration of information and adaptive management. The techniques for gathering, integrating, and using the information obtained in monitoring and evaluation are still being developed. Over time, it is anticipated that the combination of natural processes and the cumulative influence of habitat restoration efforts will increase the stability and potential productive capacity of the Wolf Creek watershed. Watershed analysis does provide a useful tool for designing restoration efforts and for evaluating outcomes of those efforts. It is expected to provide the framework for continued development, monitoring, and evaluation of those efforts.

ACKNOWLEDGMENTS

I appreciate the project support of Leo Poole of the Eugene District, BLM. I thank R. Danehy and several anonymous reviewers for their careful manuscript reviews. Sylvia Kantor assisted with editing the manuscript.

REFERENCES

Anderson, J.W. 1981. Anadromous fish projects. Pages 109–114 *in* T.J. Hassler, editor. Propagation, enhancement, and rehabilitation of anadromous salmonid populations and habitat in the Pacific Northwest. Humboldt State University, Arcata, California.

Armantrout, N.B. 1990. Index areas as population indicators. Pages 71–99 *in* Thomas J. Hassler, editor. Proceedings of the Northeast Pacific chinook and coho salmon workshop. California Cooperative Fisheries Research Unit, Arcata, California.

Armantrout, N.B. 1991. Restructuring streams for anadromous salmonids. Pages 136–149 *in* J. Colt and R.J. White, editors. Fisheries Bioengineering Symposium. American Fisheries Society Symposium 10, Bethesda, Maryland.

Armantrout, N.B. 1997. Defining aquatic habitat through remote sensing. Pages 291–301 *in* Zhou Yingqi, Hu Fuyuan, Zhou Hongqi, Cui He, Yao Chaoqi, Din Fuhui, and Lu Yi, editors. Fourth Asian Fisheries Forum, Asian Fisheries Society, Beijing, China.

Armantrout, N.B. In press. Aquatic and riparian inventories using remote sensing. UNESCO MAB Symposium: Fish and land/inland water ecotones. Zakopane, Poland.

Brown, E.R. 1985. Management of wildlife and fish habitats in forests of western Oregon and Washington. USDA Forest Service, Publication Number R6-F&WL-192-1985, Part 1—Chapter Narrative, Part 2—Appendixes. Portland, Oregon.

Dietrich, W. E., C. J. Wilson, D. R. Montgomery, and J. McKean. 1993. Analysis of erosion thresholds, channel networks, and landscape morphology using a digital terrain model. Journal of Geology 101:259–278.

FEMAT (Forest Ecosystem Management Assessment Team). 1993. Forest ecosystem management: an ecological, economic and social assessment. USDA, USDI, USDC, and EPA, Portland, Oregon.

House, R.A., J. Anderson, P. Boehne and J. Suther. 1990. Stream rehabilitation manual. Oregon Chapter, American Fisheries Society, Corvallis.

House, R.A., and P.L. Boehne. 1987. Evaluation of instream enhancement structures for salmonid spawning and rearing in coastal Oregon streams. North American Journal of Fisheries Management 5:283–295.

Jacobs, S. E., and C. X. Cooney. 1997. Oregon coastal salmon spawning surveys, 1994 and 1995. Oregon Department of Fish and Wildlife, Portland.

ODFW (Oregon Department of Fish and Wildlife). 1997. Draft Siuslaw River salmon management plan. Oregon Department of Fish and Wildlife, Florence.

Regional Ecosystem Office. 1994. A Federal agency guide for pilot watershed analysis. Portland, Oregon.

Strahler, A.N. 1957. Quantitative analysis of watershed geomorphology. Trans. American Geophysical Union 38:913–920.

Teensma, P.D.A. 1987. Fire history and fire regimes of the central western Cascades of Oregon. Doctoral dissertation, University of Oregon, Eugene.

USDA and USDI (USDA Forest Service and USDI Bureau of Land Management). 1994. Record of decision for amendments to Forest Service and Bureau of Land Management planning documents within the range of the northern spotted owl. Portland, Oregon.

USDI (United States Department of the Interior). 1995a. Wolf Creek watershed analysis. U.S. Department of Interior, Bureau of Land Management, Eugene District, Eugene, Oregon.

USDI (United States Department of the Interior). 1995b. Record of decision and resource management plan. USDI, Bureau of Land Management, Eugene District, Eugene, Oregon.

Weyerhaeuser Corporation. 1996. Upper Siuslaw watershed analysis. Weyerhaeuser Corporation, Springfield, Oregon.

38 Watershed Restoration in Deer Creek, Washington— A Ten-Year Review

James E. Doyle, Greta Movassaghi, Michelle Fisher, and Roger Nichols

Abstract.—In 1984 the Deer Creek Group, a coalition of local landowners and managers, state agencies, fishing groups, and local Native American tribes, formed in response to concerns about declining fish populations as a result of timber harvesting activities in the Deer Creek Watershed of the North Cascades Mountains in Washington State. The group addressed these concerns through inventorying, monitoring, and identification of restoration opportunities throughout the watershed and then developed a restoration strategy. The focus of federal restoration efforts in Deer Creek over the last 10 years has been to reduce the impact of management activities and to promote the return of natural hydrologic and erosion processes. Restoration objectives included reducing coarse sediment delivery, promoting stream channel recovery and natural revegetation of riparian and floodplain areas, and restoring fish habitat by mechanically stabilizing hill slopes and streambanks. The Deer Creek watershed restoration program has been considered successful for a variety of reasons, and it demonstrates the role watershed restoration can play in a regional or provincial sustainable fisheries strategy to protect and restore west coast salmon and steelhead populations.

INTRODUCTION

Deer Creek, a tributary to the North Fork Stillaguamish River in northwestern Washington State, once supported one of the largest populations of native, wild summer-run steelhead *Oncorhynchus mykiss* in the Pacific Northwest, in addition to providing habitat for coho salmon *O. kisutch* and native char *Salvelinus malmo* populations. The loss or degradation of aquatic habitat and subsequent decline and near-extinction of the native fish populations was mostly attributed to concentrated timber management activities on landslide-prone areas throughout the watershed as well as a number of major storm events over the past 30 to 40 years, including several major landslides and flood events since the early 1980s. Deer Creek has been the focus of comprehensive watershed planning and restoration since 1984. Although approximately two thirds of the watershed is in federal ownership, all landowners began to coordinate short- and long-term timber harvesting operations within the watershed. For example, the U.S. Forest Service (USFS) deferred further timber harvesting in the Deer Creek watershed because of the critical decline in fish populations and the loss/decline in fish habitat conditions. Watershed restoration was initiated on federal land in 1984 with the goal of restoring the watershed's physical and biological processes that form and maintain the favorable aquatic habitat conditions of the past. Initially, restoration efforts emphasized reducing mass soil movement (landslides and debris torrents) from logging roads, old harvest units, and naturally unstable areas in addition to upgrading and stormproofing the existing road system. Subsequent efforts focused on controlling bedload transport and deposition as well as streambank stability by restoring structure to the stream channel network. Preliminary assessments of watershed damage resulting from two large flood events in November 1995 showed that 10 years of comprehensive watershed restoration and hydrologic recovery in Deer Creek has protected critical aquatic habitat and prevented major road and landslide failures. In 1994 more than 460 adult steelhead returned to Deer Creek, a significant increase from the past 9 to 10 years when less than 100 returned to spawn (C. Kramer, Washington Department of Fish and Wildlife, unpublished data). For the third year in a row, juvenile steelhead densities are increasing over previous years. By protecting and restoring aquatic habitat conditions in Deer Creek, anadromous fish populations are rebounding in this watershed.

In 1994, with implementation of President Clinton's Forest Ecosystem Plan. Deer Creek—as a key watershed—was targeted as a top priority watershed in the Western Washington Cascades Province for continued watershed restoration. Working with the State of Washington, both a federal and a state "jobs for the woods" restoration program was put into action in this watershed. This chapter reviews watershed management efforts by the USFS and other land managers as we pass the 10-year mark and look toward the future.

THE DEER CREEK WATERSHED—BACKGROUND

The Deer Creek watershed encompasses approximately 67 square miles, of which approximately half is National Forest land and the rest is state and privately owned. The major land use is forestry. Most of the watershed is classified as a temperate evergreen forest dominated by a western hemlock *Tsuga heterophylla* and silver fir *Abies amabilis* vegetation series. Within these two vegetation series, dominant tree species include western hemlock, silver fir, western red cedar *Thuja plicata*, and Douglas fir *Pseudotsuga mensiezii* (Henderson et al. 1992). The topography is characterized by formerly glaciated valleys separated by sharp ridges, with elevations ranging from 1,600 to 4,900 ft in the headwaters to near 200 ft at the mouth. The main stem of Deer Creek is 24 mi with 23 individual tributaries totaling an additional 56 mi of stream channel. Approximately two thirds of this stream network is accessible to anadromous fish. The effect of glaciation, in particular the deposition of lacustrine clays and silts, has strongly influenced the morphology and sediment production of the watershed. These glacial sediments in the steep lower slopes of the watershed's valleys are prone to mass wasting and erosion (Collins et al. 1994). Average annual precipitation ranges from 75 inches in the lower watershed to 110 inches or more at higher elevations. Precipitation occurs throughout the year, with 75% falling between October and March. In the North Cascade Mountains, elevation influences whether winter precipitation occurs as rain or snow. Middle watershed elevations (1600–2600 ft) are transitional *rain-on-snow* zones—snow may build up and melt several times during a winter. Winter storms, often accompanied by heavy rains and wind, may melt snow at these elevations, causing high flows and subsequent flood damage.

About two thirds of the Deer Creek stream channel network is accessible by anadromous fish and historically has supported runs of steelhead trout, coho salmon, and native char. Based on historical accounts, Deer Creek fish habitat consisted of a variety of riffles and high quality pools formed by a multitude of huge boulders with deep, clear, cold water. The wild steelhead run in Deer Creek has evoked strong emotions among past and present generations of anglers and writers. In 1918, the famous western novelist, Zane Gray, fished Deer Creek while passing from Seattle on his way to the Campbell River in British Columbia. During his second day on the stream, he hooked his first steelhead—an experience that began a long association with steelhead fishing about which Gray wrote extensively. In 1937 the Washington State Game Commission closed Deer Creek to all fishing to protect and maintain the natural production of steelhead within the watershed. In 1943 the North Fork Stillaguamish River, downstream from Deer Creek, was designated for fly fishing only—probably the first time a western river was restricted to fly fishing (Raymond 1973).

Major timber harvesting began in the watershed in the 1950s. By the mid-1980s the Deer Creek fishery was in decline due to the cumulative effects of timber harvesting: increased land slides, channel sedimentation, and loss of fish habitat. Timber has been harvested over the entire lower watershed and on one third of the federal land in the upper watershed.

THE DEER CREEK GROUP

Declining fish populations, coupled with habitat loss and degradation in Deer Creek, made protection and restoration of the remaining habitat a high management priority. In 1984 the Deer Creek Group, a coalition of local landowners and managers, state agencies, fishing groups, and local Native American tribes, formed in response to these concerns. Members of the original Deer Creek Group were:

Landowners
>Georgia Pacific
>Scott Paper
>Department of Natural Resources
>USFS-Mt. Baker Snoqualmie National Forest
>Small landowners coalition

Resource Agencies
>Washington Department of Game
>Washington Department of Fisheries
>Snohomish County-Department of Public Works

Treaty Tribes
>Stillaguamish Tribe
>Tulalip Tribe

Environmental Organizations
>Federation of Fly Fishers
>Audubon Society
>Washington Environmental Council

Mixed land ownership in the watershed and conflicting mandates of resource management agencies made it imperative that a forum be created where all concerned parties could begin a dialog to address resource issues. The objectives of the Deer Creek Group were to

- develop understanding and communication among participants,
- coordinate, review, and provide recommendations on forest management plans and activities,
- identify, prioritize, and implement restoration measures,
- collect and exchange technical information and ideas,
- educate participants and the public on complexities of natural resource management, and
- resolve forest road related problems.

Nonetheless, these efforts could not stop or reverse the decline in the adult and juvenile summer-run steelhead populations in Deer Creek. In the fall of 1993 Washington Trout formally petitioned the National Marine Fisheries Service to consider Deer Creek summer-run steelhead as a federally endangered species.

WATERSHED ASSESSMENT AND RESTORATION STRATEGY DEVELOPMENT

In 1984 the USFS and other resource management agencies and tribes of the Deer Creek Group began to address concerns about Deer Creek through inventorying, monitoring, and identification of restoration opportunities throughout the watershed. This major assessment and planning effort led to the identification of the natural landscape interactions (climate, hydrologic response, erosion processes) that had been altered by management activities and, subsequently, to the development of a restoration strategy. This watershed-scale assessment was focused mainly on the aquatic ecosystem and was not as broad in scope as present federal watershed analysis efforts.

This assessment identified various land management activities throughout the watershed, which contributed over time to a loss of historical aquatic habitat and were the main factors responsible for the decline of the native fish runs (including the watershed's famous summer-run steelhead). In particular, concentrated timber harvest and associated road building modified the hydrologic regime which, in turn, resulted in increased mass wasting (landslides) and channel degradation. Such degradation was most evident where high road density altered the natural drainage pattern in steep, first- and second-order channels that had been clear cut. These channels flushed repeatedly during storms and became a chronic source of coarse sediment input to the downstream channel network.

Over the past 10 years, the focus of federal restoration efforts in Deer Creek has been to reduce the impact of management activities throughout the watershed and to promote the return of natural hydrologic and erosion regimes. The ultimate goal has been to restore major portions of the historic aquatic habitat. The restoration strategy was two-fold: to employ long-term aquatic resource protection, followed by a comprehensive restoration program. Resource protection—in particular timber harvest and road building prescriptions and standards—was an early key step of the restoration strategy. Timber harvesting and planning continued during the early phases of watershed assessment and initial restoration implementation. Recognition of the cause and effect relationships of management activities and resource conditions led watershed specialists to identify and study ways to quantify these relationships. The concept of hydrologic cumulative effects was employed to assess and explain conditions across the watershed that contributed to stream channel degradation and fish habitat loss. Thresholds were developed using these cumulative effects models to enhance management decisions. Based on such thresholds, the USFS in 1986, and again in 1990, deferred any future timber harvesting on federal land in the watershed until aquatic habitat conditions were improved. With natural watershed recovery now possible through these management decisions, the restoration component of the strategy could be developed and implemented.

The restoration component focused on erosion processes and sedimentation. The general objective was to reduce coarse sediment delivery and to promote stream channel recovery, natural revegetation of riparian and floodplain areas, and restored fish habitat by mechanically stabilizing hill slopes and streambanks. The long-term goal was to recover depressed native fish stocks, particularly summer-run steelhead trout. The restoration program focused on implementing three categories of restoration treatments (road, hill slope, and inchannel) over multiple years. The restoration treatment objectives were to

- reduce coarse sediment input from road and hill slopes to the downstream channel system,
- reduce the risk of catastrophic failure (major landslides from roads and hill slopes),
- reduce sediment recruitment from the stream channel banks and,
- promote restoration of the natural channel sediment regime (transport and deposition).

Projects were developed from the 1984 watershed assessment and coordinated with other agencies, tribes, and landowners through the Deer Creek Group. In particular, USFS projects were prioritized by first treating the sediment source areas (current and potential). Individual sites were prioritized by an informal risk assessment of the probability of slope failure and the potential resource impacts if failure occurred. Other factors considered were accessibility to the site, success potential of the treatment, and project cost. This prioritization scheme emphasized implementation of road and hill slope projects before most inchannel work.

WATERSHED RESTORATION IMPLEMENTATION

Once a restoration strategy was developed through watershed assessment, the implementation team designed and implemented specific treatments throughout the watershed that were intended to meet the overall restoration goals through more narrowly defined project site treatment objectives. The initial step in stabilizing hill slope sites was to identify the sources of sediment (often road-related) and to prescribe treatments for reducing further sedimentation. Once the sediment supply was reduced, treatments were designed to demobilize coarse sediment at sites of deposition. Generally, hill slope treatments were intended to stabilize and revegetate bare, eroding slopes.

Design factors and treatments for roads:

Design factors
> Size of the drainage area
> Future use of the road
> Previous slope failure history at the site
> Adequacy of culvert condition, size, and spacing
> Drainage ditch condition and dimensions
> Evidence of road prism cracking or instability
> Position of the road prism on the hill slope
> Hydrologic regime of the hill slope
> Surface water dispersion on the road prism and banks.

Treatments (road obliteration, road storm proofing, and/or road upgrading)
> Installation of improved culverts, or hardened, dipped crossings (concrete fords or open box culverts)
> Installation of bridges to replace ineffective culverts
> Removal of culverts from inactive roads and restoration of natural drainage
> Construction of effective drainage ditches and insloping/outsloping of roads to direct water and reduce road fill saturation that causes road fill failure
> Installation of waterbars to intercept water and provide a controlled flow in a drainage ditch
> Removal of sidecast or settling road fill materials to reduce the risk of mass wasting
> Revegetation of cut banks, fill slopes, and/or obliterated road beds to reduce surface erosion and to stabilize the soil and ground cover

Design factors and treatments for sediment stabilization:

Design factors
> Hill slope gradient at the site
> Size of the drainage area, nature of the sediment (coarse or fine)
> Natural features and vegetation at the site
> The hydrologic regime at the site.

Treatments
> Installation of retaining structures such as sediment fences and check dams
> Revetments, drains, and trenches on unstable areas to reduce the risk of mass wasting
> Seeding, mulching, and planting of native trees and shrubs

Inchannel treatments focused on the location and position of woody debris. The majority of the treatments featured either repositioning wood on the lateral margins of the active channel or decreasing the mobility or rapid transport of this wood out of the active channel. Large woody debris complexes were installed along designated channel margins to reduce sediment recruitment from the stream banks to the channel. These treatments were also designed to add structural diversity (roughness) missing from the channel system and to then allow time for the channel to adjust. Little mechanical reshaping or regrading of the channel was employed.

Besides these restoration efforts by the USFS, others in the Deer Creek Group have allocated considerable resources toward Deer Creek restoration. For example, the Stillaguamish Tribe has spent nearly $500,000 in assessments, watershed analysis, prescription formulation, and restoration in the watershed. The Tribe is presently embarking on a riparian restoration project in 70 different locations throughout the watershed (P. Stevenson, Stillaguamish Tribe, personal communication).

MONITORING AND MAINTENANCE

When concerns were raised in the early 1980s about the effects of management activities on fish, it was clear that an integrated strategy to monitor key watershed parameters on national forest land as well as other downstream lands was needed. Prior to 1984 only a few limited habitat and fish population surveys were conducted in Deer Creek. In the summer of 1984, the USFS initiated the development of a watershed-scale, interagency, multi-resource monitoring program. The main objective of this monitoring was to determine avenues for correcting current aquatic resource problems and improving future resource management decisions. Most of the inventories and surveys were coordinated by the USFS and included state agency and tribal fishery personnel. These data collection efforts included stream surveys, fish species abundance surveys, spawning gravel and stream channel morphology assessments, stream temperature monitoring, and road and landslide inventories. After the first year of monitoring, an initial report recommended (1) the monitoring effort be continued and expanded to include the whole watershed; (2) pilot restoration projects be identified; and (3) a hydrologic cumulative effects assessment be conducted for the watershed.

The most intensive long-term monitoring effort in Deer Creek has been the juvenile fish population relative abundance effort led by the Washington Department of Fish and Wildlife. Seven index sites located throughout the watershed have been evaluated each year (summer low-flow) since 1984. In 1995 and 1996 the USFS took on the task of adding seven additional index sites to augment the original seven sites. These additional sites were added to areas of the watershed suspected as fish *refugia* areas. Fish refugia are suspected or designated areas providing high quality fish habitat, currently or in the future. Refugia are cornerstones of most species conservation strategies. Although fragmented areas of suitable habitat are important in Deer Creek, Moyle and Sato (1991) argue that to recover aquatic species, refugia should be focused at the watershed scale.

Another intensive monitoring effort in Deer Creek (USFS) has been on specific project effectiveness. A regular review of effectiveness of site treatments aided the scheduling of annual project maintenance, thus allowing existing projects to be modified and future project designs to be adjusted. Project modifications were common, particularly in the early years of the restoration program when techniques had to be adjusted to local site conditions or unfamiliar methods required minor redesign. Many project sites (road and inchannel) have required annual maintenance since implementation.

RESULTS AND CONCLUSIONS

Implementation of watershed restoration in Deer Creek has been a major undertaking for all landowners. The treatment results listed here are general because, even with detailed or statistical analysis, it would be difficult to obtain definitive results without conducting cost-prohibitive, long-term studies. Even with such studies, the complexity and dynamic nature of aquatic ecological interactions at a watershed scale would make it difficult to link the observed results with management actions. Most federal watershed restoration programs, as well as any other agency or tribal restoration effort, including the Deer Creek restoration, have restricted monitoring budgets; monitoring is usually restricted to comparing qualitative data. This type of monitoring, known as *implementation and effectiveness* monitoring, is limited to producing trend information. For example, at the watershed scale, this level of monitoring is intended to show the trend or change in the amount of landslide activity, coarse sediment transport and deposition, road failures, stream channel stability, habitat loss or degradation, stream temperatures, and ultimately, fish population status.

Nonetheless, a number of general results following 10 years of watershed restoration and protection in Deer Creek are noteworthy:

- Forty-eight out of 68 mi of federal roads have been treated since 1984 with little or no failure on treated roads.
- Coarse sediment input to the upper watershed stream system has been reduced, despite the continuing course sediment input from the DeForest Creek slide to the main stem channel.
- Hill slope sediment fences have effectively controlled coarse sediment transport.
- Inchannel structures have been effective in storing and transporting coarse sediments in upper Deer Creek.
- Fish habitat conditions have improved in the upper watershed since 1990.
- Anchoring with cable or rebar has been effective for immobilizing inchannel structures.
- Mechanical stabilization of streambanks has resulted in more stable bank revegetation.
- Restricting timber harvest has promoted natural hydrologic recovery (on federal land).
- Fish populations, especially summer-run steelhead, have increased significantly since 1993.

It may be difficult to view this work as ecological restoration because the intent was not actually to restore the watershed to its pre-management condition. The multiple use mandate of the USFS required that restoration strategies strike a balance between land use and resource protection and recovery. The emphasis of restoration in Deer Creek has been to accelerate natural recovery through a scope of work that was accessible, affordable, and achievable.

The watershed restoration program in Deer Creek has been considered successful because:

- The program assisted in the development of a watershed restoration strategy now embodied as a component in the Northwest Forest Plan aquatic conservation strategy.
- The USFS was the initial partner in developing and maintaining working relationships within the Deer Creek Group which became a model for the Washington State Timber, Fish, and Wildlife Program.
- Deer Creek has remained a federal restoration priority for 10 years and a revised restoration strategy including assessment and planning is underway.
- With USFS taking the lead, and with support from other members of the Deer Creek Group, restoration treatments have been implemented on state and on private lands in the watershed beginning in the early 1990s.
- The USFS has responded to many regional and international requests to share knowledge and skills through numerous field trips and workshops.
- Since the USFS declared a moratorium on timber harvest in 1986, significant revegetation of timber stands has occurred, resulting in partial recovery of the watershed's natural hydrologic function.
- Restoration treatments survived the 1990 and 1995 flood events, some estimated to be 50-year events.
- Relatively high quality fish habitat has been maintained in the upper watershed and is currently serving as *refugia* habitat for summer-run steelhead, coho salmon, and native char.
- Summer low-flow juvenile fish population densities in Deer Creek have been maintained or have slightly increased since 1994, a pivotal year in a decade-long trend in declining population densities.

Deer Creek watershed restoration has been successful, largely because of cooperative efforts, both within the USFS and with external partners in the Deer Creek Group. Efforts in Deer Creek may serve as a model for development of strategies and techniques necessary to implement a successful multi-year watershed restoration program in other Pacific Northwest watersheds. Watershed restoration over multiple river basins is important for any sustainable fisheries strategy designed to protect and restore west coast salmon and steelhead populations.

ACKNOWLEDGMENTS

In developing this chapter, we thank Curt Kramer, Washington Department of Fish and Wildlife, for his valuable years of insight, experience, and monitoring efforts on fish utilization in Deer Creek and the aquatic staffs of both the Stillaguamish and Tulalip Tribes in forming and carrying out the various partnerships employed to assess, evaluate, and monitor Deer Creek. Thanks to the other members of the Deer Creek Group, especially the other landowners for their restoration efforts, and to the Federation of Fly Fishers for their dedication to obtain outside funding for Deer Creek restoration work. We would also like to thank Sylvia Kantor (Sustainable Fisheries Foundation) for her editing.

REFERENCES

Collins, B.D., and six coauthors. 1994. Watershed assessment and salmonid habitat restoration strategy for Deer Creek, North Cascades of Washington. Report to the Stillaguamish Tribe and Washington Department of Ecology. 10,000 Years Institute, Seattle.

Henderson, J.A., R.D. Lescher, D.H. Peter, and D.C. Shaw. 1992. Field guide to the forested plant associations of the Mt. Baker Snoqualmie National Forest. USDA Forest Service PNW Technical Paper R6-ECOL-TP028-91.

Moyle, P.B., and G.M. Sato. 1991. On the design of preserves to protect native fishes. Pages 155–169 *in* W.L. Minckley and J.E. Deacon, editors. Battle against extinction: native fish management in the American west. University of Arizona Press, Tucson.

Raymond, S. 1973. The year of the angler. Winchester Press, Piscataway, New Jersey.

39 Integrating History into the Restoration of Coho Salmon in the Siuslaw River, Oregon

Matthew Booker

Abstract.—Any discussion of sustainable fisheries for Pacific salmon should consider the experience gained from more than a century of management and thousands of years of Native American subsistence fishing. Efforts to manage salmon fisheries sustainably can profit from an examination of the causes for declining salmon populations. This chapter traces the history of coho salmon *Oncorhynchus kisutch* in the Siuslaw River, Oregon from before Euro-American contact to the present. Siuslaw River coho have declined to a fraction of previous levels. A complex of factors is responsible, but watershed modification resulting in loss of spawning and rearing habitat and overharvest are the primary reasons. The basin likely has a lowered carrying capacity for coho salmon. Reduced spawner escapements have also contributed to the declines.

Attempts to sustain coho production in the Siuslaw River have been ongoing since shortly after Euro-American settlement began more than a century ago. Managers have tried hatcheries, restrictions on commercial and recreational harvest, and habitat improvement strategies. Where managers once sought maximum yields, current restoration efforts reflect the need to stabilize coho populations at sustainable levels. Salmon populations depend on numerous interrelated factors, including human harvest and land use. If a sustainable fishery is one in which future generations can also participate, then ongoing habitat protection and restoration efforts offer hope for a brighter future.

INTRODUCTION

Mass media attention on the decline of salmon in the Pacific Northwest in recent years has focused on the big dams on the Columbia River. Journalists have presented the crisis to the public in compelling terms: the big hydropower dams in the Columbia basin mean inevitable death for the salmon (see, for example, Dietrich 1995; Cone and Ridlington 1996). However, dams are not the only reason for the decline of once-abundant salmon runs. Along the entire Oregon Coast, coho salmon *Oncorhynchus kisutch* have been federally listed as a threatened species despite a general lack of hydropower dams in their home rivers (Weitkamp et al. 2000; Portland Oregonian 1998).

The Siuslaw River on the central Oregon Coast was once a prime producer of coho salmon. Despite the fact that there are no hydro dams in the basin, Siuslaw coho salmon populations have declined from an average of about 209,000 fish in 1889–1896 to an average of only 3,828 adult spawners in 1990–1995 (ODFW 1997). The story of Siuslaw coho is a case study of the decline of coastal coho throughout the Pacific Northwest. A complex of factors caused the disappearance of coho, and a number of issues must be addressed in order to restore coho populations. By reviewing the history of this once-productive watershed, I hope to shed light on how to approach restoration and provide an example for other watersheds.

INFORMATION SOURCES

Information about prehistoric fish populations and habitat is available from a number of sources. This chapter relies on recent archaeological data, ethnographic accounts, agency documents from throughout the 20th century, newspaper stories, and local histories from the Siuslaw basin. Personal communications with basin residents and management agency personnel also contributed to this study.

Federal agencies manage most of the lands within the Siuslaw basin. The Northwest Forest Plan (USDA and USDI 1994) stipulated that watershed analyses be completed as part of its aquatic conservation strategy for Federal lands. Watershed analyses bring together relevant geological, geomorphic, social, historical, wildlife and fisheries data for each sub-basin. Since 1995, the U.S. Forest Service (USFS) and Bureau of Land Management (BLM) have released watershed analyses covering the entire Siuslaw watershed except the lower mainstem and estuary (H. Plumley, USFS, personal communication).

Determining past abundance of salmon populations is a difficult task. Estimates of prehistoric and historic coho populations are dependent on limited data. The first data for Siuslaw coho salmon are found in cannery records from the late 19th century (Mullen 1981). A reported one-time harvest of 462,000 coho in 1899 by Siuslaw River canneries is considered exceptional (W. Beidler, ODFW, personal communication); extrapolating from cannery records, researchers place the annual coho run at the end of the 19th century between 218,750 (Sedell and Luchessa 1982) and 438,000 fish (Lichatowich and Nicholas 1991).

HISTORY OF SIUSLAW COHO SALMON

THE WATERSHED

The Siuslaw River watershed encompasses an area of about 773 square miles, most of which is in Lane County, Oregon (State Water Resources Board 1965). Over two thirds of the land in the Siuslaw basin is now public land managed by federal agencies, mostly forested, and in a checkerboard pattern with private ownership (Board of Commissioners 1952; USDI 1995a; USDI 1995b; USDI 1996).

The Siuslaw River watershed once provided excellent coho habitat, in part because of its geology. The upper part of the basin slowly tilts eastward away from the coast, resulting in low gradients throughout most of the river's uplands (USDI 1995a). With few exceptions, natural barriers to fish migration are rare. A long and biologically rich estuary stretches inland for miles. Massive western red cedar *Thuja plicata*, Douglas fir *Pseudotsuga menziesii*, and Western hemlock *Tsuga heterophylla*, grew along the streams. Fires, landslides and windstorms brought these large trees into the streams (USFS 1996). The prehistoric Siuslaw River and its tributaries were jammed with logs, which trapped the generally fine-grained sediments of the basin. Numerous beaver dams added to the diversity of wetland and off-channel habitats (USDI 1996).

All these factors combined to make the Siuslaw a perfect habitat for coho salmon which require food and shelter for their first year spent in fresh water and for cutthroat trout *O. clarki*. Chinook salmon *O. tshawytscha* were also present in large numbers, as were smaller numbers of steelhead trout *O. mykiss*. The Siuslaw may once have had the largest run of coho salmon south of the Columbia River (T. C. Dewberry, Pacific Rivers Council, personal communication), and was claimed to have "the best sea-run cutthroat trout fishing in the Northwest" as recently as the 1970s (Division of State Lands 1973).

ABORIGINAL PEOPLES AND SALMON

Prior to Euro-American contact, the Siuslaw Indians occupied or utilized natural resources throughout the Siuslaw basin. As many as 50 villages and seasonal camps may have existed within the watershed and along the adjacent coast (Skinner 1971; Zenk 1990). An estimated 900–2,100 Siuslaw Indians were present at the time of Euro-American contact in the early 19th century (Boyd 1990; Zenk 1990).

The "seasonal round" of the Siuslaw people involved migration to areas within the watershed to take advantage of various seasonally abundant resources. Winter villages were apparently located near the estuary, where shellfish and other tidewater foods were available year-round. Dried salmon and fern roots were important winter foods (Beckham et al. 1982). From spring through autumn, the Siuslaw people moved upriver to take advantage of plants, berries, and anadromous fish, including Pacific lampreys *Lampetra tridentatus* (Beckham et al. 1982).

Salmon were an important subsistence food for the Siuslaw people. Fishing methods were designed to take advantage of the fishing site topography; for example, dams or weirs were constructed in upstream areas, or salmon could be speared in shallow water (Thomas 1991). In the lower river, extensive weirs were built of stakes hammered into the mud flats, which were designed to trap fish stranded by tidal changes. Intertidal weirs provided a steady supply of numerous fish species on a daily basis, year-round (Byram 1995). Surveys by University of Oregon archaeologists from 1993 to 1995 identified several weir sites in the Siuslaw estuary that were radiocarbon-dated to 900 years before present, with use continuing into the early 20th century (Byram 1995).

Intertidal weirs were used by aboriginal people from Southeast Alaska to Northwestern California. Wooden intertidal weirs have been interpreted as technologies that allowed intensive salmon fishing for preservation and storage of a critical winter subsistence food (Moss et al. 1990). Radiocarbon dates of wood stake weirs in Southeast Alaska and British Columbia indicate that intertidal weirs have been in use continuously for more than 3,000 years (Moss et al. 1990). It is possible that weirs—and salmon fishing—have been in use even longer on the Oregon coast, but sea level and geological changes, due to earthquake subsidence and tsunamis, may have eradicated evidence of earlier weirs (M. Moss, University of Oregon, personal communication).

Siuslaw Indian impacts on salmon due to harvest and land management have not been studied in depth, but it is clear that salmon were abundant and an important food for the Siuslaw people before and after Euro-American contact. Weirs in the Siuslaw River estuary indicate that salmon harvest has occurred in the Siuslaw River for at least 1,000 years. This continuous harvest of salmon suggests that the Siuslaw basin does have a history of sustainable salmon fisheries.

EARLY SETTLEMENT AND INDUSTRY

The first non-Indian settlement in the Siuslaw basin occurred in the upper reaches of the basin near the Willamette Valley in the 1850s (USDI 1996). Most of the lower Siuslaw River watershed and the sand dunes at its mouth remained in the sole possession of Native inhabitants until the 1870s (Beckham et al. 1982). In 1875 Congress removed the Alsea sub-agency—including most of the Siuslaw basin—from the Siletz reservation, opening land for settlement by Euro-Americans (Beckham 1990).

In 1876 the first salmon cannery and fishery opened on the shifting sands where Florence now stands (Beckham et al. 1982). In 1879 the cannery and companion sawmill closed because of sand blowing into the machinery (Lomax 1971). Reports of "huge runs of spawning salmon in the Siuslaw River" attracted others who built canneries in the estuary (Eugene Register Guard 1966).

In 1884 William Kyle opened a cannery and store in Florence, and a cooperative cannery was built near the mouth of the North Fork of the Siuslaw River in 1887. These canneries imported Chinese laborers to process the fish and prepare the cans, bringing in all the necessary equipment and supplies from Astoria and San Francisco. The Kyle cannery packed about 350 cases of salmon a day (Lomax 1971). From 1887 to 1892, 68,762 cases of salmon were canned and shipped from the Siuslaw River (Lomax 1971).

The early industrial activity on the lower river was accompanied by a flood of settlers who poured into the Siuslaw basin looking for fertile farmland. Many of the early land claimants may have been Coos and Siuslaw Indians who registered for land grants (Beckham et al. 1982; Zenk 1990). The first non-Indian farmers on the lower river settled near the head of tidewater on the Siuslaw's long estuary. Settlement in the fertile bottomlands occurred quickly as early migrants sought out the best farm land (Skinner 1971).

Prior to Euro-American settlement, the Siuslaw River and its tributaries contained vast quantities of large woody material, including numerous debris jams. Sedell and Luchessa (1982) cited one early explorer's inability to navigate the lower reaches of the Siuslaw River due to the large quantities of wood. Farmers cleared logjams to reduce rainy-season ponding of floodwaters and salvage operators removed logs from stream channels and floodplains.

"Splash dams" on Siuslaw tributaries allowed early loggers to move logs downriver for processing and shipping. Splash dams destroyed instream structure and scoured streambeds, causing great damage to fish habitat (Sedell and Luchessa 1982; USDI 1995a). Splash dams were operated until at least 1910 on several Siuslaw tributaries (USDI 1995b; USFS 1996). By 1893, the impact on salmon numbers from over-fishing and habitat loss was alarming enough to inspire the building of a fish hatchery at Mapleton (Eugene Register Guard 1959).

Fish exports were the key to the early economy of the lower Siuslaw basin. Connected to the Willamette Valley only by a marginal road through densely forested mountain ranges, settlers in the lower basin were dependent on ocean shipping to bring in supplies. Small steamships brought parts for the canneries, food, and all the necessary accoutrements of the late 19th century, and then were loaded with cases of salmon for sale in the boomtown of San Francisco (Eugene Register Guard 1959).

Despite a steady demand for goods, shipping was held back by hazardous conditions. Ocean-going vessels faced snags, shifting channels, and shoals in the Siuslaw estuary and at the mouth. Marine underwriters refused to insure Siuslaw-bound craft due to the lack of beacons and unimproved conditions (Lomax 1966). Actual work on jetties at the river mouth began in 1893, but was not completed until 25 years later. The jetties completed in 1918 provided a 13-ft deep channel to shipping across the bar and prevented the river mouth from drifting north and south, a chronic problem for shippers (Lomax 1966). The ecological impact of modifications to the Siuslaw estuary has never been studied, but snagging and dredging often remove valuable habitat in estuaries (Maser and Sedell 1994).

SAWDUST

As fishermen swept the estuary and settlers cleared the huge cedars and logjams out of the valley floors for farmland and to reduce flooding, logging began in earnest in the Siuslaw basin. Difficulties in transporting raw logs or lumber to markets restrained large-scale logging in the Siuslaw watershed until the beginning of World War I. With ocean shipping to California and overseas markets a chronic problem because of the hazardous bar entrance and shallow estuary, logging boosters looked to an overland railroad connection to the populous Willamette Valley.

A rail line was finally completed to the top of tidewater in 1914. A publicist for the Willamette Pacific Railroad told an audience of prospective investors why the railroad had been built.

> You see real timber on this road as you do not see on any other line that I know of. The railroad passes through some of the finest trees seen anywhere in the world, and most of it is as yet uncut, but occasionally you see a tract that has been logged off. I tell you, this road is piercing one of the richest countries in the entire world (Eugene Morning Register 1914).

By the end of World War II, it had become clear to many Siuslaw River residents that the way to make a living was not on the river or at sea, but in the forests (Eugene Register Guard 1975). Small mills sprang up throughout the region to mill the huge trees, and the logging boom was on.

A retired logger recalled that 18 small mills in the 1940s employed between 400 and 500 men in the Siuslaw watershed (Teeuwen 1980).

In 1952 the Board of Commissioners of the Port of Siuslaw advertised that "slightly more than two-thirds of the [Douglas] fir volume [in the Siuslaw basin] is in old-growth trees, the remainder in second-growth timber," but the ancient forests were disappearing fast (Board of Commissioners 1952). In 1925 there were only 48 mills in all of Lane County and 286 million board feet of timber was cut, but by 1949 there were 200 mills and over 1.3 billion board feet of timber was logged in the county (Board of Commissioners 1952). By 1965 the State Water Resources Board was referring to "the decline of the timber industry" on the mid-Oregon Coast, and predicting tourists and retirees would become the important economic base of the area (State Water Resources Board 1965).

FISHERIES DECLINE

Even before the tremendous old growth forests of the Siuslaw basin were heavily logged in the 1920s and 1930s, coho were in decline. By 1920 the distinct early run of wild coho was almost gone (T. C. Dewberry, Pacific Rivers Council, personal communication). The gillnet fishery in the estuary selectively targeted the early coho that filled up the estuary in late summer, waiting for the first rains of autumn to raise the stream levels. Perhaps the clear water and lower flows aided the fishermen, who drifted their nets with the tide at night (T. C. Dewberry, Pacific Rivers Council, personal communication). One Siuslaw River gill-netter remembered boating 100 fish in a single night during the 1920s, and coho averaging 15 pounds. During the Depression years, his was one of 60 gillnet boats on the river (Eugene Register Guard 1975).

Annual catches in the river gillnet fishery varied, but after the 1920s the trend was downward (Board of Commissioners 1952). To counteract the downward spiral, the Oregon State Fish Commission reduced fishing intensity, restricted gillnet mesh-size, and reduced the commercial gillnet season from 189 days in 1939 to just 81 days in 1949 (Board of Commissioners 1952).

Awareness of the decline of Pacific salmon has existed since early in the 20th century. The Oregon Territorial Constitution of 1848 prohibited dams on any river or stream with salmon runs unless the dam was "constructed so as to allow salmon to pass freely up and down such rivers and streams" (Eugene Register Guard 1993). Despite this prescient law, splash dams for log transport and hydroelectric dams have severely diminished salmon populations. As historian Richard White (1995) has pointed out, more than 60 years ago all of the present threats to Pacific salmon were known and discussed, a fact that has not prevented the dramatic decline of salmon in the interim. In the Siuslaw River, the greatest declines in the coho population probably occurred near the turn of the century and again in the 1960s, following the roading and logging of much of the remaining old growth in the basin and subsequent loss of habitat (T. C. Dewberry, Pacific Rivers Council, personal communication). Currently, coho are estimated to be at less than 1% of past levels in the Siuslaw River (Dewberry 1995).

TROLLING

With the decline in the estuary fishery and increasing competition, Siuslaw fishermen began to look for new opportunities. The first salmon troller in the ocean off the mouth of the Siuslaw appeared in 1941; the owner/operator told a journalist that he first began to fish in the ocean because the river gillnet fishery was too competitive (Eugene Register Guard 1975).

Though logging became the mainstay of the local economy, ocean fishing remained good enough off the mouth of the Siuslaw to support salmon trollers. Fresh salmon were iced, sorted, and weighed at the Columbia River Packers' Association plant in Florence and trucked to Astoria, where they were sent to the retail market, either fresh or canned. One troller in 1949 reported a 4-day catch of 1,778 pounds, mostly chinook, for a cash value of $413.23, an average haul (Eugene

Register Guard 1949). In 1953 slightly more than 200,000 pounds of salmon were caught by Florence-based trollers. In 1955 trollers had a record year to date; over 400,000 pounds of salmon were landed, and as many as 82 boats unloaded at the Columbia River Packers' Association plant daily (Eugene Register Guard 1955).

However, the decline of salmon—especially coho—was apparent even as trollers landed record catches. In 1951, the first year that state fishery officials began to conduct spawning surveys of coho, an estimated 17,000 coho spawned in the Siuslaw River. In the 3-year period from 1958 to 1960, fewer than 3,000 coho made it to the spawning grounds (ODFW 1997). The last cannery on the Siuslaw closed in 1956, the same year Oregon voters passed an initiative banning commercial river fishing statewide (ODFW 1997; Eugene Register Guard 1993).

As historian Jay Taylor has shown, the closure of the river gillnet fishery did not reduce fishing pressure but simply pushed fishermen out of the rivers and into the ocean. Oregon trollers landed about 167,000 coho in 1950, and they landed almost 675,000 coho in 1957, a year after the river fisheries were closed (Taylor 1997). But the boom was short-lived. Troll landings of coho decreased by 80% from 1963 to 1967, and by 1969, only 38 commercial fishing boats were registered in Florence (Bureau of Governmental Research and Service 1969). Once again, concern about declining salmon runs resulted in further limits to commercial fishing seasons.

The coho salmon catch, backbone of the Oregon troll fishery, fluctuated during the 1970s and 1980s but showed a decided downward trend. The average annual catch statewide from 1971 to 1975 was 981,000 coho. From 1977 to 1981 the catch averaged 551,000 coho, and the troll fishery caught 320,000 coho in 1983, only 14,000 coho in 1984, and 84,000 coho in 1985 (PFMC 1983; PSC 1994). Oregon Department of Fish and Wildlife (ODFW) biologists estimated only 44,265 coho spawned in all of Oregon's coastal rivers north of Cape Blanco in 1993 (Jacobs and Cooney 1995). This compares unfavorably to estimates of over 200,000 coho returning annually to the Siuslaw River alone in the late 19th century (ODFW 1997).

The Siuslaw River also sustained a sizeable sport fishery. In 1949, over 27,000 fishing licenses were sold in all of Lane County (Board of Commissioners 1952). A study conducted in 1971 surveyed anglers who fished in the Siuslaw estuary from March to October of 1971. About 4,200 boat angler trips and 22,500 shore angler trips occurred during this single season, in the estuary alone (Gaumer et al. 1974). During the 1980s, river anglers caught an average of about 900 coho per year (ODFW 1997). The sport fishery for Siuslaw coho was closed in 1993, a year of very poor returns (ODFW 1997). Despite clearly depleted stocks, the coastal sport and commercial coho fishery continued to be an important part of the coastal economy into the early 1990s.

Harvest Management

For many years, fisheries managers responded to declining salmon runs by reducing harvest. They started by limiting commercial fishing. Managers first regulated the river gillnet fishery in 1899, requiring licenses to fish commercially. Later regulations enlarged the mesh sizes of the nets and shortened the fishing season. The same philosophy was applied to ocean fishing as it developed. Ocean net fisheries were banned and trollers operated with gear restrictions, minimum fish sizes, and ever-shorter seasons. Finally, fisheries managers turned to the sport fishery, banning sportfishing in the Siuslaw River in 1993, again with the goal of increasing spawning escapement.

Since 1951, ODFW personnel have counted spawning salmon in the same coastal Oregon stream locations every year, and coast-wide salmon populations are extrapolated from the survey results (ODFW 1997). Until 1990, ODFW surveys were carried out in areas that were more densely populated with coho than average streams, making estimates of the overall coho run artificially high. ODFW estimates of coho numbers in the Siuslaw basin were consistently higher than estimates made by the BLM from its own surveys, more than twice as high from 1986 to 1988 (Armantrout 1990). In 1990, dissatisfaction with salmon escapement survey results prompted the ODFW to revise its stream survey methods (Jacobs and Cooney 1995).

Beginning with the 1990 season, ODFW spawning surveys have included random sampling of coho spawning streams. The result has been "significantly lower" estimates of coho densities on the Oregon Coast, and correspondingly lower estimates of overall run size (Cooney and Jacobs 1995). In the Siuslaw River, for example, the reaches of stream traditionally surveyed had about two and a half times as many spawning coho as randomly selected stream sections (ODFW 1997). It now appears that Oregon's coastal coho have been over-counted for many years, obscuring the enormity of their decline (Portland Oregonian 1995).

HATCHERIES

An important aspect of contemporary fishery management in the Pacific Northwest is artificial enhancement. The first hatchery on the Siuslaw was authorized in 1893, and operated fitfully through at least 1898. Located near Mapleton, in the upper Siuslaw estuary, the hatchery was in a difficult location and the egg-take relied on fish landed by seiners in the estuary, most of which died in transit (Taylor 1997). Later attempts to enhance Siuslaw salmon runs relied for the most part on fish from outside the basin. Thousands of surplus adults from hatcheries outside the basin were planted from 1964 to 1988 (ODFW 1997).

ODFW's management plan for coho salmon in the Siuslaw has emphasized natural production since 1982. Formerly, fisheries managers stocked coho fry and pre-smolts in the Siuslaw Basin, but "stocking of pre-smolt coho was discontinued when it was found that the hatchery fish were displacing wild juvenile coho" (ODFW 1997).

On the Oregon Coast generally, hatchery production of coho fry and then smolts reached into the millions as early as the first decade of the 20th century (Lichatowich and Nicholas, in press). As managers realized that survival increased if the fish were larger, smolt production supplanted fry and fingerling plantings. Coastal hatchery production peaked in the late 1970s around 60 million coho smolts (Lichatowich and Nicholas, in press). Hatchery production increased coast-wide as natural production decreased; the result was that wild coho were replaced by hatchery fish, allowing continued harvest of the mixed stocks and increasing the danger of genetic intermingling in streams.

Hatcheries have seriously damaged wild salmon populations. Young hatchery fish compete with wild fish for limited food in the mainstem and estuary. Lichatowich and Nicholas (in press) demonstrated that smolt releases from hatcheries and total adult production (wild and hatchery combined) on the Oregon Coast followed an almost inverse relationship from 1975 to 1990. As annual smolt releases climbed from 30 million to over 60 million in the 1980s, combined wild and hatchery production for the entire Oregon Coast dropped from over 4 million adult coho in 1975 to less than a million adult coho in 1990 (Lichatowich and Nicholas, in press).

Adult hatchery coho are targeted for high harvest throughout their range. Wild fish intermingling with hatchery fish are harvested incidentally. Some hatchery coho may spawn with wild fish, ultimately threatening the genetic integrity of the population (Weitkamp et al. 1995). Hatchery fish can mask problems with freshwater habitat by maintaining high coho returns while native coho populations decline (PRC 1995). Most importantly, large numbers of hatchery fish can obscure the severity of declines in wild runs, reducing the impetus for action (Lichatowich and Nicholas, in press).

HABITAT MANAGEMENT

In the second half of the 20th century, managers have also tried to increase wild salmon runs by improving instream habitat for coho salmon. Some investigators believed the Siuslaw River was too warm and had too little flow in summer to support coho, which spend a year in fresh water before heading to sea. The State Water Resources Board (1965) identified numerous sites for water storage reservoirs in the Siuslaw basin and suggested water releases from the reservoirs would improve juvenile rearing in the river (State Water Resources Board 1965). Whether these reservoirs would have functioned as intended, or would have caused other problems is unknown; none were built, and water temperatures remain a serious issue (USDI 1995b).

Another way agencies improved instream habitat was by opening up new spawning areas. Efforts to build a fish ladder from Lake Creek to Triangle Lake began in the early 1960s, and hatchery coho were stocked above the falls beginning in 1964 (State Water Resources Board 1965; ODFW 1997). The ladder was completed in 1989, and more than 350,000 hatchery smolts were planted from 1990 to 1997 in hopes of establishing a self-sustaining run above Triangle Lake (ODFW 1997). As of January 1998, ODFW biologists were confident that coho had naturalized above Triangle Lake, although it remained unclear what percentage of the returning adults had been spawned naturally (George Westphal, ODFW, personal communication).

Agencies worked hard to remove what they perceived as barriers to coho migration. The State Water Resources Board (1965) reported that "The [Oregon State Fish and Game] Commission has engaged in extensive stream clearance to open up additional spawning and rearing in tributary areas." In fact, until 1980 a significant portion of state habitat money for the Siuslaw River went to blasting log jams and beaver dams, perhaps as much as 90% of the local habitat budget (Sedell and Luchessa 1982; T. C. Dewberry, Pacific Rivers Council, personal communication). More recently, research has shown that it is precisely the pools behind log jams, beaver dams, and other large woody debris that provide the critical rearing habitat and winter refugia juvenile coho require (ODFW 1997). Restoration efforts in recent years have focused on increasing the quantity and quality of pools and refuge areas for coho, and early reports indicate some success (Armantrout 1991; T. C. Dewberry, Pacific Rivers Council, personal communication).

INTEGRATING HISTORY INTO RESTORATION

Coho salmon in the Siuslaw River have undergone a precipitous decline in the past century. That decline was the result of a complex of factors, including forces outside the basin such as ocean conditions and offshore fisheries.

Native peoples apparently harvested Siuslaw coho for millennia without compromising the sustainability of the runs. Commercial fishing began in the late 19th century and grew quickly. Unsustainably high harvest rates in the 1880s continued until coho and chinook populations crashed in the early 20th century. Thereafter, increasing regulation on fisheries was unable to guarantee sufficient coho escapement to allow runs to rebuild. ODFW now believes that 22,800 coho must return to the Siuslaw in order to fully "seed" the basin. This represents about 44 fish spawning in each of the Siuslaw's estimated 514 miles of suitable habitat (ODFW 1997). This level of escapement has not occurred since stream surveys began in 1950. Nonetheless, commercial and sport fisheries targeting Siuslaw coho were allowed until 1993 (ODFW 1997).

Large numbers of coho salmon released from hatcheries have resulted in possible alteration of wild stock fitness. Harvests targeted on hatchery fish may have been excessive for commingled wild fish. Even when state agencies forbade harvest, poaching of spawning fish has been a problem. A basin resident recalled poachers wantonly destroying salmon in Deadwood Creek in the 1910s, and illegal harvest continues to be a threat to salmon even today in several basin streams (Ball 1980; Armantrout 1990).

Coho spawning and rearing habitat loss in the Siuslaw basin resulted from logging and its byproducts, splash-damming, and road-building; from agricultural practices and residential development; and from damage to the estuary (USDI 1995a; USDI 1996; USFS 1996). Large woody debris was directly removed from streams throughout the basin and replacement capacity was lost due to removal of riparian trees. This has resulted in lethally high summer stream temperatures and downcutting of streambeds (Armantrout 1991; Dewberry 1995; USFS 1996; USDI 1996). Complex channels dominated by large masses of wood have been replaced by simplified bedrock channels that provide far less refuge for salmonids and cannot retain sediments (ODFW 1997).

The Siuslaw estuary was once a highly productive part of the watershed, offering young salmon excellent habitat during the transition to their saltwater phase (Division of State Lands 1973). The estuary was dredged to maintain a shipping channel and gillnetters cleared snags from the estuary to protect their gillnets (Lomax 1966; Eugene Register Guard 1975; USFS 1996). Upstream activities such as loss of large woody debris supply and increased fine sediment loading may have also negatively affected the estuary.

Dewberry (1996) believes the root of the coho problem is that the Siuslaw Basin's capacity to rear juvenile coho is only 1 or 2% of its level a century ago. As evidence, he points to Siuslaw River chinook. According to Dewberry (Pacific Rivers Council, personal communication), Siuslaw chinook are experiencing strong returns because they do not spend much of their life cycle in freshwater. In contrast, steelhead, cutthroat and coho populations—all dependent upon healthy freshwater ecosystems—are in serious decline. This at times has likely been exacerbated by too few spawners utilizing the remaining habitat.

This analysis ties the coho problem to the complete transformation of coastal rivers and their tributaries. Without downed trees in the stream and valley floors that capture and retard sediments and organic material washed in during storms, streams are unable to retain nutrients and invertebrate populations have declined (Dewberry 1996). The absence of logjams means that streams lose the pools that sheltered young coho in their first year of life.

The snowballing effects of the absence of logjams in the Siuslaw basin demonstrate how actions—such as removing jams to ease salmon passage—can have unexpected results. The challenge for restorationists is to avoid the unknown dangers that have arisen for every other effort to bring back coho.

REACHING CONSENSUS ON RESTORATION

Many groups are working to rebuild Siuslaw coho, including private timberland owners, commercial fishermen, conservationists, and state and federal agencies. Until the 1980s, restoration efforts focused on hatchery supplementation of the salmon runs and removal of log jams, which were considered to be barriers to coho migration. Fisheries managers later came to believe that the major obstacle facing wild coho salmon was lack of suitable spawning and rearing habitat (ODFW 1997). Stream restoration efforts, led by government agencies, that emphasized the placement of in-stream structures have proved to be very expensive (ODFW 1997).

Restoration projects have increasingly focused on increasing the amount of large woody debris in streams, though efforts have also included blasting pools in bedrock, constructing boulder weirs, building gabions to create pools, adding spawning gravel, and planting streamside vegetation (Armantrout 1991; USDI 1995a). Restorationists have pointed out that, because few large trees exist in Siuslaw basin riparian areas, recruitment of new wood is limited (Armantrout 1991; Dewberry 1995). Stream restoration efforts attempt to maintain sufficient instream habitat until large streamside conifers reach sufficient size to once again supply large woody debris to streams (Armantrout 1991; Dewberry 1995).

Whether sufficient rearing habitat for coho exists or not, too few adult coho are returning from the ocean to fully seed the Siuslaw basin. In some basin streams, spawner densities from 1986 to 1994 averaged about 24 coho per mile of stream surveyed, only half the density the ODFW considers adequate (USDI 1995b; ODFW 1997). Many Siuslaw basin streams have had much lower densities of spawning coho (Armantrout 1990). If insufficient coho successfully reproduce, quality and quantity of freshwater habitat becomes moot. Rehabilitation of coho populations must include both providing high quality freshwater habitat and sufficient numbers of wild fish to use that habitat.

The decline of coho salmon in the Siuslaw River is the result of the intersection of many factors, from overfishing to habitat destruction. It is a complicated history that suggests no simple solutions exist. However, certain lessons can be drawn that may lead to a more sustainable future for Siuslaw coho and the people who depend on them.

History has shown that actions taken to benefit coho salmon populations—gear restrictions, shortened seasons, fishery closures, hatchery augmentation, fish passage enhancement via blasting logjams—have often failed to achieve their objective and some have even harmed the salmon. In the same way that harmful activities are evaluated and discontinued, habitat restoration projects should also be evaluated. Restoration projects need to include long-term monitoring to determine whether projects actually benefit coho. A variety of restoration techniques are currently being used, and some may be far less effective—and cost effective—than others.

Restoration should also focus on the best habitat. As one study put it in 1990, "a high percentage of the run utilizes a small percentage of the available habitat" (Armantrout 1990). Efforts to preserve and restore the best habitat are likely to result in greater benefit to coho than spreading many projects throughout the basin. Now completed for most of the Siuslaw basin, watershed analyses (USDI 1995a; USDI 1995b; USDI 1996; USFS 1996) are a first step toward management that takes local conditions, including habitat quality, into account.

Finally, restoration must confront the legacy of history. The events that have led to the decline of Siuslaw coho were for the most part unpredictable. Management actions had unexpected and unintended consequences. Nature acted too, in ways that sometimes worked against coho. In many ways, what scientists have learned is what does not work. Unpredictable change is likely to continue to be a feature of salmon fisheries.

To the extent that sustainability implies stability and predictability, it may be a difficult challenge for managers. Salmon populations naturally fluctuate a great deal, and habitats within the Siuslaw system are no longer capable of supporting runs of a quarter million coho annually. Sustainability today means ensuring coho stocks do not slide downward to extinction, as many have already (Nehlsen et al. 1991). Restoration then has very different goals than past attempts at enhancement. Rather than focusing on abundance and harvest, managers must prioritize conditions that permit and enhance coho survival. Restoration is a holding action to sustain coho while waiting for populations to rebuild. It is an entirely different perception of resources than has been characterized by much of the history of the past century, but it is one that reflects current realities.

In the century or more that must pass before large conifers once again contribute instream habitat and shade in the Siuslaw watershed, habitat restoration may be the only way that freshwater habitat can be maintained for coho. Projects installed by government agencies and by cooperative agreements among landowners and agencies are attracting use by adult and juvenile coho (Armantrout 1991; Dewberry 1995). Programs to protect what remains of salmon watersheds, such as the salmon refugia proposed by Lichatowich et al. (2000), as well as habitat restoration, may be key to maintaining and enhancing conditions supporting optimal freshwater survival while populations rebuild.

Changes in the coastwide salmon management system may also facilitate restoration. A harvest management system de-emphasizing mixed stock troll fisheries and emphasizing near-shore or terminal fisheries, where more of the harvest benefits would accrue to local communities, is more likely to engender incentive for local watershed restoration and protection. This shift in harvest patterns will also make it easier to protect weaker runs.

Hopefully, these management alternatives can preserve the remnant Siuslaw River coho salmon until human decisions and natural conditions once again favor abundance. Because the historical patterns of Siuslaw resource exploitation and habitat degradation have been similar in watersheds up and down the coast, it is likely that some of the recommendations for rebuilding Siuslaw coho are applicable to restoration and ultimate sustainability in other watersheds.

ACKNOWLEDGMENTS

I am grateful to Scott Byram and Madonna Moss of the Department of Anthropology, University of Oregon; Will Beidler, Oregon Department of Fish and Wildlife; Matt Klingle, Department of History, University of Washington; and the reviewers of this chapter: Neil Armantrout of the BLM, stream ecologist Charley Dewberry, and Dennis Todd of the University of Oregon Honors College.

REFERENCES

Armantrout, N. B. 1990. Index areas as population indicators. Pages 71–99 *in* T. J. Hassler, editor. Northeast Pacific chinook and coho salmon workshop. California Cooperative Fishery Resource Unit, Arcata, California.

Armantrout, N. B. 1991. Restructuring streams for anadromous salmonids. Pages 136–149 *in* J. Colt and R. J. White, editors. Fisheries Bioengineering Symposium. American Fisheries Society Symposium 10. American Fisheries Society, Bethesda, Maryland.

Ball, L. B. 1980. The story of Deadwood Creek. Siuslaw Pioneer 1980:88–93. Siuslaw Pioneer Museum, Florence, Oregon.

Beckham, S. D. 1990. History of Western Oregon since 1846. Pages 180–188 *in* W. Suttles, editor. Handbook of the North American Indians: Northwest Coast, Volume 7. Smithsonian Institution, Washington D. C.

Beckham, S. D., K. A. Toepel, and R. Minor. 1982. Cultural resource overview of the Siuslaw National Forest, Western Oregon. United States Department of Agriculture, Pacific Northwest Region, Portland.

Board of Commissioners. 1952. Review report on survey of Siuslaw River and bar entrance. Board of Commissioners, Port of Siuslaw Commission, Florence, Oregon.

Boyd, R. T. 1990. Demographic History, 1774–1874. Pages 135–148 *in* W. Suttles, editor. Handbook of the North American Indians: Northwest Coast, Volume 7. Smithsonian Institution, Washington, D.C.

Bureau of Governmental Research and Service. 1969. Central Oregon Coast: commercial fishing and fish processing, Vol. V. Central Lincoln People's Utility District and Lincoln, Lane, Douglas and Coos Counties, Newport, Oregon.

Byram, S. 1995. Fishing technologies from intertidal wet sites on the Oregon Coast. Unpublished research paper presented at the Hidden Dimensions Conference, 25 April 1995, Vancouver, British Columbia.

Cooney, C. X., and S. E. Jacobs. 1995. Oregon coastal salmon spawning surveys, 1993. Oregon Department of Fish and Wildlife, Portland.

Cone, J., and S. Ridlington. 1996. The Northwest salmon crisis: a documentary history. Oregon State University Press, Corvallis.

Dewberry, T. C. 1995. Knowles Creek report. Pacific Rivers Council, Eugene, Oregon.

Dewberry, T. C. 1996. Knowles Creek report 1996: lessons learned from the flood of February 1996. Pacific Rivers Council, Eugene, Oregon.

Dietrich, W. 1995. Northwest passage: the great Columbia River. Simon & Schuster, New York.

Division of State Lands. 1973. Oregon estuaries. State of Oregon, State Land Board, Division of State Lands, Salem.

Eugene Morning Register. May 16, 1914. Lane County Historical Museum collections, Eugene, Oregon.

Eugene Register Guard. September 22, 1949. Register Guard Publishing, Eugene, Oregon.

Eugene Register Guard. February 27, 1955. Register Guard Publishing, Eugene, Oregon.

Eugene Register Guard. February 22, 1959. Register Guard Publishing, Eugene, Oregon.

Eugene Register Guard. June 19, 1966. Register Guard Publishing, Eugene, Oregon.

Eugene Register Guard. March 23, 1975. Register Guard Publishing, Eugene, Oregon.

Eugene Register Guard. August 3, 1993. Register Guard Publishing, Eugene, Oregon.

Gaumer, T., D. Demory, and L. Osis. 1974. 1971 Siuslaw River estuary resource use study. Fish Commission of Oregon, Division of Management and Research, Portland.

Jacobs, S. E., and C. X. Cooney. 1995. Improvement of methods used to estimate the spawning escapement of Oregon coastal natural coho salmon. Annual Progress Report 1993–1994, Fish Research Project. Oregon Department of Fish and Wildlife, Portland.

Lichatowich, J. A., and J. W. Nicholas. In press. Oregon's first century of hatchery intervention in salmon production: evolution of the hatchery program, legacy of a utilitarian philosophy and management recommendations. Proceedings of international symposium on biological interactions of enhanced and wild salmonids, Nanaimo, British Columbia.

Lichatowich, J. A., G. R. Rahr III, S. M. Whidden, and C. R. Steward. Sanctuaries for Pacific salmon. Pages 675–686 *in* E. E. Knudsen, C. R. Steward, D. D. MacDonald, J. E. Williams, and D. W. Reiser, editors. Sustainable fisheries management: Pacific salmon. Lewis Publishers, Boca Raton, Florida.

Lomax, A. L. 1966. Early port development on the lower Siuslaw River. Lane County Historian 11: 35–39. Lane County Historical Society, Eugene, Oregon.

Lomax, A. L. 1971. Early shipping and industry in the lower Siuslaw valley. Lane County Historian 16: 32–38. Lane County Historical Society, Eugene, Oregon.

Maser, C., and J. R. Sedell. 1994. From the forest to the sea: the ecology of wood in streams, rivers, estuaries and oceans. St. Lucie Press, Delray Beach, Florida.

Moss, M., J. Erlandson, and R. Stuckenrath. 1990. Wood stake weirs and salmon fishing on the Northwest Coast: evidence from Southeast Alaska. Canadian Journal of Archaeology 14:143–158.

Mullen, R. E. 1981. Oregon's commercial harvest of coho salmon, *Oncorynchus kisutch* (Walbaum), 1892–1960. Oregon Department of Fish and Wildlife. Information Report 81-3, Portland.

Nehlsen, W., J. E. Williams, and J. A. Lichatowich. 1991. Pacific salmon at the crossroads: stocks at risk from California, Oregon, Idaho, and Washington. Fisheries 16(2): 4–21.

ODFW (Oregon Department of Fish and Wildlife). 1997. Siuslaw River basin fish management plan. Oregon Department of Fish and Wildlife, Portland.

Portland Oregonian. January 5, 1995. Oregonian Publishing Company, Portland.

Portland Oregonian. August 4, 1998. Oregonian Publishing Company, Portland.

PFMC (Pacific Fishery Management Council). 1983. Draft proposed framework plan for managing the ocean salmon fisheries off the coasts of Washington, Oregon, and California commencing in 1984. Pacific Fishery Management Council, Portland.

PRC (Pacific Rivers Council). 1995. Coastal salmon and communities at risk: the principles of coastal salmon recovery. Pacific Rivers Council, Eugene, Oregon.

PSC (Pacific Salmon Commission). 1994. Ninth annual report 1993/1994. Pacific Salmon Commission, Vancouver, British Columbia.

Sedell, J. R., and K. J. Luchessa. 1982. Using the historical record as an aid to salmonid habitat enhancement. Pages 210–223 *in* N. B. Armantrout, editor. Proceedings of a symposium on acquisition and utilization of aquatic habitat inventory information. American Fisheries Society, Western Division, Portland.

Skinner, M. L. 1971. Florence, the "fir-clad" city. Lane County Historian 16: 25–31. Lane County Historical Society, Eugene, Oregon.

State Water Resources Board. 1965. Mid-coast basin report. Oregon State Water Resources Board, Salem.

Taylor, J. E. 1997. Making salmon: economy, culture, and science in the Oregon fisheries, precontact to 1960. Doctoral dissertation. University of Washington, Seattle.

Teeuwen, R. C. 1980. Oregon's coastal people: their lives and livelihoods–a pictorial essay. J. Rasmussen, editor. Oregon Coastal Zone Management Association, Newport, Oregon.

Thomas, R. 1991. Life in the Siuslaw valley prior to European settlement: glimpses of tribal lifestyle at the confluence of Siuslaw, Kalapuya, Yoncalla and Lower Umpqua tribal domain. Unpublished research paper, Oregon Collection archives, University of Oregon, Eugene.

USDA and USDI (USDA Forest Service, USDI Bureau of Land Management). 1994. Record of decision for amendments to Forest Service and Bureau of Land Management planning documents within the range of the Northern spotted owl. Portland, Oregon.

USDI (United States Department of the Interior). 1995a. Wolf Creek watershed analysis. U.S. Department of the Interior, Bureau of Land Management, Eugene District, Eugene, Oregon.

USDI (United States Department of the Interior). 1995b. Lake Creek watershed analysis. U.S. Department of the Interior, Bureau of Land Management, Eugene District, Eugene, Oregon.

USDI (United States Department of the Interior). 1996. Siuslaw watershed analysis. U.S. Department of the Interior, Bureau of Land Management, Eugene District, Eugene, Oregon.

USFS (United States Forest Service). 1996. Indian/Deadwood watershed analysis. USDA Forest Service, Siuslaw National Forest, Corvallis, Oregon.

Weitkamp, L. A., and six coauthors. 1995. Status review of coho salmon from Washington, Oregon and California. NOAA Tech. Memo. NMFS-NWFSC-24. U.S. Department of Commerce, Seattle.

Weitkamp, L. A., T. C. Wainwright, G. J. Bryant, D. J. Feel, and R. G. Kope. 2000. Review of the status of coho salmon from Washington, Oregon, and California. Pages 111–118 *in* E. E. Knudsen, C. R. Steward, D. D. MacDonald, J. E. Williams, and D. W. Reiser, editors. Sustainable fisheries management: Pacific salmon. Lewis Publishers, Boca Raton, Florida.

White, R. 1995. The organic machine. Hill and Wang, New York.

Zenk, H. B. 1990. Siuslawans and Coosans. Pages 572–579 *in* W. Suttles, editor. Handbook of the North American Indians: Northwest Coast, Volume 7. Smithsonian Institution, Washington, D.C.

Section VIII

Toward Sustainability

40 Community Education and Cooperation Determine Success in Watershed Restoration: the *Asotin Creek Model Watershed Plan*

Angela Thiessen and Linda Vane

Abstract.—Asotin Creek, located in the southeastern corner of Washington State, is similar in many ways to other salmon-bearing streams in the lower Snake River system. The watershed has been significantly affected by human activities and catastrophic natural events, such as floods and droughts. It supports only remnant salmon and trout populations compared to earlier years and will require protection and restoration of its fish habitat and riparian areas in order to increase its salmonid productivity. The *Asotin Creek Model Watershed Plan* was the first plan in the state of Washington to be concerned specifically with habitat protection and restoration for salmon and trout and, therefore, is a useful model for salmon recovery efforts elsewhere. The planning process was guided by a plan of work developed jointly by a Landowner Steering Committee (local landowners) and a Technical Advisory Committee (local volunteers and agency representatives). Trust, credibility, commitment, and active communication between the two committees were key to the successful conclusion of the planning process. Much of the information presented herein was abstracted from the April 1995 final report, the *Asotin Creek Model Watershed Plan* (LSC 1995).

INTRODUCTION

The Asotin County Conservation District (ACCD) undertook a 2-year water quality monitoring study in Asotin Creek in 1991. The study provided baseline data, helped identify water quality problems, and was instrumental in the subsequent development of a watershed restoration plan. The study involved land users and the public in an extensive public information campaign in which signs were installed, pamphlets were distributed, and public meetings were held.

In 1993 the ACCD was selected by the Northwest Power Planning Council (NPPC) to develop a Model Watershed Plan (Plan) in which the ACCD would identify resource problems, formulate and evaluate alternatives, and ultimately recommend a salmon recovery and watershed management strategy. In the 2 years that followed, the ACCD instituted a highly successful public planning process that engaged local landowners, community volunteers, government agencies, and scientists in a cooperative venture. Leadership was provided by a number of ranchers and farmers with property along Asotin Creek and by landowners on upland ridges throughout the watershed. This chapter describes a successful model of how local organizations, landowners, and citizens can work together toward the sustainability of salmonid habitats.

THE WATERSHED SETTING

A RURAL COMMUNITY

Asotin County and the Asotin Creek watershed are located in rural southeastern Washington State. The county is sparsely populated (18,900 residents). Fewer than 200 people reside within the 208,257-acre Asotin watershed. Although most of the local residents live outside the watershed, extensive agricultural and forestry activities occur within its boundaries. The watershed comprises wetlands, forested land, cropland, pastureland, and rangeland. Rangeland and pastureland occupy about 43% of the watershed, while cropland occupies about 26%. The watershed is host to at least 142 farms/ranches, with only 73 of them managed by full time owner-operators. Crops are harvested on about 115 of these establishments, which produce primarily wheat and barley. Eighty-one farms/ranches raise mainly beef cattle, while at least one ranch raises hogs. Cattle graze on both private and public lands.

Most of the cropland in Asotin County is classified as highly erodible land by the federal government and is farmed using methods designed to reduce erosion, as required under the 1985 United States Department of Agriculture (USDA) Farm Bill. Farmers in the watershed have long recognized the need for good soil conservation practices. The first terrace system was established in the 1940s. At present 78% of the cultivated cropland is terraced.

In spite of the care farmers have taken to reduce erosion, significant amounts of sediment enter Asotin Creek from agricultural runoff. The Natural Resources Conservation Service (NRCS) estimated that over 23,000 tons of soil from cropland are lost annually and washed into non-farmed areas such as grassed waterways, terraces, roadside ditches, and drainage ways. Much of this sediment eventually finds its way into the stream system (LSC 1995).

ASOTIN CREEK

The Asotin Creek drainage, with its headwaters in the Blue Mountains, comprises 360 miles of perennial and intermittent stream channels. The watershed is home to anadromous populations of chinook salmon *Oncorhynchus tshawytscha*, steelhead *O. mykiss*, bull trout *Salvelinus confluentus*, and Pacific lamprey *Lampetra tridentata*.

At higher elevations, Asotin Creek drains the northeast sector of the Umatilla National Forest. Waters within the national forest routinely meet Washington State water quality standards for temperature, turbidity, and fecal coliform bacteria. However, water quality standards for temperature and bacteria levels were not met in water samples taken in 1990 and 1993 from lower reaches of Asotin Creek.

Most of the Asotin Creek mainstem and portions of its major tributaries have been straightened, diked, or relocated as the result of flooding, flood-proofing for property protection, and road construction. These and other changes have caused the Asotin Creek channel to become wider, straighter (less sinuous), and more confined (less floodplain), with steeper banks and generally faster streamflow than in the past.

Asotin Creek's historic stream channel was flatter, more sinuous, and less entrenched than it is today. Once characterized by alternating point bars and well-developed pools and riffles, Asotin Creek has since transformed into a less complex riffle/glide system. The loss of well-defined geomorphic features, coupled with changes in the historic flow and sediment regimes, has led to widespread habitat degradation and reductions in fish numbers (LSC 1995).

A COMMUNITY PLANS FOR RESTORED SALMON HABITAT

Having been selected by the NPPC to develop a Model Watershed Plan, the ACCD took steps to involve the community from the earliest stages of the planning process. It was understood by the ACCD and other participating agencies that the local landowners' cooperation and participation would be critical to the success of the project. As a first step, the ACCD held public meetings to inform the community and solicit feedback about the project. Landowners from throughout the county, as well as local citizens and community groups, were invited to attend several public meetings. At first reluctant to engage, landowners eventually began to voice their concerns and thereafter became central to the planning process. For its part, the ACCD and cooperating agencies gave a clear signal that they wanted, needed, and valued the landowners' opinions.

From the interest generated at these meetings, the ACCD and the NRCS decided to form the seven-member Landowner Steering Committee (LSC). The LSC was designed to assist the ACCD in developing a comprehensive conservation plan while representing the views and needs of the community and taking a leading role in the planning process.

To assist the LSC with fulfilling their role, the ACCD also established a Technical Advisory Committee (TAC) comprised of local volunteers and representatives from government agencies with an interest in the management of the Asotin Creek watershed. The primary duties of the TAC were to conduct a resource inventory of the watershed, identify resource problems, develop well-considered alternatives, and recommend salmon recovery measures. Represented on the TAC were:

- USDA Natural Resources Conservation Service
- USDA Forest Service
- Washington Department of Fisheries and Wildlife
- Washington Department of Ecology
- Washington Department of Natural Resources
- Washington State University Cooperative Extension Service
- Bonneville Power Administration
- Clearwater Power Company

Members of the TAC included individuals representing the fields of fisheries, biology, wildlife management, range and soil conservation, and stream geomorphology. It was their job to ensure that all scientific and biological measures were considered and implemented.

A work plan was developed by the LSC and the TAC. Both groups met regularly throughout the process to review progress and exchange information. Trust, credibility, commitment, and active communication between the two committees were of central importance to the success of the process.

RESULTS OF THE PLANNING PROCESS

A commitment to cooperation, which was the governing principle of the Asotin Creek planning process, was reflected in the project mission statement developed by local landowners and volunteers with the assistance of agency staff. The statement adjured participants to "*complete and implement an integrated plan for the Asotin Creek watershed which will meet landowner objectives and agency acceptance, in order to protect and enhance all resource bases with concern for long-term sustainability.*"

With its mission in mind the planning team identified project objectives. Drawing upon the results of a formal watershed analysis, the team identified four primary problems to be addressed by the Plan. After two years of data collection, analysis, interpretation, and discussion, the *Asotin Creek Model Watershed Plan* was completed in April of 1995. The final *Asotin Creek Model Watershed Plan* recommends that salmon recovery measures be undertaken in four key areas.

Stream and Riparian Areas.—To include riparian native woody plantings, wetland enhancements, reconnecting off-channel rearing sites, stream meander reconstruction, placement of instream habitat structures, and fencing.

Forestland.—To include stock water and/or fish and wildlife ponds, critical area planting, and tree planting.

Rangeland.—To include stock trails and walkways, noxious weed control, well development, and fencing.

Cropland.—To include permanent grass cover, grassed waterways, terraces, filter strips, and sediment basins.

Specific strategies to implement these measures are outlined in the Plan. The ACCD Watershed Coordinator will be responsible for the implementation of the Plan and will continue to coordinate with landowners, agencies, committees, and contractors as necessary. Although long-term funding for implementing the Plan has not been secured, the ACCD, with the support of the LSC, will identify funding sources and will develop specific project plans, obtain permits where needed, and implement the salmon recovery measures contained in the Plan. What follows is a description of the process by which those measures were developed, and insights gained by agency personnel who participated in the process.

KEYS TO SUCCESS

The development of the *Asotin Creek Model Watershed Plan* was a locally based process that reflected community concerns and benefited from the technical and scientific expertise of several agencies. Members of the community at large and local landowners, in particular, were brought into the decision making process. Scientists and government staff provided organizational assistance and guidance on technical issues. Together, project participants developed a proactive and highly successful approach to resource management that was reflected in the completed Plan. Key factors in the success of the planning project were (1) a highly effective watershed coordinator responsible for project management and the recruitment of community participants; (2) cooperation from local landowners; (3) support from government agencies; and (4) an aggressive outreach and education program.

WATERSHED COORDINATOR

Because he was an area resident with a farming and ranching background, the Asotin Watershed Coordinator was well-suited to act as a liaison between local residents and government agencies. It has been our experience that citizens often feel intimidated by agency personnel and in some cases even refuse to deal with government employees. Because landowner input and support was sought, the Watershed Coordinator spent much of his time meeting and talking with landowners. He assured them that their participation was important to the entire process, from salmon recovery planning through implementation. Having engaged the landowners in the planning process, the Watershed Coordinator proved to be highly effective in helping to communicate landowners' perspectives on sensitive issues to project participants who were unfamiliar with the farmers' and ranchers' way of life. As a local resident he was already familiar with other watershed residents and shared their concerns for the future of the watershed.

LANDOWNERS

When asked to participate in the watershed planning process, landowners were eager to learn more about the watershed, participate in public meetings, and serve on committees. Overall, the development of the Plan was marked by splendid landowner support, cooperation, and participation. The conservation district staff members believed they were fortunate to work with, and enjoyed the trust of, a very special group of individuals. The success experienced would not have been possible if landowners had not played such an important role in developing the Plan.

Obtaining landowner cooperation, acceptance, and permission is sometimes perceived as an extraordinary and often insurmountable challenge. This can be due to underestimating the willingness of landowners to cooperate. Because the landowners' livelihood is dependent upon the availability of natural resources, they are often genuinely interested in preservation and restoration. We found that landowners, when asked to participate, became valuable contributors to the planning process.

Landowners not only demonstrated a commitment to protecting the integrity of the watershed's natural resources, they proved to be very willing and able to educate themselves. Many local farmers and ranchers attended seminars and conferences to learn about the latest techniques and research findings in the agricultural world. Most subscribe to journals, newspapers, and magazines to further educate themselves. Oftentimes landowners and other local residents proved to be more knowledgeable about methods and best management practices than were agency representatives.

The landowners in the Asotin Creek watershed share a unique bond in having been born and raised in Asotin County. Most went to school together, showed animals in 4-H and FFA together, and often fished the same streams together. Since early childhood, they have developed strong ties with each other and with the land. As individuals whose livelihoods depend on the productivity of the soil and forests, and the availability of clean water, Asotin Creek landowners proved more than willing to take steps to protect and enhance their natural resources.

FORMATION OF THE LANDOWNER STEERING COMMITTEE

In deciding who should serve on the LSC, the ACCD considered many factors, knowing it would be important to involve people from all sectors of the community. They wanted landowners who lived along the creek, as well as upland farmers and ranchers, to participate. The ACCD believed it was important to include a representative from a fishing or environmental group, but unfortunately neglected to nominate a member from such a group. Non-landowners were not well represented on the LSC. Consequently, some people thought the selection process and composition of the LSC was biased. The LSC was perceived by some as a "good old boy" network comprised of the more influential ranchers and farmers from the area.

The skepticism was somewhat justified in that three of the biggest landowners who lived along Asotin Creek were chosen for the LSC. One of these individuals had just purchased a large ranch after making his living as a commercial fisherman in Alaska for 40 years. In considering this individual for the LSC, the selection committee believed this was an opportunity to bring a diverse background and a unique perspective to the committee. The other two ranchers along the creek had showed a great deal of interest at the public meetings, were involved in other organizations in the county, and had large herds of cattle that they wintered, calved, and grazed along the creek. The remaining four members were ranchers and/or farmers on upland ridges throughout the watershed.

One of the criteria given the most weight in the selection process was the potential committee members' commitment to protecting and enhancing the watershed for the benefit of all. Group unity is difficult to attain if members are interested in participating only to protect their own land and assets. Each member chosen had, in his or her own way, genuine concern for the environment. Frequently, the topic of discussion at committee meetings focused on the benefits and joy of cool clean water, clean air, and lush green mountains that all of the committee members personally valued in their lives.

Most of the LSC members believed that implementing recommendations to improve water quality, deliver less sediment to streams, and lower water temperatures would ultimately benefit the landowners themselves. Their acceptance of this basic truth is very important to the ultimate success of the Plan. Landowners necessarily hold their own economic well-being as a high priority. Their lands and pastures are extremely valuable to the production of livestock and hence to themselves. Generally, a landowner will not relinquish his water rights or let anyone fence off a stream unless he sees a personal benefit. If it can be shown that the practice will increase streambank vegetative cover and reduce loss of pasture from eroding streambanks, while at the same time providing better conditions for fish, landowners will certainly be willing to cooperate.

AGENCY SUPPORT

The support and cooperation of various agencies also contributed to the successful completion of the Plan. Agency representatives took a back seat role and made themselves available for necessary technical assistance. This approach worked well because landowners did not feel pressured or forced by regulatory agencies to participate in the planning process. The Assistant State Conservationist for the NRCS stated that, *"The intended purpose of the Asotin Creek Model Watershed Plan was to demonstrate that you could take a local watershed group and provide technical assistance that would provide a technically sound plan that can be implemented by the local landowners."* (F. Easter, NRCS, personal communication)

Agency support and acceptance increased our success dramatically. Agency representatives took part in every phase of the planning process in the Asotin Creek watershed. They surveyed, conducted studies, collected data, completed the watershed analysis, attended many meetings, and later even served as construction officers during project implementation. The landowners gained a high degree of respect for the agency representatives, which helped keep communication lines open among those involved in developing the Plan.

PUBLIC OUTREACH AND EDUCATION

Public outreach and education proved to be an efficient and cost-effective means of exposing large numbers of people to the watershed and the planning effort. The ACCD produced a quarterly newsletter, prepared several news articles for local papers and the local radio station, installed demonstration signs at project sites, and conducted watershed tours for local citizens, government officials, legislators, and various interest groups. Tree cuttings and plantings along streambanks attracted a variety of individuals and groups such as Boy and Girl Scouts, FFA, 4-H clubs, and many local students.

Landowners and members of the LSC participated actively in the outreach effort by allowing demonstration sites on their land that fenced off portions of Asotin Creek. These demonstration projects involved installing samples of alternative watering devices, installing windbreaks and dormant stock plantings, and even a gutter system on a wintering area hay shed to reduce soil saturation and the delivery of contaminated runoff to the stream.

It was especially useful to have respected landowners in the community, such as the LSC members, take an active role in promoting salmon recovery efforts and serving as role models for other cooperators. When these individuals experienced success and talked to other landowners about the success, there was much less skepticism and much greater enthusiasm. Many people toured the demonstration projects and observed firsthand the improvements made to fish and fish habitat. When they learned that significant benefits could be gained with minimal effort and financial commitment by the landowner, most people began to champion the resource objectives and methods set forth in the Plan.

A critical part of any public outreach and education program is education in the school system. We believe that many children learn best through hands-on activities and will retain information much longer when they can put it to practical use. The ACCD made a point to speak at school staff meetings to educate teachers and administrators about the Plan, explain how it was developed, and discuss the importance of our watershed. We are fortunate in that Asotin Creek is only three blocks away from the local elementary, junior high, and high schools. Having the stream almost literally in one's own back yard provided numerous educational opportunities. Students participated in a number of field days on Asotin Creek during which they experienced their watershed through hands-on experiments and activities. Students learned to conduct a number of water quality tests, such as measuring dissolved oxygen, sampling macroinvertebrates, and measuring sediment deposition. They also learned about the life cycle of salmon and took tours to better understand the purpose and methods for instream and riparian habitat restoration.

DISCUSSION

By working through an interdisciplinary team and using an "ecosystem approach" to planning, the watershed analysis identified factors that collectively contribute to the degradation of fish habitat in Asotin Creek. The *Asotin Creek Model Watershed Plan* identified a level of restoration that was deemed both necessary and "doable" by laymen and experts alike. The planned level of restoration was thought to be reasonable for this particular watershed because (1) resources in the watershed are still at a "treatable" level; (2) recommended treatment is considered cost-effective; (3) funding expectation is high on the part of the ACCD; (4) the ACCD's administrative and technical infrastructure are in place; and (5) there is widespread support for the Plan's recommendations within the local community.

Public support for the Plan is reflected by the strong involvement of local residents in our programs, even though participation is completely voluntary. The people who participate do so because they want to, not because they have to. One local rancher and landowner who sat on the LSC, is a good example. Early in the process, he voiced his enthusiasm and willingness to cooperate when he said, "We need to step up and take the initiative and help form the policies, rather than react to them after they are in place" (NRCS 1994).

Education and cooperation underlie the success of any watershed restoration effort. The *Asotin Creek Model Watershed Plan* is no exception. When the public is informed, landowners are involved in the planning process, and government agencies work cooperatively with affected parties, the goal of sustaining the natural resources that affect our lives is within our collective reach.

ACKNOWLEDGMENTS

This study evolved from a request by Stu Trefry, then Assistant to the Director of the Washington State Department of Agriculture, who wanted others to learn of the success experienced in the Asotin Creek watershed. We would like to thank Mr. Trefry for the opportunity to share our story. In addition, we would like to thank the local landowners and all of the individuals and organizations who were involved in the development and implementation of the *Asotin Creek Model Watershed Plan*. Particular thanks go to Mr. Brian Sangster, the Asotin County Watershed Coordinator. Primary project funding was provided by the Bonneville Power Administration and the Washington State Conservation Commission. We would also like to thank Dick Stone and two anonymous reviewers for their helpful reviews of the draft manuscript.

REFERENCES

LSC (Landowner Steering Committee). 1995. Asotin Creek model watershed plan. Asotin County Conservation District, Clarkston, Washington.

NRCS (Natural Resources Conservation Service). 1994. A place to come home to: recovering salmon habitat in the west. Natural Resources Conservation Service, U.S. Department of Agriculture, Washington, D.C.

41 Spring-Run Chinook Salmon Work Group: A Cooperative Approach to Watershed Management in California

Nat Bingham and Allen Harthorn

Abstract.—Spring-run chinook salmon populations in California's Central Valley have been depressed by a wide range of habitat problems, but particularly water project operations. Listing of the spring-run chinook under the Federal and California State Endangered Species Acts (ESA) was proposed by environmental groups in 1992. The Pacific Coast Federation of Fishermen's Associations (PCFFA), a commercial fishermen's group, proposed and initiated an alternative recovery strategy. The Spring-run Chinook Salmon Work Group was organized by a coalition of fishermen, environmental groups, and government agency biologists. Initial efforts of the Work Group led to extensive community outreach activity by the University of California Sea Grant Program. The Work Group also supported the formation of local watershed conservancies in several key watersheds which had remnant spring-run populations. As a result of this collaborative process, a broad range of habitat protection and restoration measures have already been implemented. Thanks to Work Group habitat restoration activities, a 5-year ESA listing reprieve was obtained during which local environmental groups agreed to suspend their listing petition process. In the meantime, the Pacific Rivers Council petitioned NMFS to list the spring-run in 1996. As of publication, listing of spring-run chinook has occurred in California under the state ESA and the run was proposed for federal listing March 9, 1998 (63 Fed. Reg. 11482). A final decision on federal listing of the spring-run is expected in September of 1999. Successful community-based habitat restoration activities in the meantime may have helped mitigate the impact of such a listing or shortened the time between listing and recovery.

INTRODUCTION

This narrative will tell the story of how and why spring-run chinook salmon declined in California and how that led to pressure for state and federal listings, which in turn led to the active involvement of commercial salmon fishermen in a collaborative recovery effort. It will describe how the fishermen got involved with cattle ranchers, farmers, and environmentalists to build a coalition dedicated to recovering spring-run salmon. The difficult process the farmers, ranchers, and fishermen went through to begin to understand each others' true interests and needs relative to watershed management is well worth following. As society's demands for water and development continues to increase, to the detriment of salmon and their habitats, so will the need for such interactions. The social and political process followed for potentially revitalizing the Central Valley spring-run chinook salmon may serve as a model for restoring sustainability of other depleted Pacific salmon runs.

HISTORICAL BACKGROUND

Four races of chinook salmon inhabit California's Central Valley during the freshwater phase of their anadromous life history. One race, the spring-run chinook salmon, migrates into San Francisco Bay from the Pacific Ocean primarily as 3- and 4-year-old adults. They move up the Sacramento River in March, entering their natal tributary streams from mid-February to June. Spring-run salmon typically ascend the tributary streams draining higher elevation, northern Sierra Nevada mountains where they find cool, deep pools in well-shaded canyons and "hold" without feeding throughout summer.

Triggered by decreasing water temperatures in the fall, some spring-run chinook salmon migrate to higher elevations, while others remain near their holding areas to spawn. Some instances of spawning migration from the holding areas to lower elevations, where there is unusually cold water available at temperatures comparable to streams at higher elevations, have also been reported, particularly in Butte Creek (Yoshiyama et al. 1996). As early as December, the salmon fry begin to emerge from the gravel. Some remain in their natal stream for up to a year, migrating downstream as yearlings, while others move downstream as young-of-the-year fry.

Before the extensive damming and diversion of the Central Valley tributary streams, spring-run chinook populations were robust. Spring-run salmon seem to be totally dependent on free-flowing, pristine, alpine streams. Historically the spring-run in both the Sacramento and San Joaquin rivers have numbered in the hundreds of thousands of spawning adults (Fisher 1994). Before the turn of the 20th century, the spring-run chinook dominated the commercial fishery. The spring-run passed through Suisun Bay in April, where they were harvested in an inriver drift gillnet fishery. Spring-run chinook salmon were preferred over the fall-run due to their bright ocean-fresh condition. Catches in this fishery ranged as high as 12 million lbs. from 1870 to 1920 (Clark 1940).

Hydraulic mining and dam construction degraded habitat conditions for salmon. Mining destroyed spawning habitat in many Sierra streams, where placer gold deposits were found and exploited. As the easily accessible deposits in streambeds were exhausted, the miners discovered that rich placer deposits were buried deep under layers of unconsolidated sediment in ancient riverbeds adjacent to the foothill canyons. A new technology quickly evolved. Streams were diverted into flumes high in the watersheds and channeled into high-pressure pipelines. Water cannons were then used to wash the sediment overstory downstream. Millions of yards of debris choked the rivers, causing major floods in the Sacramento Valley (Kelley 1989). Some of the first environmental protection laws in the nation were enacted to resolve disputes, including litigation between farming, fishing, and mining interests. In 1893, these laws halted hydraulic mining except where it could be shown that it would not materially injure navigable streams or adjacent lands (this ban is still codified as California Public Resources Code §3981). Later, many of the miners' water diversion systems and flumes were taken over and converted to hydropower projects. Many of these old systems are still in use today, and continue to impact fish habitat. The miners' doctrine of "first claim, first use" then became the basis of California's appropriative water rights law.

Beginning with the damming of the Tuolumne River at La Grange in 1890, high dams were constructed without fish passage on all but a few minor tributaries to the Central Valley, blocking the upstream migration of spring-run chinook. Shasta Dam, on the mainstem of the Sacramento River, blocked access to the upper Sacramento, Pit, and McCloud rivers for both the spring- and the unique winter-run chinook salmon. Friant Dam completely diverted all flows from the San Joaquin river. By the 1950s, the spring run had entirely disappeared from the San Joaquin basin. In the Sacramento River Basin, the remaining spring run was largely confined to four tributaries— Big Chico, Mill, Deer, and Butte creeks—which still had productive habitat. In Mill and Deer creeks, this was largely due to the absence of gold-bearing sediments, the inaccessibility of the canyons, and the ownership of the watersheds by ranchers (and later the U.S. Forest Service), who acted as protective stewards of habitat. Nonetheless, illegal taking or poaching of summering adult spring-run salmon became a problem along Deer Creek with the construction of State

Highway 32 which facilitated access to the holding areas (Warden Gayland Taylor, CDFG, personal communication).

In contrast, Butte Creek was extensively impacted by hydraulic mining. A large network of water diversions into flumes were constructed, which were later converted into hydroelectric power facilities. Large areas of mine tailings are still present in the Butte Creek watershed. Although Mill and Deer creeks did have water diversions constructed on them for irrigating pastures and orchards, these dams were low-head structures laddered for fish passage. On Butte, Mill, and Deer creeks, water rights adjudications in the 1920s appropriated the available instream flows. Adequate flows for the upstream migration of spring-run salmon below and around the diversion dams were available in average water years, but insufficient in dry or drought years. Flow issues on Butte Creek remain a problem with illegal diversions of water occurring beyond appropriative rights, particularly in the Butte Sink area. In Big Chico Creek, a large pumping station (M&T Ranch) 1/4 mile upstream of the confluence with the Sacramento River had caused reverse flows up from the Sacramento River. This was resolved by relocating the pumping plant to the Sacramento River. The other problem in the lower canyon was caused by slides of large blocks of volcanic caprock into the canyon, creating a migration barrier to fish.

RECENT DECLINES IN SPRING-RUN CHINOOK ABUNDANCE

Native spring-run chinook salmon populations have now been extirpated in all Central Valley tributaries except Mill, Deer, Butte, and Big Chico creeks, where they had also declined through 1950–70. The operation of the state and federal pumping plants at Tracy, which export large volumes of water from the south Sacramento-San Joaquin Delta to southern California for agricultural and urban use, has also caused extensive losses of emigrating salmon.

On the Feather River, a population of spring-run salmon had been maintained at the hatchery, which was constructed to mitigate for the loss of habitat above the dam at Oroville. However, this sub-population has subsequently been hybridized with fall-run chinook salmon as a result of hatchery spawning practices (Ward 1998). By 1990 runs which had averaged 4,000–5,000 in each of the four "healthy" creeks had declined to a few hundred (Campbell and Moyle 1990). As a result, fisheries activists (particularly the Cal-Neva Chapter of the American Fisheries Society and the Natural Heritage Institute (NHI)) had become concerned enough about the status of the spring-run chinook salmon to propose a formal listing under the Federal ESA in 1992.

HARVEST ISSUES

Overharvest also contributed to the declines of Central Valley fisheries, particularly in the last century before there were adequate state and federal controls. Run sizes are also dependent on many environmental factors beyond the control of harvest managers.

Commercial salmon fisheries are dependent on access to healthy and abundant salmon populations. Therefore, it is not surprising that the inriver commercial gillnet fishery on the Sacramento River declined progressively as spring-run chinook decreased in abundance. By 1953 the gillnet fishery was closed. Today the commercial and recreational salmon fisheries off California are largely hook and line ocean troll fisheries. These fisheries primarily harvest fall-run chinook, many of which originate from the public hatcheries built to mitigate the impacts of dam construction. The National Marine Fisheries Service (NMFS) manages the ocean fishery, acting on advice of the Pacific Fishery Management Council. Any federal ESA issues for protecting or managing potential or listed threatened or endangered salmon species in the fishery are addressed through Section 7 consultation within NMFS. Since spring-run salmon may intermingle in the ocean with far more abundant (mostly hatchery-produced) fall-run stocks, an ESA listing may well result in major closures of ocean fisheries to prevent even incidental take of ESA-listed fish.

LAYING THE GROUNDWORK FOR COOPERATION

The commercial fishery, represented by the Pacific Coast Federation of Fishermen's Associations (PCFFA), and the passenger fishing vessel fleet (charter boats), represented by the Golden Gate Fishermen's Association (GGFA), have a long history of working together to increase salmon populations. For example, they have sponsored and operated the Salmon Stamp program, which collects an additional state fishing license fee. Proceeds from the Salmon Stamp fund, totaling about $1 million a year, are used to fund a variety of salmon restoration and enhancement projects in California. PCFFA has committed to, and follows, a strong habitat protection advocacy agenda which often involves coalitions with environmental groups.

The increasing concern over the status of spring-run chinook drove several environmental organizations to propose listing of this run under the ESA. However, due to the positive history of cooperation with fishing groups, the environmental organizations consulted with PCFFA prior to deciding whether to petition for a listing of spring-run chinook. At a meeting in early 1992, attended by representatives from PCFFA, GGFA, NHI, and the Sierra Club Legal Defense Fund (now Earthjustice Legal Defense Fund (ELDF)), PCFFA proposed an alternative strategy to facilitate the recovery of spring-run populations. PCFFA pointed out that listing under ESA might result in the entire ocean salmon fishery being severely curtailed, which would cause major economic impacts as well as an unpredictable political backlash.

PCFFA's proposal was to organize an outreach effort in the key watersheds where spring-run chinook were still present. The objective of the outreach effort would be to simulate the formal recovery process by implementing similar actions to those that would be mandated under the federal ESA. However, participation in the process would be voluntary. Critical for the success of the proposed strategy was the acquisition of sufficient funding for habitat protection and restoration projects. After discussion, the environmental groups agreed to the strategy. PCFFA took the lead on the outreach programs in the tributary watersheds while NHI assumed the lead of an initiative in the Delta.

PCFFA staff and volunteers recruited participation by the Department of Wildlife, Fisheries, and Conservation Biology at the University of California. Its Sea Grant outreach extension project provided staff and administrative support. PCFFA provided facilitation services. A small grant to fund the project was provided from the fishermen's Salmon Stamp Program.

ROLE OF THE SPRING-RUN CHINOOK SALMON WORK GROUP

Sea Grant organized and facilitated the first meeting of what became the Spring-run Chinook Salmon Work Group in Red Bluff, on the Sacramento River, in October 1992. The participants at the first meeting were largely fishery biologists from state and federal agencies, representatives from sport fisheries on the Sacramento River, and rice farming groups that had been working with fishermen on issues related to the screening of the water diversion at the Glenn-Colusa Irrigation District pumping plant.

At the first meeting, the Work Group agreed to operate on an open, collaborative basis, with decisions made by group consensus. The Work Group identified 38 issues and limiting factors impacting spring-run salmon populations. To address these issues and concerns, the Work Group developed a broad programmatic goal and a recovery strategy to support progress toward the goal. A basic program goal, "Restore Sacramento River System native spring-run chinook salmon runs and their habitat," was adopted.

The first priority in the recovery strategy was to protect the relatively strong spring-runs in Mill, Deer, Butte, and Big Chico creeks. This would be followed by habitat restoration and reintroduction of spring-run chinook in adjacent watersheds where populations had been lost or severely reduced. The Work Group explicitly recognized that the fisheries problems related to entrainment of juvenile salmon by the massive Bay-Delta water export pumping facilities were a primary causative factor in the spring-run chinook declines but that this issue was being addressed in other forums; the Work Group agreed to stay informed on Delta issues.

Three major limiting factors were identified in Mill, Deer, and Butte Creeks, with the primary issue being maintenance of sufficient instream flows to allow up- and downstream migration. On Mill Creek, a program begun before the initiation of the Work Group was already underway. A cooperative program had been worked out between the California Department of Water Resources (CDWR), the California Department of Fish and Game (CDFG), and the Los Molinos Mutual Water Co. to exchange well water for instream flows on Mill Creek. During the critical weeks when spring-run salmon were attempting to migrate up Mill Creek, instream flows would be provided by Los Molinos Mutual Water Co. To compensate, early growing season irrigation needs were met by pumping from the wells. Funding for well construction was provided by the Salmon Stamp Fund and the CDWR "Four Pumps Agreement" mitigation fund. Restored instream flows may be a big factor in recent improved returns.

The second problem that was identified, and given high priority for action, was poaching. Because of the long summer instream holding period, spring-run adult salmon are particularly vulnerable to illegal take. Enhanced funding for CDFG wardens' overtime was identified as the best solution to this problem.

The third issue was the need for habitat protection in the upper watersheds of Deer, Mill, Big Chico, and Butte creeks. While the condition of the holding, spawning, and rearing habitat was generally good, sediment-related watershed impacts, caused by road construction and forestry, were identified. Ownership in the headwaters is divided between public (Lassen National Forest, Bureau of Land Management), industrial private timberlands (small portion of ownership), and rural residential subdivisions (mostly in the Butte Creek watershed). To encourage the participation of local stakeholders, the Work Group began to hold monthly public meetings in locations near Mill, Deer, Big Chico, and Butte creeks. Residents and landowners from these watersheds were contacted and invited to the meetings by project staff.

RESTORATION ACTIVITIES

The following first-year accomplishments were achieved by members of the Work Group:

- Funding was secured for additional patrol hours by CDFG wardens to increase protection of adult spring-run salmon holding in Mill, Deer, Big Chico, and Butte creeks.
- Lassen National Forest, managers of much of upper Mill and Deer watersheds, agreed to protect aquatic habitats in those creeks relative to erosion control.
- The recreational trout fishing regulations for the area were changed to catch and release only by the California Fish and Game Commission. Butte Creek was completely closed to fishing.
- The planting of artificially propagated trout was terminated. This was designed to reduce habitat competition in the spring-run salmon rearing areas.
- Workshops were organized to teach University of California Agricultural Extension agents about salmon habitat needs.

WATERSHED CONSERVANCIES

Along with the potential of the ESA listing of spring-run chinook salmon, Mill and Deer creek landowners were faced with another challenge from the environmental and fisheries community. The Friends of the River were petitioning the State to have Mill and Deer creeks declared as Wild and Scenic Rivers under a statutory process which would mandate no further construction of dams, protection of riparian corridors, and provision of public access to the creeks. Of these, the public access mandate most concerned ranchers. Their perception was that the land and the salmon were better protected by limiting public access to the creeks through their lands.

After extended discussion within their own communities, the landowners on Mill and Deer creeks decided to develop their own watershed organizations, which came to be known as the Deer and Mill Creek Watershed Conservancies. A core group of concerned and dedicated community members volunteered many hundreds of hours to organize the conservancies. They wanted to share their sense of stewardship of the land and work on a cooperative basis with the agencies responsible for the protection of natural resources. The primary voting membership in the conservancies was limited to landowners in the watershed, but other parties of interest, such as government agency staff, environmentalists, and fishermen were also welcome to participate in the meetings. The Deer Creek Watershed Conservancy adopted the mission statement: "Dedicated to preserving natural resources, private property rights, and responsible land stewardship." Shortly after its organization, the Deer Creek Watershed Conservancy sponsored a legislative alternative to Wild and Scenic Rivers legislation. A bill, AB 1413, was introduced in the California Legislature which prohibited further dam construction on Mill and Deer creeks, but would not mandate the onerous public access requirements contained under the Wild and Scenic designation. This legislation was supported by the environmental and fisheries groups and was signed into law in late 1994 as the "Mill and Deer Creek Protection Act of 1995" (now Cal. Public Resources Code §5093.70).

The Mill and Deer Creek Watershed Conservancies went on to begin the process of watershed planning for their watersheds. In addition, two more conservancies were organized in nearby Butte and Chico watersheds. A watershed conservancy was organized in 1995 in Butte Creek, which has the most extensive drainage basin, the most complex hydrology, eleven water diversion dams, four bypass weirs, and fragmented land ownership. The City of Chico, which owns a large portion of the Big Chico Creek watershed, organized a task force which is performing many of the same functions as the conservancies. As a testament to its success, several on-the-ground habitat restoration projects have already been initiated using CAL-FED Bay Delta Category III funds and the Central Valley Project Restoration Fund. Implementation of watershed restoration projects requires significant human and financial resources.

RESTORATION FUNDING

One of the Work Group members, representing the inland recreational fishing community, successfully lobbied Congress on behalf of the Work Group for an add-on line item appropriation of $300,000 to the U.S. Bureau of Reclamation (BOR) to fund community-based projects for restoring spring-run chinook salmon runs in the Central Valley. The Work Group was asked to assist the National Fish and Wildlife Foundation (NFWF—the granting agency for the funds), and the BOR in the selection of projects to be funded from this special source. This proved to be a difficult and time-consuming process for the Work Group. Since the Work Group was ad-hoc, it was difficult to define representation from meeting to meeting. After much discussion, a project selection technical committee was appointed. This committee was charged with developing a ranked list of project proposals to the full Work Group for their consensus recommendation to the NFWF. The funding process placed a noticeable strain on the unity and collaborative spirit of the Work Group. Almost all the project proposals submitted for funding were submitted by Work Group members. Anxiety over which project proposals would be funded caused tension among Work Group members. In spite of the difficulties, consensus was reached on a list of projects:

- A watershed inventory of potential road-related sediment sources in the Mill and Deer Creek basins by Meadowbrook Associates;
- A water temperature study in Butte Creek by the CDWR;
- Administration and continued facilitation of the Spring-run Chinook Salmon Work Group by the University of California Sea Grant Program;

- Evaluation of the habitat potential of the Upper Butte Creek watershed by The Nature Conservancy;
- An instream flow study on Mill Creek by D.W. Alley & Associates;
- Control of bamboo at several sites on Deer Creek by the Deer Creek Conservancy;
- Installation of livestock exclusion fencing by the Deer Creek Conservancy; and
- Involvement in the Bay-Delta long-term planning process by the NHI.

Most of these projects are now completed or near completion. Having gone through the difficult and time-consuming project-selection process, the Work Group decided that it did not want to continue selecting projects. The line item appropriation for spring-run salmon projects has been continued by Congress, but projects are now being selected directly by the NFWF. In addition, two large-scale funding programs are now reaching into the watersheds through the Federal Central Valley Project Improvement Act and the CAL-FED Bay-Delta Ecosystem Restoration Program. The Work Group continues its involvement in funding restoration projects by helping the conservancies and local groups access these funds.

During its first year, the Work Group also organized and participated in several field trips to obtain a first-hand perspective on habitat problems affecting spring-run chinook salmon. On one of these trips, the Work Group visited Deer Creek Meadows (a large meadow area where cattle grazing has eliminated riparian vegetation), which is located at the upper end of the creek's spawning reaches. The owner of a portion of Deer Creek Meadows (and a timber company) shared the sense of stewardship his family had developed over four generations on the land. He asked the group what was lacking in terms of fish habitat at the site. Although the site is located higher in the watershed than spring-run salmon migrate, Work Group members described the absence of riparian vegetation, trees, and lack of stream bank definition caused by cattle grazing. Almost a year later this landowner announced at a meeting of the Deer Creek Conservancy that he was fencing along the creek to provide a riparian habitat corridor and building cattle bridges across the stream. He had decided to take this action voluntarily, to help restore the creek. This example points to the results that are achievable through the voluntary collaborative approach made possible by the educational activities of the Spring-run Chinook Salmon Work Group.

A CITIZENS' ACTION SUCCESS STORY

The Spring-run Chinook Salmon Work Group began as a planning group to identify the factors limiting spring-run salmon and to recommend actions to resolve the identified problems. It then passed through a transitional stage where its members secured funding for restoration projects and supported the self-organization of the watershed conservancies. As the local watershed conservancies became functional, they took over the task of watershed assessment and planning, and are presently undertaking on-the-ground restoration projects.

The Work Group then evolved into its present form and function: a conduit for information exchange, reporting on activities in the watersheds; and an education and communication forum between landowners and government. It should be understood that, without the initial effort by the Work Group to facilitate the process of collaborative cooperation between landowners, fishermen, and environmentalists, it is doubtful that so much could have been accomplished without a listing under the ESA. Only time will tell whether these efforts have helped stem the spring-run salmon declines but, at the very least, the process has stimulated and supported communication and restoration activities among diverse groups. Another spin-off benefit of Work Group involvement is that, in the event of a listing, the pro-active approach already taken in the upper watersheds will likely reduce the federal ESA recovery onus on upper watershed landowners and emphasize further restoration activities on the lower portions of affected watersheds.

So far these conservation strategies do seem to be paying off. Preliminary counts of spring-run chinook returning as spawners indicate that more than 10,000 of this run have returned to Butte

Creek alone as of mid-June 1998, one of the highest returns in recent history. This is especially significant given recent hostile ocean conditions generated by El Niño weather patterns. Only time will tell whether this represents a real trend or not.

Lessons Learned

The formation of the Spring-run Chinook Salmon Work Group from a coalition of diverse interest groups has led to notable progress in important habitat restoration. The acquisition of significant restoration funding was stimulated by Work Group activities. Formation of local watershed conservancies, resulting from Work Group involvement, set the stage for actual restoration projects. Private landowners have successfully been encouraged to increase their level of habitat stewardship with a number of beneficial results. Work Group education and communication programs continue to encourage and support grass-roots habitat restoration.

The Work Group staff have learned that an adaptive approach to the management of the group's activities is critically important. The format and function of any similar citizens' action group must be allowed to evolve as its members wish it to. As local, grass-roots organizations (conservancies, resource conservation districts, watershed councils, etc.) organize and take on the task of watershed assessment and planning, it is essential for ad hoc entities such as the Spring-run Chinook Salmon Work Group to gracefully make the transition into a low-profile, support and educational role.

As of the publication date, spring-run chinook have been listed under the State of California ESA, resulting in changes in instream water management and Delta water exports. Another result is that the CAL-FED Operations Group has developed a Spring-run Protection Plan. The National Marine Fisheries Service will announce whether it will federally list the spring-run chinook in September 1999. Regardless of whether the listing occurs, the voluntary, collaborative actions taken in the tributary watersheds should go a long way towards early recovery of these magnificent fish, and may well make a difference whether the spring-run is ultimately listed as "threatened" or in the much more restrictive "endangered" category as well as its chances for early delisting and recovery. The successes of bringing diverse groups together to take positive and proactive action for benefiting seriously depleted Central Valley spring-run chinook salmon can and should serve as a model for citizens' activism supporting sustainability of Pacific salmon. Recent returns of spring-run chinook have been greater than in many years, holding promise that this strategy is working.

ACKNOWLEDGMENTS

We wish to thank Chris Leninger of the Deer Creek Conservancy, Paul Ward (CDFG), Dr. Peter B. Moyle, Glen Spain, and two anonymous reviewers for reviewing and commenting on the text.

REFERENCES

Campbell, E.A., and P.B. Moyle. 1990. Historical and recent population sizes of spring-run chinook salmon in California. Pages 155–216 *in* T. Hassler, editor. Proceedings, 1990 Northeast Pacific Chinook and Coho Salmon Workshop. American Fisheries Society and Humboldt State University, Arcata, California.

Clark, F. H. 1940. California salmon catch records. California Fish and Game 26:49–66.

Fisher, F. 1994. Past and present status of Central Valley chinook salmon. Conservation Biology 8:870–873.

Kelley, R. 1989. Battling the Inland Sea: American Political Culture, Public Policy and the Sacramento Valley 1850–1986. University of California Press, Berkeley.

Ward, P. 1998. Interview in "Western Water." Water Education Foundation. January/February 1998.

Yoshiyama, R. M., E. R. Gerstung, F. W. Fisher, and P. B. Moyle 1996. Historical and present distribution of chinook salmon in the Central Valley Drainage of California. Sierra Nevada Ecosystem Project Final Report to Congress, Volume 3. Center for Water and Wildland Resources, University of California at Davis (Available on the Internet at: http://ceres.ca.gov/snep/pubs).

42 Creating Incentives for Salmon Conservation

Rodney M. Fujita and Tira Foran

Abstract.—Many of the well-known factors that harm wild salmon populations (habitat degradation, hydropower development, some aspects of hatchery management, and overfishing) are symptoms of deeper, underlying causes. The loss or diminution of our connection to nature may be one such fundamental cause. Society's general emphasis on economic growth (an increase in simple wealth) instead of true economic development (an increase in well-being) may be another. The prevailing paradigm that guides most natural resource management decisions appears to be that of a struggle to strike a balance between economic growth and environmental protection. In this struggle, environmental protection is seen as a burden that retards growth and economic growth is seen as inimical to environmental protection, suggesting that environmental protection goals must be "balanced" (through compromise) with economic growth goals. In this chapter, we discuss some problems inherent in this paradigm, as it applies to salmon ecosystems in the U.S. Pacific Northwest. We also recommend a shift in policy toward a "hard on the goals, soft on the people" approach: an approach that adopts rigorous environmental goals, while granting the flexibility that industry and others need to develop cost-effective ways to achieve those goals. In addition, we recommend some policy and institutional changes that emphasize the creation of economic incentives for sustainable development—wise stewardship of resources resulting in a stream of benefits over the long term, grounded in comprehensive protection of ecosystem health.

INTRODUCTION

Salmon support both ecosystems and economies. Salmon-based economies span the gamut from small coastal villages that may derive the majority of their income from commercial salmon fishing (e.g., communities around Bristol Bay, Alaska) to larger cities where salmon fishing may be a much smaller, but nonetheless distinct component of the local economy. In addition, while some communities might not derive a significant portion of their income from salmon today, fishing for salmon may have represented a greater component of the economic and cultural history of these communities in the past. Enhanced salmon fishing and related activities might also be important components of a community's local economic development strategy for the future.

The threats to salmon and salmon-based economies from hydropower, hatcheries, habitat degradation, and overfishing are well known (Nehlsen et al. 1991). Basically, these threats resulted initially from a lack of balance between different types of economic activity. Eventually, as fisheries and the economies associated with them declined, attempts were made to strike a balance between economic activities that damaged salmon ecosystems and other economic activities that depended on healthy fish populations and ecosystems (e.g., fishing and tourism). A great deal of money has been spent on attempts to mitigate the damage caused by economic activities that harmed salmon. Mitigation tended to emphasize interventions, such as the construction of hatcheries and fish ladders, rather than the protection or restoration of habitats and the physical processes necessary for habitat maintenance and enhancement.

Environmental standards, based on estimates of how much damage ecosystems can sustain, represent another attempt to balance damaging activities with environmental protection and eco-system-dependent economic activities. Water quality standards, instream flow standards, forest practice codes, harvest targets, and myriad regulations designed to meet these standards were promulgated as environmental values gained in importance. This "command-and-control" approach tends to specify actions that can or cannot be taken, with the aim of meeting environmental goals which may be based on a limited understanding of what constitutes ecosystem health. Regulations, developed by bureaucrats, often forced industries to take actions that were seen as cumbersome, inefficient, and needlessly costly. Such regulations created strong incentives to cheat or find loop-holes, and did not address other strong incentives to harvest resources unsustainably, to pollute, or to degrade habitat to the detriment of salmon and the economies that depend on them.

New approaches that replace such perverse incentives with economic incentives for wise resource use and comprehensive ecosystem protection and restoration are gaining wider acceptance. These approaches are based on the premise that the best way to balance environmental protection and the sustainability of ecosystem-dependent economic activities with potentially harmful activities is to clearly state rigorous ecological goals, and then provide businesses the flexibility to develop their own ways of meeting the goals ("hard on the goals, soft on the people"). We believe policies based on this idea are more likely to result in wise, consensus-based solutions that reduce costs while achieving environmental goals. In this chapter, we present ideas on how to further develop and implement such approaches, with emphasis on salmon ecosystems of the U.S. Pacific North-west. We focus on creating economically sustainable fisheries and on protecting fish habitat.

CREATING SUSTAINABLE FISHERIES

Modern fisheries often suffer from incentive systems that do not favor sustainable resource use, vaguely articulated (and sometimes conflicting) management objectives, and inadequate risk man-agement (Hilborn and Walters 1992; Ludwig et al. 1993; Fujita et al. 1998). We begin this section by elaborating on the currently unmet need for ecologically sound harvest policies. We then discuss various ways management institutions could be reformed to create incentives for sustainable development.

ECOLOGICALLY SOUND HARVEST POLICY

Salmon harvest goals and regulations are currently aimed at achieving the limited "conservation" goal of ensuring that enough fish escape harvest each year to support maximum sustained yield over time (Chan and Fujita, 1994a). Harvest rate targets were derived based on data on the relationship between numbers of spawners and number of young fish that "recruit" to the fishery. As a result, exploitation rates (the percentage of the population that is harvested) have been quite high. For example, exploitation rates for Oregon wild coho salmon within the jurisdiction of the Pacific Fishery Management Council ranged from 60% (from 1953–1964) up to 87% (in 1976) (Overholtz 1994). They have since fallen to about 18% as a result of fishery closures in recent years. Oregon coastal chinook harvest rates within the Council's jurisdiction have recently ranged from 39% to 55% (Overholtz 1995). To rebuild and sustain the fishery, exploitation rates of many salmon stocks should be reduced and more conservative parameter estimates should be used in stock-recruitment models to hedge against natural variation and scientific uncertainty, and to account for the ecological roles of salmon (Knudsen 2000).

There are many sources of scientific uncertainty that reduce the accuracy of salmon abundance forecasts. Overly optimistic forecasts lead to overfishing, because managers set harvest rates and quotas too high (Knudsen 2000). Improved salmon spawner surveys indicate that previous surveys grossly overestimated abundance of Oregon wild coho (Jacobs and Cooney 1993). The spawner-recruitment relationships which form the basis of escapement goals (targets for how many fish should escape harvest each year) are highly uncertain due to natural variability and to the lack of data at high levels of spawner abundance (Hilborn and Walters 1992).

Striving to achieve maximum sustainable yield (MSY) appears to be an inherently risky strategy, since small increases in harvest mortality over the MSY target lead to relatively large decreases in recruitment (Overholtz 1994). Alternatives to MSY include maximizing economic yield (Clark 1985), as well as policies that decrease the exploitation rate as the spawning biomass falls below a danger threshold or that increase the probability of large recruitment events (Hilborn and Walters 1992; Overholtz 1994). For example, to rebuild the coho fishery, and maintain it at a sustainable level, Overholtz (1994) recommended a moratorium on coho harvest for a few years, followed by harvest at much lower rates based on the goal of increasing the probability of good recruitment to rebuild the population.

Studies that incorporate the effects of natural variability and trends in habitat quality indicate that a return to coho harvest rates characteristic of years prior to the recent ban on directed harvest will result in continued decline, even if habitat quality is substantially improved (Overholtz 1994). Therefore, a thorough review of escapement goals (targets for how many fish should escape the fishery each year) is needed, with a view toward reducing harvest rates on wild stocks. This strategy will also provide a hedge against natural variations in ocean productivity, habitat quality, and other factors.

Further modifications of escapement goals and harvest rates should be made to account for the fact that, in addition to supporting a fishery, salmon play other roles in ecosystems. Recent studies indicate that the carcasses of spawners contribute significantly to freshwater stream productivity and serve as an important food source for fish and wildlife (Bilby et al. 1995). Salmon also serve as a food source for terrestrial and marine mammals, although the importance of salmon in marine mammal diets is uncertain. Salmon may have other roles that have not yet been identified. For example, salmon may regulate populations of prey organisms to some extent, especially when salmon abundance is high and/or concentrated (Shortreed et al. 2000). Chan and Fujita (1994a) suggest a modest 10% buffer be applied to escapement goals to provide enough salmon to natural ecosystems to fulfill these various ecological roles, until enough is known to develop better escapement goals.

More research is needed to reduce scientific uncertainty in abundance projections and in setting escapement goals and harvest rate targets (Knudsen 2000). The effects of environmental variables such as ocean productivity (or its proxies, such as sea surface temperature, upwelling index, etc.), predation, and freshwater flows must be better understood if harvest management is to be fine-tuned to maximize harvests within the constraints imposed by natural variability and the need to protect ecosystem health. While such research would be valuable, it is likely to require many years and substantial resources to achieve a sufficient understanding of how such variables affect salmon distribution and abundance. Therefore, it is prudent to adopt a precautionary approach—that is, conservative harvest rates, escapement goals, and assumptions—until and unless better information becomes available.

Lack of steady and adequate funding for research is a serious impediment to improving fishery management. A landings tax could be used to fund research, if the research is properly directed and supervised by impartial scientists. Such a tax could be separate from or integrated with other taxes targeted for other purposes (e.g., to support habitat restoration, a topic discussed later with respect to "salmon utilities"). It will be important to ensure that fishermen are substantively involved in the expenditure of the tax revenues, and that the tax be gradually imposed to minimize economic hardship. It will also be important to ensure tax policy as a whole provides a level playing field for wild-caught and farmed salmon. In addition, a focused research program conducted in cooperation with fishermen could greatly improve the information base needed for good harvest management, if proper oversight is provided and if the data are analyzed in a timely manner.

Reforming Institutions to Create Incentives for Sustainable Development

In the past, natural resource agencies in California, Oregon, and Washington have had some trouble resisting pressure to allow unsustainable harvests (Chan and Fujita 1994a). A common agency response to declining fisheries has been to augment natural production with expensive hatcheries. Hatcheries, however, can create adverse economic, ecological, genetic, and management impacts (Chan and Fujita 1994b). These state agencies have now developed (or are developing) "wild fish policies" aimed at protecting and restoring wild stocks of salmon and other species.

We believe the current structure of many existing fishery management agencies and regional fishery management councils appears to create more incentives for unsustainable harvests than for long-term sustainability and ecosystem protection. The regional fishery management councils are essentially quasi-democratic: the majority rules, but not all stakeholders are represented. Participants do not have well defined rights to harvest fish; allocations are negotiated and voted on. As a result, even fishermen who are conservation-minded are reluctant to reduce their harvests, because the economic returns of such behavior to these individuals are uncertain. Accountability to the greater public interest is usually quite limited (although accountability is improving due to the increased involvement of environmental organizations in fishery management). Enforcement is often inadequate because of funding shortfalls.

To countervail traditional pressures for unsustainable resource use, the general public's environmental restoration and protection interests must rise to the same level as the interests of traditional resource users. Existing public education efforts should be augmented. It may also be necessary to broaden the revenue base of natural resource management agencies, which may have somewhat closer ties to the interests of the commercial and sportfishing industries than to the interests of the public at large. The strong influence of industry on the agencies may be due, in part, to the partial reliance of the agencies on taxes and license fees paid by commercial fishers, sport fishers, and hunters.

Recently, a coalition of fishing and environmental organizations proposed a mechanism to reduce the disproportionate influence of the commercial, sportfishing, and hunting industries by broadening the client base of the California Department of Fish and Game (CDFG), while simultaneously providing a badly needed increase in funding to the agency. The coalition proposed that CDFG receive an amount equivalent to $1 per California resident (approximately $30 million) from existing state general funds without imposing new taxes (California Assembly Bill 1315). Although the measure failed to pass in 1997, it illustrates a potentially effective short-term method to provide needed resources to existing agencies, while increasing accountability to the general public.

Likewise, a retail tax on consumers of wild salmon might increase accountability to a broader spectrum of public values and increase the amount of funding available for resource management. The tax would have to be designed to minimize disruptive effects on wild salmon markets and to avoid violating free-trade agreements, which may make the imposition of the tax complex and difficult. Wild salmon markets are already in trouble due to slow market response to changes in fish abundance and to competition from farmed salmon; a tax on wild salmon may depress these markets further. On the other hand, a tax on farmed salmon would likely be interpreted as a barrier to free trade. The path of least resistance will largely depend on regional and local politics. For example, a consumer fish tax may result in opposition from consumer groups and either harvesters, processors, or fish farmers, depending on how the tax is structured. A re-allocation of general tax revenues to salmon management and conservation may result in opposition from a wide variety of groups, depending on how large the proposed re-allocation is.

Another way to achieve greater accountability would be to create a broad constituency for better salmon harvest management by organizing salmon habitat restorationists, local economic stakeholders (e.g., chambers of commerce of coastal towns), and other interested parties into effective advocacy coalitions. Because many traditional community-based institutions have been successful at maintaining sustainable harvests, ideas such as community-based management, self-governance, and co-management have generated considerable interest (Pinkerton 1994; Townsend 1994). Ideally, local communities would develop and administer their own fishery regulations, creating buy-in and incorporating the broadest possible spectrum of values and knowledge. Many different kinds of institutional structures can be envisioned to achieve this goal. The challenge becomes one of choosing a structure that creates effective incentives for stewardship and ecosystem protection.

Monitoring and enforcement by a wide spectrum of community residents could both improve compliance with regulations and improve the regulations themselves to reflect economic and social concerns in the wider community. Several accountability mechanisms set up by fishermen or fishing communities exist. For example, Maine lobstermen hold each other accountable to conservation rules with a system of escalating warnings and social sanctions (Wilson 1994).

One way to create a broader constituency for good harvest management *and* good habitat management may be to grant authority to manage a share of the allowable harvest to a group made up of fishermen, scientists, environmentalists, local business people, watershed groups, and other stakeholders. This would establish a quasi-property right that would be expected to create a stake in better management for an entire community. Community development quotas (CDQs) represent a move in this direction. CDQs have already been employed as a form of economic development assistance for coastal villages in Alaska (BSFA 1994). They are allocations of fish harvest privileges granted to a number of villages, which can either lease or fish them. Since the benefits of wise management of the CDQ would accrue to a specific community (not diffused over a large region), CDQs may create incentives for stewardship.

Individual transferable quotas, or ITQs, also show promise as tools for sustainable fish harvest management (Fujita et al. 1996a, b; Fujita et al. 1998). ITQs are transferable allocations of fishing privileges (usually as percentages of the total allowable catch) to individuals. The conservation benefits of ITQs depend strongly on the ability to set a protective total allowable catch (TAC) level. Ideally, the TAC does not fluctuate too much over time, but rather increases fairly steadily as stocks rebuild. A fisherman's shares of the TAC would thus increase in size and value in proportion to the success of conservation and rebuilding efforts, creating a direct economic incentive for sustainability. However, salmon abundance and distribution varies substantially from year to year, sometimes without a clear relationship to harvest policies, and often varying more in response to environmental variables (e.g., ocean productivity) (e.g., Cramer 2000). Any workable ITQ program for salmon must be sensitive to these issues, and must also reflect community values (e.g., limit the aggregation of shares to acceptable levels, to avoid dominance of the fishery by a few individuals or firms) if it is to foster conservation and gain sufficient support to be workable over the long term. The incentives for stewardship created by individual quota approaches may compensate, to some degree, for less than perfect enforcement and accountability mechanisms.

Another way to create positive incentives for long-term stewardship of fishery resources may be to clearly define the allocation of costs and benefits within the fishery management institution (Townsend 1994). In such an institution, votes on management measures could be proportional to the number of shares held by a participant (e.g., the number of individual quota share units) Participants could also sell the future stream of benefits associated with the fishing quota by selling shares.

Townsend (1994) argues that if fishery management institutions provided participants with management influence (votes) in proportion to the shares they held and with well-defined harvest privileges, conservation-minded fishermen could buy shares from participants who do not favor, for example, a reduction in total allowable catch to allow for stock rebuilding. In this way, some fishermen would assume all of the economic risk of conservation in exchange for an anticipated future benefit that would accrue only to them. Other participants would be compensated financially for lost fishing opportunities.

Townsend (1994) also contends that free transferability of shares is essential, so that far-sighted investors will have an economic incentive to buy a controlling interest in the management institution. He provides an example of an environmental organization that could purchase harvest privileges at a low price when the resource is depleted (reflecting inefficiencies owing to overcapitalization and other factors). The environmental organization could reduce total harvest in this way and then lease harvest rights back to fishermen once stock recovery is accomplished.

This type of institution has some advantages in theory; however, some practical considerations suggest potential problems. For example, majority shareholders could face other kinds of incentives for unsustainable harvests, such as a heavy debt load. Another consideration is that interceptions must be minimized so that benefits of foregoing harvest accrue to those with a financial stake in stock rebuilding, and not to mixed stock fishers who contribute little or nothing to stewardship initiatives. Another problem is that since harvest shares often become very costly in an open market, "far-sighted" investors must also be wealthy investors, a description that fits few environmental groups or smaller fishing operations.

If, however, shares are not granted in perpetuity but rather for a specific term only, shareholders could choose between bidding for the right to harvest today vs. bidding for the right to harvest in the future. We would expect commercial fishing interests to bid up the price of shares of the current harvest, because of current debt burdens and other short-term economic pressures. This tendency to discount the future would reduce pressure on share prices of future harvests (Leuthold et al. 1989), allowing far-sighted fishermen and investors an opportunity to invest in a larger share of future harvests. Such a far-sighted investment strategy could prove potentially lucrative in highly variable fisheries, such as the Pacific sardine, whose biomass appears to fluctuate naturally on a 30–60 year cycle (Baumgartner et al. 1992), or Pacific salmon. A fishery that included futures contracts could potentially spread fishing effort out over time. Fishermen and other investors would also have a vested interest in improving their understanding of the ecological factors and natural cycles influencing abundance and distribution. For example, if salmon abundance off Alaska decreases next year, and remains low for several years, share prices may decrease. An investor who could predict the next cycle of high productivity (for example, on the basis of the cycle of oceanic regime shifts in the Pacific) could purchase shares in the harvest of the year 2008 (which could be valuable if productivity actually increases) at low prices in 1998.

PROTECTING SALMON HABITAT

Habitat damage has been the major factor in the extinction of many wild fish, and salmon are no exception. Although the data are spotty, at least 106 major Northwest salmonid populations are already thought to have gone extinct, primarily as a result of dams, water diversions, logging, hatcheries, and other kinds of habitat degradation (Nehlsen et al. 1991). Because salmon habitat is distributed all over the region—on federal, state, and private lands—salmon have suffered from poor environmental practices promoted or allowed by governments at all levels. Almost every conceivable form of environmental insult, originating everywhere from the highest ridge tops to the coastal ocean, harms salmon. These include poor forest practices, spraying of herbicides, water diversions for agriculture and urban development, habitat modification for development, and pollution inputs to waterways.

New Standards for Ecosystem Health

Even the most progressive policies and institutions cannot protect the environment or natural resources essential for sustaining economic development if environmental standards do not accurately reflect "the biological bottom line." This bottom line is the threshold beyond which natural ecosystem structure and function must not be altered.

Environmental standards usually apply to single attributes of single species or of single chemicals (e.g., the concentration of a particular toxic compound that results in 50% mortality of the test organisms). These standards continue to be important, and should be refined. However, there is a need to develop new standards to ensure that ecosystem health is protected at large spatial scales.

Ecosystem health must be defined in an operational, quantitative way if this concept is to be implemented (CCME 1996). Operational definitions that can be quantified have been developed to guide a number of attempts to restore large natural systems, including Germany's Rhine River and Florida's Everglades/Florida Bay/Coral Reef system. An ongoing process to establish ecological health goals and indicators for California's San Francisco Bay-Delta system has resulted in some preliminary operational definitions (Levy et al. 1996). Ecological health has been defined in this process as the structural (e.g., extensive wetlands that exhibit some degree of connectivity) and functional (e.g., peak flows and sufficient instream flows) attributes that allow the system to maintain itself without unusually high extinction rates, population declines, or other adverse changes. The overall goal of San Francisco Bay-Delta-River ecosystem restoration efforts has been expressed as "the re-establishment of a healthy, functional system that supports a diversity of habitat types along with their resident communities of plants and animals, and is self-sustaining (requiring minimal intervention) and resilient to stresses" (Levy et al. 1996). Specific restoration objectives include various biodiversity parameters and the continued provision of ecosystem services (e.g., sustainable fisheries, aesthetics, recreational opportunities, etc.) (Levy et al. 1996; Delta Protection Act of 1992, California Public Resources Code Section 21080.22 and Division 19.5). The restoration paradigm emerging from this definition of health emphasizes the restoration and/or maintenance of certain key processes that have shaped and maintained the system over time (e.g., peak flows, sediment transport, etc.).

Ecosystem health goals, as well as more traditional environmental standards, should be responsive to new scientific information, but should not be compromised to accommodate economic or social interests. Rather, these interests should be accommodated (to the extent that they are valid and valued by the public) with policies that are flexible enough to allow regulated communities to come up with ways to meet the goals at minimum cost. Ideally, policies would create strong economic incentives (supplementary to existing penalties) to encourage the regulated community to meet goals. There are a number of policy tools available to create such incentives. In some cases, institutional reform or entirely new institutions will be needed.

New Tools for Creating Economic Incentives for Conservation

Once ecosystem protection goals are clearly defined, a number of tools can be used to achieve them. Some of these tools have the potential for eliminating incentives to damage habitat, creating incentives to conserve habitat, and fostering voluntary participation in conservation efforts. Furthermore, new tools are necessary for solving some of the problems plaguing efforts to restore populations of endangered species. It is especially important to stop penalizing private landowners, whose land includes habitat for endangered species, with inflexible regulations and perverse economic incentives (Bean and Wilcove 1996). Nowhere is this more important than in salmon ecosystems, which incorporate vast amounts of private land. For example, many small private forest landowners in the Pacific Northwest engaged in "panic cutting" in response to the listing of the spotted owl as an endangered species. These landowners, fearing inflexible regulations associated with the Endangered Species Act, cut trees that could have become owl habitat before any owls took up residence, because they feared that stringent land use restrictions would be imposed if owls were found on their property (Bean and Wilcove 1996). Such fears are also generated by recent and proposed listings of salmonid populations under state and federal endangered species acts.

Creating "Safe Harbors."—One solution to this problem is to enter into "safe harbor" agreements with landowners, in which the landowner agrees to protect habitat in exchange for assurances from a government agency that no further restrictions will be required in the future (Bean and Wilcove 1996). While such an agreement would eliminate the agency's ability to strengthen protections on the land in response to new scientific information, this cost may be outweighed in many circumstances by the need to induce landowners to protect habitat. This risk could be reduced by insisting on quite stringent and comprehensive restrictions (preferably based on indicators and standards of ecosystem health, as discussed above) in the agreement.

Creating Tax Incentives.—Federal law imposes a 37 to 55% tax on estates. It is widely recognized that this creates pressure on landowners, who otherwise would like to pass the land on to the next generation, to instead sell land or liquidate natural resources (e.g., harvest timber) so that they can pay the tax. Many examples of this behavior have been documented (Bean and Wilcove 1996). One solution would be to defer estate taxes on land for which a conservation agreement between the landowner and the government is in effect. The tax would come due if heirs or subsequent owners violated or discontinued the agreement.

Environmental, timber, mining, and other landowner interests came together during a Keystone Center dialogue to endorse estate tax reform and several other ideas for creating tax incentives for protecting habitat on private lands. These included federal tax credits for conservation practices, tax credits for property taxes paid on land subject to a conservation agreement, tax deductions for costs associated with protecting habitat, and a habitat trust fund that would generate revenues from surplus lands with little or no ecological value (Keystone Center 1995).

Accommodating Economic Interests While Achieving Environmental Benefits.—Agreements between willing sellers and buyers that provide both economic and environmental benefits should be pursued. For example, the Environmental Defense Fund recently brokered a path-breaking water leasing project. This lease permits the Bonneville Power Administration (BPA) to transfer up to 16,000 acre-feet of water annually for 3 years from a ranch in eastern Oregon back into the Snake River to assist the recovery of threatened and endangered anadromous fish. This agreement is the largest water transfer agreement ever consummated in the Western U.S. for the purposes of improving instream flows. The ranch will earn more in the long run by selling water rather than using it to farm with. BPA will be able to enhance hydropower production by acquiring water at a lower cost than the oil and gas now used to generate electricity during peak periods (EDF Letter 1993).

TRANSFERABLE RESOURCE PRIVILEGES

Transferable resource privileges provide another tool for introducing the flexibility that is needed to accommodate economic interests with environmental protection. For example, permission to log lands with relatively poor fish and wildlife habitat value (e.g., in a national forest) could be granted in exchange for assurances that the landowner avoid logging a riparian forest on his or her land which is providing high habitat values for salmon.

NEW INSTITUTIONS FOR PROTECTING AND RESTORING HABITAT

New institutions, or institutional reforms, may be needed to create positive economic incentives and funding for protecting and restoring habitat. Some of the promising reforms are Area Quotas, Land and Water Trusts, Fisheries and Habitat Management Compacts, and Salmon Utilities.

Area Quotas.—An Area Quota would go beyond the basic CDQ and constitute a grant of salmon harvest privileges to be managed by communities within the watershed that support salmon harvested there. Stakeholders (fishermen, ranchers, loggers, environmentalists, and other residents of these communities) would be responsible for either making harvest and habitat protection decisions themselves, or for making recommendations to a government agency (or multi-agency group) with the appropriate jurisdiction. In theory, this type of institution would create incentives to engage in habitat restoration/protection efforts and responsible harvest practices to provide a long-term revenue stream to society as represented by the Area Quota. There would also be a need to conform harvest and habitat practices with the greater community's values, which presumably would be broader and longer term than the interests of any one sector.

One of the greatest challenges standing in the way of efforts to link salmon harvest decisions more closely to the watersheds that produce salmon is the fact that much of the harvest occurs in waters distant from the watershed of origin, often by fishers who are not part of one of the communities within the watershed. One way to resolve this problem would be to require the Pacific Fishery Management Council to ensure larger minimum escapements to each watershed, so that local stakeholders can realize more of the economic benefits associated with good watershed stewardship. Another way would be to impose a landing tax on salmon harvested outside of the watersheds of origin, and distribute the revenue to watersheds in proportion to their production of salmon.

Land and Water Trusts.—Land trusts are organizations that acquire land and either manage it to maintain habitat and/or ecological services, or sell it to States for the same purpose. Land trusts can facilitate the acquisition of important habitat land from willing sellers. Economic analyses of the costs and benefits of current land uses, and the removal of subsidies, may induce more landowners to sell if large tracts of land must be acquired. For example, agriculture in former wetland or deltaic areas may be shown to be unsustainable and uneconomic because of land subsidence and costs associated with levee maintenance, flood control, disaster relief, treatment of agricultural discharge, etc.

Water trusts can acquire instream water rights to protect and restore aquatic habitat. Because water rights issues in the West are often highly charged, creative ways of accommodating the economic interests of landowners with the need for increased instream flows to protect fish are needed. The Oregon Water Trust recently concluded an innovative agreement with a cattle rancher, in which the Trust delivered hay to the rancher in exchange for the water the rancher would have used to grow his own hay. The increased instream flow is critically needed to preserve one of the best steelhead spawning tributaries of the Deschutes River in Oregon (Oregon Water Trust 1996).

Fisheries and Habitat Management Compacts.—The Environmental Defense Fund and the Confederated Tribes of the Warm Springs Reservation of Oregon analyzed the need for establishing a new management entity for Oregon's Deschutes River Basin (Moore et al. 1995). They concluded that no existing institution had sufficient regulatory authority, financial resources, or institutional capacity to implement ecosystem management in the Basin. Several options for establishing a new entity to fill this gap were investigated. The goals of the new institution were clearly articulated, and include the development of market-based mechanisms to protect and restore fish habitat, such as water leases, water transfers, water conservation, and pollution caps coupled with tradable discharge permits.

The proposed structure of the new basin-wide management institution is that of a Congressionally authorized compact or intergovernmental agreement between the State of Oregon and the Warm Springs Tribes. The compact would define formal roles for all of the region's stakeholders, including irrigation districts, farmers, ranchers, environmentalists, urban interests, recreational interests, hydropower interests, and federal and state land management agencies.

Although it may be difficult to design and implement such a compact, this approach offers many advantages over the other options considered. A compact could have a great deal of authority both to manage the various natural resources in the Basin, including land and water resources, and to design and implement management plans across jurisdictional lines, including on federal lands. This authority and the breadth of the compact's mission could dramatically improve prospects for coordinated resource management. Furthermore, a compact would explicitly and formally recognize both the sovereignty of the Warm Springs Tribes and the Tribes' existing legal rights to use the natural resources within their ancestral domain.

Salmon Utilities.—As another way to create appropriate incentives for habitat protection and restoration, Hawken (1993) has proposed the establishment of fish utilities, which are modeled on electric power utilities. Such utilities may be able to incorporate a broader suite of social and economic values into fishery management while providing economic incentives to protect and restore habitat. Because utilities are governed by public utility commissions, they are regulated by their constituencies. While some utilities in the past have made unwise investments (e.g., nuclear power plants), many are now investing in energy conservation (because it is much less costly than building new power generating capacity).

Hawken (1993) envisions a utility whose charter would be to increase stocks of wild salmon through habitat protection and restoration. The governing board of the fish utility could be elected to provide for public accountability. The utility would be authorized to collect fees on salmon landed. It would be guaranteed a profit depending on its success as measured by appropriate indicators of health of salmon populations and their ecosystems. Its charter would mandate the investment of a large portion of its revenue into habitat restoration. Because the landing fees would provide a fairly steady revenue stream, the utility could also issue bonds at low interest rates to help pay for long-term restoration projects that otherwise might be difficult to fund. Increased salmon landings resulting from stock rebuilding (a function of improved habitat, increased salmon abundance, and fewer constraints imposed on the fishery to protect sensitive stocks) would further increase revenues and investments in habitat restoration. Cooperative arrangements with relevant agencies (e.g., Memoranda of Understanding) would be necessary, since many different agencies have jurisdiction over various parts of the range of wild salmon. Hawken (1993) envisions many other benefits of a salmon utility, including a pooled pension fund for fishermen, increased employment, and enhanced education in resource management.

Many questions remain concerning the implementation of fish utilities. For example, what would the relationship of the utility be to harvest management bodies? Would the utility complement existing management authorities, or replace them? Fishermen in some areas (e.g., Washington and California) already pay fees on the salmon they land for habitat enhancements. Would the utility's dependence on landing taxes create an incentive for unsustainable harvest levels? Can a utility be structured to create positive incentives for sustainable harvest management? How could the utility maintain investments if landing fees dropped due to poor ocean conditions, constraints imposed by salmon runs listed under state or federal endangered species acts, or market gluts resulting from farmed fish production or other factors? The role of fishermen in the utility must be defined; are they to be elected board members, serve as employees, or fulfill some other role? Would board members be elected by fishermen or the wider community? Landings taxes would likely be strongly resisted unless fishermen were granted substantial influence over the expenditure of revenues. Nevertheless, the fish utility is an interesting model institution that perhaps has the potential for facilitating and sustaining habitat restoration and protection over the long time horizons often required.

The scale of participation in any management system, including CDQs, Area Quotas, Fisheries and Habitat Management Compacts, and Salmon Utilities, is a very important factor determining feasibility and success. The inclusion of an extremely large number of stakeholders would be expected to result in relatively little participation by most, as well as logistical difficulties. In addition, consensus-building would probably be very difficult. On the other hand, the granting of management authority or quasi-property rights to very small groups of stakeholders raises other issues, such as accountability to science-based performance standards and acceptability of decisions to the broader community. Many other factors must also be carefully considered. For example, the structure of the management body would also be very important; the optimal structure would probably vary from area to area, depending on pre-existing relationships among stakeholders and other factors.

CONCLUSIONS

In this chapter, we identified the major, fundamental causes of the degradation of salmon populations and the economies that depend on them. These are (1) the loss or diminution of our connection to nature; (2) society's general emphasis on short-term economic gain; and (3) the lack of strong incentives for conservation. We have also described several new approaches for addressing these challenges, and for harmonizing environmental protection with economic development. These approaches can augment or replace more traditional command-and-control approaches, so as to create incentives for long-term stewardship and the protection of ecological integrity. Some of these approaches, such as the building of community-based institutions with area quotas, are intended to forge a closer link between salmon harvesting and salmon habitat protection. Such institutions might also begin to heal the rift between human communities and natural ecosystems.

If salmon economies and ecosystems are to achieve sustainability, we must prevent overfishing, address threats to salmon habitat, and create incentives for conservation. We offer the following recommendations.

- Reduce harvest rates until weak populations are rebuilt.
- Improve the scientific basis for harvest management.
- Protect and restore salmon habitat.
- Establish stable funding mechanisms for salmon management and habitat protection, such as a consumer fish tax or reallocation of general tax revenues.
- Consider establishing new institutions in which stakeholders can vote their shares for conservation, and be assured that benefits from conservation measures will accrue to them.
- Consider the establishment of Area Quotas to be managed by groups made up of fishermen, fishing community members, watershed community members, environmentalists, and others to create incentives for both sustainable harvest management and habitat protection and restoration.
- Increase the constituency for good harvest management through education and a broader revenue base for natural resource management agencies.
- Define ecosystem health goals and indicators, including quantitative targets for each indicator.
- Pursue creative ways to accommodate economic interests and create incentives for conservation, while securing environmental benefits, such as water transfers, transferable resource use rights, "safe harbor" agreements, conservation agreements, estate tax reform, and tax credits for conservation.
- Consider the establishment of Salmon Utilities to implement habitat restoration efforts.

ACKNOWLEDGMENTS

The authors wish to thank Evelyn Pinkerton and Bonnie McCay for helpful comments on an earlier draft of this study.

REFERENCES

Baumgartner, T., A. Soutar, and V. Ferreirabartrina. 1992. Reconstruction of the history of Pacific sardine and Northern anchovy populations over the past 2 millenia from sediments of the Santa Barbara Basin, California. California Cooperative Oceanic Fisheries Investigations Reports 33:24–40.

Bean, M.J., and D.S. Wilcove. 1996. Ending the impasse. The Environmental Forum 13(4):22–28.

BSFA (Bering Sea Fishermen's Association). 1994. The CDQ Report. June 1994. Bering Sea Fishermen's Association, Anchorage, Alaska.

Bilby, R., B.R. Franson, and P.A. Bisson. 1995. Incorporation of nitrogen and carbon from spawning coho salmon into the trophic system of small streams: evidence from stable isotopes. Canadian Journal of Fisheries and Aquatic Science 53:164–173.

CCME (Canadian Council of Ministers of the Environment). 1996. A framework for developing ecosystem health goals, objectives, and indicators. Tools for ecosystem management. Prepared by the Water Quality Guidelines Task Group, Winnipeg, Manitoba.

Chan, F., and R.M. Fujita. 1994a. Ocean salmon fishery management: Reforms for recovering our biological and economic heritage. Environmental Defense Fund Report. Environmental Defense Fund, Oakland, California.

Chan, R., and R.M. Fujita. 1994b. The impacts of hatcheries on wild salmon populations in the Pacific Northwest. Environmental Defense Fund Report. Environmental Defense Fund, Oakland, California.

Clark, C.W. 1985. Bioeconomic modeling and fisheries management. Wiley, New York.

Cramer, S.P. 2000. The effect of environmentally driven recruitment variation on sustainable yield from salmon populations. Pages 485–503 in E.E. Knudsen, C.R. Steward, D.D. MacDonald, J.E. Williams, and D.W. Reiser, editors. Sustainable fisheries management: Pacific salmon. Lewis Publishers, Boca Raton, Florida.

EDF Letter. 1993. Western Salmon Get Help in Fighting Upstream Battle. Environmental Defense Fund Newsletter 24(6):1–3. Environmental Defense Fund, New York.

Fujita, R.M., D. Hopkins, and W.R. Willey. 1996a. Creating incentives to curb overfishing. Forum for Applied Research and Public Policy 11(2):29–34.

Fujita, R.M., J. Philp, and D. Hopkins. 1996b. The conservation benefits of Individual Transferable Quotas (ITQs). Environmental Defense Fund Report. Environmental Defense Fund, Oakland, California.

Fujita, R.M., I. Zevos, and T. Foran. 1998. Innovative approaches for fostering conservation in marine fisheries. Ecological Applications 8(1) Supplement:139–150.

Hawken, P. 1993. The Ecology of Commerce. HarperCollins, New York.

Hilborn, R., and C.J. Walters, 1992. Quantitative fisheries stock assessment: choice, dynamics, and uncertainty. Chapman & Hall, New York.

Jacobs, S.E., and C.X. Cooney. 1993. Improvement of methods used to estimate the spawning escapement of Oregon Coastal Natural coho salmon. Oregon Department of Fish and Wildlife, Division of Fish Research Project F-145-R-1, Annual Progress Report, Portland.

Keystone Center. 1995. Keystone Dialogue on Incentives to Protect Endangered Species on Private Lands. Keystone Center, Keystone, Colorado.

Knudsen, E.E. 2000. Managing Pacific salmon escapements: The gaps between theory and reality. Pages 237–272 in E.E. Knudsen, C.R. Steward, D.D. MacDonald, J.E. Williams, and D.W. Reiser, editors. Sustainable fisheries management: Pacific salmon. Lewis Publishers, Boca Raton, Florida.

Leuthold, R.M., J. Junkus, and J.E. Cordier. 1989. The theory and practice of futures markets. Lexington Books, Lexington, Massachusetts.

Levy, K., R.M. Fujita, T.F. Young, and W. Alevizon. 1996. Restoration of the San Francisco Bay-Delta-River ecosystem: Choosing indicators of ecological integrity. Report from the "Restoration of the San Francisco Bay-Delta-River Ecosystem: Choosing indicators of ecological integrity" workshops. Environmental Defense Fund, Oakland, California.

Ludwig, D., R. Hilborn, and C. Walters. 1993. Uncertainty, resource exploitation, and conservation: lessons from history. Science 260:17–36.

Moore, D., Z. Willey, and A. Diamant. 1995. Restoring Oregon's Deschutes River: Developing Partnerships and Economic Incentives to Improve Water Quality and Instream Flows. Environmental Defense Fund and the Confederated Tribes of the Warm Springs Reservation. Environmental Defense Fund, New York.

Nehlsen, W., J.E. Williams, and J.A. Lichatowich, 1991. Pacific salmon at the crossroads: Stocks at risk from California, Oregon, Idaho, and Washington. Fisheries 16(2):4–20.

Oregon Water Trust. 1996. Working Cooperatively to Benefit Oregon's Streamflows and Water Rights Holders. Pamphlet available from Oregon Water Trust, Portland, Oregon.

Overholtz, W.J. 1994. Oregon Coastal Natural coho: Prognosis and strategies for stock rebuilding. Oregon Trout, Portland, Oregon.

Overholtz, W.J. 1995. Exploitation rate goals, relative yields, and other important management information for Oregon coastal chinook stocks. Oregon Trout, Portland, Oregon.

Pinkerton, E.W. 1994. Local fisheries comanagement: a review of international experiences and their implications for salmon management in British Columbia. Canadian Journal of Fisheries and Aquatic Sciences 51:2363–2378.

Shortreed, K.S., J.M.B. Hume, and J.G. Stockner. 2000. Using photosynthetic rates to estimate the juvenile sockeye salmon rearing capacity of British Columbia lakes. Pages 505–521 in E.E. Knudsen, C.R. Steward, D.D. MacDonald, J.E. Williams, and D.W. Reiser, editors. Sustainable fisheries management: Pacific salmon. Lewis Publishers, Boca Raton, Florida.

Townsend, R.E. 1994. Fisheries self-governance: corporate or cooperative structures? Marine Policy 19:39–45.

Wilson, J. 1994. Self-governance in the Maine lobster fishery. Pages 241–258 in Karyn L. Gimbel, editor. Limiting access to marine fisheries: keeping the focus on conservation. World Wildlife Fund and the Center for Marine Conservation. Center for Marine Conservation, Washington, D.C.

43 Long-Term, Sustainable Monitoring of Pacific Salmon Populations Using Fishwheels to Integrate Harvesting, Management, and Research

Michael R. Link and Karl K. English

Abstract.—This chapter describes how using fishwheels to integrate assessment, research, and harvesting activities on the Nass River, British Columbia, has provided an opportunity to create an effective, long-term monitoring program for Nass area salmon stocks. Four fishwheels have been operated on the Nass River for several years and we have demonstrated the usefulness of live-capture technology to (1) obtain inseason abundance estimates that can be used by fisheries managers to manage fisheries; (2) obtain annual, postseason escapement estimates and biological information from upriver-bound salmon; (3) provide a reliable supply of fish for use in basic research; and (4) selectively harvest abundant species or stocks while fish from less productive species or stocks are returned unharmed to the river. We describe several practical and economic benefits that have accrued from the fishwheel program on the Nass River and discuss how these benefits may be applicable to other river systems. The fishwheel program has improved the management of Nass area salmon stocks and the entire cost of the annual fishwheel program could be paid for by the revenue from harvesting a portion of the annual catch of salmon in the fishwheels. By integrating assessment programs with species-selective harvesting, long-term and sustainable monitoring of fish stocks may become achievable where it has not been possible in the past.

INTRODUCTION

A fundamental requirement of sustainable salmon fisheries management is a means of accurately assessing the abundance and related health indices of the populations. It is impossible to adequately evaluate the effects of our management actions without rigorous, long-term stock assessment programs. Armed with population status and trend data, fisheries agencies can more efficiently allocate their limited resources to a wide range of potential restoration and management initiatives. Unfortunately, rigorously designed, long-term stock assessment programs are rare in salmon fisheries management. Short-term funding initiatives, administrative obstacles, shifting political and public priorities, shrinking government budgets, and other factors have hindered the establishment of long-term salmon stock assessment programs (Walters 1995; PFTT 1997).

We describe a fisheries management program based on the application of fishwheel technology that serves as the foundation for an effective, long-term assessment program for salmon stocks in the Nass River, British Columbia (B.C.). The underpinning research was conducted as part of an agreement reached between the Nisga'a people, a First Nation on the North Coast of B.C., and the governments of Canada and B.C. to share the salmon resources of the Nass River. In previous papers, we described the essential features of this Fisheries Agreement (AIP 1996; Link and English 1998). In this chapter, we document the benefits of a salmon assessment program that includes the ability to harvest some of the healthier fish stocks that are being assessed. Although the system described below has evolved rapidly since 1992, it has proven reliable and, consequently, we are optimistic that it will allow for sustainable management of Nass area salmon stocks. Because the Nass River and its salmon stocks are similar to various other Pacific Coast drainages, insights presented here may be successfully applied elsewhere.

STUDY AREA

The Nass River watershed comprises 20,000 km^2 of northwestern B.C. (Figure 43.1). The Nass River (K'alii Aksim Lisims) flows through the heart of the Nisga'a traditional territory and supports a rich salmon resource that has allowed the Nisga'a and their culture to flourish for thousands of years. The average annual returns of all species of salmon *Oncorhynchus* spp. to the Nass range from 1.5 to 2.5 million fish. About 1 million fish are harvested in commercial ocean fisheries annually with a landed valued of $5 to $7 million (Cdn).

There are about 2,500 Nisga'a who live in four communities in the Nass Valley (Figure 43.1) and about 3,000 Nisga'a who live outside the valley. In addition to actively participating in the general commercial ocean gillnet and seine fisheries, the Nisga'a annually harvest from 30,000 to 50,000 sockeye salmon *O. nerka*, from 5,500 to 8,500 chinook salmon *O. tshawytscha*, and smaller numbers of the other species of salmon in an inriver gillnet fishery.

FISHWHEELS

The integration of assessment, research, and harvesting activities has been made possible on the Nass River by the introduction of live-capture fishwheels (Figure 43.2; Meehan 1961; Donaldson and Cramer 1971; Link et al. 1996). Although not part of their traditional culture, the Nisga'a have embraced fishwheels as an efficient and useful fisheries management and harvesting tool.

The Nisga'a Tribal Council (NTC) operates four aluminum fishwheels on the Nass River from June to September of each year (Link et al. 1996; Link and English 1996; Link and English 1997; Link and Gurak 1997; Link, in press a,b). The fishwheels are operated 24 h per day and capture all species of salmon (Table 43.1). Two fishwheels are located in a canyon near the village of Gitwinksihlkw (40 km upstream from the ocean) and two are a further 16 km upstream in a canyon near Grease Harbour (Figure 43.1). The fishwheels catch a portion of the adult sockeye, chinook, coho salmon *O. kisutch*, pink salmon *O. gorbuscha*, chum salmon *O. keta*, and steelhead trout *O. mykiss* that ascend the mainstem river on their spawning migration. Most of the fish captured in the Gitwinksihlkw fishwheels are tagged with external tags. The upstream Grease Harbour fishwheels are used to determine the proportion of the sockeye salmon population tagged at the Gitwinksihlkw fishwheels. The ratio of tagged to untagged sockeye salmon in the Grease Harbour fishwheels is used to estimate the daily sockeye salmon escapement within the season.

INTEGRATED ACTIVITIES

Fishwheels are currently used on the Nass River to (1) calculate daily inseason estimates of sockeye salmon abundance that are used to manage commercial fisheries; (2) obtain annual, postseason escapement estimates for sockeye, chinook, and coho salmon and biological data on all species of salmon; (3) provide a ready supply of individual fish of all species of salmon for basic research purposes; and (4) selectively harvest the comparatively abundant sockeye and release unharmed the less productive species caught by the fishwheels. These activities comprise a single integrated management program that is made possible through fishwheel technology.

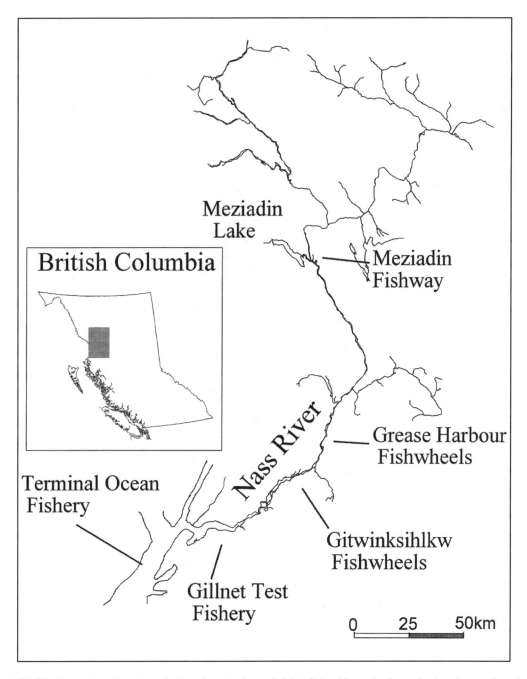

FIGURE 43.1 The Nass River, its location relative to British Columbia, and relevant landmarks mentioned in the text.

Inseason estimates of the sockeye salmon abundance based on fishwheel catches provide the Department of Fisheries and Oceans, Canada, with the information they need to manage the ocean commercial fisheries on Nass River sockeye salmon on a day-to-day basis. A combination of historical and inseason estimates of catchability (portion of the run captured per unit of effort) is used to estimate inriver sockeye salmon abundance. Daily estimates of abundance are also possible for chinook and coho salmon, but are of lesser value since the commercial harvest of these stocks is not managed inseason (Link and English 1996).

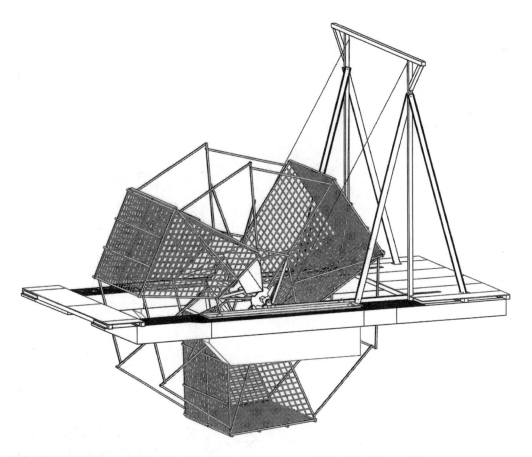

FIGURE 43.2 Diagram of the Nass River fishwheel. The overall length is 13 m and the tower is 7 m high. The river current flows from left to right and the fishwheel revolves in a counterclockwise direction at about two revolutions per minute. Live tanks are fitted into the center of the pontoons.

TABLE 43.1
Average annual numbers of adult salmon captured and tagged in the Nass River fishwheels, and the average abundance of fish passing the Grease Harbour fishwheel sites, 1994–1997. Three fishwheels were operated in 1994 and four thereafter. The estimated abundance was derived from mark-recapture experiments using recoveries at the spawning grounds.

Species	Average number of fish captured in all fishwheels combined	Average number of fish tagged	Number of fish "excess" to tagging needs	Average estimated abundance of fish passing the fishwheels
Sockeye	24,100	7,800	16,300	275,000
Chinook	2,400	1,400	1,000	19,000
Coho	5,200	1,836	3,364	61,000
Pink	14,500	0	14,500	n/a
Chum	240	0	240	n/a
Steelhead	320	115	205	n/a

n/a—not available (no estimate made)

Postseason estimates of chinook, sockeye, and coho spawning escapement are based on the proportion of tags recovered during spawning ground surveys and from an upstream fishway (Figure 43.1). Scale and length data collected from fishwheel-caught salmon enable estimates of annual age and size frequency distributions of returning salmon. Sockeye salmon scales taken from the fishwheel catch are also used to determine the contribution of Nass sockeye salmon to ocean fisheries in southeast Alaska and northern B.C.

In addition to helping to fulfill management needs, the fishwheels provide a ready, low-cost supply of adult salmon for basic research. For example, since 1992, over 700 adult chinook salmon, 100 sockeye salmon, and 30 coho salmon captured in the fishwheels have been fitted with radio transmitter tags (Koski et al. 1996a,b; Link and Gurak 1997) and monitored as they continued their upstream migrations. Much of this research would have been prohibitively expensive were it not for the ample supply of adult salmon provided by the fishwheels.

Several other research projects have benefited from the ongoing fishwheel program. Recently, steelhead catch data have been used to compute preliminary annual escapement estimates of steelhead for comparison with estimates of the potential escapement based on habitat capacity. Data on the timing and abundance of Pacific lamprey *Lampetra tridentata* in the Nass River have also been obtained from the fishwheel catches and simple mark-recapture experiments (M. Link, unpublished data). Steelhead, chinook, and chum salmon tissues have been collected and analyzed to examine the utility of microsatellite DNA analysis for determining stock composition in mixed-stock fisheries. These and other research initiatives have been undertaken at very little additional cost to the fishwheel program.

In recent years, the NTC has harvested "surplus" sockeye from the fishwheels, while releasing unharmed the less abundant steelhead, chinook, coho, pink, and chum salmon (Link and English 1996; Link and Gurak 1997). The harvested fish were deemed surplus to escapement requirements in years when the sockeye escapement target was exceeded. Fish captured in the fishwheels are held in large, water-filled live tanks that are fitted inside the fishwheel pontoons. The tagging/harvesting crew visited the fishwheels several times a day to check the live tanks and either harvest or release captured fish. In each of the years 1993–95, some of the sockeye salmon captured in the fishwheels were sold under a special license for an average of $12.00 (Cdn) per fish. Under a final treaty currently being negotiated, the Nisga'a will be entitled to a share of the Nass salmon stocks which varies with the abundance of fish returning to Canadian waters (AIP 1996). Under the final treaty, and only when there are other fisheries targeting Nass sockeye salmon, the Nisga'a will be able to sell fish caught in the fishwheels that are in excess of the stock assessment program.

IMPROVEMENTS IN FISHERIES MANAGEMENT

A gillnet test fishery was conducted near the mouth of the Nass River from 1966 until 1993 (Figure 43.1; Southgate et al. 1990). Large variability (fivefold range over the period of record) in the annual catchability (proportion of the entire run caught) of the gillnet gear resulted in imprecise and unreliable estimates of the sockeye salmon escapement (Link et al. 1996; Link and Peterman 1998). In 1994, the Department of Fisheries and Oceans requested that the NTC provide inseason sockeye salmon escapement estimates using the fishwheels. The use of fishwheels to sample and tag sockeye salmon has led to more precise and accurate inseason estimates of sockeye abundance than were obtained previously.

The proportion of the total annual sockeye run captured by the fishwheels is 10 to 20 times greater than was caught in the historical gillnet test fishery. This, in conjunction with low interannual variability (twofold range to date) in catchability, has increased confidence in fishwheel-based escapement estimates. The reliability of these estimates was further enhanced by application of mark-recapture data to calculate fishwheel catchability for relatively short time intervals within each season. In the past, because the gillnet test fishery killed the fish it captured, it was not possible to determine or account for inter- and intra-season variation in catchability (e.g., by using a mark-recapture experiment) (Link and English 1996).

Link and Peterman (1998) used a simulation model to quantify the increase in yield from the Nass sockeye salmon stock due to switching from the gillnet to the fishwheel test fishery. They found that although the fishwheel program cost considerably more to operate than the gillnet test fishery, it could be expected to increase yields from the fishery by $180,000 over the added costs.

Although the methodology is still being refined, fishwheel-based assessment techniques have led to more effective implementation of a fixed-escapement policy by the Department of Fisheries and Oceans compared to earlier management approaches (Les Jantz, Department of Fisheries and Oceans, personal communication). Managers are now more confident when calling for fishery restrictions designed to protect weak runs and/or increase the harvest of abundant runs than in the past.

LOWERING THE MARGINAL COST OF ASSESSMENT

One means of developing a cost-effective and sustainable monitoring program lies in lowering the marginal cost of assessment and research activities by integrating them with harvesting. Alone, the current fishwheel program is expensive. The field program, inseason and postseason data analysis, and reporting cost approximately $180,000 (Cdn) annually including overhead (i.e., labor, payroll overhead, capital equipment maintenance and replacement, office building, and support staff). However, the fishwheel program has produced an average of 16,300 sockeye salmon excess to the tagging needs each year over the last 4 years (Table 43.1). Under the proposed treaty, the Nisga'a will be entitled to harvest and sell these fish, worth about $195,000 ($12 Cdn per fish). Thus, the entire fishwheel program could be paid for using income from the sale of sockeye salmon alone. By keeping the marginal cost (income minus cost of the program and harvesting) of the stock assessment program low (essentially zero), the potential for maintaining the program over the long run is high. When one considers the full complement of benefits (e.g., greater yields from the fishery, protection of weak stocks, selective harvest, increased involvement in fisheries management and local employment for the Nisga'a), the case for a fishwheel-based stock assessment program becomes very compelling.

DISCUSSION

Effective stock assessment techniques combined with a low-cost and biologically defensible harvest regime are required if salmon populations and associated fisheries and management systems are to be sustained. Our experience on the Nass River suggests that there may be opportunities elsewhere to use fishwheels and/or other live-capture techniques (traps, weirs and selective seine fishing) as cornerstones of rational and sustainable stock assessment programs.

We do not see fishwheel-based monitoring and assessment as a panacea for creating sustainable salmon assessment programs in the Pacific Northwest. On the contrary, their use is limited to river systems with adequate depth, flow, and river morphology as well as moderately abundant salmon populations. Despite the proven effectiveness of fishwheels in the Nass River, they may be poorly suited to conditions in other river systems. For fishwheels to be a cost-effective harvesting tool, salmon must be relatively abundant. Since many rivers today have reduced or modified flows, and salmon populations in many areas are depressed and may not warrant the cost of a fishwheel-based monitoring program, the approach described here may have limited utility. However, in those situations where rivers or salmon populations are inappropriate for fishwheels, we see traps or weirs as opportunities for providing similar benefits to what we have described for fishwheels on the Nass River. We believe that live-capture gear combined with rigorous stock assessment methods and selective harvesting offer great potential for sustainably managing salmon populations.

Although our analysis of the marginal cost of the Nass River fishwheel program is based on empirical data, the NTC has not yet sold enough fish to cover the cost of the program. Aside from selling limited numbers of "surplus" sockeye salmon from the fishwheel catch once the escapement target has been met (in 3 of the last 6 years), the NTC is not currently in a position to sell fish excess to the assessment program (Table 43.1). However, under the treaty that is currently being negotiated, the NTC will be granted an entitlement that in most years would encompass the numbers of sockeye and chinook salmon excess to the tagging program (Table 43.1). Whether the sale of fish may or may not occur in the future, the economics of our analysis hold; the value of excess fish captured in the fishwheels is greater than the cost to conduct the monitoring program and the subsequent analysis and reporting.

Some could argue that we are only proposing to tax the existing fishery by harvesting fish in the fishwheels instead of using existing fishing gear elsewhere in the salmon migratory routes. There is some truth to this conjecture. Fish harvested by the fishwheels could easily be harvested by existing ocean fishers. A landing tax could be imposed on these fishers and the tax could be used to fund management. However, the cost of fishwheel-based harvesting is considerably lower than traditional net harvesting technologies used today (Jaeger 1997). By using fishwheels to capture these fish that otherwise could be captured in the existing fisheries, it will be possible to capture much of the potential resource rent (revenue from the harvest above a normal return from the harvesting expenditures) from the salmon harvest. A more efficient harvesting procedure means that more value is obtained from these "transferred" fish than if they were caught in the existing ocean net fishery. Put another way, fewer dollars are taken from fishers' income through the transfer of fish compared with the transfer of a landing tax. This newly created wealth is available to pay for the assessment program. Additionally, taxing existing salmon fishing fleets appears to be even less tenable when the net value of the fishery is negative, as is the case in many areas at this time (ICF 1988).

Along with their potential economic benefits, fishwheels offer a means of improving the quality of monitoring and assessment programs (Link and English 1998; Jaeger 1997). The Nass River, like many salmon-bearing rivers, supports multiple species of salmon with similar run timing and differing productivity. The less productive species, like the Nass River steelhead, are vulnerable to overharvesting. Mixed-species harvest rates are often set to capture the benefits of the more abundant and productive species. In these situations, the weaker stocks cannot sustain themselves and are driven to low abundance or extinction.

Live-capture fishwheels enable species-selective harvesting and, therefore, represent an improvement over most traditional net gear in mixed-species fisheries. In river systems where hatchery fish have been externally marked so that they can be distinguished from wild fish, live-capture fishing gear may be used to increase the "stock-specific" selectivity of harvest, thereby increasing yields while protecting less productive wild stocks (Jaeger 1997). Live-capture fishing technology such as fishwheels offer tools to conduct low-cost, species-specific harvesting that is long overdue in salmon fisheries management (Walters 1995).

Rigorous, long-term stock assessment programs will be difficult to implement and sustain now and in the future. In addition, we acknowledge the magnitude of other obstacles that hinder the long-term survival and sustainable use of Pacific salmon in North America (this volume). At times, the "doom and gloom" prognosis for Pacific salmon can discourage progress by exposing the enormity of the problems (this volume; Walters 1995). Monitoring the health and abundance of salmon stocks is often difficult but is the fundamental first step toward action to improve the situation for salmon populations. By integrating stock assessment programs with species-selective harvesting, long-term monitoring of fish stocks may become achievable where it has not been possible in the past.

ACKNOWLEDGMENTS

We are grateful to Bill Griffiths, Bill Jaeger, Dave Peacock, and Gary Searing for providing constructive criticism that improved this study. Funding to prepare the manuscript was provided by the Nisga'a Tribal Council and LGL Limited.

REFERENCES

AIP (Agreement-In-Principle). 1996. Nisga'a Treaty Negotiations, Agreement-in-Principle. Ministry of Aboriginal Affairs, Communications Branch, Victoria, British Columbia.

Department of Fisheries. 1958. Results of Nass River biological surveys for the years 1956 and 1957, including a preliminary assessment of the possible effects of the proposed hydro-electric project. Department of Fisheries, Canada, Vancouver, British Columbia.

Donaldson, I. J., and F. K. Cramer. 1971. Fishwheels of the Columbia. Binfords and Mort Publishers, Portland, Oregon.

ICF Technologies Inc. 1988. Economic impacts and net economic values associated with non-Indian salmon and sturgeon fishers: a report to the State of Washington, Department of Community Development. ICF Technologies Incorporated, Redmond, Washington.

Jaeger, W. K. 1997. Better salmon management in the Pacific Northwest by combining technical and institutional innovations. Natural Resources Journal 37:785–808.

Koski, W. R., M. R. Link, and K. K. English. 1996a. Distribution, timing, fate and numbers of chinook salmon returning to the Nass River watershed in 1992. Canadian Technical Report of Fisheries and Aquatic Sciences 2129.

Koski, W. R., R. F. Alexander, and K. K. English. 1996b. Distribution, timing, fate and numbers of chinook salmon returning to the Nass River watershed in 1993. Canadian Manuscript Report of Fisheries and Aquatic Sciences 2371.

Link, M. R. In press, a. The 1996 Fishwheel Project on the Nass River, B.C. Canadian Manuscript Report of Fisheries and Aquatic Sciences.

Link, M. R. In press, b. The 1997 Fishwheel Project on the Nass River, B.C. Canadian Manuscript Report of Fisheries and Aquatic Sciences.

Link, M. R., K. K. English, and R. C. Bocking. 1996. The 1992 fishwheel project on the Nass River and an evaluation of fishwheels as an inseason management and stock assessment tool for the Nass River. Canadian Manuscript Report of Fisheries and Aquatic Sciences 2372.

Link, M. R., and K.K. English, 1996. The 1993 fishwheel project on the Nass River and an evaluation of fishwheels as an inseason management and stock assessment tool for the Nass River. Canadian Technical Report of Fisheries and Aquatic Sciences 2130.

Link, M. R., and K. K. English. 1997. The 1994 Fishwheel Project on the Nass River, B.C. Canadian Manuscript Report of Fisheries and Aquatic Sciences. No. 2421.

Link, M. R., and K. K. English. 1998. A fisheries agreement with the aboriginal Nisga'a nation: building an institution to promote a sustainable fishery and fishery management system. Pages 149–161 in T.J. Pitcher, P.J.B. Hart, and D. Pauly, editors. Reinventing Fisheries Management. Chapman & Hall, New York.

Link, M. R., and A. C. Gurak. 1997. The 1995 Fishwheel Project on the Nass River, B.C. Canadian Manuscript Report of Fisheries and Aquatic Sciences. No. 2422.

Link, M. R., and R. M. Peterman. 1998. Estimating the value of in-season estimates of abundance of sockeye salmon (*Oncorhynchus nerka*). Canadian Journal of Fisheries and Aquatic Sciences 55:1408–1418.

Meehan, W. R. 1961. Use of a fishwheel in salmon research and management. Transactions of the American Fisheries Society 90:490–494.

PFTT (Pacific Fisheries Think Tank). 1997. British Columbia salmon: a fishery in transition. Pacific Fisheries Think Tank. Report No. 1. Institute of Fisheries Analysis, Simon Fraser University, Burnaby, British Columbia.

Southgate, D. R., B. Spilsted, and L. Jantz. 1990. A review of the Nass River test fishery biological program for 1989. Canadian Data Report of Fisheries and Aquatic Sciences 805.

Walters, C. J. 1995. Fish on the line: the future of Pacific fisheries. David Suzuki Foundation, Vancouver, British Columbia.

44 Sanctuaries for Pacific Salmon

*James A. Lichatowich, Guido R. Rahr, III,
Shauna M. Whidden, and Cleveland R. Steward*

Abstract.—Over 100 years ago, a plea was made by salmon conservationist Livingston Stone to create a system of refuges that would protect wild salmonids from the effects of development and a burgeoning Northwest population. His plea went unheeded, and artificial propagation of salmon through hatchery operations was used instead to mitigate for habitat loss. This chapter traces the history of the fish refuge concept over the last century, explores its application to current salmonid recovery efforts, and examines the challenges of implementation. We present a definition of a salmon refuge that creates, for the first time, a blueprint for the protection of salmonids and their habitats that is uniquely suited to the specific needs and complex life histories of anadromous species. We propose a three-step process to identify, prioritize, and establish refuges that will protect terrestrial and freshwater habitats and associated populations of salmon.

INTRODUCTION

In the early spring of 1892, the American Fisheries Society gathered in New York City for its 21st annual meeting. One of the speakers was Livingston Stone, a retired Unitarian Minister and one of the country's early advocates for fish conservation. Stone (1892) delivered a passionate speech predicting the demise of salmon *Oncorhynchus* spp. and calling on the Society to support the establishment of "salmon parks." Stone's words were both moving and prophetic:

> Provide some refuge for the salmon, and provide it quickly, before complications arise which may make it impracticable, or at least very difficult…If we procrastinate and put off our rescuing mission too long, it may be too late to do any good. After the rivers are ruined and the salmon gone they cannot be reclaimed…all the power of the United States cannot restore salmon to the rivers after the work of destruction has been completed.

Stone proposed the establishment of refuges, or protected areas, for Pacific salmon. Stone's plea went unheeded, as did several other proposals for salmon refuges over the next 100 years. The salmon refuge or sanctuary remains today the road not taken in salmon management.

It is true that some watersheds are protected today from significant habitat degradation because they lie partly or wholly in designated wilderness areas, national parks, or other protected lands. But the protection they offer to salmon, steelhead, and trout is by default rather than design. We have largely failed to protect salmon habitat, and now, after more than 150 years of habitat degradation, salmon throughout the Pacific Northwest are in serious decline (Nehlsen et al. 1991). They have been extirpated from about 40% of their historic range in Washington, Oregon, Idaho and California. In other parts of their range the surviving salmon populations are threatened or endangered (NRC 1996).

Although the overall decline of each salmon stock can be attributed to a combination of factors including overharvest, hatchery programs, and variable ocean conditions, the largest single factor has been habitat loss (Nehlsen et al. 1991). Management without salmon parks or refuges has not adequately protected habitats. Clearly, there is the need to develop a new approach, one that includes sanctuaries for salmon.

The purpose of this chapter is to resurrect Livingston Stone's basic idea and to initiate a discussion of the design, utility, and feasibility of salmon refuges. This chapter is divided into three parts: a history of the concept; current need and design considerations; and a suggested process for establishing salmon refuges.

SALMON REFUGES—HISTORY OF AN IDEA

In the 1870s, within 10 years of the birth of the salmon canning industry in the Columbia River, the economic value of the fishery and the canned product had grown tenfold. Concerned about maintaining the long-term viability of this new and profitable industry, the Oregon legislature petitioned the U.S. Fish Commission for help. Spencer Baird, the U.S. Fish Commissioner, responded to the request. Baird reported that threats to the salmon industry came from three sources: overharvest, dams, and habitat changes (Baird 1875). For a $15,000 to $20,000 investment, Baird said, artificial propagation of salmon would make salmon so abundant that there would be no need to worry about enforcing protective regulations. Baird's report pinpointed the birth of a myth— that salmon did not require healthy rivers and wise stewardship to remain abundant and that humans could maintain the productivity of salmon through controlled, artificial means.

In 1889, after 17 years of experience with the artificial propagation of salmon, Livingston Stone appeared to be having second thoughts. In that year, Stone joined a group of scientists who were sent to Alaska by the U.S. Fish Commission to investigate the Alaskan salmon and their fisheries. Stone's job was to look for opportunities to introduce artificial propagation to Alaska. When he returned, Stone mulled over what he saw and compared Alaskan rivers and their abundant salmon runs to the rivers and wild salmon runs in the Pacific Northwest, and then published his ideas in *Transactions of the American Fisheries Society* (Stone 1892). In his article, Stone made a passionate plea for the protection of salmon. The vehicle for that protection was a salmon park or refuge. Stone argued forcefully that the nation needed to set aside some rivers exclusively for the salmon, and that there was no other way to prevent the rapid development of the Pacific Northwest from destroying the salmon and its habitat.

In 1892 salmon were still abundant in Oregon, Washington, and Idaho, and Stone admitted that it sounded foolish to call for salmon parks. But he reminded his audience that the buffalo were once abundant. Yellowstone National Park was created just in time to protect the last few animals from extinction.

Stone (1892) argued the salmon were caught between the greed of fishermen and the destructive forces of development and civilization. He reminded his colleagues of the fate of Atlantic salmon *Salmo salar* on the east coast:

> It was the mills, the dams, the steamboats, the manufacturers injurious to the water, and similar causes, which, first making the stream more and more uninhabitable for the salmon, finally extermi- nating them altogether.

Stone was forced to admit that not even the hatchery could overcome the destructive effects of a growing population and advancing civilization.

Stone specifically recommended designating the Uganik River, on Afognak Island (near Kodiak Island in Alaska), as the Nation's first salmon park. He was an effective salesman. On Christmas Eve of 1892, President Benjamin Harrison signed a proclamation designating Afognak Island and Uganik River a salmon and forest reserve. The proclamation protected the river from development and closed commercial and subsistence fisheries both in the river and in the surrounding marine waters (Harrison 1892).

Although Stone understood the destructive effects of development, he was not ready to give up on the conventional wisdom of the late 19th century that hatcheries were a better refuge for salmon than natural habitat in a wild river. Even in the Afognak salmon sanctuary, he believed that man could still improve on nature. The sanctuary would include a salmon hatchery with the capacity to incubate 78 million eggs. The sanctuary and its fishery restrictions were put into place not to ensure an adequate escapement of salmon to maintain healthy, natural production, but to ensure a supply of eggs to the hatchery. In his campaign for the establishment of salmon parks, Stone apparently did not detect the contradiction in his argument. If hatcheries were effective and more productive than wild rivers, as most early fish culturists like Stone believed (Bottom 1996; Lichatowich et al. 1996), why did the hatchery need a protected, wild river to maintain a supply of eggs?

The fishing restrictions around Afognak Island and in the Uganik River were gradually lifted and with the passage of the White Act in 1923, they were abolished altogether. In the short space of 30 years the salmon sanctuary reverted to simply a hatchery program. The hatchery was closed in 1934 for economic reasons and because there were questions about the value of the operation. The buildings were turned over to the Navy to be used as a recreation center (Roppel 1982, 1986). So ended the Nation's first and only salmon refuge.

In the same year that President Harrison established the Afognak Salmon and Forest Reserve, Commander J. J. Brice, U.S.N. was ordered to survey military and other government reservations and identify potential sites for Pacific salmon hatcheries. His job was to "formulate a plan to restore salmon in their original numbers" (Brice 1895). Brice's plan recommended construction of four central hatcheries and 20 auxiliary stations from California to Alaska. However, buried in his plan was a single, easily overlooked paragraph that contained a recommendation not tied to artificial propagation. According to Brice, the region needed to set aside a river as a fish preserve. He recommended the Klamath River of northern California as the first candidate (Brice 1895). Brice's recommendation was not implemented.

During the first decade of the 20th century a new national word, "conservation," emerged. This word originated in the west and had a different meaning in 1910 than it does today. Conservation referred to the control of the flow of western rivers for irrigation or other purposes (Hays 1959). Conservation meant the maximum use of natural resources with maximum efficiency. Its meaning was strictly utilitarian.

In 1910 the President of the American Fisheries Society endorsed the strictly utilitarian approach to conservation. In his address to fishery professionals at their 40th annual meeting, Seymour Bower explained what conservation meant to fishery management:

> I believe that we should take the broad general ground and consider the food problem as a whole and not solely in the interests of those minor fisheries. By this I mean, for illustration—and perhaps some will cry treason—that if a section of timber must be left standing indefinitely and the lands withheld from agriculture, in order to preserve a few trout streams as such, then the timber and the trout should go. On the other hand, timber belts bordering such splendid trout waters as the Au Sable and other rivers in Michigan and elsewhere, should be preserved or replaced, for here the conditions are reversed. The soil for the most part being poor, the actual food value of these streams is much greater than the potential food value of the lands through which they flow (emphasis added; Bower 1911).

Two years after Bowers presented his views on conservation, Henry Ward addressed the 42nd annual meeting of the American Fisheries Society and took a position in direct opposition to Bower. At that time refuges and game parks were being set aside for mammals and birds in increasing numbers, and Ward believed fishes deserved similar treatment and that river reaches or even entire watersheds should be set aside for the protection of native fish fauna. What was most unusual about Ward's talk was his concern for the small, noncommercial native fishes. He warned his colleagues that native fishes were disappearing at a rapid rate from eastern rivers and that they needed the protection of fish refuges (Ward 1912). In the era when conservation was a vehicle for development and maximum use, Ward's plan for the preservation of national rivers and the protection of small native fishes fell on deaf ears.

Prolonged drought, changing ocean conditions, excessive harvest, and continuing habitat degradation sent the Pacific salmon into decline in the 1930s. Persistent depletion of native stocks in the face of increasing hatchery production made it painfully obvious to some biologists that hatcheries were not capable of preventing or reversing the decline. John Cobb, Dean of the College of Fisheries at the University of Washington, went so far as to list artificial propagation as one of the threats to the fishing industry (Cobb 1930). The salmon's decline and the failure of artificial propagation rekindled interest in salmon refuges. Charles Pollock, Washington State Supervisor of Fisheries, viewed the establishment of sanctuaries as a matter of grave concern. In his annual report for 1932, Pollock stated that there was "The need of definite legislation mandatory on all state departments involved to set aside certain watersheds as permanent fish sanctuaries to guarantee both commercial and recreational fisheries of this state in the future" (Pollock 1932).

In Oregon the concept of fish refuges gained support from a different source—the people. Oregon citizens have often used the initiative petition to bring about change in fisheries management, including at least four petitions for salmon sanctuaries. In 1928, the voters of Oregon were asked to decide whether the Deschutes Basin (except for White River), the Umpqua River from the South Fork to the coast, the North Fork of the Umpqua River, the Rogue River and its tributaries, and the McKenzie River and its tributaries should be set aside as fish sanctuaries. The petitions would have stopped all water developments in those basins and, as stated in the preamble to the Deschutes petition,

> That the maintenance, so far as is still possible, in the natural condition and free of encroachments by commercial interests, of the Deschutes River and its tributaries except White River, for food and game fish propagation and for recreational purposes, shall be and is hereby declared a beneficial use of the waters thereof (Oregon Voter 1928).

Petitions to establish preserves on the other rivers had the same language. Unfortunately, a majority of Oregonians did not endorse fish refuges. All four measures were defeated by a margin of about two to one (Cone and Ridlington 1996).

In 1938, Oregon Governor Charles Martin asked the Oregon State Planning Board to study the declining salmon fishery in the Columbia River and report back to him with recommendations. Among other recommendations, the Board stated that there was a need to study the feasibility of establishing salmon preserves by setting aside some of the tributaries to the Columbia exclusively for salmon spawning (OSPB 1938). We could find no evidence that the recommendation was implemented.

In the 1940s, massive hydroelectric development in the Columbia and Snake rivers renewed interest in the concept of salmon refuges. The original Lower Columbia River Fisheries Development Program (LCRFDP) contained six recommendations including the establishment of salmon refuges in all the tributary streams entering the Columbia River below the proposed McNary Dam (Laythe 1948). To implement the refuges, Oregon and Washington had to pass enabling legislation. The State of Washington passed a Sanctuary Bill by overwhelming majorities of 42 to 1 in the Senate and 79 to 18 in the House. However, the Oregon House of Representatives voted 41 to 18 on March 3, 1949 to build power dams on the Deschutes River, effectively killing the sanctuary provisions of the LCRFDP.

Finally, in 1959 Assistant Secretary of the Interior Ross Leffler made a last minute plea to declare the Snake River a salmon sanctuary instead of proceeding with the planned hydroelectric development that would inundate and obstruct miles of prime salmon habitat (Leffler 1959). Leffler failed, the hydroelectric projects were built, and Snake River salmon populations plummeted. Today the Snake River sockeye and chinook salmon are listed under the federal Endangered Species Act.

Ironically, while the concept of fish refuges was unable to gain momentum, the federal wildlife refuge system was rapidly expanding to provide refuges for migratory waterfowl (Gabrielson 1943). So while the development of a nationwide system of refuges helped to conserve and recover declining waterfowl populations, the parallel development and operation of fish hatcheries failed to meet salmon and steelhead production objectives and instead contributed to the decline of native stocks (NRC 1996).

Not until the environmental movement of the 1960s challenged the fundamental assumptions of the conservation movement, was the efficiency and desirability of a dependence on hatcheries seriously questioned. By then it was too late. The rivers were politically fragmented into mixed public and private ownership and physically fragmented by more than a century of dams and habitat degradation.

CURRENT NEED FOR SALMON SANCTUARIES

In the 107 years since Livingston Stone called for the establishment of salmon parks, the need for salmon refuges has grown stronger. Yet our ability to establish true refuges for native salmon has diminished. The remaining productive habitat is disappearing rapidly. However, no factor is as important to the persistence of wild salmonids as the presence of good habitat for spawning, rearing, cover, and migration. Habitat loss or degradation is a major factor causing the decline of 175 of 195 salmon and steelhead stocks described as "at risk" or of "special concern" in the Pacific Northwest (Nehlsen et al. 1991). At least 95% of the remaining healthy native stocks of anadromous salmonids in California, Oregon, and Washington are threatened to some degree by habitat degradation (Huntington et al. 1996).

Many of the remaining, robust, naturally spawning stocks of salmon and steelhead occur in watersheds that not only contain relatively intact habitat but also receive some degree of protection from human activities. Stocks in this category include species that spend lengthy periods as juveniles and adults in freshwater, and are therefore more susceptible to freshwater habitat degradation, such as summer and winter steelhead trout *O. mykiss*, coho salmon *O. kisutch*, and spring chinook salmon *O. tschawytscha*. For example, the Bogachiel and Hoh rivers on Washington's Olympic Peninsula are partially protected by the Olympic National Park and support healthy populations of coho, spring chinook, and winter steelhead. The North Umpqua River and North Fork of the John Day River in Oregon, which both share headwaters on federal lands protected through either Wilderness Area or Key Watershed designations, and mainstem reaches protected through the National Wild and Scenic Rivers System, support healthy summer steelhead and spring chinook stocks in the Pacific Northwest (Huntington et al. 1996).

Within the last 5 years, a growing number of technical reports and proposals have echoed Livingston Stone's words of 107 years ago, that we need to identify and protect the most productive and diverse remaining assemblages of native fish and their habitats (FEMAT 1993; Frissell 1996; Williams et al. 1996). We believe that, if an honest commitment and real protection of salmon habitats is not forthcoming within the next few years, depletion and extinction of northwest salmon will reach massive proportions.

IMPLEMENTING THE SALMON REFUGE CONCEPT

> Conservation must nurture the whole life history, not focus inordinate attention on elusive 'bottle-necks' to production. (Healey 1994)

We define a salmon refuge as a watershed that meets the following minimum criteria.

1. It must contain habitats compatible with salmon spawning, rearing and migration, and the biophysical processes to create and maintain those habitats over time (Stanford et al. 1996).
2. It must contain native populations of Pacific salmon which are capable of expressing a major part of their life history diversity.
3. It must be managed to ensure that salmon populations and essential habitats are maintained in the future, and it will receive greater protection than is provided by existing laws and regulations.

Our definition asserts that habitat and salmon are inexorably linked and that watersheds constitute the basic management unit (Healey and Prince 1995). This perspective departs from traditional salmon management, which tends to manage salmon and their habitats separately, and often at inappropriate scales.

To be considered for selection as a salmon refuge, a watershed must support diverse, locally adapted populations of salmon and the habitats necessary to sustain them. A watershed that currently does not meet these criteria can qualify later for refuge status if steps are taken to restore the physical (habitat) and biological (salmon population and life histories) properties necessary to satisfy the first two requirements listed above. Given the precarious status of existing salmon populations and habitats, these requirements currently limit the number of watersheds that might qualify for refuge status.

The restrictiveness of our definition is not meant to imply that critical salmonid habitats at the sub-watershed level habitat should not be protected. These needs can and should be met through complementary management approaches as described more thoroughly below. Moreover, if properly protected, these areas can serve as important foci for the recovery of watershed processes and depressed populations of salmon.

Although the focus of this chapter is on freshwater refuges, a larger network of interconnected freshwater, estuarine, and marine refuges is needed for the recovery and conservation of salmon populations in the Pacific Northwest. Some of the landscape elements that might be included in a refuge system are:

- Small, as well as large, streams
- Alluvial channels and channel migration zones
- Critical off-channel habitats
- Main stem habitats
- Wetlands
- Floodplains
- Riparian areas
- Estuaries
- Bays, reefs, and other nearshore habitats, and
- Critical marine habitats

In the following sections we propose a three-step process for identifying, evaluating, and protecting watersheds as salmon refuges.

Step 1—Identify and Prioritize Watersheds To Be Designated as Salmon Refuges

Threshold, evaluation, and ranking criteria (*sensu* Weiner et al. 1997) are needed to identify candidate watersheds and to select those to manage as salmon refuges. Threshold criteria are minimum objectives or standards that must be met for a watershed to receive further consideration. We propose that the requirements specified in our refuge definition (see above) be applied as threshold criteria to identify candidate refuges.

Some of the preliminary work necessary to identify candidate refuges has already been done or is underway. For example, Huntington et al. (1996) identified watersheds with healthy native salmon and steelhead stocks; the Oregon Chapter of the American Fisheries Society mapped aquatic diversity areas in most major river basins in Oregon (Li et al. 1995); the Forest Ecosystem Management Assessment Team identified key watersheds on federal lands in northwest California, western Oregon, and Washington (FEMAT 1993); the Interior Columbia Basin Ecosystem Management Project has identified aquatic strongholds in the interior Columbia River Basin in Oregon, Washington, and Idaho (Quigley and Cole 1997), the Oregon Biodiversity Project has identified aquatic "Conservation Opportunity Areas" throughout the state of Oregon (OBP 1998); and Thurow et al. (2000) identified core areas of healthy salmon and steelhead populations in the Columbia River basin. A review of these studies would yield a list of watersheds that are prime candidates for salmon refuges.

After candidate refuges have been identified, they should be evaluated further to determine which ones would make the best salmon refuges. Several evaluation and ranking criteria should be considered, including:

- Number, sizes, and trend of salmon populations within the proposed refuge;
- Importance to the recovery of threatened or endangered populations of salmon, or the protection of genetically or ecologically unique populations of salmon; and
- Quantity, quality, and connectedness of key salmon habitat within the designated area, including habitats used for spawning, rearing, and migration.

So that the greatest ecological benefits are obtained, the evaluation should also consider the spatial distribution of refuges, their biological diversity or uniqueness, the ecological services which they render, and their susceptibility to degradation. If properly designed and managed, salmon refuges would not only protect salmon, they would conserve the diverse species and biophysical patterns and processes typical of healthy ecosystems.

Our emphasis on salmon populations, habitat, and watersheds presents some methodological difficulties. Techniques for mapping salmon life histories onto habitat templates and for monitoring salmon life histories will have to be developed (e.g., Jones and Moore 2000). The relative and overall importance of different habitat and biological components of the watershed will need to be determined on a case-by-case basis.

Criteria for determining the physical size of the refuge need to be developed. What watershed area or stream order constitutes the minimum size for a salmon refuge? How much preference should be given to headwater areas? Clearly, upland landscapes and headwater streams perform critical functions that affect downriver conditions, and therefore must be protected. But lowland streams, mainstem habitats, wetlands, and floodplains are also essential to salmon populations; they, too, should be protected. These areas would be among the most difficult to acquire and manage as refuges because they tend to be situated in heavily impacted areas.

In addition to refuge size, managers need to consider features that affect the movement of salmon and the flow of materials and energy across refuge and ecosystem boundaries. The specific location of a refuge within the landscape and the presence of dams or other impediments may influence refuge effectiveness.

Similarly, the location of a salmon refuge relative to other refuges or salmon populations will affect the overall success of salmon conservation efforts. Salmon populations were historically distributed over a large geographic area, isolated to a degree by distance and adaptations to local conditions, but joined together by constant but low levels of straying, interbreeding, and genetic exchange. This pattern of reproductive isolation and exchange is key to the long-term health of all salmonid species. Therefore, a regional network of salmon refuges comprising viable, diverse, and interconnected populations of salmon and essential habitats is needed to accommodate the evolutionary and ecological needs of the species.

In defining areas to designate as salmon refuges, managers should account for environmental uncertainty and the human tendency to err. Ideally, refuges should be free from external control and constraints, requiring minimal intervention by managers. They should be large enough to accommodate natural disturbances operating at both small and large scales (Reeves et al. 1995). They should also be capable of dampening the effects of unanticipated or adverse environmental conditions occurring outside their designated boundaries.

We believe that some good candidate refuges already fit the criteria. Examples include the Cedar and Tolt Rivers in western Washington, which supply drinking water for Seattle, drain watersheds where human activities are severely restricted, and support significant populations of salmon. The 51-mile long Hanford Reach, which contains the last free-flowing, productive mainstem chinook salmon spawning habitat on the Columbia River, has been proposed for federal designation as a Wild and Scenic River (Whidden 1996). The Hanford Reach should be protected as a salmon refuge. We also note that many existing protected areas (including national parks, wilderness areas, Wild and Scenic Rivers, national wildlife refuges, recreation areas, and monuments) were selected primarily for aesthetic, recreational, and socio-economic reasons, and therefore may be inadequate as salmon refuges.

STEP 2—PROTECT CANDIDATE REFUGES FROM FURTHER DEGRADATION

It is essential that the important habitats within each candidate refuge receive prompt and special attention. These habitats have been described as "biological hotspots" (Frissell 1996), and can range from stream reach to the entire watershed in size. In many cases, the priority habitats exist in headwater reaches of river systems on federal lands (Li et al. 1995), and may be readily acquired and protected. In other cases, they may be on private lands where cooperative management agreements, conservation easements, land exchanges, or purchase of riparian land will be necessary. For example, in King County, Washington, "biological hotspots" on private and public lands have been identified as Significant Resource Areas and protected through implementation of county Basin Plans (Anonymous 1995; Fresh and Lucchetti 2000).

While the approach to protecting "hotspots" must be individually tailored to the unique opportunities within each watershed, a variety of different tools are available. For instance, there may be federal funds available to acquire interests in land or water for inclusion in the National Wildlife Refuge, Forest or Parks systems. Further, if funds are available, areas may also be protected as new or expanded state or local county parks. In many areas, coordinated cooperative agreements and interagency memorandums of agreements are being drafted to provide interim protections to key habitat areas. Sometimes a local land trust can help in these transactions by providing technical assistance, expertise, and funding. In some states, there are mechanisms available to protect critical aquatic habitat from degradation using the Outstanding Resource Water designation under the Clean Water Act. This can be a powerful tool to protect the water quality values that support salmon, steelhead, and trout. Every candidate watershed should be assessed to determine the appropriate strategy for protecting the watershed's biological hotspots and ensuring that the overall quality of the watershed does not deteriorate pending a more formalized designation.

Once protected, the biological "hotspots" will serve as the sources of the native aquatic and terrestrial fauna and flora needed to recolonize recovering or restored habitat elsewhere in the watershed (Frissell 1996). These protected areas are the seeds from which the actual watershed-level salmon refuge will grow.

STEP 3—DESIGNATE AND MANAGE REFUGES TO CONSERVE SALMON, HABITAT, AND BIOLOGICAL DIVERSITY

A watershed will not move from candidate to actual salmon refuge status until a network of protected habitats has been established, management programs for their protection implemented, and criteria

1 and 2 of our definition are met. Once the biological and physical criteria for refuge status have been met, and effective management implemented, we propose that an official designation process be initiated. However, it is evident from the long history of the salmon refuge idea that nothing will happen unless state, federal, and tribal management agencies make the salmon refuge concept a priority. Therefore, we suggest that the official designation process be initiated by one of these agencies, that public funds be used to acquire and maintain salmon refuges, and that salmon refuges be integrated and managed over the long term by appropriate federal, state, or tribal resource agencies.

Most of the candidate refuges include a combination of land ownerships and management jurisdictions and will require a range of acquisition and development strategies. For example, habitat protection on public lands will require the full implementation of existing and proposed state and federal aquatic conservation strategies. For some species, such as chum salmon *O. keta* and coho salmon, critical spawning and rearing habitat occur primarily on private lands, so a number of regulatory and voluntary mechanisms should be employed by state governments and local landowners to protect and restore habitat. They could include

a. public-private land exchanges to consolidate public ownership within the candidate salmon refuge;
b. salmon land trusts within each basin to build local capacity to acquire instream flows and priority habitat;
c. conservation easement programs to protect habitats from degradation; and
d. education of private landowners through small watershed planning and protection groups.

We feel strongly that land acquisition is the best vehicle for protecting riparian areas and adjacent lands. Voluntary measures, although praiseworthy, will not by themselves afford the protection needed to maintain viable salmon populations.

A good model for establishing a network of native salmon protected habitats is the system of protected and restored wetlands created through the North American Waterfowl Management Plan (NAWMP 1986). Signed by government and private partners in 1986, the NAWMP has assisted in the protection or enhancement of over four million acres of waterfowl habitat in North America. Habitats have been successfully protected and restored through a range of voluntary landowner conservation agreements and acquisitions by state and federal agencies as well as private groups like The Nature Conservancy and Ducks Unlimited (Carey Smith, U.S. Fish and Wildlife Service, personal communication).

Another potential avenue for refuge establishment is the Endangered Species Act, which provides money from the Land and Water Conservation Fund and other federal sources for the acquisition of interests in land and water to conserve threatened and endangered species. As of September 1998, 56 refuges ranging from 14 to 116,585 acres have been acquired under authority of the Endangered Species Act for the benefit of threatened and endangered species. Examples include Crystal River (manatees), Oklahoma Bat Caves (bats), and Hakalau Forest (Hawaiian birds).* A similar program of land acquisition and refuge establishment could be implemented for threatened and endangered salmon populations in the Pacific Northwest.

MAKING LIVINGSTON STONE'S DREAM A REALITY

It has taken only 100 years to push most of our formerly abundant populations of wild salmon to the brink of extinction. During this same period, the human population of the Pacific Northwest has grown exponentially and will only continue to increase pressure on the watersheds that still

---

* This information was obtained October 16, 1998, from the U.S. Fish and Wildlife Service Internet website on endangered species, located at http://refuges.fws.gov/NWRSFiles/WildlifeMgmt/EndangeredSpeciesMgmt.html.

produce wild salmon (NRC 1996). If salmon, steelhead, trout, and associated native species are to be part of our future, bold steps must be taken today to protect their last strongholds. This proposal is intended to spark discussion that can lead to the land, water, and fish management reforms needed to prevent further decline of Pacific Northwest salmon. The identification and protection of candidate refuges will not happen overnight, and it will always be controversial. The refuge strategy must be integrated into fish conservation programs at all scales—from local watershed councils to state and federal land reform programs. With public support for salmon restoration at an all-time high, and salmon conservation programs being developed throughout the Northwest, now is a perfect time to create a refuge system for Pacific salmon. Livingston Stone had the right idea. We must now make it happen.

ACKNOWLEDGMENTS

The authors would like to acknowledge the many contributions that have made this study possible. First, we would like to thank Oregon Trout for funding Guido Rahr and Shauna Whidden during most of the research and writing stages. We would also like to thank the Fish Refuge Working Group members who have contributed countless hours toward the development and advancement of the fish refuge concept. And finally, we are indebted to Ray Hubley for his tireless vigilance in keeping the fish refuge idea alive over the last decade or more and his direct and constant assistance to our current efforts to make the refuge idea a reality.

REFERENCES

Anonymous. 1995. Protection of significant resource areas in King County. Unpublished manuscript, King County Department of Natural Resources, Seattle.

Baird, S. 1875. The salmon fisheries of Oregon. Oregonian, March 3, 1875, Portland.

Bottom, D. L. 1996. To till the water: a history of ideas in fisheries conservation. Pages 569–597 in D. J. Stouder, P. A. Bisson, and R. J. Naiman, editors. Pacific salmon and their ecosystems. Chapman & Hall, New York.

Bower, S. 1911. Fishery conservation. Transactions of the American Fisheries Society 40:95–100.

Brice, J. J. 1895. Establishment of stations for the propagation of salmon on the Pacific coast. Miscellaneous Documents, U.S. House of Representatives, 53rd Congress, Washington, D.C.

Cobb, J. N. 1930. Pacific salmon fisheries. U.S. Dept. of Commerce, Bureau of Fisheries Document 1092, Washington, D.C.

Cone, J., and S. Ridlington. 1996. The Northwest salmon crisis: a documentary history. Oregon State University Press, Corvallis.

FEMAT (Forest Ecosystem Management Assessment Team). 1993. Forest ecosystem management: an ecological, economic and social assessment. USFWS, NOAA, BLM, USFS, NPS and the EPA, Portland, Oregon.

Frissell, C. 1996. A new strategy for watershed protection, restoration and recovery of wild native fish in the Pacific Northwest. Pages 1–25 in R. Doppelt, project manager. A guide to the restoration of watersheds and native fish in the West. Pacific Rivers Council, Eugene, Oregon.

Fresh, K. L., and G. Lucchetti. 2000. Protecting and restoring the habitats of anadromous salmonids in the Lake Washington watershed, an urbanizing ecosystem. Pages 525–543 in E. E. Knudsen, C. R. Steward, D. D. MacDonald, J. E. Williams, D. W. Reiser, editors. Sustainable fisheries management: Pacific salmon. Lewis Publishers, Boca Raton, Florida.

Gabrielson, I. 1943. Wildlife refuges. The Macmillan Company, New York.

Harrison, B. 1892. Afognak forest and fish culture reserve. A proclamation. Record Group 22 U.S. Fish and Wildlife Service, National Archives, Washington, D.C.

Hays, S. 1959. Conservation and the gospel of efficiency. Atheneum Press, New York.

Healey, M. C. 1994. Variation in the life history characteristics of chinook salmon and its relevance to conservation of the Sacramento winter run of chinook salmon. Conservation Biology 8:876–877.

Healey, M. C., and A. Prince. 1995. Scales of variation in life history tactics of Pacific salmon and the conservation of phenotype and genotype. Pages 176–184 *in* J. L. Nielsen and D. A. Powers, editors. Evolution and the aquatic ecosystem: defining unique units in population conservation. American Fisheries Society Symposium 17, Bethesda, Maryland.

Huntington, C., W. Nehlsen, and J. Bowers. 1996. A survey of healthy native stocks of anadromous salmonids in the Pacific Northwest and California. Fisheries 21(3):6–14.

Jones, K. K., and K. M. S. Moore. 2000. Habitat assessment in coastal basins in Oregon: implications for coho salmon production and habitat restoration. Pages 329–340 in E. E. Knudsen, C. R. Steward, D. D. MacDonald, J. E. Williams, D. W. Reiser, editors. Sustainable fisheries management: Pacific salmon. Lewis Publishers, Boca Raton, Florida.

Laythe, L. L. 1948. The fishery development program in the Lower Columbia River. Transactions of the American Fisheries Society 68:43–55.

Leffler, R. 1959. The program of the Fish and Wildlife Service for the anadromous fish of the Columbia River. Bulletin of the Oregon State Game Commission. October. Pages 3–7.

Li, H. and 12 coauthors. 1995. Safe havens: refuges and evolutionarily significant units. Pages 371–380 *in* J. L. Nielsen and D. A. Powers, editors. Evolution and the aquatic ecosystem: defining unique units in population conservation. American Fisheries Society Symposium 17, Bethesda, Maryland.

Lichatowich, J. A., L. E. Mobrand, R. J. Costello, and T. S. Vogel. 1996. A history of frameworks used in the management of Columbia River chinook salmon. Bonneville Power Administration, DOE/BP-33243-1, Portland, Oregon.

NRC (National Research Council). 1996. Upstream: salmon and society in the Pacific Northwest. Committee on Protection and Management of Pacific Northwest Anadromous Salmonids, National Academy of Science, Washington, D.C.

Nehlsen, W., J. E. Williams, and J. A. Lichatowich. 1991. Pacific salmon at the crossroads: stocks at risk from California, Oregon, Idaho and Washington. Fisheries 16(2):4–21.

NAWMP (North American Waterfowl Management Plan). 1986. U.S. Department of Interior, U.S. Fish and Wildlife Service, Washington, D.C.

OBP (Oregon Biodiversity Project). 1998. Oregon's living landscape: strategies and opportunities to conserve biodiversity. Defenders of Wildlife, Lake Oswego, Oregon.

Oregon Voter. 1928. Save rivers for fish. 54(2):21–24, Portland.

OSPB (Oregon State Planning Board). 1938. A study of commercial fishing operations on the Columbia River. Report submitted to the Governor of Oregon, Salem.

Pollock, C. R. 1932. Fortieth and forty-first annual reports of the State Department of Fisheries and Game. Olympia, Washington.

Quigley, T. M., and H. Bigler Cole. 1997. Highlighted scientific findings of the Interior Columbia Basin ecosystem management project. U.S. Department of Agriculture, Forest Service, Pacific Northwest Research Station and U.S. Department of Interior, Bureau of Land Management, General Technical Report PNW-GTR-404, Portland, Oregon.

Reeves, G. H., L.E. Benda, K.M. Burnett, P.A. Bisson, and J.R. Sedell. 1995. A disturbance-based ecosystem approach to maintaining and restoring freshwater habitats of evolutionarily significant units of anadromous salmonids in the Pacific Northwest. Pages 334–349 *in* J. L. Nielsen and D. A. Powers, editors. Evolution and the aquatic ecosystem: defining unique units in population conservation. American Fisheries Society Symposium 17, Bethesda, Maryland.

Roppel, P. 1982. Alaska's salmon hatcheries 1891–1959. Alaska Historical Commission Studies in History 20, Anchorage.

Roppel, P. 1986. Salmon from Kodiak: an history of the salmon fishery of Kodiak Island, Alaska. Alaska Historical Commission Studies in History 216, Anchorage.

Stanford, J., and six coauthors. 1996. A general protocol for restoration of regulated rivers. Regulated Rivers: Research and Management 12:391–413.

Stone, L. 1892. A national salmon park. Transactions of the twenty-first Annual Meeting of the American Fisheries Society 21:149–162.

Thurow, R. F., D. C. Lee, and B. E. Rieman. 2000. Status and distribution of chinook salmon and steelhead in the interior Columbia River basin and portions of the Klamath River basin. Pages 133–160 *in* E. E. Knudsen, C. R. Steward, D. D. MacDonald, J. E. Williams, D. W. Reiser, editors. Sustainable fisheries management: Pacific salmon. Lewis Publishers, Boca Raton, Florida.

Ward, H. 1912. The preservation of the American fish fauna. Transactions of the American Fisheries Society 42:157–170.

Weiner, A., and five coauthors. 1997. The *Exxon Valdez* oil spill: habitat protection as a restoration strategy. Restoration Ecology 5(1):44–55.

Whidden, S. 1996, The Hanford Reach: protecting the Columbia's last safe haven for salmon. Environmental Law 25:265–297.

Williams R., and 11 coauthors. 1996. Return to the river: restoration of salmonid fishes in the Columbia River ecosystem. Independent Scientific Group, Portland, Oregon.

45 One Northwest Community— People, Salmon, Rivers, and the Sea: Toward Sustainable Salmon Fisheries

Donald D. MacDonald, Cleveland R. Steward, and E. Eric Knudsen

Abstract.—Pacific salmon management is in crisis. Throughout their range, salmon and steelhead populations are being adversely affected by human activities. Without coordinated, effective, and timely action, the future of the Pacific salmon resource is most certainly in doubt. To address the challenges that are currently facing salmon management, concerned citizens representing a diverse array of government agencies and non-governmental organizations have agreed to cooperate in the development of a Sustainable Fisheries Strategy for west coast salmon and steelhead populations. The Strategy builds on the contents of this book, resulting from the Sustainable Fisheries Conference and subsequent community- and watershed-based citizen forums. This chapter presents the key elements of the Strategy including a common vision for the future, a series of guiding principles, and specific strategies for supporting sustainable fisheries. As such, the Strategy embraces an ecosystem-based approach to managing human activities, rather than the traditional egocentric approach to managing salmonid populations and associated habitats. A system of community-based, watershed-oriented councils, including all stakeholders and agency representatives, is proposed for effective transition to ecosystem-based salmon and steelhead management. It is our hope that everyone involved in Pacific salmon management will embrace both the spirit and the specific elements of the Sustainable Fisheries Strategy as we face the difficult challenges ahead.

INTRODUCTION

Humans have maintained a bond with Pacific salmon for thousands of years. Since the earliest residents of the Northwest crossed the Siberian land bridge, salmon have been an integral element of the economy, society, and culture that has developed on the west coast of North America. While significant social and economic changes have occurred since western people arrived some 200 years ago, the importance of salmon to the region remains essentially undiminished. In many ways, salmon continue to define the Pacific Northwest and its people, both in terms of our culture and societal values. One has difficulty considering the plight of the salmon without experiencing a personal sense of loss; our fate seems inextricably linked to theirs. The extinction of even one stock of salmon or steelhead diminishes our stature as humans and stewards of the resource.

ARTICULATING OUR COMMON VISION

People throughout the Northwest share a common desire to see salmon and steelhead populations returned to health. This overwhelming concern over the fate of the salmon has given rise to a plethora of salmon protection, enhancement, and restoration plans and initiatives. However, in spite of our best efforts, we have not succeeded and are unlikely to restore salmon until these disparate efforts are effectively integrated into an ecosystem-based framework that encompasses the entire North American coast from California to Alaska. For this reason, concerned citizens representing a diverse array of government agencies and non-governmental organizations from throughout the Pacific Northwest have agreed to cooperate in the development of a Sustainable Fisheries Strategy for west coast salmon and steelhead populations (see Table 45.1).

The Sustainable Fisheries Strategy presented below was largely derived from the results of the "Toward Sustainable Fisheries" conference held in Victoria, British Columbia April 26–30, 1996 and attended by more than 500 delegates. The preceding chapters in this book, the results of several Forums on Sustainable Fisheries, as well as abundant other writings (particularly NRC 1996 and Stouder et al. 1997), provided additional background information for refining the Sustainable Fisheries Strategy described in this chapter. A full description of the Sustainable Fisheries Strategy is provided in SFF (1996) and on the Sustainable Fisheries Foundation home page (http://www.island.net/~sff-mesl).

Conference delegates shared technical information and engaged in facilitated discussions on five key issues related to fisheries sustainability, including:

- Developing harvest management strategies for sustainable fisheries;
- Protecting and restoring freshwater, estuarine, and marine habitats;
- Integrating communities into sustainable fisheries management;
- Increasing the production of salmon and steelhead; and
- Developing institutional and regulatory structures that favor sustainable fisheries.

To support the development of the Sustainable Fisheries Strategy, conference delegates were asked to establish a long-term vision for the future, determine which factors are constraining our ability to achieve this vision, and identify the actions that are needed to overcome these constraints. As a result, a shared vision of sustainable fisheries and a series of principles were developed that would support this vision. While the precise wording of the vision varied, it was generally agreed that:

> Our long-term vision for the Pacific Northwest includes healthy, diverse, and productive ecosystems, viable aboriginal, sport, and commercial salmon and steelhead fisheries, and vital and stable communities throughout the historical range of Pacific salmonids.

This vision clearly articulates a desired future condition that recognizes the inherent linkages between Pacific salmon, our continuing use of these resources, and the health of the environment and communities in which we live. It also encompasses a range of social, cultural, economic, and ecological values, thereby reflecting the interests and needs of virtually every resident of the region. The vision also reflects an ecosystem-based approach which views salmon and people as inextricably linked and recognizes that decisions affecting the well-being of the salmon will also influence our economy, culture, and communities—both now and in the future (Knudsen et al. 2000). The following guiding principles can be used to focus environmental management and salmon restoration programs in the future.

- People have the right to use Pacific fisheries resources for social, cultural, and economic purposes. With this right goes the responsibility to maintain and restore the abundance and diversity of salmon and steelhead populations throughout their range.
- Ecosystem integrity and the economic well-being of communities throughout the Northwest are interdependent; therefore, achieving and protecting ecosystem integrity is an essential part of economic and social development in the region.

- The concept of sustainable use of renewable natural resources provides a basis for addressing societal needs for a healthy ecosystem and a healthy economy.
- Resource management decisions must be ecologically based, conservation-driven, scientifically sound, and risk-averse.
- The watershed is the fundamental management unit for Pacific salmonids; salmon populations are the product from healthy watersheds. However, estuarine and marine habitats through which the salmon migrate must also be considered in management decisions.
- Freshwater, estuarine, and marine habitats must be effectively protected and, where necessary, restored.
- Watersheds must be managed to support a range of compatible land and water uses while protecting downslope and downstream resources.
- Humans must be effectively integrated into the ecosystem through cultural and community awareness, education, ethical responsibility, and adequate legislation and enforcement.
- Communities of place and communities of interest should be empowered in the decision-making process through the implementation of ecosystem-based approaches to fisheries and watershed management.
- The biological and genetic diversity of salmon and steelhead populations must be maintained.
- Habitat protection and restoration initiatives and production-based (rather than allocation-based) harvest management strategies should be used in conjunction with conservation aquaculture and other tools, to restore the production of Pacific salmonids.
- Specific management objectives and quantifiable indicators of progress toward sustainable fisheries should be developed, monitored, and reported.
- Emphasis on mixed-stock fisheries must be reduced where necessary through the development of terminal and stock-specific fisheries which utilize selective harvesting methods.
- Harvest management decisions must be based on high quality information about the productivity of salmon and steelhead stocks.
- Fundamental research into the nature of the relationship between salmon and their environment, and the effects of human activities on that relationship, must be fully supported.
- Institutional structures, legislation, and regulations must be reformed to better support sustainable fisheries.
- An integrated hierarchy of institutions should be established in which the institutions concerned with natural resource management are organized and/or operate at watershed, regional, and coast-wide levels.
- Institutions must have the authority to make decisions, the resources necessary to ensure their implementation, and be held accountable for the results of their actions or inactions.
- Compliance with and enforcement of applicable statutes must be improved.
- A comprehensive set of economic and/or tax incentives for industry and private landowners should be offered to increase compliance with regulations and to foster responsible environmental stewardship.
- Perverse disincentives (e.g., subsidies for irrigation and river transportation) for responsible stewardship and effective management must be eliminated.
- Communication among all of the groups involved in salmon management must be enhanced.

KEY ELEMENTS OF THE SUSTAINABLE FISHERIES STRATEGY

In addition to a long-term vision and guiding principles, the Sustainable Fisheries Strategy also includes a series of strategic actions that should be implemented to support the transition to sustainable fisheries management. The recommended strategic actions are organized into five topic areas, including habitat protection and restoration, harvest management, salmon and steelhead

TABLE 45.1
Organizations participating in the development of the Sustainable Fisheries Strategy.

The organizations that are participating in the development of the Sustainable Fisheries Strategy and Implementation Plan include:

- Acres International Ltd.
- American Fisheries Society (also including: Alaska, Idaho, Humboldt, North Pacific International, and Portland chapters; Canadian Aquatic Resources Section, Marine Fisheries Section; Western Division)

- B.C. Hydro
- B.C. Hydro and Power Authority—Kootenay Generation Area
- B.C. Ministry of Agriculture, Fisheries, and Food
- B.C. Ministry of Environment Lands and Parks
- B.C. Ministry of Forests
- B.C. Wildlife Federation
- Boeing Company
- Bonneville Power Administration
- Browning-Ferris Industries
- Bullitt Foundation
- Brainerd Foundation
- Bureau of Reclamation
- Canadian Consulate
- Canadian Department of Transport
- Canadian Forest Service
- Celgar Ltd.
- City of Castlegar
- Coastal Salmon Restoration Group
- Columbia Basin Trust
- Columbia Power Corporation
- Columbia River Inter-Tribal Fish Commission
- Cominco Ltd.
- Commercial Fishing Industry Council
- Confederated Tribes of the Colville Reservation
- Environment Canada
- Fisheries and Oceans Canada

- For The Sake of the Salmon (including: National Marine Fisheries Service; U.S. Fish and Wildlife Service; U.S. Environmental Protection Agency; Natural Resources Conservation Service; U.S. Forest Service; Bureau of Land Management; Bureau of Indian Affairs; Governor Kitzhaber; Governor Lowry; Governor Wilson; Northwest Indian Fisheries Commission; Coos County (Oregon); City of Portland; King County (Washington); Mendocino County (California); Oregon Forest Industry Council; Washington Forest Protection Association; Portland General Electric; Pacific Power and Light; Seattle City Light; California Forestry Association; Public Power Council; Grant County Public Utility District; Washington Association of Conservation Districts; Northwest Sportfishing Industry Association; Oregon Outdoors Association; Oregon Charterboat Association; Westport Charter Association; Pacific Coast Federation of Fishermen's Associations; Salmon For All; United Anglers of California; Northwest Steelheaders; Trout Unlimited; Pacific Rivers Council; Oregon Trout; Long Live the Kings; California Trout; Friends of the River; Oregon Wildlife Heritage Foundation)

- Forest Alliance of British Columbia
- Forest Renewal British Columbia
- Greenpeace
- Harder Foundation
- Hatfield Consultants Ltd.
- Idaho Department of Fish and Game
- Institute for Fisheries Resources
- MacMillian Bloedel Ltd.
- National Marine Fisheries Service
- National Roundtable on the Environment and the Economy
- Northwest Ecosystem Institute
- Northwest Power Planning Council
- Oregon Department of Fish and Game
- Pacific Coast Federation of Fishermen's Association
- Pacific Energy Institute
- Pacific Salmon Alliance
- Pacific Salmon Foundation
- Powerex
- Salmon Enhancement and Habitat Advisory Board
- Salmonid Enhancement Task Group

- Save Our Wild Salmon Coalition (including: Alaska Trollers Association; American Rivers; Antioch Living Systems Collective; Association of Northwest Steelheaders; Boulder-White Clouds Council; Clearwater Forest Watch; Coalition for Salmon and Steelhead Habitat; Coast Range Association; Defenders of Wildlife; Federation of Fly Fishers; Friends of the Earth; Idaho Conservation League; Idaho Rivers United; Idaho Steelhead and Salmon Unlimited; Idaho Wildlife Federation; Institute of Fisheries Resources; Long Live The Kings; The Mountaineers; Natural Resources Defense Council; Northwest Conservation Act Coalition; Northwest Environmental Defense Center; Northwest Resource Information Center; Northwest Sportfishing Industry Association; Oregon Natural Desert Association; Oregon Natural Resources Council; Oregon Outdoors Association; Oregon Trout; Pacific Coast Federation of Fishermen's Association; Pacific Rivers Council; Purse Seine Vessel Owners Association; Rivers Council of Washington; Salmon For All; Salmon For Washington; Sawtooth Wildlife Council; Sierra Club; Sierra Club Legal Defense Fund; Trout Unlimited; Washington Kayak Club; Washington Trollers Association; Washington Wilderness Coalition; Water Watch of Oregon; Western Ancient Forest Campaign); The Wilderness Society
- Seattle Aquarium Society
- U.S. Environmental Protection Agency

TABLE 45.1 (continued)
Organizations participating in the development of the Sustainable Fisheries Strategy.

- Selkirk College
- Shuswap Nation Fisheries Commission
- Simpson Timber
- Squamish River Watershed Committee
- State of Alaska Governor's Office
- Steelhead Society of British Columbia
- Sustainable Fisheries Foundation
- Triton Environmental Consultants Ltd.

- U.S. Geological Survey
- Washington Department of Fish & Wildlife
- Washington Trout
- West Kootenay Power
- Western Forest Products
- Weyerhaeuser Company Foundation
- Wild Salmon Watch

production, community-based fisheries management, and institutional and regulatory structures. The specific recommendations that relate to each topic are presented in the following sections.

I. **Habitat Protection and Restoration**—Sustainable habitats are key to conserving native populations of salmon and other aquatic organisms. The following definition of sustainable habitat provides a basis for considering specific strategies for protecting and restoring freshwater, estuarine, and marine habitats.

> Sustainable habitat is the physical space and collection of biotic and abiotic processes and entities that constitute a properly functioning ecosystem capable of maintaining itself within the bounds and patterns produced by natural disturbance processes.

Sustainable habitat is self-perpetuating, resilient, and permits the full expression of biological diversity (e.g., life history types, phenotypic and genotypic characteristics, trophic structures, and species assemblages) and production potential, and is spatially linked to other habitats (e.g., a stream entering an estuary, which is linked with the ocean). Habitat management should be based on an ecosystem perspective, that is, managers should address the needs of the larger biological community rather than focus exclusively on individual species. It is more ecologically and economically effective to protect habitat than to restore it, so special emphasis should be placed on establishing salmon habitat protected areas before further damage occurs, including both freshwater and marine refugia. Habitat restoration efforts should rely largely on natural processes and assist the natural regenerative capacity of the ecosystem.

Many general and specific actions needed to support sustainable freshwater, estuarine, and marine habitats in the Pacific Northwest are included in the Sustainable Fisheries Strategy. These recommended actions, which form the basis of a comprehensive strategy for protecting and restoring salmon and steelhead habitats, are summarized below.

I-1 **Increase societal awareness of ecosystem processes and foster the aquatic stewardship ethic.** Several specific actions are needed to achieve this goal, including educating the public and policy makers; advocating a stewardship ethic; forming partnerships between landowners and government; developing clear action plans to foster public support for habitat protection initiatives, and holding politicians accountable for their decisions; and, using appropriate methods for determining the non-market values of fisheries resources and conveying this information to the public.

I-2 **Increase scientific understanding about ecosystem processes, habitat requirements by life stage, and linkages among the physical, chemical, and biological**

components of the ecosystem. Some of the actions that are needed to achieve this goal include evaluating and, if necessary, improving methods and criteria for data collection; conducting more studies to obtain the required knowledge, particularly for watershed analysis; standardizing information reporting procedures; sharing data and information among interested parties; defining the parameters for "healthy habitat" (and compare it to the public's perception of "healthy habitat"); conducting a coastwide assessment of habitats to prioritize protection and restoration initiatives; and, evaluating past/current restoration projects to determine those that achieve the best results.

I-3 **Implement an ecosystem-based approach to habitat management throughout the Pacific region.** Some of the actions that support this goal include defining an overarching goal for sustainable habitat; developing a hierarchy of habitat management objectives at the local, regional, and international levels; developing strategic plans as a means of prioritizing habitat protection and restoration initiatives; developing the capacity to understand and manage whole ecosystems (i.e., develop systems analysis techniques); managing for ecological diversity; coordinating planning and management activities within communities, between agencies, and across jurisdictions; standardizing information reporting and dissemination procedures; and evaluating current habitat management practices, using the best as models for sustainable habitat management.

I-4 **Establish management structures that facilitate meaningful local involvement in habitat management decisions.** Specifically, watershed councils should be established and given the authority to make habitat management decisions, provided that minimum standards are met. In this way, community-based solutions to habitat issues can be encouraged and supported.

I-5 **Reform institutional and regulatory structures and processes so that they support sustainable habitat initiatives.** The specific actions that are recommended include setting up a process to establish minimum standards for habitat quality and quantity; periodically reviewing and, as necessary, revising minimum standards; using watersheds and watershed aggregates (i.e., major river basins) as the primary units for habitat management (a total watershed perspective is needed in which ecological, economic, and social interests are considered); establishing the protection of healthy fish populations and habitat as a top priority, with restoration initiatives considered subsequently; passing stronger legislation to protect all life stages of Pacific salmonids; regulating land exchanges to ensure that no harm comes to fish or their habitat; providing tax incentives to property owners who dedicate land to conservation; using zoning to control development; implementing buy-back programs to obtain lands adjacent to critical salmonid habitats (a fund should be set up by each appropriate government for this); identifying preferred development areas that do not overlap with critical fish habitats; establishing sanctuaries or refugia for habitat conservation and restoration; and, changing the British Columbia Water Act to include provisions for riparian zone protection and protection of water quality on crown lands.

I-6 **Protect salmonid habitats by increasing enforcement and compliance monitoring activities.** Additional funding will be needed to achieve this goal. Societal attitudes will need to change to favor increased enforcement while disincentives for habitat protection need to be eliminated.

I-7 **Restore degraded habitats in accordance with established priorities.** Some of the actions that are recommended to restore habitats include amending the Clean Water Act to facilitate restoration activities that improve water quality (e.g., reducing agricultural run-off); stabilizing upslope and upstream habitats to reduce erosion;

recreating essential spawning and rearing habitats where these have been degraded or destroyed; removing dams where possible and appropriate; restructuring dams and dam operations to facilitate fish passage; seeking alternative energy sources that are more fish friendly; and, restructuring or operating dams to assure that water storage and release operations can maintain downstream ecosystem structure and function for the benefit of salmon and steelhead.

II. **Harvest Management**—Viable tribal and aboriginal, commercial, and sport fisheries are essential for maintaining the unique culture, communities, and economies of the Northwest. Yet, declines in the abundance of salmon have necessitated a significant abatement of fishing effort in many areas. As a result, many salmon-dependent communities and economies have been severely affected and will continue to be harmed until salmon and steelhead stocks are restored to their full abundance.

Rebuilding and/or sustaining salmon and steelhead populations, and viable fisheries, will require increasing spawner abundance, as well as restoring the habitats upon which they depend, in many areas throughout the Northwest. It is unlikely that our goal of sustainable salmonid populations, sustainable fisheries, and sustainable communities can be achieved within the current harvest management framework. A major philosophical shift is required in harvest management, from emphasizing fishing to emphasizing healthy escapements. The transition to more effective harvest management systems will require a number of other changes, including adoption of an ecosystem-based management system in which the watershed is the essential stewardship unit and local watershed interests are directly involved in fisheries and environmental management decisions. Salmon escapement goals (i.e., how many fish must return to a stream to sustain the run) must be established for each population using a habitat-based conservation approach (i.e., *not* maximum sustained yield for aggregated populations) and must consider non-human uses of salmon and the ecosystem (i.e., biodiversity, wildlife, etc.). Resource managers must adopt a risk averse approach to harvest management, which minimizes the potential for not achieving escapement goals. Restructuring of salmon fisheries is also needed in many cases to reduce mixed-stock fisheries, increase terminal fisheries, emphasize selective harvesting methods, and efficiently employ stock-specific fisheries. Some of the specific strategies that are recommended to address these needs include the following.

II-1 **Practice conservation-based rather than exploitation-based harvest management**. The first step in this process is to establish "good" escapement goals (i.e., goals that fully utilize the habitat, support biological sustainability, and genetic variability). These goals should be reviewed periodically and refined to reflect changing conditions. Management should be responsive to population vulnerability and explicitly minimize the risks associated with harvesting activities. Importantly, management systems should recognize and respond to the variability of the environment and, hence, fluctuations in salmonid populations.

II-2 **Separate allocation decisions from other harvest issues**. For harvest management strategies to be successful, decisions on how many fish to catch must be based on good science (i.e., they must not be affected by political pressure). Once these decisions have been made, then it is possible to decide how those fish are to be shared among the various user groups. The Alaska Board of Fisheries provides a good example of an effective fish allocation system.

II-3 **Adopt community-based approaches to fisheries management**. Communities must be empowered to participate effectively in the fisheries management process. To this end, multi-stakeholder or multi-shareholder processes must be developed in which participants share in the identification of the problems and creation of innovative solutions. These processes must be open and transparent to foster the honest discussions among the shareholders on motivations, values, knowledge, limitations, and uncertainties. Importantly, linkages among community involvement, empowerment, and accountability must be established, with specific authority devolved to the local level and more general authority retained at the regional level. Minimum standards should be established at higher government levels for habitat protection and other related issues to assure that local decisions on land and water management are consistent with sustainable fisheries and ecosystem management goals. In addition, governmental oversight of community-based decisions is needed to assure that the minimum standards are met. There are several excellent models for integrating communities into the fisheries and environmental management process, including the Alaska Community Advisory Committees (there are 80 committees in all), who advise the Alaska Board of Fisheries, and the Skeena Watershed Committee.

II-4 **Develop harvest management strategies that support sustainable salmonid production**. Some of the actions needed to support sustainable production of salmon and steelhead include establishing and empowering independent organizations to mediate disputes between jurisdictions on salmonid harvest; increasing population-specific management capability by reducing the use of mixed-population fisheries; moving toward terminal fisheries and selective harvesting methods; de-emphasizing the need to harvest all hatchery fish at the expense of commingled wild fish; collecting real-time information on harvest and improve enforcement through shareholder participation; and, restricting harvest by all sectors to increase escapements and ultimately salmonid production.

II-5 **Educate the public on the trade-offs associated with resource management decisions**. It is important that the public recognize that the number of salmon available to harvest depends on many other land and water management decisions. For example, there are direct trade-offs between salmon production and hydroelectric power generation or between housing development and local watershed health. The public needs to be informed about these relationships so they can make informed decisions about the management of watersheds throughout the Pacific Northwest.

III. **Salmon and Steelhead Production**—In response to decreasing natural production resulting from overexploitation and habitat loss, fisheries managers began to turn to artificial production techniques as a means of satisfying the growing demand for Pacific salmon. While releases of juvenile salmonids from production hatcheries, spawning channels, and other facilities dramatically increased survival, the fisheries on these enhanced runs have adversely affected both the numbers and genetic biodiversity of commingled wild salmonid populations. Maintaining high levels of genetic diversity is fundamentally important for achieving sustainable fisheries objectives, particularly in the face of changing climatic conditions on a global scale (that is, seemingly unimportant populations at the edge of the geographic range for each species, may turn out to be the important populations for colonizing altered habitats in the future). For this reason, salmonid production should focus on the conservation and restoration of wild stocks throughout their historic range. The following are key strategies that should be pursued to achieve our goal of sustainable salmonid populations.

III-1 **Increase the natural production of salmon and steelhead.** Some of the specific actions that should be pursued to support this goal include establishing priorities by comparing existing production with production capacity on a stream by stream basis; improving artificial propagation methods and strategies for tactically applying the improved technology to support sustainable fisheries objectives; and, recognizing that protecting and restoring habitat, while ensuring that optimal numbers of adults spawn in the habitat, is the key to realizing sustainable production goals. The concept of optimal spawner abundance must also include consideration of how the number of carcasses influences freshwater production capacity.

III-2 Ensure genetic diversity of salmon and steelhead populations. Managers, harvesters, and the public must be educated about the importance of preserving genetic biodiversity. In addition, harvest managers should fully consider inherent differences in productivity rates among individual populations whenever management decisions are made that affect population aggregates.

III-3 Refocus hatchery programs where necessary. Instead of focusing on production for harvest, artificial propagation of salmon and steelhead should be directed at helping to restore depressed fish populations (i.e., assist in the realization of conservation goals). Hatchery programs should be modified so they do not compromise either the abundance or genetics of wild stocks. Small-scale, temporary hatcheries should replace the large production hatcheries that have been used in the past. Further refocus efforts to produce quality, rather than quantity of, hatchery fish. Production hatcheries that increase catches without harming wild stocks should be used, at least in the near term, to ease the transition to more sustainable fisheries.

III-4 Enhance communication and education. There is a critical need to deliver the message of sustainability more effectively to the public. It is also important to articulate the long-term implications of unsustainable resource management practices to the public and decision-makers, in a way that will capture their attention and stimulate their involvement. By fostering a better understanding of the issues, an atmosphere can be created where people share the responsibility for the problem and work cooperatively toward creative solutions, potentially including alternative governance systems.

III-5 Create an atmosphere that supports sustainable populations of salmonids. Existing laws and regulations should be assessed and reformed, as needed, to support sustainable fisheries. Communities must be involved in this process to facilitate the development of cooperative solutions and a sense of stewardship. Public education can foster and instill a sense of sustainability from one generation to the next. One of the approaches that can be used to achieve this objective involves integrating these concepts in educational curricula in schools, media forums, and political institutions.

III-6 Reconnect people with natural systems. The absence of support for sustainable resource management practices and our inability to recognize the intrinsic value of ecosystems is linked very strongly to the lack of connectivity between society and the natural environment. Some of the opportunities for reconnecting people with the natural systems include the establishment of nature camps for education and creation of viewing areas at important migration corridors, spawning grounds, and rearing habitats.

III-7 Conduct research on sustainability. There is a need for ongoing research on sustainable agriculture, forestry, mining, and fishing. For such research to generate tangible benefits, however, it must be coordinated among the various jurisdictions involved, conducted by multidisciplinary teams, and be adequately funded. Opportunities should be pursued to develop and communicate to the general public and decision-makers the linkages between local economies and ecosystem health. Community forums and other meetings should be convened to encourage non-scientists to participate in the identification of critical information gaps, research priorities, and more effective decision-making processes. Importantly, decision-making processes must evolve to recognize the inherent uncertainty and variability associated with the natural environment and, hence, natural fluctuations in salmonid production.

III-8 Recognize and reflect the intrinsic linkages between production, habitat, and harvesting in fisheries management strategies. Importantly, political and institutional structures must evolve to facilitate the implementation of strategies that explicitly recognize these linkages. These strategies must be adaptive to accommodate the variability of natural systems.

IV. **Community-Based Fisheries Management**—Community-based fisheries management is an essential element of the coastwide Sustainable Fisheries Strategy. When asked why communities should be involved in the management of salmon and steelhead populations and associated habitats, conference participants responded that:

> We must move toward community-based fisheries management to re-establish the link between the authority to manage the resource, the obligation to do it well, and the consequences of failing to do so.

In other words, salmon-dependent communities in both coastal and interior regions are usually most affected by changes in the abundance of the salmonid resource. Therefore, these communities should have a means of participating effectively in decision-making processes related to the management of salmon and steelhead populations and associated habitats—their future depends on it. The following specific strategies are recommended to facilitate a transition toward community-based fisheries management.

IV-1 Convene community workshops to develop a common vision for ecosystem management. Such multi-stakeholder processes will provide a means of defining functional production units and developing partnerships among salmon-based interest groups. These partnerships will be essential for implementing strategies that are designed to support the goals and objectives that are established for each production unit (i.e., using an ecosystem-based approach). In addition, these workshops can be used to gather and disseminate information on each production unit. Such workshops should be convened as soon as possible.

IV-2 Develop effective mechanisms for collecting, maintaining, and disseminating information to community groups and the public. Communities will be empowered by gaining access to information that is related to fisheries and watershed management (including local, traditional, and scientific information). Implementation of this recommendation will require two steps. First, communities should work with agencies to identify their information needs. Then, agencies and communities should work together to identify the most effective means of transferring the necessary information. Development of such mechanisms should be initiated in the near-term.

IV-3 Foster the establishment of environmental ethics and a commitment to responsible stewardship. In addition to effective public education programs, this goal should be advanced through demonstration projects that involve the public and schools. These projects should emphasize ecosystem-based natural resource management and identify the linkages between individuals and the salmonid resource. This is a long-term goal that should be supported by near-term action.

IV-4 **Establish mechanisms that facilitate the equitable distribution of the costs and benefits associated with the sustainable management of fisheries resources.** That is, communities that bear the costs associated with habitat protection and restoration, resource enhancement, or foregoing the harvest of resources should reap the benefits associated with increased returns of salmon in subsequent years. Similarly, those that derive financial benefits by harvesting and processing Pacific salmon should contribute to the costs associated with their management. Therefore, mechanisms need to be developed that facilitate the fair distribution of costs and benefits. Such mechanisms should consider the costs of freshwater production and the role of pasturage in the ocean. It was recognized that the development of such mechanisms would be a long-term goal. In the near-term, the principle of equitability can be advanced by moving toward terminal and more selective fisheries.

IV-5 **Establish mechanisms to provide sustainable funding for community-based fisheries and watershed management initiatives.** To be sustainable, resource management must be self-funding and operated on a cost-recovery basis. That is, the true costs associated with resource use must be passed on to the users of land, water, and fish resources. The main potential sources of funding include royalties from harvesters and processors (with costs passed on to the consumer), land users, water users, other resource users, and sportfishing interests (lodges and anglers). However, start-up funding from government agencies will also be needed in the near term to facilitate the development of community-based management systems (e.g., by funding watershed coordinators, etc.).

IV-6 **Establish incentives to support sustainable salmonid production.** Tax incentives and/or penalties provide an effective means of changing the behavior of land and water users. These should be pursued and, where necessary, reformed. In addition, habitat report cards should be created to evaluate progress toward habitat protection and restoration goals and to identify stakeholders who are contributing to the success of these initiatives. Eco-labeling programs should also be established to identify fish-friendly harvesters, industries, farms, and utilities. Public awareness campaigns, shareholder pressure, boycotts, and legislation should also be used to encourage land and water users to "do the right thing." Opportunities for joint management between local landowners and other community groups should be encouraged through the use of economic incentives.

IV-7 **Let others see and respond to the "Common Vision" for sustainable fisheries.** This can be accomplished through advertising, including web pages, brochures, ads on TV and radio, magazine articles, and other means. The advertising should include an invitation to participate in the development and implementation of cooperative solutions. Adaptive management workshops should also be convened, which provide computer models as tools to explore "what if" scenarios. A variety of demonstration projects and sites should also be developed. It is critical that we develop some examples of effective, ecosystem-based management, so we can point to them and talk with those who have experienced success.

IV-8 **Build broadly-based partnerships that involve all interested environmental agencies, institutions, foundations, and watershed stakeholder groups in this "Common Vision."** Most of these groups have existing communication networks. Better communication could be ensured by establishing a list of groups that need to be involved and then developing a strategy to involve them. All interested parties should be invited to participate.

V. **Institutional and Regulatory Structures**—There is general agreement and ample evidence that our institutions have failed to prevent the decline of Pacific salmon and the deterioration of our watersheds. Therefore, these structures must be reformed if we are to effectively manage our natural resources. This realization underlines the need to identify new mandates and policy objectives by which institutions will be guided, the structural and functional attributes of effective institutions, and the types of institutions best suited to managing human activities at local, regional, and international scales. These goals are reflected in the following broad vision for the future:

> Institutions and an institutional framework must sustain diverse, interjurisdictional fish populations in the future.

By interjurisdictional populations, we mean runs that migrate between jurisdictions within which management authority is exercised (e.g., local management areas, and state, provincial, federal, or international waters). Salmon managers have the difficult responsibility of balancing the conservation and use of the resource, ensuring that fish and fishing opportunities are available to people, and at the same time maintaining the health of the stocks and ecosystems upon which they depend. Working under the premise that existing institutional structures may be incapable of meeting today's challenges, a number of alternative institutional structures for governance of salmon restoration efforts are recommended, including:

V-1 **Reform institutional structures, legislation, and regulations to better support sustainable fisheries**. Current institutional structures and associated arrangements cannot adequately address the challenges that are facing fisheries managers. This is primarily because many activities affecting habitat and harvest management are outside the purview of existing salmon management agencies. Therefore, the existing decision-making structure must be changed to achieve sustainable fisheries. Reformed management structures must be more consistent with biological realities and must more effectively involve shareholder groups. While consensus decisions are the preferred option, an over-riding authority must have with the power to convene meetings of stakeholders and arbitrate final decisions. Importantly, reformed institutional structures must provide a means of separating conservation decisions (e.g., escapement goals, harvestable surplus, etc.) from allocation decisions. A good example is found in Alaska, where sustained yield is mandated in the state constitution, the Department of Fish and Game sets biologically based escapement goals, and The Board of Fisheries decides allocation issues with input from the Department, public advisory committees, and concerned citizens.

V-2 **Alternative management structures should provide an effective means of devolving management authority to the communities.** For joint management structures to work, it is important that the responsibilities of participating organizations and the consequences of failing to live up to those responsibilities be clearly identified in associated agreements. It is recommended that reforms in institutional structures take place at three levels to support sustainable fisheries, including the watershed, regional, and coastwide levels. Specific recommendations for new structures are discussed below.

V-3 **Watershed Councils should be formed for the specific purpose of developing scientifically credible management plans and environmental standards which apply directly to the local watershed.** One of their objectives will be to instill a sense of ownership in the watershed management plan and a sense of responsibility in stakeholders whose activities in the watershed affect salmon survival. It is important that people recognize their role in restoring and protecting the local environment. The prospects of achieving sustainable fisheries will be greatest if people, businesses, and governments work together at the local level. Such partnerships are the key to success of the new management structures.

Equating watersheds to management units has both theoretical and practical value because salmon populations often segregate by drainage and, more fundamentally, the movement of water and matter is usually confined by gravity to the watershed. This watershed-based approach makes it easier to monitor population abundance, assess terrestrial, riparian, and aquatic impacts, and develop appropriate management prescriptions.

Watershed Councils should include local citizens, landowners, representatives of major stakeholder groups, and representatives of appropriate government agencies. Resource specialists should assist the Councils in developing goals, strategies, and standards for the watershed. The Councils should be able to avail themselves of a "tool box" of resources and approaches to guide their efforts.

Along with the development of a comprehensive management plan, Watershed Councils should be responsible for the preparation and analysis of baseline inventories, the specification of appropriate performance standards (indicators), evaluation of proposed actions and progress toward management goals, coordination with other Watershed Councils, and the integration of their plans with higher institutional levels (e.g., Major Basin Councils; see below). The Councils also need to consult with other institutions on the adequacy of the plans and their joint implementation. The Watershed Councils should prescribe and oversee monitoring and evaluation activities to ensure that sufficient progress is being made or that actions are taken to correct deficiencies. The process would be a dynamic one in which new information gained through monitoring and evaluation would be continually fed into the evolving knowledge base. Council members must be willing to use this information to evaluate the consequences of their actions and to change, if necessary, earlier decisions that were made without benefit of complete knowledge, (i.e., using adaptive management approaches).

V-4   Major River Basin Councils should be established for larger watersheds that have many tributary watersheds—such as the Fraser, Skeena, Columbia, Sacramento, and Yukon rivers—to guide management efforts. As used here, a basin is an aggregation of smaller watersheds to which are applied common management goals, standards, and prescriptions. The designation of a Basin Council is particularly relevant when managing groups of salmon populations whose range spans more than one watershed. The Basin Council should facilitate interjurisdictional cooperation on issues involving transboundary stocks and human activities that transcend individual watershed boundaries.

Establishment of a Basin Council is recommended when the whole can be more effectively managed than the parts (given the options available to the manager), or when management coverage by existing Watershed Councils is inadequate. Examples of areas not likely to be fully addressed by Watershed Councils include main stem reaches of larger rivers and their estuaries.

Basin Councils should include representatives of both government agencies and non-governmental organizations. The overarching purpose of Basin Councils would be to develop and administer management plans that would coordinate the activities and authorities of Watershed Councils and reduce the potential for conflict, duplication, and confusion. As such, the Basin Councils should review and recommend local adaptations and modifications of watershed-based approaches and projects, provide a source of technical and financial support for watershed management initiatives, and help coordinate and review monitoring and evaluation activities across watersheds. The Watershed Councils should also ensure that the relevant information is incorporated into the management process and should define appropriate performance standards that apply basinwide.

V-5 **A North Pacific Conservation Council should be established as a higher order institution with responsibility to coordinate among watershed and basin councils and resolve disputes.** The North Pacific Council should be comprised equally of Canadian and U.S. representatives, including both independent scientists and members of specific sectors, including environmental groups, commercial fishing interests, aboriginal organizations, and the academic community. It is important that the representatives on such a North Pacific Council be apolitical and non-partisan to avoid the problems that have historically plagued interjurisdictional management of Pacific salmon. The Council should be funded jointly by federal, state, and provincial governments as part of their ongoing commitment to conservation.

Some of the key responsibilities of a North Pacific Council would be to coordinate the activities of the basin and watershed councils; prepare a "state of the stocks" report annually; prepare a "state of the habitat" report annually; recommend focused research and monitoring programs to address transboundary fisheries management issues (e.g., ocean conditions, offshore harvest, etc.); resolve transboundary disputes; encourage standardization of data collection procedures; and, facilitate access to information. It is likely that the North Pacific Council would also need to be supported by a Secretariat and/or Advisory Committee to assist the Council in fulfilling its mandate.

We recognize that some of the proposed institutional changes are drastically different from existing institutional structures. Significant questions remain regarding the effectiveness of the proposed council-based approach to institutional structures for managing watersheds and fisheries. Success will depend not only on the creativity and energy of stakeholders in the watershed councils, but the willingness of existing fisheries management and research organizations to remain flexible in working with developing councils. Some of the important challenges facing new institutional structures include: establishing new processes for developing annual harvest regulations; maintaining a strong technical basis for decision-making; and ensuring that fisheries research will be responsive to the needs of the councils. Most importantly, successful implementation will require creativity, risk, and courage among political leaders, government agencies, and shareholder groups.

SUMMARY AND CONCLUSIONS

One of the clear messages that has emerged in the information contained in this book, from other sources, and from the in-depth discussions that occurred during the Victoria conference and thereafter, is that Pacific salmon management is an extremely complex undertaking. Salmon and steelhead populations are influenced by a wide range of human activities. In turn, the social, cultural, and spiritual well-being of the people of the Pacific Northwest are inextricably linked to the fate of salmon. Therefore, it is incumbent upon all of us to protect and restore salmon and steelhead populations throughout their range and, in so doing, assure that future generations can fully appreciate this priceless resource.

Most people agree that the existing fisheries and environmental management systems have generally failed to protect salmon and steelhead populations, particularly south of Alaska. Therefore, significant reform of our institutional and regulatory structures is needed to support sustainable fisheries goals. To be effective, these new management structures will need to be based on broad partnerships and embrace input from many different sources. Importantly, effective solutions to the complex challenges that we are currently facing in the management of Pacific salmon and steelhead populations will require implementation of ecosystem-based approaches to natural resource management. We hope that the Sustainable Fisheries Strategy and this book will provide government agencies, First Nations and Tribal organizations, and conservation and community groups with a framework for sustainable management of our shared fisheries resources. However, its success depends on effective implementation, planning, and cooperative action among all of the partners involved in the process.

While the road to salmon recovery will undoubtedly be long and hard, we can achieve our common vision for the future of salmon and steelhead populations, if we choose to work together in the spirit of cooperation and common purpose. We must—if our children and theirs are to enjoy the benefits associated with sustainable fisheries that we have come to treasure.

ACKNOWLEDGMENTS

The Sustainable Fisheries Strategy would not be possible without the creativity, suggestions, comments, critiques, and discussions with countless individuals who participated in the Toward Sustainable Fisheries conference and follow-up community-based conferences. We appreciate the dedication and perseverance of all the authors contributing to this book. Thanks to C. Burger, K. Hyatt, and J. Williams for their reviews and comments on this manuscript.

REFERENCES

Knudsen, E.E., D.D. MacDonald, and C.R. Steward. 2000. Setting the stage for a sustainable Pacific salmon fisheries strategy. Pages 3–13 in E.E. Knudsen, C.R. Steward, D.D. MacDonald, J.E. Williams, and D.W. Reiser, editors. Sustainable fisheries management: Pacific salmon. Lewis Publishers, Boca Raton, Florida.

NRC (National Research Council). 1996. Upstream: Salmon and society in the Pacific Northwest. National Academy Press, Washington, D.C.

Stouder, D.J., P.A. Bisson, and R.J. Naiman, editors. 1997. Pacific salmon and their ecosystem status and future options. Chapman & Hall, New York.

SFF (Sustainable Fisheries Foundation). 1996. Towards sustainable fisheries: Building a cooperative strategy for balancing the conservation and use of westcoast salmon and steelhead populations. Preliminary Draft for Review and Comment. Ladysmith, British Columbia and Bothell, Washington.

Index

A

Abies amabilis (silver fir), 618
Abies grandis (grand fir), 608
Aboriginal Fisheries Strategy (AFS), 76
Aboriginal fishing, B.C.
    allocation proposal, 76
    canning industry impact, 76
    conservation techniques, 80, 81
    equipment used, 80
    federal prohibitions against, 80–81
    fishing rights distribution, 85–86
    government suppression of rights, 81–82
    history of, 79–80
    lack of employment opportunities, 85
    native reliance on, 75–76
Absention principle, 81–82, 224
ACCD (Asotin County Conservation District), 639
Acer spp. (maple), 527, 608, 609
Acipenser transmontanus (white sturgeon), 53
ACS (Aquatic Conservation Strategy), 605
ADF&G (Alaska Department of Fish and Game), 162, 183,
    274, 584
Adult equivalent (AEQ) fishing mortalities, 459
Afognak Island (Alaska), 676–677
AFS (Aboriginal Fisheries Strategy), 76
Agriculture impact on habitats, 16, 45
Alaska, *see also* Southeast Alaska salmon
    Alsek River
        fisheries, 278
        habitat studies, 274
        location and geography, 274
        research programs, 288, 290, 291
        stock programs and status, 286–288
    Bristol Bay fishery, 8
    cost–benefit analysis, hatchery programs, 401
    fisheries management, statewide, 6
        allocation and conservation separation, 65
        habitat protection policies, 64, 182–183
        leadership in, 65–66
        production improvements, 16–17
        sustainable communities commitment, 65
        sustained yield requirement, 63–64
    global marketplace position, 399–400
    habitat protection, Tongass study
        discussion and recommendations, 593–597
        methods, 585–587
        results, 587–593
        team charter, 584–585
    protection policies, 183
    ranching programs
        disease dissemination, 417
        ecological interactions, 417
        Northern Southeast, 414–415
        overview, 407–409
        Prince William Sound, 409–413
        production, 409
        Southern Southeast, 413–414
        straying and genetic risks, 416
    salmonid status, 584
    Stikine River
        fisheries, 275–277
        habitat studies, 274
        location and geography, 274
        research programs, 288, 290, 291
        stock programs and status, 278–282
    Taku River
        fisheries, 278
        habitat studies, 274
        location and geography, 274
        research programs, 288, 290, 291
        stock programs and status, 282–286
Alaska Board of Fisheries, 64, 65, 183
Alaska Department of Fish and Game (ADF&G), 162, 183,
    274, 584
Alaska Forest Resources and Practices Act, 182
Alaska Genetic Policy, 416
Alder (*Alnus rubra*), 300, 527, 606
Aleutian low pressure field, 40, 471, 473, 474
Alsek River (Alaska, Canada)
    fisheries, 278
    habitat studies, 274
    location and geography, 274
    research programs, 288, 290, 291
    stock programs and status
        chinook salmon, 286–287
        coho salmon, 288
        sockeye salmon, 287–288
American Fisheries Society, 24, 676, 677
Anotopterus pharao (daggertooth), 177
Anthropogenic factors
    accounted for in climate cycles, 469, 470–471
    adverse effects on salmon habitats, 6
    cumulative effects, 470–471
    habitats effected by, 35–37
    logging, 182–183
    risk reduction techniques, 597
    road location and design, 589–590, 594–595
    urbanization
        adverse effects on salmon habitats, 6
        impact on sustainability, 37
Appendices
    escapement goal-setting methods summary, 262–265
    fisheries resource conservation, 59–62
    management units and stocks, 272

populations estimated, 264–267
populations for estimated type, 268–271
Appropriations Act for Interior and Related Agencies
(1994), 584
Aquaculture, *see also* Hatcheries
marketing and, 399–400
net-pen reared salmon
cost of regulatory compliance, 399
market for, 397
opposition in U.S., 398
production history, 396
Aquatic Conservation Strategy (ACS), 605
Aquatic Science Team, 134
Area quotas for conservation, 662–663
Artificial production, *see also* Hatcheries
mitigation vs. enhancement, 381
net-pen reared salmon
cost of regulatory compliance, 399
market for, 397
opposition in U.S., 398
production history, 396
ranching programs, Alaska, 396
disease dissemination, 417
ecological interactions, 417
Northern Southeast, 414–415
overview, 407–409
Prince William Sound, 409–413
production, 409
Southern Southeast, 413–414
Artificial reefs compensation measures, 345
Asotin County Conservation District (ACCD), 639
Asotin Creek Model Watershed Plan (WA)
agency support, 644
community plans for restoration, 640–641
ecosystem approach, 645
landowner steering committee, 643
mission and objectives, 641–642
public outreach and education, 644
successful plan elements
landowners involvement, 642–643
watershed coordinator selection, 642
watershed setting, 640
Atlantic cod (*Gadus morhua*), 6
Atlantic salmon (*Salmo salar*), 394, 676
Atlantic Zone Fisheries Resource Conservation Council, 52
Australia farmed salmon industry, 397

B

Baird, Spenser, 676
Bankfull width (BFW), 547
Barkley Sound (B.C.), 302
Basin Councils and sustainable fisheries strategy, 699–700
BCF (British Columbia Ministry of Fisheries), 51
Beaver (*Castor canadensis*), 604
Beaver dams, 609, 610, 626
Beverton-Holt model, 485–486
BFW (bankfull width), 547
Big Chico Creek and spring-run chinook, 648
Big leaf maple (*Acer macrophyllum*), 608

Biological Restoration and Compensation Program
(BRCP), *see* Panther Creek BRCP
Biological Review Team (BRT), 104, 111, 119; *see also*
National Marine Fisheries Service
Blackbird Mine, 567, 568, 579
BLM (Bureau of Land Management), 602, 603, 626
Board of Fisheries (Alaska), 64, 65, 183
Bonneville Dam, 135, 136
Bonneville Power Administration (BPA), 382, 662
Bottleneck effects, 556
Bower, Seymour, 677
Brachyramphus marmoratus (marbled murrelets), 604
BRCP (Biological Restoration and Compensation
Program), *see* Panther Creek BRCP
Brice, J.J., 677
Bristol Bay fishery, 8
British Columbia (B.C.), *see also* Canada
Aboriginal fishing
allocation proposal, 76
canning industry impact, 76
conservation techniques, 80, 81
equipment used, 80
federal prohibitions against, 80–81
fishing rights distribution, 85–86
government suppression of rights, 81–82
history of, 79–80
lack of employment opportunities, 85
native reliance on, 75–76
catch and release policy, 70
estuary management, Fraser River
assessment of results, 345
compensation projects, 343–345
habitat loss, 342
mitigation referral system, 343
policy definitions, 342
recommendations, 346
experimental watershed project, Carnation Creek
adult chum returns, 302–303, 306
adult coho returns, 302–303, 305–306
chum smolts marine survival, 315
coho smolts marine survival, 312–317, 320
data collection, 300–301
design limitations, 322
environmental and population variables, 300
juvenile chum, 307–308
juvenile coho, 308–311
location and description, 298–299
objectives and purpose of, 297–298
steelhead and cutthroat trout, 320–322
streamside forest-harvest treatments, 299–300
summary, 323–24
farmed salmon industry, 71–72
fishery management
management policy, 69–70, 73–74
operational measures, 70–72
public policy, 68–69, 72–73
nursery lakes, 507–508
principal watersheds, 77
rearing capacity estimates, PR model
assumptions, 514, 516
assumptions in earlier models, 506

biologically feasible production estimates, 519
measurements correlation to fish yield, 506
model development, 509–510
nursery lakes in B.C., 507–508
predictions for B.C. lakes, 517
productivity variables measurement, 508–509
testing of, 511, 514
uses as a model, 507
versatility and uses, 519
salmon management
authority for, 86–87
commercial fishery impact, 82
economic considerations, 83–85
mixed-stock problem, 78–79
British Columbia Forest Practices Code (1996), 69, 298, 356
British Columbia Lands Act, 357
British Columbia Ministry of Fisheries (BCF), 51
Brook lamprey (*Lampetra richardsoni*), 602
BRT (Biological Review Team), 104, 111, 119; *see also* National Marine Fisheries Service
Buffers, stream
management deficiencies, 594
status of, Tongass, 588–589
Bull trout (*Salvelinus confluentus*), 568, 640
Bureau of Land Management (BLM), 602, 603, 626
Butte Creek spring-run chinook history, 648–649

C

CAFSAC (Canadian Atlantic Fisheries Scientific Advisory Council), 52, 59–60
California (CA)
coastal steelhead ESUs, 128–129
coho salmon status, 114
high pressure field, 40
steelhead trout status, 128
watershed management, spring-run chinook
harvest issues, 649
history of, 648–649
outreach effort, 650
recent declines in abundance, 649
Work Group role, 650–654
California Department of Fish and Game (CDFG), 658
Canada, *see also* British Columbia
Alsek River
fisheries, 278
habitat studies, 274
location and geography, 274
research programs, 288, 290, 291
stock programs and status, 286–288
cost–benefit analysis, hatchery programs, 400–401
farmed salmon industry in, 397
federal legislation for fishery management, 51
fishing agreements with U.S., 224–226
Stikine River
fisheries, 275–277
habitat studies, 274
location and geography, 274

research programs, 288, 290, 291
stock programs and status, 278–282
Taku River
fisheries, 278
habitat studies, 274
location and geography, 274
research programs, 288, 290, 291
stock programs and status, 282–286
Canadian Atlantic Fisheries Scientific Advisory Council (CAFSAC), 52, 59–60
Canadian Fisheries Act (1888), 342, 356
Canadian Fisheries Act (1967), 51
Canning industry
history in B.C., 80
impact on native fishing, B.C., 76
Carbon dioxide (CO_2) in aquatic habitats, 40, 41, 42
Carez lyngbyei (sedge), 343, 344
Carnation Creek Experimental Watershed Project, *see* Carnation Creek study
Carnation Creek study (British Columbia)
adult chum returns
declines due to forest-harvesting, 306
population collapse predictions, 306
population distributions observed, 302
reasons for decline in, 302–303
adult coho returns
declines due to forest-harvesting, 306
fishery impact on, 305–306
population collapse predictions, 306
population distributions observed, 302–303
reasons for decline in, 303, 305
response to conservation measures, 305
chum smolts marine survival, 315
coho smolts
forest harvesting, 312
marine survival, 313–317, 320
data collection
populations surveys, 301
techniques used, 300–301
variables studied, 300
design limitations
absence of control watershed, 322
atypical road network and, 322
environmental and population variables, 300
juvenile chum
forest harvesting effects on, 307–308
reduced ocean survival, 308
juvenile coho
decline in habitat complexity, 310–311
forest harvesting effects on, 308–310
sediment migration effects, 311
location and description, 298–299
objectives and purpose of, 297–298
steelhead and cutthroat trout, 320–322
streamside forest-harvest treatments, 299–300
summary, 323–24
Castor canadensis (beaver), 604
Catastomus macrocheilus (suckers), 602
CDFG (California Department of Fish and Game), 658
CDQs (community development quotas), 659
Cedar River, 528, 529, 531

Cedar (*Thuja plicata*), 527, 608, 618
Central Valley California, *see* Spring-run chinook
    watershed management
CERCLA (Comprehensive Environmental Response,
    Compensation and Liability Act) (U.S.), 567
Cervus elaphus (elk), 604
CFCs (chloro-fluorocarbons) in aquatic habitats, 40
CFFG (Coastal Fisheries-Forestry Guidelines), 298
CH_4 (methane) in aquatic habitats, 40
Channelization of terrestrial habitat, 344
Char (*Salvelinus malmo*), 617
Cherry salmon (*Oncorhynchus masu*), 396
Chile farmed salmon industry, 397
Chinook salmon (*Oncorhynchus tshawytscha*)
    annual harvests, 166, 171
    distribution, 170–171
    economic feasibility of enhancement programs,
        385–388
    escapement trends, 171–172
    exploitation rate reduction, Snake River
        harvest strategies, 458–459
        management decisions, 463–465
        PSC model, 455–457, 464
        results and discussion, 460, 461
    recruitment variations
        effects on sustainable yield, 491
        ocean harvest rate, 488–489
        ocean survival, 489–490
        past recruitment rate estimations, 488
        recruitment rate indices, 486, 488
    reintroduction and restoration, Panther Creek
        donor stock selection, 571–576
        stock supplementation methods, 576
    in Siuslaw River watershed, 626
    status and distribution, Columbia and Klamath rivers
        classification criteria, 137
        current status, 148–151
        database use, 138
        emphasis area identification and importance,
            153–155
        factors influencing, 151–153
        geographic study area, 134–135
        implications for sustainability, 155–156
        known status, 147
        potential historical range, 138, 146–147
        predictive models, 138–142
        predictive models results, 147–148
        subspecies studied, 136
        subwatershed divisions, 137–138
    stock assessment and management
        annual harvests, 166, 171
        escapements, 171–172
    stock programs and status
        Alsek River, 286–287
        Stikine River, 278–279
        Taku River, 282–283
    threatened species status, 452
    watershed-based management, spring-run
        harvest issues, 649
        history of, 648–649
        outreach effort, 650

recent declines in abundance, 649
    Work Group role, 650–654
Chloro-fluorocarbons (CFCs) in aquatic habitats, 40
Chub mackerel (*Scomber japonicus*), 317
Chum salmon (*Oncorhynchus keta*)
    annual harvests, 178–180
    distribution, 178
    escapements, 180–181, 182
    habitat assessment, Carnation Creek
        adult declines due to forest-harvesting, 306
        juveniles and forest harvesting, 307–308
        population collapse predictions, 306
        population distributions observed, 302
        reasons for decline in, 302–303
        smolt marine survival, 315
    production of, 181–182
    stock assessment and management, 182
        annual harvests, 178–180
        distribution, 178
        escapements, 180–181, 182
        production of, 181–182
    stock losses, 45
    stock programs and status
        Stikine River, 281–282
        Taku River, 286
Classification trees, 138–139
Climate and sustainability, *see* Decadal climate cycles
Clupea harrengus (herring), 53, 317
CO_2 (carbon dioxide), 41, 42
CO_2 (carbon dioxide) in aquatic habitats, 40
Coastal Fisheries-Forestry Guidelines (CFFG), 298
Coastrange sculpins (*Cottus aleuticus*), 301
Coho salmon (*Oncorhynchus kisutch*)
    annual harvests, 165–167
    distribution, 165
    economic feasibility of enhancement programs,
        385–388
    escapement trends, 167–168
    habitat assessment, Carnation Creek
        adult declines due to forest-harvesting, 306
        fishery impact, 305–306
        juveniles and forest harvesting, 308–310
        marine survival of smolts, 313–317, 320
        population collapse predictions, 306
        population declines, 303, 305
        population distributions observed, 302–303
        response to conservation measures, 305
        sediment migration effects, 311
        smolts and forest harvesting, 312
    habitat assessment in Oregon
        field methods for data collection, 332
        fish productivity and habitat conditions, 335–338
        habitat data analyses, 333
        habitat quality assessments, 333–335
        restoration strategy development, 338
        survey design, 330–332
    marine survivals, 170
    production of, 168–169
    recruitment variations
        ocean harvest rate, 488–489
        ocean survival, 489–490

past recruitment rate estimations, 488
recruitment rate indices, 486, 488
sustained yield effected by, 491
restoration in Siuslaw River watershed, 633–634
ecosystem current state, 632–633
history of, 626–630
restoration projects, 602, 633–634
status review, ESA
ESU identification, 112–113
extinction risk assessment, 114–115, 116–117
listing status, 115
stock assessment and management
annual harvests, 165–167
distribution, 165
escapements, 167–168
marine survivals, 170
production, 168–169
stock decline history, 96
stock programs and status
Alsek River, 288
Stikine River, 281
Taku River, 285
Columbia River, 32
chinook and steelhead status and distribution
classification criteria, 137
current status, 148–151
database use, 138
emphasis area identification and importance,
153–155
factors influencing, 151–153
geographic study area, 134–135
implications for sustainability, 155–156
known status, 147
potential historical range, 138, 146–147
predictive models, 138–142
predictive models results, 147–148
subspecies studied, 136
subwatershed divisions, 137–138
coastal steelhead ESUs, 126
conservation efforts, 66
economic evaluation, 384
historic populations of salmonids, 4, 133–134
history of enhancement propagation, 383–384
history of population declines, 678
inland steelhead ESUs, 129–130
pre-development salmon population, 16
stock extinctions, 96
Columbia River Fish Management Plan, 454
Columbia River Terminal Fishery Project (TFP), 382
Commercial fishery, see also Harvest management
annual harvests
chinook, 166, 171
chum, 178–180
coho, 165–167
pink, 176
sockeye, 166, 173
fleet reduction and licensing, B.C., 69–70
impact on salmon management, B.C., 82
troll regulations
bilateral agreements, U.S. and Canada, 224–225
history of, 223–224

international agreements, 224
reciprocal fishing agreements, U.S. and Canada,
225
Commercial Fishery Entry Commission (CREC), 441
Community development quotas (CDQs), 659
Community efforts toward sustainability
Alaska fisheries, 65
Asotin Creek plan
agency support, 644
community plans for restoration, 640–641
ecosystem approach, 645
landowner steering committee, 643
mission and objectives, 641–642
public outreach and education, 644
successful plan elements, 642–643
watershed setting, 640
community-based management, 696–697
Deer Creek Group, 619
institutional reform, 658–659
public education, 644, 658
Spring-run Work Group role
lessons learned, 654
program goal, 650
recovery strategy, 650–651
restoration funding, 652–653
restoration progress, 651
results and current function, 653–654
watershed conservancies, 651–652
Compensation measures
artificial reefs, 345
bank construction, 345
channelization, 344
compensation for loss definition, 342
vegetation transplanting, 343–344
Comprehensive Environmental Response, Compensation
and Liability Act (CERCLA) (U.S.), 567
Conference on the Law of the Sea Treaty, U.N., 82, 225
Conservation of stock, see Stock conservation
Convention for Antarctic Exploitation, 60–62
Copepods (Cyclops columbianus), 173
Coquille River (Oregon), 488
Cottus sp. (sculpins), 301, 602
CREC (Commercial Fishery Entry Commission), 441
CRiSP model, 478
Cutthroat trout (Oncorhynchus clarki clarki), 95, 96–97,
320–322
Cyclops columbianus (copepods), 173

D

Dace (Rhinichthys sp.), 602
Daggertooth (Anotopterus pharao), 177
Dams
anthropogenic impact of, 479–480
habitat degradation from, 648
Decadal climate cycles
anthropogenic effects accounted for, 469, 470–471
climate pattern shifts
ecosystem mechanisms, 474
fish and, 471

hydrosystem, 475–476
latitudinal patterns, 473–474
ocean/climate mechanisms, 471–472
plankton response, 472–473
PNI and Columbia River salmon, 475
regime shift indicator, 474–475
salmon response, 473
climatic variations effects, approaches for, 468–469
human and climatic impacts, cumulative effects, 470–471
interaction of processes
quantifying climate and anthropogenic processes, 479–480
salmon survival and climate patterns, 477–479
stock recruitment equation, 468
Deer Creek (WA) watershed assessment and restoration
background, 618–619
Deer Creek Group, 619
monitoring and maintenance, 622
restoration implementation, 621–622
results and discussion, 622–623
spring-run chinook history, 648–649
strategy development, 619–620
Deer (*Odocoilesu hemionus*), 609
Department of Fisheries and Oceans (DFO) (Canada)
Fraser River sockeye studies, 210
management responsibility, 51, 52, 59–60
native land claims authority, 87
"no net loss" principle, 68, 357
operational measures for fishery management, 70–72, 76
transboundary rivers studies, 274
Deschutes Basin (Oregon), 135, 678
DFO, *see* Department of Fisheries and Oceans (Canada)
Diking
habitat impacts from, 355
impact on sustainability, 39
Division of Commercial Fisheries Management and Development (Alaska), 183
Division of Sport Fish (Alaska), 183
Donor stock selection
candidate drainages, Panther Creek, 573, 576
genetic outbreeding depression, 573
length of migration, 572
minimum stock sizes for genetic diversity, 573
RASP analysis, 571–572
run-timing, 572–573
similarity of drainage basins, 572
Douglas fir (*Pseudotsuga menziesii*), 527, 608, 618, 626
Drainage basins, 530, 572

E

Earthjustice Legal Defense Fund (ELDF), 650
Ecological reporting units (ERUs), 135
Economic considerations
commercial value to communities, 655
conservation incentives, 661–662
consumer fish tax considerations, 658
cost–benefit analyses

enhancement programs, 400–401
fishwheels, 672, 673
ranched salmon, 394, 400–401
riparian restoration, 578
economic value of fisheries, Alaska, 161
enhancement programs
contributions, 387–390
cost summary, 386
feasibility summary, 390–391
feasibility variables, 382–383
hatchery costs per fish, 385
production costs, 384–385
production costs vs. ex-vessel value, 387
stock survival rates, 385
farmed salmon production, global, 394, 399
funding sources for research, 657
harvest size and market value, 399
institutional reform for positive incentives
area quotas, 662–663
land and water trusts, 663
management compacts, 663
utilities, fish, 663–664
landed value of Nass River harvest, 668
LWD rehabilitation costs, 558
marketing management, 402–404
Panther Creek restoration, 579
restoration funding, spring-run chinook, 652–653
salmon management, B.C.
Aboriginal employment opportunities, 85
commercial grade quality, 84–85
harvest production, benefits of increased, 85
selective trap fisheries, 84
stock identification, 84
stock-specific management techniques, 83
salmon markets
annual harvests, 393–394
global, 396–399
market price, 382
net-pen reared salmon, 397
price decreases, 394
stock decreases, 394
value of commercial vs. recreational catches, 72
watershed-based management, 403
Ecosystem-based management
Asotin Creek Model Watershed Plan, 645
characteristics of, 9
ecosystem health goal adoptions, 660–661
functional elements, 9
habitat protection in LWW
adaptive management, 537–538
agency change, 537
conservation area establishment, 538–540
context and scale considerations, 535–536
ecologically relevant management units, 534–535
goals and objectives, 536–537
restoration and protection emphasis, 536
integration of views, 9
LWD rehabilitation, 546
need for, 8
in Siuslaw River watershed, 634
sustainable fisheries strategy, 691–693

Eelgrass (*Zostera marina*), 344
EEZ (exclusive economic zone), 224, 225, 227
ELDF (Earthjustice Legal Defense Fund), 650
Elk (*Cervus elaphus*), 604
El Niño, 302, 308, 472, 500
Elwha River (WA), 108
Endangered Species Act (ESA)
    chinook status, 452
    coho salmon status review
        ESU identification, 112
        extinction risk assessment, 114–115, 116–117
        listing status, 115
    endangered species definition, 106–107
    listed salmon stocks, 4, 5, 95–96, 228, 452
    Panther Creek BRCP, 568–569
    pink salmon status review, 103
        endangered species definition, 106–107
        even-year ESU, 104, 106
        even-year extinction risk, 107–108
        odd-year ESU, 104–106
        odd-year extinction risk, 108–109
    refuge establishment using, 683
    steelhead trout status review
        coastal ESU identification, 120, 126–129
        coastal ESU risk categories, 121–125
        inland ESU identification, 129–130
        inland ESU risk categories, 125
        listing status, 130–131
    threatened species definition, 114
Enhancement programs, *see* Artificial production;
    Hatcheries
Environmental Review Process (Canada), 69
Environmental Variability scalars, 457
Equipment for salmon fishing
    Aboriginal fishing techniques, 80, 83
    fishwheels, 80, 83
        contribution to fisheries management, 671–672
        cost–benefit analysis, 672, 673
        harvesting procedure efficiencies, 673
        limits to use, 672
        population estimates, 669–671
        research opportunities, 671
        uses of, 668
    gillnet fisheries, 80
        Alsek, Stikine, and Taku rivers, 276–278
        Columbia River, 387
        high mortalities with, 84
        Nass River, 671
    weirs, 83, 627
ERUs (ecological reporting units), 135
ESA, *see* Endangered Species Act
Escapements management
    chinook escapement trends, 171–172
    coho escapement trends, 167–168
    deficiencies in, 238–239
    definitions for, 165, 238
    downward evolution of goals, 239
    escapement estimations table, 264–267
    escapement estimations type table, 268–271
    goal-setting methods summary table, 262–265
    management recommendations

        management actions, 255–256
        population health maintenance, 254–257
        public education and expectations, 257–258
        technological advancements needed, 251–253
    management units and stocks table, 272
    mixed population fisheries
        escapement goals and, 239–240
        harvest priority, 250–251
        number of populations, 249–250
        run management effects, 251
    production capacity determination, 238
    spawner/recruit models, 239, 243–244
    sustainability and, 238
    theoretical methods assessment, 240
        data collection methods quality, 246–248
        escapements estimation accuracy, 241–243, 246
        estimate type quality, 248–249
        goal-setting methods classifications, 240–241
        goal-setting methods evaluation, 245
        gradual goal reduction examples, 245–246
        gradual goal reduction possibility, 241
        mixed population fisheries occurrence, 243
        spawner/recruit model weaknesses, 243–245
Estuary management, Fraser River (B.C.)
    assessment of results, 345
    compensation projects, 343–345
    habitat loss, 342
    mitigation referral system, 343
    policy definitions, 342
    recommendations, 346
ESU, *see* Evolutionarily Significant Unit
Eulachon (*Thaleichthys pacificus*), 53
Euphotic volume (EV) model, 505, 509
Euphotic zone depth (EZD), 509
Evolutionarily Significant Unit (ESU)
    coho salmon identification, 112–113
    concept of, 4, 5, 24–25
    conservation of stock, 56
    data review process, 112
    pink salmon, 103, 104–106
        endangered species definition, 106–107
        even-year, 104, 106
        even-year extinction risk, 107–108
        odd-year, 104–106
        odd-year extinction risk, 108–109
    qualities of, 103
    steelhead trout
        coastal identification, 120, 126–129
        coastal risk categories, 121–125
        inland identification, 129–130
        inland risk categories, 125
        listing status, 130–131
Exclusive economic zone (EEZ), 224, 225, 227
Exploitation rate reduction, Snake River
    harvest strategies
        adult equivalent fishing mortalities, 459
        status quo policy, 458–459
    management decisions
        alternate analyses, 464–465
        assumptions considerations, 464

objectiveness of harvest analysis, 464
    policy selection, 463
PSC Chinook Model
    allocation issues, 454
    assumptions, 457
    background, 455–456
    calibration, 457
    spawner/progeny relationship, 457
    uses of, 464
results and discussion
    relative legal catches, 461
    single-region policies, 460, 461
    status quo policy predictions, 460
EZD (euphotic zone depth), 509

F

FAO (Food and Agriculture Organization), 6
Farmed salmon
    in British Columbia, 71–72
    economic effects, 394, 399
    international, 397
Faroe Islands farmed salmon industry, 397
FEMAT (Forest Ecosystem Management Assessment
        Team), 96
FHAP (Fish Habitat Assessment Procedure) (B.C.), 555
First Nations, 72; *see also* Aboriginal fishing, B.C.
Fisheries Act (1888) (Canada), 76, 356
Fisheries Act (1967) (Canada), 51, 68
Fisheries Rehabilitation, Enhancement, and Development
        Division (FRED), 396
Fishery Resource Assessment Model (FRAM), 424
Fishery resource conservation, *see also* Stock conservation
    definitions of, 52
    ecosystem perspective, 55–56
    federal legislation, Canada, 51
    keystone species preservation
        ecosystem linkage, 56–57
        stock management, 57, 62
    operational objectives
        biomass vs. biodiversity management, 53–55
        evolution of perspective, 53, 55
        human impact and, 56–57
        management principles, 58–62
Fish Habitat Analysis Team, 584
Fish Habitat Assessment Guidebook, 556
Fish Habitat Assessment Procedure (FHAP) (B.C.), 555
Fish utilities, 663–664
Fishwheels on Nass River
    contribution to fisheries management, 671–672
    cost–benefit analysis, 672, 673
    harvesting procedure efficiencies, 673
    limits to use, 672
    population estimates, 669–671
    research opportunities, 671
    uses of, 668
Floods
    hydrologic regimes and, 530–531
    LWD rehabilitation consideration of, 556
    rehabilitation following, 553

Flows, instream, *see also* Compensation measures; Large
        woody debris
    rehabilitation
        dams impact on, 479–480, 648
        diking impact on, 39, 355
        hydroelectric power impact on, 6, 45
        logging impact on, 182–183, 628–629
        stream buffers and, 588–589, 594
Food and Agriculture Organization (FAO), 6
Forest Ecosystem Management Assessment Team
        (FEMAT), 96
Forest management activities
    adverse effects on salmon habitats, 5
    forest harvesting effects, Carnation Creek
        adult chum returns, 302–303, 306
        adult coho returns, 302–303, 305–306
        chum smolts marine survival, 315
        coho smolts marine survival, 312–317, 320
        data collection, 300–301
        design limitations, 322
        environmental and population variables, 300
        juvenile chum, 307–308
        juvenile coho, 308–311
        location and description, 298–299
        objectives and purpose of, 297–298
        steelhead and cutthroat trout, 320–322
        streamside forest-harvest treatments, 299–300
        summary, 323–324
    impact on sustainability, 37
    interdependence with habitats, 45
    logging
        impact on habitats, 182–186
        in Siuslaw River watershed, 628–629
Forest Practices Act (Alaska), 64
Forest Practices Code (Canada), 69, 298, 356
FRAM (Fishery Resource Assessment Model), 424
Fraser River (British Columbia), 32
    Aboriginal fishing history, 80–81
    estuary management
        assessment of results, 345
        compensation projects, 343–345
        habitat loss, 342
        mitigation referral system, 343
        policy definitions, 342
        recommendations, 346
    human population growth impact, 33–35
    land and water use impact, 37–39
    sockeye salmon
        fishing plan, 210, 212
        in-season data analysis, 211–212
        in-season data collection, 210–211
        management plan objectives, 212–215, 216
        population history, 207–209
        production database, 210
        responsible parties, 209–210, 212
        run-size estimation process, 212
        stock status, 215–216
    stock management, 78
    urban streams, lower valley
        early development, 350–351
        fishery resource, 350

location and description, 349–350
population growth and land use, 351–355
protection of fish habitat, 356–358
recommendations for, 358–359
Fraser River Estuary Management Program (FREMP), 343
Fraser River Panel, 210, 212
Frazer Lake (Alaska), 23–24
FRED (Fisheries Rehabilitation, Enhancement, and
      Development Division), 396
FREMP (Fraser River Estuary Management Program), 343

G

Gadus morhua, 6
Game Creek (Alaska) watershed
      description, 586
      headwater condition assessment, 591–592
      instream condition assessment, 590
      riparian areas, 593
Gasterosteus aculeatus (stickleback), 516
GCFA (Golden Gate Fishermen's Association), 650
Genetic diversity
      artificial production as risk to, 7
      vs. biomass management, 53–55
      donor stock selection, 573
      maintenance importance, 56
      maintenance of, 23, 24
      phenotype differences and, 155
      straying risk and, 416
Gillnet fisheries
      Alsek, Stikine, and Taku rivers, 169, 276–278
      Columbia River, 387
      high mortalities with, 84
      Nass River, 671
Gitxsan-Wet'suwet'en land claims, 87
Global salmon markets
      aquaculture and
            marketing impact, 399–400
            net-pen reared salmon, 396–399
      implications for PNW
            marketing management trends, 402
            watershed-based management, 402–403
      international markets, 396–399
      public and private ranching
            cost–benefit analysis, 400–401
            impact of, 400
      U.S. market position, 399–400
Global warming
      freshwater habitats impacted by, 42
      mean air temperature increases, 40
      salmon migration impacted by, 41–42
      salmon production impacted by, 40
      surface water temperature increases, 42–43
Golden Gate Fishermen's Association (GCFA), 650
Government agencies
      Board of Fisheries (Alaska), 64, 65, 183
      British Columbia Ministry of Fisheries, 51
      Bureau of Land Management (U.S.), 602, 603, 626
      California Department of Fish and Game, 658

Department of Fish and Game (Alaska), 162, 183, 274,
      584
DFO (Canada), *see* Department of Fisheries and
      Oceans
Food and Agriculture Organization (U.N.), 6
Fraser River Panel (B.C.), 210, 212
MELP (Canada), 51
NMFS (U.S.), *see* National Marine Fisheries Service
Oregon Department of Fish and Wildlife, 384, 630–632
PFMC (U.S.), 221, 226, 454
PMFC (U.S.), 223
PSC (U.S.), *see* Pacific Salmon Commission
U.S. Forest Service, 584, 626
Grande Ronde River (Oregon), 96
Grand fir (*Abies grandis*), 608
Gravel mining operations impact on sustainability, 37
Grease Harbour (B.C.) fishwheels, 668
Greenhouse gases, 40
Growth Management Act (1990) (WA), 538

H

Habitat assessment and management
      basin-wide management institutions, 663
      benchmark values considerations, 334
      coho salmon in Carnation Creek
            adult declines due to forest-harvesting, 306
            fishery impact, 305–306
            juveniles and forest harvesting, 308–310
            marine survival of smolts, 313–317, 320
            population collapse predictions, 306
            population declines, 303, 305
            population distributions observed, 302–303
            response to conservation measures, 305
            sediment migration effects, 311
            smolts and forest harvesting, 312
      coho salmon in Oregon streams
            field methods for data collection, 332
            fish productivity and habitat conditions, 335–338
            habitat data analyses, 333
            habitat quality assessments, 333–335
            restoration strategy development, 338
            survey design, 330–332
      estuary management, Fraser River
            assessment of results, 345
            compensation projects, 343–345
            habitat loss, 342
            mitigation referral system, 343
            policy definitions, 342
            recommendations, 346
      freshwater degradation and, 152
      life cycle link to, 33
      maintenance difficulties, 32
      preservation importance, 24
      protection policies in Alaska, 64, 182–183
      protective legislation, B.C., 68–69
      requirements, 20
      in Siuslaw River watershed, 631–632
      sustainable fisheries strategy, 691–693

urban streams in LFV
    development impacts, 352–353
    estuarine losses, 353
    future of, 358
    population growth impacts, 351–352
    protection of fish habitats, 355–358
    recommendations, 358–359
Habitat protection and restoration, *see also* Stock
        conservation
    BRCP, Panther Creek
        assumptions for, 568
        baseline restoration program, 577–578
        categories of restoration options, 569
        chinook reintroduction strategies, 571–576
        compensation of interim losses, 578
        costs of program, 579
        engineering concepts used, 569–571
        feasibility of program, 579–580
        history of salmon, 565–567
        measures considered, 569
        statutes and policies, federal and state, 568–569
    coho salmon, Siuslaw River
        ecosystem approach, 634
        large woody debris rehabilitation, 602, 633
    compensation measures
        artificial reefs, 345
        bank construction, 345
        channelization, 344
        vegetation transplanting, 343–344
    ecosystem approach, LWW
        adaptive management, 537–538
        agency change, 537
        conservation area establishment, 538–540
        context and scale considerations, 535–536
        ecologically relevant management units, 534–535
        goals and objectives, 536–537
        restoration and protection emphasis, 536
    effectiveness of, Tongass evaluation
        discussion and recommendations, 593–597
        methods, 585–587
        results, 588–593
        study team charter, 584–585
    large woody debris rehabilitation
        candidate stream assessments, 553, 555–557
        in Carnation Creek, 299–300
        costs of projects, 558
        definition of LWD, 546
        documenting fish presence, 557–558
        ecological role of LWD, 310, 547–549
        ecosystem approach, 546
        philosophy of, 546–547
        recommendations, 558–559
        site selection, 550–553
        in Siuslaw River, 602, 633
        successful design factors, 549–550
    legislation
        Alaska, 64, 182–183
        British Columbia, 68–69
    loss and degradation, 4–6
    stock conservation
        economic incentives for conservation, 661–662

ecosystem health standards, 660–661
        institutional reform, 662–664
        transferable resource privileges, 662
    sustainable fisheries strategy, 691–693
urbanization impacts, LWW
    elimination of stream habitat, 530
    littoral zone changes, 533–534
    physical habitat quality, 530–532
    stream biota affected by, 532–533
    water quality changes, 532, 534
watershed analysis, Wolf Creek
    aquatic habitat restoration projects, 604–606,
        613–614
    background, 602
    climate, 608
    fish, 610–611
    geology and geomorphology, 607
    hydrology, 609–610
    land use, 612
    methods for, 603–604
    monitoring and evaluation, 607, 614–615
    stream channel, 610
    synthesis of results, 612–613
    vegetation, 608–609
    wildlife, 609
watershed assessment, Deer Creek
    background, 618–619
    Deer Creek Group, 619
    monitoring and maintenance, 622
    restoration implementation, 621–622
    restoration treatment objectives, 620
    results and discussion, 622–623
    spring-run chinook history, 648–649
    strategy development, 619–620
Harvest management, 6; *see also* Commercial fishery;
        Hatcheries
    accuracy of abundance counts, 656–657
    exploitation rate reduction, 656
    fisheries regulations methods, 630–631
    fishwheels for monitoring, Nass River
        contribution to fisheries management, 671–672
        cost–benefit analysis, 672, 673
        harvesting procedure efficiencies, 673
        limits to use, 672
        population estimates, 669–671
        research opportunities, 671
        uses of, 668
    goals and rates modifications needed, 657
    status declines, 153
    sustainable fisheries strategy, 693–694
Hatcheries, *see also* Aquaculture; Harvest management
    cost–benefit analysis, Alaska, 394, 400–401
    enhancement programs economic analysis
        contributions, 387–390
        cost summary, 386
        feasibility summary, 390–391
        feasibility variables, 382–383
        hatchery costs per fish, 385
        history in Columbia River, 383–384
        production costs, 384–385
        production costs vs. ex-vessel value, 387

stock survival rates, 385
farmed salmon
  in British Columbia, 71–72
  economic effects, 394, 399
  international, 397
harvest priority in escapement management, 250–251
history of, 16
impact on Pacific salmon populations, 222
impact on wild stocks, 153
management policy in Alaska, 185–186
overescapement, 256
releases in LWW, 528
residuals management, 375
role in population restoration, 25
selective marking, B.C., 70–71
in Siuslaw River watershed, 631
wild vs. hatchery fish study, *see* Hatchery study,
  Yakima River
Hatchery study, Yakima River
guidelines for impact reduction
  release location, 375–376
  release numbers, 374–375
  release size, 374
  release timing, 376–377
  residuals management, 373–374, 375
incidence of disease, 372
population abundance, 372
population release, 366–367
residuals, 368
sampling, 367–368
water temperature effects, 372
wild vs. hatchery
  displacements, 368–369
  interaction rates, 370–372
  numbers and biomass, 369–370
  types of interactions, 368
Hemlock (*Tsuga heterophylla*), 527, 608, 618, 626
Herring (*Clupea harrengus pallasi*), 317
Hidden Falls project, 415
Human activities, *see* Anthropogenic factors
Hydroacoustic methods for escapement estimates
  Fraser River, 210
  PSC program, sockeye salmon, 211
  sockeye salmon, 213
Hydroelectric power
  adverse effects on salmon habitats, 6, 45
  habitat protection policies, Alaska, 64
  salmon conservation efforts, B.C., 68
  status declines due to, 152

I

ICBEMP (Interior Columbia River Basin Ecosystem
  Management Project), 134
Idaho habitat restoration, *see* Panther Creek BRCP
Individual transferable quotas (ITQs), 659
Industrial development effects on habitats, 6
Institutional reform for sustainable fisheries
  area quotas, 662–663
  community involvement, 658–659

harvest privileges and futures, 659–660
increased tax funding, 658
land and water trusts, 663
management compacts, 663
public education, 658
shortcomings of, 8, 656
stakeholder interests balanced, 658
strategies, 698–700
utilities, fish, 663–664
Integrated fisheries management model
  conclusions, 448
  fishing regime scenarios, 441–442
  fixed harvest-rate regime results, 445–446, 448
  goal of modeling exercise, 447
  interception-balancing fishing regime results, 446, 448
  model application, 440
  model structure, 437–439
  MSY fishing regime results, 447, 448
  parameter estimation, 440–441
  random variability option, 447
  results, 442–445
Interceptions definition, 436
Interior Columbia River Basin Ecosystem Management
  Project (ICBEMP), 134
International Convention for the High Seas Fisheries, 224
International Pacific Salmon Fisheries Commission
  (IPSFC), 209, 225
Introduced species impact on status, 152–153
Ireland farmed salmon industry, 397
Iron and phytoplankton production, 473
ITQs (individual transferable quotas), 659

J

Japanese Aquatic Resources Conservation Act, 396
Japan farmed salmon industry, 396

K

Kadake Creek (Alaska) watershed
  description, 587
  instream condition assessment, 590
  landslides in, 592
  riparian areas, 593
K'alii Aksim Lisims (Nisga'a), 83, 668
Kamchatka (Russia) steelhead trout
  anadromous/nonanadromous forms, 202
  distribution, 196
  intraspecific diversity, 202
  life strategies, 197–201
  study program
    life history, 201
    methods, 197
    objectives of, 196
    population structures, 197–200
    site, 197
  uniqueness of, 195–196
Klamath Mountains Province coastal steelhead ESUs, 127
Klamath Resource Information System, 99

Klamath River chinook and steelhead status and
　　distribution
　　classification criteria, 137
　　current status, 148–151
　　database use, 138
　　emphasis area identification and importance, 153–155
　　factors influencing, 151–153
　　geographic study area, 134–135
　　implications for sustainability, 155–156
　　known status, 147
　　potential historical range, 138, 146–147
　　predictive models, 138–142
　　predictive models results, 147–148
　　subspecies studied, 136
　　subwatershed divisions, 137–138
Kokanee salmon, 22, 516
Kvachina River (Russia), 197

L

Lake Washington Watershed (LWW)
　　anadromous salmonids in
　　　life history and status, 528–530
　　　life history diversity, 530
　　　species, 527
　　description, 526–527
　　ecosystem approach
　　　adaptive management, 537–538
　　　agency change, 537
　　　conservation area establishment, 538–540
　　　context and scale considerations, 535–536
　　　ecologically relevant management units, 534–535
　　　goals and objectives, 536–537
　　　restoration and protection emphasis, 536
　　urbanization impacts
　　　elimination of stream habitat, 530
　　　littoral zone changes, 533–534
　　　physical habitat quality, 530–532
　　　on stream biota, 532–533
　　　water quality, 532, 534
Lampetra richardsoni (brook lamprey), 602
Lampetra tridentatus (Pacific lamprey), 602
　　in Asotin Creed, 640
　　in Nass River, 671
　　in Siuslaw River watershed, 627
Land and water trusts, 663
Landslides, 591–592
Large woody debris (LWD) rehabilitation
　　candidate stream assessments
　　　channels, 555–556
　　　design and project success, 557
　　　factors to consider, 555
　　　fish habitat, 556
　　　hydraulics, 556
　　　LWD assessment, 553, 555
　　in Carnation Creek, 299–300
　　costs of projects, 558
　　definition of LWD, 546
　　documenting fish presence, 557–558
　　ecological role of LWD

　　　biological function, 548–549
　　　clearance and logging effects, 310, 548
　　　physical function, 547–548
　　ecosystem approach, 546
　　philosophy of, 546–547
　　recommendations, 558–559
　　site selection considerations, 549–550
　　　candidacy determination, 553
　　　floodplain rehabilitation, 553
　　　large wood function in upslope areas, 550
　　　riparian zone rehabilitation and protection, 550,
　　　　552
　　in Siuslaw River watershed, 602, 633
　　Tongass National Forest study, 590
LCRFDP (Lower Columbia River Fisheries Development
　　Program), 678
Legislation, federal and state
　　CERCLA (U.S.), 567
　　Environmental Review Process (Canada), 69
　　ESA, *see* Endangered Species Act
　　Fisheries Act (Canada), 51, 342, 356
　　Forest Practices Act (Alaska), 64
　　Forest Practices Code (Canada), 69, 298, 356
　　Growth Management Act (WA), 538
　　Japanese Aquatic Resources Conservation Act, 396
　　Magnuson Fisheries Conservation and Management
　　　Act (U.S.), 436
　　Municipal Act (Canada), 356
　　Oceans Act (Canada), 342, 346
　　Pacific Salmon Treaty, *see* Pacific Salmon Treaty
　　Tongass Timber Reform Act (Alaska), 182
　　White Act (U.S.), 187, 677
Life cycle steps, 20
Limited entry regulations, 161–162
Littoral zones, 533–34
Local-stock fisheries, 164, 184
Logging, *see also* Forest management activities
　　impact on habitats, 182–183
　　in Siuslaw River watershed, 628–629
Longfin smelt (*Spirinchus thaleichtys*), 516
Lower Columbia River Fisheries Development Program
　　(LCRFDP), 678
Lower Columbia River/SW Washington coast
　　coho salmon ESU risks, 114–115
　　steelhead ESUs, 126
Lower Fraser River Valley (B.C.) urban streams
　　early development, 350–351
　　fishery resource, 350
　　location and description, 349–350
　　population growth and land use, 351–352
　　　impacts on fish habitat, 352–353
　　　impacts on streams and rivers, 353–355
　　protection of fish habitat
　　　jurisdictional issues, 357
　　　jurisdiction coordination needed, 358
　　　legislation, 356
　　　policy and guidelines development, 356–357
　　　stewardship programs, 357–358
　　　strategic planning for, 358
　　recommendations, 358–359
LWD, *see* Large woody debris

LWW, *see* Lake Washington Watershed
Lyon's Ferry Hatchery (LYF), 453

M

Magnuson Fisheries Conservation and Management Act
    (1976) (U.S.), 436
Management compacts, 663
Management unit definition, 238
Maple (*Acer* spp.), 527, 608–609
Marbled murrelets (*Brachyramphus marmoratus*), 604
Marine survival
    chum salmon, 308, 315
    coho salmon, 170
        migration timing and, 315
        predation intensity changes, 317, 320
        relationship to logging, 316
        sea-surface salinities and, 315–316
Markets for salmon, *see also* Economic considerations
    annual harvests, 393–394
    aquaculture and, 399–400
    global
        aquaculture and, 396–400
        implications for PNW, 402–403
        international markets, 396–399
        public and private ranching, 400–401
        U.S. position, 399–400
    management of, 402–404
    market price, 382
    net-pen reared, 397
    price decreases, 394
    stock decreases, 394
Marsh benches, 343
MELP (Ministry of Environment, Lands and Parks)
    (Canada), 51
Merluccius productus (Pacific hake), 317
Methane (CH_4) in aquatic habitats, 40
Micropterus dolomieu (smallmouth bass), 534
Middle Columbia River inland steelhead ESUs, 129
Migration, *see also* Proportional migration model
    behavior, 22–23
    global warming impact on, 41–42
    life history and, 17–18
Mill Creek, 648
Minimum sustainable escapements (MSE), 254
Mining, salmon declines due to, 567
Ministry of Environment, Lands and Parks (MELP)
    (Canada), 51
Mitigation
    artificial production used for, 381
    conservation and, 655
    definition of (Canada), 342
Mixed stock fisheries
    avoidance of interactions, 414–415
    definition of, 164
    escapements management
        escapement goals and, 239–240
        harvest priority, 250–251
        number of populations, 249–250
        run management effects, 251

hatchery impacts, 153, 222
management problem in B.C., 78–79
managing for lower harvest rates, 501–502
overharvest risk management, 184, 412–414
proportional migration model example, 432
Modeling approaches, *see also* Predictive models
    decadal climate cycles
        anthropogenic effects accounted for, 469, 470–471
        climate pattern shifts, 471–476
        climatic variations effects, approaches for,
          468–469
        human and climatic impacts, cumulative effects,
          470–471
        interaction of processes, 477–480
        stock recruitment equation, 468
    integrated (multi-stock) fisheries model
        conclusions, 448
        fishing regime scenarios, 441–442
        fixed harvest-rate regime results, 445–446, 448
        goal of modeling exercise, 447
        interception-balancing fishing regime results, 446,
          448
        model application, 440
        model structure, 437–439
        MSY fishing regime results, 447, 448
        parameter estimation, 440–441
        random variability option, 447
        results, 442–445
    photosynthetic model, rearing capacity
        assumptions, 514, 516
        assumptions in earlier models, 506
        biologically feasible production estimates, 519
        measurements correlation to fish yield, 506
        model development, 509–510
        nursery lakes in B.C., 507–508
        predictions for B.C. lakes, 517
        productivity variables measurement, 508–509
        testing of, 511, 514
        uses as a model, 507
        versatility and uses, 519
    proportional migration model
        fisheries comparison, 430–431
        fishery abundance, 428
        implications for mixed-stock fisheries, 432
        inputs and model specification, 428–429
        management proposal, 429–430
        methodology, 425
        stock reconstruction, 425–426
        time step, first, 426–428
        time steps, subsequent, 428
        useability of model, 431–432
    PSC chinook model, 455–457, 464
    recruitment variation effects model
        density-dependent models shortcomings, 485–486
        overharvest due to MSY goal, 486
        stock-recruitment analyses, 492, 498–502
        variation determination methods, 486, 488–491
        wide variation in simulation, 492
    restoration program, chinook BRCP, 577–578
    spawner/recruit models
        assumptions inaccuracies, 243

constant carrying capacity assumption, 244
data inaccuracies, 243
in escapements management, 238, 239
habitat degradation unaccounted for, 244
theoretical weaknesses of, 243–244
Morone saxatilis (striped bass), 253
Mortality
adult equivalent, 459
in models, 424–425
non-catch, 430
proportional migration model, 429
MSE (minimum sustainable escapements), 254
Multi-stock fisheries model
integrated fisheries management model
conclusions, 448
fishing regime scenarios, 441–442
fixed harvest-rate regime results, 445–446, 448
goal of modeling exercise, 447
interception-balancing fishing regime results, 446, 448
model application, 440
model structure, 437–439
MSY fishing regime results, 447, 448
parameter estimation, 440–441
random variability option, 447
results, 442–445
management objectives for PST, 435–436
socioeconomic factors incorporated into, 436
types of, 436
Municipal Act (Canada), 356
Myriophyllum spicatum (watermilfoil), 533
Myxobolus articus, 211

N

N₂O (nitrous oxide) in aquatic habitats, 40
Nass River (B.C.) watershed, 83
description, 668
fishwheel use
contribution to fisheries management, 671–672
cost–benefit analysis, 672, 673
harvesting procedure efficiencies, 673
limits to, 672
population estimates, 669–671
research opportunities, 671
National Habitat Policy (Canada), 51, 57
National Marine Fisheries Service (NMFS)
coho salmon status review
ESU identification, 112
extinction risk assessment, 114–115, 116–117
listing status, 115
listings under ESA, 4, 5, 452
pink salmon status review, 103
ESU identification, 104–106
extinction risk assessment, 106–109
policy on ESUs, 103
Snake River recovery plan, 568–569
spring-run chinook, Central Valley, 649
steelhead status review
ESU identification, 120

listing status, 130–131
transboundary rivers studies, 274
National Wild and Scenic Rivers System, 679
Natural production, 7; *see also* Artificial production
NAWMP (North American Waterfowl Management Plan), 683
Nehalem River (Oregon), 488
Net-pen reared salmon
cost of regulatory compliance, 399
market for, 397
opposition in U.S., 398
production history, 396
New Zealand farmed salmon industry, 397
Nisga'a, 83, 668
Nisga'a Tribal Council (NTC), 668
Nitrous oxide (N₂O) in aquatic habitats, 40
NMFS, *see* National Marine Fisheries Service
"No Net Loss" policy (Canada), 68, 357
Non-point pollution sources, 532
North American Waterfowl Management Plan (NAWMP), 683
Northern spotted owl (*Strix occidentalis*), 96, 604
North Pacific Anadromous Fish Commission, 65
North Pacific Conservation Council, 700
North Pacific Fisheries Convention, 224
Northwest Power Planning Council (NPPC), 639
Norway farmed salmon industry, 396, 397, 399
NPPC (Northwest Power Planning Council), 639

O

Oceanographic conditions, 471
Oceans, *see also* Global warming
adverse effects of high temperature, 6
climatic conditions impact on, 40
Oceans Act (1997) (Canada), 342, 346
ODFW (Oregon Department of Fish and Wildlife), 384, 630–632
Odocoilesu hemionus (deer), 609
Old Franks Creek (Alaska) watershed
description, 587
instream condition assessment, 590
landslides in, 592
riparian areas, 593
Olympic National Park, 679
Olympic Peninsula
coastal steelhead ESUs, 126
coho salmon ESU risks, 115
Oncorhynchus
clarki clarki (cutthroat trout), 95, 96, 320–322
gorbuscha, *see* Pink salmon
keta, *see* Chum salmon
kisutch, *see* Coho salmon
mykiss, *see* Rainbow trout; Steelhead trout
nerka, *see* Sockeye salmon
tshawytscha, *see* Chinook salmon
Optimum escapement definition, 505
Oregon Department of Fish and Wildlife (ODFW), 384, 630–632
Oregon (OR)

coastal steelhead ESUs, 127
coho restoration in Siuslaw River watershed
    ecosystem current state, 632–633
    history of, 626–630
    restoration projects, 602, 633–634
coho salmon ESU risks, 114
cost–benefit analysis, hatchery programs, 401
fish refuge concept in, 678
Snake River
    exploitation rate reduction, 455–462
    history of population declines, 678
    inland steelhead ESUs, 130
    stock extinctions, 96
    URB stock, 453
stream survey, coho salmon
    field methods for data collection, 332
    fish productivity and habitat conditions, 335–338
    habitat data analyses, 333
    habitat quality assessments, 333–335
    restoration strategy development, 338
    survey design, 330–332
watershed analysis, Wolf Creek
    aquatic habitat restoration projects, 604–606,
        613–614
    background, 602
    methods for, 603–604
    monitoring and evaluation, 607, 614–615
    results and discussion, 608–613
Oregon Production Index (OPI) harvest rate, 488

P

PACFISH, 584
Pacific Coast Federation of Fishermen's Associations
        (PCFFA), 650
Pacific Fishery Management Council (PFMC) (U.S.)
    environmental impact statement, 221
    Snake River chinook harvesting, 454
    sport fishery regulations, 226
Pacific hake (Merluccius productus), 317
Pacific herring (Clupea harengus), 53
Pacific lamprey (Lampetra tridentatus)
    in Asotin Creek, 640
    in Nass River, 671
    in Siuslaw River watershed, 602, 627
Pacific Marine Fisheries Commission (PMFC) (U.S.), 223
Pacific Northwest Index (PNI), 475, 478–479
Pacific Northwest ocean fisheries, see also individual
        species; Pacific salmonids
    ESA listed stocks, 228–229
    hatcheries impact on populations, 222
    history of exploitation, 219–220
    impacts on salmon, 221–222
    knowledge of salmon, 220–221
    management alternatives, 229–230
    management capacity development
        Indian fishing rights, 227
        individual stock as basic unit, 227
        information availability, 227–228
        international cooperation, 227

technological advancements, 228
        reciprocal fishing agreements, U.S. and Canada,
            225–226
        sport regulations, 226
    sustainability management approaches
        adaptive, 231
        collaborative, 231
        issues concerning, 232–233
        risk averse, 231–232
    troll regulations
        bilateral agreements, U.S. and Canada, 224–225
        history of, 223–224
        international agreements, 224
        reciprocal fishing agreements, U.S. and Canada,
            225
Pacific Salmon Commission (PSC), 209
    chinook model for exploitation rate reduction
        allocation issues, 454
        assumptions, 457
        background, 455–456
        ecosystem approach, 454
    Snake River chinook harvesting, 454
Pacific salmonids (Oncorhynchus spp.), see also individual
        species; Pacific Northwest ocean fisheries;
        Southeast Alaska salmon
    anadromous behavior, 32–33
    evolutionary units, 24–25
    habitat and population restoration
        fish hatcheries role, 25
        strategy for, 26
        watershed approach, 25
    habitat complexity, 21
    habitat requirements, 20–21
    history
        fishery declines, 16
        hatcheries appearance, 16
        populations, 133–134
    human population growth impact, 33–35
        global activity effects, 40–44
        lack of clear management goals, 39
        local activity effects, 35–37
        regional activity effects, 37–39
    interdependence of environmental impacts, 45–47
    keystone species preservation, 56–57
    life history
        life cycle, 19, 33
        migration of juveniles, 17–18
        spawning activity, 18–19
    listings under ESA, 134
    local adaptation
        genetic basis of differences, 23
        genetic diversity maintenance, 23, 24
        habitat preservation importance, 24
        intraspecies distinctions, 22
        migrational behavior, 22–23
        spawning and temperature, 22
    Pacific Northwest arena, 32
    spawning population decline impact on, 21
    stock conservation, see Stock conservation
    stock losses, 44–45
    sustainability and integrated management, 47

water temperature and spawning, 22
Pacific Salmon Treaty (PST), 66, 68
    allocation issues, 257
    harvest sharing provisions, 435–436
    stock rebuilding goal, 455
Pacific Stock Assessment Review Committee (PSARC), 52,
        59–60
Panther Creek BRCP (Idaho)
    assumptions for, 568
    baseline restoration program, 577–578
    categories of restoration options, 569
    chinook reintroduction strategies
        donor stock selection, 571–576
        stock supplementation methods, 576
    compensation of interim losses, 578
    costs of program, 579
    engineering concepts used, 569–571
    feasibility of program, 579–580
    history of salmon, 565–567
    measures considered, 569
    statutes and policies, federal and state, 568–569
Passing-stock fishery, 164, 184–185
PCFFA (Pacific Coast Federation of Fishermen's
        Associations), 650
PFMC, *see* Pacific Fishery Management Council
Phenotype diversity and sustainability, 155
Phosphorus loadings, 534
Photosynthetic rate (PR) model, rearing capacity
    assumptions
        energy transfer efficiency constancy, 516
        low planktivore competition, 514, 516
        rearing capacity as smolt production control, 514
        spawner-to-fry relationship components, 514
        zooplankton and grazing pressure, 516
    assumptions in earlier models, 506
    biologically feasible production estimates, 519
    measurements correlation to fish yield, 506
    model development
        EV model, 509
        PR units in revision, 509–510
    nursery lakes in B.C., 507–508
    predictions for B.C. lakes, 517
    productivity variables measurement, 508–509
    testing of
        correlation to total fish production, 511
        insufficient B.C. data, 511, 514
        lake surface area variation considerations, 511
        optimum escapements predictions, 514
    uses as a model, 507
    versatility and uses, 519
Phylonema oncorhynchi, 211
Picea sitchensis (Sitka spruce), 112
Pickleweed (*Salicornia virginica*), 344
Pikeminnow (*Ptychocheilus oregonensis*), 602
Pink salmon (*Oncorhynchus gorbuscha*)
    annual harvests, 176
    distribution, 176
    escapement trends, 176–177
    fishery harvests, 177–178
    population risk factors, 109
    status review, 103

endangered species definition, 106–107
    even-year ESU, 104, 106
    even-year extinction risk, 107–108
    odd-year ESU, 104–106
    odd-year extinction risk, 108–109
    stock assessment and management, 177
    stock programs and status
        Stikine River, 281–282
        Taku River, 285–286
Placer mining impact on sustainability, 39
PMFC (Pacific Marine Fisheries Commission) (U.S.), 223
PNI (Pacific Northwest Index), 475, 478–479
Policy for the Management of Fish Habitat (Canada), 51,
        342
Pollution, *see* Anthropogenic factors; Urbanization
Population definition, 238
Population per management unit definition, 249–250
Potential historical range
    chinook salmon study results, 146–147
    steelhead trout study results, 143
PR, *see* Photosynthetic rate (PR) model
Predictive models, *see also* Modeling approaches
    chinook salmon study results, 147–148
    classification trees, 138–142
    steelhead trout study results, 145
Prickly sculpins (*Cottus asper*), 301
Priest Rapids (Oregon), 453
Prince William Sound (Alaska) ranching program
    overharvest risk for wild stocks, 411–412
    overharvest risk reduction techniques, 412–413
    run failures history, 409–410
    wild declines vs. hatchery abundance, 410–411
Prince William Sound Aquaculture Corporation (PWSAC),
        410
Principles of Cooperation for Salmon Conservation, 66
Proportional migration model
    description
        fishery abundance, 428
        inputs and model specification, 428–429
        methodology, 425
        stock reconstruction, 425–426
        time step, first, 426–428
        time steps, subsequent, 428
    example
        fisheries comparison, 430–431
        implications for mixed-stock fisheries, 432
        management proposal, 429–430
        usability of model, 431–432
PRO Salmon, 103, 108–109
Prosopium williamsoni (whitefish), 568
PSARC (Pacific Stock Assessment Review Committee), 52,
        59–60
PSC Chinook Model
    exploitation rate reduction
        allocation issues, 454
        assumptions, 457
        background, 455–456
        calibration, 457
        spawner/progeny relationship, 457
    uses of, 464
PSC (Pacific Salmon Commission), 209

Pseudotsuga menziesii (Douglas fir), 527, 608, 618, 626
PST, *see* Pacific Salmon Treaty
Ptychocheilus oregonensis (pikeminnow), 602
Puget Sound/Strait of Georgia
    coastal steelhead ESUs, 120, 126
    coho salmon ESU risks, 115
Pulp mills impact on sustainability, 39
PWS, *see* Prince William Sound ranching program
PWSAC (Prince William Sound Aquaculture Corporation),
    410

Q

Quercus garryana (white oak), 112

R

Rainbow trout (*Oncorhynchus mykiss*), 568; *see also*
    Hatchery study, Yakima River
Ranching programs, Alaska, *see also* Artificial production;
    Enhancement programs
    cost–benefit analysis, 394, 400–401
    disease dissemination, 417
    ecological interactions, 417
    economic effects, 394
    Northern Southeast, 414–415
    overview, 407–409
    Prince William Sound
        overharvest risk for wild stocks, 411–412
        overharvest risk reduction techniques, 412–413
        run failures history, 409–410
        wild declines vs. hatchery abundance, 410–411
    production, 409
    Southern Southeast, 413–414
    straying and genetic risks, 416
RASP (Regional Assessment of Supplementation Project),
    571–572
Rearing capacity model (PR model)
    assumptions
        energy transfer efficiency constancy, 516
        low planktivore competition, 514, 516
        rearing capacity as smolt production control, 514
        spawner-to-fry relationship components, 514
        zooplankton and grazing pressure, 516
    assumptions in earlier models, 506
    biologically feasible production estimates, 519
    measurements correlation to fish yield, 506
    model development
        EV model, 509
        PR units in revision, 509–510
    nursery lakes in B.C., 507–508
    predictions for B.C. lakes, 517
    productivity variables measurement, 508–509
    testing of
        correlation to total fish production, 511
        insufficient B.C. data, 511, 514
        lake surface area variation considerations, 511
        optimum escapements predictions, 514
    uses as a model, 507

versatility and uses, 519
Recreational fishery
    limited entry, B.C., 71, 72
    sport fishery regulations, PNW, 226
Recruitment variation and sustainable yield
    density-dependent models shortcomings, 485–486
    natural and anthropogenic factors in equation, 468
    overharvest due to MSY goal, 486
    stock-recruitment analyses
        density independent variation in recruitment, 501
        implications of lognormal distribution of
            survivals, 492, 498
        lower harvest rate policy needed, 501–502
        realized vs. calculated values, 499
        risk-averse strategy needed, 500
        variation in recruitment, 499–500
        variation in survival from environmental factors,
            499
    variation determination methods
        effects on sustainable yield, 491
        ocean harvest rate, 488–489
        ocean survival, 489–490
        past recruitment rate estimations, 488
        recruitment rate indices, 486, 488
    wide variation in simulation, 492
Red alder trees (*Alnus rubra*), 300, 606
Redd (spawning nest), 18–19
Redwoods (*Sequoia sempervirens*), 112
Reedsport conference (Oregon), 96
Reefs, artificial, as compensation measure, 345
Refuges for salmon
    concept implementation
        candidate protection, 682
        criteria for a refuge, 680
        official designation process, 682–683
        system elements, 680
        watershed candidate evaluation, 680–682
    conservation area establishment, 538–540
    current need for, 679
    historic call for, 675–676
    history of
        artificial production endorsement, 676
        artificial production failure, 678
        conservation, utilitarian approaches to, 677
        first park established, 676–677
        lack of public support, 678
    refugia areas, 622
Regeneration rates in natural populations, 56
Regional Assessment of Supplementation Project (RASP),
    571–572
Restoration, *see* Habitat protection and restoration;
    Watershed assessment and restoration
Rhinichthys sp. (dace), 602
Richardsonius balteatus (shiners), 602
Ricker curve
    integrated (multi-stock) fisheries model, 437
    lack of fit to plotted curves, 486
    model concepts, 239
    recruitment variation model, 491
    shortcomings of, 485–486
    spawner/recruit models, 457

Riparian zones, 531
    cost–benefit analysis, chinook restoration, 578
    habitat conservation strategy, 593
    habitat drainage and sustainability, 39
    protection of, 550, 552
    recommendations for management of, 596
    restoration in Wolf Creek, 606
Robertson Creek (British Columbia), 305–306
Rubus spectabilis (salmonberry), 300
Russia
    farmed salmon industry, 396
    Kamchatka steelhead trout
        anadromous/nonanadromous forms, 202
        distribution, 196
        intraspecific diversity, 202
        life strategies, 197–201
        study program, 197–201
        uniqueness of, 195–196

S

Sacramento-San Joaquin River, 32
Salicornia virginica (pickleweed), 344
Salix sp. (willow), 609
Salmonberry (*Rubus spectabilis*), 300
Salmonid Enhancement Program (SEP) (Canada), 78
Salmon National Forest, 565
Salmo salar (Atlantic salmon), 394, 676
Salvelinus confluentus (bull trout), 568, 640
Salvelinus malmo (char), 617
Sanctuaries, *see* Refuges for salmon
Saprolegnia, 372
Sardines (*Sardinops* spp.), 471, 473
Sawtooth Hatchery, 577
Scomber japonicus (chub mackerel), 317
Scotland farmed salmon industry, 397
Sculpins (*Cottus* sp.), 301, 602
Sea-run cutthroat trout (*Oncorhynchus clarki clarki*), 95, 96–97, 320–322
Sedge (*Carez lyngbyei*), 343, 344
Selective fisheries modeling approaches
    current model description, 424–425
    limitations to harvest, 423–424
    proportional migration model description
        fishery abundance, 428
        inputs and model specification, 428–429
        methodology, 425
        stock reconstruction, 425–426
        time step, first, 426–428
        time steps, subsequent, 428
    proportional migration model example
        fisheries comparison, 430–431
        implications for mixed-stock fisheries, 432
        management proposal, 429–430
        useability of model, 431–432
    single pool modeling, 424
Selective mark fisheries, 70–71
SEP (Salmonid Enhancement Program) (Canada), 78
Sequoia sempervirens (redwoods), 112
SFF (Sustainable Fisheries Foundation), 10

SFS (Sustainable Fisheries Strategy), 10
Shiners (*Richardsonius balteatus*), 602
Shoshone Bannock Indian Tribes, 568
Shuswap Lake (B.C.), 43–44
Sierra Club Legal Defense Fund, 650
Single pool modeling, 424
Sitka Salmon Summit, 66
Sitka spruce (*Picea sitchensis*), 112
Siuslaw Indians, 626–627
Siuslaw River (OR) coho salmon, *see also* Wolf Creek
        watershed analysis
    ecosystem current state, 632–633
    history of
        aboriginal peoples, 626–627
        annual run, 626
        early settlement and industry, 627–628
        fisheries decline, 629, 630
        habitat management, 631–632
        harvest management, 630–631
        hatcheries, 631
        logging, 628–629
        trolling, 629–630
        watershed description, 626
    restoration projects
        ecosystem approach, 634
        LWD rehabilitation, 602, 633
Skagit River (WA), 528
Skeena River (B.C.) stock management, 78
Skeena Watershed Committee, 72
Smallmouth bass (*Micropterus dolomieu*), 534
Snake River (Idaho), 568
Snake River (Oregon), 454
    exploitation rate reduction
        harvest strategies, 458–459
        management decisions, 463–465
        PSC model, 455–457, 464
        results and discussion, 460, 461
    history of population declines, 678
    inland steelhead ESUs, 130
    stock extinctions, 96
    URB stock, 453
Snatolvayam River (Russia), 197
Snohomish River (Washington), 104
Social welfare policy (Canada), 69
Sockeye salmon (*Oncorhynchus nerka*)
    annual harvests, 166, 173
    conservation of stock, 53
    distribution, 172–173
    escapement trends, 173–174
    fisheries, contribution to, 174–176
    genetic analysis of, 23–24
    genetic basis of ecological differences, 23
    local adaptation, 22
    management in Fraser River
        fishing plan, 210, 212
        in-season data analysis, 211–212
        in-season data collection, 210–211
        management plan objectives, 212–215, 216
        population history, 207–209
        production database, 210
        responsible parties, 209–210, 212

run-size estimation process, 212
   stock status, 215–216
migration and water temperature, 41–42
nursery lakes in B.C., 507–508
rearing capacity estimates model
   assumptions, 514, 516
   assumptions in earlier models, 506
   biologically feasible production estimates, 519
   measurements correlation to fish yield, 506
   model development, 509–510
   nursery lakes in B.C., 507–508
   predictions for B.C. lakes, 517
   productivity variables measurement, 508–509
   testing of, 511, 514
   uses as a model, 507
   versatility and uses, 519
spawning activity, 18–19
stock assessment and management, 174
   annual harvests, 166, 173
   distribution, 172–173
   escapements, 173–174
   fishery harvests, 174–176
   losses, 45
stock programs and status
   Alsek River, 287–288
   Stikine River, 279–281
   Taku River, 283–284
Southeast Alaska salmon, *see also individual species*;
   Pacific salmonids
chinook
   annual harvests, 166, 171
   distribution, 170–171
   escapements, 171–172
   stock assessment and management, 172
chum
   annual harvests, 178–180
   distribution, 178
   escapements, 180–181, 182
   production of, 181–182
   stock assessment and management, 182
coho
   annual harvests, 165–167
   distribution, 165
   escapements, 167–168
   marine survivals, 170
   production of, 168–169
   stock assessment and management, 169
data sources and limitations, 162–164
definitions, 164–165
economic value of fisheries, 161
enhancement and hatchery management, 185–186
environmental impacts on escapements, 187–188
fisheries management
   escapement assessment, 185
   governing bodies, 183
   harvest rate reductions, 185
   local-stock fisheries, 184
   mixed-stock fisheries, 184
   passing-stock fisheries, 164, 184–185
geography and topography, 162
habitat protection, 182–183

overfishing history, 186–187
pink
   annual harvests, 176
   distribution, 176
   escapements, 176–177
   fishery harvests, 177–178
   stock assessment and management, 177
sockeye
   annual harvests, 166, 173
   distribution, 172–173
   escapements, 173–174
   fishery harvests, 174–176
   stock assessment and management, 174
stock future, 188–189
Southern Southeast Regional Aquaculture Association
   (SSRAA), 413
Sparrow decision, 76
Spawner/recruit models
   assumptions inaccuracies, 243
   constant carrying capacity assumption, 244
   data inaccuracies, 243
   in escapements management, 238, 239
   habitat degradation unaccounted for, 244
   theoretical weaknesses of
      assumptions inaccuracies, 243
      constant carrying capacity assumption, 244
      data inaccuracies, 243
      habitat degradation unaccounted for, 244
Spawning
   genetic basis of differences, 23
   habitat complexity needs, 21
   life cycle steps, 20
   life history and, 18–19
   nutrients from carcasses, 20–21
   population decline impact on, 21
   temperature relationship, 22
Spirinchus thaleichtys (longfin smelt), 516
Sport fisheries, *see also* Commercial fisheries
Sport fisheries regulations, Pacific Northwest, 226
Spring-run chinook watershed management (CA)
   harvest issues, 649
   history of, 648–649
   outreach effort, 650
   recent declines in abundance, 649
   Work Group role
      lessons learned, 654
      program goal, 650
      recovery strategy, 650–651
      restoration funding, 652–653
      restoration progress, 651
      results and current function, 653–654
      watershed conservancies, 651–652
SSRAA (Southern Southeast Regional Aquaculture
   Association), 413
Status of stock, *see individual species*; Stock conservation
Steelhead trout (*Oncorhynchus mykiss*)
   catch and release policy, B.C., 70
   ecotypes, 120
   habitat assessment, Carnation Creek, 320–322
   hatchery impact study, *see* Hatchery study, Yakima
      River

Kamchatka study
    anadromous/nonanadromous forms, 202
    distribution, 196
    intraspecific diversity, 202
    life strategies, 197–201
    study program, 197–200, 201
    uniqueness of, 195–196
listings under ESA, 134
in Siuslaw River watershed, 626
status and distribution, Columbia and Klamath rivers
    classification criteria, 137
    current status, 145–146, 151
    database use, 138
    emphasis area identification and importance,
        153–155
    factors influencing, 151–153
    geographic study area, 134–135
    implications for sustainability, 155–156
    known status, 143, 147
    potential historical range, 138, 143, 146–147
    predictive models, 138–142
    predictive models results, 145, 147–148
    subspecies studied, 135
    subwatershed divisions, 137–138
status review
    coastal ESU identification, 120, 126–129
    coastal ESU risk categories, 121–125
    inland ESU identification, 129–130
    inland ESU risk categories, 125
    listing status, 130–131
stock losses, 45
Stickleback (Gasterosteus aculeatus), 516
Stikine River (Alaska, Canada)
    fisheries, 275–277
    habitat studies, 274
    location and geography, 274
    research programs, 288, 290, 291
    stock programs and status
        chinook salmon, 278–279
        chum salmon, 281–282
        coho salmon, 281
        pink salmon, 281–282
        sockeye salmon, 279–281
Stock conservation, see also Fishery resource conservation
    definition of stock, 97, 164
    degradation causes, 664
    economic value to communities, 655
    extinctions, 96–97
    habitat protection
        economic incentives for conservation, 661–662
        ecosystem health standards, 660–661
        institutional reform, 662–664
        transferable resource privileges, 662
    information gathering techniques, 279
    mixed-stock management problem, 78–79
    populations at risk, 95–96
    recommendations, 665
    regulatory shortcomings, 8, 656
    stock-level distinctions, 97
    stock-specific management, B.C., 71
    stock status

coastwide tracking and reporting system, 99
communication with public needed, 99–100
factors affecting, 98
sustainable fisheries
    ecologically sound harvest policy needed,
        656–657
    institutional reform, 658–660, 662–664
watershed approach needed for, 98
wild fish, B.C., 69
Stone, Livingston, 675
Strait of Juan de Fuca (WA), 108, 109
Stream buffers
    management deficiencies, 594
    status of, Tongass, 588–589
StreamNet, 99
Striped bass (Morone saxatilis), 253
Strix occidentalis (northern spotted owl), 96, 604
Suckers (Catastomus macrocheilus), 602
Sustainable fisheries
    definition, 4
    ecological signals of unsustainability, 6
    ecosystem approach principles, 688–689
    ecosystem-based management, 8–9, 691–693
    key elements, 10
        community-based management, 696–697
        habitat protection and restoration, 691–693
        harvest management, 693–694
        institutional and regulatory reform, 698–700
        salmon and steelhead production, 694–696
    long-term vision statement, 688
    obstacles to, 4–7
    participants in, 688, 690–691
Sustainable Fisheries Foundation (SFF), 10
Sustainable Fisheries Strategy (SFS), 10

T

TAC (total allowable catch), 211
Taku River (Alaska, Canada)
    fisheries, 278
    habitat studies, 274
    location and geography, 274
    research programs, 288, 290, 291
    stock programs and status
        chinook salmon, 282–283
        chum salmon, 286
        coho salmon, 285
        pink salmon, 285–286
        sockeye salmon, 283–284
Teanaway River (Washington), see Hatchery study, Yakima
        River
TEC (Transboundary Technical Committee), 274
Temperature effects
    forest harvesting, 312
    global warming
        freshwater habitats impacted by, 42
        impact on salmon migration, 41–42
        impact on salmon production, 40
        mean air temperature increases, 40
        surface water temperature increases, 42–43

hatchery study, Yakima River, 372
    spawning relation, 22
    water, 32, 372
Terminal fishery, definition of, 164
Terminal Fishery Project (TFP), Columbia River, 382
Terrestrial habitat, lowering as compensation measure, 344
TFP (Terminal Fishery Project), Columbia River, 382
Thaleichthys pacificus (eulachon), 53
Thuja plicata (cedar), 527, 608, 618, 626
Timber, Fish, and Wildlife Ambient Monitoring Manual,
        555
Timber industry, *see* Forest management activities
Tongass habitat management evaluation
    discussion and recommendations
        additional protection needed, 595–597
        existing procedure effectiveness, 593–595
    methods
        expert field review, 585–586
        literature review, 585
        watershed analyses, 586–587
    results
        expert field review, 588–590
        literature review, 587–588
        watershed analyses, 590–593
    study team charter, 584–585
Tongass National Forest (Alaska), *see* Tongass habitat
        management evaluation
Tongass Timber Reform Act, 182
Total allowable catch (TAC), 211
Transboundary river salmon populations, *see* Alsek River;
        Stikine River; Taku River
Transboundary Technical Committee (TEC), 274
Tree Point/Nakat Inlet, 413
Troll regulations
    bilateral agreements, U.S. and Canada, 224–225
    history of, 223–224
    history of trolling in Siuslaw River, 629–630
    international agreements, 224
    reciprocal fishing agreements, U.S. and Canada, 225
Trout, *see* Steelhead trout
Tsuga heterophylla (hemlock), 112, 527, 608, 618, 626

U

Uganik River (Russia), 676–677
Ultraviolet radiation impact on salmon, 40, 42
Umpqua River (Oregon), 488
United Nations Convention on the Law of the Sea, 82, 225
United States (U.S.), *see also specific states*
    bilateral agreements with Canada, 224–225
    farmed salmon industry in, 397
    reciprocal fishing agreements with Canada, 225–226
Upriver bright (URB) stock, 453
Urbanization
    adverse effects on salmon habitats, 6
    impact in LWW
        littoral zone changes, 533–534
        physical habitat quality, 530–532
        stream biota effected by, 532–533
        water quality changes, 532, 534

impact on sustainability, 37
    interdependence with habitats, 45
    streams in LFV
        early development, 350–351
        fishery resource, 350
        location and description, 349–350
        population growth and land use, 351–355
        protection of fish habitat, 356–358
        recommendations, 358–359
URB (upriver bright) stock, 453
Urkholok River (Russia), 197
U.S. Endangered Species Act, *see* Endangered Species Act
U.S. Forest Service (USFS), 584, 626
U.S. National Marine Fisheries Service (NMFS), *see*
        National Marine Fisheries Service
Utilities, fish, 663–664

V

Vegetation transplanting as compensation measure,
        343–344
Vine maple (*Acer circinatum*), 609

W

Ward, Henry, 677
Washington (WA)
    Asotin Creek watershed plan, *see* Asotin Creek Model
        Watershed Plan
    coastal steelhead ESUs, 126
    coho salmon status, 114–115
    Deer Creek, *see* Deer Creek watershed assessment and
        restoration
    hatchery study, Yakima River
        guidelines for impact reduction, 374–377
        incidence of disease, 372
        population abundance, 372
        population release, 366–367
        residuals, 368
        sampling, 367–368
        water temperature effects, 372
        wild vs. hatchery, 368–370
    LWW, *see* Lake Washington Watershed
    pink salmon status, 103
    steelhead trout status, 126
Water, *see also* Watershed headings
    diversion impact on sustainability, 38–39
    temperature effects, *see* Temperature effects, 533
Watermilfoil (*Myriophyllum spicatum*), 533
Watershed analysis, Oregon
    stream survey, coho salmon
        field methods for data collection, 332
        fish productivity and habitat conditions, 335–338
        habitat data analyses, 333
        habitat quality assessments, 333–335
        restoration strategy development, 338
        survey design, 330–332
    Wolf Creek, *see* Wolf Creek watershed analysis

Watershed assessment and restoration
    Asotin Creek plan
        agency support, 644
        community plans for restoration, 640–641
        ecosystem approach, 645
        landowner steering committee, 643
        mission and objectives, 641–642
        public outreach and education, 644
        successful plan elements, 642–643
        watershed setting, 640
    Deer Creek
        background, 618–619
        Deer Creek Group, 619
        monitoring and maintenance, 622
        restoration implementation, 621–622
        restoration treatment objectives, 620
        results and discussion, 622–623
        spring-run chinook history, 648–649
        strategy development, 619–620
Watershed-based management, *see also* Ecosystem-based
    management
    analysis needed prior to disruption, 595
    analysis procedures, 586–587
        headwater condition assessment, 591–592
        instream condition assessment, 590–591
        riparian management areas, 593
    expert field review ratings, Tongass, 588
    marketing management in PNW, 402–404
    refuge candidate evaluation
        criteria, 681
        good candidate identification, 682
        implementation considerations, 681
        threshold criteria, 680–681
    refuge management unit, 680
    spring-run chinook salmon
        harvest issues, 649
        history of, 648–649
        outreach effort, 650
        recent declines in abundance, 649
        Work Group role, 650–654
Watershed Councils in sustainable fisheries strategy,
    698–699
Watershed Forums, 535
Water temperature effects, *see* Temperature effects
Weirs, 83, 627
Western hemlock (*Tsuga heterophylla*), 112
Western red cedar (*Thuja plicata*), 527, 618, 626
White Act (1923) (U.S.), 677
White Act (1924–1959) (U.S.), 187
Whitefish (*Prosopium williamsoni*), 568
White oak (*Quercus garryana*), 112
White sturgeon (*Acipenser transmontanus*), 53
Wild chinook salmon (*Oncorhynchus tshawytscha*), 45;
    see also Chinook salmon
Willamette Valley logging history, 628–629

Willow (*Salix* sp.), 609
Windthrow, 590–591, 593
Winter chinook (*Oncorhynchus tshawytscha*), 95;
    see also Chinook salmon
Wolf Creek watershed analysis
    aquatic habitat restoration projects
        activities used, 605–606
        background, 604–605
        channel restoration, 606
        criteria for developing, 605
        efforts and results, 613–614
        riparian restoration, 606
    background, 602
    methods for, 603–604
    monitoring and evaluation, 607, 614–615
    results and discussion
        climate, 608
        fish, 610–611
        geology and geomorphology, 607
        hydrology, 609–610
        land use, 612
        stream channel, 610
        synthesis of results, 612–613
        vegetation, 608–609
        wildlife, 609

Y

Yakima River (WA) hatchery study
    displacements, 368–369
    guidelines for impact reduction, 373–377
    incidence of disease, 372
    population abundance, 372
    population release, 366–367

    sampling, 367–368
    water temperature effects, 372
    wild vs. hatchery
        interaction rates, 370–372
        numbers and biomass, 369–370
        types of interactions, 368
Yaquina River (Oregon) habitat assessment
    field methods for data collection, 332
    fish productivity and habitat conditions, 335–338
    habitat data analyses, 333
    habitat quality assessments, 333–335
    restoration strategy development, 338
    survey design, 330–332
Yukon River, 32

Z

Zostera marina (eelgrass), 344